Orogenic Processes: Quantification and Modelling in the Variscan Belt

Special Publication reviewing procedures

The Society makes every effort to ensure that the scientific and production quality of its books matches that of its journals. Since 1997, all book proposals have been refereed by specialist reviewers as well as by the Society's Publications Committee. If the referees identify weaknesses in the proposal, these must be addressed before the proposal is accepted.

Once the book is accepted, the Society has a team of series editors (listed above) who ensure that the volume editors follow strict guidelines on refereeing and quality control. We insist that individual papers can only be accepted after satisfactory review by two independent referees. The questions on the review forms are similar to those for *Journal of the Geological Society*. The referees' forms and comments must be available to the Society's series editors on request.

Although many of the books result from meetings, the editors are expected to commission papers that were not presented at the meeting to ensure that the book provides a balanced coverage of the subject. Being accepted for presentation at the meeting does not guarantee inclusion in the book.

Geological Society Special Publications are included in the ISI Science Citation Index, but they do not have an impact factor, the latter being applicable only to journals.

More information about submitting a proposal and producing a Special Publication can be found on the Society's web site: www.geolsoc.org.uk

It is recommended that reference to all or part of this book should be made in one of the following ways.

FRANKE, W., HAAK, V., ONCKEN, O. & TANNER, D. (eds) 2000. *Orogenic Processes: Quantification and Modelling in the Variscan Belt*. Geological Society, London, Special Publications, **179**.

GERDES, A., WÖRNER, G. & FINGER, F. 2000. Hybrids, magma mixing and enriched mantle melts in post-collisional Variscan granitioids: the Rasterberg pluton, Austria. *In*: FRANKE, W., HAAK, V., ONCKEN, O. & TANNER, D. (eds). *Orogenic Processes: Quantification and Modelling in the Variscan Belt*. Geological Society, London, Special Publications, **179**, 415–431.

GEOLOGICAL SOCIETY SPECIAL PUBLICATION NO. 179

Orogenic Processes: Quantification and Modelling in the Variscan Belt

EDITED BY

WOLFGANG FRANKE
(Universität Giessen)

VOLKER HAAK
(GeoForschungsZentrum, Potsdam)

ONNO ONCKEN
(GeoForschungsZentrum, Potsdam)

and

DAVID TANNER
(Universität Freiburg)

2000
Published by
The Geological Society
London

THE GEOLOGICAL SOCIETY

The Geological Society of London was founded in 1807 and is the oldest geological society in the world. It received its Royal Charter in 1825 for the purpose of 'investigating the mineral structure of the Earth' and is now Britain's national society for geology.

Both a learned society and a professional body, the Geological Society is recognized by the Department of Trade and Industry (DTI) as the chartering authority for geoscience, able to award Chartered Geologist status upon appropriately qualified Fellows. The Society has a membership of 9099, of whom about 1500 live outside the UK.

Fellowship of the Society is open to persons holding a recognized honours degree in geology or a cognate subject, or not less than six years' relevant experience in geology or a cognate subject. A Fellow with a minimum of five years' relevant postgraduate experience in the practice of geology may apply for chartered status. Successful applicants are entitled to use the designatory postnominal CGeol (Chartered Geologist). Fellows of the Society may use the letters FGS. Other grades of membership are available to members not yet qualifying for Fellowship.

The Society has its own Publishing House based in Bath, UK. It produces the Society's international journals, books and maps, and is the European distributor for publications of the American Association of Petroleum Geologists (AAPG), the Society for Sedimentary Geology (SEPM) and the Geological Society of America (GSA). Members of the Society can buy books at considerable discounts. The Publishing House has an online bookshop (http://bookshop.geolsoc.org.uk).

Further information on Society membership may be obtained from the Membership Services Manager, The Geological Society, Burlington House, Piccadilly, London W1V 0JU (E-mail: enquiries@geolsoc.org.uk; tel: +44 (0) 207 434 9944).

The Society's Web Site can be found at http://www.geolsoc.org.uk/. The Society is a Registered Charity, number 210161.

Published by The Geological Society from:
The Geological Society Publishing House
Unit 7, Brassmill Enterprise Centre
Brassmill Lane
Bath BA1 3JN, UK
Orders: Tel. +44 (0)1225 445046
Fax +44 (0)1225 442836
Online bookshop: http://bookshop.geolsoc.org.uk

British Library Cataloguing in Publication Data

A catalogue record for this book is available from the British Library.

ISBN 1-86239-073-8

Typeset by Alden Multimedia, Westonzoyland, UK

Printed by The Alden Press, Oxford, UK

Distributors

USA
AAPG Bookstore
PO Box 979
Tulsa
OK 74101-0979
USA
Orders: Tel. +1 918 584-2555
Fax +1 918 560-2652
E-mail bookstore@aapg.org

Australia
Australian Mineral Foundation Bookshop
63 Conyngham Street
Glenside
South Australia 5065
Australia
Orders: Tel. +61 88 379-0444
Fax +61 88 379-4634
E-mail bookshop@amf.com.au

India
Affiliated East-West Press PVT Ltd
G-1/16 Ansari Road, Daryaganj,
New Delhi 110 002
India
Orders: Tel. +91 11 327-9113
Fax +91 11 326-0538
E-mail affiliat@nda.vsnl.net.in

Japan
Kanda Book Trading Co.
Cityhouse Tama 204
Tsurumaki 1-3-10
Tama-shi
Tokyo 206-0034
Japan
Orders: Tel. +81 (0)423 57-7650
Fax +81 (0)423 57-7651

Contents

Orogenic processes: quantification and modelling in the Variscan belt

WOLFGANG FRANKE[1], VOLKER HAAK[2], ONNO ONCKEN[2] & DAVID TANNER[3]

[1]*Institut für Geowissenschaften, Universität Giessen, Senckenbergstrasse 3, D-35390 Giessen, Germany*

[2]*GeoForschungsZentrum Potsdam, Telegrafenberg, D-14473 Potsdam, Germany*

[3]*Geologisches Institut, Universität Freiburg, Albertstrasse 23b, D-79104 Freiburg, Germany*

Research into the orogenic processes that shaped the continental crust of Europe has a long-standing tradition. Why the need to quantify and model? It is not just satisfactory to identify ancient examples of subduction zones, accretionary prisms, island arcs, extensional collapse and other standard items of the geodynamic menu. Such interpretations need to be quantified: what was the extent and composition of subducted crust, angle and speed of subduction, amount and composition of melts produced, heat sources for metamorphism? All such interpretations have to conform to first principles, and also to stand the test of quantitative balancing—a concept first developed for the conservation of length or volume in tectonic cross sections. Also in other fields, the correlation of causes and effects and the internal consistency of dynamic models requires a numerical approach.

Modelling a Palaeozoic orogen, for some people, may look like a hopeless task from the very beginning. The fossil record of ancient orogenic crust is limited by erosion from above, and re-equilibration of the Moho from below. The Variscan basement had already been eroded and buried in late Permian and Mesozoic time, and was later exhumed again by Alpine foreland compression, uplifted on the shoulders of grabens or above a mantle plume. Good outcrops are scarce. Living orogens are more attractive in many respects, but it was not advisable to have our group dispersed over the planet: the project needed regional focus and interdisciplinary collaboration. The European Variscides were the obvious target, since a wealth of basic information was available from the beginning, and projects such as the German seismic reflection programme DEKORP and EUROPROBE had provided important observations. More important still: deeply eroded ancient orogens expose crustal levels which are inaccessible in modern mountain belts. The origin and exhumation of high-grade metamorphic rocks was one of the most important aspects of our work.

With the papers presented in this volume, our understanding of the Variscan orogen has grown substantially. However, quantification and modelling are still fragmentary in several aspects and regions. Many contributions are still descriptive, or present only qualitative constraints on the geodynamic evolution. Quantification and modelling require, amongst other things, precise chronological control. In the largely unmetamorphosed externides, this prerequisite had already been achieved over decades of biostratigraphic dating. Dating of magmatic and metamorphic events in crystalline rocks is much more sophisticated and expensive, especially since it has to be combined with a thorough assessment of geochemistry, metamorphic petrology and structural analysis. Our project has produced, among other things, an extensive set of isotopic data. Only upon this base, has it been possible to advance to the level of numerical modelling even in crystalline rocks.

The present volume combines review articles with reports on recent progress in our project. The foldout map helps to locate the main areas of outcrop and tectonostratigraphic units, and a time-table compiled by **McKerrow & van Staal** allows the correlation of tectonic, metamorphic and magmatic events with the sedimentary record of the upper crust.

A first group of papers deals with the tectonic subdivision, palaeogeographic evolution and a geodynamic overview of the Variscan belt. In an invited contribution, **McKerrow *et al.*** present a palaeogeographic review mainly based upon biogeographic data. **Tait *et al.*** summarize the palaeomagnetic evidence. These two perspectives are largely, but not completely coherent: while

From: Franke, W., Haak, V., Oncken, O. & Tanner, D. (eds). *Orogenic Processes: Quantification and Modelling in the Variscan Belt.* Geological Society, London, Special Publications, **179**, 1–3. 0305-8719/00/$15.00

biogeography supports the assembly of Pangaea by the early Devonian, palaeomagnetism appears to require a wide Devonian ocean between Laurussia (including the newly accreted Armorican Terrane Assemblage and Avalonia) and Gondwana. This controversy can only be resolved by ongoing palaeomagnetic studies in the eastern parts of Gondwana. The contribution by **Franke** summarizes the evolution of the mid-European segment of the Variscides and presents a plate tectonic model. **Franke & Zelazniewicz** trace the mid-European units eastwards. Terrane correlation across the Elbe Fault Zone permits the reconstruction of a disrupted orocline, which forms the eastern termination of the Variscides against the East-European Craton. Further contributions deal with terranes on the eastern and southeastern flank of the Bohemian Massif, as revealed by K–Ar ages of detrital micas (**Belka *et al.***), U–Pb ages (Brunovistulian or Moravo-Silesian: **Finger *et al.***) or combined Nd and U–Pb studies (**Hegner & Kröner**).

The terranes which now make up the Variscan orogenic collage were detached by rifting, during the Cambrian and early Ordovician, from the northern Gondwana margin. The evolution of the best preserved example of an early Palaeozoic rift basin, the Saxo-Thuringian Belt, and its Cadomian basement, is summarized by **Linnemann *et al.*** on the basis of new zircon ages from magmatic rocks. **Floyd *et al.*** review the geochemical characteristics of magmatic rocks from the Saxo-Thuringian belt in the mid-European segment of the Variscides and in the Sudetes adjacent to the east. This paper stresses the continuity of palaeogeographic elements across the Elbe Fault Zone, and the concept of early Palaeozoic crustal extension. **Kröner *et al.*** present evidence for this process from lower crustal rocks now exposed in the Staré Mesto Belt at the southeastern margin of the Bohemian Massif. This contribution also demonstrates the numerous ambiguities inherent in any palaeogeographic interpretation of metamorphosed rocks: instead of one Saxo-Thuringian narrow ocean, whose trace appears on both flanks of a disrupted orocline (**Franke & Zelazniewicz**), **Kröner *et al.*** propose, for the same area, three belts of thinned crust: one ocean, one back-arc basin and one rift.

A second group of papers deals with the evolution of the Rheno-Hercynian externides on the northern flank of the Variscan Belt. **Oncken *et al.*** assess the Rheno-Hercynian evolution from the passive margin of Laurussia into the foreland fold and thrust belt. **Seyferth & Henk** ambitiously attempt to combine the effects of deformation and metamorphism in a numerical model for the collision of the passive margin and the active, southern margin of the Rheno-Hercynian Belt, the Mid-German Crystalline High. **Littke *et al.***, on the basis of thermal maturity and fission track data, have modelled heat flow during the extensional and compressional stages and reconstructed the highly variable eroded thickness over the Variscan Belt. In addition, constraints on post-Variscan re-heating are discussed which coincide with tectono-magmatic events in Europe. The Mid-German Crystalline High, has sourced a well-preserved foreland basin, whose evolution is treated by **Ricken *et al.*** Backthrusting has carried the Crystalline High toward the southeast, over the low-grade Palaeozoic sequences of the Saxo-Thuringian Belt. This retro-wedge has been reconstructed by **Schäfer *et al.***

The Saxo-Thuringian Belt is probably the most complex part of the European Variscides, because it contains high-grade metamorphic rocks and fossiliferous sedimentary rocks in close juxtaposition. As was already been proposed by F. E. Suess in 1912, some of the high-grade units represent tectonic klippen. However, eclogites and HP-granulites also occur in antiformal positions, where they emerge from under the floor of the Saxo-Thuringian foreland basin. One of these antiforms, the Saxonian Granulite Dome, is the type locality of the granulite facies. Isotopic dating has revealed that the high-grade rocks were formed at the same time as the early Carboniferous flysch which now overlies them. This apparent paradox has been the target of important activities. A joint venture of the DEKORP seismic profiling programme and the Orogenic Processes Group produced new refraction and reflection seismic lines. It is demonstrated that high-grade rocks underlie the entire Saxo-Thuringian foreland (**Krawczyk *et al.***). **Werner & Lippolt** present a regional survey of Ar/Ar ages from Saxo-Thuringian high-pressure rocks and adjacent units. They deduce a complex sequence of subduction, exhumation and renewed tectonic stacking. **Franke & Stein** summarize the available constraints on the exhumation of the high-grade rocks, and derive a model, in which low-viscosity rocks formed in the Saxo-Thuringian orogenic root are displaced by hydraulic forces into the foreland crust. This concept is substantiated in a numerical model presented by **Henk**.

Temperatures derived for the Saxonian granulites are in excess of 1000 °C. At the same time, the gneisses of the Moldanubian were subject to >700 °C at less than 4.5 kbar. This is discussed in detail by **O'Brien** and introduces the problem of the heat source for high-temperature metamorphism and magmatism. **Henk *et al*.** and **Wittenberg *et al*.** assess the alternative modes of heat input into the crust—the role of mantle wedges, and the ascent of asthenospheric material by delamination or slab break-off. **Gerdes *et al*.** demonstrate the influence of the mantle on the formation of granitoids in the Bohemian Massif.

In an invited paper, **Eisele *et al*.** use geochemical and sedimentological evidence to reconstruct a magmatic arc in the relatively unknown Vosges segment of the Moldanubian Zone.

This research was financed, from 1992 to 1999, by Deutsche Forschungsgemeinschaft (DFG) in the form of a priority programme. The participants and co-ordinators are grateful for this very substantial support. Special thanks are due to the reviewers of our projects, and especially to Dietrich Maronde as the representative of DFG, for his unrelenting and competent advice. The project was co-ordinated by a multi-disciplinary team of earth scientists.

We also thank U. Bayer, C. Beaumont, H. Becker, J. Behrmann, G. Bergantz, D. Bernoulli, P. Blümel, M. Brown, T. McCann, Q. Crowley, H. Downes, A. Dudek, S. Egan, E. Eide, S. Ellis, R. Feist, M. Fernàndez, W. Fielitz, F. Finger, P. Floyd, H. Fritz, R. Gover, F. Henjes-Kunst, P. Jakeš, R. Kay, T. Korja, J.-M. Lardeaux, B. Leveridge, R. Oberhänsli, F. Patocka, T. Pharaoh, A. von Quadt, F. Neubauer, J.-L. Mansy, P. Matte, M. Meschede, J. von Raumer, T. Reischmann, M. Roberts, P. Sachs, D. Snyder, M. Sosson, H. Thybo, T. Torsvik, R. Van der Voo, S. Wdowinski, O. Werner, S. Willett, J. Zalasiewicz and seven anonymous reviewers for reviewing the manuscripts thoroughly.

The Palaeozoic time scale reviewed

W. S. McKERROW[1] AND C. R. VAN STAAL[2]
[1]*Department of Earth Sciences, Parks Road, Oxford OX1 3PR, UK*
[2]*Geological Survey of Canada, 601, Booth Street, Ottawa, Ont., K1A 0E8, Canada*

Abstract: A new revised time scale for the Palaeozoic era is presented on the basis of a compilation of existing U–Pb age data of volcanic rocks associated with zonal fossil-bearing sedimentary rocks.

Datable rocks are rarely found conveniently at period or stage boundaries. Age dates of rocks that impose constraints on the time limits of biostratigraphic units therefore allow more precise estimates of geological events and rates of evolution. Particularly, the revision of the time scale on the basis of one precise dating method is highly desirable, as it eliminates systematic errors that may exist as a result of comparing different age dating methods with different decay constants. The decay constant uncertainty varies between the various isotope systems and may be compounded by additional uncertainties introduced by errors in the standards used (e.g. $^{40}Ar/^{39}Ar$, Renne *et al.* 1998).

Since the revision by Tucker & McKerrow (1995), the correlation of Palaeozoic stratigraphy with ages from isotopes has become much more precise at several period and stage boundaries. This has been due largely to a number of new U–Pb ages on zircons from tuffs interbedded or associated with zonal fossils, and, to a lesser extent, to the development of more precise palaeontological correlations.

The base of the Cambrian succession, which can be recognized by either the *Phycodes pedum* Zone or the *Sabellidites cambriensis* Zone (compare Rowland *et al.* (1998, fig. 4) with Landing *et al.* (1998, fig. 1)) is considered to be at *c.* 543 Ma (Grotzinger *et al.* 1995) on the basis of some well-constrained dates from Namibia. The ages of the succeeding Early Cambrian stages can be extrapolated from the data given by Bowring & Erwin (1998) and Landing *et al.* (1998), which amend parts of previous scales considerably. The older scales used some SHRIMP dates; recent work appears to confirm earlier suggestions (Roddick & Bevier 1995; Tucker & McKerrow 1995) that these analyses generally give dates that are too young. The start of the Tommotian age 531 ± 1 Ma is similar to Tucker & McKerrow (1995), but the start of the Atdabanian age, when trilobites first became abundant, has to be *c.* 522 Ma or slightly younger, and corresponding adjustments must then be made to the ages of the later stages in the Early and Mid-Cambrian period (see Table 1). The end of Early Cambrian time is constrained by a zircon age of 511 ± 1 Ma of upper Branchian rocks in southern New Brunswick (Landing *et al.* 1998).

An age of 505.1 ± 1.3 Ma from a Middle Cambrian sample from Antarctica (Encarnación *et al.* 1999) suggests that the start of the Late Cambrian age lies at *c.* 500 Ma. This fits well with the more precise data for the start of the Ordovician period, where Davidek *et al.* (1998) reported an age of 491 ± 1 Ma for a sample near the end of the Late Cambrian age of Wales, indicating that the start of the Ordovician period lies close to 490 Ma.

The Tremadoc–Arenig boundary probably lies soon after the 483 ± 1 Ma age for the latest Tremadoc time (Landing *et al.* 1997) and we suggest an age close to 480 Ma, following the re-evaluation of the 471 ± 3 Ma SHRIMP age to *c.* 476 Ma for the *D. deflexus* Subzone, Llyfnant flags (Compston & Williams 1992) as proposed by Landing *et al.* (1998).

Similarly, the Arenig–Llanvirn boundary should lie close to 465 Ma instead of close to 470 Ma (Tucker & McKerrow 1995), on the basis of the chronostratigraphic correlations of Mitchell *et al.* (1997) and the zircon age 464 ± 2 Ma reported by Huff *et al.* (1997) for uppermost Arenig (*Undulograptus austrodentus*) rocks in the Precordilleran Terrane of Argentina. This reassessment is consistent with a zircon age of $471 + 3/-1$ Ma for tuffaceous rocks in northern New Brunswick (Sullivan & van Staal 1996) that contain mid- to late Arenig brachiopods (Neuman 1984) and conodonts (Nowlan 1981).

From: FRANKE, W., HAAK, V., ONCKEN, O. & TANNER, D. (eds).
Orogenic Processes: Quantification and Modelling in the Variscan Belt.
Geological Society, London, Special Publications, **179**, 5–8. 0305-8719/00/$15.00

Table 1.

Base of	Permian		290 Ma?
	Stephanian	305	
	Westphalian	315	
	Namurian	325	
	Viséan	334	
Base of	Carboniferous—Tournaisian		362
	Famennian	376.5	
	Frasnian	382.5	
	Givetian	387.5	
	Eifelian	394	
	Emsian	409.5	
	Pragian	413.5	
Base of	Devonian–Lochkovian		418
	Přídolí	420	
	Ludlow	424	
	Wenlock	429	
Base of	Silurian–Llandovery		443
	Ashgill	449	
	Caradoc	459	
	Llanvirn	465	
	Arenig	480	
Base of	Ordovician–Tremadoc		490
	Late Cambrian	500	
	Middle Cambrian	511	
	Toyonian	513	
	Botomian	518	
	Atdabanian	522	
	Tommotian	531	
Base of	Cambrian—Nemakit-Daldynian		543

Landing *et al.*'s suggestion that conodonts purported to represent the Arenig–Llanvirn boundary interval in north–central Newfoundland associated with volcanic rocks dated at 473–469 Ma (Dunning & Krogh 1991) are better interpreted to indicate a Late Arenig age is consistent with this interpretation. The revision of the stratigraphic definition of the base of the Caradoc Series so that it now includes all of the *Nemagraptus gracilis* Zone (Fortey *et al.* 1995), means that start of the Caradoc epoch should now be older than formerly. It should be younger than the 460.4 ± 2.2 Ma age of a Llandeilian (not Llandeilo because it is now a stage in the Lanvirn Series) ash at Llandrindod Wells, Wales, and younger than the 456.1 ± 1.8 Ma age of the lower Caradoc Llanwrtyd volcanic series (see Tucker & McKerrow (1995, items 15 and 17); we suggest an age of 459 Ma for the Llanvirn–Caradoc boundary. Alkali basalts and comendites dated at 459 ± 3 Ma (Sullivan & van Staal 1996) immediately underlying limestone containing conodonts of the *Prioniodus variabilis* Subzone of the *Amorphognathus tvaerensis* Zone (Nowlan 1981) in northern New Brunswick are consistent with this interpretation.

Kaljo *et al.* (1998, p. 308) suggested an age of 429 Ma (instead of 428 Ma) for the start of Wenlock time on the basis of an age of 430.1 ± 2.4 Ma for an uppermost Llandovery ash, but otherwise there has been no major revision of the Tucker & McKerrow (1995) scale for the Late Ordovician and Silurian periods.

Tucker *et al.* (1998) have produced some more precise dates for the Devonian period. A K-bentonite thought to be in the upper part of the *Icriodus woschmidti* Zone has an age of 417.6 ± 1.0 Ma, which suggests that the start of the Devonian period cannot be much younger than 418 Ma. The Late Ludlow age is dated at 420.2 ± 3.9 Ma (see Tucker & McKerrow 1995), which leaves very little time for the Přídolí epoch. According to Kriz (1998, p. 187), the Přídolí succession has six graptolite zones compared with the eight in the Ludlow sequence, and we therefore think it is probable (contrary to Tucker *et al.* (1998) that the start of the Přídolí age is at least as old as 420 Ma.

Tucker *et al.* (1998) suggested small increases to the age of the start of the Lochkovian and Pragian epochs, where the dates of the stages are fairly closely constrained. In the same paper, the start of Emsian five is extended downwards by 9 Ma to 409 Ma on the basis of a 408.3 ± 1.9 Ma from a K-bentonite in New York that is probably of early Emsian age. The Mid- and Late Devonian stages are not so well constrained. The start of the Carboniferous period has to lie between 364 and 356 Ma, and we follow Tucker *et al.* (1998) in placing it at 362 Ma.

In the Carboniferous period, the start of Viséan time is close to the 334 ± 1 Ma age from a felsic tuff from the *Gnathodus texanus* conodont Zone (Trapp *et al.* 1998). Ar/Ar ages of tonstein and sanidine-bearing tuffs indicate an age of *c.* 325 Ma for the start of the Namurian age (Hess & Lippolt 1986). The bases of the Westphalian and Stephanian ages are respectively dated as 305 Ma and 315 Ma from Ar/Ar ages on sanidine from ash falls in goniatite-bearing strata from the Ruhr (Burger *et al.* 1997).

We are grateful to B. Tucker and T. Skulski for comments.

References

BOWRING, S. A. & ERWIN, D. H. 1998. A new look at evolutionary rates in deep time: uniting paleontology and high-precision geochronology. *GSA Today*, **8**, 1–8.

BURGER, K., HESS, J. C. & LIPPOLT, H. J. 1997. Tephrochronologie mit Kaolin–Kohlentonsteinen: Mittel zur Korrelation paralischer und limnishcer Ablagerungen des Oberkarbons. *Geologisches Jahrbuch*, **A147**, 3–39.

COMPSTON, W. & WILLIAMS, J. L. 1992. Ion-probe ages for the British Ordovician and Silurian stratotypes. *In*: WEBBY, B. D. & LUARIE, J. R. (eds) *Global Perspectives on Ordovician Geology*. Balkema, Rotterdam.

DAVIDEK, K., LANDING, E., BOWRING, S. A., WESTROP, S. R., RUSHTON, A. W. A., FORTEY, R. A. & ADRAIN, J. M. 1998. New uppermost Cambrian U–Pb date from Avalonian Wales and age of the Cambridge–Ordovician boundary. *Geological Magazine*, **135**(3), 305–309.

DUNNING, G. R. & KROGH, T. E. 1991. Stratigraphic correlation of the Appalachian Ordovician using advanced U–Pb zircon geochronology techniques. *In*: BARNES, C. R. & WILLIAMS, S. H. (eds) *Advances in Ordovician Geology*. Geological Survey of Canada, Papers, **90-9**, 85–92.

ENCARNACIÓN, J., ROWELL, A. J. & GRUNOW, A. M. 1999. A U–Pb age for the Cambrian Taylor Formation, Antarctica: implications for the Cambrian time scale. *Journal of Geology*, **107**, 497–504.

FORTEY, R. A., HARPER, D. A. T., KINGHAM, J. K., OWEN, A. W. & RUSHTON, A. W. A. 1995. A revision of Ordovician series and stages from the historical type area. *Geological Magazine*, **132**(1), 15–30.

GROTZINGER, J. P., BOWRING, S. A., SAYLOR, B. Z. & KAUFMAN, A. J. 1995. Biostratigraphic and geochronologic constraints on early animal evolution. *Science*, **270**, 598–604.

HESS, J. C. & LIPPOLT, H. J. 1986. $^{40}Ar/^{39}Ar$ ages of tonstein and tuff sanidines: new calibration points of the upper Carboniferous time scale. *Chemical Geology (Isotope Geosciences Series)*, **59**, 143–154.

HUFF, W. D., DAVIS, D., BERGSTRÖM, S. M., KREKELER, M. P. S., KOLATA, D. R. & CINGOLANI, C. A. 1997. A biostratigraphically well-constrained K-bentonite U–Pb zircon age of the lowermost Darriwilian Stage (Middle Ordovician) from the Argentine Precordillera. *Episodes*, **20**(1), 29–33.

KALJO, D., KIIPLI, T. & MARTMA, T. 1998. Correlation of carbon isotope events and environmental cyclicity in the East Baltic Silurian. *In*: LANDING, E. & JOHNSON, M. (eds) *Silurian Cycles: Linkages of Dynamic Stratigraphy with Atmospheric, Oceanic, and Tectonic Changes*, 297–312.

KRIZ, J. 1998. Recurrent Silurian–Lowest Devonian cephalopod limestones of Gondwanan Europe and Perunica. *In*: LANDING, E. & JOHNSON, M. (eds) *Silurian Cycles: Linkages of Dynamic Stratigraphy with Atmospheric, Oceanic, and Tectonic Changes*, 183–198.

LANDING, E., BOWRING, S. A., DAVIDEK, K. L., WESTROP, S. R., GEYER, G. & HELDMAIER, W. 1998. Duration of the Early Cambrian: U–Pb ages of volcanic ashes from Avalon and Gondwanaland. *Canadian Journal of Earth Science*, **35**, 329–338.

——, ——, FORTEY, R. A. & DAVIDEK, K. L. 1997. U–Pb zircon date from Avalonian Cape Breton Island and geochronologic calibration of the Early Ordovician. *Canadian Journal of Earth Science*, **34**, 724–730.

MITCHELL, C. E., XU, C., BERGSTRÖM, S. M., YUAN-DONG, Z., ZHI-HAO, W., WEBBY, B. D. & FINNEY, S. C. 1997. Definition of a global boundary stratotype for the Darriwilian Stage of the Ordovician System. *Episodes*, **20**(3), 158–166.

NEUMAN, R. B. 1984. Geology and paleobiology of islands in the Ordovician Iapetus Ocean: review and implications. *Geological Society of America Bulletin*, **94**, 1188–1201.

NOWLAN, G. S. 1981. *Some Ordovician conodont faunules from the Miramichi Anticlinorium, New Brunswick*. Geological Survey of Canada Bulletin, **345**.

RENNE, P. R., KARNER, D. B. & LUDWIG, K. R. 1998. Absolute ages aren't exactly. *Science*, **282**, 1840–1841.

RODDICK, J. C. & BEVIER, M. L. 1994. U–Pb dating of granites with inherited zircon: conventional and ion-microprobe results from two Paleozoic plutons, Canadian Appalachians. *Chemical Geology*, **119**, 307–329.

ROWLAND, S. M., LUCHININA, V. A., KOROVNIKOV, I. G., SIPIN, D. P., TARLETSKOV, A. I. & FEDOSEEV, A. V. 1998. Biostratigraphy of the Vendian–Cambrian Sukharikha River section, northwestern Siberian

Platform. *Canadian Journal of Earth Science*, **35**, 339–352.

Sullivan, R. W. & van Staal, C. R. 1996. Preliminary chronostratigraphy of the Tetagouche and Fournier groups in northern New Brunswick. *In*: *Radiogenic Age and Isotopic Studies*. Report 9; Geological Survey of Canada, Current Research, **1995-F**, 43–56.

Trapp, E., Zellmer, H., Baumann, A., Mezger, K. & Wachendorf, H. 1998. Die Kieselgesteins-Fazies des Unterkarbons im Harz-Biostratigraphie, U–Pb-Einzelzirkon-Alter, Petrographie und Sedimentologie. *Terra Nostra*, **98**(3), V368.

Tucker, R. D. & McKerrow, W. S. 1995. Early Paleozoic chronology: a review in light of new U–Pb zircon ages from Newfoundland and Britain. *Canadian Journal of Earth Science*, **32**, 368–379.

——, Bradley, D. C., ver Straeten, C. A., Harris, A. G. Ebert, J. R. & McCutcheon, S. R. 1998. New U–Pb zircon ages and the duration and division of Devonian time. *Earth and Planetary Science Letters*, **158**, 175–186.

The Late Palaeozoic relations between Gondwana and Laurussia

W. S. McKERROW[1], C. MAC NIOCAILL[1], P. E. AHLBERG[2], G. CLAYTON[3], C. J. CLEAL[4] & R. M. C. EAGAR[5]

[1]*Department of Earth Sciences, University of Oxford, Parks Road, Oxford OX1 3PR, UK (e-mail: stuartm@earth.ox. ac.uk)*

[2]*Department of Palaeontology, Natural History Museum, Cromwell Road, London SW7 5BD, UK*

[3]*Department of Geology, Trinity College, Dublin, Ireland*

[4]*National Museums and Galleries of Wales, Cathays Park, Cardiff CF10 3NP, UK*

[5]*The Manchester Museum, University of Manchester, Manchester M13 9PL, UK*

Abstract: Reconstructions based on biogeography, palaeomagnetism and facies distributions indicate that, in later Palaeozoic time, there were no wide oceans separating the major continents. During the Silurian and Early Devonian time, many oceans became narrower so that only the less mobile animals and plants remained distinct. There were several continental collisions: the Tornquist Sea (between Baltica and Avalonia) closed in Late Ordovician time, the Iapetus Ocean (between Laurentia and the newly merged continents of Baltica and Avalonia) closed in Silurian time, and the Rheic Ocean (between Avalonia and Gondwana and the separate parts of the Armorican Terrane Assemblage) closed (at least partially) towards the end of Early Devonian time. Each of these closures was reflected by migrations of non-marine plants and animals as well as by contemporary deformation. New maps, based on palaeomagnetic and faunal data, indicate that Gondwana was close to Laurussia during the Devonian and Carboniferous periods, with fragments of Bohemia and other parts of the Armorican Terrane Assemblage interspersed between. It follows that, after Early Devonian time, the Variscan oceans of central Europe can never have been very wide. The tectonic evolution of Europe during Devonian and Carboniferous time was thus more comparable with the present-day Mediterranean Sea than with the Pacific Ocean.

Palaeozoic palaeocontinents

During the Palaeozoic era, there were several events when major continents rifted apart and others when they collided. No continent remained as a distinct entity throughout this era; so each palaeocontinent can only be defined after consideration of its history. The major Palaeozoic continents that are represented in Europe are Laurentia, Baltica, Avalonia and Gondwana, with smaller continental fragments of the Armorican Terrane Assemblage, including Bohemia and Armorica.

Laurentia, the Palaeozoic continent based on North America, did not include the Mesozoic and Cenozoic additions to the Western Cordillera, nor is it defined (Scotese & McKerrow 1990) as including the terranes accreted to the east coast in the Devonian and Carboniferous periods. Much of the northern Appalachians (eastern Newfoundland, New Brunswick, Nova Scotia and coastal New England) were parts of Avalonia until Silurian time. The Piedmont and Florida were parts of Gondwana until Carboniferous time. But Laurentia did include (as indicated by their Cambrian and Early Ordovician faunas) north-west Ireland, Scotland, Greenland, much of Svalbard, the North Slope of Alaska and the Chukotsk Peninsula of eastern Siberia (Scotese & McKerrow 1990). Palaeomagnetic data (Mac Niocaill & Smethurst 1994) and the prevalence of warm-water carbonate facies show that Laurentia remained close to the Equator throughout Palaeozoic time.

Baltica consists of Scandinavia and the East European Platform, as far east as the Urals. Its southern margin extends from the North Sea to the Black Sea along the Trans-European Suture, a line much complicated by later deformation. In

From: FRANKE, W., HAAK, V., ONCKEN, O. & TANNER, D. (eds).
Orogenic Processes: Quantification and Modelling in the Variscan Belt.
Geological Society, London, Special Publications, **179**, 9–20. 0305-8719/00/$15.00

Cambrian time, Baltica was at high southern latitudes (Torsvik *et al.* 1996); its trilobite faunas were similar to those of Laurentia but less diverse. The faunas were also similar to those of Avalonia, which were even less diverse, thus indicating that, before Ordovician time, Baltica was situated close to Avalonia in a position off the Florida or Venezuelan margins of Gondwana (McKerrow *et al.* 1992). Subsequently, Baltica moved slowly northwards; eventually warm-water carbonates appear in Late Ordovician time, and Baltica straddles the equator in Silurian and Devonian time.

Avalonia, consists of the eastern parts of the Northern Appalachians (from Connecticut to the Avalon Peninsula in Newfoundland), southern Ireland, Wales, England, Belgium and the Rheno-Hercynian belt of northern Germany (Cocks *et al.* 1997). In eastern Germany, Avalonian crust extends across the Elbe Line (Finger & Steyrer 1995; contra Cocks *et al.* 1997); an isolated fragment may also be present east of Bohemia in the Moravo-Silesian terrane.

During Cambrian time, Avalonia was attached to Gondwana not far from the South Pole. At this time it had clastic sedimentary facies, which suggest it was attached to the northern margin of South America rather than to Morocco where different Cambrian faunas are present in carbonates (McKerrow *et al.* 1992; Torsvik *et al.* 1996). Avalonia shows evidence of latest Precambrian arc rocks assigned to Cadomian time; its basement is thus not distinguishable from those parts of northern Gondwana now seen in the Armorican Terrane Assemblage of central and southern Europe; similar arc rocks also occur in Morocco (Piqué 1981). Avalonia became detached from Gondwana in Early Ordovician time and moved rapidly northwards to collide with Baltica in latest Ordovician time and with Laurentia in Silurian time (Cocks & Fortey 1982; McKerrow *et al.* 1991; Trench *et al.* 1992).

Gondwana was a large continent that originated in latest Precambrian time by amalgamation of parts of Africa, South America, India, Antarctica and Australia (Torsvik *et al.* 1996). In Cambrian time, it included all of Europe south of Avalonia and Baltica. In Late Ordovician time the South Pole was in North Africa and, although the terranes of southern Europe had separated from Gondwana, they were still far enough south to carry periglacial deposits (Havlíček 1989). Palaeomagnetic data show that, while Gondwana moved northwards during Silurian time, several south European terranes (including parts of Bohemia and Armorica) moved more rapidly (Tait *et al.* 1995; Torsvik *et al.* 1996).

The narrowing oceans

As two continents approach, it becomes progressively more difficult to use palaeomagnetic or sedimentological data to distinguish between them, but there are some organisms that remain useful until there is a continuous shallow-water or terrestrial connection (McKerrow & Cocks 1976; McKerrow 1978).

Marine organisms can be classified, according to their mode of life, into the following groups: (a) free-swimming and floating (or pelagic) nekton and plankton, such as the Early Palaeozoic graptolites and the Late Palaeozoic ammonoids and some fishes (including some Palaeozoic sharks). (b) Bottom-dwellers (benthos) that had a pelagic larval stage, and that include most marine invertebrates (e.g. trilobites and brachiopods). Some fishes (including placoderms and primitive lungfish) appear to have been benthic in coastal waters (Janvier 1996), but as it is not known whether they had pelagic larvae, they may or may not fall into this category. (c) A few marine organisms, such as some ostracodes, which hatch out from eggs on the sea floor and have no pelagic growth stage. Benthic ostracodes cannot cross deep water. (d) Aquatic organisms that are restricted to non-marine (rivers and lakes) or brackish environments. Many Devonian fish (for example, holoptychiids, phyllolepids and the more advanced lungfishes) are known only from Old Red Sandstone sediments, which are either non-marine or which have a limited marine influence. They appear to have been unable to cross narrow stretches of open sea. (e) Terrestrial organisms (plants with large seeds; post-Devonian tetrapods) that could only spasmodically cross open water, and thus are normally restricted in range until after (sometimes long after) continental collision.

In Cambrian and Early Ordovician time, when many continents were isolated by wide oceans, only the plankton (and a very few other forms) could cross all oceans (Cocks & Fortey 1990; Scotese & McKerrow 1990; McKerrow & Cocks 1995). Plankton normally show a geographical distribution related to climate, whereas most other organisms serve to distinguish individual continents (or groups of adjacent continents). In Late Ordovician time, many brachiopods and trilobites (benthos with pelagic larvae) start to have global distributions as the major continents become closer; until in Silurian time the only common fossils to distinguish many continents are ostracodes and fish. By the end of Early Devonian time, many ostracodes became widely distributed (Berdan 1990). The Devonian fish faunas show parallel changes: the fish of Laurussia (the continent resulting from the fusion of Laurentia, Avalonia and Baltica) started to merge with those of Gondwana and became increasingly similar through Mid- and Late Devonian time (Young 1990). There is no indication of any permanent barrier to migration across the Variscan orogen after Early Devonian time.

European suture zones

In central Europe sutures of five Palaeozoic oceans have been recognized (Fig. 1).

The Iapetus suture

The Iapetus suture in the British Isles extends from the Shannon estuary in western Ireland to the mouth of the River Tweed in northeast England. The Iapetus Ocean opened between Laurentia and Baltica at around 600 Ma. Closure between Laurentia and Laurussia (the fused Baltica–Avalonia) occurred in Mid-Silurian time (McKerrow *et al.* 1991; van Staal *et al.* 1998), when turbidites derived from Laurentia were deposited in a rapidly depressed foreland basin covering central Ireland in Wenlock time (Hutton & Murphy 1987) and the English Lake District in Ludlow time (Leggett *et al.* 1983). This foreland basin developed along the northwest margin of Avalonia as the continent was subducted below the active margin of Laurentia. It was persistent and continuous enough to prevent the migration of ostracodes across the Iapetus suture until Devonian time (Berdan 1990).

The Tornquist suture

The Tornquist suture (equivalent to the Trans-European Suture Zone of some workers) extends eastwards from the mouth of the Tweed, to the north of Hamburg, and passes east of the Görlitz area (in easternmost Germany) and Silesia (in southern Poland) not far from the course of the River Odra (Oder). It lies distinctly south of the Tornquist Line, after which it was named (Cocks & Fortey 1982).

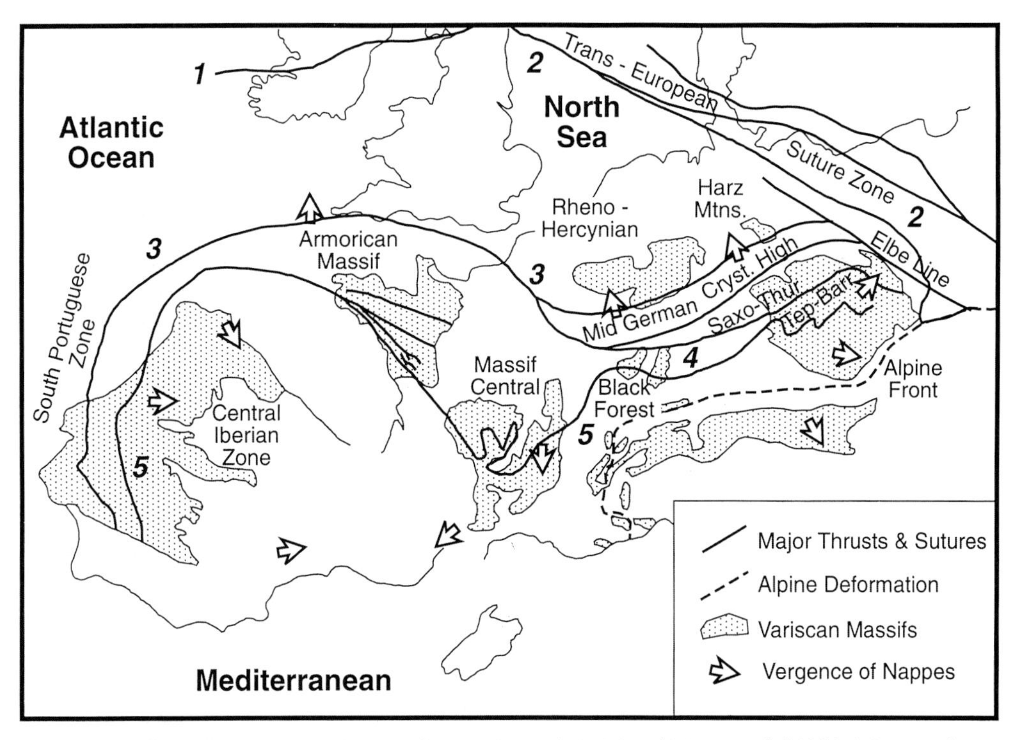

Fig. 1. Palaeozoic sutures in Europe (in part after Franke *et al.* (1995) and Matte *et al.* (1990)). 1, Iapetus Ocean; separating Avalonia from Laurentia. 2, Tornquist Sea; separating Avalonia from Baltica. 3, Rheic Ocean; separating Avalonia from the Armorican Terrane Assemblage. 4, Ophiolites south of the Saxo-Thuringian terrane. 5, Ligerian–Massif Centrale–Moldanubian ophiolites.

The Tornquist Sea (between Avalonia and Baltica) may have originated at the same time as the Iapetus Ocean (*c.* 600 Ma); Baltica was certainly separated from Gondwana (which included Avalonia until Early Ordovician time) throughout the Cambrian period (McKerrow *et al.* 1992). The Tornquist Sea probably closed in Late Ordovician time (Van Staal *et al.* 1998), at the time of the Shelveian Orogeny of western England (Toghill 1992); certainly the Early Silurian benthic ostracode distributions (Berdan 1990) show that there was no continuous deep water between Baltica and Avalonia.

The arc related to closure of the Tornquist Sea is seen in the Ordovician igneous rocks of the English Lake District (Kokelaar *et al.* 1984) and eastern England (Pharaoh *et al.* 1993). The Ordovician and Silurian volcanic series in eastern England (Pharaoh *et al.* 1993), the Ardennes and the Northern Phyllite Belt (Stillman 1988; Franke 1989) erupted between the Tornquist Sea and the Rheic Ocean; they may have been related to the subduction of either ocean (McKerrow et al., 1991; Franke et al., 1995).

The Rheic suture

The Rheic suture (Fig. 1) extends eastwards from southwest England to the south of the Ardennes and to the south of the Rheno-Hercynian terrane and the Harz Mountains of northern Germany (Franke *et al.* 1995; Cocks *et al.* 1997).

There is still some discussion about the position of the Rheic suture east of Dresden and in the Czech Republic and southern Poland (Belka *et al.* this volume; Finger *et al.* this volume). If the Mid-German Crystalline High and the Saxo-Thuringian terrane continue to the northeast of the Elbe line (Linnemann *et al.* 1998), then the Rheic suture (lying between Avalonia and Baltica to the north and Barrandia and other Armorican terranes to the south) may extend northeast to the vicinity of the Odra Fault, contrary to the conclusions of Cocks *et al.* (1997). The Rheic suture clearly may then merge with the Tornquist suture in Silesia. The eastward continuation of the Rheic suture is further complicated by the possible presence of an isolated remnant of Avalonia in Moravo-Silesia (Finger & Steyrer 1995). As Avalonia, Bohemia and Saxo-Thuringia were all originally parts of Gondwana, latest Precambrian (600–550 Ma) Cadomian arc rocks occur on many different terranes, and it is not possible to recognize the Rheic suture unless Palaeozoic sediments crop out.

The Rheic Ocean opened in Early Ordovician (Arenig) time, when Avalonia split off from the South American parts of Gondwana (Cocks & Fortey 1982; Scotese & McKerrow 1990), and closed (at least locally) in Emsian time (Van Staal *et al.* 1998). The closure of the Tornquist Sea (in Late Ordovician time) and of the Iapetus Ocean (in Wenlock time) left the Rheic Ocean, separating Bohemia and Armorica from Laurussia, as the principal surviving major long-lived ocean in Late Palaeozoic time (Fig. 2).

The Giessen, Harz and Lizard ophiolites are overlain in some sections by thin, condensed Emsian–Eifelian cherts and shales. This Rheno-Hercynian basin appears to have opened in late Emsian time, very close to the line of the Rheic suture (Franke, this volume). These sequences have been thrust over the southern margins of Avalonia in the Lizard (of SW England), Giessen (north of Frankfurt) and the Harz Mountains (Franke 1989, p. 72, and this volume); they appear to have originated from a line very close to the Rheic suture. If the Rheic Ocean closed in Emsian time, these ophiolites appear to have developed in a recently opened ocean nearby.

The suture south of the Saxo-Thuringian terrane

This suture has been inferred from allochthonous Lower Palaeozoic mid-ocean ridge basalt (MORB) type rocks, which are now present between the Saxo-Thuringian terrane and the Teplá–Barrandian of Bohemia (Franke *et al.* 1995, pp. 581–582; Linnemann *et al.* this volume), and from palaeomagnetic data showing that Saxo-Thuringia and Bohemia rotated independently during Silurian and Devonian time (Tait *et al.* 1995, this volume). The Saxo-Thuringian basin has thick clastic deposits and bimodal volcanic rocks which indicate a Cambro-Ordovician rift basin (Franke *et al.* 1995, p. 581); it is thus possible that the Saxo-Thuringian terrane separated from Gondwana around this time. Allochthonous latest Cambrian (500 Ma) eclogites derived from MORB-type rocks south of the Saxo-Thuringian terrane may represent fragments of an ocean, which, although long-lived, does not appear to have ever been a barrier of biogeographical significance.

As the Saxo-Thuringian terrane dies out towards the Rhine Valley, this suture merges west with the Rheic suture (Fig. 1).

The Ligerian (south Brittany), Massif Central, Moldanubian (south Bohemia) suture

This suture marks the separation of several distinct microcontinents (including Bohemia and Brittany) from Gondwana during Ordovician

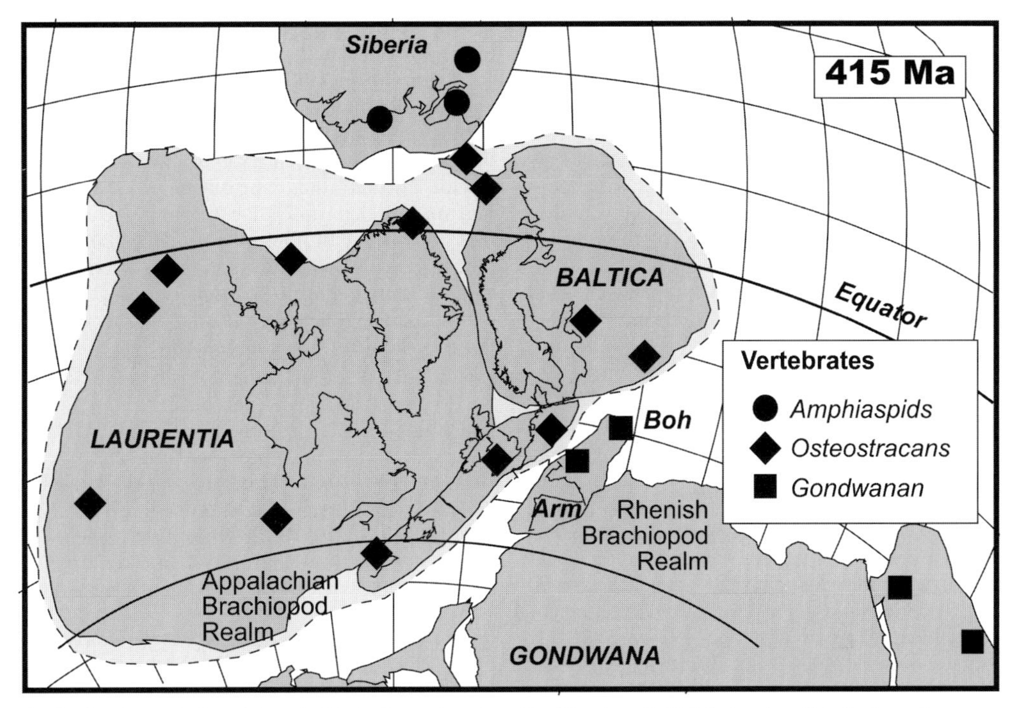

Fig. 2. Continental dispositions in Early Devonian time (Lochkovian, 415 Ma) showing distribution of non-marine (and ?coastal) fish (after Young 1990). It should be noted that different families occurred in Siberia, Laurussia and in Gondwana, and that Armorica and Bohemia appear to be attached to Gondwana. The boundary of the Appalachian and Rhenish (marine) brachiopod realms during Emsian time (after Boucot *et al.* 1969) is also shown; this now appears to be related to climate, as it lies parallel to latitude. Its position helps to align Gondwana with the northern continents.

time. It should perhaps be distinguished as several distinct sutures, but until we know just how many different terranes are present in central and southern Europe, it is convenient to consider them together (Matte *et al.* 1990; Franke *et al.* 1995).

The northern edge of Gondwana extended from South America (with Avalonia attached before Ordovician time), through Florida, West Africa and southern Europe to Turkey, Afghanistan to South China (McKerrow *et al.* 1992). This margin was active until the completion of Cadomian arc-related igneous activity around 550 Ma, close to the start of the Cambrian period (Tucker & McKerrow 1995; Landing *et al.* 1998). Extension on this margin is recorded by Cambrian subsidence in Wales, central Armorica, Saxo-Thuringia and Bohemia, but the earliest evidence of separation from Gondwana is seen in Avalonia, which probably rifted in Early Ordovician time (Cocks & Fortey 1982; Van Staal *et al.* 1998); subsequently there is palaeomagnetic evidence that Bohemia had separated from Gondwana by Late Ordovician time (Tait *et al.* 1995, this volume).

Biogeography

Ordovician and Silurian time

During Early Ordovician time, wide oceans separated Laurentia, Baltica and Gondwana so that the distribution of most benthic faunas reflected geographical isolation (Cocks & Fortey 1990); it was only the pelagic forms (such as graptolites and some conodont animals) that were able to migrate freely around the world (Berry & Wilde 1990). In the Early Palaeozoic era, we estimate that the majority of animals with pelagic larvae (such as trilobites and brachiopods) could traverse oceans up to 1000 km wide (McKerrow & Cocks 1986). In Late Ordovician time, trilobites and brachiopods have widespread geographical distributions related to latitude (Cocks & Fortey 1990). We conclude that, by this time, there were no oceans wider than 1000 km that separated the major continents without intermediate island stop-overs, such as, for example, the various components of the Armorican Terrane Assemblage situated between Gondwana and Laurussia.

During Silurian time, benthic animals with pelagic larval stages (such as like brachiopods and trilobites), could migrate freely between the major continents such that only climatic factors, rather than isolation, played a major role in their biogeography (Cocks & Fortey 1990). Less diverse high-latitude assemblages can be recognized in some benthos. Geographical isolation was present in only two groups of less mobile animals, which remained distinct: benthic ostracodes, which had (and have) no pelagic larval stage, and certain fish taxa, which were mainly restricted to non-marine and coastal environments (Berdan 1990; Young, 1990). These less mobile forms permit recognition of the Rheic Ocean through much of Early Devonian time.

Early and Mid-Devonian time

The biogeographical evidence for a narrow Rheic Ocean during Early Devonian time (Fig. 2) is reinforced by palaeoclimatic and lithological data (Scotese & Barrett 1990) and by palaeomagnetic data (Van der Voo 1993). Available palaeomagnetic poles from Gondwana for Late Silurian and Early Devonian time fall in South America, and reconstruction of Gondwana places the northern margins (north Africa and Arabia) in the tropics. Hence the Rheic Ocean cannot have been more than a few hundred kilometres wide.

The situation is made more complicated by the fact that several oceanic sutures are present between Gondwana and Baltica (Fig. 1). In Early Devonian time (Lochkovian and Pragian time) the ostracodes (which were marine but had no pelagic larval stages) crossed from Laurussia to Bohemia and Armorica, but did not reach Africa (Berdan 1990). This palaeontological evidence supports palaeomagnetic data (Tait *et al.* 1994, 1995) indicating that these microcontinents had rifted from Gondwana before Silurian time and were close to Laurussia at the start of Devonian time.

Our new reconstruction, which incorporates palaeomagnetic data with climatic, lithological and biogeographical data, now permits orientation of the the major continents so that new explanations can now be postulated for certain biogeographical observations. In particular, the Emsian brachiopod distributions first described 30 years ago (Boucot *et al.* 1969) are now seen to be related to latitude (Fig. 2). The Rhenish province of North Africa and Europe (with a few sporadic occurrences in Nova Scotia, Quebec and Nevada) developed in low latitudes, whereas the Appalachian province of the USA lies more than 40° south of the Equator. The boundaries between the two provinces on opposing margins of Gondwana with Laurussia allow the two continents to be matched with some confidence, and suggest that Morocco was opposed to Nova Scotia as the ocean closed.

Before Emsian or Givetian time, the Rheic Ocean affected the distribution of only non-marine (and coastal?) fish (Young 1990); after this time, the main factors governing biogeographical distributions were increasingly related to climate and to the disposition of suitable local environments (lakes, rivers, deltas, etc.) rather than to isolation by wide oceans.

Late Devonian time

In Late Devonian time, many non-marine and marginal marine fishes developed world-wide distributions (Fig. 3). Among the placoderm fishes, for example, the antiarchs *Bothriolepis* (mostly restricted to the Old Red Sandstone, but with some marine occurrences) and *Remigolepis* (entirely restricted to the Old Red Sandstone) both originated in Gondwana, but spread across Laurussia (Greenland, Scotland and Russia) during Late Devonian time (Young 1984). Similarly, the phyllolepid arthrodires first appeared in Givetian time in Australia and occur throughout much of Laurussia in Famennian time (Long 1984).

Some groups of sarcopterygian fish also became widespread in Late Devonian time. The tristichopterids appeared first in Givetian time in Laurussia, and soon spread to Gondwana, but generic differences between the two continents indicate that they did not migrate freely (Ahlberg & Johanson 1997); they appear to have been largely non-marine, although a marine occurrence is recorded in Morocco (Lelièvre *et al.* 1993). The Late Devonian holoptychiid *Holoptychius* was widespread in Laurussia throughout Frasnian and Famennian time, but crossed to Gondwana (Australia) during Famennian time (Young 1993). The earliest rhizodont sarcopterygians occurred in Gondwana in Late Devonian time (Johanson & Ahlberg 1998) and reached North America in Famennian time (Andrews & Westoll 1970); they spread to Europe in the Early Carboniferous period. The proximity of Laurentia and Gondwana during Late Devonian time is further illustrated by the lungfish *Soederberghia*, which had skeletal adaptations for air-breathing (cranial ribs, stalked parasphenoid) and which is represented by a single species in the Famennian succession of Greenland, North America and Australia (Ahlberg *et al.* 2000).

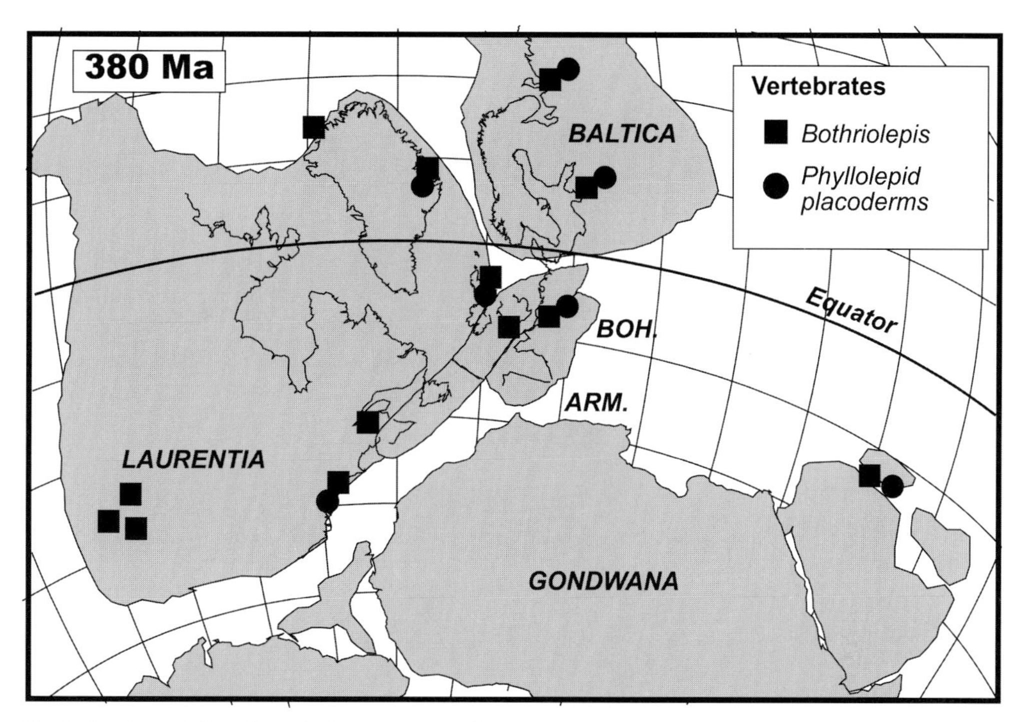

Fig. 3. Continental dispositions in Late Devonian time (Frasnian, 380 Ma) showing distribution of some non-marine (?coastal) fish (after Young 1990). By this time many fish families freely cross the suture of the Rheic Ocean, which was thus closed in places.

Tetrapods, which were initially more or less aquatic and tied to Old Red Sandstone environments, appear to have originated in late Frasnian time in Laurussia, where closely related (Givetian–early Frasnian) fish are also present (Ahlberg & Clack 1998). These animals crossed to Gondwana in Famennian time, when they reached Australia (Campbell & Bell 1977; Young 1993; Ahlberg & Clack 1998).

Devonian vegetation consisted mainly of spore-producing plants and shows little global provincialism (Meyen 1987; Edwards & Berry, in Cleal 1991). For example, the Devonian floras of northern Venezuela (Berry *et al.* 1993) are broadly similar to those of New York State (Banks *et al.* 1985); and the tree *Calloxylon*, previously known from the Late Devonian rocks of North America, also occurs in Morocco (Meyer-Berthaud *et al.* 1977) These examples support the evidence for no great separation across the Rheic suture.

Early Carboniferous time

In Early Carboniferous time, many plant distributions were dominated by climate (Fig. 4), but some floras in Venezuela (Pfefferkorn 1977) and Morocco and Algeria (Lejal-Nicol, in Wagner *et al.* 1985, p. 387) include lycophytes (e.g. *Lepidodendron*, *Lepidodendropsis*) and some seed plants (*Triphyllopteris*, *Sphenopteridium*) characteristic of the European Palaeoarea floras (*sensu* Cleal & Thomas, in Cleal 1991). Such floras are well known throughout southern Laurentia and Avalonia; they include the classic floras of the Price, Pocono and Mauch Chunk Formations in the central Appalachians (Read 1955), the Horton Group in Nova Scotia (Bell 1960), Teilia in North Wales and the oil shales of southern Scotland (Kidston 1923–1925). The other European records of these floras (in northern and eastern Germany and Silesia) were reviewed by Meyen in Vakhrameev *et al.* (1978).

While Latest Devonian miospores had cosmopolitan distributions, those in Early Carboniferous (Viséan) time show strong latitudinal correlation (Sullivan 1965, 1967; Clayton 1985, 1996). The *Monilospora* Microflora occurs in northwest Canada, Svalbard and the northern parts of Baltica. The *Grandispora* Microflora is well documented from the South Portuguese Zone, the British Isles, northern Germany (Rügen), northern Poland (Pomerania) and

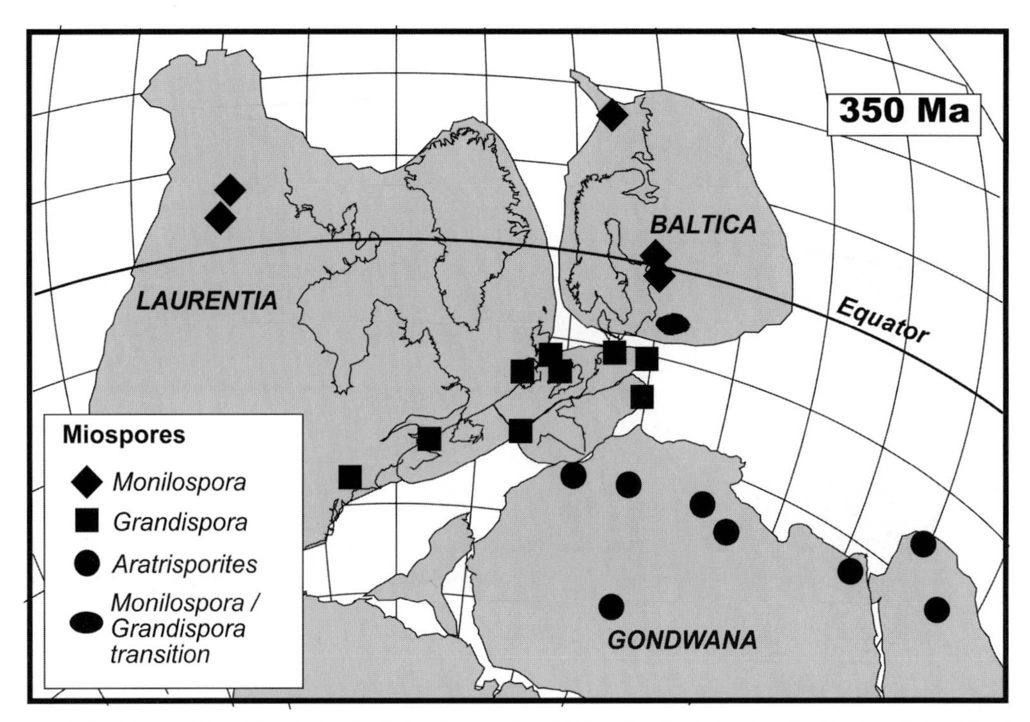

Fig. 4. Continental dispositions in Early Carboniferous time (Viséan, 350 Ma) showing distribution of three floral regions (after Sullivan 1965, 1967; Clayton 1985, 1996). They are associated with climatic belts. It should be noted that Avalonia and the Armorican Terrane Assemblage share the *Grandispora* Microflora.

Romania (Moesian Platform). A third microflora (*Aratrisporites saharaensis*) is known from North Africa (Morocco to Egypt), Niger, Syria and Saudi Arabia. Whereas the *Monilospora* and *Grandispora* Microfloras show some transitional associations, no such gradations have yet been described between these and the *Aratrisporites saharaensis* Microflora, except possibly in South America. During Early Carboniferous time, the northern margin of Africa was apparently a barrier to plant migration; perhaps some plants were more sensitive to minor changes in climate than others.

Carboniferous tetrapods are considerably more advanced than those in the Devonian period. They seem to be the product of a basal Carboniferous monophyletic radiation, which is widely represented in Laurussia, but only sporadically in Gondwana (Thulborn *et al.* 1996).

Late Carboniferous time

During Late Carboniferous time, paralic coalfields extended across Laurussia from North America across Britain and the Rheno-Hercynian region of Germany, and thence to Silesia and the Donetz area. This extensive region suffered several marine incursions, and clearly most of the Westphalian deposition (largely non-marine) occurred close to sea level (Paproth 1991) on the northern (passive margin) side of the Rheic suture; areas formerly composing Avalonia and Baltica (compare Fig. 1 with Trueman (1946, fig. 7)). By contrast, the Stephanian intermontane limnic basins developed mainly to the south of the suture on the various components of the Armorican Terrane Assemblage.

Coal-forming environments were present south of the Rheic suture in the Saar–Lorraine basin during late Westphalian time (possibly earlier) and continued until early Stephanian time (Weingart 1976); then after a short hiatus, they returned in mid-Stephanian time (Germer *et al.* 1968; Cleal 1984). In France, intermontane basins appeared in early Stephanian time (Bouroz *et al.* 1990), whereas in Spain, some coalfields originated in early Westphalian time, though they did not become widespread until late Westphalian time (Wagner & Alvarez-Vázquez 1991). Some of these internal basins yield non-marine bivalves that include species known in the Appalachians and northern Europe (Eagar 1983, 1984, 1985, 1994). It is thus evident that, by latest

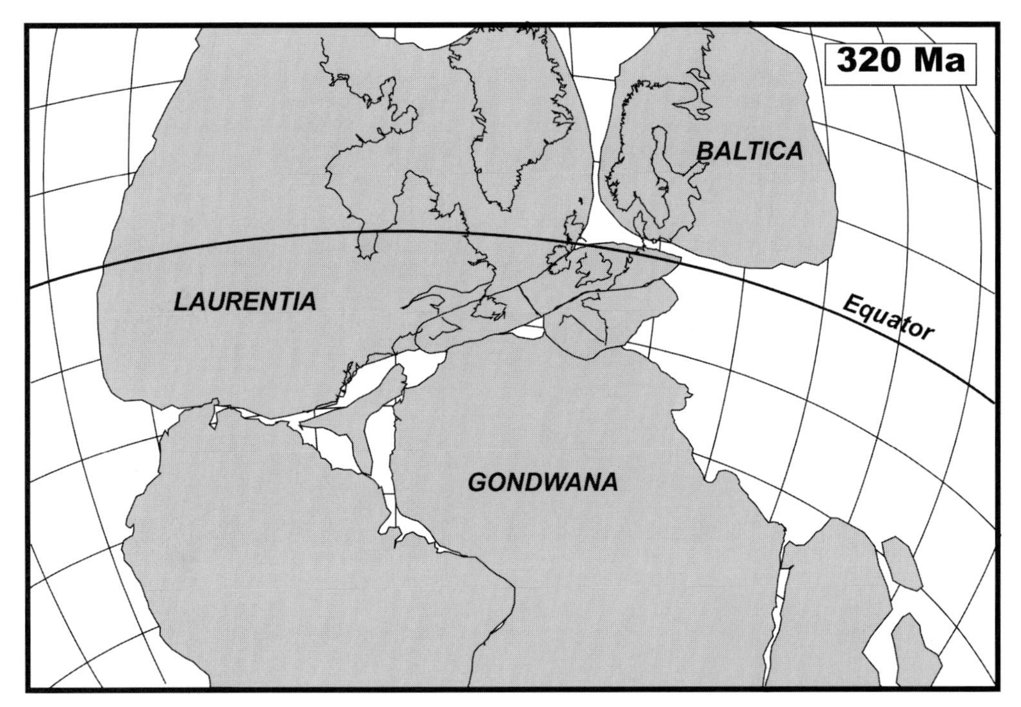

Fig. 5. Continental dispositions in Late Carboniferous time (Westphalian, 320 Ma). Apart from climatic influences, all the non-marine floras and faunas crossed freely from Gondwana to the northern continents.

Carboniferous time, non-marine mollusc faunas had crossed all the Variscan sutures in central Europe.

The Variscan orogen south of the Rheic suture (Fig. 1) was sufficiently uplifted before Westphalian D times, so that no Namurian or early Westphalian plant remains are known. However, well-documented late Westphalian floras occur in the Djerada Basin of Morocco (Jongmans 1952) and the Bechar Basin of Algeria (Jongmans & Deleau 1951; Migier 1982), which are very similar to the floras of the paralic basins in Britain and the Ruhr; they include *Paripteris pseudogiganta*, which Laveine *et al.* (1993) used to argue for strong links between North Africa and Europe during later Westphalian time. Further west, in northern Venezuela, Odreman & Wagner (1979) reported a fossil flora that is very similar to Westphalian D floras of Laurussia; these include material from the Upper Allegheny Formation of the Appalachians (Wagner & Lyons 1997), the Sydney coalfield of Nova Scotia (Zodrow 1986) and the Pennant Formation of Wales (Cleal 1997).

Late Carboniferous palynological data are not as definite as those of the Early Carboniferous period, but there is close similarity between Westphalian miospore assemblages from Europe and Africa, suggesting absence of the barrier to migration postulated in Viséan time.

The Westphalian plant distributions indicate that northern Gondwana was close to Laurasia (Fig. 5). Some of the best indicators of this proximity are the medullosalean pteridosperms, whose seeds could reach lengths of 100 mm (Gastaldo & Marten 1978) and can be transported by wind over only a few metres (Laveine *et al.* 1999). Postulating a wide ocean between these continents would require seed dispersal mechanisms that are unknown in Palaeozoic gymnosperms.

Conclusions

It can be demonstrated from both palaeontological and palaeomagnetic evidence that, after the Rheic Ocean closed in Emsian time, there were no wide oceans associated with the sutures in the European Variscides. In particular, the distributions of plants and freshwater faunas appear to be controlled by climate rather than by isolation. We conclude that the Variscides of Europe were more analogous to the present-day Mediterranean Sea than to the modern western Pacific.

We thank R. Feist and another referee for comments.

References

AHLBERG, P. E. & CLACK, J. A. 1998. Lower jaws, lower tetrapods—a review based on the Devonian genus *Acanthostega. Transactions of the Royal Society of Edinburgh: Earth Sciences*, **89**, 11–46.

—— & JOHANSON, Z. 1997. Second tristichopterid (Sarcopterygii, Osteolepiformes) from the Upper Devonian of Canowindra, New South Wales, Australia, and phylogeny of the Tristichopteridae. *Journal of Vertebrate Paleontology*, **17** (4), 653–673.

——, —— & DAESCHLER, E. B. 2000. The Late Devonian lungfish *Soederberghia* (Sarcopterygii, Dipnoi) from Australia and North America, and its biogeographical implications. *Journal of Vertebrate Paleontology*, in press.

ANDREWS, S. M. & WESTOLL, T. S. 1970. The post-cranial skeleton of rhipidistian fishes excluding Eusthenopteron. *Transactions of the Royal Society of Edinburgh*, **68**, 391–489.

BANKS, H. P., GRIERSON, J. D. & BONAMO, P. M. 1985. The flora of the Catskill clastic wedge. *In*: WOODROW, D. L. & SEVON, W. D. (eds) *The Catskill Delta*, Geological Society of America, Special Paper, **201**, 125–141.

NELKA, Z., AHRENDT, H., FRANKE, W. & WEMMER, K. 2000. The Baltica–Gondwana suture in central Europe: evidence from K/Ar ages of detrital muscovites and biogeographical data. *This volume*.

BELL, W. A. 1960. *Mississippian Horton Group of type Windsor–Horton District, Nova Scotia*. Geological Survey of Canada, Memoir, **215**.

BERDAN, J. M. 1990. The Silurian and Early Devonian biogeography of ostracodes in North America. *In*: MCKERROW, W. S. & SCOTESE, C. F. (eds) *Palaeozoic Palaeogeography and Biogeography*. Geological Society, London, Memoir, **12**, 223–231.

BERRY, C. M., CASAS, J. E. & MOODY, J. M. 1993. Diverse Devonian plant assemblages from Venzuela. *Documents du laboratoire de Geologie de Lyon*, **125**, 29–42.

BERRY, W. B. N. & WILDE, P. 1990. Graptolite biogeography: implications for palaeogeography and palaeoceanopgraphy. *In*: MCKERROW, W. S. & SCOTESE, C. R. (eds) *Palaeozoic Palaeogeography and Biogeography*. Geological Society, London, Memoir, **12**, 129–137.

BOUCOT, A. J., JOHNSON, J. G. & TALENT, J. A. 1969. *Early Devonian Zoogeography*. Geological Society of America, Special Papers, **119**.

BOUROZ, A., GRAS, H. & WAGNER, R. H. 1970. A propos de la limite Westphalian–Stéphanian et du Stéphanian inferieur. *Les Congrès et Colloques de l'Université de Liége*, **55**, 205–225.

CAMPBELL, K. S. W. & BELL, M. W. 1977. A primitive amphibian from the Late Devonian of New South Wales. *Alcheringa*, **1**, 369–381.

CLAYTON, G. 1985. Dinantian miospores and intercontinental correlation. *Comptes Rendus Dixième Congrès International de Stratigraphie et de Géologie du Carbonifère, Madrid 1983*, **4**, 9–23.

—— 1996. Mississipian miospores. *In*: JANSONIUS, J. & MCGREGOR, D. C. (eds) *Palynology, Principles and Applications*. American Association of Stratigraphic Palynologists Foundation, **2**, 589–596.

CLEAL, C. J. 1984. The Westphalian D floral biostratigraphy of Saarland (Fed. Rep. Germany) and a comparison with that of South Wales. *Geological Journal*, **19**, 327–351.

—— (ed.) 1991. *Plant Fossils in Geological Investigation: the Palaeozoic*. Ellis Horwood, Chichester.

—— 1997. The palaeobotany of the upper Westphalian and Stephanian of southern Britain and its geological significance. *Review of Palaeobotany and Palynology*, **65**, 882–890.

COCKS, L. R. M. & FORTEY, R. A. 1982. Faunal evidence for oceanic separations in the Palaeozoic of Britain. *Journal of the Geological Society, London*, **139**, 465–478.

—— & —— 1990. Biogeography of Ordovician and Silurian faunas. *In*: MCKERROW, W. S. & SCOTESE, C. F. (eds) *Palaeozoic Palaeogeography and Biogeography*. Geological Society, London, Memoir, **12**, 97–104.

——, MCKERROW, W. S. & VAN STAAL, C. R. 1997. The margins of Avalonia. *Geological Magazine*, **134**, 627–636.

EAGAR, R. M. C. 1983. The non-marine bivalve fauna of the Stephanian C of north Portugal. *In*: LEMOS DE SOUSA, J. J., OLIVEIRA, J. T. (eds) *The Carboniferous of Portugal*. Lisbon, **29**, 179–185.

—— 1984. Late Carboniferous–Early Permian non-marine bivalve faunas of northern Europe and eastern North America. *Neuvième Congrès International de Stratigraphie et de Géologie du Carbonifère (Urbana 1979)*, **2**, 559–576.

—— 1985. The stratigraphical and palaeoecological distribution of non-marine bivalves in the coal-fields of north-west Spain. *Dixième Congrès International de Stratigraphie et de Géologie du Carbonifère, Madrid*, **2**, 455–476.

FINGER, F. & STEYRER, H. P. 1995. A tectonic model for the eastern Variscides: indications from a chemical study of amphibolites in the south-eastern Bohemian Massif. *Geologica Carpathia, Bratislava*, **46**(3), 137–150.

——, HANŽL, P., PIN, CH., VON QUADT, A. & STEYRER, H. P. 2000. The Brunovistulian: Avalonian Pre-cambrian sequence at the eastern end of the Central European Variscides? *This volume*.

FRANKE, W. 1989. Tectonostratigraphic units in the Variscan belt of Central Europe. *In*: DALLMEYER, R. D. (ed.). *Terranes in the Circum-Atlantic Palaeozoic Orogens*. Geological Society of America, Special Papers, **230**, 67–90.

—— 2000. The mid-European Segment of the Variscides: tectono-stratigraphic units, terranes, boundaries and plate evolution. *This volume*.

——, DALLMEYER, R. D. & WEBER, K. 1995. Geodynamic evolution. *In*: DALLMEYER, R. D., FRANKE, W. & WEBER, K. (eds) *Pre-Permian Geology of Central and Eastern Europe*. Springer, Berlin, p. 579–593.

GASTALDO, R. A. & MARTEN, L. C. 1978. *Trigonocarpus leeanus*, a new species from the Middle Pennsylvanian of southern Illinois. *American Journal of Botany*, **65**, 882–890.

GERMER, R., KNEEUPER, G. & WAGNER, R. H. 1968. Zur Westfal/Stefan-Grenze und zur Frage der asturischen Faltungsphase im Saarbrucker Hauptsattel. *Geologica et Palaeontologica*, **2**, 59–71.

HAVLÍČEK, V. 1989. Climatic changes and development of benthic communities through the Mediterranean Ordovician. *Sbornik Geologickych Ved, Geologie*, **44**, 79–116.

HUTTON, D. H. W. & MURPHY, F. C. 1987. The Silurian of the Southern Uplands and Ireland as a successor basin to the end-Ordovician closure of Iapetus. *Journal of the Geological Society, London*, **144**, 765–772.

JANVIER, P. 1996. *Early Vertebrates*. Oxford Science, Oxford.

JOHANSON, Z. & AHLBERG, P. E. 1998. A complete primitive rhizodont from Australia. *Nature*, **394**, 569–572.

JONGMANS, W. J. 1952. *Note sur la flore du terrain Carbonifère de Djerada (Maroc oriental)*. Notes et Mémoires du Service Géologique du Maroc, **91**.

—— & DELEAU, P. C. 1951. *Les bassins houillers du Sud-Oronais. Livre II. Contribution à l'&[fa]acute;etude paléontologique*. Bulletin du Service de la Carte Géologique de l'Algérie, 1re Série, Paléontologie, **13.**

KIDSTON, R. 1923–1925. *Fossil Plants of the Carboniferous Rocks of Great Britain*. Memoir of the Geological Survey of Great Britain, Palaeontology, **2**, 1–670.

KOKELAAR, B. P., HOWELLS, M. F., BEVINS, R. E., ROACH, R. A. & DUNKEY, P. N. 1984. The Ordovician marginal basin of Wales. *In*: KOKELAAR, B. P. & HOWELLS, M. F. (eds) *Marginal Basin Geology*. Geological Society, London, Special Publications, **16**, 245–269.

LANDING, E., BOWRING, S. A., DAVIDEK, K. L., WESTROP, S. R., GEYER, G. & HELDMAIER, W. 1998. Duration of the early Cambrian: U–Pb ages of volcanic ashes from Avalon and Gondwana. *Canadian Journal of Earth Sciences*, **35**, 329–338.

LAVEINE, J.-P., LEMOIGNE, Y. & ZHANG, S. 1993. General characteristics and paleobiogeography of the Parispermaceae (genera *Paripteris* Gothan and *Linopteris* Presl), pteridosperms from the Carboniferous. *Palaeontographica, Abteilung B*, **230**, 81–139.

LAVEINE, J.-P., ZHANG, S., LEMOIGNE, Y. & RATANASTHIEN, B. 1999. Paleogeography of east and south-east Asia during Carboniferous times on the basis of paleobotanical information; some methodological comments and additional results. *In*: RATANASTHIEN, B. & Rieb, S. L. (eds) *Proceedings of the International Symposium on Shallow Tethys (ST) 5, Chiang Mai, Thailand*. Chiang Mai University, Thailand, 55–72.

LEGGETT, J. K., MCKERROW, W. S. & SOPER, N. J. 1983. A model for the crustal evolution of southern Scotland. *Tectonics*, **2**, 187–210.

LELIÈVRE, H., JANVIER, P. & BLIECK, A. 1993. Silurian–Devonian vertebrate biostratigraphy of western Gondwana and related terranes (South America, Africa, Armorica–Bohemia, Middle East). *In*: LONG, J. A. (ed.) *Palaeozoic Vertebrate Biostratigraphy and Biogeography*. Belhaven, London, 139–173.

LINNEMAN, U., GEHMLICH, M., TICHOMIROWA, M. & BUSCHMANN, B. 1998. Introduction to the Pre-symposium Excursion (part I): the Peri-Gondwanan basement of the Saxo-Thuringian Composite Terrane. *Schr. Staatl. Mus. Min. Geol., Dresden*, **9**, 7–13.

——, ——, ——, *et al.* 2000. From Cadonian subduction to Early Palaeozoic rifting: the evolution of Saxo-Thuringia at the Margin of Gondwana in the olight of single ziron geochronology and basin development (Central European Variscides, Germany). *This volume*.

LONG, J. A. 1984. New phyllolepids from Victoria and the relationships of the group. *Proceedings of the Linnaean Society of New South Wales*, **107**, 263–308.

MAC NIOCAILL, C. & SMETHURST, M. A. 1994. Palaeozoic palaeogeography of Laurentia and its margins: a reassessment of palaeomagnetic data. *Geophysical Journal International*, **116**, 715–725.

MATTE, Ph., MALUSKI, H., RAJLICH, P. & FRANKE, W. 1990. Terrane boundaries in the Bohemian Massif: results of large scale Variscan shearing. *Tectonophysics*, **177**, 151–170.

MCKERROW, W. S. 1978. *The Ecology of Fossils*. Duckworth, London.

—— & COCKS, L. R. M. 1976. Progressive faunal migration across the Iapetus Ocean. *Nature*, **263**, 304–306.

—— & —— 1986. Oceans, island arcs and olistostromes: the use of fossils in distinguishing sutures, terranes and environments around the Iapetus Ocean. *Journal of the Geological Society, London*, **143**, 185–191.

—— & —— 1995. The use of biogeography in the terrane assembly of the Variscan Belt of Europe. *Studia Geophysica et Geodetica, Prague*, **39**, 269–275.

——, DEWEY, J. F. & SCOTESE, C. F. 1991. The Ordovician and Silurian development of the Iapetus Ocean. *In*: BASSETT, M. G., LANE, P. & EDWARDS, D. (eds) *The Murchison Symposium*. Special Paper in Palaeontology, **44**, 165–178.

——, SCOTESE, C. R. & Brasier, M. D. 1992. Early Cambrian reconstructions. *Journal of the Geological Society, London*, **149**, 599–606.

MEYEN, S. V. 1987. *Fundamentals of Palaeobotany*. Chapman & and Hall, London.

MEYER-BERTHAUD, B., WENDT, J. & GALTIER, J. 1977. First record of a large *Callixylon* trunk from the late Devonian of Gondwana. *Geological Magazine*, **134,** 847–853.

MIGIER, T. 1982. Profil utworów Westfalu C–D Basenu Mezarif w Algierii. *Biuletyn Instytutu Geologicznego*, **338**, 23–70.

ODREMAN, R. O. & Wagner, R. H. 1979. Precisiones sobre alguanas floras Carboniferas y pérmicas de los Andes Venezolanos. *Boletin de Geologia, Caracas*, **13.25**, 77–79.

PAPROTH, E. 1991. Carboniferous palaeogeographic development in Central Europe. *Comptes Rendus Onzième Congrès International de Stratigraphie et de la Géologie Carbonifère (Beijing 1987)*, **1**, 177–186.

PFEFFERKORN, H. W. 1977. Plant megafossils in Venzuela and their use in geology. *Memoria V Congresso Geologico Venzeolano*, **1**, 407–414.

PHARAOH, T. C., BREWER, T. S. & WEBB, P. C. 1993. Subduction-related magmatism of late Ordovician age in eastern England. *Geological Magazine*, **130**, 647–656.

PIQUÉ, A. 1981. Northwestern Africa and the Avalonian plate: relations during late Precambrian and late Paleozoic time. *Geology*, **9**, 319–322.

READ, C. B. 1955. *Floras of the Pocono Formation and Price Sandstone in parts of Pennsylvania, Maryland and West Virginia.* US Geological Survey, Professional Papers, **263**.

SCOTESE, C. R. & BARRETT, S. F. 1990. Gondwana's movement over the South Pole during the Palaeozoic: evidence from lithological indicators of climate. *In*: MCKERROW, W. S. & SCOTESE, C. R. (eds) *Palaeozoic Palaeogeography and Biogeography.* Geological Society, London, Memoir, **12**, 75–85.

—— & MCKERROW, W. S. 1990. Revised world maps and introduction. *In*: MCKERROW, W. S. & SCOTESE, C. R. (eds) *Palaeozoic Palaeogeography and Biogeography.* Geological Society, London, Memoir, **12**, 1–21.

STILLMAN, C. J. 1988. Ordovician to Silurian volcanism in the Appalachian–Caledonian Orogen. *In*: HARRIS, A. L. & FETTES, D. J. (eds) *The Caledonian–Appalachian Orogen.* Geological Society, London, Special Publications, **38**, 275–290.

SULLIVAN, H. J. 1965. Palynological evidence concerning the regional differentiation of Upper Mississipian floras. *Pollen et Spores*, **7**, 539–563.

—— 1967. Regional differences in Mississipian spore assemblages. *Review of Palaeobotany and Palynology*, **1**, 185–192.

TAIT, J., BACHTADSE, V. & SOFFEL, H. C. 1994. Silurian paleogeography of Armorica: new paleomagnetic data from Central Bohemia. *Journal of Geophysical Research*, **99**, 2897–2907.

——, —— & —— 1995. Upper Ordovician paleogeography of the Bohemian Massif: implications for Armorica. *Geophysical Journal International*, 211–218.

——, SCHÄTZ, M., BACHTADSE, V. & SOFFEL, H. 2000. Palaeomagnetism and Palaeozoic palaeogeography of Gondwana and European terranes. *This volume.*

THULBORN, T., WARREN, A., TURNER, S. & HAMLEY, T. 1996. Early Carboniferous tetrapods in Australia. *Nature*, **381**, 777–780.

TOGHILL, P. 1992. The Shelveian event, a late Ordovician tectonic episode in southern Britain (Eastern Avalonia). *Proceedings of the Geologists' Association*, **103**, 31–35.

TORSVIK, T. H., SMETHURST, M. A., MEERT, J. G. *et al.* 1996. Continental break-up and collision in the Neoproterozoic and Palaeozoic—a tale of Baltica and Laurentia. *Earth-Science Reviews*, **40**, 229–258.

TRENCH, A., TORSVIK, T. H. & MCKERROW, W. S. 1992. The palaeogeographic evolution of Southern Britain during early Palaeozoic times: a reconciliation of palaeomagnetic and biogeographic evidence. *Tectonophysics*, **201**, 75–82.

TRUEMAN, A. E. 1946. Stratigraphical problems in the Coal Measures of Europe and North America. *Quarterly Journal of the Geological Society, London*, **102**, li–xciii.

TUCKER, R. D. & MCKERROW, W. S. 1995. Early Paleozoic chronology: a review in light of new U–Pb zircon ages from Newfoundland and Britain. *Canadian Journal of Earth Sciences*, **32**, 368–379.

VAKHRAMEEV, V. A., DOBRUSKINA, I. A., MYEN, S. V. & ZAKLINSAJA, E. D. 1978. *Paläozoische und mesozosche Floren Eurasiens und die Phytogeographie dieser Zeit.* Fischer, Jena.

VAN DER VOO, R. 1993. *Palaeomagnetism of the Atlantic, Tethys and Iapetus Oceans.* Cambridge University Press, Cambridge.

VAN STAAL, C. R., DEWEY, J. F., MAC NIOCAILL, C. & MCKERROW, W. S. 1998. Cambrian–Silurian tectonic evolution of the northern Appalachians and British Caledonides: history of a complex, west and southwest Pacific-type segment of Iapetus. *In*: BLUNDELL, D. J. & SCOTT, A. C. (eds) *Lyell; the Past is the Key to the Present.* Geological Society, London, Special Publications, **143**, 199–242.

WAGNER, R. H. & ALVAREZ-VÁZQUEZ, C. 1991. Floral characterisation and biozones of the Westphalian D Stage in NW Spain. *Neues Jahrbuch für Geologie und Paläontologie, Abhandlungen*, **183**, 171–202.

—— & LYONS, P. C. 1997. A critical analysis of the higher Pennsylvanian megafloras of the Appalachian region. *Review of Palaeobotany and Palynology*, **95**, 255–283.

——, WINKLER PRINS, C. F. & MARTINEZ DIAZ, C. (eds) 1985. *The Carboniferous of the World II. Australia, Indian subcontinent, South Africa, South America and North Africa.* IUGS Publication **20**.

WEINGART, H. W. 1976. Das Oberkarbon in der Tiefbohrung Saar 1. *Geologisches Jahrbuch*, **A27**, 399–408.

YOUNG, G. C. 1984. Comments on the phylogeny and biogeography of antiarchs (Devonian placoderm fishes) and the use of fossils in biogeography. *Proceedings of the Linnean Society of New South Wales*, **107**, 443–473.

—— 1990. Devonian vertebrate distribution patterns and cladistic analysis of palaeogeographic hypotheses. *In*: MCKERROW, W. S. & SCOTESE, C. F. (eds) *Palaeozoic Palaeogeography and Biogeography.* Geological Society, London, Memoir, **12**, 243–255.

—— 1993. Middle Devonian macrovertebrate biostratigraphy of eastern Gondwana. *In*: LONG, J. A. (ed.) *Palaeozoic Vertebrate Biostratigraphy and Biogeography.* Belhaven, London, 208–251.

ZODROW, E. L. 1986. Succession of paleobotanical events: evidence for mid-Westphalian D floral changes, Morien Group (Late Pennsylvanian, Nova Scotia). *Review of Palaeobotany and Palynology*, **47**, 292–326.

Palaeomagnetism and Palaeozoic palaeogeography of Gondwana and European terranes

JENNY TAIT, MICHAEL SCHÄTZ, VALERIAN BACHTADSE & HEINRICH SOFFEL

Institut für Allgemeine und Angewandte Geophysik, Ludwig-Maximilian-Universität, Theresienstraße 41, D-80333 München, Germany

Abstract: Neoproterozoic to Late Palaeozoic times saw the break-up of the supercontinent Rodinia, and the subsequent construction of Pangaea. The intervening time period involved major redistribution of continents and continental fragments, and various palaeogeographical models have been proposed for this period. The principal differences between these models are with regard to the drift history of Gondwana, the timing of collision between northern Africa and Laurussia, and formation of Pangaea. Palaeomagnetic evidence provides basically two contrasting models for the Ordovician to Late Devonian apparent polar wander (APW) path for Gondwana involving either rapid north and southward movement of this continent, or gradual northward drift throughout Palaeozoic time. In contrast, palaeobiogeographical models suggest contact between Laurussia and Gondwana as early as mid-Devonian time with the continents basically remaining in this configuration until break-up of Pangaea in the Mesozoic era. This is in conflict, however, with most palaeomagnetic data, which demonstrate that in Late Devonian time, north Africa and the European margin of Laurussia were separated by an ocean of at least 3000 km width. This is also in agreement with the geological record of present-day southern Europe, which argues against any collision of northern Africa with Europe in Devonian time. With regard to formation of Laurussia, however, palaeobiogeographical and palaeomagnetic data are in excellent agreement that by mid-Devonian time the oceanic basins separating Baltica, Laurentia, Gondwana-derived Avalonia and the Armorican Terrane Assemblage (ATA) had all closed. Palaeomagnetic and geological data are also in agreement that the Palaeozoic basement rocks of the European Alpine realm formed an independent microplate, which was situated to the south of Laurussia. In Late Silurian times it was separated by an ocean of *c.* 1000 km, and by Late Devonian time was approaching the southern Laurussian margin. According to palaeomagnetic data, the northern margin of Gondwana was still further to the south in Late Devonian time, and according to the geological record in southern Europe, the main continent–continent collision of northern Africa with European Laurussia and closure of the intervening ocean occurred in Late Carboniferous times. Location of this suture is situated to the south of the Palaeozoic alpine units (e.g. the Greywacke zone, Carnic Alps, Sardinia and Sicily), but has been obscured by younger deformational events and cannot be precisely positioned. Assessing available evidence and as discussed in the text, it is proposed that the most likely scenario is that the northern margin of Gondwana drifted gradually northwards from Ordovician to Late Carboniferous times when it collided with Laurussia, resulting in formation of Pangaea.

All reconstructions and models for the drift history of the Palaeozoic plates presented in this paper are based primarily on palaeomagnetic evidence. The advantage of palaeomagnetism over other palaeogeographical methods is that it depends only upon the hypothesis that the time-averaged Earth's magnetic field can be described as a geocentric axial dipole field (i.e. magnetic North corresponds to geographical North). Although recently questioned by some workers (Kent & Smethurst 1998), this hypothesis has been demonstrated to hold for at least the last 800 Ma and one of its most significant features is the direct relationship between magnetic inclination and geographical latitude. Palaeomagnetic data, therefore, can provide quantitative information for the (palaeo-) latitude at which rocks were formed and is independent of any external factors. This is in contrast to the other most commonly used techniques, such as palaeobiogeography, which may be strongly influenced by external factors such as global

From: Franke, W., Haak, V., Oncken, O. & Tanner, D. (eds).
Orogenic Processes: Quantification and Modelling in the Variscan Belt.
Geological Society, London, Special Publications, **179**, 21–34. 0305-8719/00/$15.00

climate variations and eustatic oscillations, which can influence faunal distributions and interchange. Also, dispersal of larvae is directly related to the duration of larval development and the ocean-current velocity and palaeo-circulation patterns, all of which are extremely hard to quantify for geological time periods. These questions regarding faunal distribution patterns have been a matter of debate and discussion for many years. Details of the various aspects are beyond the scope of this paper and the reader is referred to Cloud (1961), Ross (1975), Klapper & Johnson (1980), Schallreuter & Siveter (1985), Fortey & Cocks (1988), Jokiel (1990), Oclzon (1990), Schönlaub (1992), Benedetto (1998) and McKerrow *et al.* (this volume), for further discussion.

The break-up of Rodinia

The main elements controlling and driving Palaeozoic palaeogeography are Laurentia, Baltica, Gondwana and Siberia, which formed as a result of the break-up of the Proterozoic supercontinent Rodinia. This was a fairly long-lived process (Dalziel 1992), and the final break-up of Rodinia is constrained as occurring between 590 and 550 Ma from continental margin subsidence and sedimentation rates (Bond *et al.* 1984; Van Staal *et al.* 1998). Laurentia and Baltica separated from Amazonia and Africa and drifted towards the equator, thus opening the Iapetus Ocean. By Early to mid-Cambrian time, Baltica had a diverse trilobite fauna indicating lower palaeolatitudes and warmer-water environment. Formation of Gondwana into its Palaeozoic configuration occurred during break-up of Rodinia, and consolidation of this large continental mass was completed by latest Cambrian time (see discussion by Van Staal *et al.* (1998). During this time the northern margin of Gondwana was an active margin, resulting in the Cadomian deformation event. Cadomian basement rocks have been identified in the European Palaeozoic blocks of Iberia, Bohemia, the Armorican massif and also within basement rocks of the Alpine realm, thus demonstrating that they were all situated on the northern margin of Gondwana, adjacent to north Africa in Late Cambrian times. Other European and North American Gondwana-derived terranes and microplates, such as Avalonia, the Gander terrane and Florida, were probably adjacent to northern South America (see also Van Staal *et al.* (1998) and references therein for further discussion of Rodinia and its break-up).

Formation of Pangaea

Most palaeogeographical models are in good agreement with regard to the Palaeozoic drift history of Laurentia, Baltica and Gondwana-derived Avalonia and Armorican Terrane Assemblage (ATA). Much discussion and debate surrounds the drift history of Gondwana itself, however. Many recent palaeobiogeographical models (see, e.g. McKerrow *et al.* this volume) tend to favour collision of northern Africa with the European plates in Early or mid-Devonian times, and from the palaeomagnetic data at least two widely contrasting models have been proposed. What is clear is that at least by Late Carboniferous times the ocean between northern Gondwana and the northern continents had closed. Timing of collision, however, is vital for a full understanding of the orogenic processes involved within Variscan Europe.

In the following sections various models for Gondwana will be discussed, together with arguments for and against these models and their implications for Palaeozoic palaeogeography within the overall tectonic framework for the development of the Caledonian and Variscan orogenies. A brief description of the drift histories for Baltica, Laurentia and Avalonia will also be given; for more details regarding these plates the reader is referred to Van der Voo (1993) and various more recent reviews as mentioned in the text. A more in-depth discussion of the Palaeozoic drift histories and palaeogeography of the ATA and Alpine domains leading up to formation of Pangaea is given.

Gondwana

Gondwana, which comprised South America, Africa, Madagascar, Antarctica, Australia and India, dominated Palaeozoic palaeogeography and accounted for more than two-thirds of the entire continental crust. The drift history of this continent in Palaeozoic times, however, remains the least well constrained of all the Palaeozoic plates, and is a matter of much debate within the geoscience community. From the existing palaeomagnetic database alone essentially two contrasting models for the Palaeozoic apparent polar wander (APW) path of Gondwana have been proposed (Fig. 1). They indicate either gradual northward movement of Gondwana from Late Ordovician to Early Carboniferous times, or, alternatively, a much more complex path implying collision of northern Africa with Laurussia in the Late Silurian–Early Devonian time and then again in Carboniferous times.

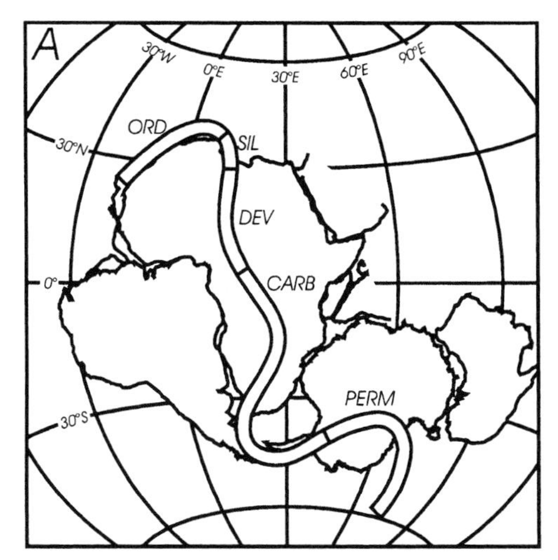

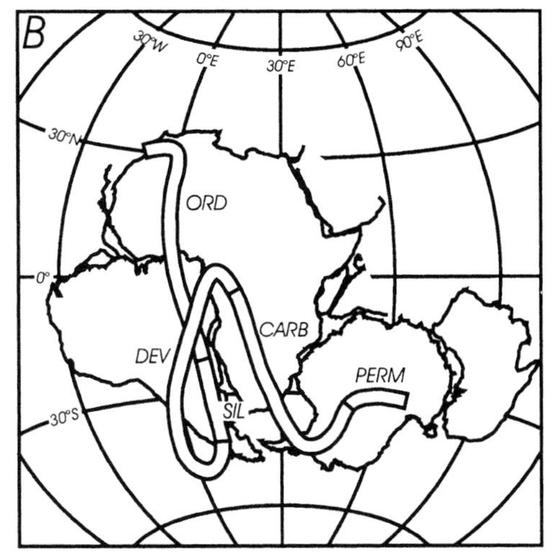

Fig. 1. Alternative APW curves for the Palaeozoic of Gondwana, (**a**) path X after Bachtadse & Briden (1991); (**b**) path Y modified after Schmidt *et al.* (1990), taking into account more recently published radiometric age for the AIR pole as mentioned in the text and listed in Table 1.

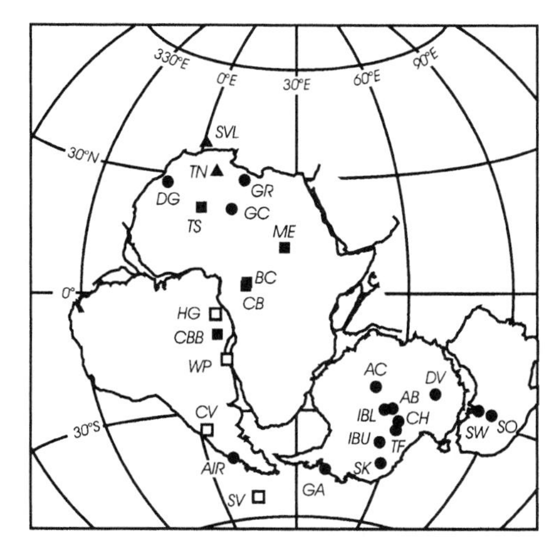

Fig. 2. Selected paleopole positions ($Q > 3/7$) for Ordovician to Permian time from Australia (■), Antarctica (▲) and Africa (●) as listed in Table 1. Pole positions from the Lachlan fold belt (Australia) are shown as open squares. Gondwana is reconstructed using the rotation parameters of Lawver & Scotese (1987). Africa is in present-day co-ordinates.

Obviously, these different models have important implications for the geodynamic evolution of the Caledonian and Variscan orogenic systems in Europe and North America as well as our understanding and interpretation of the geological record in Gondwana itself.

Both models agree that in Early Ordovician times the south pole was situated in northwest Africa, for the Late Devonian both paths are anchored by the small cluster of high-quality palaeopoles in central Africa (poles BC, CB and CBB in Fig. 2 and Table 1), and by Late Carboniferous to Permian times the south pole was clearly located in Antarctica. The differences between the APW paths are how these three groups are connected. The simpler model involves gradual and continuous northward movement of Gondwana, and simply connects these three tie points (Fig. 1a). Such a path was first suggested by McElhinny (1968), subsequently termed path X by Morel & Irving (1978), and more recently was redefined by Bachtadse & Briden (1991) on the basis of new, although poorly constrained, palaeomagnetic data from South Africa (pole GC, Fig. 2). The alternative model, originally termed path Y by Morel & Irving (1978), and subsequently modified by Schmidt *et al.* (1990), is more complex with at least one distinct loop in Silurian times (Fig. 1b), and requires rather high drift rates for Gondwana in the Palaeozoic era. The Silurian and Devonian segments of the original Schmidt *et al.* (1990) APW path were based on palaeomagnetic results from the Aïr complex of Niger (Hargreaves *et al.* 1987), northern Africa, which yield a palaeopole position in south Chile (pole AIR, Fig. 2). At that time, these rocks were thought to have an Early Silurian (*c.* 435 Ma) emplacement age. More recent radiometric ages, however, yield an Early Devonian age (409 Ma, Moreau *et al.* 1994). The Siluro-Devonian sector of the APW path of Schmidt *et al.* therefore, needs to be modified accordingly, as shown in Fig. 1. The Devonian sector of this APW path is

Table 1. *Ordoricean to Permian palaeopoles with quality factors >3/7*

Formation	Symbol	Age	Lat. (°N)	Long. (°E)	k	α_{95} (dp, dm)	Reference
Africa							
Damara granites*	DG	458 ± 8[2]	27	351	–	5	(Corner & Nethorn 1978)
Graafwater Formation	GR	Ol[5]	28	14	25	9	(Bachtadse *et al.* 1987)
Aïr intrusive rocks	AIR	409[4]	−43	9	50	6	(Hargreaves *et al.* 1987)
Gilif Hills C	GC	377 ± 5[2]	26	12		(20, 21)	(Bachtadse & Briden 1991)
Mount Eclipse	ME	Cl[5]	12	27		(8.7, 8.7)	(Chen *et al.* 1994)
Ain Ech Chebbi, Algeria*	AC	Cu[5]	−25	55		(35, 3.5)	(Daly & Irving 1983)
Illizi Basin (lower formation)	IBL	Cu[5]	−29	56	161	3.5	(Henry *et al.* 1992)
Dwyka varves	DV	P-C[5]	−25	67	–	12	(Opdyke 1999)
Tiguentourine formation	TF	P-C[5]	−34	61	170	4.1	(Derder *et al.* 1994)
Illizi Basin (upper formation)	IBU	P-C[5]	−38	57	837	2.8	(Henry *et al.* 1992
Galula red beds, Tanzania	GA	P-C[5]	−46	40		(8.4, 8.4)	(Opdyke 1964)
Abadla Red Beds, Algeria	AB	Pl[5]	−29	58	–	(2, 2)	(Merabet *et al.* 1998)
Songwe–Ketawaka red beds	SO	Pl[5]	−28	90	59.8	12	(Opdyke 1964)
Chougrane red beds	CH	Pl[5]	−32	64	20	5	(Daly & Pozzi 1976)
Songwee–Kiwira red beds	SW	Pl-u[5]	−27	85	84.2	5.3	(Nyblade *et al.* 1993)
Madagascar							
Sakoa Group glacial deposits	SK	Cu[5]	−43	60	50	8	(McElhinny *et al.* 1976)
East Antarctica							
Southern Victoria Land	SVL	475 ± 10[2]	37	360		(5.9, 5.9)	(Grunow 1995)
Teall Nunatak	TN	475 ± 4[2]	31	5	163	7.2	(Lanza & Tonarini 1998)
Australia							
Tumblagooda Sandstone*	TS	Ou[5]	24	7	556	3	(Schmidt & Embleton 1990)
Canning Basin*	CB	Du[5]	2	16	62	8	(Hurley & Van der Voo 1987)
Canning Basin	CBB	Du[5]	−12	9		(14.6, 14.6)	(Chen *et al.* 1995)
Brewer Conglomerate	BC	Du[5]	3	17		(6.4, 6.4)	(Chen *et al.* 1993)
Hervey Group[1]	HG	Du[5]	−5	8	–	(8.4, 16.2)	(Li *et al.* 1988)
Snowy River volcanic rocks[1]	SV	Dl[5]	−54	15	–	(10.9, 14.5)	(Schmidt *et al.* 1987)
Comerong volcanic rocks[1]	CV	Dm[5]	−34	2	–	(7.2, 7.2)	(Schmidt *et al.* 1986)
Worange Point[1]	WP	Du[5]	−17	12		(10.9, 10.9)	(Thrupp *et al.* 1991)

Selected palaeopoles (quality factor > 3/7 including good age control, acceptable statistics and adequate demagnetization; see van der Voo (1990) for Ordovician to mid-Permian time for Africa, Madagascar, Australia and Antarctica in African co-ordinates (rotation parameters after Lawver & Scotese (1987)). α_{95}, semi-cone of 95% confidence for the palaeopole; dp, dm, semi-axes of oval of 95% confidence of the palaeopole. Lat., Long., latitude and longitude of the palaeomagnetic palaeopole. k, precision parameter after Fisher (1953). O, Ordovician; S, Silurian; D, Devonian; P-C, Permo-Carboniferous. l, lower; m, middle; u, upper, * Palaeopoles used in the palaeogeographical reconstructions in Fig. 3.

[1]Palaeopoles derived from the Lachlan fold belt (LFB) (allochthonous?).[2] Rb/Sr age.[3] K/Ar age.[4] $^{40}Ar/^{39}Ar$ age. [5]Stratigraphic age.

then based on Latest Silurian to Late Devonian results from the Lachlan fold belt of southeast Australia. These data describe a swathe of pole positions (SV, CV, WP and HG in Fig. 2) from southern Chile in Siluro-Devonian time to the Late Devonian cluster of palaeopoles in central Africa (Schmidt *et al.* 1986, 1987; I *et al.* 1991; Thrupp *et al.* 1991).

In recent years, the more complex APW path for Gondwana has gained significantly in popularity and is generally accepted as the more reliable model. However, many controversies surround the acceptance of this APW path. The main problems are the lack of field tests to unambiguously constrain the age of magnetization of the Aïr complex, and that the model relies heavily on palaeomagnetic data from the Lachlan fold belt (LFB) of southeast Australia. Whether or not Siluro-Devonian data from this region can be used to constrain Gondwana remains highly controversial. The LFB of eastern Australia can be subdivided into a number of different tectono-stratigraphic terranes or zones (Collins & Vernon 1922; Scheibner & Basden 1996; Gray *et al.* 1997). In some models, these different blocks are considered to be (para)autochthonous to cratonic Australia, with only minimal differential movement of the blocks during Palaeozoic time (Powell 1983, 1984; Coney 1992). According to this hypothesis, any subduction zone must have been outboard of the present-day eastern margin of Australia. Alternatively, the presence of ophiolites along the major fault zones, different Ordovician faunal characteristics, and different Silurian to Early Devonian tectono-metamorphic histories of juxtaposed blocks have led to the interpretation that these faults represent Palaeozoic subduction zones or major tectonic discontinuities, with potentially large-scale differential movement of the individual blocks between at least late Ordovician and Late Devonian–Early Carboniferous time (Collins & Vernon 1992; Glen 1993; Gray *et al.* 1997; Glen *et al.* 1998; B. Webby, pers. comm.)

While reasonable doubt remains as to the tectonic evolution of SE Australia, it is the current authors' opinion that palaeomagnetic data from this region should not be used to defined the APW path of Gondwana. The only palaeopole then supporting the complex APW path is the Aïr pole. The main arguments against using this palaeopole to constrain the Devonian segment of the APW path are: (a) the age of magnetization in these rocks is not well constrained; (b) it would require high drift rates of Gondwana with repeated changes in direction of motion; (c) it implies either Early Devonian collision of Gondwana with the northern continents (for which there is no geological evidence in the alpine domains as discussed below), or it would invoke a complex tectonic regime for the northern Gondwana–southern European realm to avoid collision but allow for the equatorial position of northern Africa. Taking all factors into consideration, the complex APW is not considered reliable. Adopting the more conservative APW path, therefore, palaeomagnetic data indicate that in Early Devonian time, north Africa was still separated from the European sector of Laurussia by an ocean of *c.* 2500 km, but which was considerably narrower at its eastern and western extremities. This APW path is similar to that proposed by Scotese *et al.* (1999), which is based on palaeoclimatic indicators.

For mid-Devonian time, some palaeobiogeographical models support an Emsian–Eifelian connection between the African margin of Gondwana and Laurussia, with continued faunal interchange and proximity until final consolidation of Pangaea in Late Carboniferous time. This model is based on fossil fish similarities in Northern Australia and Laurentia (McKerrow *et al.* this volume), and the similarity of Emsian ostracode faunas in Algeria and Baltica (Berdan 1991). It is in apparent conflict, however, with late Devonian palaeomagnetic data for Gondwana. There are three palaeopoles for the Frasnian–Famennian and one for the Visean of Godwana (Hurley & Van der Voo 1987; Chen *et al.* 1993, 1994, 1995), which are grouped in central Africa (Fig. 2), thus indicating major separation of Gondwana from Laurussia (Fig. 3e). These palaeomagnetic data are all from cratonic Australia, and it is difficult to find grounds for their rejection; either our concept of the configuration of Gondwana itself could be argued, or local rotation of the areas studied could be inferred. Both of these solutions, however, are unlikely. Alternatively, the magnetization identified may represent a younger remagnetization event. Although possible, this would also seem unlikely given the quality of the palaeomagnetic data in question and the implications such an argument would involve (i.e. at some time after the Devonian period the south pole was situated in central Africa). Thus there remains a major discordance between palaeomagnetic and palaeobiogeographical data for this time period. What is clear, however, is that, in southern Europe there is little evidence for a Devonian continental collision between Gondwana and Laurasia as Devonian sequences in the European Alpine realm reflect a period of continuous sedimentation in a passive margin environment until Late Carboniferous times (Stampfli 1996). Devonian faunas of the southern Alps are also significantly different from those of Gondwana (Schönlaub 1992). Early Devonian deformation is seen, however, in the Appalachians and Britain (Acadian Orogeny). The causes of this event remain ambiguous, but are frequently attributed to collision with Gondwana, thus apparently supporting the complex APW path and, specifically, the Aïr palaeopole, which places the northern margin of Africa at the equator in Early Devonian time. Given this scenario, it is difficult to envisage why the effects of collision between Gondwana and Laurasia should be manifest within the Laurentian–Avalonian realm but not in southern Europe.

Formation of Pangaea is constrained by palaeomagnetic data as having occurred in Latest Palaeozoic time. Data from Algeria (AC and IBL, Fig. 2) demonstrate that by Late Carboniferous times the northern margin of Africa was approaching the southern margin of Europe and final collision probably occurred in latest Carboniferous to Early Permian times.

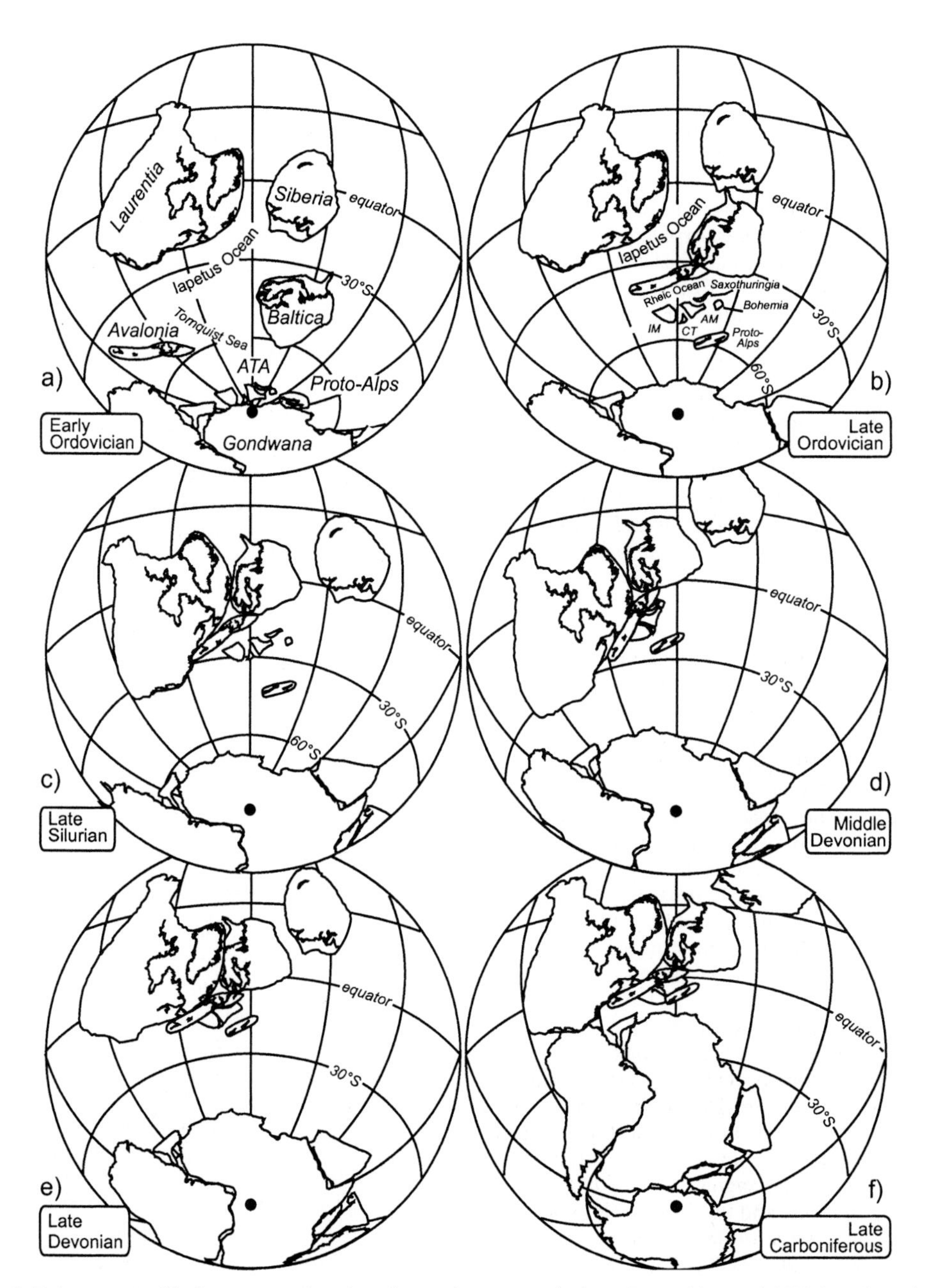

Fig. 3. Palaeogeographical reconstructions based on palaeomagnetic data. Proto-Alps and ATA based on data mentioned in the text (for Avalonia, see data compilation of Torsvik *et al.* 1993); for Baltica, see review by Torsvik *et al.* 1992; for Laurentia, see MacNiocaill & Smethurst (1994); for Siberia, see review by Smethurst *et al.* (1998). The palaeopoles used for Gondwana are indicated (*) in Table 1 (Late Silurian and Mid-Devonian palaeopole are interpolated from the Late Ordovician and Late Devonian palaeopoles). IM, Iberian Massif; AM, Armorican Massif; CT, Catalan terrane.

Laurentia

The APW path for Laurentia (comprising most of present-day North America, Greenland and Scotland), is now well constrained for the whole of the Palaeozoic era. Faunal, lithological and palaeomagnetic data are in good agreement that after the break-up of Rodinia, Laurentia drifted northwards towards the equator thus opening the Iapetus Ocean along its present-day eastern and southern margins. Throughout most of the Palaeozoic era, this continent remained in equatorial latitudes (Fig. 3) with the development of warm-water carbonates, and is characterized by the classic warm-water Edgewood faunas. For further discussion of the Palaeozoic palaeogeography of Laurentia, the reader is referred to MacNiocaill & Smethhurst (1994), for a review and compilation of previously published palaeomagnetic data, and to Witzke (1990) and McKerrow *et al.* (1991).

Baltica

The Palaeozoic APW path for Baltica is also fairly well constrained. In Early Ordovician times, Baltica was rotated with respect to its present-day orientation and was located at intermediate to high palaeolatitudes (Torsvik *et al.* 1990, 1992, 1996; Perroud *et al.* 1992). Ordovician faunas of Baltica are significantly different from those of Gondwana and Laurentia, and the lithological record also demonstrates a gradual northward movement of Baltica throughout Palaeozoic time. Palaeobiogeographical indicators show a change at the Ordovician–Silurian boundary from subtropical and temperate faunas, to equatorial warm-water faunas (Owen *et al.* 1991). After collision with Avalonia and closure of the Tornquist Sea in Late Ordovician–Early Silurian time, and final closure of the Iapetus Ocean between Baltica and Laurentia in Siluro-Devonian time, Baltica remained in equatorial palaeolatitudes until the end of the Palaeozoic era. For further details concerning palaeomagnetic data for Baltica, its palaeogeography and the Palaeozoic evolution of the Iapetus Ocean the reader is referred to McKerrow *et al.* (1991), Torsvik *et al.* (1996) and Van Staal *et al.* (1998).

Avalonia

In earliest Ordovician time Avalonia (i.e. New England, New Brunswick, Nova Scotia, the Avalon Peninsula, Southern Britain and continental Europe north of the Rheno-Hercynian zone and south of Scandinavia) was still contiguous with the northern margin of Gondwana (Trench *et al.* 1992; Torsvik *et al.* 1993). Its faunas were coincident with those from Bohemia and the Armorican Massif, but in marked contrast to those of Baltica and Laurentia (Cocks & Fortey 1982; Cocks *et al.* 1997). By Llanvirn (Fig. 3a) times it was drifting northwards away from Gondwana. The Rheic Ocean (*sensu* McKerrow *et al.* 1991) opened in its wake, and Early Ordovician subduction-related volcanism in Southern Britain indicates the change from a passive to an active margin environment (Kokelaar *et al.* 1984; Cooper *et al.* 1993). In Llandeilo times the shallow-water faunas of Avalonia were endemic (Lockley, 1983; Cocks 1993), and palaeomagnetic data demonstrate a palaeolatitude of 45° S (Trench & Torsvik 1991); thus Avalonia was separated from Laurentia by the southern Iapetus Ocean, and from Baltica by the narrowing Tornquist Sea. Timing of closure of the Tornquist Sea is not clear from palaeomagnetic evidence alone, but from geological evidence (Pharaoh *et al.* 1993; Cocks *et al.* 1997) it is thought to have closed by Late Ordovician–Early Silurian times (Fig. 3b).

Armorican Terrane Assemblage

The Gondwana-derived European Palaeozoic massifs and terranes located south of the Rheno-Hercynian zone, but north of the Alpine realm, are referred to as the Armorican Terrane Assemblage. These units (i.e. Bohemia, Saxo-Thuringia, Iberia and Armorican Massif) were previously considered to have comprised a coherent microplate (the Armorican microplate). It is now clear from palaeomagnetic and geological evidence, however, that the Variscan fold belt is a complex tectonic collage of terranes (Franke *et al.* 1995; Tait *et al.* 1997; Tait 1999).

The presence of Cadomian basement in all these terranes shows that they are all Gondwana derived, and in Early Ordovician times they were still adjacent to the northern margin of Gondwana (Fig. 3a). Palaeomagnetic data demonstrate that by Late Ordovician times the Teplá–Barrandian zone of the Bohemian Massif was located at 40° S (Fig. 3b), and thus independent of Gondwana. No unambiguous palaeomagnetic data are available from Armorica or Iberia (see Perroud *et al.* (1983), Perroud & Van der Voo (1985) and Tait *et al.* (1997) for discussion). Strong faunal and lithological similarities in Ordovician to Devonian successions of the Armorican Massif and Central Iberia indicate similar ecological conditions and demonstrate that they were both part of the same domain

(Robardet 1996). From Ordovician data alone, it remains equivocal as to whether they, and Saxo-Thuringia, remained adjacent to northern Gondwana in Late Ordovician time, or if they rifted away in mid-Ordovician times (Perroud & Van der Voo 1985) with the Teplá–Barrandian. Faunal and lithological similarities, however, would tend to support the latter hypothesis. Palaeobiogeographical models have generally assumed that all terranes remained at high palaeolatitudes, adjacent to the northern margin of Gondwana until at least Late Ordovician time. This is based primarily on the presence of Ashgillian glacial deposits throughout the ATA. However, the lower palaeolatitude demonstrated by palaeomagnetic data is not in conflict with the presence of glacial sediments, at least in the Bohemian Massif, as it has been shown that they were deposited from seasonal or floating ice during the Late Ordovician period of global cooling which also allowed the colonization of cold-water faunas in previously warmer-water realms (Brenchley *et al.* 1991; Owen *et al.* 1991).

Late Silurian to Early Devonian palaeomagnetic data from the Teplá–Barrandian, Saxo-Thuringia and the Armorican Massif (Tait *et al.* 1994; Kößler *et al.* 1996; Tait 1999) demonstrate that they were at palaeolatitudes of between 20° and 30° S (Fig. 3c), thus continuing to move northwards towards the southern margin of Baltica and Avalonia. This is in good agreement with geological evidence for gradual closure of the Rheic Ocean, which separated Avalonia and the ATA in Ordovician and Silurian times (Sommermann *et al.* 1992). Significantly, however, these results also demonstrate that the Teplá–Barrandian and Saxo-Thuringia did not belong to a coherent microplate. Whereas the magnetic inclinations obtained from these regions are in good agreement, the declination values differ significantly, demonstrating major tectonic discontinuity between the two terranes (see Tait *et al.* (1997) for further discussion).

Palaeomagnetic data are now available for the Siluro-Devonian of the eastern Pyrenees and Catalunia of northeastern Spain (the Catalan terrane). These data indicate palaeolatitudes of 30° S (Tait *et al.* 2000). The Catalan terrane, therefore, was at similar palaeolatitudes to the ATA in latest Silurian time (Fig. 3c). Whether or not this holds for the Iberian Massif is not clear, as correlation of basement rocks between these regions is extremely difficult because of strong reactivation during the Alpine Orogeny, and the Catalan terrane may have formed an independent tectonic unit in Palaeozoic time (Robardet & Gutierrez-Marco 1990).

In summary, the similarity of latest Silurian brachiopod and trilobite faunas of Iberia and Armorica, and ostracode faunas of Bohemia and the Armorican Massif (Kriz & Paris 1982), do not allow for any oceanic separation between these terranes. These faunal data, therefore, support the interpretation from palaeomagnetic studies that all these terranes of the ATA, including the Catalan terrane, had similar drift histories throughout Palaeozoic time, and in late Silurian times were located within the latitudinal band of 20–30° S and proximal to the southern Baltica–Avalonia margin (Fig. 3c). Before Emsian–Givetian times, however, vertebrate and invertebrate faunal assemblages were significantly different between Bohemia and southern Britain, thus final closure of the Rheic Ocean did not occur until the mid-Devonian (Fig. 3d).

Proto-Alps

The Palaeozoic palaeogeography and tectonic evolution of the Alpine realm is extremely complex as a result of the poly-metamorphic and poly-deformational events suffered by these rocks. Nevertheless, a number of different Palaeozoic tectonostratigraphic terranes can be recognized, particularly in the Eastern Alps. Metamorphic and deformational signatures vary greatly throughout the various basement units of the Alpine realm. The different characteristics of the pre-Alpine metamorphic evolution have been interpreted to result from stepwise accretion of these terranes onto the active Laurussian margin from Devonian to Permian times (Frisch and Neubauer 1989; Schönlaub 1992; Neubauer *et al.* 1999; and references therein). Their Gondwana origin is clear from basement rocks in which a Neoproterozoic to Cambrian deformation event can be recognized (i.e. 'Cadomian' age U–Pb crystallization and deformation). 'Cadomian' ages obtained from $^{40}Ar^{39}Ar$ dating of detrital micas in Ordovician sediments throughout the Upper Austroalpine and Southalpine units also support this hypothesis (Dallymeyer & Neubauer 1994; Antonitsch *et al.* 1994; Handler *et al.* 1997; Thóni 1999).

In the external units of the central and western Alps, Silurian to Devonian magmatic and metamorphic events have been recognized. In contrast, no pre-Carboniferous orogenic events have been observed in the Eastern Alps (Upper Austroalpine and Southalpine units: Noric–Bosnian terrane of Neubauer & von Raumer (1993)) where there was continuous sedimentation from Ordovician to Late Carboniferous times in a passive margin-type environment (Schönlaub 1992; Stampfli 1996; Frey *et al.* 1999; Neubauer *et al.* 1999). Palaeozoic sequences in the Noric–Bosnian terrane document Ordovician crustal thinning and rifting, deposition in a black-arc basin environment, followed by development of a passive margin with shallow-water marine sedimentation, which persisted until Late Carboniferous time (Neubauer *et al.* 1999).

Upper Ordovician fossils (including brachiopods and ostracodes) from the Noric–Bosnian terrane are more closely related to coeval warm-water faunas of Baltica and Avalonia than to those of Africa (Schönlaub & Heinisch 1993). The onset of rift-related volcanism and magmatism in mid-Ordovician times, and faunal associations, therefore, suggest separation of the Noric–Bosnian terrane from Gondwana in mid-Ordovician times (Loeschke & Heinisch 1993), and available faunal and lithic data (Schönlaub 1992) suggest a Late Ordovician palaeolatitude of 40–50° S (Fig. 3b). As yet, there are no palaeomagnetic data to confirm this. Using the nomenclature of Stampfli (1996), the so-called Palaeotethys ocean, which opened up behind the alpine terranes as they drifted away from Gondwana, closed by northward-directed subduction and collision of Gondwana with Laurussia in Carboniferous time (Stampfli 1996).

For Late Silurian times, trilobite faunas of the Alps are similar to those of Bohemia, and new palaeomagnetic data from the Greywacke zone (part of the Upper Austroalpine nappe complex) yield a palaeolatitude of 45–50° S (Fig. 3c) (Schätz *et al.* 1996). This is in general agreement with the position estimated from palaeobiogeographical evidence (Schönlaub 1992) and documents the gradual northward drift of the Proto-Alps, and, importantly, illustrates that the Alpine terranes were distinct from the ATA, which was at significantly lower palaeolatitudes in Late Silurian time (Fig. 3c).

The Devonian successions of the Alps are characterized by abundant shelly fossils, reef development and condensed pelagic cephalopod limestones (Burchette (1981) and others). There was distinct exchange of faunas with Siberia, Kazakhstan, the Urals, Australia, Avalonia and the ATA, but remarkably poor similarity of faunas between the Alps and northern Africa. Reasons for such a widespread distribution of these Devonian fossil groups in the northern hemisphere are puzzling and have been a matter of much debate. Effects such as Coriolis forces and the presence of equatorial gyres, which would aid fossil dispersion, have been proposed; for further discussion the reader is referred to Cloud (1961), Ross (1975), Klapper & Johnson (1980), Oclzon (1990) and Schönlaub (1992). Palaeomagnetic data from the Noric–Bosnian terrane (Schätz *et al.* 1996; Tait *et al.* 1998) indicate that it was positioned at 25° S in mid-Devonian time (Fig. 3d), and the sedimentary record describes a passive margin environment with development of substantial carbonate reefs in this southern Alpine realm. The more northerly units of the Alps (External Massifs) had already accreted with the active southern margin of Laurussia (von Raumer 1998).

The Carboniferous period saw drastic changes in the Alps. Restricted flysch-type sedimentation was initiated in Early Namurian time and the climax of deformation occurred between the Namurian and Late Westphalian stages as a result of consolidation of the alpine terranes and blocks with the Laurussian margin. Open marine sedimentation occurred along the southern margin of Laurussia (in the southern Alps) with development of shallow-water carbonates, marking the northern border of the Palaeotethys (von Raumer 1998). This marine connection persisted all along the southern margin of the Alpine terranes until at least Late Carboniferous time, thus arguing against any major continental collision with Gondwana before this (Stampfli 1996). Lower to mid-Carboniferous calc-alkaline magmatism documents oblique northward-directed subduction of the Palaeotethys beneath southern Europe with active subduction continuing into late Carboniferous time (Ziegler 1986; Mercolli & Oberhánsli 1988) with final collision and formation of Pangaea (Fig. 3f) by late Carboniferous to Early Permian times (Frisch & Neubauer 1989).

Conclusions

Laurentia, Baltica, Avalonia and the ATA; formation of Laurussia

Closure of the Tornquist Sea separating Baltica and Avalonia, and the northern Iapetus Ocean between Laurentia and Baltica, which resulted in the Caledonian orogenic event are now well constrained, and have been dealt with in some detail by McKerrow *et al.* (this volume) and so are not repeated here. Palaeomagnetic data from various elements of the ATA provide a number of new insights into the geodynamic evolution of the Variscan fold belt. They indicate that all elements of the ATA had drifted away from the northern margin of Gondwana by Late Ordovician times (Fig. 3b). By Late Silurian–Early Devonian time Bohemia, Saxo-Thuringia, the Armorican Massif and the Catalan terrane were all proximal to the southern margin of Baltica–Avalonia, and formed a chain of continental blocks (ATA) within the latitudinal belt of 20–30° S (Fig. 3c). Faunal evidence demonstrates that closure of the Rheic Ocean between the ATA and Avalonia did not occur until mid-Devonian time (Fig. 3d). Final suturing and consolidation of the ATA, closure of the Saxo-Thuringian basin, and rotation of Bohemia into its present-day orientation had occurred by Late Devonian times (Fig. 3e) (Tait *et al.* 1997). Thus from lithological, faunal and palaeomagnetic data, the accretion of Gondwana-derived terranes to the southern margin of Laurussia was essentially a continuing process throughout the Palaeozoic era.

Laurussia, the Proto-Alps and Gondwana; formation of Pangaea

Palaeomagnetism, structural geology and lithostratigraphic data from the Alps indicate that the first and main collision between the European margin of Laurussia and Gondwana occurred in Mid- to Late Carboniferous times. Stepwise accretion and amalgamation of the Alpine terranes onto the southern margin of Laurussia occurred from Devonian to Mid-Carboniferous times (Fig. 3). Late Carboniferous collision of Gondwana was diachronous along the suture zone, with sinistral transpressive movement in the western regions accompanied by shearing and eastward transport of smaller terranes and blocks (von Raumer 1998). The Devonian–Carboniferous deformation in central Europe may, therefore, have been driven by collision of Gondwana-derived terranes with the southern active margin of Laurussian. The first evidence for any major continent collision from the south was in late Carboniferous time when Gondwana collided with Laurussia, resulting in formation of Pangaea (Fig. 3f).

The authors gratefully acknowledge financial support from the DFG.

References

ABOUS-DEEB, J. M. & TARLING, D. H. 1984. Upper Palaeozoic palaeomagnetic results from Algeria and Tunisia. *Tectonophysics*, **101**, 143–157.

ANTONITSCH, W., NEUBAUER, F. & DALLMEYER, R. D. 1994. Paleozoic evolution within the Gurktal Nappe Complex, Eastern Alps. *Journal of the Czech Geological Society*, **39**, 2–3.

BACHTADSE, V. & BRIDEN, J. C. 1991. Palaeomagnetism of Devonian ring complexes from the Bayuda Desert, sudan—new constraints on the apparent polar wander path for Gondwanaland. *Geophysical Journal International*, **104**, 635–646.

——, VAN DER VOO, R. & HAELBICH, I. W. 1987. Paleozoic paleomagnetism of the western Gondwana. *Earth and Planetary Science Letters*, **84**, 487–499.

BENEDETTO, J. L. 1998. Early Palaeozoic brachiopods and associated shelly fauna from western Gondwana: their bearing on the geodynamic history of the pre-Andean margin. *In*: PANKHURST, R. J. & RAPELA, C. W. (eds) *The Proto-Andean Margin of Gondwana*. Geological Society, London, Special Publications **142**, 57–83.

BERDAN, J. M. 1991. The Silurian and early Devonian biogeography of ostracodes in North America. *In*: MCKERROW, W. S. & SCOTESE, C. R. (eds) *Palaeozoic Biogeography and Palaeobiogeography*. Geological Society, London Memoir, **12**, 223–231.

BOND, G. C., NICKESON, P. A. & KOMINZ, M. A. 1984. Breakup of a supercontinent between 625 Ma and 555 Ma: new evidence and implications for continental histories. *Earth and Planetary Science Letters*, **70**, 325–345.

BRENCHLEY, P. J. M., ROMANO, T. P., YOUNG, G. C. & STORCH, P. 1991. Hirnantian glaciomarine diamicitites—evidence for the spread of glaciation and its effect on Upper Ordovician faunas. *In*: BARNES, C. R. & WILLIAMS, S. H. (eds) *Advances in Ordovician Geology*. Geological Survey of Canada, 325–336.

BURCHETTE, T. P. 1981. European Devonian reefs: a review of current concepts and models. *In*: MOONEY, D. F. (ed.) *European Fossil Reef Models*. SEPM Special Publications, **30**, 85–142.

CHEN, Z., LI, X., POWELL, M. & BALME, B. E. 1993. Palaeomagnetism of the Brewer Conglomerate in central Australia, and fast movement of Gondwanaland during the Late Devonian. *Geophysical Journal International*, **115**, 564–574.

——, ——, —— & —— 1994. An Early Carboniferous paleomagnetic pole for Gondwanaland: new results from the Mount Eclipse Sandstone in the Ngalia Basin, central Australia. *Journal of Geophysical Research*, **99**, 2909–2924.

——, —— & —— 1995. Paleomagnetism of the Upper Devonian reef complexes, Canning Basin, Western Australia. *Tectonics*, **14**, 154–167.

CLOUD, P. E. 1961. Paleobiogeography of the marine realm. *American Association for the Advancement of Science Publications*, **67**, 151–200.

COCKS, L. R. M. 1993. Triassic pebbles, derived fossils and the Ordovician to Devonian palaeogeography of Europe. *Journal of the Geological Society, London*, **150**, 219–226.

—— & FORTEY, R. A. 1982. Faunal evidence for oceanic separations in the Palaeozoic of Britain. *Journal of the Geological Society, London*, **139**, 465–478.

——, MCKERROW, W. S. & VAN STAAL, C. R. 1997. The margins of Avalonia. *Geological Magazine*, **134**, 627–636.

COLLINS, W. J. & VERNON, R. H. 1992. Palaeozoic arc growth, deformation and migration across the Lachlan Fold Belt, southeastern Australia. *Tectonophysics*, **214**, 381–400.

CONEY, P. J. 1992. The Lachlan belt of eastern Australia and Circum-Pacific tectonic evolution. *Tectonophysics*, **214**, 1–25.

COOPER, A. H., MILLWARD, D., JOHNSON, E. W. & SOPER, N. J. 1993. The early Palaeozoic evolution of northwest England. *Geological Magazine*, **130**, 711–724.

CORNER, B. & NETHORN, D. I. 1978. *Results of a palaeomagnetic survey undertaken in the Damara Mobile Belt, South West Africa, with special reference to the magnetisation of the Uraniferous Pegmatitic Granites*. PEL-260, Atomic Energy Board, Pretoria.

DALLMEYER, R. D. & NEUBAUER, F. 1994. Cadomian $^{40}Ar/^{39}Ar$ apparent age spectra of detrital muscovites from the Eastern alps. *Jounral of the Geological Society, London*, **151**, 591–598.

DALY, L. & IRVING, E. J. 1983. Paléomagnetisme des roches carbonifères du Sahara central; analyse des aimantation juxtaposées; configuration de la Pangée. *Annales Geophysicae*, **1**, 207–215.

—— & POZZI, J. P. 1976. Résultats paléomagnetiques du Permien inférieur et du Trias Marocain: comparaison avec les données africaines et sud-americaines. *Earth and Planetary Science Letters*, **29**, 71–80.

DALZIEL, I. W. D. 1992. Antarctica: a tale of two supercontinents. *Annual Review of Earth and Planetary Sciences*, **20**, 501–526.

DERDER, M. E., HENRY, B., MERABET, N. E. & DALY, L. 1994. Palaeomagnetism of the Stephano-Autunian Lower Tiguentourine formations from stable Sharan craton (Algeria). *Geophysical Journal International*, **116**, 12–22.

EMBLETON, B. J. J. 1972. The palaeomagnetism of some Palaeozoic sediments from central Australia. *Journal and Proceedings, Royal Society of New South Wales*, **105**, 86–93.

FISHER, R. A. 1953. Dispersion on a sphere. Proceedings of The Royal Society of London, Series A, **217**, 295–305.

FORTEY, R. A. & COCKS, L. R. M. 1988. Arenig to Llandovery faunal distributions in the Caledonides. *In*: HARRIS, A. L. & FETTES, D. J. (eds) *The Caledonian–Appalachian Orogen*. Geological Society, London, Special Publications, 233–246.

FRANKE, W., DALLMEYER, D. & WEBER, K. 1995. Geodynamic evolution. *In*: DALLMEYER, R. D., FRANKE, W. & WEBER, K. (eds) *Pre-Permian Geology of Central and Eastern Europe*. Springer, Berlin, 579–593.

FREY, M., DESMONSSS, J. & NEUBAUER, F. 1999. The new metamorphic map of the Alps: introduction. *Schweizerische Mineralogische und Petrographisophe Mitteitungen*, **79**(1), 1–4.

FRISCH, W. & NEUBAUER, F. 1989. Pre-Alpine terranes and tectonic zoning in the eastern Alps. *In*: DALLMEYER, R. D. (ed.) *Terranes in the Circum-Atlantic Palaeozoic Orogens*. Geological Society of America, Special Papers, **230**, 91–100.

GLEN, R. A. 1993. Palaeomagnetism and terranes in the Lachlan Orogen. *Exploration Geophysics*, **24**, 247–256.

——, WALSHE, J. L., BARRON, L. M. & WATKINS, J. J. 1998. Ordovician convergent-margin volcanism and tectonism in the Lachlan sector of east Gondwana. *Geology*, **26**(8), 751–754.

GRAY, D. R., FOSTER, D. A. & BUTCHER, M. 1997. Recognition and definition of orogenic events in the Lachlan Fold Belt. *Australian Journal of Earth Sciences*, **44**, 489–501.

GRUNOW, A. 1995. Implications for Gondwana of new Ordovician paleomagnetic data from igneous rocks in southern Victoria Land. *Journals of Geophysical Research*, **100**(B7), 12589–12603.

HANDLER, R., DALLMEYER, R. D. & NEUBAUER, F. 1997. $^{40}Ar/^{39}Ar$ ages of detrital white mica from Upper Austroalpine units in the Eastern Alps, Austria evidence for Cadomian and contrasting Variscan sources. *Geologische Rundschau*, **86**, 69–80.

HARGRAVES, R. B., DAWSON, E. M. & VAN HOUTEN, F. B. 1987. Paleomagnetism and age of mid Palaeozoic ring complexes in Niger, West Africa, and tectonic implications. *Geophysical Journal of the Royal Astronomical Society*, **90**, 705–729.

HENRY, B., MERABET, N., YELLES, A., DERDER, M. M. & DALY, L. 1992. Geodynamical implications of a Moscovian palaeomagnetic pole from the stable Sahara craton (Illizi basin, Algeria). *Tectonophysics*, **201**, 83–96.

HURLEY, N. F. & VAN DER VOO, R. 1987. Paleomagnetism of Upper Devonian reefal limestones, Canning Basin, Western Australia. *Geological Society of America Bulletin*, **98**, 138–146.

JOKIEL, P. L. 1990. Long-distance dispersal by rafting: reemergence of an old hypothesis. *Endeavour*, **14**, 66–73.

KENT, D. V. & SMETHURST, M. A. 1998. Shallow bias of paleomagnetic inclinations in the Paleozoic and Precambrian. *Earth and Planetary Science Letters*, **160**(3–4), 391–402.

KLAPPER, G. & JOHNSON, J. G. 1980. Endemism and dispersal of Devonian conodonts. *Journal of Palaeontology*, **54**, 400–455.

KOKELAAR, B. P., HOWELLS, M. F., BEVINS, R. E., ROACH, R. E. & DUNKLEY, P. N. 1984. The Ordovician marginal basin of Wales. *In*: KOKELAAR, B. P. & HOWELLS, M. F. (eds) *Marginal Basin Geology*, Geological Society, London, Special Publications, **16**, 245–269.

KÖBLER, P., TAIT, J., BACHTADSE, V., SOFFEL, H. C. & LINNEMAN, U. 1996. Palaeomagnetic investigations of lower Palaeozoic rocks of the Thueringisches Schiefergebirge. *Terra Nostra*, **96**(2), 114–116.

KRIZ, J. & PARIS, F. 1982. Ludlovian, Pridolian and Lochkovian in La Meignanne (Massif Armoricain): biostratigraphy and correlations based on bivalvia and chitinozoa. *Geobios*, **15**(3), 391–421.

LANZA, R. & TONARINI, S. 1998. Palaeomagnetic and geochronological results from the Cambro-Ordovician Granite Harbour Intrusives inland of Terra Nova Bay (Victoria Land, Antarctica).

Geophysical Journal International, **135**, 1019–1027.

Lawver, L. A. & Scotese, C. R. 1987. A revised reconstruction of Gondwanaland. *In*: McKenzie, G. D. (ed) *Gondwana Six: Stratigraphy, Sedimentology and Paleontology*. Geophysical Monography, American Geophysical Union, **40**, 17–23.

Li, Z. X., Powell, C. M., Embleton, B. J. J. & Schmidt, P. W. 1991. New palaeomagnetic results from the Amadeus Basin and their implications for stratigraphy and tectonics. *In*: Korsch, R. J. & Kennard, J. M. (eds) *Geological and Geophysical Studies in the Amadeus Basin, Central Australia*. Bureau of Mineral Resources, Canberra, Australia, 349–360.

——, Schmidt, P. W. & Embleton, B. J. J. 1988. Paleomagnetism of the Hervey Group, central New South Wales, and its tectonic implications. *Tectonics*, **7**, 351–367.

Lockley, M. G. 1983. A review of brachiopod dominated palaeocommunities from the type Ordovician. *Palaeontology*, **26**, 111–145.

Loeschke, J. & Heinisch, H. 1993. Palaeozoic volcanism of the Eastern Alps and its palaeotectonic significance. *In*: Raumer, J. von & Neubauer, F. (eds) *The Pre-Mesozoic Geology in the Alps*. Springer, Berlin, 441–455.

MacNiocaill, C. & Smethurst, M. A. 1994. Palaeozoic palaeogeography of Laurentia and its margins: a reassessment of palaeomagnetic data. *Geophysical Journal International*, **116**, 715–725.

McElhinny, M. W. 1968. Palaeomagnetic directions and pole positions—VIII, pole numbers 8/1 to 8/186. *Geophysical Journal of the Royal Astronomical Society*, **15**, 409–430.

—— & Opdyke, N. D. 1968. Paleomagnetism of some Carboniferous glacial varves from Central Africa. *Geophysical Journal of the Royal Astronomical Society*, **73**, 689–696.

——, Embleton, B. J. J., Daly, L. & Pozzi, J. P. 1976. Paleomagnetic evidence for the location of Madagascar in Gondwanaland. *Geology*, **4**, 455–457.

McKerrow, W. S., Dewey, J. F. & Scotese, C. F. 1991. The Ordovician and Silurian development of the Iapetus Ocean. *Special Paper, in Palaeontology*, **44**, 165–178.

——, MacNiocaill, C., Ahlberg, P. E., Clayton, G., Cleal, C. J. & Eager, R. M. C. 2000. The late Palaeozoic relations between Gondwana and Laurussia. *This volume*.

Merabet, N., Bouabdallah, H. & Henry, B. 1998. Paleomagnetism of the Lower Permian redbeds of the Abadla Basin (Algeria). *Tectonophysics*, **293**(1–3), 127–136.

Mercolli, I. & Oberhänsli, R. 1988. Variscan tectonic evolution in the Central Alps: a working hypothesis. *Schweizerische Mineralogische and Petrographische Mitteilungen*, **68**, 491–500.

Moreau, C., Demaiffe, D., Bellion, Y. & Boullier, A.-M. 1994. A tectonic model for the location of Palaeozoic ring complexes in Aïr (Niger, West Africa). *Tectonophysics*, **234**, 129–146.

Morel, P. & Irving, E. 1978. Tentative paleocontinental maps for the early Phanerozoic and Proterozoic. *Journal of Geology*, **86**, 535–561.

Neubauer, F. & Raumer, J. von 1993. The Alpine basement: a linkage between Variscides and East-Mediterranean mountain belt. *In*: Raumer, J. von & Neubauer, F. (eds) *The Pre-Mesozoic Geology in the Alps*. Springer, Berlin, 641–663.

——, Hoinkes, G., Sassi, F. P., Handler, R., Höck, V., Koller, F. & Frank, W. 1999. Pre-Alpine metamorphism of the Eastern Alps. *Schweizerische Mineralogische Petrograchische Mitteilungen*, **79**(1), 41–62.

Nyblade, A. A., Lei, Y., Shive, P. N. & Tesha, A. 1993. Paleomagnetism of Permian sedimentary rocks from Tanzania and the Permian paleogeography of Gondwana. *Earth and Planetary Science Letters*, **118**, 181–194.

Oclzon, M. S. 1990. Ocean currents and unconformities: the North Gondwana Middle Devonian. *Geology*, **18**, 509–512.

Opdyke, N. D. 1964. The paleomagnetism of the Permian redbeds of southwest Tanganyika. *Journal of Geophysical Research*, **69**, 2477–2487.

Owen, A. W., Harper, D. A. T. & Jia-yu, R. 1991. Hirnantian trilobites and brachiopods in space and time. *In*: Barnes, C. R. & Williams, S. H. (eds) *Advances in Ordovician Geology*. Geological Survey of Canada, 179–190.

Perroud, H. & Van der Voo, R. 1985. Palaeomagnetism of the Late Ordovician Thouars Massif, Vendée province, France. *Journal of Geophysical Research*, **90**, 4611–4625.

——, Bonhommet, N. & Van der Voo, R. 1983. Palaeomagnetism of the Ordovician dolerites of the Crozon Peninsula (France). *Geophysical Journal of the Royal Astronomical Society*, **72**, 307–319.

——, Robardet, M. & Bruton, D. L. 1992. Palaeomagnetic constraints upon the palaeogeographic position of the Baltic Shield in the Ordovician. *Tectonophysics*, **201**, 97–120.

Pharaoh, T. C., Brewer, T. S. & Webb, P. C. 1993. Subduction-related magmatism of late Ordovician age in eastern England. *Geological Magazine*, **130**, 647–656.

Powell, C. M. 1983. Tectonic relationship between the Late Ordovician and Late Silurian palaeogeographies of south eastern Australia. *Journal of the Geological Society of Australia*, **30**, 353–373.

—— 1984. Silurian to Mid-Devonian. *In*: Veevers, J. J. (ed.) *Phanerozoic Earth History of Australia*, Oxford University Press, Oxford, 309–328.

Robardet, M. & Gutierrez-Marco, J. C. 1990. Sedimentary and faunal domains in the Iberian peninsula during lower Paleozoic times. *In*: Dallmeyer, R. D. & Martinez-García, E. (eds) *Pre-Mesozoic Geology of Iberia*. Springer, Berlin.

——, Verniers, J., Feist, R. & Paris, F. 1994. Le Paléozoique anté-varisque de France, contecte paléogéographique et géodynamique. *Géologie de la France*, **3**, 3–31.

Ross, R. J. 1975. Early Paleozoic trilobites, sedimentary facies, lithospheric plates, and ocean current. *Fossils and Strata*, **4**, 307–329.

Schallreuter, R. E. L. & Siveter, D. J. 1985. Ostracodes across the Iapetus Ocean. *Palaeontology*, **28**(3), 577–598.

Schätz, M., Bachtadse, V., Tait, J., Soffel, H. C. & Heinisch, H. 1996. New palaeomagmatic results from the southern flank of the European Variscides from the Northern Greywacke Zone, E-Alps. *Terra Nostra, Alfred-Wegener-Stiftung*, **96/2**, Köln, 165–168.

Scheibner, E. & Basden, H. 1996. *Geology of New South Wales—Synthesis. Volume 1. Structural Framework*. Geological Survey of New South Wales, Memoir, Geology, **13**(1).

Schmidt, P. W. & Embleton, B. J. J. 1990. The palaeomagnetism of the Tumblagooda Sandstone, Western Australia: Gondwana Palaeozoic apparent polar wandering. *Physics of the Earth and Planetary Interiors*, **64**, 303–313.

——, Cudahy, T. J. & Powell, C. M. 1986. Prefolding and premegakinking magnetizations from the Devonian Comerong volcanics, New South Wales, Australia, and their bearing on the Gondwana pole path. *Tectonics*, **5**, 135–150.

—— & Palmer, H. C. 1987. Pre- and post-folding magnetizations from the early Devonian Snowy River Volcanics and Buchan Caves Limestone, Victoria. *Geophysical Journal of the Royal Astronomical Society*, **91**, 155–170.

——, Powell, C. M. Li, Z. Z & Thrupp, G. A. 1990. Reliability of Palaeozoic palaeomagnetic poles and APWP of Gondwanaland. *Tectonophysics*, **184**, 87–100.

Schönlaub, H. P. 1992. Stratigraphy, biography and paleoclimatology of the Alpine Paleozoic and its implications for plate movements. *Jahrbuch der Geologischen Bundesanstalt*, **135**(1), 381–418.

—— & Heinisch, H. 1993. The classic fossiliferous Palaeozoic units of the Eastern and Southern Alps. *In*: Raumer, J. von & Neubauer, F. (eds) *The Pre-Mesozoic Geology in the Alps*. Springer, Berlin, 395–422.

Scotese, C. R., Boucot, A. J. & McKerrow, W. S. 1999. Gondwanan palaeogeography and palaeoclimatology. *Journal of African Earth Sciences*, **28**(1), 99–114.

Smethurst, M. A., Khramov, A. N. & Torsvik, T. H. 1998. The Neoproterozoic and Palaeozoic palaeomagnetic data for the Siberian Platform: from Rhodinia to Pangaea. *Earth-Science Reviews*, **43**, 1–24.

Soffel, H. C., Saradeth, S., Briden, J. C., Bachtadse, V. & Rolf, C. 1990. The Sabaloka Ring Complex revisited: palaeomagnetism and rock magnetism. *Geophysical Journal of the Royal Astronomical Society*, **102**, 411–420.

Sommermann, A.-E., Meisl, S. & Todt, W. 1992. Zirkonalter von drei verschiedenen Metavulkaniten aus dem Südtaunus. *Geologisches Jahrbuch Hessen*, **120**, 67–76.

Stampfli, G. M. 1996. The intra-alpine terrain: a paleotethyan remnant in the Alpine Variscides. *Eclogae Geologicae Helvetiae*, **89**(1), 13–42.

Tait, J. 1999. New Early Devonian paleomagnetic data from NW France: paleogeography and implications for the Armorican microplate hypothesis. *Journal of Geophysical Research*, **104**(B2), 2831–2839.

Tait, J. A., Bachtadse, V. & Dinarès-Turell, J. 2000. Paleomagnetism of Siluro-Devonian sequences, NE Spain. *Journal of Geophysical Research*, **105**, **B10**, 23595–23604.

——, ——, Franke, W. & Soffel, H. C. 1997. Geodynamic evolution of the European Variscan fold belt: palaeomagnetic and geological constraints. *Geoligische Rundschau*, **86**, 585–598.

——, —— & Soffel, H. 1994. Silurian paleogeography of Armorica: new paleomagnetic data from central Bohemia. *Journal of Geophysical Research*, **99**(B2), 2897–2907.

Tait, J., Schätz, M., Bachtadse, V. & Soffel, H. 1998. Paleogeography of paleozoic terranes in the Variscan and Alpine fold belts. *Schriften des Staatlichen Museums für Mineralogie und Geologie zu Dresden*, **9**, 192–193.

Thöni, M. 1999. A review of geochronological data from the Eastern Alps. *Schweizeriehe Mineralogische and Petrographische Mitterlungen*, **79**(1), 209–230.

Thrupp, G. A., Kent, D. V., Schmidt, P. W. & Powell, C. M. 1991. Palaeomagnetism of red beds of the Late Devonian Worange Point Formation, SE Australia. *Geophysical Journal International*, **104**, 179–201.

Torsvik, T. H., Oleson, O., Ryan, P. D. & Trench, A. 1990. On the palaeogeography of Baltica during the Palaeozoic: new palaeomagnetic data from the Scandinavian Caledonides. *Geophysical Journal International*, **103**, 261–279.

——, Smethurst, M. A., Meert, J. G. *et al.* 1996. Continental break-up and collision in the Neoproterozoic and Palaeozoic: a tale of Baltica and Laurentia. *Earth-Science Reviews*, **40**, 229–258.

——, van der Voo, R., Trench, A., Abrahamsen, N. & Halvorsen, E. 1992. BALTICA—a synopsis of Vendian–Permian palaeomagnetic data and their palaeotectonic implications. *Earth-Science Reviews*, **33**, 133–152.

——, Trench, A., Svensson, I. & Walderhaug, H. J. 1993. Palaeogeographic significance of mid-Silurian palaeomagnetic results from southern Britain: major revision of the apparent polar wander path for eastern Avalonia. *Geophysical Journal International*, **113**, 651–668.

Trench, A. & Torsvik, T. H. 1991. A revised Paleozoic apparent polar wander path for south Britain (eastern Avalonia). *Geophysical Journal International*, **104**, 227–233.

——, ——, Denith, M. C., Walderhaug, H. & Traynor, J.-J. 1992. A high southerly palaeolatitude for Southern Britain in Early Ordovician times: palaeomagnetic data from the Treffgarne Volcanic Formation SW Wales. *Geophysical Journal International*, **108**, 89–100.

VAN STAAL, C. R., DEWEY, J. F., MACNIOCAILL, C. & MCKERROW, W. S. 1998. The Cambrian–Silurian tectonic evolution of the Northern Appalachians and British Caledonides; history of a complex, west and southwest Pacific-type segment of Iapetus. *In*: BLUNDELL, D. J. & SCOTT, A. C. (eds) *Lyell; the Past is the Key to the Present*. Geological Society, London, Special Publications, **143**, 199–242.

VAN DER VOO, R. 1990. The reliability of paleomagnetic data. *Tectonophysics*, **184**, 1–10.

—— 1993. *Paleomagnetism of the Atlantic, Tethys and Iapetus Oceans*. Cambridge University Press, Cambridge.

VON RAUMER, J. F. 1998. The Palaeozoic evolution in the Alps: from Gondwana to Pangea. *Geologische Rundschau*, **87**, 407–435.

WITZKE, B. J. 1990. Paleoclimatic constraints for Paleozoic paleolatitudes of Laurentia and Euramerica. *In*: MCKERROW, W. S., SCOTESE, C. R. (eds) *Palaeozoic Biogeography and Palaeobiogeography*. Geological Society, London, Memoir, **12**, 57–73.

ZIEGLER, P. A. 1986. Geodynamic model for the Palaeozoic crustal consolidation of Western and Central Europe. *Tectonophysics*, **126**(2–4), 303–328.

The mid-European segment of the Variscides: tectonostratigraphic units, terrane boundaries and plate tectonic evolution

W. FRANKE

Institut für Geowissenschaften der Universität, D-35 390 Giessen, Germany
(e-mail: wolfgang.franke@geolo.uni-giessen.de)

Abstract: The mid-European segment of the Variscides is a tectonic collage consisting of (from north to south): Avalonia, a Silurian–early Devonian magmatic arc, members of the Armorican Terrane Assemblage (ATA: Franconia, Saxo-Thuringia, Bohemia) and Moldanubia (another member of the ATA or part of N Gondwana?).

The evolution on the northern flank of the Variscides is complex. Narrowing of the Rheic Ocean between Avalonia and the ATA occurred during the late Ordovician through early Emsian, and was accompanied by formation of an oceanic island arc. By the early Emsian, the passive margin of Avalonia, the island arc and some northern part of the ATA were closely juxtaposed, but there is no tectonometamorphic evidence of collision. Renewed extension in late Emsian time created the narrow Rheno-Hercynian Ocean whose trace is preserved in South Cornwall and at the southern margins of the Rhenish Massif and Harz Mts. Opening of this 'successor ocean' to the Rheic left Armorican fragments stranded on the northern shore. These were later carried at the base of thrust sheets over the Avalonian foreland. Closure of the Rheno-Hercynian Ocean in earliest Carboniferous time was followed by deformation of the foreland sequences during the late lower Carboniferous to Westphalian.

Closure of narrow oceanic realms on both sides of Bohemia occurred during the mid- and late Devonian by bilateral subduction under the Bohemian microplate. In both these belts (Saxo-Thuringian, Moldanubian), continental lithosphere was subducted to asthenospheric depths, and later partially obducted. Collisional deformation and metamorphism were active from the late Devonian to the late lower Carboniferous in a regime of dextral transpression. The orthogonal component of intra-continental shortening produced an anti-parallel pair of lithospheric mantle slabs which probably joined under the zone of structural parting and became detached. This allowed the ascent of asthenospheric material, with important thermal and rheological consequences. The strike slip displacements were probably in the order of hundreds of kilometres, since they have excised significant palaeogeographic elements.

This paper presents a survey of the main tectono-stratigraphic units of the Variscan Belt in central Europe, their palaeogeographic affinities, and plate kinematic evolution. It is intended to provide a thematic frame and database for the other contributions in this volume. Wherever possible, it refers to these contributions, or to recent review papers. More detailed treatments are necessary in regions with controversial interpretations, especially in metamorphic terrains. Palaeogeographic affinities are discussed at the end of the descriptions of tectono-stratigraphic units. The timing of extension, subduction and collision is assessed in a separate chapter. Continuation of the mid-European units eastward into the Sudetes is treated in Franke & Żelaźniewicz (this volume).

The Variscan orogen in extra-Alpine Europe is a terrane collage resulting from subduction/collision in three major belts, most of which were already been recognized by Kossmat (1927): the Rheno-Hercynian, the Saxo-Thuringian and the Moldanubian (see enclosure at back and Fig. 1). Modern tectonic concepts, from Bard *et al.* (1980) over Matte (1986), Matte *et al.* (1990), to Franke (1989, Franke *et al.* 1995*a*) agree in a bilateral symmetry with grossly southward subduction on the north flank (Rheno-Hercynian, Saxo-Thuringian) and northward subduction on the south flank (Moldanubian).

Rheno-Hercynian Belt (RH)

The RH is exposed in the Rhenish Massif to the west and east of the Rhine, a small area of outcrop in the Werra Grauwackengebirge, in the Harz Mts and in the Flechtingen horst (see enclosure). Summaries of the palaeogeographic

From: FRANKE, W., HAAK, V., ONCKEN, O. & TANNER, D. (eds). *Orogenic Processes: Quantification and Modelling in the Variscan Belt.* Geological Society, London, Special Publications, **179**, 35–61. 0305-8719/00/$15.00

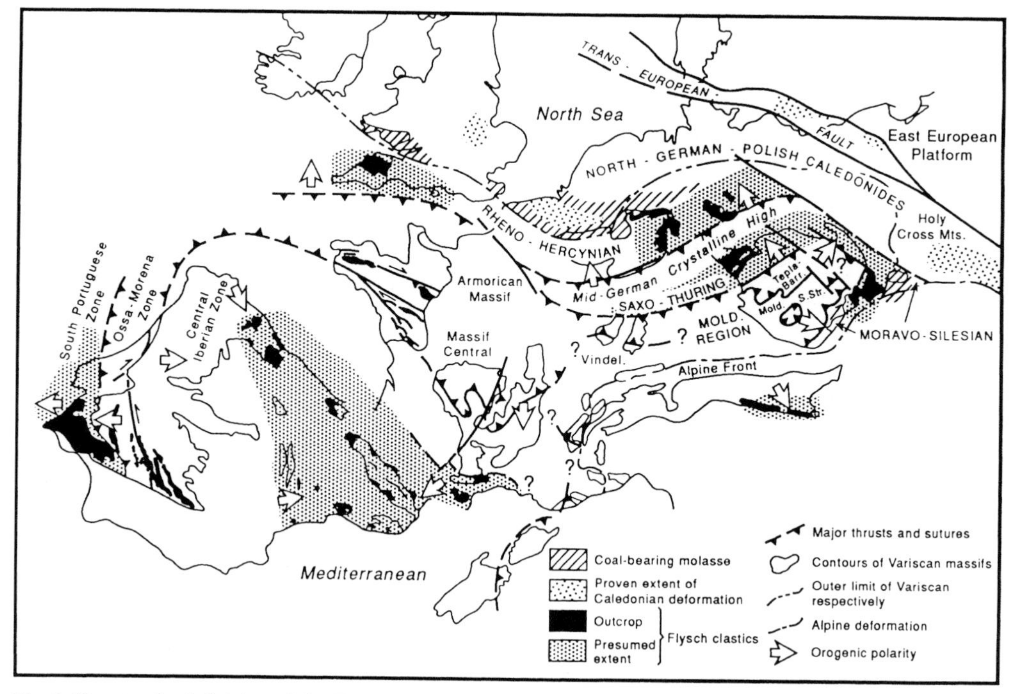

Fig. 1. Structural subdivision of the European Variscides (from Franke 1992).

evolution have been published by Meischner (1971), Franke *et al.* (1978), Walliser (1981), Engel *et al.* (1983), and Franke (1995). Westwards, the RH Belt can be traced into south Cornwall, and possibly into southern Portugal (Franke & Engel 1982; Holder & Leveridge 1986; Olivera & Quesada 1998). A former eastern continuation of the RH is seen in the Moravo-Silesian Belt (Engel & Franke 1983; Finger & Steyrer 1995; Franke & Żelaźniewicz this volume). However, this applies only to the Devonian extension and volcanism, Early Carboniferous flysch and Late Carboniferous coal-bearing molasse. Palaeogeographic affinities during the early Palaeozoic and late Proterozoic are still controversial (Belka *et al.* this volume).

All segments of the RH have a very similar sedimentary record, which is dominated by Devonian clastic shelf sediments (Fig. 2). These were derived from Caledonian sources to the north and deposited at the southern, passive margin of the Old Red Sandstone Continent (Laurussia). An originally coherent set of thin-skinned thrust sheets (the Giessen–Werra–Harz nappes, Fig. 3) contains Devonian MORB-type metabasalts and pelagic sediments, and a complex association of rock types derived from the distal margin of the Old Red continent. The Harz Mts and the Rhenish Massif mainly differ in the level of erosion: in the latter, most of the allochthon has been eroded.

Parautochthon: basement and breakup unconformity

The south Hunsrück, at the southeastern margin of the Rhenish Massif, west of the Rhine, exposes a narrow strip of Cadomian paragneisses (Wartenstein Gneiss, 574 ± 3 Ma Pb–Pb zircon evaporation: Molzahn *et al.* 1998, 550 ± 20 Ma, K–Ar and Ar/Ar on hornblende and mica: Meisl *et al.* 1989).

The basement is overlain by late Gedinnian clastic sediments (Dittmar 1996). Further to the NW, in the Ardennes, the late Gedinnian overlies folded and weakly metamorphosed Cambrian and Ordovician rocks of a ‘Caledonian’ (Acadian) basement (Pharaoh *et al.* 1993; van Grootel *et al.* 1997). During continued crustal extension, the marine transgression advanced towards the NW, and reached the southern margin of the Brabant Massif in Frasnian time (Kasig & Wilder 1983; Walter *et al.* 1985). In the eastern part of the Rhenish Massif, the basement is unknown. The Ebbe and Remscheid Anticlines expose Ordovician and Silurian shaly sequences without any indications of a ‘Caledonian’ event (Gruppenbericht 1981).

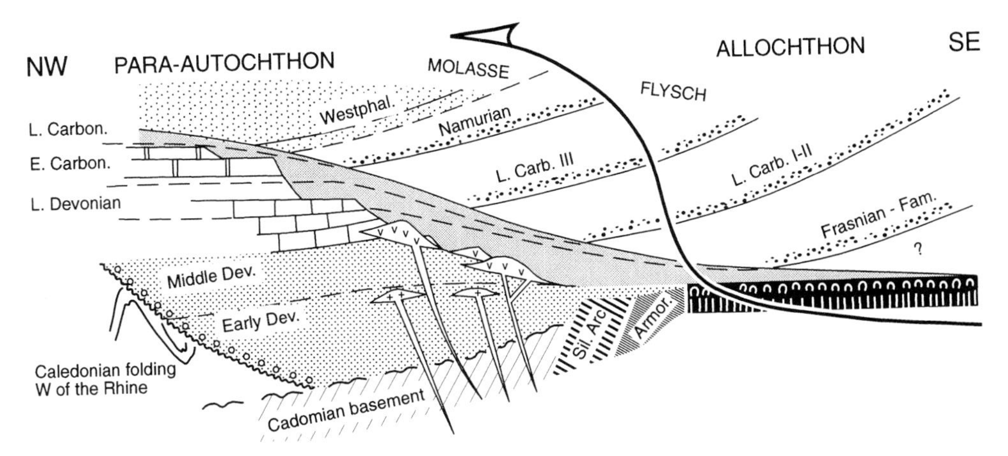

Fig. 2. Dynamic stratigraphy of the Rheno-Hercynian Belt.

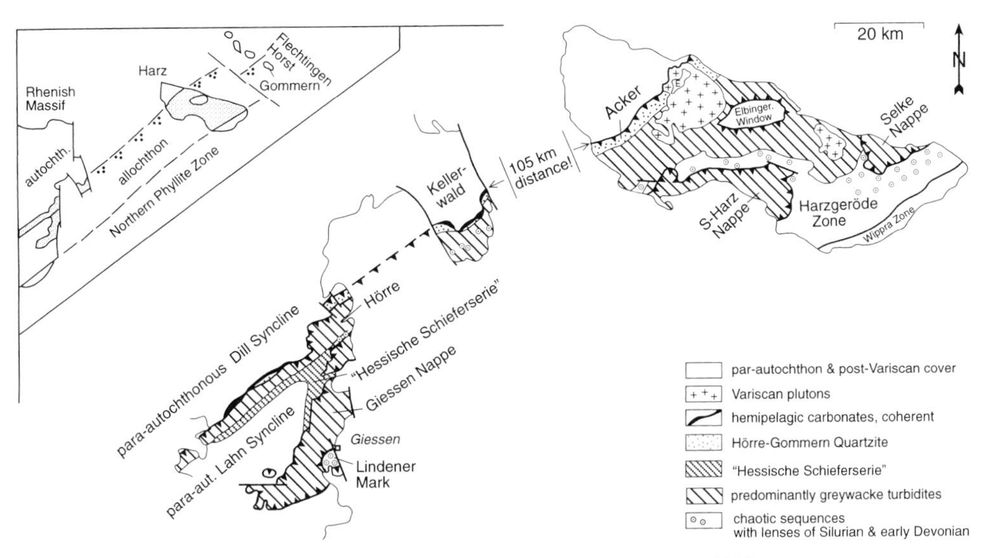

Fig. 3. Allochthonous units in the Rhenish Massif and Harz Mts (after Franke 1995).

These are overlain by thick clastic sediments of early Gedinnian and younger age. In the Taunus Mts, at the southeastern margin of the Rhenish Massif, east of the Rhine, the oldest deposits known so far are Gedinnian arkoses (Hahn 1990). It appears, that sedimentation in the Rhenish Massif commenced in earliest Devonian time (i.e., around 415 Ma—see McKerrow & van Staal this volume).

Parautochthon: Devonian rifting

A renewed pulse of extension occurred in Late Emsian time. It is documented in widespread felsic volcanism in southeastern parts of the Rhenish Massif (Kirnbauer 1991), in high heat flow derived from vitrinite reflectance studies (Littke *et al.* this volume), and by formation of oceanic crust in the future allochthon (see below).

The evolution during the mid- and late Devonian is characterized by clastic sedimentation on a rapidly subsiding shelf, which receded northwestwards in time (Langenstrassen 1983), thus giving way to hemipelagic sedimentation. Siliciclastic debris was derived from Caledonian sources to the NW, which is documented in K–Ar ages of detrital micas ranging from 459 to 409 Ma (Huckriede *et al.* 1998; Küstner *et al.* 1998). The U–Pb systematics of detrital zircons reveal Laurentia or Baltica related sources (Haverkamp *et al.* 1991). Mid-Devonian to

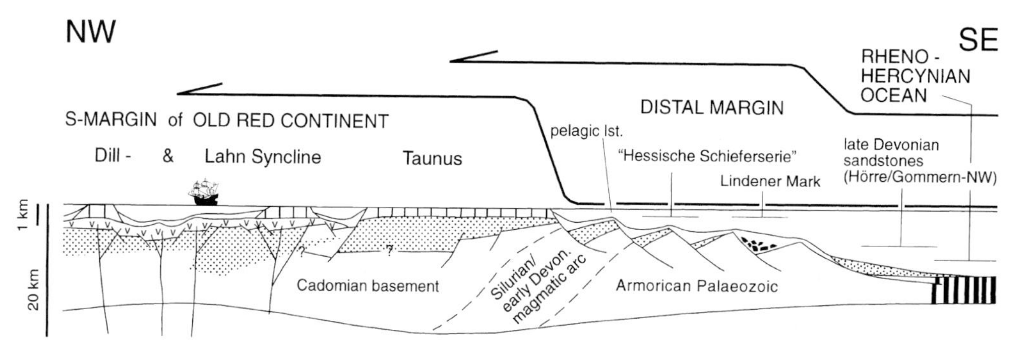

Fig. 4. Palinspastic restoration of allochthonous units in the Rhenish Massif.

Frasnian reef carbonates were formed during intervals of reduced clastic supply. After the terminal clastic input in the basal Carboniferous (Gattendorf stage; cd I), the clastic shelf was transformed into the Carboniferous Limestone Platform which extended westwards into the central British Isles.

Volcanism in post-Emsian time was mainly basic, with pronounced peaks in the Givetian and Tournaisian/early Viséan. It occurred along one minor, east–west-trending lineament in the northern part of the Rhenish Massif ('Haupt-Grünsteinzug' in the Ostsauerland Anticline, see enclosure at back, only Givetian), and in a broad belt with mid-Devonian to Viséan volcanism, which extends from the Lahn/Dill area of the Rhenish Massif into the northwestern part of the Harz Mts. The southeastern margin of the Rhenish Massif (Taunus Mts) is devoid of volcanic rocks. These relationships suggest a pinch and swell geometry of the extended crust (Fig. 4). The chemical composition of the basic volcanic rocks indicates a within-plate environment (Floyd 1995).

Closure of the Rheno-Hercynian Basin is recorded in greywacke turbidites derived from an active margin to the SE (the Mid German Crystalline High, see below). The clastic wedge reached the parautochthonous foreland of the Harz Mts in the late Tournaisian (Stoppel & Zscheked 1971) and propagated northwards with time (Engel & Franke 1983; Franke & Engel 1986). By Namurian C time, sedimentation outpaced subsidence, and sedimentation continued in the paralic, coal-bearing molasse environment of the Ruhr District in the northern part (and north of) the Rhenish Massif.

The development of the Rhenish Massif, west of the Rhine, differs in several respects: siliciclastic shelf sedimentation is followed by platform carbonates already at the base of the middle Devonian (east of the Rhine: Givetian), synsedimentary magmatism is negligeable. These features are best explained by a reduced amount of synsedimentary extension with respect to the areas east of the Rhine.

Allochthons

Allochthonous units in the Rheno-Hercynian Belt, already recognized by Kossmat (1927), are derived from the southeastern, deeper parts of the Rheno-Hercynian Basin. The present review expands upon the view of Engel *et al.* (1983), who proposed an allochthonous position not only for the Giessen–Südharz–Selke Unit, but also for the Hörre–Gommern and closely related units adjacent to the NW. A detailed discussion of the problem goes beyond the scope of this paper (see Meischner 1991 for an alternative view). The allochthon can be subdivided into NE-trending sub-belts, most of which can be traced from the Rhenish Massif into the Harz Mts. These are briefly characterized in order from the NW to the SE (Fig. 3).

Pelagic limestones. A condensed sequence of Eifelian to late Famennian pelagic carbonates forms a narrow, discontinuous belt at the northwestern front of the allochthon in the Rhenish Massif, east of the Rhine (Fig. 3). Similar carbonates occur in the 'Hessische Schieferserie' to the SE of the Hörre–Gommern Zone, and in the south-easternmost part of the parautochthon in the South Hunsrück (west of the Rhine, Müller & Stoppel 1981).

Hörre–Gommern Zone, NW part. A conspicuous belt of quartz arenites extends over almost 300 km from the Rhenish Massif into the Flechtingen Horst ('Kellerwald-Quarzit', 'Hörre–Gommern-Quarzit', Meischner 1991; Jäger & Gursky 1998). It is intercalated with pelagic shales and radiolarian cherts of Famennian to Viséan age. Detrital micas from the arenites have yielded Caledonian K–Ar ages

(454–417 Ma, H. Huckriede pers. comm. 1999). The siliciclastic material was probably derived from Pomerania at the western margin of the Baltic shield, where time-equivalent shallow-water sandstones have been proved in boreholes (Zelichowski 1995). Since they are not detectable in the parautochthon, they probably entered the Rheno-Hercynian Basin from the East. Tectonic lenses of the 'Hörre–Gommern-Quarzit' have also been detected within the 'Hessische Schieferserie' and on top of the parautochthon of the Lahn Syncline (Wierich & Vogt 1997). These relationships strongly argue for an allochthonous position of the Hörre–Gommern Zone.

Hörre–Gommern Zone, SE part. Middle Devonian to early Tournaisian limestone turbidites are intercalated with shales and one Tournaisian sandstone member with Caledonian mica ages (Huckriede *et al.* 1998). These are overlain by Famennian to early Viséan greywackes (Bender & Homrighausen 1979).

'Hessische Schieferserie'. This unit occupies a structural position between the parautochthon of the Lahn Syncline and the Giessen Nappe. It contains early Devonian sandstones overlain by middle to late Devonian hemipelagic shales and limestones, and Tournaisian radiolarian cherts (Bender 1965; Kegel 1933). Volcanic rocks are scarce.

Armorican fragments. The Giessen–Werra–Südharz–Selke Unit (see below) carries at its base a tectonic mélange of sedimentary rocks, whose faunas show clear Armorican affinities (see discussion in Oczlon 1994; Franke & Oncken 1995; Plusquellec & Jahnke 1999). This association comprises early Ordovician quartzites, Silurian shales and limestones, as well as Lochkovian to early Emsian limestones, sandstones, conglomerates and local debris-flow deposits.

Giessen Werra Südharz/Selke Nappe. This unit carries, at its base, isolated tectonic slices of MORB-type as well as some intraplate metabasalts (Wedepohl *et al.* 1983; Grösser & Dörr 1986; Floyd 1995; Ganssloser *et al.* 1995). The ocean floor rocks are overlain by an extremely condensed sequence of Emsian to late Devonian shales and radiolarian cherts, indicating an Emsian or slightly older age of the metabasalts. These condensed pelagic sediments are overlain by greywacke turbidites with metamorphic debris. The onset of this flysch-type sedimentation in the Frasnian (Rhenish Massif) or early Famennian (Harz Mts) reflects early orogenic activities in the Mid German Crystalline High adjacent to the SE.

Remnants of the Giessen–Harz Unit are exposed in few isolated outcrops and some boreholes in a narrow belt between the Northern Phyllite Zone and the Mid German Crystalline High, thus indicating the root zone of the allochthonous units (e.g., Oncken *et al.* 1995*a*, *b*). Hence, the transport distance amounts to at least 60 km. The parautochthon was shortened by *c.* 50%, with deformation increasing toward the SE (Behrmann *et al.* 1991; Hollmann & von Winterfeld 1999; Oncken *et al.* 1999, this volume).

Palinspastic restoration of the allochthons, as based upon their present-day tectonic sequence, reveals a non-volcanic extensional regime which is transitional between the Rheno-Hercynian parautochthon to the NW and the Devonian ocean floor to the SE (Fig. 4).

Northern Phyllite Zone (NPZ)

The Northern Phyllite Zone is a narrow belt of rocks situated between the main, very low grade parts of the Rhenish Massif and Harz Mts to the NW, and the Mid German Crystalline High to the SE. The common feature of the 'phyllites' is pressure-dominated metamorphism (3–6 kbar–300 °C), which took place in Carboniferous time (c. 325 Ma, Oncken *et al.* 1995*a*).

The NPZ contains a very complex association of protoliths (Anderle *et al.* 1995): in the South Hunsrück, west of the Rhine, it is possible to identify metamorphosed equivalents of the Rheno-Hercynian shelf sequences. In the South Taunus, east of the Rhine, the NPZ consists of meta-volcanic rocks, which can be attributed to a Late Silurian–Early Devonian magmatic arc (Table 1; Meisl 1995). In addition, there are Ordovician metapelites, whose microflora indicates a cold-water realm (Reitz *et al.* 1995), and, hence, has affinities to the Armorican Terrane Assemblage or Gondwana. In the Harz Mts, the NPZ contains metamorphosed equivalents of the Harzgerode Allochthon (with exotic blocks of Silurian and Devonian rocks), and, further to the SE, Ordovician clastic metasediments and within-plate metavolcanic rocks.

Since westerly parts of the NPZ contain rocks derived from the Rheno-Hercynian Shelf and, further east, portions of Armorica, the NPZ apparently contains a segment of the 'Rheic' Suture (Cocks & Fortey 1982) between Avalonia and the Armorican terrane Assemblage (Fig. 5). The Rheic Suture is marked by the arc-derived metavolcanic rocks of the South Taunus, which

Table 1. *U–Pb zircon ages of late Silurian–early Devonian magmatic arc rocks in the Northern Phyllite Zone and in the Mid German Crystalline High*

	Reference	Age (Ma)
South Cornwell	Dörr *et al.* (1999)	
Pebble in ?Frasnian conglomerate		
U–Pb zircon		422 ± 4
Rhenish Massif		
Saar 1 well	Sommermann & Satir (1993)	
Albitgranit		
Zircon, single grain, $^{207}Pb/^{206}Pb$		444 ± 22
Krausaue, Bingen	Sommermann *et al.* (1994)	434 + 34/−22
South Taunus	Sommermann *et al.* (1992)	
Zircon, U–Pb multigrain		
Serizitgneis		426 ± 14/−15
Felsokeratophyr		433 ± 9/−7
Grünschiefer		442 ± 22
Mid German Crystalline High		
Pfalz	Reischmann & Anthes (1996)	
Pb/Pb zircon, evap.		
Albersweiler Gneis 1 xenocryst		*c.* 433
Spessart	Dombrowski *et al.* (1995)	
Rotgneis		
Pb/Pb zircon, evap.		418 ± 18
Rb/Sr WR		439 ± 15
Haibach gneiss		
Pb/Pb zircon, evap.		410 ± 18
Rb/Sr WR		407 ± 14
Ruhla	Zeh *et al.* (1997)	
Pb/Pb zircon, evaporation		
Steinbacher Augnengneis		398 ± 3
Liebensteiner Gneis		412 ± 4
Schmalwassersteingneis		413 ± 9

record consumption of the Rheic Ocean. The greywackes and metabasalts SE of the NPZ probably mark the Rheno-Hercynian Suture Zone. This younger element and the fault at the northwestern boundary of the NPZ truncate the internal structure of the NPZ (the Rheic suture included) at an acute angle (Fig. 5, and Franke & Oncken 1995). These relationships were probably formed during late strike-slip displacements.

Mid German Crystalline High (MGCH)

The evolution of the MGCR has recently been summarized by Oncken (1997), Okrusch & Weber (1996), and contributions in Dallmeyer *et al.* (1995). The MGCH represents the southeastern, active margin of the Rheno-Hercynian Belt. It has sourced the Frasnian through Namurian flysch sediments, and is characterized by important arc-related plutonism especially in mid-Carboniferous time (340–325 Ma, see also Anthes 1998). Another feature of most segments of the MGCH is Silurian–early Devonian, arc-related magmatism akin to that in the Northern Phyllite Zone. However, structure, metamorphic grade and composition of the MGCH change considerably along strike.

The westernmost part of the MGCH forms the basement of the Permo-Carboniferous, intramontane Saar–Nahe Basin and has been encountered in the Saar I borehole (see foldout and Table 1). A late Ordovician/early Silurian granitoid is overlain by middle Devonian to Tournaisian unmetamorphosed sediments (Krebs 1976).

Further to the SE and east, the MGCH contains late Devonian calc-alkaline plutons. The Palatinate, west of the Rhine, probably represents a Devonian–early Carboniferous magmatic arc (Flöttmann & Oncken 1992), into which the protolith of the Albersweiler Orthogneiss intruded at 369 ± 5 Ma (Anthes & Reischmann 1997). In the Odenwald east of the Rhine, the Frankenstein Gabbro intruded an already metamorphosed association of rocks at

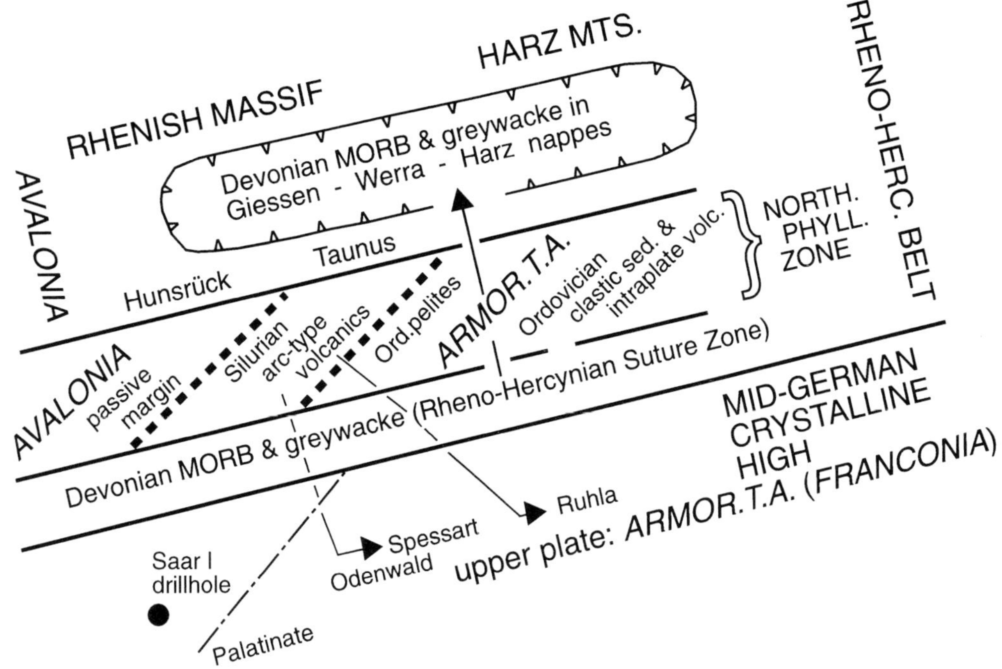

Fig. 5. Diagrammatic representation of tectonic relationships between the Rheno-Hercynian Belt, the Northern Phyllite Zone and the Mid German Crystalline High. The Silurian arc-type volcanics encased between stippled lines mark the Rheic Suture Zone. Armor.T.A.: Armorican Terrane Assemblage.

362 Ma (Kirsch *et al.* 1988). Early stages of metamorphism have been dated by K–Ar on hornblende and U–Pb on zircon at *c.* 370 Ma (Kreuzer & Harre 1975; Kirsch *et al.* 1988; Todt *et al.* 1995). According to Oncken (1997), subsidence and sedimentation in the Saar Region occurred in a forearc basin formed by subduction erosion, while time-equivalent metamorphism and plutonism were active in the Odenwald adjacent to the east and the Palatinate to the south (see Altherr *et al.* 1999 for a comprehensive treatment of plutonism in the Odenwald).

The Spessart, Kyffhäuser and Ruhla outcrops (see enclosure) exhibit amphibolite facies metamorphism with pressures up to 8–9 kbar (Nasir *et al.* 1991; Okrusch 1995; Oncken 1997; Zeh 1996). There are no indications of subduction-related HP rocks. K–Ar ages on hornblende, muscovite and biotite from the Spessart cluster between 324 and 318 Ma, indicating rapid cooling in early Late Carboniferous time (Nasir *et al.* 1991). Similar K–Ar muscovite ages (323–315 Ma) were obtained from paragneisses in boreholes east of the Spessart (Schmidt *et al.* 1986). The oldest Ar/Ar biotite ages from Ruhla range between 336 and 334 Ma (Zeh *et al.* 1997). K–Ar on biotite from the Kyffhäuser gave *c.* 340 Ma (Seim 1967). A Pb–Pb evaporation zircon age of 588 Ma from a granite-gneiss of the Kyffhäuser indicates the presence of Cadomian crust (Anthes & Reischmann 1996).

The Spessart and Ruhla areas expose, in the cores of antiformal structures, deeper structural levels, containing higher-grade equivalents of the Silurian–early Devonian magmatic arc rocks exposed in the Taunus segment of the Northern Phyllite Zone at the southeastern margin of the Rhenish Massif (Table 1). According to Oncken (1997), these relationships indicate tectonic underplating of Rheno-Hercynian crust (the passive margin of Laurussia) under the nascent active margin of the MGCH (Fig. 5).

Still further east, to the SE of the Flechtingen Horst (see enclosure), the MGCH mainly consists of granitoids, which intruded between 336 ± 4 and 327 ± 5 Ma (Pb/Pb zircon evaporation data, Anthes & Reischmann 1996).

Like in the Northern Phyllite Zone, the boundary faults of the MGCH trend ENE, while its internal subdivision is grossly oriented SW/NE (Fig. 5). The boundary faults probably represent late-Variscan strike-slip faults, which truncate an earlier, subduction-related structural pattern.

Further constituents of the MGCH have only survived as clasts in the Rheno-Hercynian flysch.

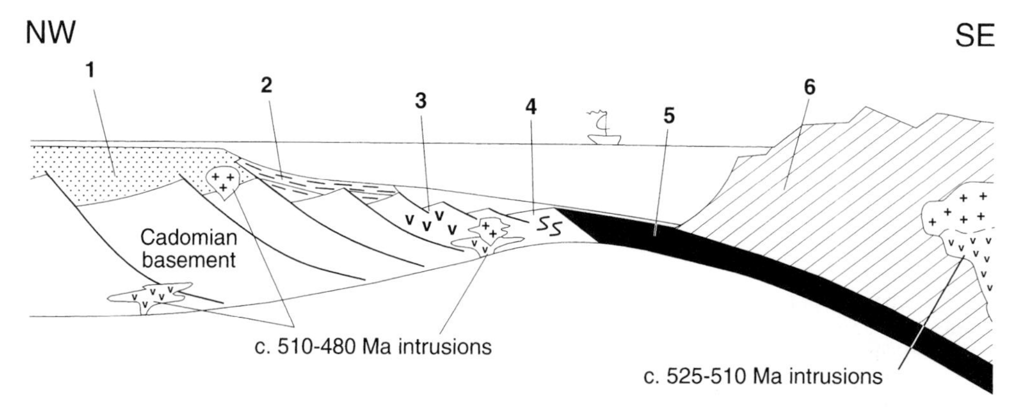

Fig. 6. Saxo-Thuringian Belt: diagrammatic palaeogeographic section through the Münchberg Klippe and adjacent parautochthon (after Franke 1995*b*). Nos. 1–6 correspond to the present-day stacking order (from bottom to top) and metamorphic grade. 1, parautochthon (fossiliferous shelf sediments); 2, fossiliferous slope deposits; 3, Cadomian basement with Devonian greenschist metamorphism; 4, Cadomian basement and early Palaeozoic magmatic rocks with Devonian metamorphism; 5, early Ordovician MOR-type basalts, *c*. 400 Ma eclogite metamorphism; 6, Cadomian basement of the Bohemian terrane, partly with Devonian amphibolite facies.

These are unmetamorphosed volcanic rocks (rhyolites, dacites, andesites), for which some imprecise, conventional U–Pb zircon data indicate a late Proterozoic age (Sommermann 1990). Besides, the Devonian through earliest Carboniferous greywackes contain detrital micas with late Proterozoic (Cadomian) ages. More frequently, especially in the younger greywackes, K–Ar mica ages cluster around 380–370 Ma (Neuroth 1997; Huckriede *et al.* 1998; Küstner *et al.* 1998). It appears that erosion of the hanging wall started with Cadomian basement rocks, followed by metamorphic rocks formed in the Rheno-Hercynian subduction zone. Since final closure of the Rheno-Hercynian Ocean did not occur before the Tournaisian (Franke & Engel 1986), these Devonian metamorphic rocks have possibly been formed by subduction erosion, as proposed by Oncken (1998).

Protolith ages and palaeogeographic derivation of the metamorphic rocks in the MGCH are largely unknown. In the lower plate assemblage, it is possible to identify the Silurian arc rocks referred to above, and one late Silurian microflora from quartzites in the Spessart segment (Reitz 1987). The upper plate is probably represented by the late Proterozoic (Cadomian) magmatic pebbles, Cadomian detrital mica ages, and the Cadomian gneiss of Ruhla, which indicate north Gondwanan affinities. The presence of Armorican rocks in the Rheno-Hercynian Belt still further north suggests that the upper plate of the MGCH likewise developed from part of the Armorican Terrane Assemblage (see Tait *et al.* this volume). This fragment is referred to below as the 'Franconia' suspect terrane.

Saxo-Thuringian Belt (STB)

Isolated outcrops of the Vesser Zone at the southeastern margin of the MGCH (foldout map) expose a greenschist-grade association of bimodal plutonic and volcanic rocks and clastic metasediments. The magmatic rocks have been dated at 518–502 Ma and are assigned to a rift environment (Kemnitz *et al.* in press).

The geological evolution of the main part of the STB has been summarized by Franke (1984*a*, *b*), Falk *et al.* (1995), as well as by Franke & Stein and by Linnemann *et al.* (this volume). The parautochthon consists of Cadomian basement and a Cambro-Ordovician rift sequence (Fig. 6), overlain by late Ordovician to Famennian pelagic sediments and Famennian to Viséan flysch sediments derived from the active Bohemian Margin to the SE.

The overlying allochthon (tectonic klippes of Münchberg, Wildenfels and Frankenberg, see enclosure) contains, in its lower part, deeper water equivalents of the Ordovician to Devonian of the parautochthon, and proximal flysch sediments. The upper allochthon exposes early Ordovician MORB-type mafic rocks metamorphosed into eclogite facies at *c*. 395 Ma (Fig. 6), and a variety of MP metamorphic rocks with zircon, hornblende and mica ages between 380 and 365 Ma (Franke *et al.* 1995*b*). The allochthon is derived from a root zone set between the Fichtelgebirge/Erzgebirge Antiform (foldout map) and the Teplá–Barrandian Unit. Equivalents of the external klippes have survived in the Marianske Lazne Unit (at the NW margin of the Teplá–Barrandian) and the Zone of Erbendorf–Vohenstrauss (ZEV), which overlies Moldanubian rocks (see enclosure). The German Continental Deep Drilling (KTB) was carried out in the northern part of the ZEV (Behr *et al.* 1997).

The deepest structural level of the STB is exposed in the Saxonian Granulite Antiform and in the Erzgebirge. These areas expose HP–MT and HP–HT rocks, which are probably derived from the orogenic root to the SE, and were intruded into the lower crust of the northwestern foreland (Henk *et al.*, Franke & Stein, Krawczyk *et al.*, this volume).

The presence of striated stones in latest Ordovician diamictites (Steiner & Falk 1981; Katzung 1999), as well as biogeographic and palaeomagnetic considerations (McKerrow *et al.* and Tait *et al.* this volume) clearly reveal that the continental crust of the Saxo-Thuringian parautochthon (Saxo-Thuringia) was at that time not very distant from Gondwana.

Although the sedimentary record and the geochemical character of the mafic magmatic rocks clearly indicate early Palaeozoic crustal extension, some observations have been taken to deduce an active margin setting in the STB and its equivalents in the Sudetes (e.g., Oliver *et al.* 1993; Kröner & Hegner 1998).

(1) The geochemical character of many Cambro-Ordovician felsic magmatic rocks is comparable with that of active margins. However, there is no sedimentary or tectonic evidence of an early Palaeozoic convergent setting. The chemical data may be explained by the waning effects of Cadomian subduction, or else by the extraction of melt from a Cadomian arc (e.g., Floyd *et al.* this volume).

(2) Late Devonian–early Carboniferous metamorphic rocks have preserved mineral ages around 480 Ma, which have been interpreted as evidence for an early Palaeozoic convergent episode. This applies to Rb–Sr muscovite ages from felsic gneisses in the Münchberg Klippe (Söllner *et al.* 1981), Ar/Ar muscovite and U–Pb titanite ages from pegmatite veins in the ZEV (Ahrendt *et al.* 1997; Glodny *et al.* 1998), zircon overgrowth in paragneisses of the ZEV (Söllner *et al.* 1997), as well as to zircons in paragneisses of the Moldanubian (Gebauer 1994). However, the main metamorphic minerals in these rocks have been dated as late Devonian and early Carboniferous, and the early Palaeozoic mineral ages are relicts of metamorphic and magmatic processes, whose P–T environment is unknown. With respect to the information derived from the sedimentary record and the geochemical affinities of the mafic magmatic rocks, the older mineral ages are best explained as products of magmatism and metamorphism in the hot lower crust of an early Palaeozoic rift. Cambro-Ordovician rift metamorphism has already been proposed by Weber & Behr (1983), and has recently been documented in the Stáre Město Belt at the eastern margin of the Bohemian Massif (Kröner *et al.* this volume), a probable eastern continuation of the Saxo-Thuringian in Germany (Franke & Żelaźniewicz this volume).

The STB underwent extreme tectonic contraction in a regime of dextral transpression with displacements top to the W. Unstacking the tectonic pile of the Münchberg Klippe leads to an orthogonal shortening component of = 180 km (Franke 1984*b*). Multiple metamorphic inversion by thrusting in the underlying Erzgebirge (Franke & Stein this volume), polyphase folding in the parautochthon (e.g., Stein 1988) and SE-vergent backfolding in the northwestern part of the belt (Schäfer *et al.* this volume) amount to at least another 100 km.

Teplá–Barrandian (TB)

The TB is the best-preserved fragment of the Cadomian Orogen in central Europe. The basement consists of late Proterozoic sediments (including latest Proterozoic flysch) and arc-related volcanic rocks, which have undergone very low-grade to amphibolite facies metamorphism and deformation at *c.* 550–540 Ma (see Zulauf *et al.* 1999 for Th–U–Pb monazite model ages, and the comprehensive treatment of the TB in Zulauf 1997). The Cadomian basement is overlain by Cambrian to middle Devonian sedimentary and volcanic rocks (Fig. 7). The geochemical characteristics of Cambrian magmatic rocks are still indicative of subduction (e.g., Patočka *et al.* 1993). However, Cambrian plutons can be shown to have intruded in a transtensional setting (Zulauf *et al.* 1997; Dörr *et al.* 1998), and the sedimentary record with early Cambrian continental deposits, mid-Cambrian marine shales and Ordovician shelf sediments documents rifting. The sedimentary and volcanic evolution of Ordovician to middle Devonian rocks reflects subsidence and extension in an intracontinental or passive margin setting. Late Givetian clastic sediments with Cadomian and Ordovician detrital mica ages record the onset of orogenic uplift, probably at the NW margin of Bohemia (Schäfer *et al.* 1997). Recent summaries of the basement are available in

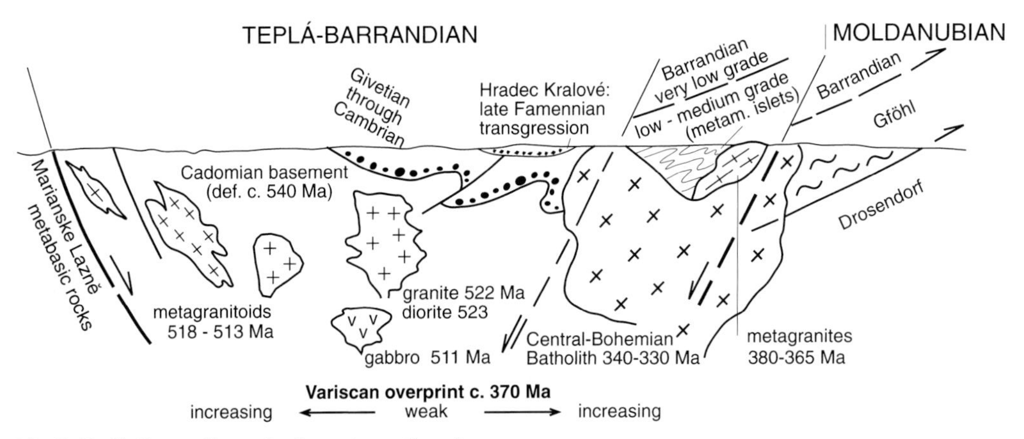

Fig. 7. Teplá–Barrandian unit: dynamic stratigraphy.

Zulauf (1997) and in Chlúpáč (1993). The Palaeozoic cover sequences have been described in detail by Chlúpáč (Chlúpáč 1994; Chlúpáč *et al.* 1998; Chlúpáč & Zdenek 1988).

The northwestern margin of the TB, as well as parts of its southeastern margin, are lined by eclogite facies and mantle rocks. These margins probably represent sutures, but were later overprinted by crustal-scale normal faulting (Zulauf 1997; Bues & Zulauf 2000). Devonian deformation and metamorphism is documented both in the northwestern and southeastern parts of the TB, with diverging tectonic polarity (Zulauf 1997). This early Variscan overprint affects continental basement and its Palaeozoic cover, i.e., the upper plate at both tectonic fronts. These rocks must have been buried either by subduction erosion, or else by out of sequence thrusting during the collisional stage. This is exemplified, e.g., in the 'Metamorphic islets' within the Central Bohemian Batholith (at the southeastern margin of the TB, see enclosure), which contain metamorphosed equivalents of the Barrandian Palaeozoic rocks and their basement.

The Staré Sedlo and Mirotice Orthogneisses in the westernmost of the 'metamorphic islets', in the roof of the Central Bohemian Pluton, have yielded U–Pb upper intercept ages of 373 ± 5 and 369 ± 4 Ma (Košler *et al.* 1993; Košler & Farrow 1994). These late Devonian plutons probably record late Devonian subduction under the southeastern margin of Bohemia. Along the northwestern margin, there is no evidence of subduction-related magmatic activities.

Variscan deformation and low-grade metamorphism has affected Givetian sediments. Folded Palaeozoic rocks encountered in boreholes beneath Cretaceous cover east of the TB type region are unconformably overlain by Famennian sediments (e.g., Chlúpáč 1994; Franke & Żelaźniewicz this volume). Hence, Variscan overprint of the TB is bracketed between the late Givetian and the late Famennian, i.e. between *c.* 385 and *c.* 365 Ma. This is backed up by Ar/Ar data (Dallmeyer & Urban 1998): in the northwestern TB, hornblende ages range between 383 and 375 Ma, and white mica between 376 and 370 Ma. One muscovite age from the southern TB yielded 362 Ma. Subsequent faulting has effected relative subsidence of *c.* 15 km of the central part of the TB, with respect to the higher grade margins of the TB and its Saxo-Thuringian and Moldanubian forelands (Zulauf 1997; Zulauf *et al.* 1998).

The TB is part of the Armorican Terrane Assemblage. This is documented in traces of the late Ordovician Saharan glaciation (Chlúpáč *et al.* 1998, p. 78), steep palaeomagnetic inclinations in Ordovician rocks (Tait *et al.* 1995, this volume) and biogeographic affinities (McKerrow *et al.* this volume). The TB has been rotated counterclockwise through at least 140° since the Silurian, and before the onset of collisional deformation in late Devonian time (Tait *et al.* 1994, this volume).

Moldanubian of the Bohemian Massif

The Teplá–Barrandian and the high-grade metamorphic rocks adjacent to the SE have been summarized, by Kossmat (1927), as the 'Moldanubian Region'. It has become customary to differentiate the high-grade, southeastern part of the Bohemian Massif as the Moldanubian *sensu stricto* (in this paper: simply 'Moldanubian'). The Moldanubian is a complex association of rocks, whose protolith ages are mostly unknown. The structure of the Moldanubian is dominated by large scale, thrust-related inversion of metamorphic facies, which has emplaced HP rocks (Gföhl Assemblage) over the parautochthonous Drosendorf Assemblage. Thrusting of the Gföhl rocks was directed toward northerly to easterly directions (Drosendorf unit: top N/NNE, Büttner & Kruhl 1997), Gföhl Gneiss: top to the ESE (Matte *et al.* 1985). Since the Gföhl allochthon extends to the SE margin of the Bohemian Massif, the distance of transport measured perpendicular to the Variscan structural trend must amount to at least 150 km. The areal distribution of the allochthon is largely controlled by updoming around the South Bohemian Batholith (see enclosure). A review of earlier papers and a first plate tectonic synthesis was published by Tollmann (1982). A recent regional overview was presented by Petrakakis (1997).

Drosendorf Assemblage

A basal part (Ostrong Unit or 'Monotonous Series') mainly contains metasedimentary and felsic meta-igneous rocks. Some rare lenses of eclogite exhibit a short-lived granulite facies overprint comparable to that observed in the overlying Gföhl Assemblage, and probably acquired at the same time (O'Brien & Vrana 1995 1997). The Monotonous Series is structurally overlain, along an important ductile fault zone, by the 'Bunte Serie' (Variegated Series), with a basal sheet of Dobra Orthogneiss and a large variety of metamorphosed sedimentary and igneous rocks. Amphibolites in the Variegated Series were derived from within-plate and ocean floor basalts (Patočka 1991).

Magmatic crystallization of the Dobra Orthogneiss has been dated at 1.38 Ga (Gebauer & Friedl 1994). The Svetlík Orthogneiss, which underlies the Gföhl Klippes of central Bohemia, has yielded a zircon age of *c.* 2.1 Ga (Wendt *et al.* 1993). The U–Pb ages of detrital or xenocrystic zircons from Drosendorf rocks in the southwestern (Bavarian) part of the Bohemian Massif cluster around 2.6–2.5, 2.0, 1.1–1.0 Ga and 600 Ma (Gebauer 1994). An early Palaeozoic metamorphic event in the Monotonous Series of Bavaria has been dated with zircon (Gebauer 1994) and Rb–Sr thin-slab measurements (474 ± 13 Ma, Köhler *et al.* 1989). An upper intercept zircon age of 358 ± 6 from an amphibolite within the 'Bunte Serie' is interpreted to represent magmatic crystallization (Friedl *et al.* 1993).

The palaeogeographic affinities of the northwestern part of the Moldanubian are uncertain. HT–LP rocks in the Oberpfalz between the Zone of Erbendorf–Vohenstrauss and the Teplá–Barrandian, contain a characteristic association of mid-Ordovician felsic metavolcanic rocks and a meta-laterite, which also occurs in the Saxo-Thuringian of the Fichtelgebirge adjacent to the NW (Mielke *et al.* 1996). It is possible that the 'Moldanubian' rocks of the Oberpfalz actually belong to the Saxo-Thuringian, which would imply a tectonic boundary against the Moldanubian proper adjacent to the east. In fact, such a boundary has been proposed by Matte *et al.* (1990), but its position is only based on speculation.

Isotopic data on metamorphism in the Drosendorf Assemblage are complex. An orthogneiss from the Kaplice Unit in the Czech Republic (equivalent of the Monotonous Unit) yielded a U–Pb age of 367 ± 20 Ma, interpreted to represent metamorphism (Kröner *et al.* 1988, p. 264). Metamorphism of the Svetlík Orthogneiss has been dated by U–Pb on titanite at 355 ± 2 Ma (Wendt *et al.* 1993). Amphibolites from the Variegated Series have yielded Ar/Ar isotope correlation ages of 341.1 ± 1.4 and 328.7 ± 3.3 Ma (Dallmeyer *et al.* 1992). Friedl *et al.* (1993) obtained 337–333 Ma by U–Pb on monazite from the Monotonous Series. Monazites from the Dobra Gneiss at the base of the Variegated Series gave 333 ± 2 Ma (Gebauer & Friedl 1994) and 336 ± 3 (Friedl *et al.* 1994). The metamorphic event at *c.* 335 Ma is dominant in the Czech part of the Drosendorf Assemblage.

A still younger HT–LP metamorphic episode appears to be confined to the Bavarian part, along the southwestern margin of the Bohemian Massif (Oberpfalz, Bavarian Forest): it is very well constrained by U–Pb zircon and monazite ages from leucosomes in anatectic gneiss from the Bavarian Forest, ranging between 326 and 322 Ma (Kalt *et al.* 2000). Concordant monazite from an anatectic segregate gave 322 Ma (Schulz-Schmalschläger *et al.* 1984). From anatectic gneisses of the Bavarian Forest, Grauert *et al.* (1990) obtained Rb–Sr thin-slab isochrons of 327–321 Ma, and concordant U–Pb monazite ages of 325–320 Ma. This younger HT–LP event (*c.* 325 Ma) is apparently not confined to the Moldanubian, but extends northwestwards into the Fichtelgebirge segment of the Saxo-Thuringian Belt and beyond (Teufel 1988; Kreuzer *et al.* 1989). It has not been detected so far in more northeasterly parts of the Saxo-Thuringian and Moldanubian Belts. Hence, it is obviously unrelated to convergent tectonics in any one of these belts, but rather represents a crosscutting, later feature.

Gföhl Assemblage

The structurally lower part of the Gföhl Unit (Raabs–Meisling Unit) contains important intercalations of amphibolite (Buschandlwand, Rehberg, Raabs). Zircons from the Buschandlwand Amphibolite have yielded a U–Pb age of 428 ± 6 Ma, taken to represent magmatic crystallization (Finger & Quadt 1995). The overlying Gföhl Orthogneiss was derived from an Ordovician granitoid (SHRIMP concordant zircon data, average 482 ± 6 Ma; Friedl *et al.* 1998). A SHRIMP II age of 469.3 ± 3.8 Ma was also obtained from a granulite in southern Bohemia (Kröner *et al.* 2000). Granitic protoliths of some granulites in central Bohemia crystallized at *c.* 370 Ma (Wendt *et al.* 1994).

Garnet peridotites enclosed in the Gföhl Assemblage were equilibrated at different peak pressures (c. 28 and 42 kbar, Medaris *et al.* 1990; 31 kbar, Carswell 1991). The UHP assemblages also contain continental crust: calcsilicate marbles have been metamorphosed at 30–40 kbar/>1100 °C (Becker & Altherr 1992). The upper part of the Gföhl Assemblage is occupied by widespread felsic granulites, partly retrogressed into amphibolite facies (Carswell & O'Brien 1992). The lower part of the Gföhl Assemblage (Raabs–Meisling Unit) has undergone pervasive medium-pressure metamorphism, which is seen to decrease downwards and is probably related to the overthrusting of hot Gföhl rocks (Friedl *et al.* 1994).

Isotopic dating of metamorphism in the Gföhl Assemblage has yielded somewhat controversial results. Sm–Nd dating of garnet–omphacite pairs from eclogites gave values of 377 ± 20 and between 342 ± 9 and 323 ± 7 Ma. The older age is taken to represent eclogite-facies metamorphism, the younger ones are referred to uplift and cooling (Beard *et al.* 1992). Sm–Nd garnet–clinopyroxene–whole rock isochrons for garnet pyroxenites yielded 370 ± 15 and 344 ± 10 Ma. Again, the older age is supposed to be closer to peak metamorphism (Carswell & Jamtveit 1990). This is corroborated by Sm–Nd data on garnet–whole rock pairs from garnet-peridotites and pyroxenites ranging between 376 and 370 Ma (Becker 1993).

Various methods of U–Pb dating of minerals from the felsic granulites and related gneisses invariably yield values around 340 Ma (concordant monazite—granulite, 340 ± 1, Schenk & Todt 1983; Mohelno granulite, 339 ± 2 Ma; Gföhl Gneiss, 337 ± 3 Ma, Vír Gneiss, 338 ± 5 Ma, van Breemen *et al.* 1982; Gföhl Gneiss, 339.9 ± 0.9 Ma, Friedl *et al.* 1994; zircons—Gföhl Gneiss, 341 ± 4 Ma, Lisov granulite, 345 ± 5 Ma; van Breemen *et al.* 1982; 338 ± 1 Ma, Aftalion *et al.* 1989; 351 ± 6, 346 ± 5 Ma; Wendt *et al.* 1994). The slightly younger monazite ages possibly represent an early stage of retrogression (van Breemen *et al.* 1982, p. 100).

It is under debate whether the zircon ages of the granulites reflect peak metamorphic conditions, especially peak pressure. Roberts & Finger (1997) have argued, on petrological grounds, that zircons in the felsic granulites were formed during uplift and decompression. Besides, the supposed age of HP granulite facies metamorphism is indistinguishable from the intrusion age of the late-tectonic Central Bohemian Batholith at 340 Ma (or even earlier, see below), which postdates emplacement of the Gföhl over the Drosendorf Assemblage. However, *in-situ* U–Pb SHRIMP II dating of zircons from the Prachatice and Blansky Les granulites has confirmed ages around 340 Ma for two distinct stages of granulite facies metamorphism and a HT–LP overprint (Kröner *et al.* 2000). Still, it cannot be excluded that at least the peak metamorphic pressures in the Gföhl Assemblage might have been achieved before 340 Ma.

The palaeogeographic affinities of the Moldanubian Units remain uncertain. Magmatic protoliths have Ordovician, Silurian and Devonian ages. Finger & Steyrer (1995) have proposed that the Dobra Gneiss of the Variegated Unit and the Bíteš Gneiss of the Moravo-Silesian (see Franke & Żelaźniewicz this volume) are equivalents, and represent the northwestern, passive margin of the Moravo-Silesian terrane. The mafic protoliths of the Raabs–Meisling Unit (lower part of the Gföhl Assemblage) are correlated to an early Palaeozoic oceanic domain between the Moravo-Silesian terrane and an unknown Moldanubian terrane to the northwest. However, the Dobra and the Bíteš Gneiss have clearly different protolith ages (1.4 Ga vs 540 Ma, respectively). Besides, the Moravo-Silesian Belt and its Rheno-Hercynian equivalent on the NW flank of the Variscides are characterized by early Devonian (not Cambro-Ordovician) rifting, and the Moravo-Silesian was juxtaposed against the Moldanubian Units along the 'Moldanubian thrust' only at a late stage of the tectonic evolution (Franke & Żelaźniewicz this volume).

The best palaeogeographic estimate (although on a still insufficient factual base) was first proposed by Matte (1986): it suggests opening of a Massif Central–Moldanubian Ocean in Ordovician time (represented by ultramafic and

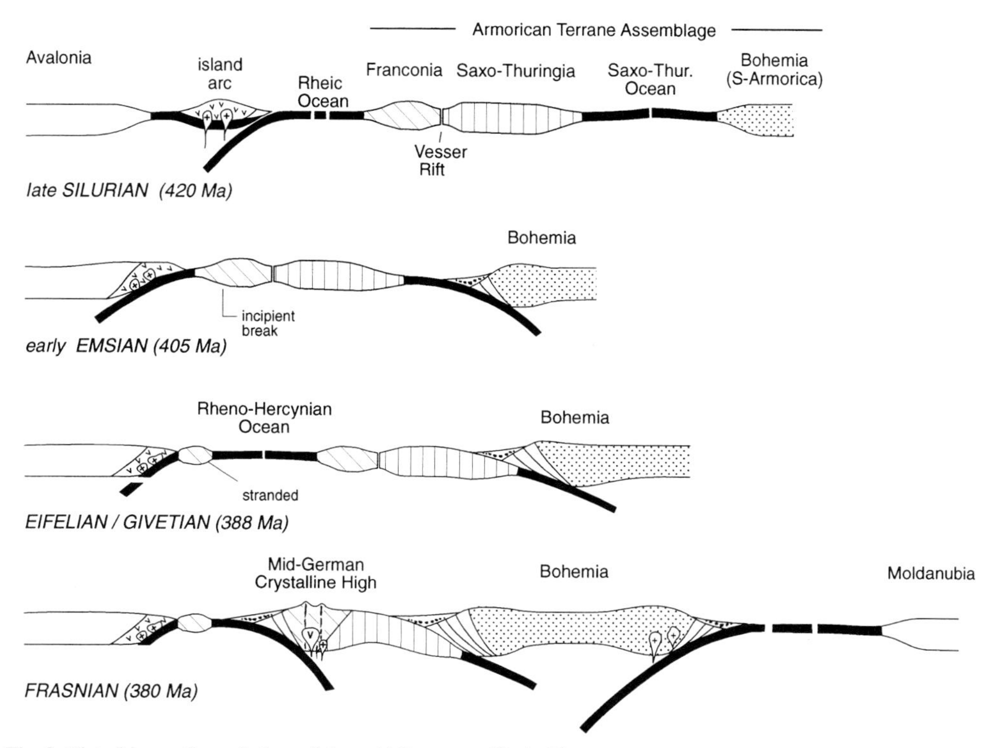

Fig. 8. Plate kinematic evolution of the mid-European Variscides.

magmatic members of the Gföhl Assemblage) and a southerly continental microplate (contained in the Drosendorf Assemblage). One of the many remaining uncertainties concerns the palaeogeographic position of the notional Drosendorf Terrane ('Moldanubia' in Fig. 8), which could either represent the northern margin of mainland Gondwana, or else another member of the Armorican Terrane Assemblage.

Granitoids in the Moldanubian and Saxo-Thuringian Belts

Apart from the late Devonian metagranitoids at the southeastern margin of the Teplá–Barrandian ('metamorphic islets' in the roof of the Central Bohemian Batholith, see above), it is possible to distinguish two age clusters of granitoids: *c*. 340 Ma and 325–305 Ma. Both post-date the tectonic emplacement of the Gföhl over the Drosendorf Assemblages, and the younger generation also post-dates the LP–HT metamorphic events (Büttner & Kruhl 1997; Finger & Clemens 1995). Schermaier *et al.* (1997) discuss possible equivalents in the Eastern Alps.

The older age group is mainly represented by the Central Bohemian Batholith, which may be referred to a volcanic arc (petrogenetic group 1 in Finger *et al.* 1997). U–Pb zircon dating yields intrusion ages around 340–330 Ma (Dörr *et al.* 1997, 1998). Zircon evaporation ages range between 351 and 340 Ma (Holub *et al.* 1997), and U–Pb dating of monazite from pegmatites gave values from 341 to 334 Ma (Novak *et al.* 1998). The Central Bohemian Batholith has been affected by the normal faulting, which has juxtaposed the Teplá–Barrandian against the Moldanubian (Zulauf 1997). The Leuchtenberg Granite of the Oberpfalz, which has undergone some tectonic deformation, has a similar age (U–Pb zircon, 338 ± 6 Ma, Köhler *et al.* 1993).

The younger age group belongs to the petrogenetic group 3 and, to a minor extent, group 4 of Finger *et al.* (1997), which are either referred to post-collisional extension in contact with mantle or mantle-derived melt (3, 4), or renewed subduction (4). Group 3 is exemplified in the South Bohemian Batholith. It comprises the Weinsberg Granite (U–Pb zircon, 318 ± 4 Ma, Quadt & Finger 1991, concordant monazite, 328 ± 5 to 323 ± 4 Ma, Friedl *et al.* 1996), the Rastenberg Granodiorite (discordant zircon, upper intercept 328 ± 10 Ma, concordant monazite 324 ± 1, 323 ± 1, Friedl *et al.* 1992; 323 ± 2, Friedl *et al.* 1993), the Eisgarn Granite (concordant monazite, 327 ± 4 Ma, Friedl *et al.*

1996) and the Freistadt Granodiorite (concordant monazite, 302 ± 2 Ma, Friedl *et al.* 1993). Equivalents are present at the southwestern margin of the Bohemian massif: Bavarian Forest, Ödwies Granite, concordant zircon 320 Ma, Regensburger Wald, Kristallgranit I, concordant zircon 315 Ma (Schulz-Schmalschläger *et al.* 1984). The younger group of granites extend northwesterwards into the Saxo-Thuringian Belt (Fichtelgebirge and western Erzgebirge, Förster *et al.* 1998, Siebel *et al.* 1997, Trzebski *et al.* 1997, Werner & Lippolt 1998).

Saxo-Thuringian and Moldanubian in SW Germany

Schwarzwald (Black Forest)

The Schwarzwald is a bivergent structure with a high-grade core (Zentralschwarzwälder Gneiskomplex, ZSG), which was emplaced by dextral transpression over low-grade units on either side: the Baden Baden Belt (BBZ) to the north and Badenweiler–Lenzkirch Belt (BLZ) to the south (Eisbacher *et al.* 1989; Krohe & Eisbacher 1988). The southernmost portion of the Black Forest is occupied again by crystalline rocks (Südschwarzwälder Gneiskomplex, SSG).

The BBZ contains a variety of greenschist-grade sedimentary and volcanic rocks, which have so far yielded only organic walled microfossils indicating an early Ordovician age (Montenari & Hanel 1998; Montenari 1999; Montenari *et al.* 2000).

The ZSG is a complex tectonic assemblage, which comprises Ordovician metasediments (Hanel *et al.* 1998) as well as metagranitoids crystallized at 510–500 Ma (Pb/Pb evaporation, Chen *et al.* 2000). Pb/Pb dating of whole rocks and garnet–feldspar pairs by Chen *et al.* (2000) reveal a HT event at *c.* 480 Ma, confirming earlier data of Hofmann & Köhler (1973), Steiger *et al.* (1973) and Kober *et al.* (1986). Ultramafic rocks and eclogites occur as small isolated bodies (Kalt & Altherr 1996). The eclogites were probably developed from Ordovician magmatic rocks (zircon upper intercept ages of 468 and 451 Ma, Kalt *et al.* 1990). In the central part of the ZSG, Sm–Nd dating on garnet, clinopyroxene and whole rock yields 327 ± 8 Ma. For similar rocks at the southern margin of the ZSG, the Sm–Nd ages vary between 337 ± 6 and 332 ± 13 Ma (Kalt *et al.* 1994). Granulite-facies overprint on eclogites in the central part of the ZSG is indistinguishable from the eclogite age (341 ± 19 Ma, Pb/Pb evaporation, mean of five zircons (Hanel *et al.* 1993). These early stages were followed by a strong HT–LP overprint at 335–330 Ma (Kalt & Grauert 1994; Kalt *et al.* 1994), and granitoid intrusions (340–330 Ma).

The evolution of the Zone of Badenweiler Lenzkirch (BLZ) has recently been summarized by Loeschke *et al.* (1998) and Güldenpfennig (1998). Ordovician–Silurian clastic metasediments are overlain by Frasnian shales and radiolarian cherts, late Devonian–Tournaisian greywacke turbidites, and Viséan felsic volcanic rocks intercalated with continental sediments. This sequence of rocks is interpreted to represent an early Palaeozoic extensional basin, later transformed into a Viséan foreland basin adjacent to a north-dipping subduction zone and overprinted by transtension (see also Echtler & Chauvet 1992).

In the Southern Schwarzwald (SSG), an allochthonous unit contains anorthosites, amphibolites and anatectic gneisses, with magmatic crystallization between 350 and 347 Ma. Zircon ages from the relative autochthon range between 343 and 341 Ma. An undeformed granite dyke dated at *c.* 342 Ma postdates thrusting (U–Pb conventional and Pb/Pb evaporation ages on zircon (Chen *et al.* 1999). According to Schaltegger (1995, 2000), deformation in the SGG is bracketed between the intrusions of the strongly sheared Schlächtenhaus granite (U–Pb monazite 334 ± 2 Ma) and the post-tectonic granites of Albtal, Bärhalde and St Blasien (U–Pb monazite and zircon, 334 ± 3 Ma–332^{+2}_{-4} Ma). The older estimate for the age of deformation (Chen *et al.* 1999) might be due to the method applied (mostly Pb/Pb evaporation ages). At least the earlier part of the tectonometamorphic evolution in the SSG is coeval with clastic sedimentation in the BLZ adjacent to the north. Since crustal stacking and marine sedimentation mutually preclude each other, the SSG must have been juxtaposed against the BLZ by later transpressional displacements.

Vosges

The Vosges have much in common with the Schwarzwald (Echtler & Altherr 1993): a northern unit (NV) with low-grade sedimentary and volcanic rocks, a central part with high-grade metamorphic units and granitoids (CV), and a southern unit with Devonian to Viséan sedimentary and volcanic sequences (SV). The tectonic structure is again bivergent with a component of dextral strike slip (Wickert & Eisbacher 1988). The Vosges have been displaced southwards by *c.* 30 km with respect to equivalent units in the Schwarzwald.

The northern Vosges, north of the Lalaye–Lubine shear zone, show a decrease of metamorphic grade and stratigraphic age from south to north. At the southern border, the phyllites of Steige and Villé have yielded acritarchs of Cambro-Ordovician to Silurian age (Reitz & Wickert 1989; Doubinger 1963). The Vallée de la Bruche sequence further north contains Devonian metabasalts, Frasnian black shales, and Famennian to early Carboniferous greywackes and conglomerates (see the latest review in Braun & Mass 1992). The conglomerates contain granite pebbles, one of which has been dated at *c.* 490 Ma (U–Pb zircon, Dörr *et al.* 1992). Post-tectonic I-type granitoids in the NW were emplaced between 336 and 330 Ma (compilation in Altherr *et al.* 2000).

The central Vosges contain garnet peridotites with peak metamorphic pressures in excess of 49 kbar, which were probably formed by subduction of subcontinental lithospheric mantle (Altherr & Kalt 1996). Many rocks underwent granulite facies metamorphism at around 335 Ma, followed by exhumation in a high heat-flow environment with granitization at *c.* 328, and granite intrusion at 326 Ma (Schaltegger *et al.* 1999).

Thrust slices at the boundary between the CV and the SV contain serpentinites and Famennian greywacke turbidites, equivalents of similar rocks in the BLZ of the Schwarzwald. Detrital zircons from the Famennian sediments have yielded concordant ages of 386 ± 2 Ma (Schaltegger *et al.* 1996). The main part of the SV is occupied by a volcano-sedimentary sequence, which starts with marine turbidites of a 'lower' and 'middle unit', overlain by terrestrial deposits (Eisele *et al.* this volume). Felsic volcanic intercalations at the top of the 'lower unit' and of the 'upper unit' have been dated at 345 ± 2 and 340 ± 2 respectively. The volcano-sedimentary sequence has been intruded by granites and diorites dated at 342 ± 1 to 339.5 ± 2.5 Ma (U–Pb zircon, Schaltegger *et al.* 1996). A severe contradiction arises from limestone turbidites in the 'middle unit', which contain foraminifera indicative of the Viséan 3b (Montenari 1999), which translates into an isotopic age of *c.* 330–325 Ma (see McKerrow & van Staal). This apparent contradiction might be resolved if the zircons from the magmatic rocks represent older xenocrysts.

The Northern Schwarzwald and Northern Vosges are conventionally regarded as a western continuation of the Saxo-Thuringian Zone of Bavaria. In the central parts of the Schwarzwald and Vosges, the presence of mantle material and HP metamorphic rocks indicates a subduction zone. Since these areas have been involved in grossly southeastward displacements, they probably correlate with the Moldanubian rocks of the Bohemian Massif. This comparison is also supported by a largely similar isotopic record of geological events, with the only exception of the 325 Ma LP event restricted to the SW margin of the Bohemian Massif. Ultramafic rocks and eclogites are usually interpreted as remnants of a Massif Central/Moldanubian Ocean, which separated the Bohemian and the Moldanubian microplates (Matte *et al.* 1990). Relicts of Ordovician metamorphism and magmatism event possibly relate to the rifting stage. However, the Variscan basement of SW Germany does not contain rocks comparable with the Teplá–Barrandian. It appears that equivalents of the Saxo-Thuringian (northern Schwarzwald and Vosges) are directly juxtaposed against the Moldanubian.

Timing of extension and convergence

With the exception of the Rheno-Hercynian 'successor ocean', all basins within the central European Variscides were formed in late Cambrian to early Ordovician time, when continental fragments became detached from the North Gondwana Margin (Franke *et al.* 1995*a*; McKerrow *et al.* and Tait *et al.* this volume). It appears that only the separation between Avalonia and the Armorican Terrane Assemblage (ATA) developed into a wide ocean, whereas the seaways between the individual Armorican Terranes, and those between the ATA and Gondwana mainland, are not detectable by palaeomagnetism and biogeography. Even in latest Ordovician time, shortly before the onset of closure of the Saxo-Thuringian Ocean, Saxo-Thuringia was still reached by the Saharan glaciation. The existence of a wide separation between the ATA and Gondwana in late Devonian time (Tait *et al.* this volume) is solely based upon palaeomagnetic data and requires further confirmation.

Plate convergence and collision are more complex. The Rheno-Hercynian Belt records two convergent episodes. The first one is the closure of the Rheic Ocean between Avalonia and the Armorican Terrane Assemblage. The bulk of magmatic arc rocks contained in the Northern Phyllite Zone and the Mid German Crystalline High are older than *c.* 410 Ma (Table 1), which indicates that subduction of oceanic crust occurred earlier, i.e., in latest Ordovician to Pragian times. The allochthonous Armorican fragments contained in the Rheno-Hercynian allochthon maintain their specific faunal character into the Early Emsian. In the Late Emsian and after, there were no biogeographic differences between Laurentia and Gondwana (McKerrow *et al.* this volume). Since Silurian–early Devonian subduction was probably directed to the north (Franke & Oncken 1995), Gedinnian extension in the Rhenish Massif might reflect back-arc spreading. The late Ordovician through early Devonian sediments of the Rheno-Hercynian Belt (the Northern Phyllite Zone included) do not reveal any trace of deformation and metamorphism related to the closure of the Rheic Ocean. If subduction was north-directed, Silurian–early Devonian deformation might have been restricted to more southerly areas, which were later overprinted by the late Devonian–early Carboniferous evolution in the Mid German Crystalline High. Alternatively, opening of the Rheno-Hercynian narrow ocean might have started, before closure of the Rheic Ocean was completed. Speculations about the geodynamic cause for this second, Rheno-Hercynian pulse of spreading are beyond the scope of this paper.

Closure of the Rheno-Hercynian Ocean occurred in late Devonian–early Carboniferous time. Incipient collision is heralded in the basal Tournaisian of the southeastern Hörre–Gommern allochthon, where northern derived sandstone turbidites with Caledonian detrital micas are seen to interfinger with greywacke turbidites derived from the Mid German Crystalline High. Ultimate closure of the oceanic realm is indicated by the first arrival of greywacke turbidites on the foreland, which is documented in the Late Tournaisian of the northwestern, autochthonous part of the Harz Mts.

In the Saxo-Thuringian Belt, collision is indicated by the rapid succession of zircon, hornblende and mica ages between *c.* 380 and 365 Ma, i.e., in late Devonian time (compilation in Franke *et al.* 1995*b*). This is corroborated by deposition of Famennian greywacke turbidites derived from the Teplá–Barrandian margin on the continental Saxo-Thuringian foreland (Franke & Engel 1986; Franke *et al.* 1992; Schäfer *et al.* 1997).

In the Moldanubian Belt, the onset of collision is less well constrained. In the Bohemian Massif, collision predates intrusion of the central Bohemian Batholith at *c.* 340 Ma as well as HP–HT metamorphism around 340 Ma in the felsic (continental) granulites. In the southern Schwarzwald and Vosges, collision must have occurred prior to the deposition of terrestrial sediments and volcanic rocks in the southern Vosges and the Badenweiler–Lenzkirch Zone in Late Viséan time.

In all three subduction zones, collision was probably preceded by subduction erosion. This is suggested by mid–late Devonian subsidence of the western Mid German Crystalline High (Oncken 1998), and the involvement of both flanks of Bohemian continental crust into late Devonian stacking and metamorphism.

Collisional deformation and metamorphism

Orthogonal component

All modern tectonic models, from Bard *et al.* (1980), through Matte (1986) to Franke (1989) and Franke *et al.* (1995*a*), have stressed the bilateral symmetry of the Variscan Belt. In central Europe, the zone of structural parting can be traced from central parts of the Vosges and the Schwarzwald (Black Forest) into the Teplá–Barrandian (Fig. 9). Thrusting on either flank is accompanied by dextral displacement, so that the overall tectonic setting is one of dextral transpression. The minimum amount of orthogonal shortening, as documented in the areal distribution of tectonic klippes and their internal architecture, amounts to at least 200 km on both flanks of the belt (e.g., Behrmann *et al.* 1991). The lithospheric mantle (and unknown portions

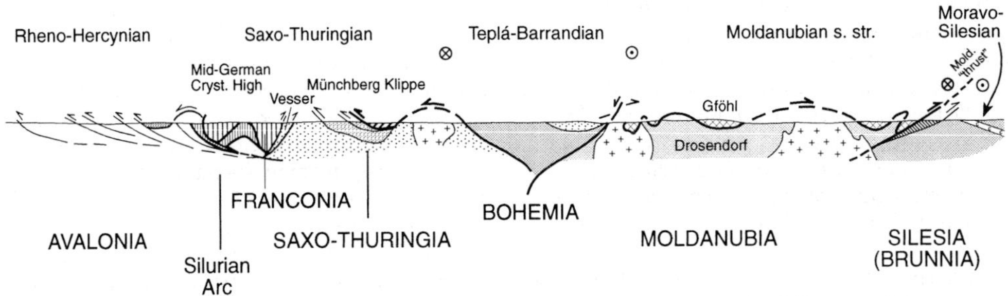

Fig. 9. Tectonic section across the mid-European Variscides, with palaeogeographic affiliations of the tectonostratigraphic units.

of the lower crust) are not incorporated into the shortening, but have been subducted. 400 km of lithospheric mantle cannot have been accommodated under the 'median massif' of the Teplá–Barrandian, because the latter is only *c.* 150 km wide. Thrusting on both flanks not only attained similar distances of transport, but also occurred more or less simultaneously, and is constrained, on the northern flank, between *c.* 380 and 330 Ma. On the southern flank, it started between the late Givetian and the late Famennian (i.e., between *c.* 380 and 365 Ma), and continued until at least *c.* 340 Ma (the time of intrusion of the late tectonic Central Bohemian Batholith). Under these circumstances, bilateral subduction of lithospheric mantle is only feasible, when the opposed mantle slabs meet at depth and descend more or less vertically into the asthenosphere.

Subduction of 150 km of lithospheric mantle, on either flank of the belt, between 380 and 365 Ma, yields a rate of orthogonal shortening of 1 cm a^{-1}: a plausible magnitude. It is speculated that this situation created slab-pull forces in the deep lithospheric root, which led to break-off of the twinned slabs, or detachment of the thermal boundary layer. In both cases, advection of asthenospheric material will have greatly enhanced heating and rheological weakening of the remaining lithospheric mantle and lower crust. Large-scale detachment of lithospheric root may also be responsible for the widespread Viséan granitoids (Schaltegger 1997, 2000; Büttner & Kruhl 1997). However, O'Brien and Henk & von Blanckenburg (this volume) discuss alternative models.

The areal distribution of crustal thickening is a matter of debate. Zulauf (1997) and Zulauf *et al.* (1998) have proposed that maximum thickening occurred under the central part of the Teplá–Barrandian, which is comparable to a Tibetan-style high plateau. After thermal softening of the orogenic root, the plateau is supposed to have collapsed, so that Teplá–Barrandian rocks subsided against the Saxo-Thuringian and Moldanubian forelands on either side. However, the late Famennian marine transgression on the Teplá–Barrandian east of the Elbe Fault Zone precludes any major elevation in this area during late Devonian–early Carboniferous time. Therefore, it is more plausible to regard the Teplá–Barrandian as the central, 'undigested' part of a microcontinental block, which was but little affected by Variscan shortening, metamorphism and granitoid intrusion. Crustal thickening and metamorphism were much more important in the northwestern and southeastern forelands of the present-day Teplá–Barrandian outcrop.

It is in deep structural levels of these more external areas, that continental subduction created high-pressure granulites. Hot and rheologically soft granulite facies rocks, formed at *c.* 340 Ma, have been expelled from the roots towards either flank: they were intruded into the Saxo-Thuringian foreland and extruded as nappes over the Moldanubian block (Henk and Stein *et al.* this volume).

A special, late phase of HT–LP metamorphism and especially prominent granitoid magmatism occurred around 325 Ma in a belt along the SW margin of the Bohemian Massif, which transsects the tectonostratigraphic zonation. This episode cannot therefore be explained by thrust-related processes within the Saxo-Thuringian or Moldanubian Belts, although metamorphism is still synchronous with ductile deformation (e.g., Stein 1988). Numerous granite bodies in this belt are oriented NW–SE and probably delineate a late tectonic fracture pattern which parallels the orientation of main horizontal stress. Since this HT belt with granitoids does not reveal higher metamorphic pressures, metamorphic temperatures must have been raised by advective heating, possibly by mafic materials (Kalt *et al.* 2000). Heating by synmetamorphic melts would require hidden plutons, since most of the exposed granite bodies clearly post-date ductile deformation. A small-scale example of these paradoxical relationships is exposed in the northwesternmost part of the SW Bohemian transverse belt: the Frankenwald Transverse Zone, a NW-trending horst zone east of the Münchberg Klippe, is likewise characterized by anomalously strong *syn*-tectonic metamorphism and NW-trending *post*-tectonic granitoids (Kunert 1999; Kosakowski *et al.* 1999).

Orogen-parallel strike-slip component

Important dextral strike-slip displacements along the orogen-parallel shear zones are indicated by severe disruptions of the palaeogeographic context at most zonal boundaries. The present-day tectonic boundaries of the Northern Phyllite Zone and the Mid German Crystalline High cut the internal palaeogeographic zonation at an acute angle (Fig. 5). The Mid German Crystalline High shed synorogenic clastic sediments into the Rheno-Hercynian foreland, but the Saxo-Thuringian flysch basin adjacent to the SE is apparently devoid of any clasts derived from this source. It appears that important parts of palaeogeography have been excised. Dextral strike-slip has also played a major role in the Schwarzwald and Vosges.

The westward termination of the Teplá–Barrandian block could be explained in a similar way: it is possible that an originally continuous 'median massif' of the Variscan Belt was segmented by large-scale strike-slip boudinage, leaving the Teplá–Barrandian and (further west) the Central/North Armorican blocks as giant tectonic clasts. These interpretations require strike-slip displacements in the order of hundreds of kilometres. The sense of shear probably corresponds to the dextral component observed in most ductile thrusts. Unfortunately, there are no markers which would constrain displacement along these mega-shears. Their activity must predate dextral transverse faulting and oroclinal bending in the eastern termination of the Variscides (Franke & Żelaźniewicz this volume).

Conclusions

In the Variscan Belt of central Europe, it is possible to identify, from north to south respectively, the following palaeogeographic elements:

- Avalonia (parautochthon of the Rheno-Hercynian Belt)
- a Siluro/Devonian island arc, formed during closure of the Rheic Ocean between Avalonia and the Armorican Terrane Assemblage, and accreted to Avalonia in the early Devonian (preserved at the southernmost margin of the Rhenish Massif)
- Rheno-Hercynian ocean floor, formed in late early Devonian time as a successor to the Rheic Ocean (base of the Lizard–Giessen–Harz Nappes)
- Armorican fragments (tectonic slices at the base of the Giessen–Harz nappes), probably once coherent with
- the Franconian suspect terrane, the northernmost member of the Armorican Terrane Assemblage (upper plate of the active margin of the Mid German Crystalline High)
- the Cambro-Ordovician Vesser Rift (basement outcrops of Vesser)
- the Saxo-Thuringian Terrane (autochthon of the Saxo-Thuringian Belt)
- the Saxo-Thuringian narrow ocean (Saxo-Thuringian upper allochthon, Mariánské Lazně mafic unit)
- the Bohemian Terrane (Teplá–Barrandian block SE of Lazne)
- the Massif Central/Moldanubian narrow ocean (remnants in the Moldanubian Gföhl Assemblage)
- the Moldanubian Terrane—part of the Armorican Terrane Assemblage, or of Gondwana mainland? (Moldanubian Drosendorf Assemblage).

Avalonia collided with Baltica in the Late Ordovian (McKerrow *et al.* this volume). The early Palaeozoic history of some SE European terranes (Silesia, Malopolska, Lysogóry) is less clear (Belka *et al.* this volume). However, Early Devonian sandstones of the Malopolska and Lysogóry as well as from the East European platform contain detrital micas with Caledonian ages comparable to those from the Rhenish Massif, and Silesia shares a common history with the Rheno-Hercynian Belt approximately since the middle Devonian.

The Rheic Ocean between Avalonia and the Armorican Terrane Assemblage was largely consumed by the late Silurian–early Devonian, and the resulting island arc was accreted to the northern continental blocks, thus completing the Old Red Sandstone Continent (Laurussia). Though Avalonia and Armorica share the same faunas from the late Emsian onwards, there is no tectonometamorphic record of collision.

Shortly after the juxtaposition of Avalonia and the Armorican Terrane Assemblage, from the Late Emsian onwards, opening of the Rheno-Hercynian narrow ocean grossly retraced the Rheic Suture. This younger basin was filled with clastic sediments derived from the Caledonides.

With the exception of the Siluro-Devonian island arc, there is no conclusive evidence of pre-Devonian subduction. Cambro-Ordovician arc-type felsic magmatic rocks may have inherited their geochemical signature from their Cadomian parent rocks, which were formed in an Andean setting. Mineral ages around 500 Ma, observed in some rocks with dominant Devonian and Carboniferous metamorphism in the Bohemian Massif and Black Forest, are relicts of pre-Variscan events. Since the sedimentary record and the geochemical character of the mafic magmatic rocks formed around 500 Ma unequivocally record crustal extension, the relict mineral ages probably reflect thermal metamorphism in the lower crust of rift basins.

In early Devonian time, oceanic crust was present in the Rheno-Hercynian, Saxo-Thuringian and Moldanubian narrow oceans. Subduction was directed grossly to the south in the two northerly basins, and to the north in the Moldanubian. Collision commenced in the earliest Carboniferous in the Rheno-Hercynian Belt, before the Famennian in the Saxo-Thuringian Basin, and is poorly constrained in the Moldanubian (pre-340 Ma). In all of the three belts metamorphism in lower parts of the upper plate suggests subduction erosion.

Collisional deformation was performed in a regime of dextral transpression. Shortening in the Rheno-Hercynian and Saxo-Thuringian reduced the width of the forelands by *c.* 50%. The overlying allochthons were transported over = 60 and 180 km, respectively. In the Saxo-Thuringian and Moldanubian Belts, continental crust was subducted to mantle depth (45 and 28 kbar, respectively). Heating to temperatures up to 1050 °C greatly reduced the viscosity of felsic materials, which were intruded into the Saxo-Thuringian, or extruded over the Moldanubian foreland (see contributions by Franke & Stein, Henk and Krawczyk *et al.* this volume).

The strike-slip component on the suture zones was important enough to disrupt facies relations, and possibly caused crustal-scale boudinage of the central Variscan terranes (Bohemia, Armorican Massif). At a later stage, displacement along dextral shear zones across the zonal boundaries and plate rotation transformed the eastern part of the belt into a disrupted orocline (Franke & Żelaźniewicz this volume).

With regard to this complex evolution, it appears that the Variscan Belt of central Europe is one of the most complex orogens in the world—tales from other belts often appear seductively simple. Or do we just know too much? Probably not: we still need much more detailed palaeogeographic resolution of the crystalline rocks, which can only be achieved by mapping of isotopic fingerprints and isotopic ages. It also remains to correlate the central European Variscides with those dispersed in the Alpine realm of the Mediterranean. Analogous to the labels on food items limiting their edible freshness, the present review should therefore be 'best consumed before 9/2005', and then rewritten.

The author gratefully acknowledges constructive reviews by B. Leveridge and Philippe Matte, and helpful suggestions by A. Gerdes, S. McKerrow, A. Kröner, D. Tanner, P. Valverde-Vaquero and A. Żelaźniewicz. Thanks is due to C. Möbus and M. Rosenau for their invaluable help with the electronic library and drawings. Facts and views presented in this paper were assembled during research projects funded by Deutsche Forschungsgemeinschaft (project nos. Fr 668/ 2, 12, 15, 18).

References

AFTALION, M., BOWES, D. R. & VRANA, S. 1989. Early Carboniferous U–Pb zircon age for garnetiferous, perpotassic granulites, Blansky Les massif, Czechoslovakia. *Neues Jahrbuch für Mineralogie, Monatshefte*, **4**, 145–152.

AHRENDT, H., GLODNY, J., HENJES-KUNST, F., HÖHNDORF, A., KREUZER, H., KÜSTNER, W., MÜLLER-SIGMUND, H., SCHÜSSLER, U., SEIDEL, E. & WEMMER, K. 1997. Rb–Sr and K–Ar mineral data of the KTB and the surrounding area and their bearing on the tectonothermal evolution of the metamorphic basement rocks. *Geologische Rundschau, Suppl. to Volume 86*, 251–257.

ALTHERR, R. & KALT, A. 1996. Metamorphic evoluton of ultrahigh-pressure garnet peridotites from the Variscan Vosges Mts. (France). *Chemical Geology*, **134**, 27–47.

——, HENES-KLAIBER, U., HEGNER, E., SATIR, M. & LANGER, C. 1999. Plutonism in the Variscan Odenwald (Germany): from subduction to collision. *International Journal of Earth Sciences*, **88**, 422–443.

——, HOLL, A., HEGNER, E., LANGER, C. & KREUZER, H. 2000. High-potassium, calc-alkaline I-type plutonism in the European Variscides: Northern Vosges (France) and northern Schwarzwald (Germany). *Lithos*, **50**, 51–73.

ANDERLE, H.-J., FRANKE, W. & SCHWAB, M. 1995. Rhenohercynian foldbelt: metamorphic units—stratigraphy. *In*: DALLMEYER, D., FRANKE, W. & WEBER, K. (eds) *Pre-Permian Geology of Central and Western Europe*. Springer, Berlin, 99–107.

ANTHES, G. 1998. *Geodynamische Entwicklung der Mitteldeutschen Kristallinschwelle: Geochronologie und Isotopengeochemie*. Dissertation, Johannes Gutenberg Universität Mainz.

—— & REISCHMANN, T. 1996. Geochronologie und Isotopengeochemie der NE Mitteldeutschen Kristallinschwelle. *Terra Nostra*, **96/2**, 9–10.

—— & —— 1997. New 207Pb/206Pb single zircon evaporation ages from the central part of the Mid German Crystalline Rise. *Terra Nostra*, **97/5**, 10–11.

BARD, J. P., BURG, J. P., MATTE, P. & RIBEIRO, A. 1980. La chaine hercynienne d' Europe occidentale en termes de tectonique des plaques. *In*: COGNE, J. & SLANSKY, M. (eds) *Géologie de l'Europe—du Précambrien aux bassins sédimentaires post-hercyniens*. 26e Congr. Intern. Géol. 1980, Colloque C6, Villeneuve, 234–246.

BEARD, B. L., MEDARIS JR, L. G., JOHNSON, C. M., BRUECKNER, H. K. & MISAŘ, Z. 1992. Petrogenesis of Variscan high-temperature Group A eclogites from the Moldanubian Zone of the Bohemian Massif, Czechoslovakia. *Contributions to Mineralogy and Petrology*, **111**, 468–483.

BECKER, H. 1993. Contrasting origin of Variscan high-p granulites and high-t peridotites (Bohemian Massif): Evidence from Sm–Nd garnet ages. *Terra Abstracts*, **5**, 380.

—— & ALTHERR, R. 1992. Evidence from ultra-high-pressure marbles for recycling of sediments into the mantle. *Nature*, **358**, 745–748.

BEHR, H. J., GRAUERT, B., OKRUSCH, M., STÖCKHERT, B., WAGNER, G. A., WEBER, K. & ZULAUF, G. (eds) 1997. The German deep drilling project, KTB. *Geol. Rundsch., suppl. to 86*, 295 pp.

BEHRMANN, J., DROZDZEWSKI, G., HEINRICHS, T., HUCH, M., MEYER, W. & ONCKEN, O. 1991. Crustal-scale balanced cross section through the Variscan fold belt, Germany: the central EGT-segment. *Tectonophysics*, **196**, 1–21.

BELKA, Z., AHRENDT, H., FRANKE, W., SCHÄFER, J. & WEMMER, K. 2000. The Baltica/Gondwana suture in central Europe: evidence from K/Ar ages of detrital muscovites. *This volume*.

BENDER, P. 1965. *Der Nordostteil der Lahnmulde zwischen Salzböde- Aar- und Biebertal*. Dissertation, Univ. Marburg, 1–140.

—— & HOMRIGHAUSEN, R. 1979. Die Hörre-Zone, eine Neudefinition auf lithostratigraphischer Grundlage. *Geologica et Palaeontologica*, **13**, 257–260.

BRAUN, A. & MASS, R. 1992. Oberdevonische Radiolarien aus dem Breuschtal (Nord-Vogesen, Elsaß) und ihr regionaler und stratigraphischer Zusammenhang. *Neues Jahrbuch für Geologie und Paläontologie, Abhandlung*, **185**, 161–178.

BUES, C. & ZULAUF, G. 2000. Microstructural evolution and geologic significance of garnet pyriclasites in the Hoher-Bogen shear zone (Bohemian Massif, Germany). *International Journal of Earth Sciences*, **88**, 808–813.

BÜTTNER, S. & KRUHL, J. H. 1997. The evolution of a late-Variscan high-T/low-P region: the southeastern margin of the Bohemian massif. *Geologische Rundschau*, **86**, 22–38.

CARSWELL, D. A. 1991. Variscan high P-T metamorphism and uplift history in the Moldanubian Zone of the Bohemian Massif in Lower Austria. *European Journal of Mineralogy*, **3**, 323–342.

—— & JAMTVEIT, B. 1990. Variscan Sm–Nd ages for the high-pressure metamorphism in the Moldanubian Zone of the Bohemian Massif, Lower Austria. *Neues Jahrbuch für Mineralogie, Abhandlung*, **162**, 69–78.

—— & O'BRIEN, P. 1992. High pressure Quartzo-Feldspathic Garnet Granulites in the Moldanubian Zone, Bohemian Massif in Lower Austria: Their P-T Conditions for Formation, Uplift History and Geotectonic Significance. *In*: *Abstracts International Workshop High Pressure Granulites—Lower Crustal Metamorphism*, Prague, 4–5.

CHEN, F., HEGNER, E., HANN, H. P. & TODT, W. 1999. Zircon Dating of a Nappe in the Moldanubian Zone of the Black Forest—Age Constraints on the Variscan Collision. *Journal of Conference Abstracts EUG 10*, **4**, 98.

——, ——, TODT, W. & HANN, H. P. 1998. Chronological constraints on Caledonian and Early Variscan events in the Moldanubian Zone of the Black Forest, Germany. *Terra Nostra*, **98/2**, 37–38.

——, —— & —— 2000. Zircon ages and Nd isotopic and chemical compositions of orthogneisses from the Black Forest, Germany: evidence for a Cambrian magmatic arc. *International Journal of Earth Sciences*, **88**, 791–802.

CHLÚPÁČ, I. 1993. *Geology of the Barrandian—A Field trip guide*. Verlag Waldemar Kramer, Frankfurt a.M.

—— 1994. Facies and biogeographic relationships in Devonian of the Bohemian Massif. *Courier Forschungsinstitut Senckenberg*, **169**, 299–317.

—— & ZDENEK, K. 1988. Possible global events and the stratigraphy of the Palaeozoic of the Barrandian (Cambrian-Middle Devonian, Czechoslovakia). *Sbornik Geologickych Ved Geologie*, **43**, 83–146.

——, HAVLÍČEK, V., KŘÍŽ, J., KUKAL, Z., STORCH, P. (eds) 1998. *Palaeozoic of the Barrandian (Cambrian to Devonian)*. Czech Geological Survey, Prague.

COCKS, L. R. M. & FORTEY, R. A. 1982. Faunal evidence for oceanic separations in the Paleozoic of Britain. *Journal Geological Society, London*, **139**, 465–478.

DALLMEYER, R. D. & URBAN, M. 1998. Variscan vs Cadomian tectonothermal activity in northwestern sectors of the Teplá–Barrandian zone, Czech Republic: constraints from 40Ar/39Ar ages. *Geologische Rundschau*, **87**, 94–106.

——, FRANKE, W. & WEBER, K. (eds) 1995. *Pre-Permian Geology of Central and Eastern Europe*. Springer, Berlin.

——, NEUBAUER, F. & HÖCK, V. 1992. Chronology of late Palaeozoic tectonothermal activity in the southeastern Bohemian Massif, Austria (Moldanubian and Moravio-Silesian zones): 40Ar/39Ar mineral age controls. *Tectonophysics*, **210**, 135–153.

DITTMAR, U. 1996. Profilbilanzierung und Verformungsanalyse im südwestlichen Rheinischen Schiefergebirge. *Beringeria*, **17**, 347.

DOMBROWSKI, A., HENJES-KUNST, F., HÖHNDORF, A., KRÖNER, A., OKRUSCH, M. & RICHTER, P. 1995. Orthogneisses in the Spessart Crystalline Complex, north-west Bavaria: Silurian granitoid magmatism at an active continental margin. *Geologische Rundschau*, **64/2**, 399–411.

DÖRR, W., FIALA, J., FRANKE, W., HAACK, U., PHILIPPE, S., SCHASTOCK, J., SCHEUVENS, D., VEJNAR, Z. & ZULAUF, G. 1998. Cambrian vs Variscan tectonothermal evolution within the Teplá–Barrandian: evidence from U–Pb zircon ages of syn-tectonic plutons (Bohemian Massif, Czech Republic). *Acta Universitatis Carolinae—Geologica*, **42**(2), 229–230.

——, ——, SCHASTOCK, J., SCHEUVENS, D., WULF, S., ZULAUF, G., AHRENDT, H. & WEMMER, K. 1997. Dating of fault related Plutons along terrane boundaries in the Bohemian Massif. *Terra Nova, Abstract supplement no. 1*, **9**, 100.

——, ——, VEJNAR, Z. & ZULAUF, G. 1998. U–Pb zircon ages and structural development of metagranitoids of the Teplá Crystalline Complex: evidence for pervasive Cambrian plutonism within the Bohemian Massif (Czech Republic). *Geologische Rundschau*, **87**, 135–149.

——, FLOYD, P. A. & LEVERIDGE, B. E. 1999. U–Pb ages and geochemistry of granite pebbles from the Devonian Menaver Conglomerate, Lizard peninsula: provenance of Rhenohercynian flysch of SW England. *Sedimentary Geology*, **124**, 131–147.

——, Piqué, A., Franke, W. & Kramm, U. 1992. Les galets granitiques du conglomérat de Russ (Dévono-Dinantien des Vosges du Nord) sont les témoins d'un magmatisme acide ordovicien. La distension crustale et le rifting saxothuringien au Paléozoique inférieur. *Comptes Rendus de l'Academie des Sciences Paris*, **315**, série II, 587–594.

Doubinger, C. 1963. Chitinozaires ordoviciens et siluriens des schistes de Steige dans les Vosges. *Bulletin du Service de Carte géologique d'Alsace Lorraine*, **16**, 125–136.

Echtler, H. P. & Altherr, R. 1993. Variscan crustal evolution in the Vosges Mountains and in the Schwarzwald: Guide to the excursion of the Swiss Geological Society and the Swiss Society of Mineralogy and Petrology (3–5 October, 1992). *Schweizer Mineralogische und Petrographische Mitteilungen*, **73**, 113–128.

Echtler, H. P. & Chauvet, A. 1992. Carboniferous convergence and subsequent crustal extension in the southern Schwarzwald (SW Germany). *Geodinamica Acta (Paris)*, 1991–1992, **5**, 1–2, 37–49.

Eisbacher, G. H., Lüschen, E. & Wickert, F. 1989. Crustal-scale thrusting and extension in the Hercynian Schwarzwald and Vosges, Central Europe. *Tectonics*, **8**, 1–21.

Engel, W. & Franke, W. 1983. Flysch sedimentation: Its relation to tectonism in the European Variscides. *In*: Martin, H. & Eder, F. W. (eds) *Intracontinental Fold Belts. Case studies in the Variscan Belt of Europe and the Damara Belt in Namibia*. Springer, Berlin, 289–321.

——, —— & Langenstrassen, F. 1983. Paleozoic sedimentation in the northern branch of the mid-European Variscides: essay of an interpretation. *In*: Martin, H. (ed.) *Intracontinental Fold Belts. Case studies in the Variscan Belt of Europe and the Damara Belt in Namibia*. Springer, Berlin, 9–42.

Falk, F., Franke, W. & Kurze, M. 1995. Saxothuringian Basin: autochthon and nonmetamorphic nappe units: stratigraphy. *In*: Dallmeyer, D., Franke, W. & Weber, K. (eds) *Pre-Permian Geology of Central and Western Europe*. Springer, Berlin, 219–234.

Finger, F. & Clemens, J. D. 1995. Migmatization and 'secondary' granitic magmas: effects of emplacement and crystallization of 'primary' granitoids in Southern Bohemia, Austria. *Contributions to Mineralogy and Petrology*, **120**, 311–326.

—— & Quadt, A. v. 1995. U/Pb ages of zircons from a plagiogranite-gneiss in the south-eastern Bohemian Massif, Austria—further evidence for an important early Paleozoic rifting episode in the easter Variscides. *Schweizer Mineralogische und Petrographische Mitteilungen*, **75**, 265–270.

—— & Steyrer, H. P. 1995. A tectonic model for the Eastern Variscides: Indications from a chemical study of amphibolites in the South-Eastern Bohemian Massif. *Geologica Carpathica*, **46**, 137–150.

——, Roberts, M. P., Haunschmid, B., Schermaier, A. & Steyrer, H. P. 1997. Variscan granitoids of central Europe: their typology, potential sources and tectonothermal relations. *Mineralogy and Petrology*, **61**, 67–96.

Flöttmann, T. & Oncken, O. 1992. Constraints on the evolution of the Mid German Crystalline Rise—a study of outcrops west of the river Rhine. *Geol. Rundsch.*, **81/2**, 515–543.

Floyd, P. A. 1995. Rhenohercynian foldbelt: autochthon and nonmetamorphic nappe units—igneous activity. *In*: Dallmeyer, D., Franke, W. & Weber, K. (eds) *Pre-Permian Geology of Central and Western Europe*. Springer, Berlin, 59–81.

Floyd, P. A., Winchester, J. A., Seston, R., Kryza, R. & Crowley, Q. G. 2000. Review of geochemical variation in Lower Palaeozoic metabasites from the NE Bohemian Massif: intracratonic rifting and plume-ridge interaction. *This volume*.

Förster, H.-J., Tischendorf, G., Seltmann, R. & Gottesmann, B. 1998. Die variszischen Granite des Erzgebirges: neue Aspekte aus stofflicher Sicht. *Zeitschrift für Geologische Wissenschaften*, **26**, 1/2, 31–60.

Franke, W. 1984*a*. Late events in the tectonic history of the Saxothuringian Zone. *In*: Hutton, D. W. H. & Sanderson, D. J. (eds) *Variscan tectonics of the North Atlantic Region*. Geological Society, London, Special Publications **14**, 33–45.

—— 1984*b*. Variszischer Deckenbau im Raume der Münchberger Gneismasse, abgeleitet aus der Fazies, Deformation und Metamorphose im ungebenden Paläozoikum. *Geotekonische Forschungen*, **68**, 1–253.

—— 1989. Tectonostratigraphic units in the Variscan belt of central Europe. *In*: Dallmeyer, R. D. (ed.) *Terranes in the Circum-Atlantic Palaeozoic Ocean*. Geological Society of America, Special Papers, **230**, 67–90.

—— 1992. Phanerozoic structures and events in central Europe. *In*: Blundell, D., Freeman, R. & Mueller (eds) *A Continent Revealed—The European Geotraverse*. Cambridge University Press, 164–179.

—— 1995. Rhenohercynian foldbelt: autochthon and nonmetamorphic nappe units—stratigraphy. *In*: Dallmeyer, D., Franke, W. & Weber, K. (eds) *Pre-Permian Geology of Central and Western Europe*. Springer, Berlin, 33–49.

—— & Engel, W. 1982. Variscan Sedimentary Basins on the Continent and relations with south-west England. *Proceedings of the Ussher Society*, **5**, 259–269.

—— & —— 1986. Synorogenic sedimentation in the Variscan Belt of Europe. *Bulletin de la Société Géologique de France*, **8, II**, 25–33.

—— & Oncken, O. 1995. Zur prädevonischen Geschichte des Rhenohercynischen Beckens. *Nova Acta Leopoldina, Neue Folge*, **71**, 53–72.

—— & Stein, E. 2000. Exhumation of high-pressure rocks in the Saxo-Thuringian belt: geological constraints and alternative models. *This volume*.

—— & Żelaźniewicz, A. 2000. The eastern termination of the Variscides: terrane correlation and kinematic evolution. *This volume*.

——, Eder, W., Engel, W. & Langenstrassen, F. 1978. Main aspects of geosynclinal sedimentation in the Rhenohercynian Zone. *Zeitschrift der deutschen geologischen Gesellschaft*, **129**, 201–216.

——, Dallmeyer, D. & Weber, K. 1995*a*. Geodynamic evolution. *In*: Dallmeyer, D., Franke, W. & Weber, K. (eds) *Pre-Permian Geology of Central and Western Europe*. Springer, Berlin, 579–593.

——, Kreuzer, H., Okrusch, M., Schüssler, U. & Seidel, E. 1995*b*. Saxothuringian Basin: Exotic metamorphic nappes: stratigraphy, structure, and igneous activity. *In*: Dallmeyer, D., Franke, W. & Weber, K. (eds) *Pre-Permian Geology of Central and Western Europe*. Springer, Berlin, 275–294.

——, Prössl, K. F. & Schwarz, J. 1992. Devonische Grauwacken im Erbendorfer Paläozoikum—Alter, tektonische Stellung und geotektonische Bedeutung. *KTB Report*, **92–4**, 213–224.

Friedl, G., McNaughton, N., Fletcher, I. R. & Finger, F. 1998. New SHRIMP-zircon ages for orthogneisses from the south-eastern part of the Bohemian Massif (Lower Austria). *Acta Universitatis Carolinae—Geologica*, **42**(2), 251–252.

——, Quadt, A. von & Finger, F. 1992. Erste Ergebnisse von U/Pb Altersdatierungsarbeiten am Rastenberger Granodiorit im Niederösterreichischen Waldviertel. *Mitteilungen Österreiches Geologischen Geselleschaft*, **137**, 131–134.

——, —— & —— 1994. 340 Ma U/Pb-Monazitalter aus dem niederösterreichischen Moldanubikum und ihre geologische Bedeutung. *Terra Nostra*, **3/94**, 43–46.

——, —— & —— 1996. Timing der Intrusiontätigkeit im südböhmischen Batholith. *Facultas-Universitätsverlag*, 6. Symposium: Tektonik—Strukturgeologie—Kristallingeologie, Salzburg, 127–130.

——, ——, Ochsner, A. & Finger, F. 1993. Timing of the Variscan orogeny in the Southern Bohemian Massif (NE-Austria) deduced from new U–Pb zircon and monazite dating. *Terra Abstracts*, **5**, 235–236.

Gansslоser, M., Vibrans, E. & Wachendorf, H. 1995. Die Metabasalte des Harzes. *Zentralblatt für Geologie und Paläontologie*, **Teil 1**, 1103–1115.

Gebauer, D. 1994. Ein Modell der Entwicklungsgeschichte des ostbayerischen Grundgebirges auf der Basis plattentektonischer Vorstellungen und radiometrischer Datierungen 1994. *Erläuterungen zur geologischen Karte von Bayern 1 : 25 000, Bl. 6439 Tännesberg*. Bayerische Geologische Landesamt.

—— & Friedl, G. 1994. A 1.38 Ga protolith age for the Dobra Orthogneiss (Moldanubian Zone of the southern Bohemian Massif, NE Austria): evidence from ion-microprobe (SHRIMP) dating of zircon. *Journal of the Czech Geological Society*, **39**, 34–35.

Glodny, J., Grauert, B., Fiala, J., Vejnar, Z. & Krohe, A. 1998. Metapegmatites in the western Bohemian massif: ages of crystallisation and metamorphic overprint, as constrained by U–Pb zircon, monazite, garnet, columbite and Rb–Sr muscovite data. *Geologische Rundschau*, **87**, 124–134.

Grauert, B., Grosse-Westermann, U., Albat, F. 1990. Interpretation konkordanter U–Th–Pb-Monazitalter moldanubischer Gneise—Schlußfolgerungen aufgrund vergleichbarer U–Th–Pb-, Sm–Nd- und Rb–Sr-Isotopenanalysen. *Beihefte European Journal of Mineralogy*, **2**, 82.

Grösser, J. & Dörr, W. 1986. MOR-Basalte im östlichen Rheinischen Schiefergebirge. *Neues Jahrbuch Geologischer und Paläontologischer Monatshefte*, **12**, 705–722.

Gruppenbericht 1981. Hamburger Arbeitskonferenz Ebbe-Antiklinorium. *Mitteilungen des geologisch-paläontologischen Institutes der Universität Hamburg*, **50**, 5–16.

Güldenpfennig, M. 1998. Zur tektonischen Stellung unterkarbonischer Grauwacken und Vulkanite der Zone von Badenweiler–Lenzkirch (Südschwarzwald). *Zeitschrift der deutschen geologischen Gesellschaft*, **149**, 213–232.

Hahn, H.-D. 1990. *Fazies grobklastischer Gesteine des Unterdevons (Graue Phyllite bis Taunusquarzite) im Taunus (Rheinisches Schiefergebirge)*. PhD Thesis, Univ. Marburg.

Hanel, M., Kalt, A., Montenari, M. & Wimmenauer, W. 1998. Bestimmung von Sedimentationsaltern hochmetamorpher Paragneise des Schwarzwaldes. *Terra Nostra*, **98/2**.

——, Lippolt, H. J., Kober, B. & Wimmenauer, W. 1993. Lower Carboniferous Granulites in the Schwarzwald Basement near Hohengeroldseck (SW-Germany). *Naturwissenschaften*, **80**, 25–28.

Haverkamp, J., Kramm, U. & Walter, R. 1991. U/Pb isotope variations of detrital zircons in sediments of the Rhenoherzynian and their significance for the Palaeozoic Geotectonic development of NW Central Europe. *Terra Abstracts*, **3**, 207.

Henk, A. 2000. Foreland-directed lower crustal flow and its implications for the exhumation of high pressure–high temperature rocks. *This volume*.

—— & von Blanckenburg, F. 2000. Syn-convergent high-temperature metamorphism and magmatism in the Variscides—a discussion of potential heat sources. *This volume*.

Hofmann, A. & Köhler, W. 1973. Whole rock Rb–Sr ages of anatectic gneisses from the Schwarzwald, SW Germany. *Neues Jahrbuch für Mineralogie Abhandlunge*, **119**, 163–187.

Holder, M. T. & Leveridge, B. E. 1986. Correlation of the Rhenohercynian Variscides. *Journal of the Geological Society, London*, **143**, 141–147.

Hollmann, G. & von Winterfeld, C. 1999. Laterale Strukturvariationen eines Vorlandüberschiebungsgürtels. *Zeitschrift der deutschen geologischen Gesellschaft*, **150**, 431–450.

Holub, F., Cocherie, A. & Rossi, P. 1997. Radiometric dating of granitic rocks from the Central Bohemian Plutonic Complex (Czech Republic): constraints on the chronology of thermal and tectonic events along the Moldanubian-Barrandian boundary. *Earth and Planetary Sciences*, **325**, 19–26.

HUCKRIEDE, H., AHRENDT, H., FRANKE, W., SCHÄFER, J., WEMMER, K. & MEISCHNER, D. 1999. Plate tectonic setting of allochthonous units in the Rhenohercynian belt: new insights from detritus analysis of synorogenic sediments. *Terra Nostra*, **99/1**, 109.

——, ——, ——, WEMMER, K. & MEISCHNER, D. 1998. Orogenic processes recorded in Early Carboniferous and Devonian clastic sediments of the Rhenohercynian Zone. *Terra Nostra*, **98/2**, 77–79.

JÄGER, H. & GURSKY, H.-J. 1998. Quarzite im Unterkarbon der Hörre–Gommern-Zone: ein paläogeographisches Problematicum im Rhenoherzynikum. *In*: *Alfred-Wegener-Stiftung*, GEO-BERLIN '98—Gemeinsame Jahrestagung DGG, DMG, GGW, Pal. Ges., Berlin, **98/3**, 148.

KALT, A. & ALTHERR, R. 1996. Metamorphic evolution of garnet-spinel peridotites from the Variscan Schwarzwald (Germany). *Geologische Rundschau*, **85**, 211–224.

—— & GRAUERT, B. 1994. Petrologie cordieritführender Gneise und Migmatite des Schwarzwaldes und des Bayerischen Waldes. *Terra Nostra*, **3/94**, 63.

——, CORFU, F. & WIJBRANS, J. 2000. Time calibration of a P-T path from a Variscan high-temperature low-pressure metamorphic complex (Bayerische Wald, Germany), and the detection of inherited monazite. *Contributions to Mineralogy and Petrology*, **138**, 143–163.

——, GRAUERT, B. & BAUMANN, A. 1994. Rb–Sr and U–Pb isotope studies on migmatites from the Schwarzwald (Germany): constraints on isotopic resetting during Variscan high-temperature metamorphism. *Journal of metamorphic Geology*, **12**, 667–680.

——, HANEL, M., SCHLEICHER, H. & KRAMM, U. 1994. Petrology and geochronology of eclogites from the Variscan Schwarzwald (F.R.G.). *Contributions to Mineralogy and Petrology*, **115**, 287–302.

——, SCHLEICHER, H. & KRAMM, U. 1990. Karbonische Eklogitbildung im Schwarzwald. *Beihefte European Journal of Mineralogy*, **2**, 120.

KASIG, W. & WILDER, H. 1983. The sedimentary development of the western Rheinische Schiefergebirge and the Ardennes (Germany/Belgium). *In*: MARTIN, H. & EDER, F. W. (eds) *Intracontinental fold belts*. Springer, Berlin, 185–211.

KATZUNG, G. 1999. Records of the Late Ordovician glaciation from Thuringia, Germany. *Zeitschrift der deutschen Geologischen Gesellschaft*, **150/3**, 595–617.

KEGEL, W. 1933. *Erläuterungen zur geologischen*. Karte von Preussen und benachbarten deutschen Ländern, Blatt Rodheim, Lieferung **317**.

KEMNITZ, H., ROMER, R. L. & ONCKEN, O. Gondwana breakup and the northern margin of the Saxothuringian belt (Variscides of Central Europe). *International Journal of Earth Sciences*, in press.

KIRNBAUER, T. 1991. Geologie, Petrographie und Geochemie der Pyroklastika des Unteren Ems/Unter-Devon (Porphyroide) im südlichen Rheinischen Schiefergebirge. *Geologische Abhandlungen Hessen*, **92**, 1–228.

KIRSCH, H., KOBER, B. & LIPPOLT, H. J. 1988. Age of intrusion and rapid cooling of the Frankenstein gabbro (Odenwald,SW-Germany) evidenced by 40-Ar/39-Ar and single-zircon 207-Pb/206-Pb measurements. *Geologische Rundschau*, **77**, 693–711.

KOBER, B., HRADETZKY, H. & LIPPOLT, H. J. 1986. Radiogenblei-Evaporationsstudien an einzelnen Zirkonkristallen zur präherzynischen Entwicklung des Grundgebirges im Zentralschwarzwald, SW-Deutschland. *Fortschitte der Mineralogie*, **64**, Beih. 1, 81.

KÖHLER, H., DODIG, G. & HÖLZL, S. 1993. Zirkondatierungen am Leuchtenberger Granit (NE Bayern). *Beihefte European Journal of Mineralogy*, **5**, 116.

——, PROPACH, G. & TROLL, G. 1989. Exkursion zur Geologie, Petrographie und Geochronologie des NE-bayerischen Grundgebirges. *Beihefte European Journal of Mineralogy*, **1**, 1–84.

KOSAKOWSKI, G., KUNERT, V., CLAUSER, C., FRANKE, W. & NEUGEBAUER, H. J. 1999. Hydrothermal transients in Variscan crust: paleo-temperature mapping and hydrothermal models. *Tectonophysics*, **306**, 325–344.

KOŠLER, J., AFTALION, M. & BOWES, D. R. 1993. Mid-late Devonian plutonic activity in the Bohemian Massif: U–Pb zircon isotopic evidence from the Staré Sedlo and Mirotice gneiss complexes, Czech Republic. *Neues Jahrbuch für Mineralogie, Monatshefte*, **9**, 417–431.

—— & FARROW, C. M. 1994. Mid-Late Devonian arc-type magmatism in the Bohemian Massif: Sr and Nd isotope and trace element evidence from the Stare Sedlo and Mirotice gneiss complexes, Czech Republic. *Journal of the Czech Geological Society*, **39**, 56–58.

KOSSMAT, F. 1927. Gliederung des varistischen Gebirgsbaues. *Abhandlungen des Sächsischen Geologischen Landesamtes*, **1**, 1–39.

KRAWCZYK, C. M., STEIN, E., CHOI, S., OETTINGER, G., SCHUSTER, K., GÖTZE, H.-J., HAAK, V., ONCKEN, O., PRODEHL, C. & SCHULZE, A. 2000. Constraints on distribution and exhumation mechanisms of high-pressure rocks from geophysical studies—the Saxothuringian case between the Bray Fault and Elbe Line. *This volume*.

KREBS, W. 1976. Zur geotektonischen Position der Bohrung Saar 1. *Geologisches Jahrbuch*, **A 27**, 489–498.

KREUZER, H. & HARRE, W. 1975. K/Ar-Altersbestimmungen an Hornblenden und Biotiten des kristallinen Odenwaldes. *Aufschluss Sonderband*, **27**, 71–77.

——, SEIDEL, E., SCHÜSSLER, U., OKRUSCH, M., LENZ, K.-L. & RASCHKA, H. 1989. K–Ar geochronology of different tectonic units at the northwestern margin of the Bohemian Massif. *Tectonophysics*, **157**, 149–178.

KROHE, A. & EISBACHER, G. H. 1988. Oblique crustal detachment in the Variscan Schwarzwald, southwestern Germany. *Geologische Rundschau*, **77**, 25–43.

KRÖNER, A. & HEGNER, E. 1998. Geochemistry, single grain zircon ages and Sm–Nd systematics of granitoid rocks from the Góry Sowie (Owl Mts), Polish west Sudetes: evidence for early Palaeozoic arc-related plutonism. *Journal of the Geological Society, London*, **155**, 711–724.

——, O'BRIEN, P. J. & PIDGEON, R. J. 2000. Zircon ages for high-pressure granulites from south Bohemia, Czech Republic, and their connections to Carboniferous high-temperature processes. *Contributions to Mineralogy and Petrology*, **138/2**, 127–142

——, STIPSKA, P., SCHULMANN, K. & JAECKEL, P. 2000. Chronological constraints on the pre-Variscan evolution of the northeastern margin of the Bohemian Massif, Czech Republic. *This volume*.

——, WENDT, I., LIEW, T. C., COMPSTON, W., TODT, W., FIALA, J., VANKOVA, V. & VANEK, J. 1988. U–Pb zircon and Sm–Nd model ages of high-grade Moldanubian metasediments, Bohemian Massif, Czechoslovakia. *Contributions to Mineralogy and Petrology*, **99**, 257–266.

KÜSTNER, W., AHRENDT, H., HANSEN, B. T. & WEMMER, K. 1998. K/Ar-Datierungen an detritischen Muskoviten und Sm–Nd-Modellalter prä- und synorogener schwach metamorpher Sedimente im Rhenherzynikum—Grundlegende Daten zur Quantifizierung orogener Prozesse am Beispiel der Varisciden. *Terra Nostra*, **98/2**, 96–98.

KUNERT, V. 1999. *Die Frankenwälder Querzone: Entwicklung einer thermischen Anomalie im Saxothuringikum*. Dissertation, Univ. Giessen.

LANGENSTRASSEN, F. 1983. Neritic sedimentation of the Lower and Middle Devonian in the Rheinisches Schiefergebirge east of the river Rhine. *In*: MARTIN H. & EDER, F. W. (eds) *Intracontinental fold belts*. Springer, Berlin, 43–76.

LITTKE, R., BÜKER, C., HERTLE, M., KARG, H., STROETMANN-HEINEN, V. & ONCKEN, O. 2000. Heat flow evolution, subsidence and erosion in the Rhenohercynian orogenic wedge of central Europe. *This volume*.

LINNEMANN, U., GEHMLICH, M., TICHOMIROWA, B., BUSCHMANN, L., NASDALA, P., JONAS, H., LÜTZNER, H. & BOMBACH, K. 2000. From Cadomian Subduction to Early Palaeozoic Rifting: The Evolution of Saxo-Thuringia at the margin of Gondwana in the light of single zircon geochronology and basin development (Central European Variscides, Germany). *This volume*.

LOESCHKE, J., GÜLDENPFENNIG, M., HANN, H. P. & SAWATZKI, G. 1998. Die Zone von Badenweiler–Lenzkirch (Schwarzwald): Eine variskische Suturzone. *Zeitschrift der Deutschen Geologischen Gesellschaft*, **149**, 197–212.

MATTE, P. 1986. Tectonic and plate tectonic model for the Variscan belt of Europe. *Tectonophysics*, **126**, 329–374.

——, MALUSKI, H. & ECHTLER, H. 1985. Tectonique. Cisaillements ductiles varisques vers l'Est—Sud-Est dans les nappes du Waldviertel (Sud-Est du Massif de Bohème, Austriche). Données microtectoniques et radiométriques 39Ar/40Ar. *Comptes Rendus de l'Academie des Sciences, Paris*, **301**, 721–726.

——, ——, RAJLICH, P. & FRANKE, W. 1990. Terrane boundaries in the Bohemian Massif: Results of large-scale Variscan shearing. *Tectonophysics*, **177**, 151–170.

MCKERROW W. S. & VAN STAAL, C. R. 2000. The Palaeozoic time scale reviewed. *This volume*.

——, MACNIOCAILL, C., AHLBERG, P. E., CLAYTON, G., CLEAL, C. J. & EAGAR, R. M. C. 2000. The late Palaeozoic relationships between Gondwana and Laurussia. *This volume*.

MEDARIS, J., L. G., WANG, H. F., MÍSAŘ, Z. & JELINEK, E. 1990. Thermobarometry, diffusion modelling and cooling rates of crustal garnet peridotite: Two examples from the Moldanubian zone of the Bohemian Massif. *Lithos*, **25**, 189–202.

MEISCHNER, D. 1971. Clastic sedimentation in the Variscan Geosyncline East of the River Rhine. *In*: MÜLLER, G. (ed.) *Sedimentology of parts of Central Europe, Guidebook VIII Int. Sed. Congress Heidelberg 1971*. Waldemar Kramer, Frankfurt, Main, 9–43.

—— 1991. Kleine Geologie des Kellerwaldes (Exkursion F). *Jahresberichte und Mitteilungen des Oberrheinischen Geologischen Vereins*, **73**, 115–142.

MEISL, S. 1995. Rhenohercynian foldbelt: Metamorphic units—igneous activity. *In*: DALLMEYER, D., FRANKE, W. & WEBER, K. (eds) *Pre-Permian Geology of Central and Western Europe*. Springer, Berlin, 118–131.

——, KREUZER, H. & HÖHNDORF, A. 1989. Metamorphose-Bedingungen und -Alter des Kristallins am Wartenstein bei Kirn/Nahe. Kurzfassungen, 5. *In*: Rundgespräch 'Geologie des europäischen Variszikums', Braunschweig, 38–39.

MIELKE, H., ROHRMÜLLER, J. & GEBAUER, G. 1996. Ein metalateritisches Denudations-Niveau als lithologisch und zeitlich korrelierbarer Bezugshorizont in Phylliten, Glimmerschiefern und Gneisen des ostbayerischen Grundgebirges. *Geologica Bavarica*, **101**, 139–166.

MOLZAHN, M., ANTHES, G. & REISCHMANN, T. 1998. Single zircon Pb/Pb age geochronology and isotope systematics of the Rhenohercynian basement. *Terra Nostra*, **98/1**, 67–68.

MONTENARI, M. 1999. Calciclastic Turbidites from the Southern Vosges Basin (Central Variscan Belt): Reconstruction of a Lost Carboniferous Carbonat Platform. *Journal of Conference Abstracts EUG 10*, **4**, 742.

—— & HANEL, M. 1998. Palynology of the metamorphic Baden-Baden-Zone: a comparison between the Schwarzwald and the Vosges. *In*: *Alfred-Wegener-Stiftung*, GEO-BERLIN '98—Gemeinsame Jahrestagung DGG, DMG, GGW, Paläontol. Ges., Berlin, **98/3**, V 233.

——, SERVAIS, TH. & PARIS, F. 2000. Palynological dating (acritarchs and chitinozoans) of Lower Paleozoic phyllites from the Black Forest/southwestern Germany. *Comptes Rendus de l'Academie des Sciences, Paris, Earth and Planetary Sciences*, **330**, 493–499.

Müller, G. & Stoppel, D. 1981. Zur Stratigraphie und Tektonik im Bereich der Schwerspatgrube 'Korb' bei Eisen (N-Saarland). *Zeitschrift der deutschen geologischen Gesellschaft*, **132**, 325–352.

Nasir, S., Okrusch, M., Kreuzer, H., Lenz, H. & Höhndorf, A. 1991. Geochronology of the Spessart Crystalline Complex, Mid-German Crystalline Rise. *Mineralogy and Petrology*, **44**, 39–55.

Neuroth, H. 1997. *K/Ar-Datierungen an detritischen Muskoviten—'Sicherungskopien' orogener Prozesse am Beispiel der Varisziden.* Göttinger Arbeiten zur Geologie und Paläontologie, **72**.

Novák, M., Cerny, P., Kimbrough, D. L., Taylor, M. C. & Ercit, T. S. 1998. U–Pb ages of monazite from granitic pegmatites in the Moldanubian Zone and their geological implications. *Acta Universitatis Carolinae—Geologica*, **42**, 309–310.

O'Brien, P. J. 2000. The fundamental Variscan problem: high-temperature metamorphism at different depths and high-pressure metamorphism at different temperatures. *This volume.*

—— & Vrána, S. 1995. Eclogites with a short-lived granulite facies overprint in the Moldanubian Zone, Czech Pepublic: petrology, geochemistry and diffusion modelling of garnet zoning. *Geologische Rundschau*, **84**, 473–488.

—— & —— 1997. The eclogites in the Monotonous Series of the Moldanubian zone and the theory of thermal pulses: a reply. *Geologische Rundschau*, **86**, 716–719.

Oczlon, M. S. 1994. North Gondwana origin for exotic Variscan rocks in the Rhenohercynian zone of Germany. *Geologische Rundschau*, **83**, 20–31.

Okrusch, M. 1995. Mid-German Crystalline High: metamorphic evolution. *In*: Dallmeyer, D., Franke, W. & Weber, K. (eds) *Pre-Permian Geology of Central and Western Europe*. Springer, Berlin, 201–213.

Okrusch, M. & Weber, K. 1996. Der Kristallinkomplex des Vorspessart. *Zeitschrifte für Geologische Wissenschaften*, **24**(1/2), 141–174.

Oliver, G. J. H., Corfu, F. & Krogh, T. E., 1993. U–Pb ages from SW Poland: evidence for a Caledonian suture zone between Baltica and Gondwana. *Journal of the Geological Society, London*, **150**, 355–369.

Olivera, J. T. & Quesada, C. 1998. A comparison of stratigraphy, structure and palaeogeography of the south Portuguese zone and south-west England, European Variscides. *Proceedings of the Ussher Society*, **9**, 141–150.

Oncken, O. 1997. Transformation of a magmatic arc and orogenic root during the oblique collision and it's consequences for the evolution of the European Variscides (Mid-German Crystalline Rise). *Geologische Rundschau*, **86**, 2–21.

—— 1998. Tectonic mass transfer prior to collision—subduction erosion versus accretion in the Mid-European Variscides. *Acta Universitatis Carolinae—Geologica*, **42**, 313–314.

——, Franzke, H. J., Dittmar, U. & Klügel, T. 1995*a*. Rhenohercynian foldbelt: metamorphic units—structure. *In*: Dallmeyer, D., Franke, W. & Weber, K. (eds) *Pre-Permian Geology of Central and Western Europe*. Springer, Berlin, 108–117.

——, Massonne, H. J. & Schwab, M. 1995*b*. Rhenohercynian foldbelt: autochthon and nonmetamorphic nappe units—metamorphic evolution. *In*: Dallmeyer, R. D., Franke, W. & Weber, K. (eds) *Pre-Permian Geology of Central and Eastern Europe*. Springer, Berlin, 82–86.

——, Plesch, A., Weber, J., Ricken, W. & Schrader, S. 2000. Passive margin detachment during arc-continent collision (central European Variscides). *This volume.*

——, von Winterfeld, C. & Dittmar, U. 1999. Accretion and inversion of a rifted passive margin—the Late Paleozoic Rhenohercynian fold and thrust belt. *Tectonics*, **18**, 75–91.

Patočka, F. 1991. Geochemistry and Primary Tectonic Environment of the Amphibolites from the Cesky Krumlov Varied Group (Bohemian Massif, Moldanubicum). *Jahrbuch für Geolologie*, **134**, 117–133.

——, Vlasimsky, P. & Blechová, K. 1993. Geochemistry of early Paleozoic volcanics of the Barrandian Basin (Bohemian Massif, Czech Republic): implications for Paleotectonic reconstructions. *Jahrbuch der Geologischen Bundesanstalt*, **136/4**, 873–896.

Petrakakis, K. 1997. Evolution of Moldanubian rocks in Austria: review and synthesis. *Journal of metamorphic Geology*, **15**, 203–222.

Pharaoh, T. C., Molyneux, S. G., Merriman, R. J., Lee, M. K. & Verniers, J (eds) 1993. The Caledonides of the Anglo-Brabant Massif. *Geological Magazine*, **130**, 561–730.

Quadt, A. von & Finger, F. 1991. Geochronologische Untersuchungen im österreichischen Teil des Südböhmischen Batoliths: U–Pb Datierungen an Zirkonen, Monaziten und Xenotimen des Weinsberger Granits. *Beihefte European Journal of Mineralogy*, **3**, 281.

—— & —— 1994. Entwicklung der mitteleuropäischen kontinentalen Kruste am Beispiel des Südböhmischen und des Tauern-Batoliths: Sr-Nd, U–Pb, REE und Spurenelement-Charakteristika. *Terra Nostra*, **3/94**, 81.

Reischmann, T. & Anthes, G. 1996. Geochronologie und Geodynamische Entwicklung der Mitteldeutschen Kristallinschwelle westlich des Rheins. *Terra Nostra*, **96/2**, 161–162.

Reitz, E. 1987. Silurische Sporen aus einem granatführenden Glimmerschiefer des Vor-Spessart, NW-Bayern. *Neues Jahrbuch für Geologie und Paläontologie, Monatsheft*, **11**, 699–704.

—— & Anderle, H.-J., Winkelmann, M. 1995. Ein erster Nachweis von Unterordovizium (Arenig) am Südrand des Rheinischen Schiefergebirges im Vordertaunus: Der Bierstadt-Phyllit (Bl. 5915 Wiesbaden). *Geologisches Jahrbuch Hessen*, **123**, 25–38.

—— & Wickert, F. 1989. Late Cambrian to Early Ordovician acritarchs from the Villé Unit, Northern Vosges Mountains (France). *Neues Jahrbuch für Geologie und Paläontologie, Monatshefte*, **6**, 375–384.

ROBERTS, M. P. & FINGER, F. 1997. Do U–Pb zircon ages from granulites reflect peak metamorphic conditions? *Geology*, **25**, 319–322.

SCHÄFER, F., ONCKEN, O., KEMNITZ, H. & ROMER, R. 2000. Retro-wedge evolution during arc-continent collision—a case study from the Variscides (Saxothuringian zone). *This volume.*

SCHÄFER, J., NEUROTH, H., AHRENDT, H., DÖRR, W. & FRANKE, W. 1997. Accretion and exhumation at a Variscan active margin, recorded in the Saxothuringian flysch. *Geologische Rundschau*, **86**, 599–611.

SCHALTEGGER, U. 1995. High-resolution chronometry of Late Variscan extensional magmatism and basin formation: examples from the Vosges, Black Forest and the Alpine Basement. *Terra Nostra*, **7/95**, 109–111.

—— 1997. Magma pulses in the Central Variscan Belt: episodic melt generation and emplacement during lithospheric thinning. *Terra Nova*, **9**, 242–245.

—— 2000. U–Pb geochronology of the southern Black Forest Batholith (Central Variscan Belt): timing of exhumation and granite emplacement. *International Journal of Earth Sciences*, **88**, 814–828.

——, FANNING, C. M., GÜNTHER, D., MAURIN, J. C., SCHULMANN, K. & GEBAUER, D. 1999. Growth, annealing and recrystallization of zircon and preservation of monazite in high-grade metamorphism: conventional and *in-situ* U–Pb isotope, cathodoluminescence and microchemical evidence. *Contributions to Mineralogy and Petrology*, **134**, 186–201.

——, SCHNEIDER, J.-L., MARIN, J.-C. & CORFU, F. 1996. Precise U–Pb chronometry of 345–340 Ma old magmatism related to syn-convergence extension in the Southern Vosges (Central Variscan Belt). *Earth and Planetary Science Letters*, **144**, 403–419.

SCHENK, V. & TODT, W. 1983. U–Pb-Datierungen an Zirkon und Monazit der Granulite im Moldanubikum Niederösterreichs (Waldviertel). *Fortschritte der Mineralogie*, **61**, Beiheft 1, 190–191.

SCHERMAIER, A., HAUNSCHMID, B. & FINGER, F. 1997. Distribution of Variscan I- and S-type granites in the Eastern Alps: a possible clue to unravel pre-Alpine basement structures. *Tectonophysics*, **272**, 315–333.

SCHMIDT, F.-P., GEBREYOHANNES, Y. & SCHLIESTEDT, M. 1986. Das Grundgebirge der Rhön. *Zeitschrift der deutschen geologischen Gesellschaft*, **137**, 287–300.

SCHULZ-SCHMALSCHLÄGER, M., PROPACH, G. & BAUMANN, A. 1984. U/Pb-Untersuchungen an Zirkonen und Monaziten von Gesteinen des Vorderen Bayerischen Waldes. *Fortschritte der Mineralogie*, **762**, 223–224.

SEIM, R. 1967. Der Para-Biotit-Plagioklasgneis des Kyffhäuserkristallins, seine metatektischen Derivate und seine Einlagerungen. *Beihefte zur Geologie*, **56**, 185.

SIEBEL, W., TRZEBSKI, R., STETTNER, G., HECHT, L., CASTEN, U., HÖHNDORF, A. & MÜLLER, P. 1997. Granitoid magmatism of the NW Bohemian Massif revealed: gravity data, composition, age relations and phase concept. *Geologische Rundschau, suppl. to volume 86*, 45–63.

SÖLLNER, F., KÖHLER, H. & MÜLLER-SOHNIUS, D. 1981. Rb–Sr-Altersbestimmungen an Gesteinen der Münchberger Gneismasse (MM), NE-Bayern—Teil 2, Mineraldatierungen. *Neues Jahrbuch für Mineralogie, Abhandlung*, **142**, 178–198.

SÖLLNER, F., NELSON, D. R. & MILLER, H. 1997. Provenance deposition and age of gneiss units from the KTB drill hole (Germany): evidence from SHRIMP and conventional U–Pb zircon age determination. *Geologische Rundschau, suppl. to volume 86*, 235–250.

SOMMERMANN, A. E. 1990. Petrographie und Geochemie der magmatogenen Gerölle in Konglomeraten des Kulms im Hinblick auf ihre Herkunft von der Mitteldeutschen Schwelle. *Geologisches Jahrbuch Hessen*, **118**, 167–197.

——, ANDERLE, H.-J. & TODT, W. 1994. Das Alter des Quarzkeratophyrs der Krausaue bei Rüdesheim am Rhein (Blatt 6013 Bingen, Rheinisches Schiefergebirge). *Geologisches Jahrbuch Hessen*, **122**, 143–157.

——, MEISL, S., TODT, W. 1992. Zirkonalter von drei verschiedenen Metavulkaniten aus dem Südtaunus. *Geologisches Jahrbuch Hessen*, **120**, 67–76.

——, SATIR, M. 1993. Zirkonalter aus dem Granit der Bohrung Saar 1. *Beih. z. Eur. J. Mineral.*, **5/1**, 145.

STEIGER, R. H., BÄR, M. T. & BÜSCH, W. 1973. The zircon age of an anatectic rock in the central Schwarzwald. *Fortschritte der Mineralogie*, **50**, Beih. 3, 131–132.

STEIN, E. 1988. Die strukturgeologische Entwicklung im Übergangsbereich Saxothguringikum/Moldanubikum in NE-Bayern. *Geologica Bavarica*, **92**, 5–131.

STEINER, J. & FALK, F. 1981. The Ordovician Lederschiefer of Thuringia. *In*: HAMBREY, M. J. & HARLAND, W. D. (eds) *Earth's pre-Pleistocene glacial record.* University Press, Cambridge, 579–581.

STOPPEL, D. & ZSCHEKED, J. G. 1971. *Zur Biostratigraphie und Fazies des höheren Mitteldevons und Oberdevons im Westharz mit Hilfe der Conodonten- und Ostracodenchronologie.* Beihefte zum Geologischen Jahrbuch, **108**.

TAIT, J., BACHTADSE, V. & SOFFEL, H. 1994. Silurian palaeogeography of Armorica: New palaeomagnetic data from central Bohemia. *Journal of Geophysical Research*, **99**, 2897–2907.

——, —— & —— 1995. Upper Ordovician palaeogeography of the Bohemian Massif: implications for Armorica. *Geophysical Journal International*, **122**, 211–218.

——, SCHÄTZ, M., BACHTADSE, V. & SOFFEL, H. 2000. Palaeomagnetism and Palaeozoic palaeogeography. *This volume.*

TEUFEL, S. 1988. Vergleichende U–Pb- und Rb–Sr-Altersbestimmungen an Gesteinen des Übergangsbereiches Saxothuringikum/Moldanubikum, NE-Bayern. *Göttinger Arbeiten zur Geologie und Paläontologie*, **35**, 1–87.

TODT, W. A., ALTENBERGER, U. & RAUMER, J. F. VON 1995. U–Pb data on zircons for the thermal peak of metamorphism in the Variscan Odenwald, Germany. *Geologische Rundschau*, 84/3, 466–472.

TOLLMANN, A. 1982. Großräumiger variszischer Deckenbau im Moldanubikum und neue Gedanken zum Variszikum Europas. *Geotektonische Forschungen*, **64**, 1–91.

TRZEBSKI, R., BEHR, H. J. & CONRAD, W. 1997. Subsurface distribution and tectonic setting of the late Variscan granites in the northwestern Bohemian Massif. *Geologische Rundschau, suppl. to volume 86*, 64–78.

VAN BREEMEN, O., AFTALION, M., BOWES, D. R., DUDEK, A., MISAR, Z., POVONDRA, P. & VRANA, S. 1982. Geochronological studies of the Bohemian Massif, Czechoslovakia, and their significance in the evolution of Central Europe. *Transactions of the Royal Society of Edinburgh, Earth Sciences*, **73**, 89–108.

VAN GROOTEL, G., VERNIERS, J., GEERKENS, B., LADURON, D., VERHAEREN, M., HERTOGEN, J. & DE VOSS, W. 1997. Timing of magmatism, foreland basin development, metamorphism and inversion in the Anglo-Brabant fold belt. *Geological Magazine*, **134**, 606–616.

WALLISER, O. H. 1981. The geosynclinal development of the Rheinische Schiefergebirge (Rhenohercynian Zone of the Variscides, Germany). *In*: ZWART, H. J. & DORNSIEPEN, U. F. (eds) *The Variscan Orogen in Europe. Geologie en Mijnbouw*. **60**, 89–96.

WALTER, R., SPAETH, G. & KASIG, W. 1985. An outline of the geological structure of the northeastern Hohes Venn area and its Northern foreland. *Neues Jahrbuch für Geologie und Paläontologie Abhandlungen*, **171**, 107–216.

WEBER, K. & BEHR, H. J. 1983. Geodynamic Interpretation of Mid-European Variscides. *In*: MARTIN, H. & EDER, F. W. (eds) *Intracontinental fold belts. Case Studies in the Variscan Belt of Europe and the Damara Belt in Namibia*. Springer, Berlin, 427–468.

WEDEPOHL, K.-H., MEYER, K. & MUECKE, G. K. 1983. Chemical composition and genetic relations of metavolcanic rocks from the Rhenohercynian Belt of Northwest Germany. *In*: MARTIN, H. & EDER, F. W. (eds) *Intracontinental Fold Belts*. Springer, Berlin, 231–256.

WENDT, J. I., KRÖNER, A., FIALA, J. & TODT, W. 1993. Evidence from zircon dating for existence of approximately 2.1 Ga old crystalline basement in southern Bohemia, Czech Republic. *Geologische Rundschau*, **82**, 42–50.

——, ——, —— & —— 1994. U–Pb zircon and Sm–Nd dating of Moldanubian HP–HT granulites from South Bohemia, Czech Republic. *Journal of the Geological Society, London*, **151**, 83–90.

WERNER, O. & LIPPOLT, H. J. 1998. Datierung von postkinematischen magmatischen Intrusionsphasen des Erzgebirges: Thermische und hydrothermale Überprägung der Nebengesteine. *Terra Nostra*, **98/2**, 160–163.

WICKERT, F. & EISBACHER, H. 1988. Two-sided Variscan thrust tectonics in the Vosges Mountains, northeastern France. *Geodinamica Acta*, **2/3**, 101–120.

WIERICH, F. & VOGT, W. 1997. Zur Verbreitung, Biostratigraphie und Petrographie unterkarbonischer Sandsteine des Hörre–Gommern-Zuges im östlichen Rhenoherzynikum. *Geologica et Palaeontologica*, **31**, 97–142.

ZEH, A. 1996. *Die Druck-Temperatur-Deformations-Entwicklung des Ruhlaer Kristallins (Mitteldeutsche Kristallinzone)*. Geotektonische Forschungen, **86**.

—— & KATZUNG, G. 1994. Zur Geologie des Kyffhäuser-Kristallins. *Neues Jahrbuch für Geologie und Paläontologie Monatshefte*, **1994/6**, 368–384.

——, BRÄTZ, H., COSCA, M., TICHOMIROWA, M. & OKRUSCH, M. 1997. 39Ar/40Ar und 207Pb/206Pb Datierungen im Ruhlaer Kristallin, Mitteldeutsche Kristallinzone. *Terra Nostra*, **97/5**, 212–215.

ZELICHOWSKI, A. M. 1995. Lithostratigraphy and sedimentologic-paleogeographic development: Western Pomerania. *In*: ZDANOWSKI, A. & ZAKOWA, H. (eds) *The Carboniferous System in Poland*. Polish Geological Institute, Warszawa, 97–100.

ZULAUF, G. 1997. Von der Anchizone bis zur Eklogitfazies: Angekippte Krustenprofile als Folge der cadomischen und variscischen Orogenese im Teplá Barrandium (Böhmische Masse). *Geotektonische Forschungen*, **89**, 302 S.

——, BUES, C., DÖRR, W., FIALA, J., KOTKOVA, J., SCHEUVENS, D. & VEJNAR, Z. 1998. Extrusion tectonics due to thermal softening of a thickened crustal root: The Bohemian Massif in Lower Carboniferous times. *Terra Nostra*, **98/2**, 177–180.

——, DÖRR, W., FIALA, J., KOTKOVÁ, J. & VEJNAR, Z. 1998. Variscan elevator-style tectonics in the Bohemian Massif—A consequence of a collapsing crustal root. *Acta Universitatis Carolinae—Geologica*, **42**(2), 363–364.

——, ——, —— & VEJNAR, Z. 1997. Late Cadomian crustal tilting and cambrian transtension in the Teplá–Barrandian unit (Bohemian Massif, Central European Variscides). *Geologische Rundschau*, **86**, 571–584.

——, SCHITTER, F., RIEGLER, G., FINGER, F., FIALA, J. & VEJNAR, Z. 1999. Age constraints on the Cadomian evolution of the Teplá–Barrandian unit (Bohemian Massif) through electron microprobe dating of metamorphic monazite. *Zeitschrift der deutschen geologischen Gesellschaft*, **150**, 627–639.

The eastern termination of the Variscides: terrane correlation and kinematic evolution

WOLFGANG FRANKE[1] & ANDRZEJ ŻELAŹNIEWICZ[2]

[1]*Institut für Geowissenschaften der Universität, D-35-390 Giessen, Germany*

[2]*Instytut Nauk Geologicznych PAN, Podwale 75, 50-449 Wrocław, Poland*

Abstract: Analysis of tectonostratigraphic units in the West Sudetes reveals the same geological events as in the areas west of the Elbe Fault Zone: a late Proterozoic (Cadomian) orogenic event, Cambro-Ordovician to Devonian rift–drift, and late Devonian to early Carboniferous subduction–collision. There is no conclusive evidence of an Ordovician orogenic event. Tectonic units in the Sudetes are shown to be related to terranes defined in western parts of the Bohemian Massif. The Lausitz–Izera Block, the Orlica–Śnieżnik Unit and the Staré Město Belt represent easterly continuations of the Saxo-Thuringian Terrane. The Rudawy Janowickie Unit and the Sudetic Ophiolite contain fragments of the Saxo-Thuringian Ocean. The protoliths of the Görlitz–Kaczawa Unit, the South Karkonosze Unit, the Góry Sowie and the Kłodzko Units either belong to the Bohemian Terrane or else were welded onto it during mid–late Devonian metamorphism and deformation. Relicts of the Saxo-Thuringian Foreland Basin are marked by flysch with olistoliths in the Görlitz–Kaczawa Unit and in the Bardo Basin. The spatial array of terranes in and around the Bohemian Massif reveals a disrupted orocline, dissected by dextral transpression along the Moldanubian Thrust. This orocline was formed when central parts of the Variscan belt were accommodated in an embayment of the southern margin of the Old Red Continent.

The Variscan basement NE of the Elbe Fault Zone is largely concealed under Permo-Mesozoic rocks. Neither its eastern termination nor its relationship with the East European Craton are well established. The largest exposed part, the West Sudetes, contains Neoproterozoic to Lower Carboniferous rocks that underwent polyphase deformation and metamorphism ranging from very low grade to eclogite facies, with peaks during Mid–Late Devonian and Viséan time (Fig. 1) Palaeomagnetic studies reveal important microplate rotation. Further complications have been caused by Permo-Mesozoic extension and basin inversion, and by extensive cover sequences.

These complexities have led to a plethora of diverging tectonic and palaeogeographical models (e.g. Matte *et al.* 1990; Aleksandrowski 1995; Wajsprych 1995), but often based only on evidence from the Sudetes, or even parts of the Sudetes (e.g. Oliver *et al.* 1993; Cymerman & Piasecki 1994; Johnston *et al.* 1994). We therefore summarize the evolution of the main tectonostratigraphic units, with their sedimentary and volcanic records, magmatic and metamorphic ages and main structural characteristics, and compare these data with those for areas to the west of the Elbe Fault Zone. On this basis, we discuss the kinematic effect of the main fault zones, which dissect the NE part of the Bohemian Massif, and outline a model of the structural evolution of the eastern termination of the exposed Variscides. We omit from this analysis Variscan units farther to the south-east concealed under the Carpathians.

Tectono-stratigraphic units

Železné hory and the basement of the North Bohemian Cretaceous Basin (ZH)

To the NE of the Elbe Fault Zone, Palaeozoic deposits and Proterozoic basement rocks are exposed in Železné hory and in boreholes that have penetrated the Upper Cretaceous cover (Fig. 1). These Palaeozoic deposits are interpreted as part of a Palaeozoic 'Centralbohemicum' domain, which also comprises the Prague Syncline of the Teplá–Barrandian Unit (Chlupáč 1994).

The Proterozoicx and Palaeozoic sequence of Železné hory ranges up-section into palaeontologically dated rocks of Silurian age, and most probably into Early Devonian time. Devonian limestone pebbles in Permian conglomerates

From: FRANKE, W., HAAK, V., ONCKEN, O. & TANNER, D. (eds).
Orogenic Processes: Quantification and Modelling in the Variscan Belt.
Geological Society, London, Special Publications, **179**, 63–86. 0305-8719/00/$15.00

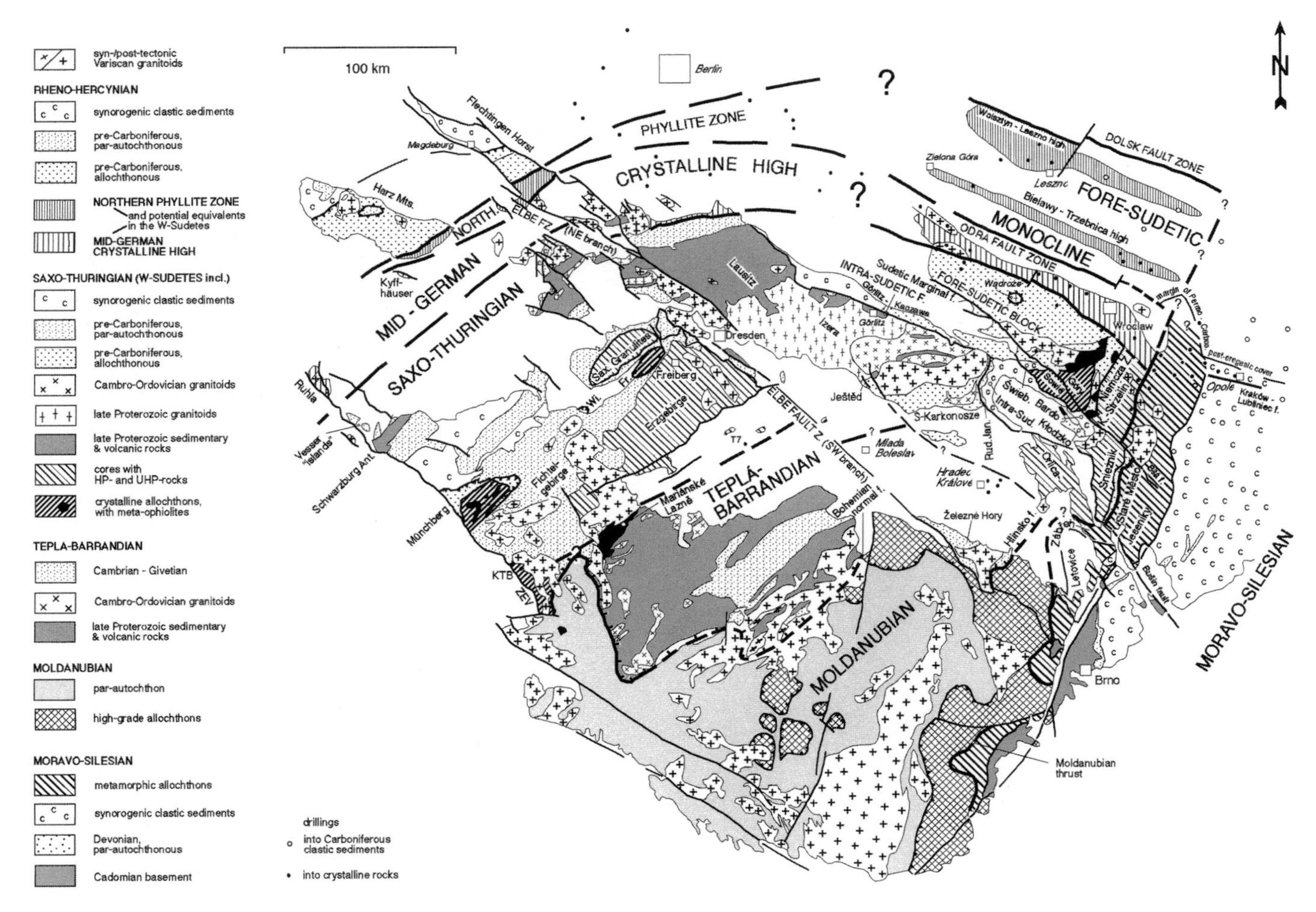

Fig. 1. Geological map of the Bohemian Massif and areas adjacent to the north and NE.

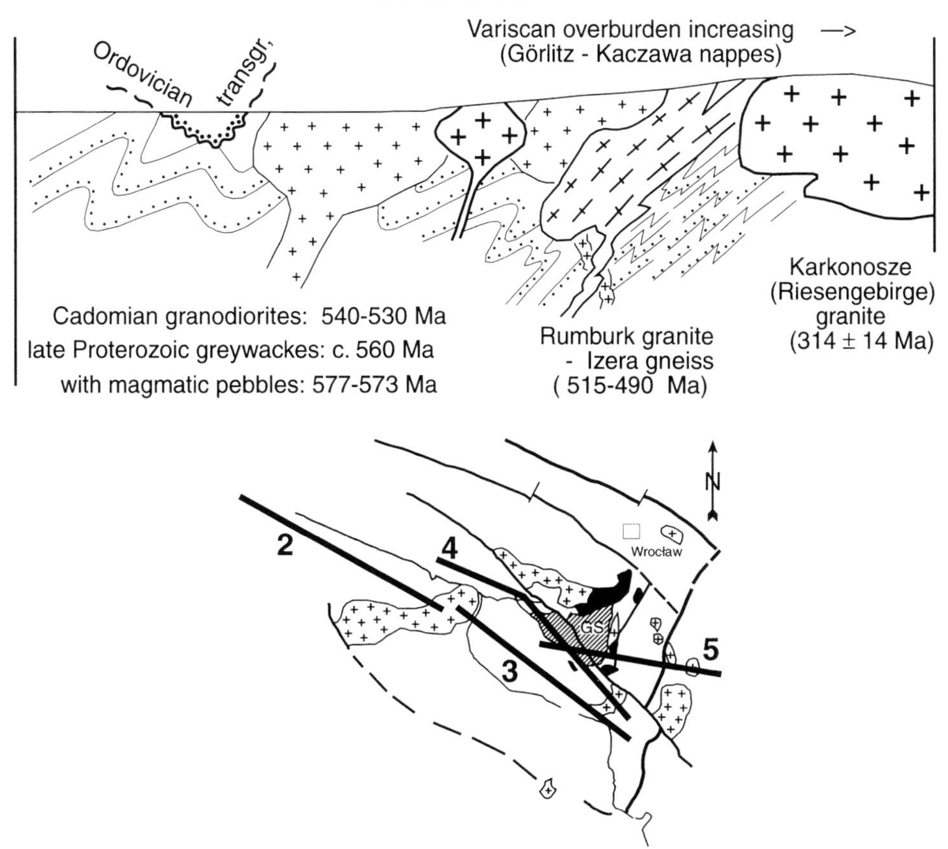

Fig. 2. Dynamic stratigraphy of the Lausitz–Izera Unit. Location of diagrammatic sections of Figs 2–5 (compare with Fig. 1). Crosses, granites.

south of Mlada Boleslav, 50 km NE of Prague, are thought to have been derived from the same unit (see enclosure at back) and Chlupáč (1994).

Near Hradec Králové, folded and slightly metamorphosed Ordovician and ?Proterozoic rocks are unconformably overlain by limestones, which span Frasnian to Early Carboniferous time (Chlupáč & Zikmundová 1976; Zukalová 1976; Čech *et al.* 1989; Chlupáč 1994). These stratigraphic constraints suggest deformation in Early or pre-Frasnian time. In the nearby Teplá–Barrandian Unit, deformation is stratigraphically constrained to Late- to post-Givetian age (Chlupáč 1993*a*, *b*; Chlupáč *et al.* 1998) and Ar–Ar isotopic dating yields values of 383–362 Ma (Dallmeyer & Urban 1998).

Lausitz (Lusatia)–Izera Unit (LI)

The oldest rocks of the Lausitz–Izera are upper Proterozoic greywackes (Figs 1 and 2). Their age is constrained by an algal flora (Burmann 1997*a*) and by zircons from a pyroclastic intercalation, dated at 562 ± 4 Ma. Magmatic pebbles from intercalated conglomerates have yielded ages of 573–577 Ma (Pb–Pb evaporation, Gehmlich *et al.* 1997). They indicate the presence of an older basement (Bielicki *et al.* 1989; Kemnitz & Budzinski 1991). These greywackes underwent latest Proterozoic (Cadomian) folding and metamorphism up to greenschist grade and were later intruded by voluminous post-tectonic I-type granodiorites (Lusatian Granodiorites). A large number of zircon data give latest Proterozoic to early Cambrian intrusion ages (535–530 Ma Pb–Pb, Gehmlich *et al.* 1997; 540 Ma U–Pb, Korytowski *et al.* 1993). Older Pb–Pb evaporation ages for the Lusatian Granodiorities, mostly between 587 and 560 Ma, have been reported by Kröner *et al.* (1994*a*). Intrusion of the granodiorites caused contact metamorphism to sillimanite grade.

During late Cambrian–Early Ordovician rifting, the Cadomian basement was extensively intruded by the Izera–Rumburk Granites between 515 and 480 Ma (U–Pb zircon: 515 + 5/−7 Ma, Korytowski *et al.* 1993; 493 ± 2 Ma, Oliver *et al.* 1993; 514 ± 5/−6 Ma, Philippe *et al.* 1995; Pb–Pb zircon: 515 ± 8 Ma, Kröner *et al.* 1994*a*; 490 ± 12 Ma, Hammer *et al.* 1997). The granites contain xenoliths of (?) Cadomian HT–HP and HT–LP rocks (Achramowicz & Żelaźniewicz 1998; Żelaźniewicz & Achramowicz 1999).

At the eastern margin of the Lausitz–Izera Unit, the Proterozoic rocks are unconformably overlain by the Lower Ordovician Dubrau Quartzite (Linnemann & Buschmann 1995; Linnemann *et al.* this volume), which, together with the Izera Granitoids and the Proterozoic rocks, underwent Variscan overprint (Ar–Ar muscovite age of *c.* 320 Ma for the Izera gneiss; Steltenpohl *et al.* (1993)). Variscan deformation and metamorphism, which are hardly recognizable in the Neoproterozoic granitoids of Lusatia, increase towards the SE, where the Izera Granites have been transformed to gneisses. Polyphase deformation (Achramowicz & Żelaźniewicz, 1998; Żelaźniewicz *et al.* 1998*a*) includes localized shearing (Czapliński 1998) at mostly steeply dipping zones with top-to-the NW dip-slip to sinistral strike-slip. Aleksandrowski *et al.* (1997) have, however, interpreted these shear zones as early flat-lying thrusts with NW-ward displacement, rotated into the present attitude during intrusion of the Karkonosze Granite at *c.* 330–310 Ma. The latest increment on these steep zones is dextral transpression, which also brought the Palaeozoic Kaczawa rocks over the Izera Footwall along the Intra-Sudetic Fault Zone (Achramowicz 1998; see Aleksandrowski 1995).

A few minor granitoid bodies within the Lausitz–Izera Unit have been dated to Late Carboniferous time (Pb–Pb 304 ± 14 Ma evaporation, Kröner *et al.* 1994; 312 ± 10 Ma, Hammer *et al.* 1997). The post-tectonic Variscan Karkonosze Granite, which forms the southern margin of the Lausitz–Izera Unit, likewise intruded during Late Carboniferous time (monzogranite from Liberec, Pb–Pb 314 ± 14 Ma; Kröner *et al.* 1994*a*; Rb–Sr whole-rock ages for 328 ± 12 Ma and 329 ± 17 Ma for porphyritic granite, 309 ± 3 for main (ridge) granite and 310 Ma for leucogranite: (Duthou *et al.* 1991).

Post-Ordovician sediments are restricted to a small, fault-bounded segment in the most southwestern part of the Lausitz–Izera Unit, SW of the Karkonosze Granite (Jítrava, NW part of the Ještěd Mts). It exposes Devonian metabasalts and limestones overlain by Lower Carboniferous shales and greywacke turbidites (Chlupáč & Hladil 1992; Chlupáč 1993*a*).

As discussed in detail by Linnemann *et al.* (this volume), the Lausitz–Izera Unit clearly represents an eastern continuation of the Saxo-Thuringian Parautochthon. The main difference is the deeper level of erosion NE of the Elbe Fault Zone, which has exposed basement with voluminous granitoids of Late Proterozoic (Cadomian), Early Ordovician and Late Carboniferous (Variscan) age.

South Krkonoše: Železný Brod–Rýchory Mts Unit (ZB)

In contrast to the mostly (meta)granitic northern flank of the Karkonosze Batholith, the southern flank is occupied by a large variety of volcanic and sedimentary rocks (Fig. 1). A granitoid from Semily dated at *c.* 540 Ma (U–Pb zircon, Dörr, pers. comm.) indicates the presence of a Cadomian basement. The low-grade metasediments with fossils of Ordovician and Silurian age (Chlupáč 1993*a*, 1994, 1997) are associated with within-plate basalt to mid-ocean ridge basalt (P-MORB) bimodal metavolcanic rocks of an intracontinental rift (Fajst *et al.* 1998). Metaporphyroids in the eastern part (Rýchory Mts) gave Rb–Sr whole-rock age of 495 ± 9 and 501 ± 8 Ma (Bendl & Patočka 1995). Sheets of porphyritic metagranite have yielded U–Pb ages of 492–481 Ma (Oliver *et al.* 1993) and Pb–Pb ages of 501 ± 10 Ma and 503 ± 6 Ma (Kröner *et al.* 1994). As the metagranitoids show the same sequence of deformational events as the surrounding metasediments and metavolcanic rocks (Grandmontagne *et al.* 1996), they probably represent subvolcanic equivalents of the rhyolites intruded into a Lower Palaeozoic sequence (Fig. 3). The polyorogenic history proposed, for example, by Chaloupský is no longer tenable (Chlupáč & Hladil 1992; Grandmontagne *et al.* 1996; Kachlík & Patočka 1998). Variscan polyphase deformation and the absence of structurally coherent radiometric ages impede reconstruction of a stratigraphic sequence and its comparison with neighbouring regions.

Locally evident initial Variscan blueschist metamorphism at 364–359 Ma (Ar–Ar phengite) was followed by regional greenschist facies retrogression between 351–344 Ma (white mica), with a last overprint dated by Ar–Ar on hornblende and muscovite at 327–320 Ma (Maluski & Patočka 1997; Marheine *et al.* 1999, pers. comm.).

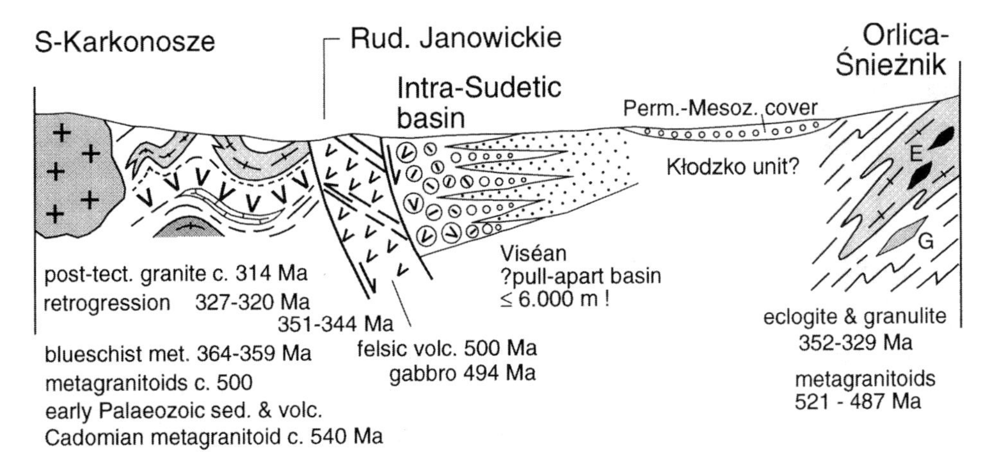

Fig. 3. Dynamic stratigraphy of the South Karkonosze and Rudawy Janowickie Units, the Intra-Sudetic Basin and the Orlica–Śnieżnik Unit (E, eclogite; G, granulite). (See Fig. 2 for location of the section.) Crosses, granites.

The early stages of deformation under the greenschist facies conditions created a penetrative foliation and a mineral lineation oriented WNW, with varying sense of shear (Grandmontagne *et al.* 1996). Mazur & Kryza (1996) extrapolated their observations in the Rudawy Janowickie and proposed for the entire Karkonosze–Izera Block a strick order of top-to-the NW motion during the blueschist and top-to-the SE motion during the greenschist phase. Such a kinematic generalization has, however, been confirmed neither for the South Krkonoše (Grandmontagne *et al.* 1996), nor for the northern part of the block (Achramowicz & Żelaźniewicz 1998; Czapliński 1998).

East Karkonosze: Rudawy Janowickie (RJ)

The Železný Brod–Rýchory Mts continue, without a major structural break, into the East Karkonosze (Lasocki Grzbiet plus Rudawy Janowickie) referred to as Rudawy Janowickie (RJ) in this paper (Fig. 1). However, the Leszczyniec Metavolcanic Unit, whose acid and mafic rocks yielded the U–Pb zircon ages of 505 ± 5 Ma and 49 ± 2 Ma (Oliver *et al.* 1993), displays mostly normal MORB (N-MORB) signature, pointing to a more advanced rift stage (Red Sea type, Patočka & Smulikowski 1997). Blueschist metamorphism terminating at 360 Ma has been reported from two places (Ar–Ar, Maluski & Patočka 1997).

Cambro-Ordovician rift-related magmatism and Devonian blueschist metamorphism is also present in the NW part of the Teplá-Barrandian Unit (Mariánské Lázně) and the Saxo-Thuringian Metamorphic Nappes, i.e. in palaeogeographical terms, the NW margin of Bohemia, the Saxo-Thuringian Ocean and the Saxo-Thuringian Foreland (Fig. 3). As the Rudawy Janowickie contain metamorphosed MORB-like metabasic rocks, they may possibly correlate with the protolith of the Münchberg Eclogite. The South Krkonoše (ZB) and Rudawy Janowickie (RJ) Complexes are probably part of the Saxo-Thuringian Terrane (Mazur *et al.* 1998).

According to Mazur (1995) the East Karkonosze represents an early stack of the WNW-verging thrust units with now inconsistent transport directions explained by later refoldings connected with the *c.* 330–310 Ma intrusion of the Karkonosze Granite and subsequent eastward extensional collapse. However, both lithotectonic and lithostratigraphic subdivisions of the Rudawy Janowickie Unit (e.g. Teisseyre 1973; Mazur 1995; Kozdrój 2000) as well as structural data (e.g. Mazur 1995; Cymerman 1996; Kozdrój 2000) are still controversial and uncertain, and so is Mazur's (1995) model.

Intra-Sudetic Basin (ISB)

The Intra-Sudetic Basin is situated to the SE of the Rudawy Janowickie Unit. It contains up to 8000 m of intra-montane fluvial to marine clastic sediments of early Carboniferous to Early Permian age, accompanied by acid to intermediate volcanism (Dziedzic & Teisseyre 1990; Awdankiewicz 1999, 2000). The earliest marine ingression is biostratigraphically dated to Late Viséan time (Go α a of the goniatite chronology; Żakowa 1963) but the underlying *c.* 6000 m thick clastic deposits possibly extend earlier to Early Viséan or even Late Tournaisian age (Teisseyre 1975). Metamorphic source areas to the north (Kaczawa) and west (Rudawy Janowickie) must have already been at the surface. Rapid sedimentation controlled by normal faulting and a supposed southeastern continuation of the strike-slip Intra-Sudetic Fault have provoked interpretations of the Intra-Sudetic Basin as a pull-apart basin (Aleksandrowski *et al.* 1997; Lankreijer 1998).

Orlica–Śnieżnik Done (OS)

The Orlica–Śnieżnik Done generally consists of a metagranitic or migmatitic core with local inserts of (U)HP granulites and eclogites derived from a variety of protoliths ranging from Fe–Ti lamprophyres and gabbros via calc-alkaline and bimodal volcanic rocks to MORB (Bakun-Czubarow 1998). This core is overlain by a middle–lower amphibolite facies series of mica schists and paragneisses with marbles and amphibolites, which is mantled, in turn, by metapelites and metabasites in greenschist facies (Figs 1 and 3). Metasediments contain doubtful microfossil evidence of the Late Proterozoić-Early Cambrian protolith age (Gunia 1996), whereas an associated leptynite yielded an Pb–Pb zircon age of *c.* 520 Ma (Kröner *et al.* 1997).

Zircons from the orthogneisses have been dated by U–Pb at 504 ± 3 to $488 \pm 4/-7$ Ma (Oliver *et al.* 1993), and by Pb–Pb at 521 ± 11 to 503 ± 4 Ma (Kröner *et al.* 1997). The wide range of ages, also revealed by U–Pb dating of abraded single zircons (Borkowska & Dörr 1998), is probably caused by inherited lead with *c.* 540 Ma and older components. Precise SHRIMP U–Pb and Pb–Pb core-to-rim data for metagranites (similar to the Izera Granite) and migmatitic gneisses point to zircon growth events at *c.* 565, 540, 500 and 340 Ma (Turniak *et al.* 1998). The *c.* 500 Ma event extensively produced new zircon crystals in both lithologies. All these data in conjunction with the presence of migmatitic (some eclogite inclusive) xenoliths within the *c.* 500 Ma granite (Żelaźniewicz 1997) and its influence upon the migmatized gneisses (Borkowska & Dörr 1998) suggest that the granite intrusion was coeval with an early migmatitization. Later metamorphic overprint at *c.* 340 Ma may have also caused some migmatization in the core of the Orlica–Snieżnik Dome (see Mazur *et al.* 1999).

Variscan metamorphism is also documented by finally tectonically emplaced eclogites, with Sm–Nd whole-rock isochrons between 352 ± 4 and 329 ± 6 Ma and a U–Pb zircon lower intercept at 337 ± 3 Ma (Brueckner *et al.* 1991). Rb–Sr dating of phengite or whole rock yielded 333–329 Ma (Bröcker *et al.* 1997). An earlier (Devonian) metamorphic event is recorded in gneisses immediately adjacent to the Miedzygórze Eclogite by a U–Pb lower intercept zircon age of 372 ± 7 Ma and a Rb–Sr thin slab whole rock isochron of 396 ± 17 Ma (Bröcker *et al.* 1997), all pointing to a complicated tectonometamorphic history of the Orlica–Śnieżnik Dome.

The Orlica–Śnieżnik Unit has much in common with the Erzgebirge Anticlinorium of the Saxo-Thuringian Belt: a lower-grade Palaeozoic mantle (lower Cambrian in the core of the Orlica–Śnieżnik Dome), Cambro-Ordovician orthogneisses set in the Cadomian rocks and, in the core, inserts of eclogite and granulite facies rocks with Early Carboniferous isotopic ages (see discussion by Franke & Stein (this volume)). This correlation probably also applies to the eastern Fore-Sudetic Block (Kamjeniec Complex), situated to the NE of the Sudetic Marginal Fault, east of the Niemcza Zone (Figs 1 and 3).

Görlitz–Kaczawa Unit (GK)

The Görlitz–Kaczawa Unit accompanies the Lausitz–Izera Unit (LI) on its NE side (Fig. 1). These units are separated by a major fault zone, termed the 'Intra-Lusatian Fault' in its northwestern segment, and 'Intra-Sudetic Fault' (or Main Intra-Sudetic Fault) to the SE. The southeastern (Kaczawa) part of the Görlitz–Kaczawa Unit consists of a Palaeozoic volcano-sedimentary succession, with conodont-bearing Ordovician clastic sediments, Silurian black shales with graptolites and Devonian siliceous shales with conodonts (Baranowski *et al.* 1990). These sediments are associated with bimodal volcanic rocks of Early Palaeozoic age, grading from WPB to N-MORB chemistry (Furnes *et al.* 1994; Floyd *et al.* this volume). The Kaczawa segment also contains chaotic assemblages (Baranowski *et al.* 1990) with Viséan limestones as the youngest components (Chorowska 1978). Such assemblages make up most of the northwestern (Görlitz) part of the Görlitz-Kaczawa Unit: Cambrian to lower Carboniferous sedimentary and volcanic rocks of very low metamorphic grade are interpreted as olistoliths in a Carboniferous flysch matrix (Thomas 1990). Metamorphism in the Kaczawa rocks has attained blueschist facies, reported only from the southeastern part of the Kaczawa Unit (Kryza *et al.* 1990; but see Smulikowski 1995), and is not in evidence towards the NW and north. Blueschist metamorphism is not precisely dated as yet, but probably occurred, as in the South Krkonoše Unit (Guiraud & Burg 1984), during late Devonian time. Northwestern and northern parts of the Kaczawa Unit show only greenschist facies. These areas, as well as the very low grade

Görlitz Assemblage, must have been deformed and metamorphosed in Carboniferous time, as they contain early Carboniferous elements.

In the northeastern part of the Kaczawa Unit, near Wądroże Wielkie, isolated outcrops of augen-gneiss emerge from under the Quaternary cover. These rocks represent a tectonic window of the Lausitz–Izera Unit under the Görlitz–Kaczawa Allochthon and probably they are part of the Cadomian basement upon which the Kaczawa Complex was laid down at the onset of rift-related deposition.

The Görlitz–Kaczawa Unit records geological events well known from the Saxo-Thuringian Belt: Cambro-Ordovician rifting on a Cadomian basement, a Silurian–Devonian pelagic interval and early Carboniferous flysch with chaotic assemblages. The juxtaposition of rocks that have undergone (?) late Devonian metamorphism with Devonian sediments is similar to the thrust stack of the Münchberg Klippe in the Bavarian segment of the Saxo-Thuringian Belt (Falk *et al.* 1995; Franke *et al.* 1995).

Świebodzice Depression (SD)

The Świebodzice Depression is a small synorogenic basin with a 4 km thick infill of very coarse-grained fluvial to shallow-marine clastic sediments of ?Frasnian and Famennian to Early Carboniferous age (Figs 1 and 4). These sediments straddle a fault system, which separates the Kaczawa basement to the north from the Góry Sowie Block to the south. The fan-delta system that filled the basin was derived from uplift to the south and southwest (Porębski 1981, 1990). The clast spectrum includes the Góry Sowie Gneisses and gabbros and minerals derived from the Sudetic Ophiolite (Otava & Sulovský 1998; Klara 1999).

Góry Sowie Block (SG)

The Góry Sowie Block consists of a paragneiss–migmatite complex with debatable amounts of metagranitoids (see Żelaźniewicz (1996) and discussion by Bröcker *et al.* (1998)), some granulites and metabasites–ultrabasites (Figs 1 and 4). Pb–Pb zircon ages around 480–490 Ma (Kröner & Hegner 1998) have been interpreted as crystallization ages of Ordovician granitoids intruded into an active plate margin, formed during convergence between Gondwana-derived terranes and Baltica. However, the coincidence of these ages with *c.* 515–480 Ma porphyritic granite intrusions and accompanying migmatization elsewhere in the Sudetes allow to interpret them as a record of the earliest migmatitic event in the Góry Sowie Complex (Żelaźniewicz 1996; Bröcker *et al.* 1998). Apparent Pb–Pb zircon ages of *c.* 440 Ma obtained for a late-tectonic nebulite and 473 Ma for a syn-tectonic granitoid vein, taken by Oliver *et al.* (1993) and Kröner & Hegner (1998) to constrain an Ordovician orogenic event, may have no geological significance (Bröcker *et al.* 1998). The same syn-tectonic granitoid vein yielded U–Pb monazite ages of 379–383 Ma and U–Pb xenotime ages of *c.* 380 Ma (Bröcker *et al.* 1998), confirmed by U–Pb monazite data of Timmermann *et al.* (2000), which point to the importance of the Mid- to Late Devonian high-temperature metamorphism and migmatization of the main tectonothermal stage in the evolution of the Góry Sowie Block (Żelaźniewicz 1990). The latter is also documented by Rb–Sr whole-rock isochron ages (thin

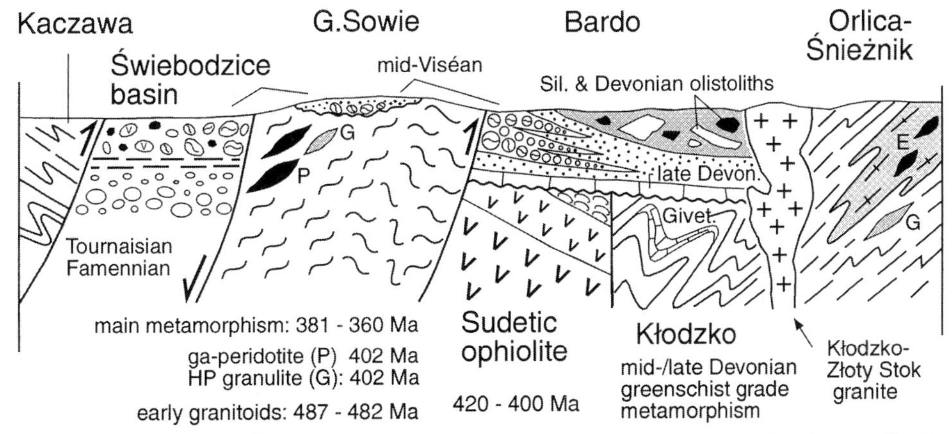

Fig. 4. Dynamic stratigraphy of the Góry Sowie Block and neighbouring units (E, eclogite; G, granulite; P, peridotite). (See Fig. 2 for location of the section.)

slab technique) of 374 ± 5 to 362 ± 8 Ma (Bröcker *et al.* 1998), as well as concordant monazite at 381 ± 2 Ma, discordant zircon at 369 ± 15 Ma, and Rb–Sr biotite and muscovite ages of 372 ± 7 and 360 ± 7 Ma obtained by Van Breemen *et al.* (1988), which all agree well with the Ar–Ar age of *c.* 360 Ma of muscovite from the above-mentioned granitoid vein. (Marheine, pers. comm.). Therefore, having integrated all field, structural, petrological, isotopic and regional data, we suggest, in agreement with Bröcker *et al.* (1998), that the 490–480 Ma important thermal event in the Góry Sowie Block does not represent supra-subduction granitoid emplacement in a 'Caledonian' magmatic arc setting, which is nowhere proved to have existed at that time in the Sudetes. On the contrary, this event probably relates to the commonly evidenced Cambro-Ordovician extension during which a large part of the Góry Sowie Complex supracrustal rocks was deeply buried, undergoing early metamorphism, migmatization and deformation.

An independent stage of metamorphism is documented in a lens of felsic granulite (Pb–Pb evaporation age of 402 ± 1 Ma, O'Brien *et al.* 1997), and in a mantle-derived garnet peridotite, which became metamorphosed under granulite facies conditions at 402 ± 3 Ma and then at *c.* 380 Ma (Sm–Nd garnet core and rim orthopyroxene–clinopyroxene–whole-rock isochrons; Brueckner *et al.* 1996). Other mantle-derived ultramafic bodies (1400 °C) were uplifted to lower-crustal depths (850 °C, 15 kbar) and then emplaced into gneisses of the Góry Sowie Complex at a time when the latter underwent progressive high-temperature migmatization of the main tectonothermal stage and the former became retrogressed (Dubińska *et al.* 1998, 1999).

The Góry Sowie Block shows some similarities with the Saxo-Thuringian Münchberg Klippe, which contains a thrust sheet composed of orthogneiss dated at *c.* 480 Ma, HP eclogites dated at 395 Ma, and amphibolite facies metamorphic rocks formed between 380 and 365 Ma, but without high-temperature overprint (see Franke *et al.* 1995; Franke this volume).

Sudetic Ophiolite (SO)

The Sudetic Ophiolite is a complex of mafic and ultramafic rocks extensively outcropping at the NE margin of the Góry Sowie Block and patchily around its SE corner (Fig. 1). Gabbros and minor sheeted dykes have N-MORB affinity with protoliths either impoverished or rich in incompatible elements (Pin *et al.* 1988; Dubińska & Gunia 1997). Predominantly harzburgitic to lherzolitic peridotites, representing mantle tectonites, and ultramafic cumulates underwent serpentinization and greenschist overprint at temperatures of *c.* 250 °C (Dubińska 1995). Higher greenschist conditions (300–370 °C, 1.7–3 kbar) and an amphibolite facies overprint are observed only in an isolated serpentinite body (Szklary) within the Niemcza Zone, east of the Góry Sowie Block (Dubińska 1997). Geochemistry of rodingites and amphibolitized dykes within the Szklary Serpentinites, suggestive of a highly depleted source and boninite-suite affinity, is taken to imply a fore-arc setting (Dubińska 1997).

U–Pb datings of abraded zircon from rocks north of the Góry Sowie Block have yielded ages of 420 + 20/−2 Ma for a gabbro (Ślęża, Oliver *et al.* 1993) and of 400 + 4/−3 Ma (single grains, concordant) for a rodingitized plagiogranite from serpentinite (Jordanów, Żelaźniewicz *et al.* 1998*b*). These findings reveal an oceanic stage attained in early Devonian time and do not support the Sm–Nd ages of *c.* 350 Ma obtained by Pin *et al.* (1988). By the end of Devonian time the ophiolite, together with the Góry Sowie Block and adjacent units (Figs 1, 4 and 5), became exposed at the surface and all delivered clasts to late Devonian synorogenic basins.

Kłodzko Unit (KU)

Clastic metasediments with marbles at the top are overlain by a volcano-sedimentary sequence with bimodal volcanic rocks and a predominantly mafic plutonic–volcanic assemblage of WPB to N-MORB character (Wojciechowska 1995) (Figs 1 and 4). A coral fauna from the marbles, formerly interpreted as being of Ludlow age (Gunia & Wojciechowska 1971), has recently been redated to Givetian time (Hladil *et al.* 1999). Metamorphism attained greenschist to lower amphibolite facies, and increases westwards. Both the Kłodzko Greenschists and the ophiolite adjacent to the north are overlain by Famennian limestones (see Haydukiewicz 1990). Hence, metamorphism and polyphase deformation are constrained between 387 and 376 Ma (see McKerrow & van Staal this volume), coeval with the HT migmatic event in the Góry Sowie Gneisses.

Bardo Basin (BB)

The Bardo Basin infill is unmetamorphosed (Fig. 1). Sedimentation started with upper Famennian to lowest Carboniferous carbonate mudstones (Haydukiewicz 1990). In the

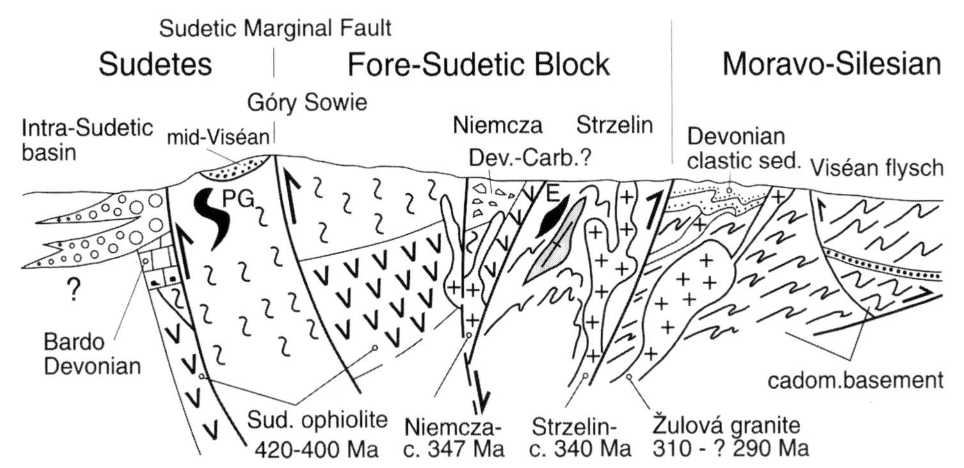

Fig. 5. Dynamic stratigraphy of the Fore-Sudetic Block NE of the Sudetic Marginal Fault and the Moravo-Silesian Belt (E, eclogite; G, granulite). (See Fig. 2 for location of the section.) Crosses, granites.

northwestern part of the basin, to the SW of the Góry Sowie Block, these carbonates overlie gabbros belonging to the Sudetic ophiolite; in the southeastern part, they are transgressive over the greenschist grade rocks of the Kłodzko Unit (Bederke 1924). Middle Viséan clastic sediments were derived from the Góry Sowie Gneisses immediately adjacent to the north, and the Kłodzko Block to the west (Fig. 4). The upper, main part of the basin fill consists of greywacke turbidites and debris flows (Wajsprych 1995). These contain clasts up to kilometre scale of upper Ordovician siliceous shales, Silurian black shales and Devonian siliceous shales dated with graptolites and conodonts (Jäger 1964; Porębska 1984; Haydukiewicz 1990, 1998). The source of these clasts is unknown. However, the Bardo Flysch with its olistoliths is very similar to time-equivalent deposits in the Görlitz–Kaczawa Unit and in the Saxo-Thuringian Belt further west.

The Sudetic Ophiolite and the Kłodzko Unit must have been juxtaposed before the deposition of the overstepping Famennian carbonate deposits and together with the adjacent Bardo Basin escaped blueschist metamorphism. Magnetic and gravimetric data suggest that the ophiolite patchily underlies the eastern part of the Góry Sowie Block (Znosko 1981), yet there is no structural evidence of a supposed thrust and all contacts are normal faults (Żelaźniewicz 1995). Gneissic clasts of late Devonian age of the Świebodzice and Bardo Basins demonstrate that the Góry Sowie Block was also incorporated into this late Devonian tectonic assemblage, which records early Variscan accretion. Evidence of the marine transgression of upper Famennian carbonate deposits over rocks folded and metamorphosed in late Devonian time is similar to the record of the boreholes near Hradec Králové (see above). It is possible that the tectonic assemblage comprising the Góry Sowie Block, the Sudetic Ophiolite and the Kłodzko Unit was accreted, in late Devonian time, to the Bohemian Terrane (Franke, this volume; 'Centralbohemicum' unit of Chlupáč (1994)).

Niemcza Zone (NZ)

The Niemcza Zone is a 5 km wide, north-trending belt immediately east of the Góry Sowie Block (Figs 1 and 5), which consists of cordierite–andalusite-bearing metagreywackes with subordinate quartzites, intruded by syntectonic granodiorities with U–Pb and Pb–Pb zircon ages of *c.* 340 Ma (Oliver *et al.* 1993; Kröner & Hegner 1998). Low-pressure metamorphism of greenschist to amphibolite facies was irregularly imposed by the intrusion of the granitoids, being constrained by an Ar–Ar hornblende age of 332 ± 2 Ma (Steltenpohl *et al.* 1993). Synmetamorphic polyphase shearing and some mylonitization imparted an asymmetric fabric to the metasediments, which have often been misinterpreted (Scheumann 1937; Cymerman & Piasecki 1994; Mazur & Puziewicz 1995) as mylonites derived from the Góry Sowie Gneisses (see Dziedzicowa 1987; Żelaźniewicz 1995; Czapliński *et al.* 1996). The dominant deformation produced E–ESE-dipping surfaces with sinistral strike-slip (Cymerman & Piasecki 1994; Mazur & Puziewicz 1995) or oblique-slip displacements. These were followed by dip-slip and dextral strike-slip to oblique-slip movements, which are also observed in the granitoids (Żelaźniewicz 1995).

Like the clastic sediments of the Bardo Basin, the greywackes contain clasts of the mylonitized Góry Sowie Gneisses. The Niemcza deposits possibly represent a metamorposed counterpart of the Bardo sequence (Oberc 1980), which was partly derived from and partly deposited upon the Góry Sowie Gneisses. Within the greywackes, there are lenses of undated bedded cherts and a small serpentinite body (Szklary). The latter is the only fragment of the Sudetic Ophiolite, which underwent metamorphic overprint at the amphibolite facies (widespread anthophyllite, >500 °C, Juskowiak 1960; Dubińska & Gunia 1997) and possibly even higher temperature conditions (720–760 °C and 3.5 kbar) as inferred from the mineralogy of cross-cutting metamorphosed mafic dykes (Dubińska 1997).

Kamieniec–Strzelin Belt (KS)

Rocks occurring in the eastern part of the Fore-Sudetic Block between the Niemcza Zone and the easternmost outcrops of crystalline 'islands' amidst Cenozoic deposits (Figs 1 and 5) are referred to in this paper as the Kamieniec Ząbkowicki–Strzelin Belt, and for the sake of brevity termed here the Kamieniec–Strzelin Belt. It contains rocks of different P–T history and probably comprises two tectonostratigraphic units separated by a major fault zone concealed by the Cenozoic cover.

The western (upper) unit consists of the Kamieniec Complex (Cwojdziński & Żelaźniewicz 1995) represented by medium-grade mica schists with minor marbles, quartzites, amphibolites and felsic to mafic metavolcanic rocks, and porphyritic metagranite dated at 504 ± 3 Ma (U–Pb zircon, Oliver *et al.* 1993). This complex also includes eclogites (Achramowicz *et al.* 1997; Bakun-Czubarow 1998) surrounded by mica schists, which likewise underwent an early HP event (Nowak 1998*a*). The Kamieniec Complex is interpreted in this paper as part of the Saxo-Thuringian Belt continuing southward into the Orlica–Śnieżnik Dome across the Sudetic Marginal Fault system (see below). A northwestward continuation into the Odra Fault Zone (Fig. 1) has been proposed (Grocholski 1986; Cwojdziński & Żelaźniewicz 1995), but requires further confirmation.

The eastern (lower) unit, referred to as the Strzelin–Otmuchów Metamorphic Unit (Cwojdziński & Żelaźniewicz 1995), is devoid of HP rocks and contains, instead, local intercalations of characteristic metasandstones interpreted as equivalent to the lower to middle Devonian quartzites, which overlie the Cadomian basement of the Moravo-Silesian Zone further to SE (Bederke 1935; Chlupáč 1989; Szczepański 1999). Consequently, crystalline rocks underlying the Devonian metasandstones may also be assigned to this Cadomian basement.

The contrasting parts of the Kamieniec–Strzelin Belt are probably separated by a north-easterly continuation of the transpressive 'Moldanubian Thrust' (see below), but with a more complex history that also includes normal faulting (Achramowicz 1999). The two units have been tectonically juxtaposed before the intrusion of the post-tectonic Strzelin Granite at 347–340 Ma (Rb–Sr isochron, Oberc-Dziedzic *et al.* 1996). Because of the poor outcrops, the subdivision and structural interpretation of the Kamieniec–Strzelin Belt are still controversial (Nowak 1998*b*; Achramowicz 1999; Mazur & Józefiak 1999; Oberc-Dziedzic 1999; Szczepański 1999).

Odra Zone (OZ)

The Odra Zone (Odra Fault Zone) is a Late Carboniferous–Permian, NW-trending horst zone in the subsurface, extending from the south-east of Wrocław toward the NW (Fig. 1). It contains amphibolite facies rocks intruded by Carboniferous granites similar in geochemistry to other late to post-orogenic Variscan granites known from the West Sudetes (Oberc-Dziedzic *et al.* 1999). Two boreholes located immediately to the NW of the Odra Zone have yielded low-grade metapelites with a microfauna of late Devonian age (Chorowska 1978).

The Odra Zone is flanked on either side by magnetic anomalies (Żelaźniewicz *et al.* 1997). Greenschist grade metavolcanic rocks drilled on the SW flank belong to the Fore-Sudetic part of the Görlitz–Kaczawa Unit. The source of the magnetic anomalies to the NE of the Odra Zone remains unknown, but they also coincide with a basement high (Bielawy–Trzebnica, Fig. 1) comprising low-grade quartz–sericite slates.

Rocks of the Odra Zone have repeatedly been correlated with the Mid-German Crystalline High (Grocholski 1986; Bankwitz *et al.* 1999), which is likewise flanked by a 'Northern' and 'Southern Phyllite Zone' (see Franke (this volume)). However, this requires confirmation, for example, by biostratigraphic and isotopic datings, especially in view of the lack in the Odra Zone of characteristic Silurian and Upper Devonian granitoids (Oberc-Dziedzic *et al.* 1999).

Staré Město Unit (SM)

The Staré Mesto Unit is situated between the Jeseník segment of the Moravo-Silesian Belt and the Orlica-Śnieżnik Unit (Fig. 1). It contains migmatite, a bimodal meta-magmatic suite ('leptyno-amphibolite series') with spinel peridotite, metagabbro dated to 514 Ma (Pb–Pb zircon, Kröner *et al.* this volume; Štípská *et al.* 1998) and a syn-tectonic tonalite sill dated to 339 Ma (Parry *et al.* 1997), whose fabrics indicate sinistral strike-slip. The metapelite series underwent granulite facies metamorphism (6–8 kbar and 800–850 °C) at 501 ± 6 Ma and are interpreted as the lower crust of a Cambro-Ordovician rift (Štípská *et al.* 1998; Kröner *et al.* this volume). Later Variscan metamorphism is constrained by an Ar–Ar amphibole age of 330.1 ± 3.4 Ma (Maluski *et al.* 1995).

Early Palaeozoic rifting is characteristic of the Armorican Terrane Assemblage and suggests affinities of the Staré Město Unit with the Saxo-Thuringian or Bohemian Terranes. With regard to the position of this unit, correlation with the Saxo-Thuringian Terrane appears more likely. The preservation of rift-related metamorphism demonstrates that afterwards the Staré Město Unit occupied a relatively external position within the orogen because it was much less affected by Carboniferous metamorphism than the Orlica–Śnieżnik rocks adjacent to the NW.

East Sudetes

The area to the SE of the Moldanubian fault zone is referred to as the 'Moravo-Silesicum' (Suess 1926), a part of the Variscan orogen with a Devonian–Carboniferous sequence deposited on the Cadomian basement extending from Bruno to Upper Silesia ('Bruno-Vistulicum' of Dudek (1980); see Finger *et al.* (this volume)).

East Sudetes: Letovice Complex (LC)

The Letovice Complex consists of metapelites and ophiolitic amphibolites (Jelínek *et al.* 1984), and represents Bohemicum, which tectonically overlies the metamorphosed (Moravian) part of the Moravo-Silesian Belt exposed in the Svratka Window. From high-grade rocks of the Moldanubian unit in the hanging wall, the Letovice Complex is separated by a 'mica schist zone' (Suess 1912). The concealed northwestern margin of the Letovice Complex as shown in Fig. 1 is based upon the magnetic map of the Czech Republic. Both the Letovice Complex and the mica schists units have undergone high-pressure metamorphism, and the Letovice Complex includes retrogressed eclogites as well as peridotite lenses (Johan & Schulmann 2000). K–Ar and Ar–Ar ages on amphiboles from the Letovice Complex scatter between 354 and 328 Ma (Macintyre *et al.* 1993).

As the Letovice Complex is juxtaposed against the Cadomian basement and Devonian–Carboniferous succession of the Moravo-Silesian Unit, which is proposed to correlate with the Rheno-Hercynian Belt, the mafic rocks might represent an equivalent of Rheno-Hercynian ocean floor formed in early Devonian time (see Franke this volume). However, early Carboniferous HP metamorphism and fragmentary meta-ophiolites have also been detected both in the Gföhl nappes of the Moldanubian region and in the Orlica–Śnieżnik Unit (probably Saxo-Thuringian Terrane). Therefore, the tectonostratigraphic and palaeogeographical affinities of the Letovice Complex remain uncertain.

East Sudetes: Moravo-Silesian (MS)

The Moravo-Silesian Belt is possibly an easterly equivalent of the Rheno-Hercynian (Engel & Franke 1983; see Dvořak & Paproth (1969) for an alternative view). Both are characterized by a Cadomian basement (see Finger *et al.* this volume), Devonian extension with bimodal magmatism, Devonian elastic sediments and shelf carbonate deposits (Chlupáč 1989), an external Carboniferous limestone platform (Belka 1987) and Lower Carboniferous flysch passing up-section and outwards into coal-bearing paralic molasse (see the review by Dvořak (1995)). The main tectonic transport is directed, in both belts, away from the Moldanubian–Saxo-Thuringian Core of the Bohemian Massif, although the western part of the Moravo-Silesian Flysch has a phase of NW-directed D2 backthrusting (Figs 1 and 5; Matte *et al.* 1990).

The foreland of the Moravo-Silesian Belt is situated to the SSE. More internal portions are exposed in the NE segment of the Moravo-Silesian Belt (Jeseníký Mts), which is characterized by stronger Devonian extension and magmatism (e.g. Hladil 1995; Schulmann & Gayer 2000). Further to the SW, in the Drahany Uplands, this deeper part of the basin has been excised by the Moldanubian Thrust, so that a more external part of the belt (Cadomian basement with epicratonic Devonian cover) abuts against the thrust.

All these units have undergone early Carboniferous metamorphism Ar–Ar cooling ages of hornblende and biotite between 311 and 301 Ma (Maluski *et al.* 1995) record late extensional exhumation with respect to the units adjacent to the NW (Schulmann & Gayer 2000). The southwestern boundary of the Jeseníký Block, the NW-trending Bušin Fault, shows likewise ductile normal faulting to the southwest.

Recent palaeomagnetic studies have revealed that the Moravo-Silesian Belt has been rotated clockwise through *c.* 90° after Devonian time (Tait *et al.* 1994, this volume). If one compares the internal palaeogeographical zonation of the Moravo-Silesian Belt with that of the Rheno-Hercynian Zone, one arrives at a rotation attaining *c.* 180° (Hladil 1995).

Main fault zones

Elbe Fault Zone (EF)

The Elbe Fault Zone is a zone of dextral strike-slip, which can be traced from a position NE of the Harz Mts down to the southeastern margin of the Bohemian Massif (Fig. 1). East of the Harz Mts, the trend of the lower Carboniferous Hörre–Gommern Quarzite (see Franke this volume) shows an offset of *c.* 30 km. This displacement must be relatively young, as it displaces units that had already undergone folding and thrusting. By comparison with the zonation of K–Ar ages on slaty cleavage obtained from equivalent parts of the Rhenish Massif, the offset of the Hörre–Gommern Unit should be younger than *c.* 315 Ma (Ahrendt *et al.* 1983).

Further south, yet still to the north of the Palaeozoic outcrop of the Bohemian Massif, the Elbe Fault Zone is constrained by several boreholes. The outcrop of Cambrian deposits contained in the North Saxonian Synform (Linnemann *et al.* this volume) displays a dextral flexure with a lateral offset of *c.* 20 km NW of Dresden. In the Dresden area, the fault zone steps over to the SW bank of the Elbe river. The Meissen pluton was intruded into this step at 330 Ma (Wenzel *et al.* 1997). These relationships provide an important constraint on the age of faulting (Mattern 1996).

The northern branch of the Elbe Fault Zone is continued, toward the SE, as the West Lausitz Fault, which forms the present-day southwestern margin of the Lausitz–Izera Unit. The West Lausitz Fault is an important late Cretaceous reverse fault, and it is uncertain whether it had an earlier increment of dextral strike-slip. In the southeastern prolongation of the Cretaceous fault, the Bušin Fault separates the Jeseniký segment from the Moravian Karst segment of the Moravo-Silesian Belt (see below). According to field observations by the authors, the Bušin Fault is a ductile normal fault. Taken altogether, there is no evidence of dextral strike-slip on the NE branch of the Elbe Fault Zone to the SE of Dresden.

The SW segment of the Elbe Fault Zone is best exposed south of Dresden, where it defines the NE boundary of the Erzgebirge Antiform. Palaeozoic sedimentary and volcanic rocks within the fault zone have undergone ductile dextral translation. A minor component of normal faulting brings the Palaeozoic rocks down against the high-grade metamorphic rocks of the Erzgebirge. Minor brittle reverse faulting occurred during late Carboniferous time and probably again during late Cretaceous time (Rauche 1992, 1994). Some post-Variscan activity of the Elbe Fault Zone is indicated by the formation of the Permian Döhlen Basin NE of the Erzgebirge, which probably represents a pull-apart basin (Mattern 1996).

Further to the southeast, the SW branch of the Elbe Fault Zone separates the Železné hory Unit (an equivalent of the Teplá–Barrandian Unit) from Moldanubian high-grade rocks to the SW. The ductile normal fault system at the SE margin of the Teplá–Barrandian Unit, largely masked by the Central Bohemian Pluton (Zulauf 1994, 1997; Scheuvens 1998), is continued east of the Elbe Fault Zone, by the normal Hlinsko Fault (Pitra *et al.* 1994). These correlations define a dextral offset of *c.* 50 km. On the basis of the displacement of regional magnetic anomalies north of Prague, Rajlich (1987) estimated this offset at some 100–120 km. Still further SE, the Elbe Fault Zone separates the Svratka Metamorphic Dome from the Moldanubian Block. The fault zone terminates abruptly at the 'Moldanubian Thrust', whose activity clearly post-dates the main (ductile) displacements along the Elbe Fault Zone (Fig. 1).

Intra-Sudetic Fault Zone (ISF)

The Intra-Sudetic Fault Zone is a NW-trending, polyphase steep fault zone separating the Lausitz–Izera from the Görlitz–Kaczawa Unit. Its continuation in either direction is hypothetical. Whereas some workers take the fault zone to be insignificant (Cymerman 1998) or of minor importance (Achramowicz 1998), Aleksandrowski *et al.* (1997) have proposed dextral displacement of 50–300 km. An age limit to the activity of the fault zone is provided by the intrusion of the Karkonosze Granite (Rb–Sr *c.* 330–310 Ma, Duthou *et al.* 1991; Pb–Pb evaporation 314 ± 414 Ma, Kröner *et al.* 1994*a*, which does not show any ductile deformation and against which the ductile parts of the fault zone do terminate (Fig. 1).

A system of SSW-trending brittle or ductile normal faults possible branches off from the Intra-Sudetic Fault to the SE of the Karkonosze Granite. It brought down the eastern portions of the Rudawy Janowickie Unit against the western ones and against the Železný Brod Unit (Mazur & Kryza 1996), yet terminated as brittle west-vergent thrust (Leszczyniec Thrust, Oberc 1960). This normal fault system largely controlled the subsidence at the Intra-Sudetic Basin during (?)pre-Viséan and Viséan times (Teisseyre 1975) and may represent the northwestern margin of a pull-apart basin. The northeastern boundary fault (possible southeastern continuation of the Intra-Sudetic Fault) would have to be sought at the southwestern margin of the Góry Sowie Block (see Aleksandrowski *et al.* 1997), and the southwestern unidentified branch of the fault template would be located at the boundary between the Intra-Sudetic Basin and Permian rocks further west. However, Cretaceous reverse faulting has overprinted all earlier displacements along these faults. Alternatively, the Carboniferous portion of the Intra-Sudetic basin can be interpreted as an intramontane, late to post-collisional trough developed over a large volcanic centre controlled by major NNW-trending fractures in its basement (Awdankiewicz 1999, 2000).

Sudetic Marginal Fault (SMF)

The Sudetic Marginal Fault is an important normal fault separating the Sudetes Mts to the SW from the Fore-Sudetic Block to the NE (Fig. 1). Its earlier increments (in Late Carboniferous and Cretaceous time) are reverse and have brought up the Fore-Sudetic Block with an estimated throw of a few kilometres, so that deeper levels are exposed NE of the fault zone. Most geological units terminate at the fault (Fig. 1), which suggests that they have been structurally thinner than the throw of the fault. The outcrop of the Góry Sowie Block is unaffected by the Sudetic Marginal Fault, because the boundary faults are steep. These relationships also prove that since Late Carboniferous time there has been no recognizable strike-slip component on the Sudetic Marginal Fault, which makes a striking contrast to the Intra-Sudetic Fault. Seismic reflection profiling reveals a listric geometry of the Sudetic Marginal Fault (Żelaźniewicz *et al.* 1997). Late, subordinate increments of its activity (since Miocene time, Dyjor 1995) have brought down the northeastern block, so that the Sudetic Marginal Fault at present forms the northeastern margin of the elevated part of the Bohemian Massif.

Moldanubian Thrust (MT)

The Moldanubian Thrust, first recognized by Suess (1912), has emplaced high-grade rocks of the Moldanubian Units over the Moravo-Silesian Belt. Structural studies have yielded clear evidence of dextral transpression (Schulmann *et al.* 1991; Cymerman 1993; Fritz & Neubauer 1993; Dallmeyer *et al.* 1995; Schulmann & Gayer, 2000) to the SE of the Elbe Fault Zone (Fig. 1). Major curvatures in the trend of the fault zone are expressed as transpressive antiformal stacks, or else as later transverse folds. The Moldanubian Thrust apparently post-dates the Elbe Fault Zone (Fig. 1; see Aleksandrowski *et al.* 1997). It is displaced, in its turn, by the southeastern continuation of the Sudetic Marginal Fault (Bilá Fault). This offset can be explained by uplift of the Fore-Sudetic Block (Oberc 1968), which demands a shift of the WNW-dipping Moldanubian Thrust toward the NW. Hence, the transpressional movements along the Moldanubian Thrust post-date the NW-trending dextral strike-slip activities and pre-date early reverse faulting along the Sudetic Marginal Fault.

The Moldanubian Thrust juxtaposes the Moravo-Silesian against Moldanubian rocks in its SW segment south of the Elbe Fault Zone and against rocks with Saxo-Thuringian affinities in the NE segment. The thrust apparently excises terranes (Figs 1 and 6), and the lateral offset must therefore amount to at least some tens of kilometres. This offset must be compensated somewhere by the SW–NE-directed shortening, probably within the concealed phyllitic basement highs in Poland.

The age of movements along the Moldanubian Thrust is constrained by the sedimentary transport of pebbles derived from Moldanubian Granulites into the Late Viséan Luleč Conglomerate of the Moravo-Silesian Flysch (Otava 1998). The Late Viséan age translates into an isotopic age of *c.* 330–325 Ma (see McKerrow & van Staal this volume).

Discussion

Terrane correlation across the Elbe Fault Zone

As proposed already by Kossmat (1927) and confirmed by Franke *et al.* (1993), at least most parts of the West Sudetes belong to the Saxo-Thuringian Zone and consist of tectonometamorphic units that can be correlated with areas west of the Elbe Fault Zone (Fig. 6, Table 1;

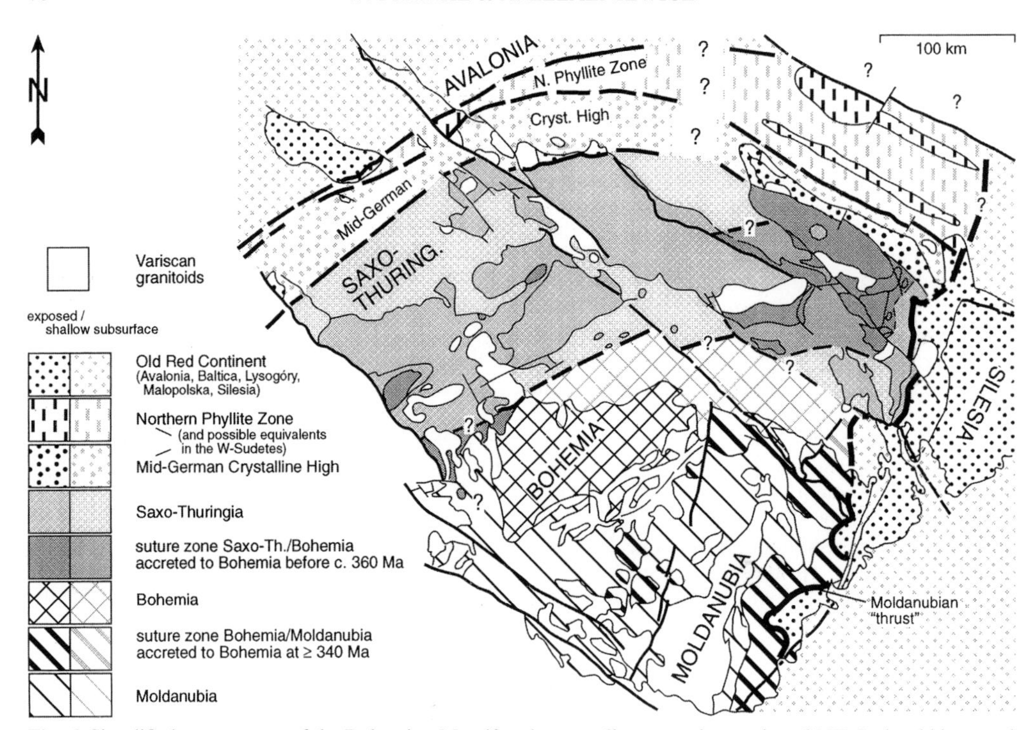

Fig. 6. Simplified terrane map of the Bohemian Massif and areas adjacent to the north and NE. It should be noted that the Northern Phyllite Zone and the Mid-German Crystalline High are tectonic units, which contain fragments of more than one terrane (see Franke this volume).

see also Franke (this volume)). These are: (1) equivalents of the Saxo-Thuringian Parautochthon and non-metamorphic nappes characterized by flysch sedimentation of early Carboniferous age and subsequent deformation (Ješted segment of the Lausitz–Izera Unit; northwestern, lower-grade part of the Görlitz–Kaczawa Unit). (2) equivalents of the *c.* 340 Ma HP ± HT rocks encountered in the Saxonian Granulitgebirge and in the Erzgebirge (Orlica–Śnieżnik Dome and probably Kamieniec Complex of the Kamieniec–Strzelin Belt); (3) equivalents of the very low to amphibolite grade rocks dated at 380–360 Ma in the Saxo-Thuringian Metamorphic Nappes and in the Teplá–Barrandian Unit (?Görlitz–Kaczawa, Kłodzko and Orlica– Śnieżnik Units); (4) equivalents of the *c.* 400 Ma HP ± HT rocks (subducted parts of the narrow Saxo-Thuringian Ocean) found in the Saxo-Thuringian Münchberg Klippe and in the Mariánské Lázně Complex (garnet peridotites and granulites in the Góry Sowie Block).

Rocks with Devonian metamorphism such as in the Kłodzko and Góry Sowie units can be assigned either to marginal parts of the Bohemian Terrane, which were incorporated into Devonian subduction–collision, or else to palaeogeographical units that were accreted to the northern margin of Bohemia in Devonian time (narrow Saxo-Thuringian Ocean and southern, distal margin of Saxo-Thuringia). Devonian metamorphism in these units hampers detailed correlation.

The last two items of Table 1 are less significant. Cadomian subduction-related magmatic rocks are widespread in all parts of the Variscan basement, and upper Cambrian–Lower Ordovician rift-related bimodal magmatic rocks are ubiquitous in the entire Armorican Terrane Assemblage as defined by Tait *et al.* (this volume).

There are also some notable differences between the West Sudetes and the areas west of the Elbe Fault Zone. First, the Devonian metamorphic assemblages west of the Elbe Fault Zone contain no blueschist facies rocks. However, this is not a major factor, as the preservation of Devonian medium-grade metamorphic rocks in the Saxo-Thuringian Belt is sporadic. Second, the Teplá–Barrandian and Saxo-Thuringian Units contain some Silurian and Lower Devonian intraplate metabasalts, but no equivalent of the 420–400 Ma ophiolite neighbouring the Góry Sowie Block. We speculate that the

Table 1. *Comparison of geological events in the West Sudetes and areas SW of the Elbe Fault Zone*

	Saxo-Thuringian non-metamorphic nappes, Para-autochthon and substratum	Saxo-Thuringian metamorphic nappes and Marianské Lazně	Teplá-Barrandian	W Sudetes
c. 330 Ma folding	+	final emplacement	–	Görlitz–NW Kaczawa, Ještěd, Bardo
Early Carboniferous flysch	+	–	–	Görlitz–NW Kaczawa, Ještěd, Bardo
340 Ma HP–HT in antiforms	Saxonian Granulites, Erzgebirge (reworked)	–	–	Orlica–Śnieżnik, Kamieniec
360–380 Ma deformation and metamorphism	–	greenschist	very low to amphibolite grade	Kaczawa, S-Karkonosze, Rudawy Janowickie, Kłodzko, Sowie (various grades)
c. 400 Ma HP $\pm$ HT	–	+	–	Sowie
480–510 MA magmatism, partly metamorphosed	+	+	+	Kaczawa, S Karkonosze, Rudawy Janowickie, Gory Sowie, Orlica–Śnieżnik, Strzelin, Staré Město
Late Cadomian magmatism	Schwarzburg Anticline (volcanic), Erzgebirge (granite)	?	+	Lausitz, S Karkonosze (granitoids)

Sudetic Ophiolite represents a late stage of extension in the Saxo-Thuringian Ocean, and was accreted to the Sudetic Thrust Stack in Late Devonian time, because it was young and hot. Equivalents of the ophiolite west of the Elbe Fault Zone might be present under the Tertiary cover of the Eger Graben, or else have been excised by Variscan extensional faulting. NW-trending faults are also present at the SW margin of the Bohemian Massif, and are possibly responsible for the formation of late Carboniferous–Permian pull-apart basins (Franke, 1989; Vollbrecht *et al.* 1989; Weber & Vollbrecht 1989). However, activity along such faults is much more prominent in the Elbe and Intra-Sudetic Fault Zones, where displacements attained the order of tens of kilometres. Third, the Devonian HT–LP metamorphism at 380–360 Ma in the Góry Sowie Block, following the *c.* 400 Ma HP event, has no visible counterpart in the Saxo-Thuringian Belt. Time-equivalent metamorphic rocks elsewhere show much lower geothermal gradients.

Orogenic events

Like their equivalents SW of the Elbe Fault Zone, the West Sudetes contain Cadomian basement which underwent important rifting in late Cambrian to early Devonian times (see also Linnemann *et al.* (this volume)). Contrary to hypothesis put forward by Oliver *et al.* (1993), Cymerman *et al.* (1997) and Kröner & Hegner (1998), there is no evidence of Early Palaeozoic ('Caledonian') plate convergence in the West Sudetes, neither are there indications of Early Palaeozoic compression-related metamorphism, deformation and coeval flysch sedimentation (Aleksandrowski 1994; Żelaźniewicz & Franke 1994; Żelaźniewicz 1997). The Lower Palaeozoic mafic magmatic rocks indicate an extensional setting (see Floyd *et al.* (this volume)), which is in perfect accord with the sedimentary record. The calc-alkaline geochemical signature of the felsic magmatic rocks can be explained by the extraction of melt from a Late Proterozoic (Cadomian) magmatic arc, or else by after-effects of Cadomian subduction. It is noteworthy that the widespread Early Palaeozoic extension has now also been demonstrated, for the first time in the Bohemian Massif, in lower-crustal rocks (Staré Město Belt, SE of the Orlica–Śnieżnik Unit: Štípská *et al.* 1998; Kröner *et al.* this volume). These findings testify to the importance of Early Palaeozoic rifting in the Cadomian basement of the Armorican Terrane Assemblage.

The Variscan convergence in the West Sudetes is documented in HP–HT metamorphic rocks formed at *c.* 400 Ma, blueschist–amphibolite facies rocks dating between *c.* 380 and 360 Ma, 340 Ma retrograde overprint and early Carboniferous flysch incorporated into the fold belt.

Oroclinal bending?

The Moravo-Silesian Belt has been rotated clockwise, with respect to its Rheno-Hercynian equivalent, through 90 ° or even 180 ° (Tait *et al.* 1994 this volume). It is uncertain whether the equivalent units were connected by a complete and intact arc structure. This problem can be assessed by the correlation of terranes across the Elbe and Intra-Sudetic Fault Zones as suggested in Fig. 6. It is possible, but by no means certain, that the Odra Fault Zone and basement highs adjacent to the NE are equivalents of the Mid-German Crystalline High, the Northern Phyllite Zone and the Rheno-Hercynian Foreland, and that the Rheno-Hercynian Foreland with its Devonian to Carboniferous deposits and volcanic rocks continues into the Moravo-Silesian Belt. These correlations will have to be tested by detailed studies on borehole cores from the basement highs in Poland. Tectonic units correlatable with the Saxo-Thuringian Belt SW of the Elbe Fault Zone can be traced around a Bohemian core region as far as the Staré Město Unit (SE of the Orlica–Śnieżnik Dome), which abuts against the Moldanubian Thrust. Further to the SW, there is no clear evidence of rocks with Saxo-Thuringian affinities, and the Moldanubian Thrust juxtaposes Moldanubian Units against the Moravo-Silesian Belt. This demonstrates that any Bohemian Arc structure is not a hairpin bend with a complete tectonic sequence on either flank. Instead, the Bohemian Massif resembles, in map view, an open 'drag fold' with a subvertical axis, disrupted by the Moldanubian Thrust. The asymmetry of the 'fold' is in accord with the component of dextral strike-slip along the Moldanubian Fault Zone. It is only the southeastern (Moravo-Silesian) limb of the 'fold' that has been 'overturned'.

The mechanics of clockwise rotation of the Moravo-Silesian Belt through 90° or even 180° during the collisional stage has not been resolved. Rotation possibly started when central parts of the present-day Bohemian Massif (the Bohemian Terrane and parts of the Saxo-Thuringian Terrane) were forced to intrude northeastwards as an indenter (Lorenz & Nicholls 1984). Such an indenter might have been confined between the Moldanubian Thrust on the SE and shear zones on either side of the

Mid-German Crystalline High (Schäfer *et al.* this volume). Once a N–S orientation of the Moravo-Silesian Block was attained, further rotation could have been achieved when basement units to the SE (now concealed by the Carpathians and the Pannonian Basin) were displaced to the NW by strike-slip along the Tornquist–Teisseyre and Trans-European Fault Zone, thus tightening the Bohemian Arc.

It is uncertain whether the more internal, Saxo-Thuringian parts of the Bohemian Arc were also affected by the clockwise rotation. The presence of Saxo-Thuringian rocks in the West Sudetes, to the NE of the concealed Teplá–Barrandian Unit around Hradec Králové, can be explained in three ways: (1) the Saxo-Thuringian units of the West Sudetes were displaced to the SE along NW-trending fault zones: the Elbe Fault Zone, a possible Variscan precursor to the Cretaceous reverse West Lausitz Fault, and the Intra-Sudetic Fault (see Aleksandrowski 1990, 1995); (2) the Saxo-Thuringian Belt was bent around the Bohemian (Teplá–Barrandian) core; (3) the Saxo-Thuringian Belt was initially thrust easterly onto the Moravo-Silesian Units and then moved northeastward by the 'Bohemian Indenter' to accommodate the dextral strike-slip along the Moldanubian Thrust Zone.

These end-member models may be distinguished by comparing the orientation of coeval early Variscan tectonic fabrics (e.g. stretching lineations and later overprints) in the various tectonic units. However, such comparisons require precise constraints on the age of the individual tectonic elements, which are at present not available.

Although the kinematic evolution of the Bohemian Arc requires further studies, it is possible to address its setting on a larger scale. It appears that the tectonic units now contained within the Bohemian Massif were moulded into a recess within the southern margin of the Old Red Continent comprising the southern margin of Avalonia to the NW, and the southwestern margin of the East European Craton to the NE. Such a model has already been proposed by Grygar & Vavro (1994).

Conclusions

All lines of evidence consistently suggest that the West Sudetes record the same geological events that also characterize the central segment of the Variscides SW of the Elbe Fault Zone: a Cadomian active margin with synorogenic greywackes and granodiorites representing a magmatic arc, Cambro-Ordovician rifting with important bimodal magmatism, Silurian and Devonian pelagic sediments, and early Carboniferous flysch with olistoliths. As in the central segment, Variscan convergence started in early to mid-Devonian time, and produced pressure-dominated metamorphic rocks. Tectonostratigraphic units in the West Sudetes are clearly correlatable with terranes known from the central and western parts of the Bohemian massif, that is with the Saxo-Thuringian Terrane, the Saxo-Thuringian Ocean and the Bohemian (Centralbohemicum) Terrane. Hence, it is not necessary to create, in the West Sudetes, additional terranes and several sutures (e.g. Cymerman *et al.* 1997). However, blueschist–greenschist facies metamorphism and intense deformation have much disrupted the stratigraphic sequences, so that it is mostly impossible to distinguish between the genuine core of the Bohemian Terrane and units accreted to it during late Devonian collision, especially because both the Bohemian Terrane and the accreted fill of the Saxo-Thuringian Basin have the same Cambro-Ordovician rift development and late Devonian deformation.

Only two important NW-trending dextral shear zones (Elbe Fault Zone and Intra-Sudetic Fault Zone) were active in late Devonian to early Carboniferous times (Aleksandrowski 1990, 1995). They have much disrupted the earlier structural pattern, and probably gave rise to the formation of fault-controlled intramontane to shallow marine basins (Świebodzice and Intra-Sudetic Basins). Strike-slip faulting and oroclinal bending have deformed the eastern part of the Variscides into a disrupted arc structure. The 'Moldanubian Thrust' at the southeastern margin of the Bohemian Massif is a late tectonic feature, and separates the slightly bent, northwestern part of the arc from its southeastern (Moravo-Silesian) part, which has been rotated through $\geqslant 190^{\circ}$. This structural pattern is explained by accommodation of Variscan units in a recess of the Old Red Continent to the north, which is defined by the NE-trending southern margin of Avalonia and the SE-trending margin of the East European Platform. The causes and the pattern of the plate-boundary forces responsible for the emplacement of the Bohemian Massif remain to be resolved in detail.

We gratefully acknowledge travel funding through the International Bureau of the German Ministry of Education and Research (BMBF) and the Polish Academy of Sciences. Thanks for important hints and critical comments are due to F. Finger, A. Kröner, K. Schulmann and P. Aleksandrowski.

References

ACHRAMOWICZ, S. 1998. Reinterpretation of the Intra-Sudetic Fault Zone. *GeoLines*, **6**, 5.

—— 1999. The Saxothuringian/Moravosilesian boundary duplex structure in the Fore-Sudetic Block. *GeoLines*, **8**, 5–6.

——, MUSZYŃSKI, A. & SCHLIESTEDT, M. 1997. The northernmost eclogite occurrence in the Saxothuringian Zone, West Sudetes (Poland). *Chemie der Erde*, **57**, 51–61.

—— & ŻELAŹNIEWICZ, A. 1998. Północna część bloku karkonosko-izerskiego: zapis dwóch orogenez. *Polskie Towarzystwo Mineralogiczne–Prace Specjalne*, **11**, 9–13.

AHRENDT, H., CLAUER, N., HUNZIKER, J. C. & WEBER, K. 1983. Migration of folding and metamorphism in the Rheinisches Schiefergebirge, deduced from K–Ar and Rb–Sr age determinations. *In*: MARTIN, H. & EDER, F. W. (eds) *Intracontinental Fold Belts*. Springer, Berlin, 232–338.

ALEKSANDROWSKI, P. 1990. Early Carboniferous strike-slip displacements at the northeast periphery of the Variscan Belt in Central Europe. *International Conference on Paleozoic Orogens in Central Europe (IGCP Program 233: Terranes in the Circum-Atlantic Paleozoic Orogens), Abstracts*, Göttingen–Giessen.

—— 1994. Discussion on U–Pb ages from SW Poland: evidence for a Caledonian suture zone between Baltica and Gondwana. *Journal of the Geological Society, London*, **151**, 1049–1050.

—— 1995. Rola wielkoskalowych przemieszczeń przesuwczych w ukształowaniu waryscyjskiej struktury Sudetów. (The significance of major strike-slip displacements in the development of Variscan structure of the Sudetes, SW Poland) *Przegląd Geologiczny*, **43**, 745–754.

——, KRYZA, R., MAZUR, S. & ŻABA, J. 1997. Kinematic data on major Variscan strike slip faults and shear zones in the Polish Sudetes, northeast Bohemian Massif. *Geological Magazine*, **34**, 727–739.

AWDANKIEWICZ, M. 1999. Volcanism in a late Variscan intramontane trough: the Intra-Sudetic Basin, SW Poland. *Geologia Sudetica*, **32**, 13–47.

—— 2000. Volcanism in a late Variscan intramontane trough: petrology and geochemistry of the Carboniferous and Permian volcanic rocks of the Intra-Sudetic Basin, SW Poland. *Geologia Sudetica*, **32**, 83–111.

BAKUN-CZUBAROW, N. 1998. Ilmenite-bearing eclogites of the West Sudetes—their geochemistry and mineral chemistry. *Archiwum Mineralogiczne*, **51**, 29–110.

BANKWITZ, P., BANKWITZ, E., WASTERNACK, J. *et al.* 1999. The Mid-German Crystalline Zone and its supposed continuation into SW-Poland. *Terra Nostra*, **99**(1), 59–60.

BARANOWSKI, Z., HAYDUKIEWICZ, A., KRYZA, R., LORENC, S., MUSZYŃSKI, A., SOLECKI, A. & URBANEK, Z. 1990. Outline of the geology of the Góry Kaczawskie (Sudetes, Poland). *Neues Jahrbuch für Geologie und Paläontologie, Abhandlungen*, **179**, 223–257.

BEDERKE, E. 1924. Das Devon in Schlesien und das Alter der Sudetenfaltung. *Fortschritte Geologie und Paläontologie*, **7**, 1–50.

—— 1935. Verbreitung und Gliederung des Devons in den Ostsudeten. *Zentralblatt für Mineraalogie, Geologie und Paläontologie*, **Abt.B**, 33–40.

BELKA, Z. 1987. The development and decline of a Dinantian carbonate platform: an example from the Moravia–Silesia Basin, Poland. *In*: MILLER, J., ADAMS, A. E. & WRIGHT, V. P. (eds) *European Dinantian Environments*. Wiley, New York, 177–188.

BENDL, J. & PATOČKA, F. 1995. The Rb–Sr isotope geochemistry of the metamorphosed bimodal volcanic association of the Rychory Mts. crystalline complex, West Sudetes, Bohemian massif. *Geologia Sudetica*, **29**, 3–18.

BIELICKI, K, HAASE, G., EIDAM, J. *et al.* 1989. Pb–Pb and Rb–Sr dating of granitoids from the Lusatian Block. *ZfI-Mitteilungen, 5th Working Meeting Isotopes in Nature Proceedings II, Leipzig*, 23–45.

BORKOWSKA, M. & DÖRR, W. 1998. Some remarks on the age and mineral chemistry of orthogneisses from the Lądek-Śnieżnik metamorphic massif—Sudetes, Poland. *Terra Nostra*, **98**(2), 27–30.

BRÖCKER, M., COSCA, M. & KLEMD, R. 1997. Geochronologie von Eklogiten und assoziierten Nebengesteinen des Orlica–Śnieżnik Kristallins (Sudeten, Poland): Ergebnisse von U–Pb, Sm–Nd, Rb–Sr und Ar–Ar Untersuchungen. *Terra Nostra*, **97**(5), 29–30.

——, ŻELAŹNIEWICZ, A. &. ENDERS, M. 1998. Rb–Sr geochronology of migmatitic gneisses from the Góry Sowie (West Sudetes, Poland): the importance of Mid–Late Devonian metamorphism. *Journal of the Geological Society, London*, **155**, 1025–1035.

BRUECKNER, H. K., BLUSZTAJN, J. & BAKUN-CZUBAROW, N. 1996. Trace element and Sm–Nd 'age' zoning in garnets from peridotites of the Caledonian and Variscan Mountains and tectonic implications. *Journal of Metamorphic Geology*, **14**, 61–73.

——, MEDARIS, J. L. G. & BAKUN-CZUBAROW, N. 1991. Nd and Sr age and isotope patterns from Variscan eclogites of the eastern Bohemian Massif. *Neues Jahrbuch für Mineralogie, Abhandlungen*, **163**, 169–196.

BURMANN, G. 1997*a*. Präkambrisches Algenbenthos aus Kieselpeliten. Teil 1: Palisadia wagneri n. gen. n. spec.—erste Makroalge aus dem Präkambrium Deutschlands. *Zeitschrift für Geologische Wissenschaften*, **25**, 41–90.

—— 1997*b*. Prákambrisches Algenbenthos aus Kieselpeliten. Teil 2: Kobylisya konzalivi n. gen. n. spec.—ein Retikulum mit Vacuolen aus der Blovice-Formation des Barrandiums/Tschechische Republik. *Zeitschrift für Geologische Wissenschaften*, **25**, 91–108.

ČECH, S., HAVLÍČEK, V. & ZIKMUNDOVÁ, J. 1989. The Upper Devonian and Lower Carboniferous in north-eastern Bohemia (based on boreholes in the

Hradec Králové area). *Vestník Ústredního Ústavu Geologického*, **64**, 65–76.

CHLUPÁČ, J. 1989. Fossil communities in the Lower Devonian of the Hruby Jesenik Mts., Czechoslovakia. *Neues Jahrbuch für Geologie und Paläontologie, Abhandlungen*, **177**, 367–392.

—— 1993*a*. Stratigraphic evaluation of some metamorphic units in the N part of the Bohemian Massif. *Neues Jahrbuch für Geologie und Paläontologie, Monatshefte*, **188**, 363–388.

—— 1993*b*. *Geology of the Barrandian: a Field-trip Guide*. Senckenbergische Naturforschende Gesellschaft & Czech Geological Survey. Waldemar Kramer, Frankfurt am Main.

—— 1994. Facies and biogeographic relationships in Devonian of the Bohemian Massif. *Courier Forschungsinstitut Senckenberg*, **169**, 299–317.

—— 1997. Palaeozoic ichnofossils in phyllites near Železný Brod, northern Bohemia. *Journal of the Czech Geological Society*, **42**, 75–94.

—— & HLADIL, J. 1992. New Devonian occurrences in the Ještěd Mts., North Bohemia. *Časopis pro Mineralogii a Geologii*, **37**, 185–191.

—— & ZIKMUNDOVÁ, J. 1976. The Devonian and Lower Carboniferous in the Nepasice borehole in East Bohemia. *Věstník Ústredního Ústavu Geologického*, **51**, 269–278.

——, HAVLÍČEK, V., KŘÍŽ, J., KUKAL, Z. & ŠTORCH, P. 1998. *Palaeozoic of the Barrandian (Cambrian to Devonian)*. Czech Geological Survey, Prague.

CHOROWSKA, M. 1978. Wizeńskie wapienie w epimetamorficznym kompleksie Gór Kaczawskich. (Viséan limestones in the metamorphic complex of the Kaczawa Mts (Sudetes).) *Rocznik Polskiego Towarzystwa Geologicznego*, **48**, 245–261.

CWOJDZIŃSKI, S. & ŻELAŹNIEWICZ, A. 1995. Podłoże krystaliczne bloku przedsudeckiego. (Crystalline basement of the Fore-Sudetic Block.) *Przewodnik LXVI Zjazdu PTG, Wrocław*, 11–28.

CYMERMAN, Z. 1993. Czy w Sudetach istnieje nasunięcie ramzowski? (Does the Ramzova overthrust in the Sudetes actually exist?) *Przegląd Geologiczny*, **41**, 700–706.

—— 1996. Transpresja i ekstensja w Rudawy Janowickie metamorphic complex (Western Sudetes).) *Przegląd Geologiczny*, **44**, 1211–1216.

—— 1998. Uskok śródsudecki a regionalne strefy ścinań podatnych w Sudetach. (Intra-Sudetic Falut and regional-scale ductile shear zones.) *Przegląd Geologiczny*, **46**, 609–616.

—— & PIASECKI, M. A. J. 1994. The terrane concept in the Sudetes, Bohemian Massif. *Geological Quarterly*, **38**, 191–210.

——, PIASECKI, M. A. J. & SESTON, R. 1997. Terranes and terrane boundaries in the Sudetes, northeast Bohemian Massif. *Geological Magazine*, **134**, 717–725.

CZAPLIŃSKI, W. 1998. Orthogneisses and metapelites from a polyphase tectonic zone—mesostructural versus microstructural evidence: an example from the Czerniawa Zdrój section (Izera–Karkonosze Block, West Sudetes). *Geologia Sudetica*, **31**, 93–104.

——, ACHRAMOWICZ, S., LOPEZ, L. A., PACHOLSKA, A. & ŻELAŹNIEWICZ, A. 1996. Major shear zones in the West Sudetes from quartz microfabric data. *Europrobe—Trans-European Suture Zone, Książ Workshop, Abstracts*.

DALLMEYER, R. D. & URBAN, M. 1998. Variscan vs Cadomian tectonothermal activity in northwestern sectors of the Teplá–Barrandian zone, Czech Republic: constraints from $^{40}Ar/^{39}Ar$ ages. *Geologische Rundschau*, **87**, 94–106.

——, FRANKE, W. & WEBER, K. (eds) 1995. *Pre-Permian Geology of Central and Eastern Europe*, VII Moravo-Silesian Zone. Springer, New York, 469–553.

DUTHOU, J. L., COUTURIE, J. P. & MIERZEJEWSKI, M. P. 1991. Oznaczenia wieku granitu Karkonoszy metodą izochronowá, rubidowo-strontową na podstawie calych próbek skalnych. (Rb/Sr age of the Karkonosze granite on the base of the whole rock method.) *Przegląd Geologiczny*, **2**, 75–79.

DUBIŃSKA, E. 1995. Rodingites of the eastern part of Jordanów–Gogołów serpentinite massif, Lower Silesia, Poland. *Canadian Mineralogist*, **33**, 585–608.

—— 1997. Rodingites and amphibolites from the serpentinites surrounding Góry Sowie block (Lower Silesia, Poland): record of supra-subduction zone magmatism and serpentinization. *Neues Jahrbuch für Mineralogie und Petrologie, Abhandlungen*, **171**, 239–279.

—— & GUNIA, P. 1997. The Sudetic ophiolite: current view on its geodynamic model. *Geological Quarterly*, **41**, 19–33.

——, ŻELAŹNIEWICZ, A. & NEJBERT, K. 1998. Subcontinental mantle-derived rocks within migmatitic gneisses of the Góry Sowie block, Sudetes. *Acta Universitatis Carolinae*, **42**, 232–233.

——, ——, —— & BYLINA, P. 1999. Ultramafic rocks from migmatitic gneisses of the Góry Sowie block, Sudetes. *Polskie Towarzystwo Mineralogiczne—Prace Specjalne*, **14**, 76–78.

DUDEK, A. 1980. The crystalline basement block of the Outer Carpathians in Moravia: Bruno-Vistulicum. *Rozpravy Československej Akademii věd, Rada Matematicko-přirodovedeckych Věd*, **90**, 1–85.

DVORÁK, J. 1995. Moravo-Silesian Zone: autochthon: stratigraphy. *In*: DALLMEYER, D., FRANKE, W. & WEBER, K. (eds). *Pre-Permian Geology of Central and Western Europe*. Springer, Berlin, 475–489.

—— & PAPROTH, E. 1969. Über die Position und Tektogenese des Rhenoherzynikum und des Sudetikum in den Mitteleuropäschen Varisziden. *Neues Jahrbuch für Geologie und Paläontologie, Monatshefte*, **127**, 65–85.

DYJOR, S. 1995. Rozwój kenozoiku na bloku przedssudeckim. Evolution of the Cainozoic on the Fore-Sudetic Block. *Przewodnik LXVI Zjazdu PTG, Wrocław*, 22–40.

DZIEDZIC, K & TEISSEYRE, A. K. 1990. The Hercynian molasse and younger deposits in the Intra-Sudetic Depression, SW Poland. *Neues Jahrbuch für*

Geologie und Paläontologie, Abhandlungen, **179**, 285–305.

DZIEDZICOWA, H. 1987. Rozwój strukturalny i metamorfizm we wschodnim obrzeżeniu gnejsów Gór Sowich. (Structural development and metamorphism in the region east of the Góry Sowie gneissic massif.) *Acta Universitatis Wratislaviensis, Prace Geologiczno-Mineralogiczne*, **10**, 221–247.

ENGEL, W. & FRANKE, W. 1983. Flysch sedimentation: its relation to tectonism in the European Variscides. *In*: MARTIN, H. & EDER, F. W. (eds) *Intracontinental Fold Belts. Case Studies in the Variscan Belt of Europe and the Damara Belt in Namibia*. Springer, Berlin, 289–321.

FAJST, M., KACHLÍK, V. & PATOČKA, F. 1998. Geochemistry and petrology of the early Palaeozoic Železný Brod volcanic complex (W Sudetes, Bohemian Massif): geodynamic interpretations. *GeoLines*, **6**, 14.

FALK, F., FRANKE, W. & KURZE, M. 1995. Saxothuringian Basin: autochthon and nonmetamorphic nappe units: stratigraphy. *In*: DALLMEYER, D., FRANKE, W. & WEBER, K. (eds) *Pre-Permian Geology of Central and Western Europe*. Springer, Berlin, 219–234.

FINGER, F., HANŽL, P., PIN, C., VAN QUADT, A. & STEYRER, H. P. 2000. The Brunovistulian: Avalonian Precambrian sequence at the eastern end of the Central European Variscides? *This volume*.

FLOYD, P. A., WINCHESTER, J. A., SESTON, R., KRYZA, R. & CROWLEY, Q. G. 2000. Review of geochemical variation in Lower Palaeozoic metabasites from the NE Bohemian Massif: intracratonic rifting and plume–ridge interaction. *This volume*.

FRANKE, W. 1989. Variscan plate tectonics in Central Europe—current ideas and open questions. *Tectonophysics*, **169**, 221–228.

—— 2000. The mid-European segment of the Variscides: tectono-stratigraphic units, terrane boundaries and plate tectonic evolution. *This volume*.

—— & ENGEL, W. 1982. Variscan sedimentary basins on the continent and relations with south-west England. *Proceedings of the Ussher Society*, **5**, 259–269.

—— & STEIN, E. 2000. Exhumation of high-grade rocks in the Saxo-Thuringian belt: geological constraints and geodynamic concepts. *This volume*.

——, KREUZER, H., OKRUSCH, M., SCHÜSSLER, U. & SEIDEL, E. 1995. Saxothuringian Basin: exotic metamorphic nappes: stratigraphy, structure, and igneous activity. *In*: DALLMEYER, D., FRANKE, W. &. WEBER, K. (eds) *Pre-Permian Geology of Central and Western Europe*. Springer, Berlin, 275–294.

——, ŻELAŹNIEWICZ, A., PORĘBSKI, S. K. & WAJSPRYCH, B. 1993. The Saxothuringian zone in Germany and Poland: differences and common features. *Geologische Rundschau*, **82**, 583–599.

FRITZ, H. & NEUBAUER, F. 1993. Kinematics of crustal stacking and dispersion in the south-eastern Bohemian Massif. *Geologische Rundschau*, **82**, 556–565.

FURNES, H., KRYZA, R., MUSZYŃSKI, A., PIN, C. & GARMANN, L. B. 1994. Geochemical evidence for progressive, rift-related early Palaeozoic volcanism in the western Sudetes. *Journal of the Geological Society, London*, **151**, 91–109.

GEHMLICH, M., LINNEMANN, U., TICHOMIROWA, M., LÜTZNER, H. & BOMBACH, K. 1997. Die Bestimmung des Sedimentationsalters cadomischer Krustenfragmente im Saxothuringikum durch die Einzelzirkon-Evaporationsmethode. *Terra Nostra*, **97**(5), 46–50.

GRANDMONTAGNE, N. E., HEINISCH, H., FRANKE, W. & ŻELAŹNIEWICZ, A. 1996. Polyphase Variscan deformation in the South Krkonoše Mts., Czechia. *Terra Nostra*, **96**(2), 88–92.

GROCHOLSKI, A. 1986. Proterozoic and Palaeozoic of southwestern Poland in a light of new data. *Biuletyn Instytutu Geologicznego*, **335**, 7–29.

GRYGAR, R. & VAVRO, M. 1994. Geodynamic model of evolution of Lugosilesian orocline of European Variscan orogeny belt. *Journal of the Czech Geological Society*, **39**, 40–41.

GUIRAUD, M. & BURG, J. P. 1984. Mineralogical and petrological study of a blueschist metatuff from the Železný Brod Crystalline Complex, Czechoslovakia. *Neues Jahrbuch für Mineralogie, Abhandlungen*, **149**, 1–12.

GUNIA, T. 1996. Problem wieku marmurów okolicy Stronia Śląskiego na podstawie mikroskamieniałości (Sudety). (The issue of marbles age from the Stronie Śląskie vicinity researched on the basis of microfossils (the Sudetes).) *Acta Universitatis Wratislaviensis, Prace Geologiczno-Mineralogiczne*, **62**, 5–48.

—— & WOJCIECHOWSKA, I. 1971. Zagadnienie wieku wapieni i fyllitów z Małego Bożkowa (Sudety Środkowe). (On the age of limestones and phyllites from Mały Bożków (Central Sudetes).) *Geologia Sudetica*, **5**, 137–164.

HAMMER, J., BRÖCKER, M. & KRAUSS, M. 1997. Alter und geologische Signifikanz von Deformationszonen im östlichen Teil des Lausitzer Granitoidkomplexes. *Terra Nostra*, **97**(5), 62–63.

HAYDUKIEWICZ, J. 1990. Stratigraphy of Paleozoic rocks of the Góry Bardzkie and some remarks on their sedimentation (Poland). *Neues Jahrbuch für Geologie und Paläontologie, Abhandlungen*, **179**, 275–284.

—— 1998. Latest Devonian conodonts from an olistolith in the northern part of the Góry Bardzkie, West Sudetes. *Geologia Sudetica*, **31**, 61–68.

HLADIL, J. 1995. Arguments in favour of clockwise block rotation in Variscides of Moravia—analyzing the Devonian facies disjunction. *Geological Research in Moravia and Silesia*, **2**, 44–48.

——, MAZUR, S., GALLE, A. & EBERT, J. R. 1999. Revised age of the Mały Bożków limestone in the Kłodzko metamorphic unit (early Givetian, late Middle Devonian) implications for the geology of the Sudetes, SW Poland. *Neues Jahrbuch für*

Geologie und Paläontologie, Abhandlungen, **211**, 329–353.

JÄGER, H. 1964. *Monograptus hercynicus* in den Westsudeten und das Alter der Westsudeten-Hauptfaltung. *Geologie*, **13**, 249–273.

JELÍNEK, E., PAČESOVÁ, M., MARTINEC, P., MISAŘ, Z. & WEISS, Z. 1984. Geochemistry of a dismembered metaophiolite complex, Letovice, Czechoslovakia. *Transactions of the Royal Society of Edinburgh: Earth Sciences*, **75**, 37–48.

JOHAN, V. & SCHULMANN, K. 2000. High pressure metamorphism at the boundary of Letovice metaophiolites, eastern Bohemian Massif. *Casopis pro Mineralogii a Geologii*, in press.

JOHNSTON, J. D., TAIT, J. A., OLIVER, G. J. H. & MURPHY, F. C. 1994. Evidence for a Caledonian orogeny in Poland. *Transactions of the Royal Society of Edinburgh: Earth Sciences*, **85**, 131–142.

JUSKOWIAK, M. 1960. Antofyllit ze Szklar koło Ząbkowic Śląskich. (Anthophyllite from Szklary near Ząbkowice Śląskie. *Kwartalnik Geologiciczny*, **4**, 311–320.

KACHLÍK, V. & PATOČKA, F. 1998. Lithostratigraphy and tectonomagmatic evolution of the Železný Brod Crystalline Unit: some constraints for the palaeotectonic development of the W Sudetes (NE Bohemian Massif). *GeoLines*, **6**, 34–35.

KEMNITZ, H. & BUDZINSKI, G. 1991. Beitrag zur Lithostratigraphie und Genese der Lausitzer Grauwacken. *Zeitschrift für Geologische Wissenschaften*, **19**, 433–441.

KLARA, G. 1999. Petrographic characteristics of mafic pebbles from the selected conglomerate horizons in Lower Silesia (SW Poland). *GeoLines*, **8**, 36–37.

KORYTOWSKI, A., DÖRR, W. & ŻELAŹNIEWICZ, A. 1993.U–Pb dating of (meta)granitoids in the NW Sudetes (Poland) and their bearing on tectono-stratigraphic correlation.) *Terra Nova*, **5**, *Abstract supplement 1*, 331.

KOSSMAT, F. 1927. Gliederung des varistischen Gebirgsbaues. *Abhandlungen des Sächsischen Geologischen Landesamtes*, **1**, 1–39.

KOZDRÓJ, W. 2000. *Ewolucja geotektoniczna krystaliniku wschodnich Karkonoszy no podstawie analizy strukturalnej oraz geochemicznej charakterystyki przeobrażonych skał magnowych*. PhD thesis, Państwowy Instytut Geologiczny, Wrocław.

KRÖNER, A. & HEGNER, E. 1998. Geochemistry, single zircon ages and Sm–Nd systematics of granitoid rocks from the Góry Sowie (Owl Mts, Polish West Sudetes): evidence for early Palaeozoic arc-related plutonism. *Journal of the Geological Society, London*, **155**, 711–124.

——, HEGNER, E., HAMMER, J., HAASE, G., BIELICKI, K.-H., KRAUSS, M. & EIDAM, W. 1994*a*. Geochronology and Nd–Sr systematics of Lusatian granitoids: significance for the evolution of the Variscan orogen in east–central Europe. *Geologische Rundschau*, **83**, 357–376.

——, —— & JAECKEL, P. 1994*b*. Pb–Pb zircon ages and Nd isotopic systematics for metamorphic rocks from the Góry Sowie Block, Sudetes, Poland, and geodynamic significance. *Journal of the Czech Geological Society*, **39**, 60.

——, —— & —— 1997. Cambrian to Ordovician granitoid orthogneisses in the Polish and Czech West Sudetes Mts. and their geodynamic significance. *Terra Nostra*, **97**(11), 67–68.

——, ŠTÍPSKÁ, P., SCHULMANN, K. & JAECKEL, P. 2000. Chronological constraints on the pre-Variscan evolution of the northeastern margin of the Bohemian Massif, Czech Republic. *This volume*.

KRYZA, R., MUSZYŃSKI, A. & VIELZEUF, D. 1990. Glaucophane-bearing assemblage overprinted by greenschist-facies metamorphism in the Variscan Kaczawa complex, Sudetes, Poland. *Journal of Metamorphic Geology*, **8**, 345–355.

LANKREIJER, A. C. 1998. *Rheology and basement control on extensional basin evolution in Central and Eastern Europe: Variscan and Alpine–Carpathian–Pannonian tectonics*. PhD Thesis, Vrije Universiteit Amsterdam.

LINNEMANN, U. & BUSCHMANN, B. 1995. Die cadomische Diskordanz im Saxothuringikum (oberkambrisch-tremadocische overlap-Sequenzen). *Zeitschrift für Geologische Wissenschaften*, **23**, 707–727.

——, GEHMLICH, M., TICHOMIROWA, M. *et al.* 2000. From Cadomian subduction to Early Palaeozoic rifting: the evolution of Saxo-Thuringia at the margin of Gondwana in the light of single zircon geochronology and basin development (Central European Variscides, Germany). *This volume*.

LORENZ, V. & NICHOLLS, I. A. 1984. Plate and intraplate processes of Hercynian Europe during the late Palaeozoic. *Tectonophysics*, **107**, 25–26.

MALUSKI, H. & PATOČKA, F. 1997. Geochemistry and ^{40}Ar–^{39}Ar geochronology of the mafic metavolcanic rocks from the Rýchory Mountains complex (West Sudetes, Bohemian Massif): palaeotectonic significance. *Geological Magazine*, **134**, 703–716.

——, RAJLICH, P. & SOUČEK, J. 1995. Pre-Variscan, Variscan and early Alpine thermotectonic history of the north-eastern Bohemian Massif: $^{40}Ar/^{39}Ar$ study. *Geologische Rundschau*, **84**, 345–358.

MACINTYRE, R. M., BOWES, D. R., HAMIDULLAH, S. & OONSTOTT, T. C. 1993. K–Ar and Ar–Ar isotopic study of amphiboles from meta-ophiolite complexes, Eastern Bohemian Massif. *Proceedings of the 1st International Conference on the Bohemian Massif, Prague 1988*. Czech Geological Survey, Prague, 195–199.

MARHEINE, D., KACHLÍK, V., PATOČKA, F. & MALUSKI, H. 1999. The Variscan polyphase tectonothermal development in the South Krkonose Complex (W-Sudetes, Czech Republic). *Journal of Conference Abstracts EUG 10*, **4**, 95.

MATTE, Ph., MALUSKI, H., RAJLICH, P. & FRANKE, W. 1990. Terrane boundaries in the Bohemian Massif: results of large-scale Variscan shearing. *Tectonophysics*, **177**, 151–170.

MATTERN, F. 1996. The Elbe zone at Dresden—a Late Paleozoic pull-apart intruded shear-zone. *Zeitschrift der Deutschen Geologischen Gesellschaft*, **147**, 57–80.

MAZUR, S. 1995. Strukturalna i metamorficzna ewolucja wschodniej okrywy granitu Karkonoszy w południowej części Rudaw Janowickich i Grzbiecie Lasockim. (Structural and metamorphic evolution of the country rocks at the contact of the Karkonosze granite in the southern Rudawy Janowicki Mts. and Lasocki Range.) *Geologia Sudetica*, **29**, 31–98.

—— & JÓZEFIAK, D. 1999. Structural record of Variscan thrusting and subsequent extensional collapse in the mica schists from vicinities of Kamieniec Ząbkowicki, Sudetic Foreland, SW Poland. *Annales Societatis Geologorum Poloniae*, **69**, 1–26.

—— & KRYZA, R. 1996. Superimposed compressional and extensional tectonics in the Karkonosze–Izera Block, NE Bohemian Massif. *In*: ONCKEN, O. & JANSSEN, C. *Basement Tectonics 11*. Kluwer, Dordrecht, 51–66.

—— & PUZIEWICZ, J. 1995. Mylonity strefy Niemczy. (Mylonites of the Niemcza zone.) *Annales Societatis Geologorum Poloniae*, **64**, 23–52.

——, ALEKSANDROWSKI, P. & AWDANKIEWICZ, M. 1998. The South and East Karkonosze metamorphic complexes (Western Sudetes): a Variscan suture zone modified by extensional collapse. *Acta Universitatis Carolina—Geologica*, **42**, 304–305.

——, TURNIAK, K. & WYSOCZANSKI, R. 1999. SHRIMP zircon geochronology of the Orlica–Śnieżnik gneisses (the Sudetes, SW Poland): tectonic implications. *Terra Nostra*, **99**(1), 146.

MCKERROW, W. S. & VAN STAAL, C. R. 2000. The Palaeozoic time scale reviewed. *This volume*.

NOWAK, I. 1998*a*. *P–T–d* path for mica schists associated with eclogites in the Fore-Sudetic Block, SW Poland. *Acta Universitatis Carolina—Geologica*, **42**, 310–311.

—— 1998*b*. Polyphase exhumation of eclogite-bearing high-pressure mica schists from the Fore-Sudetic Block, SW Poland. *Geologia Sudetica*, **31**, 3–31.

O'BRIEN, P. J., KRÖNER, A., JAECKEL, P., HEGNER, E., ŻELAŹNIEWICZ, A. & KRYZA, R. 1997. Petrological and isotopic studies on Paleozoic high-pressure granulites, Góry Sowie Mts, Polish Sudetes. *Journal of Petrology*, **38**, 433–456.

OBERC, J. 1960. Tektonika Wschodnich Karkonoszy i ich stanowisko w budowie Sudetów. (Eastern Karkonosze tectonics and their position in the Sudeten structure.) *Acta Geologica Polonica*, **10**, 1–48.

—— 1968. Granica między strukturą zachodnio- i wchodniosudecką. (The boundary between the Western and Eastern Sudetic tectonic structures.) *Rocznik Polskiego Towarzystwa Geologicznego*, **38**, 203–217.

—— 1980. Early to middle Variscan development of the West Sudetes. *Acta Geologica Polonica*, **30**, 27–52.

OBERC-DZIEDZIC, T. 1999. The metemorphic and structural development of gneisses and older schists series in the Strzelin Crystalline Massif (Fore-Sudetic Block, SW Poland). *Polskie Towarzystwo Mineralogiczne—Prace Specjalne*, **14**, 10–21.

——, PIN, C., DUTHOU, J. L. & COUTURIE, J. P. 1996. Age and origin of the Strzelin granitoids (Fore-Sudetic Block, Poland): $^{87}Rb/^{86}Sr$ data. *Neues Jahrbuch für Mineralogie, Abhandlungen*, **171**, 187–198.

——, ŻELAŹNIEWICZ, A. & CWOJDZIŃSKI, S. 1999. Granitoids of the Odra Fault Zone: late to post-orogenic Variscan intrusions in the Saxothuringian zone, SW Poland. *Geologia Sudetica*, **32**, 55–71.

OLIVER, G. J. H., CORFU, F. & KROUGH, T. E. 1993. U–Pb ages from SW Poland: evidence for a Caledonian suture zone between Baltica and Gondwana. *Journal of the Geological Society, London*, **150**, 355–369.

OTAVA, J. 1998. Material evidence of granulite uplift in Variscan siliciclastics: SE Bohemian Massif. *Acta Universitatis Carolinae—Geologica*, **42**, 315.

OTAVA, J. & SULOVAKÝ, P. 1998. Detrital garnets and chromites from the Książ formation, Świebodzice depression: implications for the Variscan evolution of the Sudetes. *GeoLines*, **6**, 49–50.

PARRY, M., ŠTIPSKA, P., SCHULMANN, K., HROUDA, F., JEŽEK, J. & KRÖNER, A. 1997. Tonalite sill emplacement at an oblique plate boundary: northeastern margin of the Bohemian Massif. *Tectonophysics*, **280**, 61–81.

PATOČKA, F. & SMULIKOWSKI, W. 1997. Petrology and geochemistry of metabasic rocks of the Rýchory Mts–Rudawy Janowickie Mts complex (West Sudetes, NE Bohemian Massif). *Polskie Towarzystwo Mineralogiczne—Prace Specjalne*, **9**, 146–150.

PHILIPPE, S., HAACK, U., ŻELAŹNIEWICZ, A., DÖRR, W. & FRANKE, W. 1995. Preliminary geochemical and geochronological results on shear zones in the Izera–Karkonosze Block (Sudetes, Poland). *Terra Nostra*, **95**(8), 122.

PIN, C., MAJEROWCZ, A. & WOJCIECHOWSKA, I. 1988. Upper Paleozoic oceanic crust in the Polish Sudetes: Nd–Sr isotope and trace element evidence. *Lithos*, **21**, 195–205.

PITRA, P., BURG, J. P., SCHULMANN, K. & LEDRU, P. 1994. Late orogenic extension in the Bohemian Massif: petrostructural evidence in the Hlinsko region. *Geodinamica Acta*, **7**, 15–30.

PORĘBSKA, E. 1984. Latest Silurian and Early Devonian graptolites from the Zdanów section, Bardo Mts., Sudetes. *Annales Societatis Geologorum Poloniae*, **52**, 89–209.

PORĘBSKI, S. J. 1981. Świebodzice succession (Upper Devonian–Lowest Carboniferous; Western Sudetes): a prograding, mass-flow dominated fan-delta complex. *Geologia Sudetica*, **16**, 101–192.

—— 1990. Onset of coarse clastic sedimentation in the Variscan realm of the Sudetes (SW Poland): an example from the upper Devonian–lower Carboniferous Świebodzice succession. *Neues Jahrbuch für Geologie und Paläontologie, Abhandlungen*, **179**, 259–274.

RAJLICH, P. 1987. Variszische duktile Tektonik im Böhmischen Massiv. *Geologische Rundschau*, **76**, 755–786.

RAUCHE, H. 1992. *Spätvariszische Spannungs- und Verformungsgeschichte der Gesteine am Südwestrand der Elbezone (Östliches Saxothuringikum, Varisziden)*. PhD thesis, Ruhr Universität Bochum.

—— 1994. Kinematics and timing of the ductile to brittle transition in mylonites from the Elbe Zone–Erzgebirge Border Shear Zone (Mid-Saxonian Fault, Eastern Saxothuringian). *Journal of the Czech Geological Society*, **39**, 88–89.

SCHÄFER, E., ONCKEN, O., KEMNITZ, H. & ROMER, R. L. 2000. Upper-plate deformation during collisional orogeny; a case study from the German Variscides (Saxo-Thuringian Zone). *This volume*.

SCHEUMANN, K. H. 1937. Zur Frage nach den Vorkommen von Kulm in der Nimptscher Kristallinzone. *Mineralogische und Petrographische Mitteilunger*, **49**, 216–240.

SCHEUVENS, D. 1998. *Die tektonometamorphe und kinematische Entwicklung im Westteil der Zentralböhmischen Scherzone (Böhmische Masse)—Evidenz für variscischen Kollaps*. PhD thesis, Johann Wolfgang Goethe-Universität, Frankfurt am Main.

SCHULMANN, K. & GAYER, R. 2000. A model for an obliquely developed continental accretionary wedge: NE Bohemian Massif. *Journal of the Geological Society, London*, **157**, 401–416.

——, LEDRU, P., AUTRAN, A. *et al.* 1991. Evolution of nappes in the eastern margin of the Bohemian Massif: a kinematic interpretation. *Geologische Rundschau*, **80**, 73–92.

SMULIKOWSKI, W. 1995. Evidence of glaucophane-schist facies metamorphism in the East Karkonosze complex, West Sudetes, Poland. *Geologische Rundschau*, **84**, 720–737.

STELTENPOHL, M. G., CYMERMAN, Z., KROGH, E. J. & KUNK, M. J. 1993. Exhumation of eclogitized continental basement during Variscan lithospheric delamination and gravitational collapse, Sudety Mountains, Poland. *Geology*, **21**, 1111–1114.

ŠTÍPSKÁ, P., SCHULMANN, K. & KRÖNER, A. 1998. Role of Cambro-Ordovician rifting in Variscan collision at the NE margin of the Bohemian Massif. *Acta Universitatis Carolinae—Geologica*, **42**, 343–344.

SUESS, F. E. 1912. Die moravischen Fenster und ihre Beziehung zum Grundgebirge des Hohen Gesenkes. *Denkschrifte Österreichische Akademie Wissenschaftlichen mathematisch-naturwissenschaftliche klasse*, **88**, 541–631.

—— 1926. *Intrusionstektonik und Wandertektonik im variszischen Grundgebirge*. Borntraeger, Leipzig, 1–212.

SZCZEPAŃSKI, J. 1999. *Mikrostrukturalna i petrologiczna charakterystyka warstw z Jegłowej w krystaliniku Wzgórz Strzelińskich*. PhD thesis, Uniwersytet Wrocławski.

TAIT, J. BACHTADSE, V., & SOFFEL, H. 1994. Silurian palaeogeography of Armorica: new palaeomagnetic data from central Bohemia. *Journal of Geophysical Research*, **99**, 2897–2907.

——, SCHÄTZ, M., BACHTADSE, V. & SOFFEL, H. 2000. Palaeomagnetism of Palaeozoic palaeogeography of Gondwana and European terranes. *This volume*.

TEISSEYRE, A. K. 1975. Sedymentologia i paleogeografia kulmu starszego w zachodniej części niecki śródsudeckiej. (Sedimentology and paleogeography of the Kulm alluvial fans in the western Intrasudetic Basin (Central Sudetes, SW Poland).) *Geologia Sudetica*, **9**(2), 5–135.

TEISSEYRE, J. 1973. Skały metamorficzne Rudaw Janowickich i Grzbietu Lasockiego. (Metamorphic rocks of the Rudawy Janowickie and Lasocki Grzbiet Ranges.) *Geologia Sudetica*, **8**, 7–118.

THOMAS, U. 1990. Unterkarbonische Wildflysch-Ablagerungen im Südteil der DDR. *Zeitschrift Angewandte Geologie*, **86**, 182–184.

TIMMERMANN, H., PARRISH, R. R., NOBLE, S. R. & KRYZA, R. 2000. New U–Pb monazite and zircon data from the Sudetes Mountains in SW Poland: evidence for a single-cycle Variscan orogeny. *Journal of the Geological Society, London*, **157**, 265–268.

TURNIAK, K., MAZUR, S. & WYSOCZANSKI, R. 1998. SHRIMP zircon geochronology for gneisses of the Orlica–Śnieżnik Dome (West Sudetes, SW Poland): evidence for an Early Palaeozoic (500 Ma) plutonism and Visean (340 Ma) HT/LP metamorphic event. *Abstracts PACE Meeting 1998, Prague*.

VAN BREEMEN, O., BOWES, D. R., AFTALION, M. & ŻELAŹNIEWICZ, A. 1988. Devonian tectothermal activity in the Sowie Góry Gneiss Block, Sudetes, southwestern Poland: evidence from Rb–Sr and U–Pb isotopic studies. *Annales Societatis Geologorum Poloniae*, **58**, 3–19.

VOLLBRECHT, A., WEBER, K. & SCHMOLL, J. 1989. Structural model for the Saxothuringian–Moldanubian suture in the Variscan basement of the Oberpfalz (Northeastern Bavaria, F.R.G.) interpreted from geophysical data. *Tectonophysics*, **157**, 123–133.

WAJSPRYCH, B. 1995. The Bardo Mts. rock complex: the Famennian–Lower Carboniferous preflysch (platform)-to-flysch (foreland) basin succession, the Sudetes. *XIII International Congress on Carboniferous–Permian, Kraków, Guide to Excursion B2*. Państwowy Instytut Geologiczny, Warszawa, 23–43.

WEBER, K. & VOLLBRECHT, A. 1989. The crustal structure at the KTB drill site, Oberpfalz. *In*: EMMERMANN, R. & WOHLENBERG, J. (eds) *The German Continental Deep Drilling Program (KTB)*. Springer, Berlin, 5–36.

WENZEL, T., MERTZ, D. F. OBERHÄNSLI, R., BECHER, T. & RENNE, P. R. 1997. Age, geodynamic setting, and mantle enrichment processes of a K-rich intrusion from the Meissen massif (northern Bohemian massif) and implications for related occurrences from the mid-European Hercynian. *Geologische Rundschau*, **86**, 556–570.

Wojciechowska, I. 1995. Geotektoniczna pozycja metabazytów metamorfiku kłodzkiego (wschodnia krawędź Bohemikum, Sudety, Polska). (Geotectonic position of metabasites of the Kłodzko Metamorphic Unit (eastern edge of the Bohemikum, Sudetes, Poland).) *Acta Universitatis Wratislaviensis Prace Geologiczno-Mineralogiczne*, **50**, 65–76.

Żakowa, H. 1963. Stratigraphy and facial extents of the Lower Carboniferous in the Sudetes. *Kwartalnik Geologiczny*, **7**, 73–94.

Żelaźniewicz, A. 1990. Deformation and metamorphism in the Góry Sowie gneiss complex, Sudetes, SW Poland. *Neues Jahrbuch für Geologie und Paläontologie, Abhandlungen*, **179**, 129–157.

—— 1995. Część przedsudecka bloku sowiogórskiego. (Fore-Sudetic part of the Góry Sowie Block.) *Przewodnik LXVI Zjazdu PTG, Wrocław*, 85–109.

—— 1996. On some controversies about the geology of the West Sudetes. *Zeitschrift für Geologische Wissenschaften*, **24**, 457–465.

—— 1997. The Sudetes as a Palaeozoic orogen in central Europe. *Geological Magazine*, **134**, 691–702.

—— & Achramowicz, S. 1999. Xenoliths of Proterozoic high-*P* and/or high-*T* metapelites within the *c.* 500 Ma Izera granite, West Sudetes, SW Poland. *Journal of Conference Abstracts, EUG 10*, **4**, 95–96.

—— & Franke, W. 1994. Discussion on U–Pb ages from SW Poland: evidence for a Caledonian suture zone between Baltica and Gondwana. *Journal of the Geological Society, London*, **151**, 1049–1055.

——, Achramowicz, S., Nowak, I. & Lorenc, M. W. 1998*a*. Northern Izero–Karkonosze Block: a mode of Variscan reworking of Neoproterozoic (Cadomian?) continental crust. *Terra Nostra*, **98**(2), 172–174.

——, Cwojdziński, S., England, R. W. & Zientara, P. 1997. Variscides in the Sudetes and the reworked Cadomian orogen: evidence from the GB-2A seismic reflection profiling in southwestern Poland. *Geological Quarterly*, **41**, 289–308.

——, Dörr, W. & Dubińska, E. 1998*b*. Lower Devonian oceanic crust from U–Pb zircon evidence and Eo-Variscan event in the Sudetes. *Terra Nostra*, **98**(2), 174–176.

Znosko, J. 1981. The problem of oceanic crust and of ophiolites in the Sudetes. *In*: Narębski, W. (ed.) *Ophiolites and initialites of Northern Border of the Bohemian Massif. Guidebook of Excursion May–June 1981*, **II**, Academy of Sciences of German Democratic Republic and Polish Academy of Sciences. 3–28.

Žukalová, V. 1976. Upper Devonian stromatoporoids, foraminifers and algae in the borehole Nepasice (eastern Bohemia). *Věstník Ústredního Ústavu Geologického*, **51**, 281–284.

Zulauf, G. 1997. Von der Anchizone bis zur eklogitfazies: Angelkippte Krustenprofile als Folge der cadomischen und variscischen Orogenese im Teplá–Barrandium (Böhmische Masse). *Geotektonische Forschungen* **89**, 1–302.

—— 1994. Ductile normal faulting along the West Bohemian Shear Zone (Moldanubian/Teplá-Barrandian boundary): evidence for late Variscan extensional collapse in the Variscan Internides. *Geologische Rundschau*, **83**, 276–292.

The Baltica–Gondwana suture in central Europe: evidence from K–Ar ages of detrital muscovites and biogeographical data

Z. BELKA[1], H. AHRENDT[2], W. FRANKE[3] & K. WEMMER[2]

[1]*Institut für Geologische Wissenschaften und Geiseltalmuseum, Martin-Luther-Universität Halle-Wittenberg, Domstr. 5, D-06108 Halle, Germany (e-mail:belka@geologie.uni-halle.de)*

[2]*Institut für Geologie und Dynamik der Lithosphäre, Universität Göttingen, Goldschmidtstr. 3, D-37077 Göttingen, Germany*

[3]*Institut für Geowissenschaften und Lithosphärenforschung, Justus-Liebig-Universität-Giessen, Senckenbergstr. 3, D-35390 Giessen, Germany*

Abstract: The Lysogory Unit, the Malopolska Massif and the Upper Silesian Massif in southern Poland are parts of a mosaic of contrasting crustal fragments separating the old Precambrian crust of the East European Platform (EEP) from the Phanerozoic mobile belts of western Europe. The geological histories of these blocks are markedly different. They have been regarded as integral parts of the palaeocontinent of Baltica (that is, the EEP), mostly because of presence of fossils typical for the Baltic realm, although geophysical and geological data and some faunal elements rather suggest linkages to the Peri-Gondwana plates. To provide additional constraints for the plate tectonic affinity of these blocks detrital muscovite grains extracted from Cambrian and Devonian clastic rocks were dated by the K–Ar method. The K–Ar cooling ages show a very complex provenance pattern for clastic material in Cambrian time. Combined with the biogeographical constraints, the new provenance data apparently show that the blocks of Lysogory, Malopolska, and Upper Silesia are in fact crustal fragments derived from the Gondwana margin, not displaced parts of the East European Craton. Thus, the Teisseyre–Tornquist Line (that is, the edge of the EEP) is the Baltica–Gondwana suture in central Europe. The combined data reveal an accretionary scenario in which the Malopolska Block was the first Gondwana-derived microplate that accreted to the margin of Baltica.

The southwestern margin of the East European Platform (EEP) is bordered by a crustal domain that is one of the most enigmatic parts of the European lithosphere. This domain separates the >850 Ma Precambrian crust of the EEP from the Variscan and Alpine mobile belts of western Europe. Previously called as 'the Epicaledonian Platform' or 'the North German–Polish Caledonides' (e.g. Ellenberger & Tamain 1980; Ziegler 1981; Franke 1989), it is currently termed the Trans-European Suture Zone (TESZ). Berthelsen (1993) introduced this term to stress the tectonic character of the domain and to encompass the pre-Variscan crust amalgamated in Early Palaeozoic time when Laurentia, Baltica, and Gondwana-derived terranes collided (see Blundell *et al.* 1992). The TESZ spans the area from the edge of the EEP (that is, the Teisseyre–Tornquist Line in central Europe) in the east to the Avalon Terrane in the west and includes basement blocks underlying the Rheno-Hercynian fold-belt in the south. Because Precambrian basement and Palaeozoic rocks in this part of Europe are mostly deeply buried under thick Mesozoic and Cenozoic cover, the structural framework and the crustal evolution of the TESZ are poorly known. Information from deep seismic experiments and boreholes shows a complex structure of the TESZ that constitutes a collage of various crustal blocks (e.g. Pozaryski 1975; Guterch *et al.* 1986, 1994; Berthelsen 1992).

In the past, however, the lack of palaeomagnetic studies and isotopic information from the basement has left plenty of room for speculations, and various models to account for the evolution of the TESZ were formulated (for reviews, see Berthelsen (1993), Dadlez (1995) and Vidal & Moczydlowska (1995)). Inspired by modern concepts of plate tectonics, a terrane hypothesis is currently favoured.

From: Franke, W., Haak, V., Oncken, O. & Tanner, D. (eds). *Orogenic Processes: Quantification and Modelling in the Variscan Belt.* Geological Society, London, Special Publications, **179**, 87–102. 0305-8719/00/$15.00

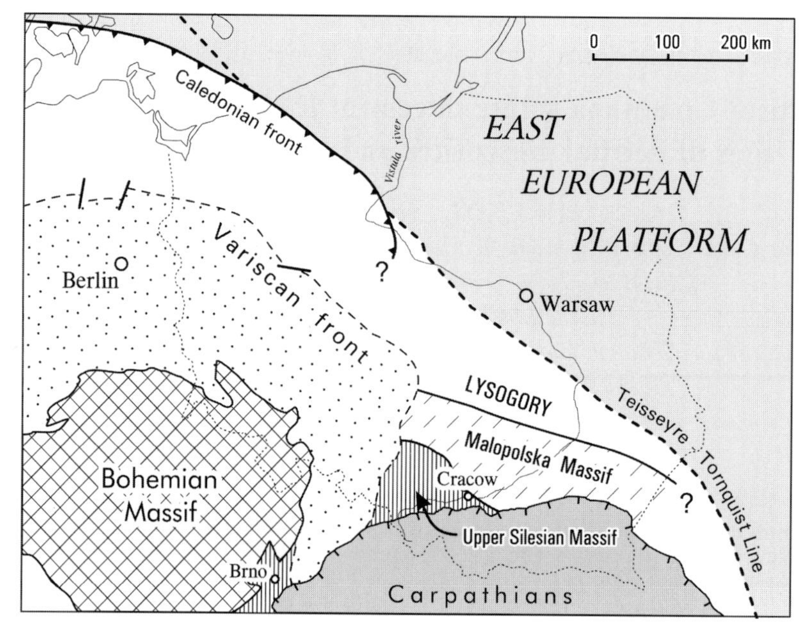

Fig. 1. Simplified structural map of central Europe showing the crustal units in the Variscan foreland of southern Poland. Dotted line indicates the Polish border.

Terranes have been differentiated on the basis of contrasting lithofacies characteristics, stratigraphy and tectonic development of the Lower Palaeozoic rocks. There are, however, strongly diverging opinions concerning the number of terranes, their outline and kinematics (e.g. Pozaryski *et al.* 1992; Franke 1995).

In southern Poland, where the post-Palaeozoic cover is relatively thin, extensive data originating from outcrops and hundreds of boreholes proved the occurrence of at least three fault-bounded crustal units. From the margin of the EEP outwards (Fig. 1), these are the Lysogory Unit, the Malopolska Massif and the Upper Silesian Massif. They are built up of fragments of continental lithosphere delimited by deep fractures recognizable in the relief of the Moho. These units followed individual tectonic and sedimentary evolutions during Early Palaeozoic time (Fig. 2). With the exception of narrow zones along the border faults, they all are characterized by a very low thermal overprint of Phanerozoic rocks (Belka 1990, 1993; Belka & Siewniak 1996). Each of them potentially is a terrane, but the time when their present positions were attained is poorly defined. The accretionary history is still a matter of controversy, as the palaeomagnetic data allow various interpretations. They suggest either a large-scale mobility of the Malopolska and Upper Silesian Massifs in Variscan time or a more 'stationary' model, positioning Upper Silesia in its present position already during Early Palaeozoic time (for discussion, see Lewandowski (1995) and Nawrocki (1995)). Traditional palaeogeographical interpretations have regarded these units as representing sedimentary realms along the Tornquist margin of Baltica at least since the beginning of the Cambrian period (e.g. Dadlez 1983; Bergström 1984; Vidal & Moczydlowska 1995). This was primarily because of records of the Early Cambrian trilobites diagnostic of the Baltic zoogeographical province in Malopolska and in Upper Silesia (Orlowski 1975, 1985). Recently, Dadlez (1995) and Pharaoh *et al.* (1996) interpreted the Lysogory Unit and the Malopolska Massif as fragments of Baltica's Precambrian crust. On the other hand, however, there are records of Cambrian brachiopods and trilobite trace fossils of certain 'Gondwanan' affinity from Lysogory (Seilacher 1983; Jendryka-Fuglewicz 1992, 1998). It should be stressed here that the Holy Cross Mountains (HCM), an area famous for numerous outcrops of fossiliferous Palaeozoic rocks, is located along the contact between Lysogory and Malopolska and comprises the marginal parts of both terranes. The two regions of the HCM differ significantly in facies development, stratigraphy, and geotectonic history. Unfortunately, this bipartite subdivision of the Holy Cross Mountains was overlooked in some palaeogeographical reconstructions (e.g. Cocks

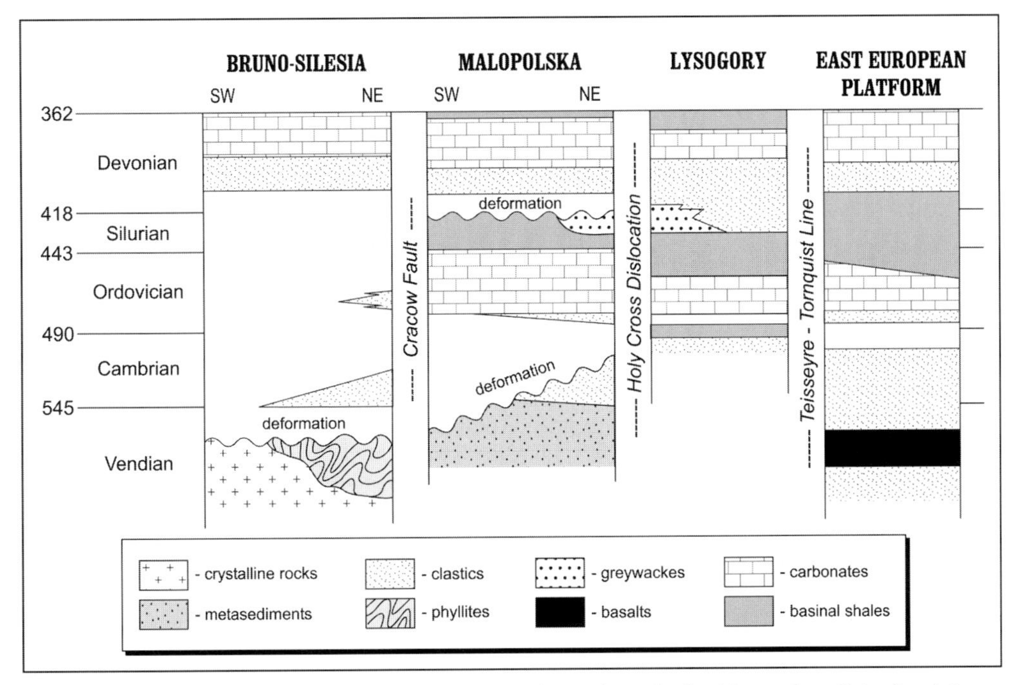

Fig. 2. Stratigraphic columns of the crustal units forming the Variscan foreland in southern Poland and the stratigraphic succession for the marginal part of the East European Platform in the Warsaw region. (Note the differences in occurrence of stratigraphic gaps (white spaces) and deformation phenomena.) The Upper Silesian Massif constitutes the NE part of Bruno-Silesia. The outline of stratigraphy of Malopolska does not include information from the San Block.

et al. 1997; Cocks & Fortey 1998) and faunal records from the Holy Cross area were considered as biogeographical data coming from a single palaeoplate.

To recognize possible plate tectonic affinities of the Palaeozoic terranes within the TESZ of southern Poland, K–Ar age determinations on detrital muscovites were carried out. Fitch *et al.* (1966) made the first attempt to use detrital micas for palaeogeographic studies in England. Horstmann (1987) and Horstmann *et al.* (1990) were the first who applied the method of dating detrital white micas by the K–Ar method in foreland basins for the indirect reconstruction of orogenic processes for an orogen itself on a regional scale. Whereas the above investigations were carried out on the Damara Orogen in Namibia, Renne *et al.* (1990) presented provenance studies by Ar/Ar laser-probe dating on muscovites and biotites from the Eocene Montgomery Creek Formation of northern California. Welzel (1991), Neuroth (1997) and Schäfer *et al.* (1997) followed with their investigations in the Mid-European Variscides. Welzel (1991) in particular, could show by intensive methodological studies the convincing usefulness and geological significance of provenance studies by dating detrital white micas with the conventional K–Ar method in sedimentary environments. She clearly proved that only white micas, in contrast to biotites, retain their original age information during erosion, transportation and sedimentation taking into account a careful preparation (see sample preparation and analytical methods).

In this paper, we provide the first provenance data from the Cambrian and Devonian clastic rocks of Lysogory, Malopolska and Upper Silesia. These data, combined with a summary of the biogeographical data, allow us to propose an accretionary scenario for these terranes that solves the seemingly incompatible faunal records.

Regional geological setting

The overall geological setting of Poland has been discussed at length elsewhere and will not be reviewed here. We focus attention upon the most fundamental features that characterize and distinguish different crustal blocks. There is no doubt that the marginal part of the EEP in Poland was physically a part of the palaeocontinent of Baltica. Its stratigraphic succession may serve therefore as a starting point and a basis for comparison of the Late Proterozoic and Early Palaeozoic geology (including the Lower Devonian succession) across crustal blocks of the TESZ in southern Poland (Fig. 2).

The East European Platform

The basement of the EEP is not exposed in Poland but it has been penetrated by more than 100 deep boreholes. It belongs to the south-western marginal part of the Fennoscandian crustal segment of the East European Craton. The Fennoscandian crust constitutes a collage of continental and arc-related terranes, which were assembled during a sequence of Proterozoic collisional events between 2.0 and 0.9 Ga (Bogdanova *et al.* 1994, 1996). Crustal growth was achieved by gradual accretion of new crust in the northwesterly direction. In its southern part, the Fennoscandian segment is composed of several SW–NE-trending belts. They are characterized by granulite or amphibolite facies metamorphism. Most workers trace these belts to the southwest, into the Polish territory, where they are truncated by the Teisseyre–Tornquist Line (TTL). This interpretation, however, is not satisfactorily constrained. Lack of modern geochronological data from Poland constitutes a general problem in correlation of crustal units distinguished in the Polish part of the East European Craton (Ryka 1982) with those of the adjacent countries.

The crystalline basement rocks are unconformably overlain by undeformed and generally flat-lying sedimentary cover. The cover rocks are not metamorphosed, generally displaying only a moderate thermal overprint. Sediment thickness varies significantly from 360 m in the east, close to the Belorussian border, to more than 5600 m (including up to 2000 m of Palaeozoic rocks) along the TTL. Remarkable differences in thickness and stratigraphy of the sequence, both corresponding to the differentiated basement topography, are also observed in the NW–SE direction. The sedimentary cover includes rocks ranging from late Precambrian to Quaternary age, but it contains numerous stratigraphic gaps. It starts with a Neoproterozoic succession of polymictic clastic rocks intercalated with a complex of volcanic and volcaniclastic rocks (Fig. 2). The onset of deposition has been inferred to occur during late Riphean time (Aren & Lendzion 1978). The volcanic rocks, which form much of the lower portion of the succession, are extensive basaltic eruptions that appear to be rift related. Recent U–Pb zircon dating of tuffs by Compston *et al.* (1995) gives an age of 551 ± 4.0 Ma. Acritarchs and other microfossils found in the succeeding clastic portion of the succession are characteristic of late Vendian time (Moczydlowska 1991).

The Upper Vendian and the overlying Cambrian siliciclastic rocks represent a transgressive–regressive depositional system developed on an epicontinental shelf, in a shallow-water area influenced by tidal currents and storms. A general facies model, as proposed by Jaworowski (1997), shows a pattern that roughly parallels the TTL, with a zone of tidal coastal sands in the east and a shelf mud zone in the west (Fig. 3). Thus, the open sea was located west of the TTL and the land in the east. The facies and thickness trends point to the westward transport of the clastic material from the surrounding basement highs in the east. Provenance of clastic material from the craton interior can also be deduced from the composition of sandstones that are very mature, usually pure quartz arenites (Sikorska 1988). White mica content is generally very low, *c.* 1%. The Cambrian sequence of the EEP is very well biostratigraphically constrained and includes predominantly Lower and Middle Cambrian rocks. The Upper Cambrian rocks, up to 1 m thick, are present only in the Peri-Baltic Depression, and are separated from the Middle Cambrian rocks by a stratigraphic gap (Lendzion 1983).

The Ordovician and the Silurian rocks in the Polish segment of the EEP are marine and show a general facies pattern similar to that of the

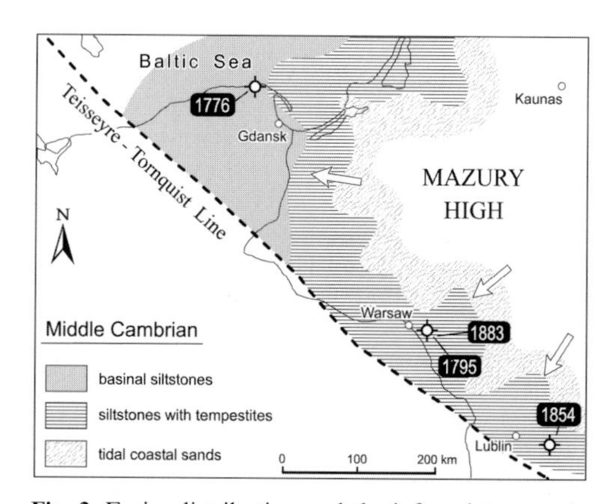

Fig. 3. Facies distribution and the inferred transport directions of clastic material in the Middle Cambrian sequence of the marginal part of the East European Platform (modified from Jaworowski (1997)). Numbers indicate K–Ar cooling ages of detrital muscovites present in the Middle Cambrian clastic rocks of the boreholes Zarnowiec IG-1, Okuniew IG-1 and Lopiennik IG-1. (For more details on samples and localities, see Table 1.)

Cambrian succession, with littoral deposition in the east and a basinal realm in the west, along the TTL. The difference, however, is that in the shallow-water environments extensive carbonate deposition prevailed and no input of clastic material from the craton interior into the shelf can be observed (for review, see Modlinski (1982) and Teller (1997)). The basinal and open-shelf sediments are therefore mainly represented by shales, often with numerous graptolites. This stable facies pattern persisted until the end of Silurian time, but during Ludlow time subsidence increased rapidly in areas along the margin of the EEP and fine-grained clastic material entered the basinal realm. According to Jaworowski (1971), the rocks represent distal turbidites deposited from currents that were generated somewhere to the west, outside of the EEP, and travelled eastwards onto its margin.

The Silurian–Devonian boundary occurs in a conformable marine succession. The Lower Devonian succession reveals a distinct regressive trend, from open-marine fine-grained clastic and carbonate rocks to lagoonal and terrestrial coarse-grained sandstones interpreted as an equivalent to the OldRed facies. In the most areas of the EEP, however, the Lower Devonian rocks have been eroded. Milaczewski (1981), who studied this succession between Warsaw and the Ukrainian border, postulated two local source area for the clastic material, one located to the northeast and another to the southwest.

The Lysogory Unit

The geology of this unit, named after a small mountain range in the Holy Cross Mountains (HCM), is based primarily on data from exposures in the northern part of this region. There is only scant information from boreholes and geophysical surveys. The tectonostratigraphic term *Lysogory* and another term also used in the literature, the *Northern Holy Cross Mountains*, are synonymous. The unit is clearly delimited only to the south, where the Holy Cross Dislocation separates it from the adjacent Malopolska Massif (Figs 2, 4 and 5). Its extent to the west, north, and southeast is still a matter of debate. Recent geophysical soundings seem to indicate that the Lysogory Unit constitutes a very narrow zone, only about 30 km wide, which is bordered to the northeast by a prominent, almost vertical deep fault (Semenov *et al.* 1998). At the moment, it is not clear whether the Lysogory Unit is directly in contact with the crust of the East European Platform or the block adjacent to the Lysogory Unit possibly represents another displaced terrane.

The crystalline basement of the Lysogory Unit has never been reached in boreholes. Geophysical data predict that it occurs at depths of 6–7 km (Semenov *et al.* 1998). The sedimentary succession of the Unit, in its lithological and facies development, displays some marked similarities to that of the EEP margin (Fig. 2). It

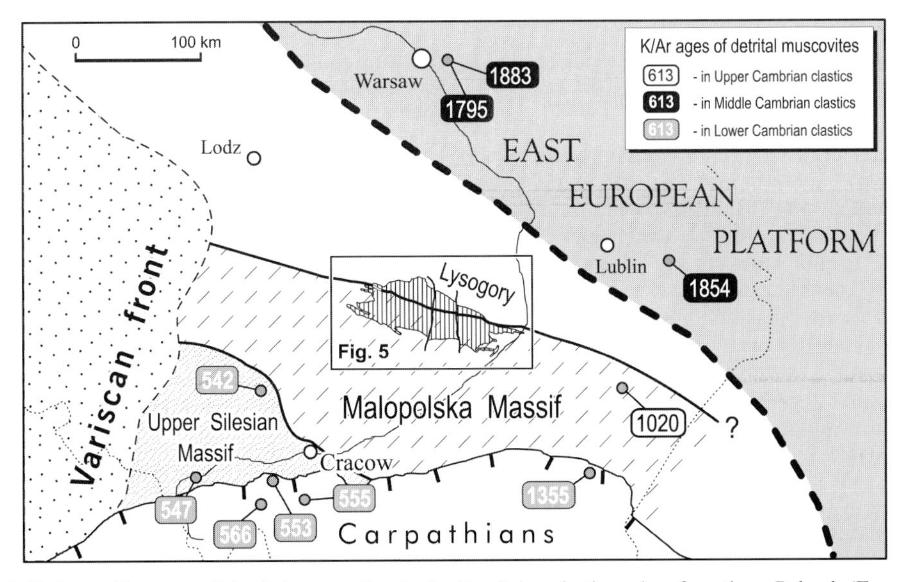

Fig. 4. K–Ar cooling ages of detrital muscovites in the Cambrian clastic rocks of southern Poland. (For more details on samples, see Table 1.) Data from the Holy Cross Mountains (boxed area) are presented in Fig. 5.

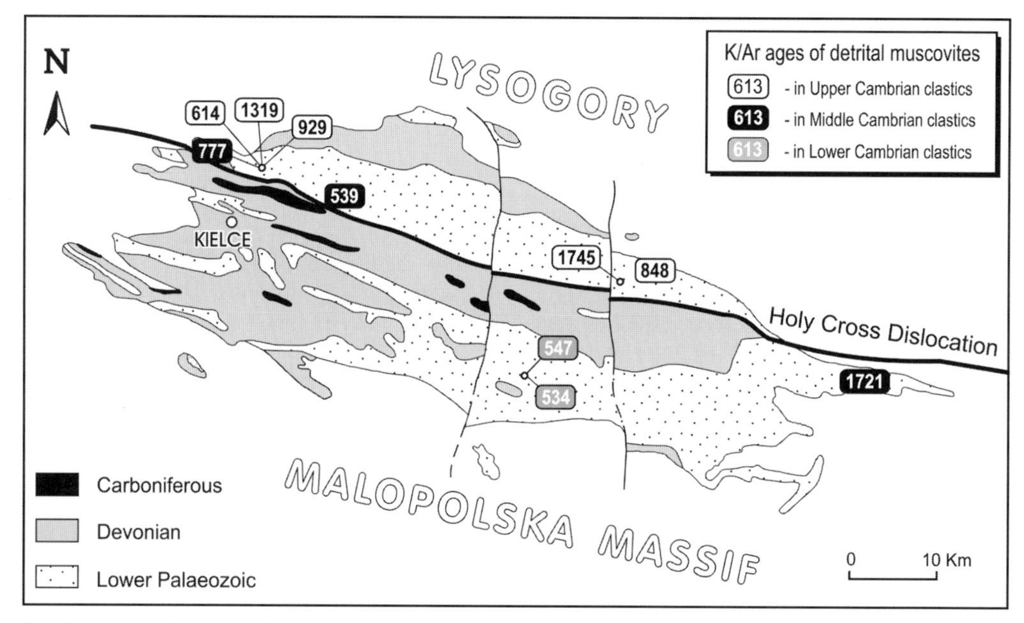

Fig. 5. K–Ar cooling ages of detrital muscovites in the Cambrian clastic rocks of the Holy Cross Mountains. (For more details on samples and localities, see Table 1.)

was, however, gently folded during Variscan deformation and, in addition, the coeval strata are thicker, at least as much as twice, as a rule. The most prominent clastic units include the Cambrian, the Upper Silurian and the Lower Devonian units. The contacts within the pre-Permian stratigraphic column are conformable (Fig. 2). The succession starts with up to 1800 m of Middle Cambrian to lower Tremadoc clastic rocks, which were deposited in a shallow-water offshore environment (Kowalczewski 1995). These rocks are predominantly sandstones yielding a very mature composition but detrital micas are extremely frequent and omnipresent in all fine-grained intercalations. Ordovician and much of Silurian time is represented by basinal graptolite shales, whereas the Upper Ludlow and the Přídolí units consist of up to 1500 m series of fine-grained greywackes with interbeds of volcaniclastic rocks. Sedimentary structures of these greywackes suggest deposition from turbidity currents. It is important to note that a similar complex of greywackes is also present in the adjacent, marginal part of the Malopolska Massif (Fig. 2). The Silurian–Devonian transition in the Lysogory Unit occurs within a complex of red polymictic clastic rocks, accumulated mainly in alluvial fans that prograded directly into a shallow marine bay (Kowalczewski *et al.* 1998). A regionally extensive siliciclastic deposition is characteristic for the entire Lower Devonian sequence. Fauna and sedimentary structures indicate an upward evolution from more continental to predominantly shallow-marine clastic rocks deposited in near-shore, storm-dominated settings (Szulczewski 1995).

The Malopolska Massif

The Malopolska Massif may be divided into two parts that exhibit individual tectonic and sedimentary evolution during Early Palaeozoic time. The western part, which we consider as the Malopolska Massif *sensu stricto*, is characterized by the pre-Ordovician deformation of Cambrian and Precambrian rocks (Fig. 2). The eastern part, which we propose to call the *San* Block, after the river that drains this area, displays a continuity in the development of the Lower Palaeozoic succession. Unfortunately, we are not able to trace precisely the tectonic boundary between these units at present. The features characteristic of the San Block can be identified in the area close to the Polish–Ukrainian border. The Malopolska Block is clearly delimited to the north and the south by the Holy Cross Dislocation and the Cracow Fault, respectively (Figs 2 and 5).

The crystalline basement of the Malopolska Block is unknown. Deep boreholes reached, in many places below the Palaeozoic sequence, a succession containing a wide spectrum of polymictic clastic rocks. Kowalski (1983) estimated its thickness to be >3000 m. These rocks are very weakly metamorphosed. A thermal overprint seems to result from deep burial and was promoted by the immature composition of clastic material. The background lithology is dominated by shales and siltstones with intercalations of greywackes and conglomerates related to turbidites and debris flows. These coarse-grained clastic deposits vary greatly in their composition. Clastic material includes acid and basic magmatic rocks, lithoclasts of sediments, and fragments of metamorphic rocks (Kowalczewski 1990). The highly immature detritus, sedimentary features, the extreme thickness, and presumably also a high rate of sedimentation suggest in our opinion deposition within a forearc–trench system. A relative large amount of arkosic material together with the presence of a both volcaniclastic and plutoniclastic detritus point additionally to provenance from uplifted arc massifs located along a continental margin. The age of the succession is poorly constrained. Rare and badly preserved acritarchs are predominantly indicative of a Vendian age. Compston *et al.* (1995) provided the first numerical age of 549 ± 3 Ma for tuffs present close to the top of the succession in the central part of the Malopolska Massif. More to the northeast, where parts of the succession are exposed in the southern Holy Cross Mountains, the age ranges up to Mid-Cambrian time. This youngest portion is not metamorphosed and locally exhibits a shallowing-upward trend associated with a much more mature clastic lithology. Consequently, white detrital micas are almost absent in the Lower and Middle Cambrian succession of the southwestern HCM.

It is important to stress that the entire Precambrian to Middle Cambrian succession of the Malopolska Massif is folded and discordantly overlain by Lower Ordovician rocks (Fig. 2). A tendency can be observed in the Malopolska Massif, moreover, that from northeast to southwest the rocks underlying the Ordovician sequence are progressively older and more highly metamorphosed. This demonstrates that the whole block of Malopolska was additionally tilted before Ordovician time and erosion removed much of the succession in the western part of the massif, close to the Cracow Fault.

The Ordovician units represent a transgressive sequence, with offshore sandstones at the base followed by more open-marine carbonate rocks and shales with graptolites. Locally, the sequence is extremely condensed and fossiliferous (Dzik & Pisera 1994). In places where argillaceous facies dominates, it is up to 150 m thick. The Silurian sequence is developed in a monotonous facies of graptolites shales. In the southern HCM, it lithologically resembles the Silurian sequence of Lysogory, but is generally thinner. Close to the Holy Cross Dislocation, however, the thickness increases remarkably and the Ludlow greywackes are about 1200 m thick.

The Lower Devonian sequence comprises clastic rocks that rest discordantly on various Lower Palaeozoic complexes (Fig. 2). Its basal surface truncates locally the substrate down to Lower Cambrian rocks (Szulczewski 1995). The rocks are chiefly very mature sandstones accumulated in alluvial and nearshore marine environments (Tarnowska 1988). In the proximity of the Holy Cross Dislocation, the Lower Devonian sequence shows a complex lithostratigraphic framework and there are several alternative explanations for the depositional character of particular complexes (e.g. Malec 1993; Szulczewski 1995; Kowalczewski *et al.* 1998). It is, however, universally accepted that the marine Lower Devonian succession oversteps the Holy Cross Dislocation.

The Precambrian and Lower Palaeozoic rocks of the San Block are known from the subsurface only. Boreholes penetrated a Precambrian basement composed of folded clastic rocks that have been affected by low-grade metamorphism. The succession is dominated by massive sandstone beds having features characteristic of deposition in a flysch basin. According to Karnkowski & Glowacki (1961), the basement is discordantly overlain by Cambrian clastic rocks. Detailed sedimentological studies are lacking but our preliminary analysis reveals the presence of shallow-water clastic packages in the lower part of the Cambrian sequence. Recent investigation using acritarchs demonstrates that these rocks were only moderately heated (Jachowicz 1998). Acritarchs also prove the presence of Lower, Middle and Upper Cambrian units in the sedimentary cover of the San Block. The stratigraphic framework, however, is not entirely clear because some clastic sequences previously attributed to Precambrian time yield Early Cambrian acritarchs. The notable difference between the Malopolska and the San Block is that the Ordovician and Silurian units in the latter area lie conformably on the Cambrian succession. Clastic and carbonate rocks at the base of the Ordovician sequence are only a few metres thick and followed by deep-water shales and siltstones that constitute the major part of the Ordovician and the Silurian succession. These rocks contain a rich graptolite fauna, the composition of which differs from those of the Malopolska Massif and the Lysogory Unit (Porebska, pers. comm.). Unlike the typical features of graptolite shales, those in the San Block are characterized by a remarkably high content of detrital micas.

The Upper Silesian Massif

Although the basement of the Upper Silesian Massif (USM) and the pre-Middle Devonian portion of its sedimentary cover are not exposed, the stratigraphic framework of this crustal block is well known as a result of extensive subsurface data (for a summary, see Bula *et al.* (1997)). It is now generally accepted that the USM constitutes the northern part of a larger block; the Bruno-vistulicum (in this paper, simply 'Bruno-Silesia' as a palaeogeographical element). The north-eastern boundary of the USM is delineated with great precision; this is the Cracow Fault, which separates this block from the Malopolska Massif (for details, see Belka & Siewniak (1996) and Bula *et al.* (1997)). Bruno-Silesia possesses an extremely thick sedimentary cover which rests on Precambrian basement of Cadomian age. The basement includes both the intrusive rocks and their metamorphic roof (Dudek 1980, 1995). A distinct polarity in the geochemical signatures of the plutonic rocks suggests that the Bruno-Silesian basement represents a fragment of an active continental margin (Finger *et al.* 1995, this volume). The Cadomian basement of the USM is discordantly overlain by a thick sequence of Cambrian clastic rocks (Fig. 2). These rocks are not deformed and display only a moderate thermal overprint (Belka 1993). The sequence varies strongly in thickness and stratigraphy (Bula & Jachowicz 1996). Sedimentary structures and trace fossils indicate deposition in a tide-dominated shallow offshore facies that grades upwards into a moderately deep offshore facies. The stratigraphic gap comprising at least Ordovician and Silurian time is a regional feature observed in the area of the USM (Fig. 2) with the exception of its northern margin, where marine Ordovician rocks were discovered (Gladysz *et al.* 1990). The Lower Devonian succession is very thin, generally less than 40 m in thickness, and poorly biostratigraphically constrained. It consists of both alluvial and shallow-water clastic rocks that truncate the Lower Palaeozoic rocks. There is, however, no angular unconformity between the Devonian and the Lower Palaeozoic units.

Biogeography

In Cambrian time, the palaeocontinent of Baltica was geographically isolated and thus the benthic faunas are different from the faunas of contemporary continents elsewhere. Among components showing this individual biotic evolution of Baltica, trilobites and brachiopods are the most important. At generic level, however, some links appear to occur between Baltica and Laurentia (Brasier 1989). The Early Cambrian trilobite faunas of Baltica are characterized by largely endemic olenellid genera such as *Holmia, Kjerulfia* and *Schmidtiellus,* and ellipsocephalids, a second group, typical of this biogeographical realm (Bergström 1984). These trilobites are widely distributed in the Baltoscandian region and in the Polish segment of the of EEP (Lendzion 1983), where they are associated with inarticulate brachiopods represented exclusively by forms that have phosphatic shells. The diversity of the brachiopod assemblages decreases in the southeasterly direction, along the TTL (Jendryka-Fuglewicz 1992, 1998). This trend may be related to climate and latitude.

By contrast, the Middle to Upper Cambrian succession of the Lysogory Unit includes fossils that are unknown from the palaeocontinent of Baltica. Inarticulate brachiopods yield strong affinity to faunas known from Avalonia (Jendryka-Fuglewicz, pers. comm.). A very rich assemblage of trilobite trace fossils is identical to those distributed throughout Gondwana and the Peri-Gondwanan microplates (Seilacher 1983). The Ordovician benthic fauna of the Lysogory Unit is not yet analysed in detail. Preliminary data indicate the presence of cosmopolitan forms amongst brachiopods.

The Malopolska Block reveals a specific faunal succession in the Lower Palaeozoic succession. Records of Baltic olenellid trilobites are known from the Lower Cambrian sequence (Orlowski 1985). The fauna is distinct and endemic at species level. In contrast, the Cambrian brachiopod faunas of Malopolska show Avalonian affinities and are dominated by forms with calcitic shells. In the Lower Cambrian sequence, for instance, there is only one Baltic species, *Westonia bottnica,* present in this Avalonian assemblage. A progressive migration of Baltic brachiopods to Malopolska can be observed during the Middle Cambrian sequence (Jendryka-Fuglewicz 1998). The Ordovician faunas, which are perfectly documented in the southern part of the Holy Cross Mountains (Dzik *et al.* 1994), belong essentially to the Baltic province, despite enlarged endemicity of ostracodes (Olempska 1994) and some links to other continents amongst conodonts (Dzik 1989).

There are only a few records of Early Palaeozoic benthic fossils in Upper Silesia. Orlowski (1975) reported the occurrence of the

Early Cambrian trilobites typical of the Baltic realm. Unfortunately, the associated brachiopod fauna has not been described until now. The conodont fauna recovered from the Middle Ordovician clastic rocks suggest, as do Cambrian trilobites, that the area was positioned within the Baltic province. It should be noted, however, that in the past, Ordovician conodont assemblages from the southwestern margin of Malopolska (vicinity of Myszkow) were regarded incorrectly as faunas of the Upper Silesian Massif (e.g. Dzik 1989).

Sample preparation and analytical procedure for K–Ar dating

K–Ar measurements were performed on detrital muscovites extracted from sandstones and siltstones of Cambrian and Devonian age. The majority of samples were taken from boreholes that pierced the Palaeozoic rocks deep in the subsurface. Sample information is presented in Table 1. After cleaning, the samples were passed twice through a jaw crusher. The loosened grain fabric was fractionated by dry sieving. The grain size with the highest amount of micas visible macroscopically was further treated to obtain at least 120 mg of pure muscovite for analysis. The micas were then enriched by applying the 'mica jet' method (Horstmann 1987). Subsequently, they were treated by different separation procedures, e.g. separation by grain shape on a self-constructed dry shaking device and/or separation by magnetic susceptibility with a Frantz magnetic separator. The muscovite concentrate was cleaned under a binocular microscope by handpicking, to prevent contamination by organic particles or unwanted minerals. Following the hand selections the muscovite concentrate was rubbed with alcohol in a porcelain mortar to remove weathered grain margins and then sieved using an 80 μm disposable sieve. A final inspection followed under a binocular microscope. After this procedure the material for analysis comprised only the fresh cores of the muscovite grains.

The argon isotopic composition was measured in a Pyrex glass extraction and purification line coupled to a VG 1200 C noble gas mass spectrometer operating in static mode. The amount of radiogenic ^{40}Ar was determined by isotope dilution method using a highly enriched ^{38}Ar spike from Schumacher, Bern (Schumacher 1975). The spike wass calibrated against the biotite standard HD-B1 (Fuhrmann *et al.* 1987). The age calculations were based on the constants recommended by the IUGS quoted by Steiger & Jäger (1977). Potassium was determined in duplicate by flame photometry using an Eppendorf Elex 63/61. The samples were dissolved in a mixture of HF and HNO_3 according to the technique of Heinrichs & Herrmann (1990). CsCl and LiCl were added as ionisation buffer and internal standard, respectively. The analytical error for the K–Ar age calculations is given with a 95% confidence level (2σ).

For detrital minerals, such as the muscovites in this study, sampling of deeply buried rocks may have a disadvantage because the mineral grains could be affected by burial diagenetic alteration. In all cases the sampled rocks were mature or slightly supramature. The highest thermal maturity we observed was in the Lower Cambrian rocks of Upper Silesia. However, they were never heated to temperatures over 180–190 $C (Belka 1993). Therefore, all obtained ages were interpreted in the sense of cooling ages of the source areas.

Results

During collection of samples we noticed striking variation in the content of detrital muscovites in the Cambrian sedimentary rocks of Poland. The tectonic boundaries between Lysogory, Malopolska and Upper Silesia are clearly visible in the distributional pattern of detrital micas as abrupt changes in the concentration of detrital mica grains in the Cambrian sequences. In the successions of Lysogory and Upper Silesia, for instance, detrital micas are large and very abundant. They are, however, very rare and extremely small in the Cambrian rocks of Malopolska. This is why we were not able to collect any samples from the southwestern Holy Cross Mountains that met requirements for K–Ar analyses. The Cambrian rocks of the EEP are also characterized by a very low content of detrital muscovites. We found only a few horizons within the Middle Cambrian units with enough material for K–Ar dating.

Table 1 summarizes the results of the K–Ar dating. They reveal a very complex provenance pattern for clastic material of the Cambrian succession. Detrital muscovites extracted from the Middle Cambrian sandstones of the EEP show a range of K–Ar cooling ages from *c.* 1775 to 1883 Ma, with approximately ± 42 Ma uncertainties. This material must have been derived from the Svecofennian crust. In one of the investigated boreholes (Okuniew IG-1), Dörr *et al.* (1998) dated migmatitic gneisses underlying the Cambrian succession and obtained a U–Pb zircon age of 1800 ± 3 Ma.These Svecofennian rocks, however, were no longer exposed at the

Table 1. *Summary of K–Ar cooling age data for detrital muscovites from Cambrian and Devonian clastic rocks of southern Poland*

No.	Sample locality	Depth (m)	Stratigraphy	Age (Ma)
Boreholes				
East European Platform				
1	Ciecierzyn	4118	Lower Devonian	415.0 ± 11.2
2	Lopiennik IG-1	1821	Lower Devonian	437.5 ± 9.8
3	Lopiennik IG-1	4487	Middle Cambrian	1853.7 ± 42.3
4	Okuniew IG-1	3662	Middle Cambrian	1882.6 ± 42.8
5	Okuniew IG-1	3695	Middle Cambrian	1795.4 ± 41.1
6	Zarnowiec IG-4	2740	Middle Cambrian	1775.6 ± 41.1
Malopolska				
7	Kanczuga 27	1540	Lower Cambrian?	1355.3 ± 29.5
8	Kostki Male 2	2690	Lower Devonian	419.9 ± 12.0
9	Krasne 27	2422	Lower Devonian	432.3 ± 10.0
10	WB-44	207	Lower Devonian	518.3 ± 10.7
11	Wola Obszanska 9	1106	Upper Cambrian	1019.7 ± 20.6
Upper Silesia				
12	Borzeta IG-1	3057	Lower Cambrian	555.2 ± 11.5
13	Goczalkowice IG-1	2745	Lower Devonian	554.8 ± 11.3
14	Goczalkowice IG-1	2977	Lower Cambrian	546.9 ± 12.5
15	Klucze 1	1940	Lower Cambrian	541.9 ± 11.0
16	Potrojna IG-1	3359	Lower Cambrian	566.5 ± 13.0
17	Wysoka 3	2456	Lower Cambrian	553.1 ± 10.6
Pomerania				
18	Unislaw IG-1	5334	Givetian	426.2 ± 10.0
Exposures				
Lysogory				
19	Bukowa Quarry	–	Emsian	467.0 ± 10.2
20	Podole Quarry	–	Emsian	425.6 ± 9.7
21	Swietomarz	–	Givetian	431.9 ± 8.7
22	Waworkow Quarry	–	Upper Cambrian	848.4 ± 19.4
23	Wisniowka Quarry	–	Middle Cambrian	777.1 ± 22.9
24	Wisniowka Quarry	–	Upper Cambrian	613.7 ± 12.6
25	Wisniowka Quarry	–	Upper Cambrian	1319.1 ± 52.1
26	Wisniowka Quarry	–	Upper Cambrian	929.4 ± 23.6
27	Jurkowice	–	Upper Cambrian	1745.3 ± 35.3
28	Wymyslona	–	Middle/Upper Cambrian	539.1 ± 14.9
Malopolska				
29	Sandomierz E	–	Middle Cambrian	1721.1 ± 37.9
30	Napekow Quarry	–	Emsian	423.3 ± 9.7
31	Kedziorka	–	Lower Cambrian	534.1 ± 18.6
32	Kedziorka	–	Lower Cambrian	547.1 ± 14.9
33	Kielce (power station)	–	Emsian	435.7 ± 9.0
34	Zbrza	–	Emsian	421.4 ± 10.2

margin of the EEP after Early Cambrian time (Fig. 3). Hence, the source was presumably the Mazury High, an area located more inwards of the EEP, from where Claesson (1995) reported several U–Pb zircon ages for the basement, ranging between 1.8 and 2.1 Ga. The absence of K–Ar cooling ages from the Mazury High, however, precludes a precise identification of source rocks.

Muscovites from the Middle–Upper Cambrian clastic rocks of the Lysogory Unit show K–Ar cooling ages indicating a provenance from more than a single source (Fig. 5). The ages of 614 ± 13 Ma and 539 ± 15 Ma are typical for a Cadomian provenance, whereas the 1745 Ma old muscovites seem to document a clastic input from the Svecofennian basement of Baltica. A strong variation in age of detrital muscovites in Lysogory suggests a mixing of different detrital mica populations. Our data, therefore, need to be controlled by Ar/Ar single-grain dating and barometric characterization of the micas. At present, we interpret other ages obtained from Lysogory as a result of mixing of Baltic and Cadomian material. Recently, a zircon population were found in the sandstones of Wymyslona (Location 28, Table 1), which is different from that of the EEP margin and must have derived from Cadomian sources (Valverde-Vaquero, pers. comm.).

In the Cambrian succession of the Malopolska Block there are also two populations of mica grains. The Lower Cambrian clastic rocks of the *Protolenus* Zone provided muscovites with Cadomian K–Ar cooling ages of 534 ± 18 Ma and 547 ± 15 Ma, respectively (Fig. 5). Thus, the former detritus is only *c*. 20 Ma older than the sedimentary age of the rock; this indicates a rapid exhumation of the source area. Upward in the section, a change in provenance from Cadomian to Baltic sources appears to occur. The Middle Cambrian sequence contains much older detritus of 1721 ± 38 Ma age. Preliminary microprobe analysis of this material shows a homogeneous chemical composition almost identical to that of muscovites recovered from the Middle Cambrian rocks of the EEP. In contrast to the Malopolska Block, the Baltic detritus appears to be already present in the Lower Cambrian rocks of the San Block. This is indicated by the 1355 ± 30 Ma age of detrital muscovites obtained from the borehole Kanczuga 27 (Fig. 4 and Table 1). It should be noted, however, that the Early Cambrian age of the investigated drill core is not well constrained stratigraphically. Moreover, further microprobe investigation will be necessary to ascertain the integrity of the mica population in the sample because the age of 1355 Ma may be the result of mixing of Cadomian and Baltic sources. The same problem must be clarified by interpreting the detritus age of 1020 ± 21 Ma from the Upper Cambrian rocks of the San Block. This age may indicate provenance from a Grenvillian source may result from a mixture of various mica populations.

Detrital muscovites extracted from the Lower Cambrian sandstones of Upper Silesia give a relatively tight range of K–Ar cooling ages from 542 to 566 Ma, with approximately ± 12 Ma uncertainties (Fig. 4). This appears to indicate that mica populations were supplied from a single source region with a Cadomian imprint. Derivation from the crystalline basement of the southern part of Bruno-Silesia, where the cooling ages of 540-590 Ma are known from the Brno Batholith (Dudek & Melkova 1975), is most probable.

Except for the area of Upper Silesia, the clastic material in the Devonian clastic rocks of Poland contains detrital muscovite grains derived from Palaeozoic sources. Contrary to the general opinion in the past, these sources were not local reworked Lower Palaeozoic rocks, but primary crystalline rocks. The very surprising result is also that, unlike in the Cambrian sequence, a unimodal mica population is widely distributed across Malopolska, Lysogory and the marginal part of the EEP. It has a narrow range of K–Ar cooling ages from 415 to 438 Ma, with approximately ± 10 Ma uncertainties (Fig. 6 and Table 1). Slightly older material of 467 ± 10 Ma was detected in the sample collected from the upper part of the Emsian sequence in the Bukowa Quarry (Lysogory). Both the Late Silurian and the Mid-Ordovician ages of detrital muscovites correspond well to the peaks of metamorphic evolution of the Scandinavian Caledonides (Dallmeyer 1988). The supply of the material from the northerly direction is additionally indicated by preliminary data we obtained from northern Poland (borehole Lebork, west of Gdansk), where the *c*. 460 Ma detritus is already present in the Upper Silurian clastic rocks.

The only area in southern Poland where the Devonian clastic rocks contain detrital muscovites with a Cadomian signature is the Upper Silesian Massif. The K–Ar cooling age of 555 ± 11 Ma suggests a provenance from the crystalline basement of Bruno-Silesia. At the moment, it is not clear to what degree Cadomian micas are mixed in the Devonian clastic rocks present in the marginal part of Malopolska, close to its contact with the USM (Fig. 6). The cooling age of 518 ± 11 Ma may be the result of mixing of Cadomian and Caledonian sources, or alternatively, the detrital micas might well be derived from 480–525 Ma granitoids, which are known from the Saxo-Thuringian belt of Germany, its equivalents in Poland, and from the western part of the Teplá–Barrandian terrane.

Discussion

Collectively, the new provenance data presented above and the results of previous biogeographical studies provide strong evidence that the crustal blocks bordering the East European Platform in southern Poland are exotic terranes. Understanding the complex history of these blocks is difficult they are not entirely suspect in regard to their palaeogeographical linkages to the palaeocontinent of Baltica. Although neither the time nor the place of their dispersion can be defined with precision at the moment, the crustal structure of these blocks, different from that of the EEP (Guterch *et al.* 1994), and the geochronological constraints (e.g. Dudek 1980; Finger

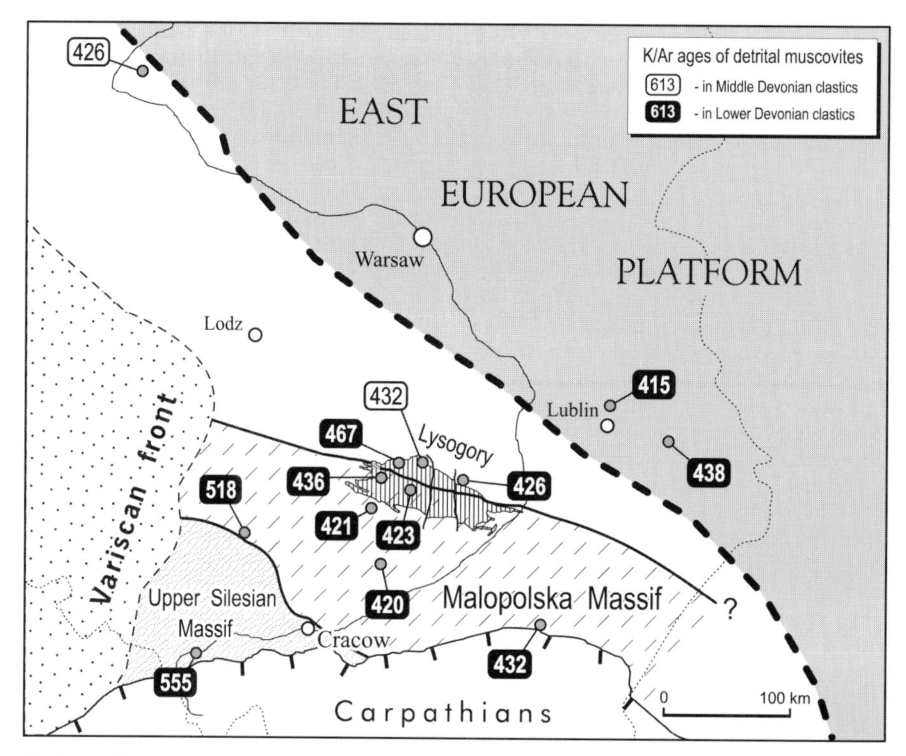

Fig. 6. K–Ar cooling ages of detrital muscovites in Devonian clastic rocks of southern Poland. (For more details on samples and localities, see Table 1). The Palaeozoic rocks of the Holy Cross Mountains are hatched.

et al. 1995) favour the interpretation that the Lysogory, the Malopolska and Bruno-Silesia are fragments derived from the Gondwana margin, not displaced parts of the East European Craton. During Early Cambrian time, these blocks were still separated from Baltica and their position in relation to each other was fundamentally different from that of today. Thus, it is now apparent that the Teisseyre–Tornquist Line is the primary suture between Baltica and Gondwana in central Europe. This boundary also formed the margin of Baltica before the first outboard Gondwanan microplates were attached to it.

One of the most important results of this study is doubtless the elucidation of seemingly incompatible faunal records from the Lower Palaeozoic sequence of Malopolska. Moreover, it seems that the Malopolska Massif is a composite terrane composed of the Malopolska and the San Blocks. The integration of the biotic succession and the provenance characteristics obtained from the Malopolska Block suggest that the accretionary history of this unit was as follows.

(1) In Early Cambrian time, this block was separated from the margin of Gondwana and had already reached a position close enough to Baltica to share Baltic province trilobites. The clastic material was still supplied from local Cadomian sources. The distance to Baltica, however, was presumably too great or ecological factors hindered transfer of brachiopod larvae from the shelf of Baltica. The co-occurrence of Baltic trilobites with inarticulate brachiopods of Avalonian affinity in Malopolska reveals that the migration potential of the Early Cambrian trilobites was greater than that of brachiopods. Thus, the latter are probably a more sensitive and useful tool in reconstructing the Cambrian palaeogeography.

(2) By Mid-Cambrian times, the Malopolska Block was closer to the Baltica margin so that more and more Baltic brachiopods settled in this realm. Another consequence was that a change in supply of clastic material from Cadomian to Baltic sources occurred.

(3) The next consequence of displacement of the Malopolska Block was the collision with the Baltic margin, which resulted in deformation of Cambrian and Precambrian rocks and in tilting of the entire block. This episode, termed in the past the ‘Sandomierz Phase’ (Samsonowicz 1926), took place after Mid-Cambrian and

before late Tremadoc time. Thus, the Malopolska Block constitutes the first Gondwana-derived microplate that accreted to the margin of Baltica. The provenance data partly confirm the earlier palaeomagnetic data, which place Malopolska close to the margin of Baltica during Early Palaeozoic time but in a more south-easterly position than at present (Lewandowski 1993). The K–Ar cooling age of detrital muscovites (*c.* 1.7 Ga) obtained from the Middle Cambrian rocks suggests that Malopolska must have accreted to the Fennoscandian segment of the East European Craton.

(4) In Early Ordovician time, the Malopolska Block was covered by a shallow sea. The more southerly palaeogeographical position of this block in relation to the Baltoscandian region was probably the reason for the enlarged endemicity of ostracode faunas (Olempska 1994).

(5) Structural and some palaeomagnetic data (Lewandowski 1993; Mizerski 1995) point to strike-slip displacement of Malopolska along the SW margin of the EEP. The facies pattern along the Holy Cross Dislocation suggests the amalgamation of Malopolska with the Lysogory Unit during late Silurian time. This is also confirmed by the unification of clastic material throughout the region in Early Devonian time.

The accretionary history of the San Block is still being studied but it seems that, like Malopolska, this unit was in the immediate proximity of Baltica already during Early Cambrian time. From the biogeographical and provenance data we can assume that Bruno-Silesia was also located closer to Baltica than to any other Peri-Gondwanan plates during Cambrian and Ordovician times. There is, however, no indication that this block was ever in a direct contact to the Baltic continent. It must have joined the Malopolska presumably during Early Devonian time. Nevertheless, Bruno-Silesia retained its faunal separate character up to Late Devonian time (Berkowski 1999).

In some respects, the Lysogory Unit is exotic to all adjacent blocks. Although it was probably also situated very close (or was even attached) to Baltica from Late Cambrian time, there is a paradox that the Cambrian fauna shows Gondwanan rather than Baltic affinities. Any kind of separation by a narrow but deep-water realm is unlikely, as there is much *c.* 1.7 Ga detritus from Baltic sources in the shallow-water environments of Lysogory. Our provenance data show clearly that Lysogory had a distinct Early Palaeozoic drift history different from those of the Avalonia or Armorica plates. We hope that future studies of the Cambrian and Ordovician biotic successions of Lysogory will provide the data necessary to constrain the place of its accretion.

Contrary to some earlier interpretations (e.g. Lewandowski 1993), the provenance data from the Devonian clastic rocks of southern Poland suggest that the final amalgamation of crustal blocks within the TESZ was largely terminated before Mid-Devonian time. During the later periods, this domain formed the foreland of the Variscan orogen. As a consequence, the TESZ in central Europe has been subjected to regional compression. The proximity of the orogenic front resulted in extensive deformation especially close to the contact faults between the Lysogory Unit, the Malopolska Massif and the Upper Silesian Massif.

Support for this study was provided by the German Research Council (DFG), grant Be 1296/5-3, and is greatly appreciated. This is a contribution to the Special Research Programme 'Orogenic Processes' funded by the DFG. We are grateful to S. Speczik (Polish Geological Institute), T. Wilczek and Z. Borys (both Polish Oil Company 'Geonafta') for access to borehole material. Although too numerous to mention each by name, we are indebted to all those geologists who have contributed to our present understanding of southern Poland and on whose work much of this paper is based. The help of Z. Bula, M. Jachowicz, Z. Kowalczewski, J. Malec, W. Moryc, Z. Obara, S. Orlowski, J. Paczesna, E. Porebska, J. Schäfer, S. Skompski, A. Zelazniewicz and K. Zychowicz is acknowledged in particular. The authors gratefully acknowledge helpful suggestions and constructive criticism by E. Eide, T. Pharaoh and D. Tanner. This paper is Contribution 302 to EUROPROBE (TESZ project). EUROPROBE is co-ordinated within the International Lithosphere Programme and sponsored by the European Science Foundation.

References

Aren, B. & Lendzion, K. 1978. Stratigraphic–lithologic characteristics of the Vendian and Lower Cambrian. *Prace Instytutu Geologicznego*, **90**, 7–41.

Belka, Z. 1990. Thermal maturation and burial history from conodont colour alteration data, Holy Cross Mountains, Poland. *Courier Forschungsinstitut Senckenberg*, **118**, 241–251.

—— 1993. Remarks on thermal maturity level in the subsurface of the Upper Silesian Coal Basin. *Acta Geologica Polonica*, **43**, 95–101.

——, Siewniak, A. 1996. Thermal maturation of the Lower Palaeozoic strata in the southwestern margin of the Malopolska Massif, southern Poland: no evidence for Caledonian regional metamorphism. *Geologische Rundschau*, **85**, 775–781.

Bergström, J. 1984. Strike-slip faulting and Cambrian biogeography around the Tornquist Zone.

Geologiska Föreningens i Stockholm Förhandlingar, **106**, 382–383.

BERKOWSKI, B. 1999. *Famennian coral faunas of southern Poland*. PhD thesis, Adam-Mickiewicz-University, Poznan.

BERTHELSEN, A. 1992. Mobile Europe. *In*: BLUNDELL, D., FREEMAN, R. & MUELLER, S. (eds) *A Continent Revealed. The European Geotraverse*. Cambridge University Press, Cambridge, 11–32.

—— 1993. Where different geological philosophies meet: the Trans-European Suture Zone. *Publications of the Institute of Geophysics, Polish Academy of Sciences*, **A20**(255), 19–31.

BLUNDELL, D., FREEMAN, R. & MUELLER, S. (eds) 1992. *A Continent Revealed. The European Geotraverse*. Cambridge University Press, Cambridge.

BOGDANOVA, S. V., BIBIKOVA, E. V. & GORBATSCHEV, R. 1994. Palaeoproterozoic U–Pb zircon ages from Belorussia: new tectonic implications for the East European Craton. *Precambrian Research*, **68**, 231–240.

——, PASHKEVICH, I. K., GORBATSCHEV, R. & ORLYUK, M. I. 1996. Riphean rifting and major Palaeoproterozoic crustal boundaries in the basement of the East European Craton: geology and geophysics. *Tectonophysics*, **268**, 1–21.

BRASIER, M. D. 1989. Towards a biostratigraphy of the earliest skeletal biotas. *In*: COWIE, J. W. & BRASIER, M. D. (eds) *The Precambrian–Cambrian Boundary*. Clarendon Press, Oxford,. 117–165.

BULA, Z. & JACHOWICZ, M. 1996. The Lower Palaeozoic sediments in the Upper Silesian Block. *Geological Quarterly*, **40**, 299–336.

——, JACHOWICZ, M. & ZABA, J. 1997. Principal characteristic of the Upper Silesian Block and Malopolska Block border zone (southern Poland). *Geological Magazine*, **134**, 669–677.

CLAESSON, S. 1995. Palaeoproterozoic age provinces in the southwestern part of the East European Craton. *In*: *Precambrian of Europe: Stratigraphy, Structure, Svolution and Mineralization*. 9ths Meeting of the Association of European Geological Societies, St Petersburg, Abstracts, 19–20.

COCKS, L. R. M. & FORTEY, R. A. 1998. The Lower Palaeozoic margins of Baltica. *Geologiska Föreningens i Stockholm Förhandlingar*, **120**, 173–179.

——, MCKERROW, W. S. & VAN STAAL, C. R. 1997. The margins of Avalonia. *Geological Magazine*, **133**, 456–466.

COMPSTON, W., SAMBRIDGE, M. S., REINFRANK, R. F., MOCZYDLOWSKA, M., VIDAL, G. & CLAESSON, S. 1995. Numerical ages of volcanic rocks and the earliest faunal zone within the Late Precambrian of east Poland. *Journal of the Geological Society, London*, **152**, 599–611.

DADLEZ, R. 1983. O koncepcji wczesnopaleozoicznych wielkich ruchow przesuwczych wzdluz krawedzi plyty laurentyjskiej i baltyckiej. *Przeglad Geologiczny*, **31**, 377–386.

—— 1995. Debates about the pre-Variscan tectonics of Poland. *Studia Geophysica et Geodaetica*, **39**, 227–234.

DALLMEYER, R. D. 1988. Polyphase tectonothermal evolution of the Scandinavian Caledonides. *In*: HARRIS, A. L. & FETTES, D. J. (eds) *The Caledonian–Appalachian Orogen*. Geological Society, London, Special Publication, **38**, 365–380.

DÖRR, W., BELKA, Z., FRANKE, W. & WISZNIEWSKA, J. 1998. Isotopic signatures from basement rocks of the East European Platform and Palaeozoic clastic sediments. PACE TMr Network Meeting, *Programme and Abstracts*, 22.

DUDEK, A. 1980. The crystalline basement block of the Outer Carpathians in Moravia: Bruno-Vistulicum. *Rozprawy Ceskoslovenské Akademie Ved*, **90**(8), 1–85.

—— 1995. Moravo-Silesian Zone—metamorphic evolution. *In*: DALLMEYER, R. D., FRANKE, W. & WEBER, K. (eds) *Pre-Permian Geology of Central and Eastern Europe*. Springer, Berlin, 508–511.

—— & MELKOVA, J. 1975. Radiometric age determination in the crystalline basement of the Carpathian Foredeep and of the Moravian Flysch. *Vestnik Ustredniho Ustavu Geologickeho*, **50**, 257–264.

DZIK, J. 1989. Conodont evolution in high latitudes of the Ordovician. *Courier Forschungsinstitut Senckenberg*, **117**, 1–28.

—— & PISERA, A. 1994. Sedimentation and fossils of the Mojcza Limestone. *Palaeontologia Polonica*, **53**, 5–41.

——, OLEMPSKA, E. & PISERA, A. 1994. Ordovician carbonate platform ecosystem of the Holy Cross Mountains. *Palaeontologia Polonica*, **53**, 1–315.

ELLENBERGER, F. & TAMAIN, A. L. G. 1980. Hercynian Europe. *Episodes*, **1980**(1), 22–27.

FINGER, F., FRASL, G., DUDEK, A., JELINEK, E. & THÖNI, M. 1995. Igneous activity (Cadomian plutonism in the Moravo-Silesian basement). *In*: DALLMEYER, R. D., FRANKE, W. & WEBER, K. (eds) *Pre-Permian Geology of Central and Eastern Europe*. Springer, Berlin, 495–507.

——, HANŽL, P., PIN, C., VON QUADT, A. & STEYRER, H. P. 2000. The Brunovistulian: Avalonian Precambrian sequence at the eastern end of the Central European Variscides. *This volume*.

FITCH, F. J., MILLER, J. A. & THOMPSON, D. B. 1966. The paleogeographic significance of age determinations on detrital micas from the Triassic of the Stockport–Macclesfield District, Cheshire, England. *Palaeogeography, Palaeoclimatology, Palaeoecology*, **2**, 281–312.

FRANKE, D. 1995. Caledonian terranes along the southwestern border of the East European Platform—evidence, speculation and open questions. *Studia Geophysica et Geodaetica*, **39**, 241–256.

FRANKE, W. 1989. Tectonostratigraphic units in the Variscan belt of central Europe. *In*: DALLMEYER, R. D. (ed.) *Terranes in the Circum-Atlantic Palaeozoic Orogens*. Geological Society of America, Special Papers, **230**, 67–90.

FUHRMANN, U., LIPPOLT, H. J. & HESS, J. C. 1987. Examination of some proposed K–Ar standards: $^{40}Ar/^{39}Ar$ analyses and conventional K–Ar-data. *Chemical Geology*, **66**, 41–51.

GLADYSZ, J., JACHOWICZ, M. & PIEKARSKI, K. 1990. Palaeozoic Acritarcha from the Siewierz vicinity (northern margin of the Upper Silesian Coal Basin). *Geological Quarterly*, **34**, 623–630.

GUTERCH, A., GRAD, M., JANIK, T. *et al.* 1994. Crustal structure of the transition zone between Precambrian and Variscan Europe from new seismic data along LT-7 profile (NW Poland and eastern Germany). *Comptes Rendus de l'Académie des Sciences, Géophysique, Série II*, **319**, 1489–1496.

——, ——, MATERZOK, R. & PERCHUC, E. 1986. Deep structure of the Earth's crust in the contact zone of the Paleozoic and Precambrian platforms in Poland (Tornquist–Teisseyre Zone). *Tectonophysics*, **128**, 251-279.

HEINRICHS, H. & HERRMANN, A. G. 1990. *Praktikum der Analytischen Geochemie*, Springer, Berlin.

HORSTMANN, U. E. 1987. Die metamorphe Entwicklung im Damara Orogen, Südwestafrika/Namibia, abgeleitet aus K–Ar-Datierungen an detritischen Hellglimmern aus Molassesedimenten der Nama Group. *Göttinger Arbeiten zur Geologie und Paläontologie*, **32**, 1–95.

——, AHRENDT, H., CLAUER, C. & PORADA, H. 1990. The metamorphic history of the Damara Orogen based on K–Ar data of detrital white micas from the Nama Group, Namibia. *Precambrian Research*, **48**, 41–61.

JACHOWICZ, M. 1998. Rola badan palinologicznych w geologii naftowej—przeszlosc i przyszlosc. *In*: *Dzien dzisiejszy przemyslu naftowego*. Konferencja Naukowo-Techniczna GEONAFTA, Wysowa, Abstracts, 66–69.

JAWOROWSKI, K. 1971. Sedimentary structures of the Upper Silurian siltstones in the Polish Lowland. *Acta Geologica Polonica*, **21**, 519–571.

—— 1997. Depositional environments of the Lower and Middle Cambrian sandstone bodies, Polish part of the East European Craton. *Biuletyn Panstwowego Instytutu Geologicznego*, **377**, 1–112.

JENDRYKA-FUGLEWICZ, B. 1992. Analiza porownawcza ramienionogow z utworow kambru Gor Swietokrzyskich i platformy prekambryjskiej w Polsce. *Przeglad Geologiczny*, **467**, 150–155.

—— 1998. Kambryjska eksplozja zycia. Najstarsze zespoly brachiopodow w profilach geologicznych Polski. *Abstracts of the XVI Palaeontological Meeting*, Wiktorowo, 18–19.

KARNKOWSKI, P. & GLOWACKI, E. 1961. O budowie geologicznej utworow podmiocenskich przedgorza Karpat srodkowych. *Geological Quarterly*, **5**, 372–416.

KOWALCZEWSKI, Z. 1990. Grubookruchowe skaly kambru na srodkowym poludniu Polski. *Prace Panstwowego Instytutu Geologicznego*, **131**, 1–82.

—— 1995. Fundamental startigraphic problems of the Cambrian in the Holy Cross Mts. *Geological Quarterly*, **39**, 449–470.

——, JAWOROWSKI, K. & KULETA, M. 1998. Klonow Beds (uppermost Silurian–?lowermost Devonian) and the problem of Caledonian deformations in the Holy Cross Mts. *Geological Quarterly*, **42**, 341–378.

KOWALSKI, W. R. 1983. Stratigraphy of the upper Precambrian and lowest Cambrian strata in southern Poland. *Acta Geologica Polonica*, **33**, 183–218.

LENDZION, K. 1983. Biostratigraphy of Cambrian deposits in the Polish part of the East European Platform. *Geological Quarterly*, **27**, 669–694.

LEWANDOWSKI, M. 1993. Paleomagnetism of the Paleozoic rocks of the Holy Cross Mts (Central Poland) and the origin of the Variscan orogen. *Publications of the Institute of Geophysics, Polish Academy of Sciences*, **A23**(265), 3–85.

—— 1995. Palaeomagnetic constraints for Variscan mobilism of the Upper Silesian and Malopolska Massifs, southern Poland—reply. *Geological Quarterly*, **39**, 283–292.

MALEC, J. 1993. Upper Silurian and Lower Devonian in the western Holy Cross Mts. *Geological Quarterly*, **37**, 501–536.

MILACZEWSKI, L. 1981. The Devonian of the south-eastern part of the Radom-Lublin area (eastern Poland). *Prace Instytutu Geologicznego*, **101**, 1–90.

MIZERSKI, W. 1995. Geotectonic evolution of the Holy Cross Mts in central Europe. *Biuletyn Panstwowego Instytutu Geologicznego*, **372**, 5–47.

MOCZYDLOWSKA, M. 1991. Acritarch biostratigraphy of the Lower Cambrian and the Precambrian–Cambrian boundary in southeastern Poland. *Fossils and Strata*, **29**, 1–127.

MODLINSKI, Z. 1982. Rozwoj litofacjalny i paleotektoniczny na obszarze platformy prekambryjskiej w Polsce. *Prace Instytutu Geologicznego*, **102**, 1–66.

NAWROCKI, J. 1995. Palaeomagnetic constraints for Variscan mobilism of the Upper Silesian and Malopolska Massifs, southern Poland—discussion. *Geological Quarterly*, **39**, 271–282.

NEUROTH, H. 1997. K/Ar-Datierungen an detritischen Muskoviten—'Sicherungskopien' orogener Prozesse am Beispiel der Varisziden. *Göttinger Arbeiten zur Geologie und Paläontologie*, **49**, 1–143.

OLEMPSKA, E. 1994. Ostracods of the Mojcza Limestone. *Palaeontologia Polonica*, **53**, 129–212.

ORLOWSKI, S. 1975 Lower Cambrian trilobites from Upper Silesia (Goczalkowice borehole). *Acta Geologica Polonica*, **25**, 377–383.

—— 1985. Lower Cambrian and its trilobites in the Holy Cross Mts. *Acta Geologica Polonica*, **35**, 231–250.

PHARAOH, T. *et al.* 1996. Trans-European Suture Zone. Phanerozoic accretion and the evolution of contrasting continental lithospheres. *In*: GEE, D. G. & ZEYEN, H. J. (eds) *EUROPROBE 1996—Lithosphere Dynamics: Origin and Evolution of Continents*. Uppsala University. 41–54.

POZARYSKI, W. 1975. Interpretacja geologiczna wynikow glebokich sondowan sejsmicznych na VII profilu miedzynarodowym. *Przeglad Geologiczny*, **23**, 163–171.

——, GROCHOLSKI, A., TOMCZYK, H., KARNKOWSKI, P. & MORYC, W. 1992. The tectonic map of Poland in the Variscan epoch. *Przeglad Geologiczny*, **40**, 643–651.

RENNE, P. R., BECKER, T. A. & SWAPP, S. M. 1990. $^{40}Ar/^{39}Ar$ laser-probe dating of detrital micas from the Montgomery Creek Formation, northern California: clues to provenance, tectonics, and weathering processes. *Geology*, **18**, 563–566.

RYKA, W. 1982. Precambrian evolution of the Polish part of the East-European Platform. *Geological Quarterly*, **26**, 257–272.

SAMSONOWICZ, J. 1926. Remarques sur la tectonique et la paléogéographie du Massif paléozoique de Swiety Krzyz. *Posiedzenia Naukowe Panstwowego Instytutu Geologicznego*, **15**, 44–46.

SCHÄFER, J., NEUROTH, H., AHRENDT, H., DÖRR, W. & FRANKE, W. 1997. Accretion and exhumation at a Variscan active margin, recorded in the Saxothuringian flysch. *Geologische Rundschau*, **86**, 599–611.

SCHUMACHER, E. (1975): Herstellung von 99,9997% ^{38}Ar für die $^{40}K/^{40}Ar$ Geochronologie. *Geochron. Chimia*, **24,** 441–442.

SEILACHER, A. 1983. Upper Paleozoic trace fossils from the Gilf Kebir–Abu Ras area in southwestern Egypt. *Journal of African Earth Sciences*, **1**, 21–34.

SEMENOV, V. Y., JANKOWSKI, J., ERNST, T., JOZWIAK, W., PAWLISZYN, J. & LEWANDOWSKI, M. 1998. Electromagnetic soundings across the Holy Cross Mountains, Poland. *Acta Geophysica Polonica*, **46**, 171–185.

SIKORSKA, M. 1988. Microlithofacies of Middle Cambrian sedimentary rocks in the Polish part of the East European Platform. *Prace Instytutu Geologicznego*, **126**, 1–47.

STEIGER, R. H. & JÄGER, E. 1977. Subcommission on Geochronology: convention on the use of decay constants in geo- and cosmochronology. *Earth and Planetary Science Letters*, **36**, 359–362.

SZULCZEWSKI, M. 1995. Depositional evolution of the Holy Cross Mts. (Poland) in the Devonian and Carboniferous—a review. *Geological Quarterly*, **39**, 471–488.

TARNOWSKA, M. 1988. Zarys historii sedymentacji osadow dewonu dolnego w poludniowej czesci Gor Swietokrzyskich. *Geological Quarterly*, **32**, 242–243.

TELLER, L. 1997. The subsurface Silurian in the East European Platform. *Palaeontologia Polonica*, **56**, 7–21.

VIDAL, G. & MOCZYDLOWSKA, M. 1995. The Neoproterozoic of Baltica—stratigraphy, palaeobiology and general geological evolution. *Precambrian Research*, **73**, 197–216.

WELZEL, B. 1991. Die Bedeutung von K/Ar-Datierungen an detritischen Muskoviten für die Rekonstruktion tektonomorpher Einheiten im orogenen Liefergebiet—ein Beitrag zur Frage der varistischen Krustenentwicklung in der Böhmischen Masse. *Göttinger Arbeiten zur Geologie und Paläontologie*, **49**, 1–61.

ZIEGLER, P. A. 1981. Evolution of sedimentary basins in North-West Europe. *In*: ILLING, L. V. & HOBSON, G. D. (eds) *Petroleum Geology of the Continental Shelf of North-West Europe*. Heyden, London, 3–39.

The Brunovistulian: Avalonian Precambrian sequence at the eastern end of the Central European Variscides?

F. FINGER[1], P. HANŽL[2], C. PIN[3], A. VON QUADT[4] & H. P. STEYRER[5]

[1]*Institut für Mineralogie der Universität, Hellbrunnerstraße 34, A-5020 Salzburg, Austria*

[2]*Czech Geological Survey, Leitnerova 22, CZ-65869 Brno, Czech Republic*

[3]*Département de Géologie, UMR 6524, Univ. B. Pascal, CNRS, 5 Rue Kessler, F-63038 Clermont Ferrand, France*

[4]*Institut für Isotopengeologie der ETH, Sonneggstraße 5, CH-8092 Zürich, Switzerland*

[5]*Institut für Geologie und Paläontologie der Universität, Hellbrunnerstraße 34, A-5020 Salzburg, Austria*

Abstract: An outline is presented of the present state of research on the Precambrian evolution history of the Brunovistulian, a large (30 000 km^2), mainly sediment covered Peri-Gondwana basement block at the eastern end of the Central European Variscides. On the basis of recent chemical, isotopic and geochronological data it is argued that the eastern half of the Brunovistulian (Slavkov Terrane) originated in an island-arc environment, documenting the rare case of Neoproterozoic crustal growth in central Europe. The western half of the Brunovistulian, the Thaya Terrane, includes more mature, recycled cratonic material and is considered to have been originally part of the Neoproterozoic Gondwana continent margin. A phase of regional metamorphism at *c.* 600 Ma, followed by extensive granitoid plutonism, probably marks the stage when the Slavkov Terrane was accreted to the Thaya Terrane by arc–continent collision. A belt of metabasites, which is intercalated between the two terranes, may represent relics of the incipient arc or a back-arc basin. A comparison of geochronological data shows that the timing of geological events recorded in the Brunovistulian does not correlate with the evolution history of the Cadomian crust in the Teplá–Barrandian zone and the Saxo-Thuringian belt. This supports the theory that the Brunovistulian is not part of Armorica but derived from a different sector of the Neoproterozoic Gondwana margin. A correlation with the Avalonian superterrane appears feasible.

During recent years, many new geological and geochronological data have become available for the Cadomian basement of the Central European Variscides (e.g. Linnemann 1995; Tichomirowa *et al.* 1997; Dörr *et al.* 1998; Zulauf *et al.* 1999; Finger *et al.* 2000; Kröner *et al.* this volume; Linnemann *et al.* this volume). These data shed light on the early evolution stages of the various allochthonous Peri-Gondwana terranes that were welded together in Palaeozoic time to establish the Central European crust. In this paper the Cadomian evolution history of the largest of the Precambrian basement blocks in Central Europe, the Brunovistulian in Moravia (Dudek 1980), is discussed and a tentative geodynamic model suggested.

With regard to the Variscan tectonic framework (see enclosure at back), there is much agreement that the Moravo-Silesian Unit with its Devonian–Carboniferous sedimentary cover and its Cadomian crystalline basement, the Brunovistulian, should be considered a continuation of the Rheno-Hercynian (Franke & Engel 1982; Dvořák 1995; Tait *et al.* 1996, 1997; Franke, this volume) and, therefore, a potential part of the Avalonian superterrane. Alternative tectonic concepts that correlated the Moravo-Silesian Unit with the southern flank of the Variscan fold belt in France (Matte *et al.* 1990) are no longer maintained (Matte, pers. comm.). For information concerning the Variscan tectonic relationship between the Moravo-Silesian and the Moldanubian, the reader is referred to Finger & Steyrer (1995) and Fritz *et al.* (1996).

Geological overview

On the basis of a systematic study of drillcores from Moravia, Dudek (1980) proposed that large amounts of well-preserved Cadomian basement

From: FRANKE, W., HAAK, V., ONCKEN, O. & TANNER, D. (eds).
Orogenic Processes: Quantification and Modelling in the Variscan Belt.
Geological Society, London, Special Publications, **179**, 103–112. 0305-8719/00/$15.00

rocks are present below the Palaeozoic to Tertiary sediments at the eastern termination of the Central European Variscides, and introduced the term 'Brunovistulicum' for this inferred basement block (Fig. 1).

At the surface these Cadomian rocks can be studied only in relatively small massifs in the western parts of Moravia and Silesia (Thaya Dome, Svratka Dome, Brno Massif, Keprník and Desná Dome). However, thanks to dense oil and gas drilling in the Carpathian foredeep between Brno and Ostrava, and on the basis of additional geophysical data, Dudek (1980) was able to compile a geological sketch map for this subcrop part of the Brunovistulian (Fig. 1). Unfortunately, little information is available about the crystalline basement beneath the thick Devonian and Carboniferous Silesian sediments north of the Neogene Molasse basin. However, as Cadomian rocks outcrop again in the Hrubý Jeseník Mountains (Desná Dome), Dudek (1980) proposed that the whole Moravo-Silesian zone rests upon the same Cadomian consolidated basement block.

As a result of the increasing Variscan deformation and metamorphism towards the Moldanubian boundary (Frasl 1970, 1991; Schulmann 1990; Fritz *et al.* 1996), there is some uncertainty about the actual western limitation of the Brunovistulian. Judging from geochronological data (see below), we consider it likely that the strongly sheared granitoid pencil gneisses of western Moravia and Silesia (Bittesch Gneiss, Keprnik Gneiss) still represent parts of the Brunovistulian, although the bodies are probably more or less dislocated from the main basement block in the east. Some workers have argued that even some bodies of granitoid gneisses of the Moldanubian Unit in Lower Austria (Dobra Gneiss, Spitz Gneiss) might represent Variscan reworked and overthrust parts of the Brunovistulian plate (Matura 1976; Finger & Steyrer 1995). This concept is, however, still a matter of controversy (see, e.g. Fuchs 1998) and the intra-Moldanubian gneisses are therefore left out of the present study.

Figure 1 illustrates that the Brunovistulian essentially consists of a large metamorphic complex with mainly paragneisses in the (north)-east, and a large granitoid complex in the (south)west (Brno Batholith, Thaya Batholith). The name Thaya Batholith is traditionally used for the granitoids in the Thaya Dome, i.e. the area between Eggenburg and Znojmo, although the separation of the Thaya Batholith from the Brno Batholith is presumably a simple result of Permian sinistral strike-slip along the Diendorf–Boskovice fault system. Near Brno, a striking narrow belt of metamorphosed basic and ultrabasic rocks passes through the Brno Batholith in a roughly north–south direction (Central Basic Belt).

The granitoids

The large abundance of Cadomian granitoids is one of the most outstanding features of the Brunovistulian block, and these rocks have been the subject of several detailed petrographical and geochemical studies (Dudek 1980, Finger *et al.* 1989; Jelínek & Dudek 1993; Leichmann 1996; Hanžl & Melichar 1997). All these studies concluded that the Brunovistulian granitoid terrane is essentially an I-type granitoid terrane. However, Finger *et al.* (1995) have additionally pointed out a pronounced east–west zoning of granite types, with the plutons east of the Central Basic Belt being relatively primitive in composition, resembling island-arc magmas (Hbl granodiorites, tonalites, quartz diorites), whereas those in the west are mainly high-K granodiorites and granites. The magmatic zonality in the Brunovistulian granitoids is very clearly expressed in the Sr and Nd isotope values (Finger *et al.* 1995; Finger & Pin 1997, in prep.). These are generally primitive in the eastern province ($^{87}Sr/^{86}Sr_{580}$ mostly 0.704–0.705, ϵNd_{580} -1 to $+3$), and mostly in the typical crustal range in the western granitoids ($^{87}Sr/^{86}Sr_{580}$ 0.708–0.710, ϵNd_{580} -4 to -7). Intermediate initial ratios (0.705–0.707, -1 to -2) have been found in the rare diorite–tonalite bodies west of the Central Basic Belt and in one distinct subalkaline granite pluton in the southern Thaya Batholith near Eggenburg (Finger & Riegler 1999). Samples of Bittesch Gneiss yielded generally very mature isotope values (ϵNd_{580} -10 to -11; see also data of Liew & Hofmann (1988)), whereas the Keprnik Gneiss has ϵNd_{580} values of around -5 to -6 (Hegner & Kröner this volume).

Regarding the magma sources it is obvious from the isotope data that the western high-K granites and granodiorites, including the Bittesch and Keprnik Gneiss, are mainly crustal derived with cratonic components involved in the melting process. Friedl *et al.* (1998) reported abundant Mesoproterozoic inherited zircon cores from the Bittesch Gneiss. However, the additional occurrence of some isotopically more primitive tonalites and diorites in the western province indicates the presence of another magma source (mantle or juvenile crust).

The granitoid magmas east of the Central Basic Belt formed either through melting of young calc-alkaline crust or contain a significant mantle component. Because of the dominance of

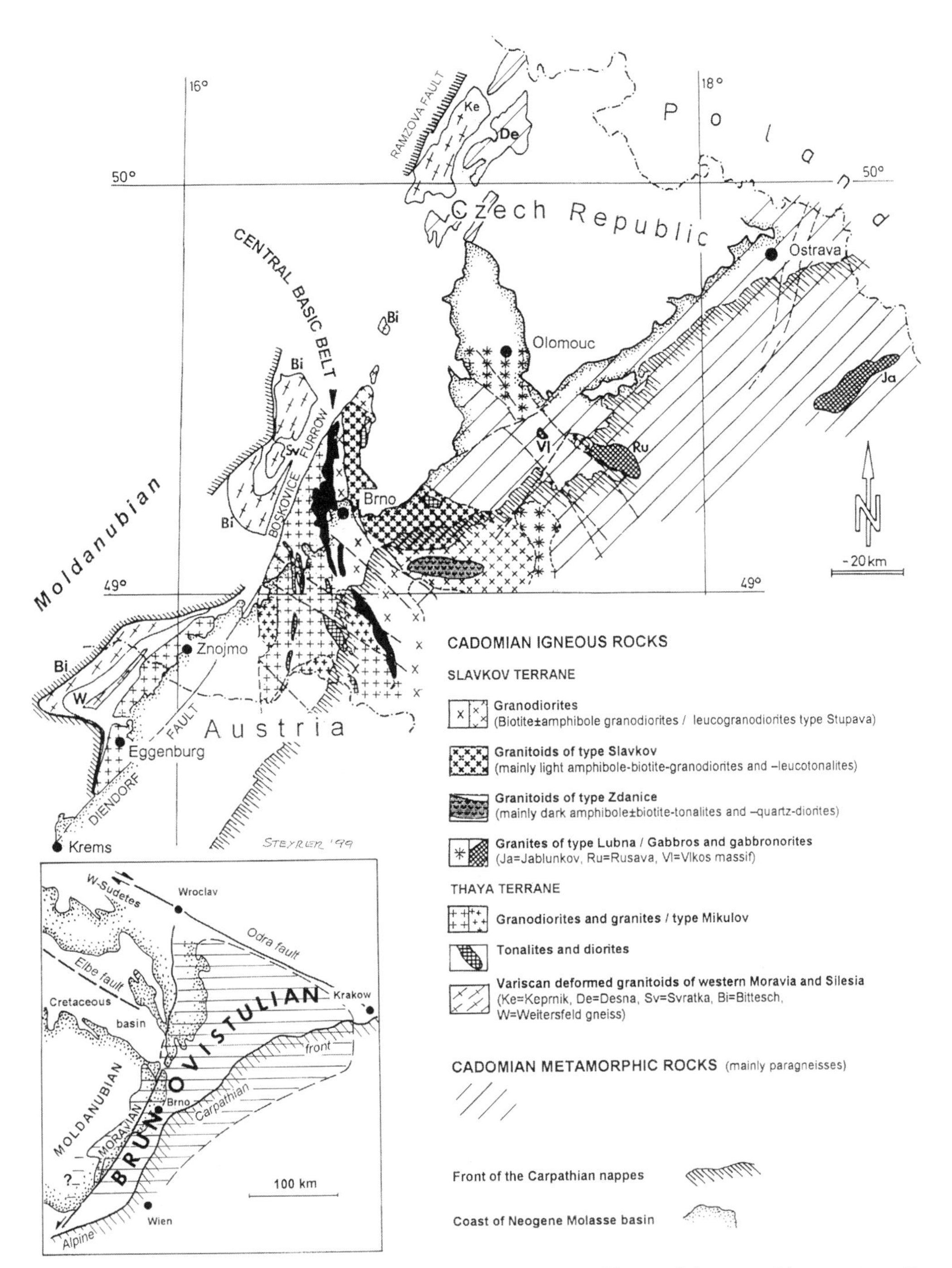

Fig. 1. Geology of the Brunovistulian block as far as is known from drillings and the exposed basement massifs. Mainly after Dudek (1980) and official maps of the Czech Republic. Inset shows the presumed extent of the Brunovistulian (horizontally hatched), according to Dudek (1980), and its position in the tectonic framework of the Central European Variscides (see enclosure at back).

Table 1. *Compilation of the currently available zircon ages from the granitoid complex of the Brunovistulian including recent data for the Bittesch and Keprnik Gneiss*

Rock type, source	Age (Ma)	Dating method	Reference
Diorite, Dolni Kounice (Brno Massif)	584 ± 5	U–Pb (zircon)	Van Breemen *et al.* (1982)
Granite, Eggenburg (Thaya Massif)	583 ± 11	U–Pb (zircon)	Friedl (1997)
Granodiorite, Gumping (Thaya Massif)	595 ± 1	Pb–Pb (zircon, evap.)	Kröner (unpub. data)
Metagranite, Korneuburg (drillcore)	587 ± 12	Pb–Pb (zircon, evap.)	M. Tichomirowa (unpub. data)
Bittesch gneiss, Messern (Thaya Dome)	586 ± 7	U–Pb (zircon)	Friedl *et al.* (1998)
Keprnik Gneiss	584 ± 8	Pb–Pb (zircon, evap.)	Kröner *et al.* (this volume)

felsic granitoids we attribute a great importance to melting processes in the crust, although in some cases (e.g. basic massifs of Rusava and Jablunkov) the presence of mantle melts seems to be clear.

Geochronological data provide increasing evidence that most of the Brunovistulian granitoids formed during a relatively short plutonic episode at around 580–590 Ma. In addition to a number of zircon ages from granitoids of the western province, which all fall in this time span (Table 1), an Ar–Ar hornblende age of *c.* 590 Ma has been recently reported for one of the eastern granitoids (Fritz *et al.* 1996). A Rb–Sr isochron of 551 ± 6 Ma, which is often cited as dating the formation of the Thaya Batholith (Scharbert & Batik 1980), is probably a mixing line and provides an invalid young age (Finger & Riegler 1999). Possible candidates for late Cadomian intrusions are the gabbroic massifs in the eastern Brunovistulian: in the contact aureole of the Jablunkov massif (Fig. 1), Th–U–total Pb monazite ages of *c.* 550 Ma have been obtained (Finger *et al.* 1999).

Because of their overall I-type nature, and on the basis of Pearce-type diagrams, the Brunovistulian granitoids were previously mostly interpreted as subduction related (Finger *et al.* 1989; Jelínek & Dudek 1993; Hanžl & Melichar 1997). On the basis of the striking east–west zoning of plutonism, a westward dipping Cadomian subduction geometry has been inferred (Finger *et al.* 1995). However, the pulsed nature of plutonism, subsequent to a phase of regional metamorphism, may suggest a collision event. As will be discussed later, we consider arc–continent collision in a Pacific-type tectonic setting as a potential model.

The metamorphic rocks

A comprehensive description of the metamorphic complex in the eastern half of the Brunovistulian was given by Dudek (1980, 1995). Most of these rocks are derived from flyschoid clastic sediments, mainly greywackes, siltstones and psammites with subordinate pelitic layers. In the area south and southeast of Olomouc, the metasediments contain abundant layers of metavolcanic rocks, mainly metabasalts and meta-andesites.

Regional metamorphism mostly reached medium-pressure amphibolite facies grade. In some cases partial melting occurred. Microfabrics sometimes provide evidence for two metamorphic phases, with a younger phase of contact metamorphism related to the intrusion of the granitoids (Dudek 1980, Hanžl *et al.* 1999). During the Variscan orogeny, parts of the Cadomian metamorphic complex were subjected to penetrative retrograde shearing under greenschist facies conditions.

Dudek (1980) has already emphasized the relatively Al_2O_3-poor character of the metasediments in the eastern Brunovistulian, implying a calc-alkaline (volcanic arc) source area. New trace element and isotope data (Finger & Pin in prep.) point in the same direction and indicate a young, chemically and isotopically little evolved island arc-type source. $^{87}Sr/^{86}Sr_{580}$ and ϵNd_{580} values of the metasediments are generally very similar to those of the adjoining granitoids (0.704–0.706; −1 to +2). This shows that not only the granitoids, but the whole continental crust in the eastern part of the Brunovistulian were primitive throughout. In the western part of the Brunovistulian, Cadomian metamorphic rocks are much less abundant. However, paragneiss relics from the roof of the Thaya Batholith and the western Brno Batholith are chemically and isotopically more evolved (ϵNd_{580} −3 to −7) than the paragneisses in the east. The detritus of these western metasedimentary units obviously derives from much older, cratonic crust (Finger & Pin in prep.).

On the basis of the evidence presented from the granitoids and paragneisses, it is unequivocal that the continental crust west of the Central Basic Belt is completely different from that in the east. The Central Basic Belt obviously marks an important terrane boundary. Finger & Pin (1997) have introduced the terms Slavkov Terrane for the eastern province and Thaya Terrane for the western province.

Kröner *et al.* (2000) and Hegner & Kröner (this volume) have recently presented Nd isotope data from a number of metasedimentary and metagranitoid rocks from the Jeseníky mountains. These data suggest that the Desná Dome can be correlated with the Slavkov Terrane, whereas the Keprnik Dome has the isotopic characteristics of the Thaya Terrane.

No precise geochronological information is available for the timing of Cadomian regional metamorphism in the Brunovistulian. K–Ar hornblende and mica ages (Dudek & Melkova 1975) and Th–U–total Pb monazite ages (Finger *et al.* 1999) broadly constrain its age between 610 and 580 Ma, suggesting that regional metamorphism and plutonism are related to one single Cadomian tectonothermal cycle. A zircon evaporation age of 599 ± 2 Ma obtained by Kröner *et al.* (2000) from retrogressed migmatites of the Desná Dome could, in our opinion, provide a reliable date for Cadomian regional metamorphism.

Little is known regarding the age of the volcanic arc material present in the metasedimentary complex of the Slavkov Terrane. The generally primitive isotope compositions suggest a Neoproterozoic age. For volcanic arc-type granitoid gneisses in the Desná Dome, Kröner *et al.* (2000) have determined formation ages of 612 ± 2 and 684 ± 1 Ma. Inherited magmatic zircons in a metagreywacke from the Desná Dome, probably representing volcanic arc detritus, gave evaporation ages of 629 62, 642 63 and 665 61 Ma (Kröner *et al.* this volume). Unlike in the Thaya Terrane, no evidence for Meso- or Palaeoproterozoic zircons has been found in the Precambrian basement of the Desná Dome.

The Central Basic Belt

The Central Basic Belt consists of metagabbros, metadiorites, some ultramafic rocks, tholeiitic metabasalts and metarhyolites. In the exposed Brno Massif, a mainly plutonic subzone in the west can be distinguished from a mainly volcanic subzone in the east. As contacts are generally tectonic, it is unclear if both formed during the same event or represent independent units. The metamorphic overprint of the Central Basic Belt was polyphase (of Cadomian and Variscan age), but did not overstep low to moderate P, T conditions (greenschist to lower amphibolite facies) (Leichmann 1996).

Petrographic and geochemical data for rocks of the Central Basic Belt have been given by Štelcl & Weiss (1986), Hanžl *et al.* (1995), Leichmann (1996), Hanžl & Melichar (1997) and Finger *et al.* (2000). Evaluation of these data shows that in the volcanic zone the melts are mainly basaltic and derived from depleted to mildly enriched mantle sources. According to Leichmann (1996), the degree of enrichment increases towards the north. The metagabbros and metadiorites of the plutonic zone have mostly cumulate compositions. They show high LREE/HREE ratios and seem to be derived from a separate mantle source (Finger *et al.* 2000).

It has been often suggested that the Central Basic Belt or parts of it could represent relics of an ophiolite (Mísař 1979; Leichmann 1996). However, the relative abundance of rhyolites and diorites distinguishes the assemblage lithologically from normal mid-ocean-ridge crust. Alternatively, the belt could be interpreted as part of an ensimatic arc or as related to back-arc basin extension.

Recently, the first geochronological data from the volcanic zone of the Central Basic Belt have been obtained. A zircon evaporation age of 725 ± 15 Ma for a tholeiitic rhyolite suggests that this zone is much older than the granitoids (Finger *et al.* 2000). Leichmann (1996) has reported field observations showing that the Central Basic Belt is intruded by granitoids of the Slavkov Terrane at its eastern margin, as well as by granitoids of the Thaya Terrane at its western margin. This means that at *c.* 580 Ma the present-day assembly of the Central Basic Belt, and the Slavkov and Thaya Terranes could have been basically established. Nevertheless, during the Variscan Orogeny the section was certainly shortened. Parts of the Central Basic Belt were thrust eastwards onto Devonian strata (Hanžl *et al.* 1999). Sinistral strike-slip tectonics led to a further disturbance of the original Cadomian relationships (Hanžl & Melichar 1997).

Discussion and conclusions

A tentative tectonic model

A geodynamic evolution model for the Brunovistulian has to accommodate the following geological observations: presence of a distinctly zoned crust with an island arc-type chemical and isotopic signature in the east (Slavkov Terrane) and a continental signature in the west (Thaya Terrane); remnants of an ensimatic basic belt between the Thaya and the Slavkov Terrane; deposition of large masses of flyschoid, arc-derived sediments in the eastern Slavkov Terrane, probably mainly between 600 and 700 Ma; Cadomian deformation and regional

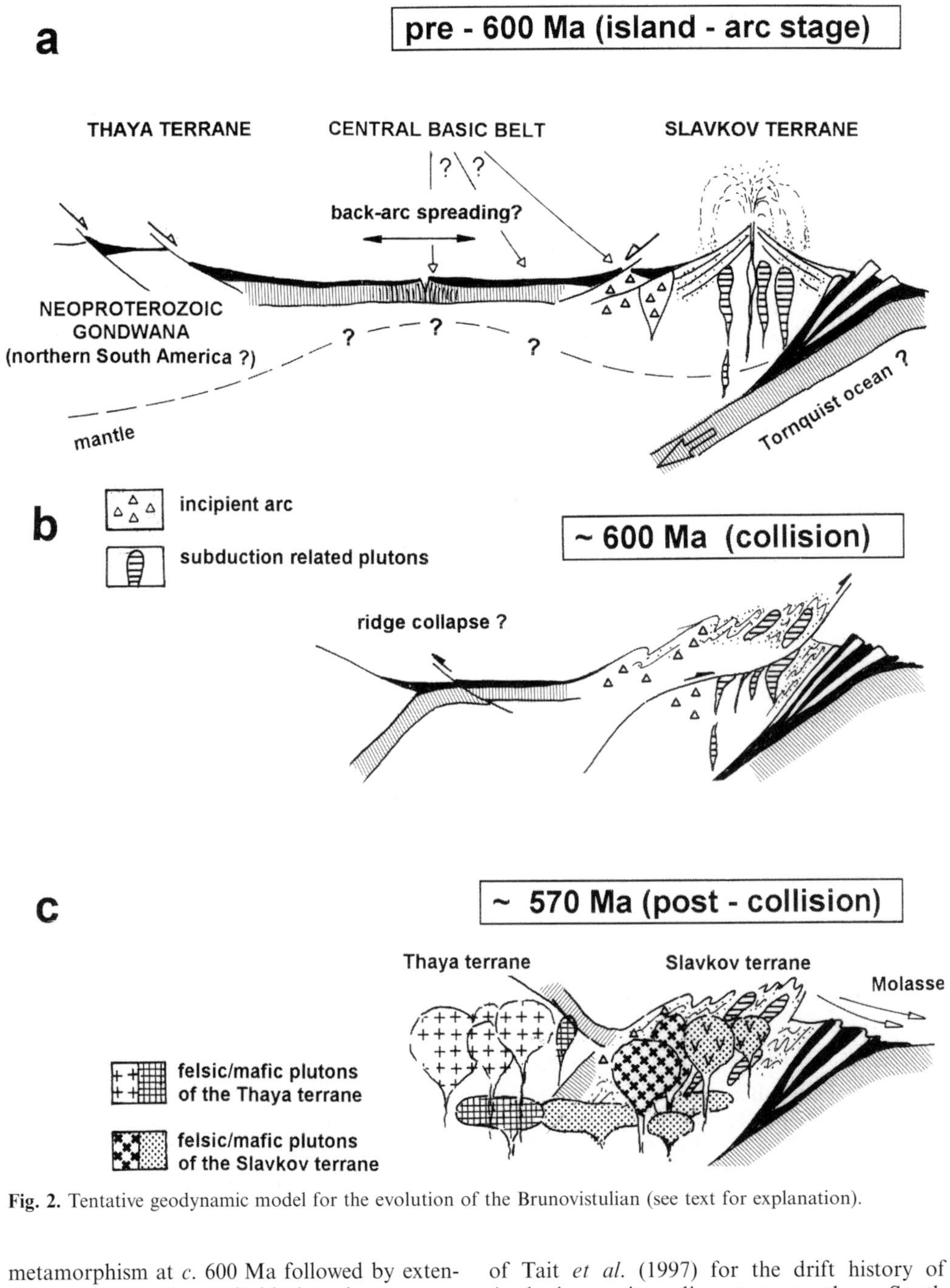

Fig. 2. Tentative geodynamic model for the evolution of the Brunovistulian (see text for explanation).

metamorphism at *c*. 600 Ma followed by extensive post-kinematic granitoid plutonism.

We assume that the starting situation was some kind of island-arc–back-arc basin setting (Fig. 2a), which had formed in Neoproterozoic time in the subduction realm of the Tornquist ocean, outboard of the Gondwana continent. Relying on the palaeomagnetic reconstructions of Tait *et al.* (1997) for the drift history of Avalonia, a site adjacent to northern South America may be visualized. Similar 'Pacific-type' orogenic settings have been proposed for most other Peri-Gondwana terranes of the Avalonian–Cadomian chain at that time (see Nance & Thompson 1996). The existence of a pre-600 Ma volcanic arc is mainly inferred on an indirect basis: first, from the sedimentation of Neoproterozoic calc-alkaline material before 600 Ma in the eastern Brunovistulian; second, from the fact that large volumes of I-type granitoids of the Slavkov Terrane probably represent a remolten meta-igneous arc-type crust. Original volcanic-arc igneous rocks seem to be preserved under the Upper Moravian Basin (metabasalts and meta-andesites; see Dudek 1980) and in the Desná Dome (e.g. Ludvikov Gneiss: 684 ± 1 Ma; Kröner *et al.* this volume).

The geological evidence for the proposed back-arc basin is still weak. As mentioned above, the rocks of the Central Basic Belt may equally well represent primitive parts of an ensimatic arc and it cannot be excluded that this arc formed more or less adjacent to the Gondwana continent margin. However, because during Neoproterozoic time, the northern Gondwana margin was a long-lived active margin (Nance & Thompson 1996), we consider the existence of island arcs and back-arc basins likely. Furthermore, it remains open to discussion whether the inferred back-arc basin between the Slavkov and the Thaya Terrane consisted of remnant crust of the Tornquist Ocean, or contained a new spreading centre. All this can only be a matter of speculation, until a much more detailed chemical and geochronological dataset is available for the rocks of the Central Basic Belt.

Because of the deformation and regional metamorphism, which affected the Brunovistulian at *c.* 600 Ma, we assume that the arc system was in a state of compression at that time (Fig. 2b). We suggest that parts of the Central Basic Belt were obducted onto the Thaya Terrane, so that they could be later intruded by crustal granitoids.

Judging from its episodic nature following regional metamorphism, the *c.* 580–590 Ma granitoid plutonism in the Brunovistulian can be rather viewed as post-collision plutonism than as normal subduction-related plutonism. We see no need to claim that subduction of the Tornquist Ocean still continued during this stage. For example, post-collisional slab break-off, or delamination of mantle lithosphere (see e.g. Henk *et al.* this volume), would also provide a suitable tectonothermal mechanism to trigger voluminous mantle and crustal melting. In fact, such a 'catastrophic' scenario would better account for the pulsed nature of plutonism than continous subduction activity and water-induced melting in the mantle wedge. The chemical zoning of plutonism may reflect simply the pre-existing crustal heterogeneity and the different source rock composition of the Slavkov and the Thaya Terranes. In our model we assume that in the Slavkov Terrane much of the previous arc building was flooded by granitoid melts or remolten at that time.

Comparison of Cadomian basement units in Central Europe

Figure 3 illustrates that the recorded Cadomian events in the Brunovistulian are significantly distinct from those in the Teplá–Barrandian Unit and the Saxothuringian.

A first important difference is that the Brunovistulian does not show evidence for Eocambrian regional metamorphism and tectonics, or for early to mid-Cambrian granitoid plutonism or related volcanism, whereas Dörr *et al.* (1998, in prep.) have pointed out that these two events are widely recorded throughout the Armorican terrane assembly, from central Europe over Brittany to Spain. In the case of the Teplá–Barrandian Unit, Zulauf *et al.* (1999) have interpreted these Eocambrian–Cambrian tectonothermal events

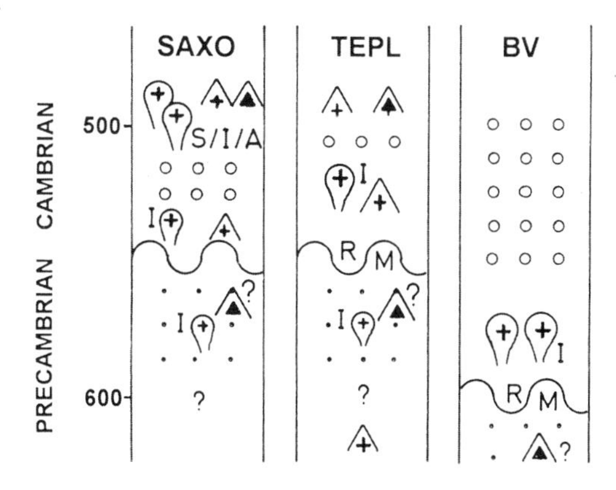

o o Molasse type sediments
• • Flysch type sediments
RM Folding / regional metamorphism
(+) granitoids (S - , I - or A - type)
volcanism
felsic / mafic to intermediate

Fig. 3. Schematic comparison of the Cadomian–Cambrian stratigraphy in the Saxo-Thuringian, Teplá–Barrandian and Brunovistulian, using data from Linnemann (1995), Tichomirowa *et al.* (1997), Dörr *et al.* (1998, 2000, pers. comm.), Linnemann *et al.* (1998), Zulauf *et al.* (1999).

as part of the process of microterrane accretion to the Armorican sector of the Gondwana margin. At the same time, a non-orogenic, broadly Molasse-like overstep sequence is documented on the Polish side of the Brunovistulian (Bula *et al.* 1997) and also in a few boreholes in south Moravia (Jachowicz & Přichystal 1997).

Second, the Brunovistulian apparently lacks evidence for the plutonic–volcanic event at *c.* 500 Ma, which is widely documented in the Armorican parts of the Bohemian Massif (von Quadt 1994; Tichomirowa *et al.* 1997; Glodny *et al.* 1998; Kröner & Hegner 1998) and mostly interpreted to mark the time when Armorica started to rift from Gondwana (Pin 1990; Tait *et al.* 1997; Floyd *et al.* this volume).

On the other hand, there is as yet no evidence for *c.* 600 Ma high-grade regional metamorphism in the Armorican parts of the Bohemian Massif. From the data available at present (Linnemann *et al.* 1998, this volume; Dörr *et al.*, in prep.) it would appear that in Neoproterozoic time, the Armorican crust was in a volcanic-arc setting with an appropriate sedimentary environment.

A comparison of the Brunovistulian with the type locality of Avalonia in Newfoundland (O'Brien *et al.* 1996) reveals broad similarities concerning both lithology and geochronology: like the Brunovistulian, the Avalonian zone (*s.s.*) comprises large amounts of juvenile Neoproterozoic arc-type crust, but also a domain where ancient source material was available (Kerr *et al.* 1995); the oldest preserved rocks are in both cases early Panafrican tholeiitic metabasite sequences (dated at *c.* 780 Ma in Avalonia and *c.* 725 Ma in the Brunovistulian); a major phase of arc magmatism at *c.* 680 Ma in Avalonia matches with the age of the Ludvikov Gneiss in the Desná Dome (684 $\pm$ 1 Ma; Kröner *et al.* this volume); the abundant Neoproterozoic calc-alkaline pyroclastics rocks and volcanogenic turbidites in Avalon could have their analogues in the metasediments of the Slavkov Terrane; arc magmatism and plutonism between 630 and 570 Ma in Avalon overlaps with the time of magmatic activity in the Brunovistulian.

In summary, a correlation of the Brunovistulian with the Avalonian superterrane seems entirely feasible on the basis of the available data.

We thank A. Dudek for constructive comments on an earlier version of this paper as well as for kindly providing sample material. A careful review of H. Fritz is also gratefully acknowledged. Furthermore, we would like to thank W. Dörr and G. Zulauf for stimulating discussion. R. Cooke and D. Tanner helped with the English; G. Riegler with the preparation of the manuscript. The study was supported by the Austrian FWF through Grant 10708.

References

BULA, Z., JACHOWICZ, M. & PŘICHYSTAL, A. 1997. Lower Palaeozoic Deposits of the Brunovistulicum. *Terra Nostra*, **97**(11), 32–38.

DÖRR, W., FIALA, J., VEJNAR, Z. & ZULAUF, G. 1998. U–Pb zircon ages and structural development of metagranitoids of the Teplá crystalline complex—evidence for pervasive Cambrian plutonism within the Bohemian massif (Czech Republic). *Geologische Rundschau*, **87**, 135–149.

DUDEK, A. 1980. The crystalline basement block of the Outer Carpathians in Moravia: Bruno-Vistulicum. *Rozpravy Československé Akademie Věd Řada Matematických a Přirodních Věd*, **90**(8), 3–85.

—— 1995. Metamorphic evolution of the Moravo-Silesian basement. *In*: DALLMEYER, R. D., FRANKE, W. & WEBER, K. (eds) *Pre-Permian Geology of Central and Eastern Europe*. Springer, Berlin, 508–511.

—— & MELKOVA, J. 1975. Radiometric age determination in the crystalline basement of the Carpathian Foredeep and of the Moravian Flysch. *Věstnik Ústředního Ústavu Geologického*, **50**, 257–264.

DVOŘÁK, J. 1995. Stratigraphy of the Moravo-Silesian zone. *In*: DALLMEYER, R. D., FRANKE, W. & WEBER, K. (eds) *Pre-Permian Geology of Central and Eastern Europe*. Springer, Berlin, 477–489.

FINGER, F. & PIN, C. 1997. Arc-type crustal zoning in the Bruno-Vistulicum, Eastern Czech Republic: a trace of the late-Proterozoic Euro-Gondwana margin. *Journal of the Czech Geological Society*, **42**(3), 53.

—— & RIEGLER, G. 1999. Der Thayabatholith und der kristalline Untergrund des Weinviertels. *Arbeitstagung Geologische Bundesanstalt 1999*, 23–31.

—— & STEYRER, H. P. 1995. A tectonic model for the eastern Variscides: indications from a chemical study of amphibolites in the south-eastern Bohemian Massif. *Geologica* Carpathica, **46**, 137–150.

——, FRASL, G., DUDEK, A., JELÍNEK, E. & THÖNI, M. 1995. Cadomian plutonism in the Moravo-Silesian basement. *In*: DALLMEYER, R. D., FRANKE, W. & WEBER, K. (eds) *Pre-Permian Geology of Central and Eastern Europe*. Springer, Berlin, 495–507.

——, ——, HÖCK, V. & STEYRER, H. P. 1989. The granitoids of the Moravian Zone of north-east Austria—Products of a Cadomian active continental margin? *Precambrian Research*, **45**, 235–245.

——, SCHITTER, F., RIEGLER, G. & KRENN, E. 1999. The history of the Brunovistulian: total-Pb monazite ages from the Metamorphic Complex. *GeoLines*, **8**, 21–23.

——, TICHOMIROWA, M., PIN, C. & HANŽL, P. 2000. Relics of an Early-Panafrican Metabasite–Metarhyolite Formation in the Brno Massif, Moravia, Czech Republic. *International Journal of Earth Sciences*, in press.

FLOYD, P. A., WINCHESTER, J. A., SESTON, R., KRYZA, R. & CROWLEY, Q. G. 2000. Review of geochemical variation in Lower Palaeozoic metabasites from the NE Bohemian Massif: intracratonic rifting and plume–ridge interaction. *This volume*.

FRANKE, W. 2000. The mid-European segment of the Variscides: tectono-stratigraphic units, terrane boundaries and plate tectonic evolution. *This volume*.

—— & ENGEL, W. 1982. Variscan sedimentary basins on the continent and relations with south-west England. *Proceedings of the Ussher Society*, **5**, 259–269.

FRASL, G. 1970. Zur Metamorphose und Abgrenzung der Moravischen Zone im niederösterreichischen Waldviertel. *Nachrichten der Deutschen Geologischen Gesellschaft*, **2**, 55–61.

—— 1991. Das Moravikum der Thaya-Kuppel als Teil der variszisch deformierten Randzone des Bruno-Vistulikums—eine Einführung. *Arbeitstagung Geologische Bundesanstalt 1991*, 49–62.

FRIEDL, G. 1997. *U/Pb-Datierungen an Zirkonen und Monaziten aus Gesteinen vom Österreichischen Anteil der Böhmischen Masse*. Phd Thesis, University of Salzburg.

——, MCNOUGHTON, N., FLETCHER, I. R. & FINGER, F. 1998. New SHRIMP-zircon ages for orthogneisses from the south-eastern part of the Bohemian Massif (Lower Austria). *Acta Universitatis Carolinae*, **42**(2), 251–252.

FRITZ, H., DALLMEYER, R. D. & NEUBAUER, F. 1996. Thick-skinned vs thin-skinned thrusting: Rheology controlled thrust propagation in the Variscan collisional belt (the southeastern Bohemian Massif, Czech Republic–Austria). *Tectonics*, **15**, 1389–1413.

FUCHS, G. 1998. Kritische Gedanken zur Neueren Geodynamischen Forschung in der Östlichen Böhmischen Masse. *Jahrbuch der Geologischen Bundesanstalt*, **141**(1), 39–43.

GLODNY, J., GRAUERT, B., FIALA, J., VEJNAR, Z. & KROHE, A. 1998. Metapegmatites in the western Bohemian massif: ages of crystallization and metamorphic overprint, as constrained by U–Pb zircon, monazite, garnet, columbite and Rb–Sr muscovite data. *Geologische Rundschau*, **87**, 124–134.

HANŽL, P. & MELICHAR, R. 1997. The Brno massif: a section through the active continental margin or a composed terrane? *Krystallinikum*, **23**, 33–58.

——, —— & LEICHMANN, J. 1999. Field excursion on Brno Massif. *Geolines*, **8**, 80–95.

——, PŘICHYSTAL, A. & MELICHAR, R. 1995. The Brno Massif: volcanites of the northern part of the metabasite zone. *Acta Universitatis Palackiane Olomucensis, Geologica*, **34**, 73–79.

HEGNER, E. & KRÖNER, A. 2000. Review of Nd isotopic data and xenocrystic and detrital zircon ages from the pre-Variscan basement in the eastern Bohemian Massif: speculations on palinspastic reconstructions. *This volume*.

HENK, A., VON BLANCKENBURG, F., FINGER, F., SCHALTEGGER, U. & ZULAUF, G. 2000. Synconvergent high-temperature metamorphis and magmatism in the Variscides–a discussion of potential heat sources. *This volume*.

JACHOWICZ, M. & PŘICHYSTAL, A. 1997. Lower Cambrian sediments in deep boreholes in south Moravia. *Vistník Českého Geologického Ústavu*, **72**(4), 329–332.

JELÍNEK, E. & DUDEK, A. 1993. Geochemistry of subsurface Precambrian plutonic rocks from the Brunovistulian complex in the Bohemian massif, Czechoslovakia. *Precambrian Research*, **62**, 103–125.

KERR, A., JENNER, G. A. & FRYER, B. J. 1995. Sm–Nd isotopic geochemistry of Precambrian to Palaeozoic granitoid suites and the deep-crustal structures of the southeast margin of the Newfoundland Appalachians. *Canadian Journal of Earth Sciences*, **32**, 224–245.

KRÖNER, A. & HEGNER, E. 1998. Geochemistry, single zircon ages and Sm-Nd systematics of granitoid rocks from the Gory Sowie (Owl Mts), Polish West Sudetes: evidence for early Palaeozoic arc-related plutonism. *Journal of the Geological Society, London*, **155**, 711–724.

——, ŠTÍPSKÁ, P., SCHULMANN, K. & JAECKEL, P. 2000. Chronological constraints on the pre-Variscan evolution of the northeastern margin of the Bohemian Massif, Czech Republic. *This volume*.

LEICHMANN, J. 1996. *Geologie und Petrologie des Brünner Massivs*. PhD thesis, University of Salzburg.

LIEW, T. C. & HOFMANN, A. W. 1988. Precambrian crustal components, plutonic associations, plate environment of the Hercynian Fold Belt of central Europe: indications from a Nd and Sr isotopic study. *Contributions to Mineralogy and Petrology*, **98**, 129–138.

LINNEMANN, U. 1995. The Neoproterozoic terranes of Saxony (Germany). *Precambrian Research*, **73**, 235–249.

——, GEHMLICH, M., TICHOMIROWA, M. *et al.* 2000. From Cadomian subduction to Early Palaeozoic rifting: the evolution of Saxo-Thuringia at the margin of Gondwana in the light of single zircon geochronology and basin development (central European Variscides, Germany). *This volume*.

——, TICHOMIROWA, M., GEHMLICH, M., BUSCHMANN, B. & BROMBACH, K. 1998. Das cadomisch–frühpalaeozoische Basement des Saxothuringischen Terranes (geotektonisches Setting, Geochronologie, Sequenzstratigraphie). *Terra Nostra*, **98**(2), 99–101.

MATTE, P., MALUSKI, H., RAJLICH, P. & FRANKE, W. 1990. Terrane boundaries in the Bohemian Massif: results of large scale Variscan shearing. *Tectonophysics*, **177**, 151–170.

MATURA, A. 1976. Hypothesen zum Bau und zur geologischen Geschichte des Grundgebirges von

Südwestmähren und dem niederösterreichischen Waldviertel. *Jahrbuch der Geologischen Bundesanstalt*, **119**, 63–74.

MísAŘ, Z. 1979. Ultrabazika jako indikátory hlubinné, blokové a vrásové stavby na příklade\e východního okraje Èeského masívu. *Zbornik Predn Konference Smolenice*, **1979**, 1191–210.

NANCE, R. D. & THOMPSON, D. (eds) 1996. *Avalonian and Related Peri-Gondwanan Terranes of the Circum-North Atlantic. Geological Society of America, Special Papers*, **304**.

O'BRIEN, S. J., O'BRIEN, B. H., DUNNING, G. R. & TUCKER, R. D. 1996. Late Neoproterozoic Avalonian and related Peri-Gondwanan rocks of the Newfoundland Appalachians. *In*: NANCE, R. D. & THOMPSON, D. (eds) Avalonian and Related Peri-Gondwanan Terranes of the Circum-North Atlantic. Geological Society of America, Special Papers, **304**, 9–28.

PIN, C. 1990. Variscan oceans: ages, origins and geodynamic implications inferred from geochemical and radiometric data. *Tectonophysics*, **177**, 215–227.

SCHARBERT, S. & BATIK, P. 1980. The age of the Thaya (Dyje) Pluton. *Verhandlungen der Geologischen Bundesanstalt*, **1980**, 325–331.

SCHULMANN, K. 1990. Fabric and Kinematic Study of the Bites Orthogneiss (Southwestern Moravia): Result of Large-Scale Northeastward Shearing Parallel to the Moldanubium/Moravian Boundary. *Tectonophysics*, **177**, 229–244.

ŠTELCL, J. & WEISS, J. 1986. *Brněnský Masív. Universita J. E. Purkyne, Brno.*

TAIT, J. A., BACHTADSE, V., FRANKE, W. & SOFFEL, H. C. 1997. Geodynamic evolution of the European Variscan fold belt: palaeomagnetic and geological constraints. *Geologische Rundschau*, **86**, 585–598.

——, —— & SOFFEL, H. C. 1996. Eastern Variscan fold belt: palaeomagnetic evidence for oroclinal bending. *Geology*, **24**, 871–874.

TICHOMIROWA, M., LINNEMANN, U. & GEHMLICH, M. 1997. Zircon ages as magmatic time marks—comparison of the crustal evolution in different units of the Saxo-Thuringian zone. *Terra Nostra*, **97**(11), 137–141.

VAN BREEMEN, O., AFTALION, M., BOWES, D. R., DUDEK, A., MISAŘ, Z., POVONDRA, P. & VRANA, S. 1982. Geochronological studies of the Bohemian massif, Czechoslovakia, and their significance in the evolution of Central Europe. *Transactions of the Royal Society of Edinburgh: Earth Sciences*, **73**, 89–108.

VON QUADT, A. 1994. U–Pb zircon data and Pb–Sr–Nd isotope chemistry from metagabbros from the KTB-borehole. *Journal of the Czech Geological Society*, **39**(1), 87–88.

ZULAUF, G., SCHITTER, F., RIEGLER, G., FINGER, F., FIALA, J. & VEJNAR, Z. 1999. Age constraints on the Cadomian evolution of the Teplá Barrandian Unit (Bohemian Massif) through electron microprobe dating of metamorphic monazite. *Zeitschrift der Deutschen Geologischen Gesellschaft*, **150**, 627–639.

Review of Nd isotopic data and xenocrystic and detrital zircon ages from the pre-Variscan basement in the eastern Bohemian Massif: speculations on palinspastic reconstructions

ERNST HEGNER[1,3] & ALFRED KRÖNER[2]

[1]*Institut für Geochemie, Universität Tübingen, Wilhelmstraße 56, D-72074 Tübingen, Germany*

[2]*Institut für Geowissenschaften, Universität Mainz, Postfach 3980, D-55099 Mainz, Germany*

[3]*Present address: Institut für Mineralogie, Petrologie und Geochemie, Universität München, Theresienstrasse 41, D-80333 München, Germany*

Abstract: Nd crustal residence ages combined with xenocrystic and detrital zircon ages of Neoproterozoic and early Palaeozoic granitoid gneisses and metasediments from the NE Bohemian Massif (West Sudetes and Erzgebirge) suggest an origin from a basement with significant Grenvillian and Svecofennian–Birimian–Amazonian age components. Large contributions from Archaean crust appear to be missing. Nd model ages of 1.4–1.7 Ga as well as Meso- to Palaeoproterozoic zircon xenocrysts ages for the Lugian Domain (Jizerské, Krkonoše, and Orlica–Snieznik Mountains) are interpreted as evidence for a predominantly Palaeoproterozoic basement that underwent rejuvenation during Meso- and Neoproterozoic–Ordovician magmatic events. Our interpretation of the Nd model ages may be relevant for the western part of the Variscan Belt, where a similar range of Nd model ages characterizes large crustal segments. Samples from the Silesian Domain (Desná Dome, Staré Město Belt, Velké Vrbno Unit) of the Brunia Microcontinent, in thrust contact with the Lugian Domain, yielded Nd crustal residence ages of 1.1–1.3 Ga, which coincide with the oldest ages of xenocrystic zircons. Brunia is interpreted as a predominantly juvenile Grenvillian-age crustal segment, possibly constituting a former arc batholith outboard from the more ensialic setting of the Lugian Domain. Nd crustal residence ages for our 77 whole-rock samples, including 41 new data, as well as 185 xenocrystic and detrital zircon ages from the northern and northeastern Bohemian Massif agree with those of Neoproterozoic rocks from northern South America and support an origin from a northwest Gondwana palaeogeographical location.

Nd isotopic data for the crystalline basement of central and western Europe are consistent with extensive reworking of older crust during the Variscan, Caledonian, and Cadomian orogenic events (Bernard-Griffiths *et al.* 1985; Michard *et al.* 1985; Liew & Hofmann, 1988; Downes *et al.* 1997; Kröner & Hegner, 1998). Liew & Hofmann (1988) interpreted the predominance of 1.4–1.7 Ga Nd mean crustal residence ages in favour of recycling of predominantly Proterozoic crust. An Archaean component, as indicated by Pb isotopic data and detrital zircon ages (e.g. Michard *et al.* 1981; Kober & Lippolt, 1985; Gebauer *et al.* 1989), was considered to be minor.

The evolution of the Variscan belt apparently contrasts with that of many Proterozoic and Phanerozoic orogenic belts that received large volumes of juvenile material through accretion of island arcs (e.g. Colorado Front Range, Sierra Nevada and Peninsular Ranges, California; DePaolo 1981; Canadian Cordillera, Samson *et al.* 1989; Ketilidian Terrane, Patchett & Bridgwater 1984; Arabian-Nubian Shield, Kröner 1985; Reischmann & Kröner 1994; Birimian Domain, West Africa, Abouchami *et al.* 1990). It can be inferred that during the Cadomian, Caledonian and Variscan events, addition of new material from the mantle to the crust was mostly confined to active continental margins.

In the eastern part of the Variscan belt, the Sudetes Mts constitute the northeastern part of the Bohemian Massif and comprise a basement of Neoproterozoic to Cambro-Ordovician age (Oliver *et al.* 1993; Kröner & Hegner 1998; Kröner *et al.* this volume). The various

From: Franke, W., Haak, V., Oncken, O. & Tanner, D. (eds).
Orogenic Processes: Quantification and Modelling in the Variscan Belt.
Geological Society, London, Special Publications, **179**, 113–129. 0305-8719/00/$15.00

lithotectonic units (terranes) of the Sudetes have been interpreted as fragments of the late Neoproterozoic to early Palaeozoic Avalonia and Armorica magmatic arcs, peripheral to West Gondwana (summary of references given by Windley (1995), Nance & Murphy (1996) and Unrug *et al.* (1999)). Rifting of segments from the active continental margin occurred in Cambro-Ordovician times. The accretion of microplates to the margin of Baltica took place during the Caledonian and Variscan Orogenies as a result of closure of the Tornquist and Rheic Oceans, respectively (Ziegler 1986; Franke 1989; Matte *et al.* 1990; Franke *et al.* 1993; Oliver *et al.* 1993; Cymerman *et al.* 1997; Kröner & Hegner, 1998; Kröner *et al.* 2000*a*). The palinspastic reconstruction of the former Neoproterozoic magmatic arcs at peri-Gondwana has only been partly successful, because of terrane dispersion as well as variable tectonic and metamorphic overprinting of the microplates. The lack of data for the composition of the basement at the former northern margin of Gondwana further has limited palinspastic reconstructions.

In this paper we place constraints on the origin and evolution of the Neoproterozoic basement blocks of peri-Gondwana now incorporated into the NE Bohemian Massif. We achieve this by the indirect evidence for the composition and mean age of the basement provided by Nd isotopic compositions of crustally derived granitoid rocks in combination with zircon xenocryst ages. We use the inferred characteristics of the basement to delineate a palaeographical position for the different components of NE Bohemia at the Neoproterozoic margin of peri-Gondwana.

Geological outline of the NE Bohemian Massif

The Bohemian Massif is composed of late Proterozoic to Palaeozoic rock assemblages that represent the easternmost lithotectonic units of the Variscan fold belt in Europe (e.g. Franke 1989; Matte 1991; enclosure at back). Along the eastern margin of the Bohemian Massif, the Moldanubian–Lugian Domain forms the axial zone of the belt. It is in thrust contact with the Neoproterozoic rocks of the Brunia Microcontinent or Silesian Domain to the SE (Fig. 1) (Matte *et al.* 1990; Matte 1991). The tectonic boundary between these major blocks is confined to the Moravo-Silesian Zone. These lithotectonic units are exposed in the western Sudetes Mountains, to the NE of the Elbe Fault Zone (Fig. 1 inset). The Lugian Domain comprises the undeformed Cadomian and minor Variscan granitoid rocks of the Lusatian Massif (LM) in the NW (Kröner *et al.* 1994). To the SE granitoid orthogneisses, blastomylonites and migmatites of Cambrian to Ordovician age extend from the Jizerské Mts through the Krkonoše Mts to the Orlica–Snieznik Dome (Oliver *et al.* 1993; Kröner *et al.* submitted). The Orlica–Snieznik Dome comprises granitic orthogneisses, migmatites as well as medium-grade schists, felsic metavolcanic rocks and amphibolites (Opletal *et al.* 1980; Pbikryl *et al.* 1996; Cymerman *et al.* 1997). Eclogites and felsic granulites occur as small lenses in tectonic contact with the gneisses.

The rocks of the Silesian Domain belong to the Brunia crustal block (Brunovistulicum of Dudek (1980)) and make up the Jeseníky Mts in the Czech East Sudetes (Fig. 2). Brunia has been interpreted as a microcontinent and is composed of Neoproterozoic gneisses and amphibolites, intruded by subordinate Cambrian granitoid rocks (Schulmann & Gayer 2000). The basement comprises a number of structural domains separated by thrusts: the Neoproterozoic–Cambrian Desná Dome, the Cadomian Cervenohorské sedlo Belt, and the Cadomian Keprník Nappe with strongly deformed Neoproterozoic orthogneisses and metasediments (Štípská & Schulmann 1995; Schulmann & Gayer 1999; Kröner *et al.* this volume).

The Moravo-Silesian Zone (Cadomian Velké Vrbno and Cambro-Ordovician Staré Město lithotectonic units) is a narrow NE-trending belt of strongly deformed and metamorphosed rocks of Brunia basement and supracrustal rocks. It defines a thrust belt and forms the SE boundary of the Lugian–Moldanubian Domain (Stípská & Schulmann 1995; Schulmann & Gayer 2000; Kröner *et al.* this volume). The Velké Vrbno Unit is composed of felsic orthogneisses, metapelites and mafic–felsic metavolcanic rocks. The Staré Město Belt consists of high-grade, lower-crustal rocks, tonalitic gneisses, granulite–facies metasediment rocks, bimodal mafic-felsic metavolcanics, ophiolitic fragments and lenses of spinel peridotite.

Analytical methods

Sm–Nd analyses were carried out on *c.* 200 mg of whole-rock powders spiked with an ^{150}Nd–^{149}Sm tracer. The powder was decomposed in a PFA beaker with a mixture of Hf and $HClO_4$, followed by decomposition in Teflon vessels heated to 180 °C for 6 days to ensure the decomposition of refractory phases. Light rare earth elements (LREE) were separated using conventional cation chromatography, and Nd was separated from Sm using HDEHP as cation exchange medium (Richard *et al.* 1976). The total procedural blanks for Nd and Sm are <30 pg and are not

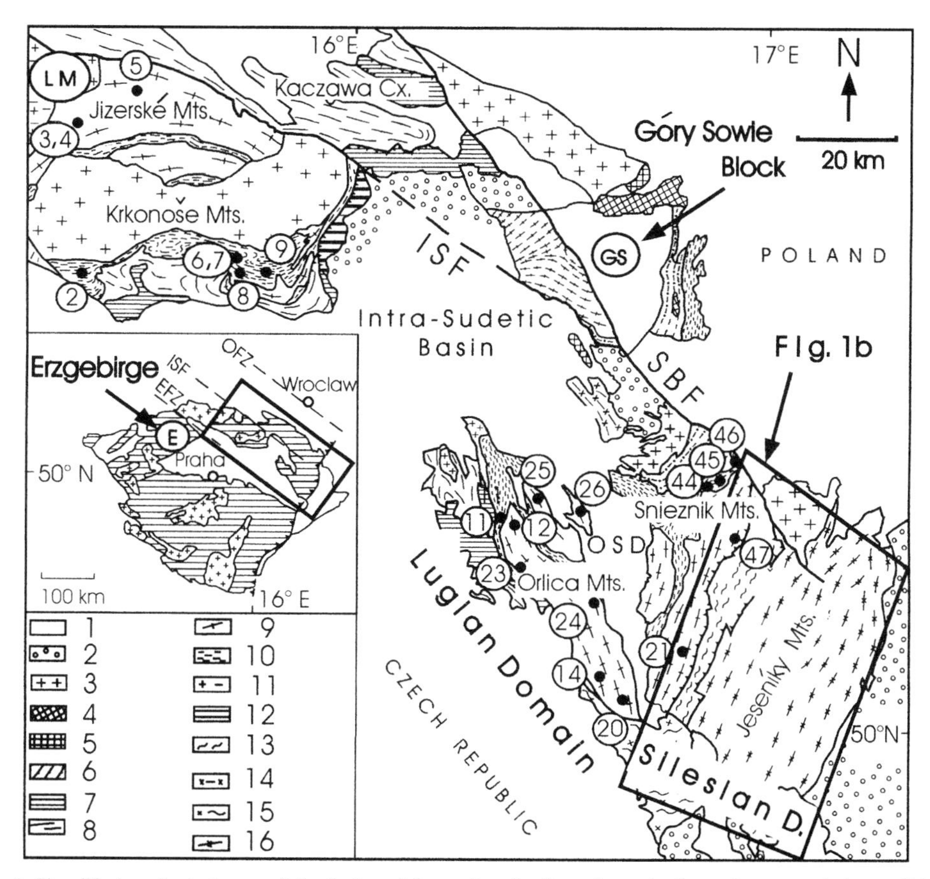

Fig. 1. Simplified geological map of the Sudetes Mountains. Outline of map is shown by rectangle in small inset map of the Bohemian Massif (see also enclosure at back). Circled numbers indicate sample localities and refer to sample numbers in Appendix 1. Sample descriptions are provided in Appendix 1. Spelling of units in Czech. Abbreviations: LM, Lusatian Massif; Kaczawa Cx., Kaczawa Complex. Inset, OFZ, Odra Fault Zone; ISF, Intra-Sudetic Fault Zone; EFZ, Elbe Fault Zone. Arrow indicates the Erzgebirge (E). Large map, SBF, Sudetic Boundary Fault; OSD, Orlica–Snieznik Dome. (For location of Góry Sowie samples (GS) see Kröner & Hegner (1998); for location of Erzgebirge samples (E) see Kröner *et al.* (1995); for location of samples from Lusatian Massif (LM) see Kröner *et al.* (1994). Lithologies in legend: 1, Post-Carboniferous sediment; 2, Devonian–Carboniferous sediment; 3, Variscan granitoid; 4, Devonian serpentinite; 5, Devonian gabbro; 6, Mylonite; 7, Palaeozoic volcanic rock; 8, Palaeozoic metavolcanic rock; 9, Cambrian–Ordovician gneiss; 10, Late Proterozoic(?) mica schist; 11, Cadomian granitoid; 12, Palaeozoic metabasite and gneiss; 13, Late Proterozoic amphibolites and micaschists; 14, Late Proterozoic metasediment and metavolcanic rock; 15, Late Proterozoic–Palaeozoic gneisses and micaschist; 16, Late Proterozoic–Devonian gneiss and metavolcanic rock.

significant for the results reported below. The ^{150}Nd–^{149}Sm tracer solution was calibrated against standards made from ultrapure metals obtained from Ames Laboratories, Iowa State University. Four determinations of the CIT Sm–Nd mixed normal solution yielded a ^{147}Sm/^{144}Nd value of 0.19650 ± 0.009% (1σ) which is only 0.08% lower than the recommended value of 0.19665 reported by Wasserburg *et al.* (1981).

Isotopic ratios were measured with a MAT 262 mass spectrometer at Tübingen University, and ^{143}Nd/^{144}Nd ratios were normalized to ^{146}Nd/^{144}Nd = 0.7219, Sm isotopic ratios to ^{147}Sm/^{152}Nd = 0.56081. During the course of this study the La Jolla Nd standard yielded ^{143}Nd/^{144}Nd = 0.511853 ± 7 (2σ, $n = 6$). Three analyses of the rock standard BCR-1 are listed in Table 1. Within-run precision ($2\sigma_{mean}$) for Nd is $<10^{-5}$. The external precision of the Nd isotopic measurements, as inferred from a large number of standard measurements, is *c.* 1.2×10^{-5}.

Results and discussion

Significance of Nd model and zircon xenocryst ages

Because the rocks analysed in this study were almost exclusively metamorphosed at medium to

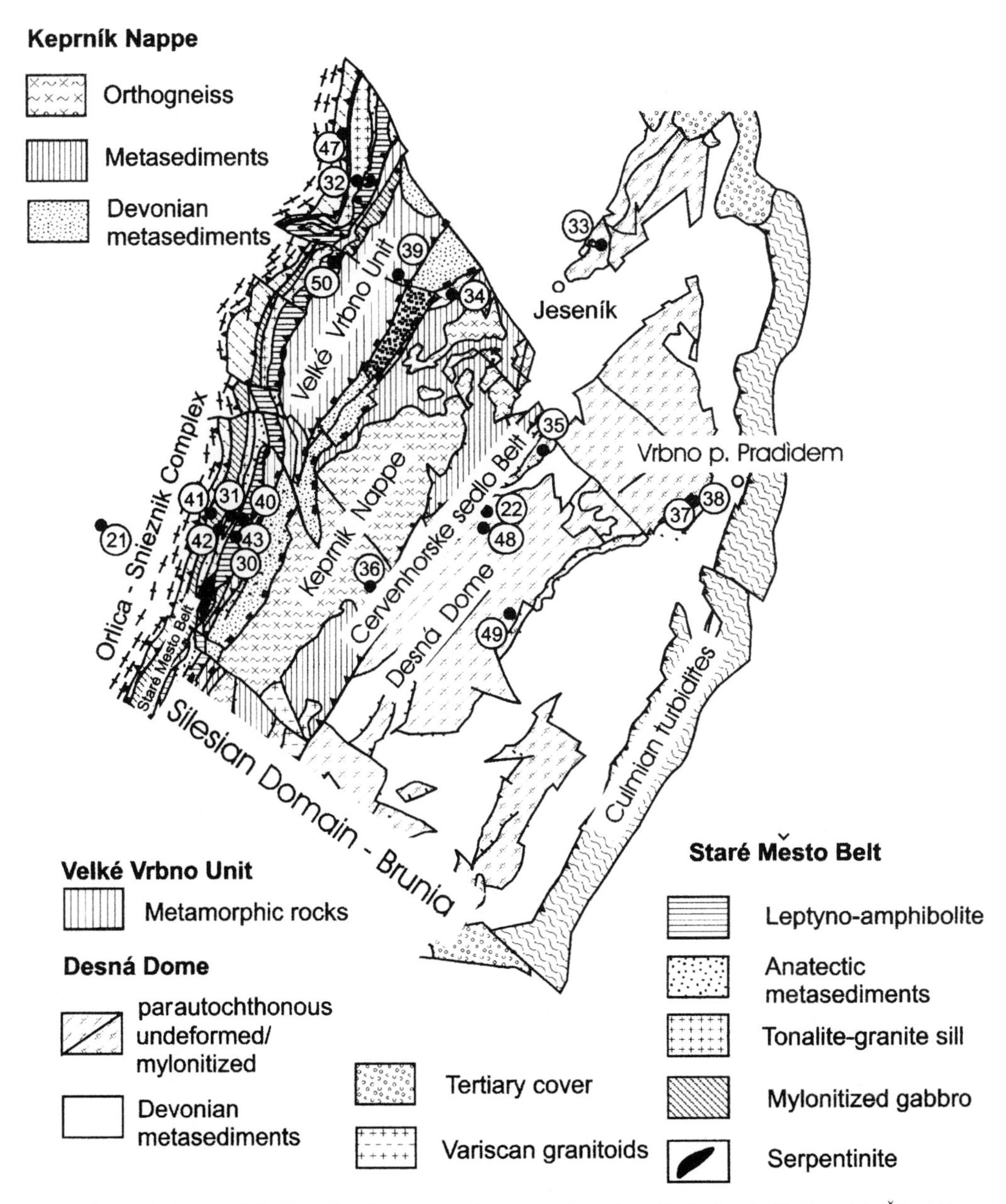

Fig. 2. Geological map of the Jeseníky Mountains, Eastern Sudetes (modified after J. Chab and P. Štípská), showing sample localities in the Desná Dome, the Staré Město Belt, Keprník Nappe, Velké Vrbno Unit and Cervenhorské sedlo Belt. Circled numbers indicate localities of samples.

high degrees, the measured Sm/Nd ratios may not necessarily reflect the original values in the protolith and may therefore lead to spurious model ages. For igneous samples, melt fractionation can also account for changes in the Sm/Nd ratio. The latter process may be responsible for the high $^{147}Sm/^{144}Nd$ ratios in some of the granitoid samples of this study (Table 1). For this reason, the Nd model ages in this paper were calculated according to the two-stage model of Liew & Hofmann (1988). This model uses the measured Sm/Nd ratio for calculation of the initial $^{143}Nd/^{144}Nd$ ratio at the time of rock formation ($\epsilon_{Nd(t)}$), and then uses a typical crustal $^{147}Sm/^{144}Nd$ ratio of 0.12 for the isotopic evolution of the protolith from the depleted

Table 1. *Sm–Nd isotopic data*

Samples	Age (Ma)	Sm	Nd	$^{147}Sm/^{144}Nd$	$^{143}Nd/^{144}Nd$	$\epsilon Nd(t)$	T_{DM}
Lugian Domain							
Jizerské Mountains							
CS 3, Granite–gneiss	506	1.44	5.010	0.1734	0.512290	−5.3	1.6
CS 4, Granite–gneiss	506	2.23	9.250	0.1458	0.512290	−5.0	1.5
CS 5, Granite–gneiss	515	5.06	27.02	0.1132	0.512151	−4.2	1.5
Krkonoše Mountains							
CS 2, Metadacite	(520)	6.73	27.81	0.1463	0.512738	5.3	0.77
CS 6, Granite–gneiss	502	8.58	42.81	0.1211	0.512151	−4.7	1.5
CS 7, Metadacite	(520)	4.45	22.82	0.1179	0.512125	−4.8	1.6
CS 8, Granite–gneiss	503	2.38	8.229	0.1751	0.512307	−5.1	1.6
CS 9, Granite–gneiss	511	1.72	5.747	0.1808	0.512379	−4.0	1.5
Orlica-Snieznik Dome							
CS 11, Metadacite	(520)	5.35	24.59	0.1316	0.512239	−3.5	1.5
CS 12, Granite–gneiss	507	1.83	6.084	0.1814	0.512256	−6.5	1.7
CS 14, Granite–gneiss	505	5.23	22.44	0.1407	0.512137	−6.2	1.7
CS 20, Granite–gneiss	503	5.39	24.16	0.1348	0.512188	−4.8	1.6
CS 21, Granite–gneiss	511	4.96	21.03	0.1426	0.512260	−3.9	1.5
CS 23, Metabasalt	(520)	4.31	16.48	0.1580	0.512804	5.8	0.73
CS 24, Granite–gneiss	(510)	3.70	13.97	0.1601	0.512254	−5.2	1.6
CS 25, Granite–gneiss	(510)	3.22	11.56	0.1685	0.512337	−4.1	1.5
CS 26, Metagreywacke	(520)	2.43	13.16	0.1114	0.511908	−8.6	1.9
CS 44a, Felsic granulite	(500)	7.34	30.93	0.1435	0.512291	−3.7	1.4
CS 44b, Felsic granulite	(500)	6.76	23.66	0.1727	0.512340	−4.4	1.5
CS 45, Felsic granulite	(500)	5.06	23.78	0.1285	0.512258	−3.5	1.4
CS 46, Granite–gneiss	502	5.85	25.94	0.1364	0.512268	−3.8	1.4
CS 47, Metadacite	523	7.36	37.40	0.1189	0.512125	−5.0	1.6
Silesian Domain							
Desná Dome							
CS 22, Granodioritic gneiss	507	4.28	26.00	0.09944	0.512303	−0.3	1.2
CS 33, Granite–gneiss	334	1.05	2.129	0.2984	0.512852	−0.2	1.1
CS 37, Granite–gneiss	685	6.76	32.20	0.1270	0.512354	0.6	1.3
CS 38, Granite–gneiss	517	5.09	24.30	0.1266	0.512178	−4.3	1.5
CS 48, Granite–gneiss	502	3.39	9.207	0.2223	0.512669	−1.0	1.3
CS 49a, Metagreywacke	598	3.98	20.41	0.1180	0.512376	0.9	1.2
CS 49b, Metagreywacke	(629)	4.14	21.98	0.1137	0.512385	1.7	1.2
Cervenohorské sedlo Belt							
CS 35, Granite–gneiss	613	3.75	20.96	0.1082	0.512303	0.4	1.2
Keprník Nappe							
CS 34, Granite–gneiss	555	3.00	17.56	0.1034	0.512059	−4.7	1.6
CS 36, Granite–gneiss	584	3.99	22.14	0.1091	0.511960	−6.7	1.8
Moravo-Silesian Transitional Zone							
Velké Vrbno Unit							
CS 39, Tonalitic gneiss	574	4.57	23.80	0.1160	0.512433	1.9	1.1
Staré Město Belt							
CS 30a, Tonalitic gneiss	503	3.69	15.12	0.1476	0.512651	3.4	0.91
CS 30 c, Melt patch	503	4.04	18.98	0.1286	0.512533	2.3	0.99
CS 31, Pelitic granulite	(520)	2.97	14.81	0.1211	0.512376	−0.1	1.2
CS 32, Tonalitic gneiss	338	7.74	40.49	0.1155	0.512259	−3.9	1.4
CS 40, Felsic metavolcanic rock	(503)	3.96	13.94	0.1720	0.512830	5.3	0.76
CS 41, Leucocr. metagabbro	504	2.68	13.60	0.1193	0.512408	0.5	1.1
CS 42, Tonalitic gneiss	345	8.83	46.14	0.1157	0.512271	−3.6	1.3
CS 50, Felsic metavolcanic rock	503	3.85	12.89	0.1804	0.512898	6.1	0.69
La Jolla ($n = 6$)					0.511853 ± 7 (2σ)		
BCR-1 ($n = 3$)		6.593	28.84	0.1382	0.512636	0.0	

$^{143}Nd/^{144}Nd$: $\leqslant 10^{-5}$ ($2\sigma_{mean}$, within-run error), external precision *c.* 1.2×10^{-5} (2σ). Nd model ages are based on a two-stage isotopic evolution of the sample (Liew & Hofmann 1988).

mantle. Given the uncertainty in the $^{143}Nd/^{144}Nd$ ratio, an error of *c*. 20 Ma may be assigned to the model age; the error in the $\epsilon_{Nd(t)}$ values is *c*. 0.3 units.

We emphasize that most of the Nd model ages reported here represent an average crustal residence times of the samples (Arndt & Goldstein 1987). This interpretation is reasonable for the model ages of the samples from the Lugian Domain, Erzgebirge and Góry Sowie Block because these samples contain xenocryst and detrital zircon with ages in excess of the Nd model age. Mixing of mantle-derived melt with old continental crust will produce Nd model ages in a given igneous rock older than that for the magmatic event generating the rock. In the extreme case of intracrustal melting without addition of juvenile material, the model age will approach that of the average age of the crust. On the other hand, Nd model ages for a number of samples from the Silesian Domain broadly agree with the xenocryst and detrital zircon ages and may thus be interpreted as crust formation ages. In this case, the Nd model age represents the time when all of the crust was derived from the mantle.

The zircon xenocryst ages summarized in this paper are the by-product of a study primarily concerned with the magmatic age of igneous rocks from the northeastern Bohemian Massif. (Kröner *et al.* 1994, 1995, this volume; Kröner & Hegner 1998). For this reason the xenocryst age spectra for each sample are probably not always representative of the entire xenocryst population. As these Pb–Pb single xenocryst ages cannot be verified to be concordant, they are of necessity minimum ages. However, in view of the fact that these ages resulted from high-temperature evaporation steps with no significant changes in their $^{207}Pb/^{206}Pb$ ratios, we are confident that the data points are concordant or nearly so. The internal precision of these measurements is *c*. 0.4% or better (Kröner & Hegner 1998), but in view of the uncertainties listed above we tentatively consider an error of about 2% to be more realistic.

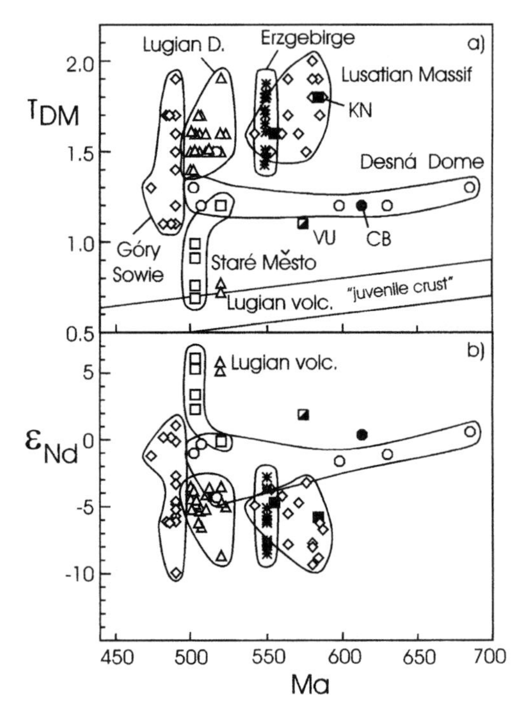

Fig. 3. (**a**) Nd model ages (T_{DM}) for samples from the Sudetes Mts, the Góry Sowie Block, the Lusatian Massif and the Erzgebirge (only rocks of Canbrian age). Region between straight lines delineates the composition of juvenile granitoid crust, allowing up to 200 Ma for the development of granitoids from mantle-derived basaltic protoliths. VU, Velké Vrbno Unit; CB, Cervenhorské sedlo Belt; Lugian D., Lugian Domain of the West Sudetes including the Orlica–Snieznik, Jizerské and Krkonoše Mts, Lugian volc., juvenile metabasalt and metadacite samples; KN, Keprník Nappe. (**b**) Initial ϵ_{Nd} values v. assumed stratigraphic ages of supracrustal rocks and crystallization ages of felsic igneous rocks from the West Sudetes Mts, Czech Republic. Data for granitoid gneisses and metasediments from the Lusatian Massif (Kröner *et al.* 1994), Erzgebirge (Kröner *et al.* 1995) and Góry Sowie Block (Kröner & Hegner, 1998) are shown for comparison. (For sources of emplacement ages, see Appendix 2.)

Variations in Nd isotopic composition and xenocrystic or detrital zircon ages

The initial ϵ_{Nd} values and corresponding Nd model ages of 41 orthogneisses, migmatites as well as metavolcanic and sedimentary rocks from the Lugian–Moldanubian and Silesian domains are listed in Table 1. Sample localities are shown in Figs 1 and 2, and sample descriptions are provided in Appendix 1. In Fig. 3, the Nd model ages and initial ϵ_{Nd} values are plotted together with our previously reported isotopic data from the Erzgebirge (Kröner *et al.* 1995), the Lusatian Massif (Kröner *et al.* 1994), and the Góry Sowie Block (Kröner & Hegner 1998) all being part of the Bohemian Massif (Fig. 1). The initial ϵ_{Nd} values for most samples refer to the crystallization ages determined on zircons (Kröner *et al.* 1994, 1995, this volume; 2000*a*; Kröner & Hegner 1998) (Appendix 2). Ages in parentheses represent values inferred from the geological context. Age uncertainties of up to 30 Ma have little effect on the calculated initial Nd isotopic ratios for rocks with typical crustal Sm/Nd ratios.

Lugian Domain (Jizerské Mts, Krkonoše Mts, Orlica–Snieznı́k Dome, Lusatia, Góry Sowie and Erzgebirge)

The predominantly mildly peraluminous granitoid rocks of the Cambro-Ordovician Lugian magmatic province (Kröner *et al.* 2000*a*) extend from the Jizerské Mts through the Krkonoše Mts to the Orlica–Snieznı́k Dome and display a moderate range in initial ϵ_{Nd} values of -3.5 to -6.5, corresponding to Nd model ages of 1.4–1.7 Ga. A metagreywacke sample (CS 26), interpreted as reflecting the average composition of this domain, has a low initial ϵ_{Nd} value of -8.6. Its provenance has an average crustal residence time of *c.* 1.9 Ga, similar to that of many Cadomian gneisses from the Lusatian Massif (Kröner *et al.* 1994).

A chemically primitive metabasalt (CS 23) from the Nové Město supracrustal unit and interpreted as rift basalt (Floyd *et al.* 1996), yielded an initial ϵ_{Nd} value of 5.8, indicating a mid-ocean ridge basalt (MORB)-like upper mantle below this segment of crust. A similarly high value of 5.3 was measured for a metadacite of the Nové Město Group (CS 2) implying that it originated by melting of underplated rift-related basaltic material.

The overall narrow range of negative ϵ_{Nd} values provides convincing evidence for an origin of the Cambrian granitoids from an isotopically uniform crustal source with a Mesoproterozoic mean crustal residence age. As the granites of the Lugian Domain are typically calc-alkaline (Kröner *et al.* 2000*a*) and contain xenocrystic zircons as old as late Archaean age, they are not likely products of melting of an underplated mafic source in an extensional setting but may have been formed in an Andean-type tectonic setting. A chemically uniform sedimentary protolith such as a metagreywacke is also appealing to explain the relatively uniform Nd model ages. Melting of greywacke prisms between accreting microplates can account for a predominantly granitic composition and isotopic homogeneity of the Lugian orthogneisses. Patchett & Bridgwater (1984) suggested that repeated recycling of crust in a subduction zone may produce similar isotopic ratios in crustal rocks, and this may be an alternative explanation. Considering the predominantly calc-alkaline composition of the orthogneisses and an absence of A-type, rifting-related granitoids, we suggest that the

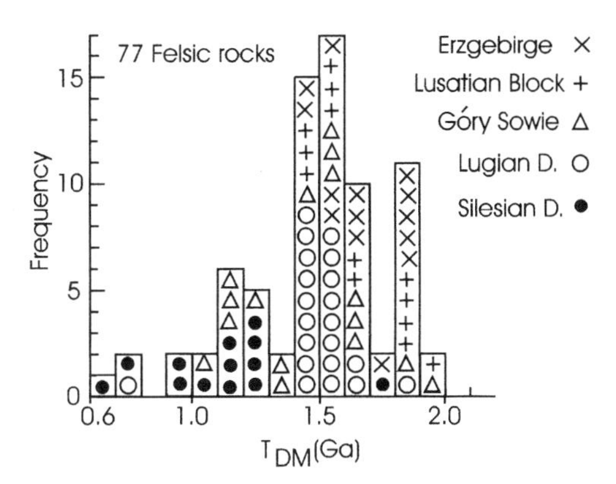

Fig. 4. Histogram showing distribution of Nd model ages from the NE margin of the Bohemian Massif (Erzgebirge, Lusatian Massif and Sudetes Mts). Data sources: this study, Kröner *et al.* 1994, 1995); Kröner & Hegner (1998). Symbols within columns denote different domains of Northern and NE Bohemia.

Cambro-Ordovician magmatic event in the Lugian Domain is probably related to subduction of the Tornquist Ocean involving recycling of Palaeo- to Mesoproterozoic crust and minor input of juvenile material from the mantle.

When comparing the Nd isotopic compositions of the Lugian samples with those for the Lusatian Massif, the Góry Sowie Block and the Erzgebirge (Fig. 1) at the northern and north-eastern margins of the Bohemian Massif, all of these regions show involvement of a Palaeoproterozoic endmember with a mean crustal residence age of *c.* 2 Ga (Figs 3 and 4). We suggest that this material represents the basement to these magmatic province and that the data spread towards younger model ages in each of these regions may be attributed to the addition of variable amounts of mantle-derived material during magmatism at continental margins. For a palinspastic reconstruction of the former site of these crustal blocks a Palaeoproterozoic basement that underwent crustal rejuvenation during Neoproterozoic, Cadomian and Cambro-Ordovician times needs to be found. Some of the gneisses of the Góry Sowie Block have higher initial ϵ_{Nd} values than those from the neighbouring units, indicating a crustal precursor with a higher proportion of juvenile material and, by implication, originating from a different palaeogeographical location.

The xenocrystic and detrital zircon grains range in age between 687 and 2620 Ma (Appendix 2) with most ages in the brackets 600–850 Ma (Cadomian–Pan-African age), 1019–1328 Ma (Grenvillian age) and 1712–2233 Ma (Svecofennian–Birimian–Amazonian age). Furthermore, there are rare Archaean grains up to 2932 Ma in age. A small but significant amount of zircons in the age range 1401–1677 Ma cannot be ascribed to any known major orogenic event in North Africa but may be related to the Rondonian and Rio Negro events (Teixeira *et al.* 1989) in the Amazon Craton of South America (see Fig. 6, below).

Silesian Domain (Desná Dome, Cervenhorské sedlo Belt and Keprník Nappe)

Neoproterozoic to Cambrian samples from the Desná Dome and the Cervenhorské sedlo Belt have similar Nd model ages of 1.2–1.3 Ga (except for one Cambrian granitic dyke sample with a model age of 1.5 Ga), suggesting that these lithotectonic units have a common origin (Table 1). The crustal residence time for the Silesian Domain is distinctly younger than the 1.4–1.7 Ga crustal residence time typical for the Lugian Domain and implies different palaeogeographical locations for these lithological units.

The small variation in the model ages suggests that Neoproterozoic as well as Cambrian rocks were derived from a similar crustal protolith with a Grenvillian crustal residence time. Grenvillian characteristics of the domain are supported by the data for two metagreywacke samples (CS 49a, b), which are interpreted as a good estimate for the overall composition of the domain. Because of the recurring Grenvillian Nd model ages in the Desná Dome and the Cervenhorské sedlo Belt, in combination with 1.0–1.1 Ga xenocrystic zircon ages (see below), we suggest that this crustal block was derived from a predominantly juvenile Grenvillian basement of northern Gondwana.

Two Cadomian orthogneisses of the Keprník Nappe are isotopically distinct from the neighbouring units (Fig. 3). They yield Nd model ages of 1.6–1.8 Ga as have been found in the Lugian Domain (Fig. 4). These data suggest that the nappe is exotic within the Silesian Domain.

The Silesian Domain is the only area of our study where only Neoproterozoic xenocrystic and detrital zircons were found. The majority of these can be ascribed to the Cadomian–Pan-African event, whereas a few grains range in age between 1019 and 1106 Ma and suggest input from a Grenvillian source.

The Moravo-Silesian Transitional Zone

A pelitic granulite sample (CS 31), interpreted to represent an average composition of the exposed upper basement of the Staré Město Belt, yielded a model age of 1.2 Ga. The Staré Město Belt may thus be interpreted as the westernmost continuation of the Silesian Domain. The belt also contains a *c.* 500 Ma bimodal mafic–felsic association that has been interpreted as an evolving rift sequence (Floyd *et al.* 1996; Kröner *et al.* this volume). The felsic metavolcanic rocks (CS 40, CS 50) have Nd model ages of 0.69–0.76 Ga indicating significant involvement of juvenile material in their generation and consistent with the assumed rift setting. Their origin may be related to underplating of basaltic melts during crustal attenuation. Two samples of *c.* 340 Ma Variscan tonalites (CS 32, CS 42) yielded model ages of 1.1–1.3 Ga that are similar to those for the *c.* 500 Ma rift-related mafic rocks in the Staré Město Belt. The Variscan tonalites may therefore represent partial melts of a mafic source formed at *c.* 500 Ma during the rift stage.

The few xenocrystic and detrital zircon ages from two samples of this zone cannot be considered as representative but suggest that a Cadomian–Pan-African source was predominant.

Origin of the pre-Variscan basement of the NE Bohemian Massif

With the inferred Nd isotopic characteristics of the basement in the NE Bohemia Massif and the ages of xenocrystic and detrital zircons in the Neoproterozoic to early Palaeozoic rocks (Appendix 2) we have important constraints for inferring possible sites of origin for the various lithotectonic units. On the basis of Nd model ages and detrital zircon ages, Nance & Murphy (1994) proposed a position for Western Avalonia near the Amazonian Craton and of Eastern Avalonia and Armorica near the West African Craton. A detailed study of detrital zircons from a paragneiss of the Moldanubian basement by Gebauer *et al.* (1989) revealed a lack of Mesoproterozoic ages, as is typical for the West African Craton. By inference, this is an unlikely palaeogeographical location for the Moldanubian basement. Below we address the ages of basement components on the basis of 77 Nd isotopic analyses obtained recently, including 41 new analyses of this study. In addition, we use 185 ages for xenocrystic and detrital zircons from the NE Bohemian Massif.

In the following we investigate the significance of the predominance of 1.4–1.7 Ga Nd model ages in the NE Bohemian Massif. In Fig. 4, the 77 Nd model ages of this study and our previously reported results for Cambro-Ordovician and Neoproterozoic granitoid gneisses and supracrustal rocks from the

Erzgebirge (Kröner *et al.* 1995) (Fig. 1, inset), the Lusatian Massif (Kröner *et al.* 1994), and the Góry Sowie (Kröner & Hegner 1998) show a range from 1 to 2 Ga with a pronounced peak at *c.* 1.4–1.7 Ga, and a minor peak at 1.0–1.2 Ga. The latter peak represents rocks from the Staré Město Belt and Desná Dome, with some of the youngest model ages of this study. There is a notable absence of average crustal residence ages >2 Ga, except for one sample from the Lusatian Massif (Kröner *et al.* 1994). Young model ages (<1 Ga), indicating juvenile crust, are sparse and represent the rift-related rocks of the Staré Město Belt and Lugian Domain. The frequency distribution of Nd model ages in Fig. 4 is similar to the range of model ages for other parts of the Bohemian Massif (Kröner *et al.* 1988; Liew & Hofmann, 1988; Liew *et al.* 1989; Nelson 1992) and the Variscan Belt in France (Bernard-Griffiths *et al.* 1985; Michard *et al.* 1985), suggesting an overall similar composition for these Gondwana-derived domains on a large scale.

To what extent the Nd model ages reflect mixing between Archaean–Proterozoic and Palaeozoic mantle-derived material may be addressed with the ages of single xenocrystic or detrital zircons from the same samples for which we report Nd model ages. These ages are shown in Fig. 5 for various regions of the northern and northeastern margin of the Bohemian Massif. In Fig. 5f, the ages for 185 xenocrystic and detrital zircons exhibit a skewed distribution with a maximum of samples yielding Neoproterozoic ages (600–900 Ma, Cadomian–Pan African). Further distinct peaks are at 1.0–1.3 Ga (Grenvillian) and 1.7–2.1 Ga (Svecofennian–Birimian–Amazonian). The number of late Archaean zircon ages and, by inference, the contribution of such sources, is apparently subordinate.

With the zircon age spectrum, the predominance of 1.4–1.7 Ga model ages for Neoproterozoic and early Palaeozoic rocks from NE Bohemia may be interpreted in a straightforward fashion. Because of the small number of Archaean zircon ages, contrasted by a large number of Palaeoproterozoic zircon ages, material with Svecofennian–Birimian–Amazonian age characteristics appears to be an important component of the basement. As it can be assumed that the basement was rejuvenated during the Grenvillian and Neoproterozoic to Ordovician magmatic events, the predominance of early Mesoproterozoic Nd crustal residence ages supports the zircon evidence for a basement with a large proportion of Palaeoproterozoic material. The amount o f juvenile material added during the young magmatic events must have been subordinate. Our interpretation of the Nd model ages may also be relevant to the western part of the Variscan Belt, where 1.4–1.7 Ga Nd model ages characterize large crustal segments (Michard *et al.* 1985; Liew & Hofmann, 1988; D'Lemos & Brown 1993).

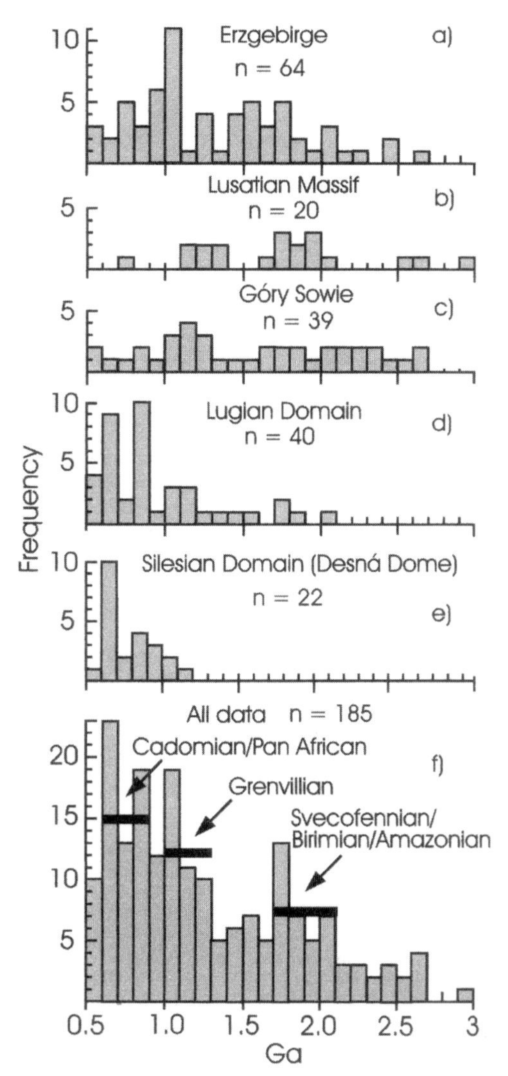

Fig. 5. Histograms showing distribution of xenocrystic and detrital zircon ages for basement rocks in the NE Bohemian Massif: Erzgebirge (**a**), Lusatian Massif (**b**), Góry Sowie Block (**c**), Lugian Domain (**d**), Desná Dome of the Silesian Domain (**e**). Summary diagram (**f**) shows coincidence of zircon ages with major orogenic events. (For data sources see Appendix 2).

With these constraints on the composition of the NE Bohemian Massif we may now look into the geological and isotopic record of Baltica and northern Gondwana as the two likely candidates for a suitable provenance from which the individual components of NE Bohemian Massif may have been derived. The Grenvillian, Svecofennian and Archaean zircon populations in the Bohemian rocks may suggest a Scandinavian provenance, but the ubiquitous presence of Cadomian–Pan-African zircons argues against this interpretation. There is no record of a major Neoproterozoic magmatic or orogenic event at 600–850 Ma in the southern and southeastern Baltic Shield (e.g. Claesson 1987; Williams & Claesson 1987; Knudsen *et al.* 1997), and Baltica is therefore an unlikely candidate for the origin of the Bohemian terranes.

We now turn to northern Gondwana as an alternative candidate. Pb isotopes and Nd model ages for the Arabian–Nubian Shield indicate that large regions were derived from the mantle during the Pan-African event between *c.* 550 and *c.* 900 Ma (Bokhari & Kramers 1981; Stacey & Stoeser 1983; Pallister *et al.* 1987; Reischmann & Kröner 1994; Stein & Goldstein 1996). An Archaean component has been detected in one relatively small area in the Arabian Shield but is subordinate and characterizes a distinct block, interpreted as a microcontinent (Stacey & Stoeser 1983; Pallister *et al.* 1987). The Neoproterozoic juvenile crustal domain of NE Africa probably extends as far east as Morocco (Cahen *et al.* 1984). From the geological record of regions along the northern margin of Gondwana as compiled by Nance & Murphy (1996) (Fig. 6) there is no evidence for major regions in northern Africa having formed during the 0.9–1.9 Ga period. This makes it unlikely that the terranes discussed in this paper have been derived from this part of Gondwana, except, perhaps, for the Silesian Domain (Brunia microcontinent), where our record shows only young xenocryst and Nd model ages (Figs. 4 and 5e). The presence of a significant population of Grenvillian-age xenocrystic zircons in most of the Bohemian terranes (Fig. 5f) is in strong contrast to the distinct lack of Grenville-age crust in North Africa (Rocci *et al.* 1991; see Fig. 6). This further corroborates our conclusion that northern and northeast Bohemia was not derived from this part of Gondwana.

In SE Mexico as well as the Ecuadorian and Colombian Andes, evidence for the existence of Grenvillian crustal component has been confirmed by whole-rock Nd model and detrital zircon ages of 1.3–1.6 Ga (Patchett & Ruiz, 1987; Noble *et al.* 1997; Restrepo-Pace *et al.* 1997), and Rb–Sr and K–Ar geochronology (Kroonenberg 1982; Priem *et al.* 1989). The range in Nd model ages in northern South America is remarkably similar to that of the Bohemian terranes, except for the Silesian Domain. Noble *et al.* (1997) reported three detrital zircon and discordia upper intercept ages, documenting the presence of detritus from Neoproterozoic (Pan-African–Cadomian), Palaeoproterozoic and Archaean crust in Ecuador. These workers tentatively suggested mixing of Archaean sources from the Amazonian Craton with juvenile Neoproterozoic–Cambrian crust similar to that form Western Avalonia. Because of the small amount of detrital zircons, recycling of 1.0–1.6 Ga crust was also considered plausible because of the proximity of Ecuador to Mesoproterozoic domains in the Amazonian Craton. The overall similarity of Nd model ages (Restrepo-Pace *et al.* 1997) and xenocrystic zircon ages between the Amazonian Craton (Fig. 6) and the Bohemian terranes discussed here support our conclusion that the Lugian Domain, the Erzgebirge, the Góry Sowie Block and the Keprník Nappe from the Silesian Domain were part of a magmatic arc situated on Grenvillian and older basement along the margin of the Amazonian Craton (Venezuela, Colombia, Ecuador). This contrasts with the proposal of Nance & Murphy (1994), who suggested derivation of Bohemia from the northern margin of West Africa.

For the Silesian Domain of the Brunia microcontinent, Grenvillian Nd model ages of predominantly 1.1–1.3 Ga indicate a different palaeotectonic position at peri-Gondwana than for the Lugian Domain with Nd model ages of 1.4–1.9 Ga. The oldest xenocrystic zircons in

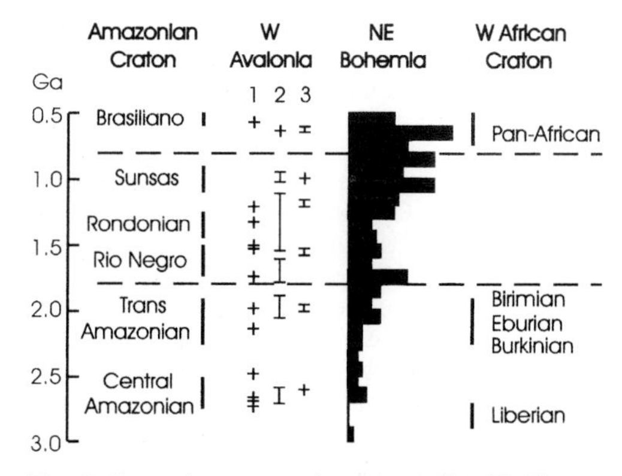

Fig. 6. Orogenic events and corresponding U–Pb ages for detrital zircons of northern Gondwana crustal components (modified from Nance & Murphy (1996)). A frequency distribution for the ages of xenocrystic and detrital zircons for the NE Bohemian Massif is shown for comparison. For Western Avalonia: column 1, ages for New England; column 2, New Brunswick; column 3, Nova Scotia.

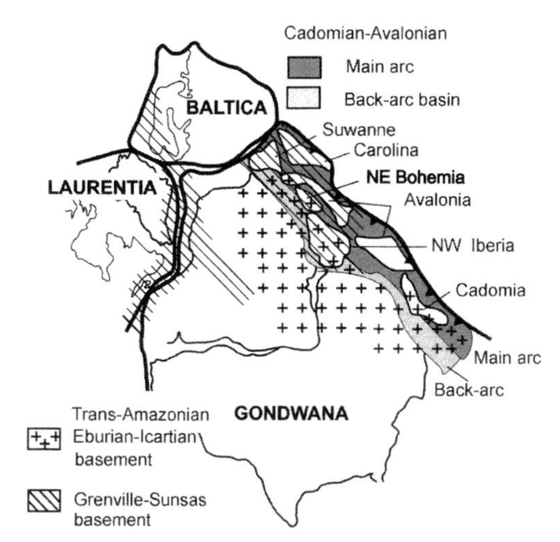

Fig. 7. Model for late Neoproterozoic distribution of terranes along the northern margin of Gondwana (modified from Fernández-Suárez *et al.* 2000) showing our preferred position of NE Bohemia within the magmatic arc–back-arc region of the Amazonian Craton.

rocks from Brunia are Grenvillian in age and coincide with most of the Nd model ages. This suggests that the Nd model ages reflect the time of crust formation, and Brunia may be considered as a fragment of predominantly juvenile Grenvillian crust. This crust could have been situated on the margin of the Amazonian Craton (Fig. 7).

Juvenile crust may be generated in island arcs that become granitic in composition after repeated melting, or it may be formed as an arc batholith with limited input of older crust. We prefer the explanation that the Silesian Domain was originally an arc batholith outboard from the more ensialic Lugian basement (Fig. 7), because a single granite from this region (CS 38) yielded a Nd model age of 1.5 Ga, indicating the presence of Mesoproterozoic material.

Conclusions

Combined Nd model ages as well as xenocrystic and detrital zircon ages of Neoproterozoic and early Palaeozoic granitoid gneisses and metasediments from the NE Bohemian Massif suggest that these rocks were derived from a basement with Grenvillian and older crustal components at the northern margin of Gondwana. Nd model ages of 1.4–1.7 Ga for the Lugian Domain, Erzgebirge, Lusatian Massif and Góry Sowie, in combination with xenocrystic and detrital zircon ages, are explained by mixing of predominantly Palaeoproterozoic material with subordinate amounts of juvenile Grenvillian age and Neoproterozoic to early Palaeozoic material. The zircon data do not allow for presence of a large proportion of Archaean crust.

The Silesian Domain of the Brunia Microcontinent yielded Nd model ages of 1.1–1.3 Ga that coincide with the oldest ages of xenocrystic zircons. We interpret these data as indicating a Grenvillian crustal segment with a predominantly juvenile composition. The Silesian Domain is interpreted as a former arc batholith outboard from the more ensialic setting of the Lugian Domain.

Our preferred position for the NE margin of the Bohemian Massif, before rifting along northern Gondwana, is on the northern margin of the Amazonian Craton in South America. An earlier proposed palaeogeographical position of Bohemia near West Africa is not supported because the lack of Grenvillian crust in this part of Gondwana.

We are grateful to P. Jaeckel, M. Opletal, K. Schulmann, P. Štípská and A. Zelazniewicz for help during sample collection and discussions. Improvements to the manuscript by F. Henjes-Kunst and T. Reischmann are appreciated. We thank J. Chab and Pavla Štípská for providing the map of Fig. 2, and G. Gutiérrez-Alonso for permission to reproduce Fig. 7. D. Tanner is thanked for editorial handling. Financial support to both authors by the Deutsche Forschungsgemeinschaft (DFG) within the Priority Programme 'Orogenic Processes' (grants Kr 590/35 and He 1857/3) is gratefully acknowledged. This is a contribution to EUROPROBE.

References

ABOUCHAMI, W., BOHER, M., MICHARD, A. & ALBARÈDE, C. J. 1990. A major 2.1 Ga event of mafic magmatism in West Africa: an early stage of crustal accretion. *Journal of Geophysical Research*, **95**(B11), 17605–17629.

ARNDT, N. T. & GOLDSTEIN, S. L. 1987. Use and abuse of crust-formation ages. *Geology*, **15**, 893–895.

BERNARD-GRIFFITHS, J., PEUCAT, J. J., SHEPPARD, S. & VIDAL, Ph. 1985. Petrogenesis of Hercynian leucogranites from the southern Armorican Massif: contribution of REE and isotopic (Sr, Nd, Pb and O) geochemical data to the study of source rock characteristics and ages. *Earth and Planetary Science Letters*, **74**, 235–250.

BOKHARI, F. Y. & KRAMERS, J. D. 1981. Island arc character and Late Precambrian age of volcanic at Wadi Shwas, Hijaz, Saudi Arabia: Geochemical and Sr and Nd isotopic evidence. *Earth and Planetary Science Letters*, **54**, 409–422.

CAHEN, L., SNELLING, N. J., DELAHL, J. & VAIL, J. R. 1984. *The Geochronology and Evolution of Africa*. Clarendon Press, Oxford.

CLAESSON, S. 1987. Isotopic Evidence for the Precambrian Provenance and Caledonian Metamorphism of High Grade Paragneisses from the Seve Nappes, Scandinavian Caledonides. I. Conventional U–Pb and SM–ND Whole Rock Data. *Contributions to Mineralogy and Petrology*, **97**, 196–204.

CYMERMAN, Z., PIASECKI, M. A. J. & SESTON, R. 1997. Terranes and terrane boundaries in the Sudetes, northeast Bohemian Massif. *Geological Magazine*, **134**, 717–725.

DEPAOLO, D.-J. 1981. A neodymium and strontium isotopic study of the Mesozoic calc-alkaline granitic batholiths of the Sierra Nevada and Peninsular Ranges, California. *Journal of Geophysical Research*, **86**(B11), 10470–10488.

D'LEMOS, R. S. & BROWN, M. 1993. Sm–Nd isotope characteristics of late Cadomian granite magmatism in northern France and the Channel Islands. *Geological Magazine*, **130**, 797–804.

DOWNES, H., SHAW, A., WILLIAMSON, B. J. & THIRLWALL, M. F. 1997. Sr, Nd and Pb isotopic evidence for the lower crustal origin of Hercynian granodiorites and monzogranites, Massif Central, France. *Chemical Geology*, **136**, 99–122.

DUDEK, A. 1980. The crystalline basement block of the outer Carpathians in Moravia. *Rozpravy Aeskoslovenské Akademie Ved*, **90**, 1–85.

FERNÁNDEZ-SUÁREZ, J., GUTIÉRREZ-ALONSO, G., JENNER, G. A., TUBRETT, M. N. 2000. New ideas on the Proterozoic–early Palaeozoic evolution of NW Iberia: insights from U–Pb detrital zircon ages. Precambrian Research, in press.

FLOYD, P. A., WINCHESTER, J. A., LEWANDWSKA, A. SZCZEPANSKI, J. & TURNIAK, K. 1996. Geochemistry of early Palaeozoic amphibolites from the Orlica–Snieznik dome, Bohemian Massif: petrogenesis and palaeotectonic aspects. *Geologische Rundschau*, **85**, 225–238.

FRANKE, W. 1989. Tectonostratigraphic units in the Variscan belt of central Europe. *In*: DALLYMEYER, R. D. (ed.) *Terranes in the Circum-Atlantic Palaeozoic Orogens*. Geological Society of America, Special Papers, **230**, 67–90.

——, ZELAZNIEWICZ, A., POREBSKI, S. J. & WAJSPRYCH, B. 1993. The Saxothuringian zone in Germany and Poland: differences and common features. *Geologische Rundschau*, **82**, 583–599.

GEBAUER, D., WILLIAMS, I. S., COMPSTON, W. & GRÜNENFELDER, M. 1989. The development of the central European crust since the Early Archaean based on conventional and ion-microprobe dating of up to 3.84 b.y. old detrital zircons. *Tectonophysics*, **157**, 81–96.

KNUDSEN, T.-L., ANDERSEN, T. & WHITEHOUSE, M. J. 1997. Detrital zircon ages from southern Norway—implications for the Proterozoic evolution of the southwestern Baltic Shield. *Contributions to Mineralogy and Petrology*, **130**, 47–58.

KOBER, B. & LIPPOLT, H. J. 1985. Pre-Hercynian mantle Pb transfer to basement rocks as indicated by Pb isotopes of the Schwarwald crystalline, SW Germany. *Contributions to Mineralogy and Petrology*, **90**, 172–178.

KRÖNER, A. 1985. Ophiolites and the Evolution of Tectonic Boundaries in the Late Proterozoic Arabian–Nubian Shield of Northeast Africa and Arabia. *Precambrian Research*, **38**, 367–380.

—— & HEGNER, E. 1998. Geochemistry, single zircon ages and Sm–Nd systematics of granitoid rocks from the Góry Sowie (Owl Mts), Polish West Sudetes: evidence for early Palaeozoic arc-related plutonism. *Journal of the Geological Society, London*, **155**, 711–724.

——, ——, HAASE, G., BIELICKY, K.-H., KRAUSS, M. & EIDAM, J. 1994. Geochronology and Nd–Sr systematics for Lusatian granitoids: significance for the evolution of the Variscan orogen in east-central Europe. *Geologische Rundschau*, **83**, 357–376.

——, ——, WILLNER, A., FRISCHBUTTER, A., HOFMANN, J. & BERGNER, R. 1995. Lates Precambrian (Cadomian) zircon ages and Nd isotopic systematics for granitoid orthogneisses of the Erzgebirge, Saxony and Czech Republic. *Geologische Rundschau*, **84**, 437–456.

——, JAECKEL, P., HEGNER, E. & OPLETAL, M. 2000*a*. Single zircon ages and whole-rock Nd isotopic systematics of granitoid gneisses from the Czech Sudetes (Jizerské hory, Krkonosze and Orlice–Snieznik Dome). *International Journal of Earth Sciences*. In press.

——, ŠTÍPSKÁ, P., SCHULMANN, K. & JAECKEL, P. 2000*b*. Chronological constraints on the pre-Variscan evolution of the northeastern margin of the Bohemian Massif, Czech Republic. *This volume*.

——, WENDT, I., LIEW, T. C. *et al.* 1988*a*. U–Pb zircon and Sm–Nd model ages of high-grade Moldanubian metasediments, Bohemian Massif, Czechoslovakia. *Contributions to Mineralogy and Petrology*, **99**, 257–266.

KROONENBER, S. B. 1982. A Grenvillian granulite belt in the Colombian Andes and its relation to the Guiana shield. *Geologie en Mijnbouw*, **61**, 325–333.

LIEW, T. C. & HOFMANN, A. W. 1988. Precambrian crustal components, plutonic associations, plate environment of the Hercynian Fold Belt of Central Europe: Indications from a Nd and Sr isotopic study. *Contributions to Mineralogy and Petrology*, **98**, 129–138.

——, FINGER, F. & HÖCK, V. 1989. The Moldanubian granitoid plutons of Austria: chemical and isotopic studies bearing on their environmental setting. *Chemical Geology*, **76**, 41–55.

MATTE, P. H. 1991. Accretionary history of and crustal evolution of the Variscan belt in Europe. *Tectonophysics*, **196**, 309-337.

——, MALUSKI, H., RAJLICH, P. & FRANKE, W. 1990. Terrane boundaries in the Bohemian Massif: result of large scale Variscan shearing. *Tectonophysics*, **177**, 151–253.

MICHARD, A., GURRIET, P., SOUDANT, M. & ALBARÈDE, F. 1985. Nd isotopes in French Phanerozoic shales: external vs internal aspects of crustal evolution. *Geochimica et Cosmochimica Acta*, **49**, 601–610.

Michard Vitrac, A., Albarède, F. & Allègre, C. J. 1981. Lead isotopic composition of Hercynian granitic K-feldspars constrains continental genesis. *Nature*, **291**, 460–464.

Miller, J. F. & Harris, N. B. W. 1989. Evolution of continental crust in the Central Andes; constraints from Nd isotopes. *Geology*, **17**, 615–617.

Nance, R. D. & Murthy, J. B. 1994. Contrasting basement isotopic signatures and the palinspastic restoration of peripheral orogens: example from the Neoproterozoic Avalonian–Cadomian belt. *Geology*, **22**, 617–620.

—— & —— 1996. Basement signatures and Neoproterozoic paleogeography of Avalonian–Cadomian and related terranes in the circum-North Atlantic. *In*: Nance, R. D. & Thompson, D. (eds) *Avalonian and Related Peri-Gondwanan Terranes of the Giraum North Atlantic*. Geological Society of America, Special Papers, **304**, 333–346.

Nelson, B. K. 1992. Crustal growth in Central Europe–isotopic evidence from the Bohemian Massif, Austria and Germany. *Geological Society of America, Abstracts with Programs*, **24**, 45.

Noble, S. R., Aspden, J. A. & Jemielita, R. 1997. Northern Andean evolution: new U–Pb geochronological constraints from Ecuador. *Geological Society of America Bulletin*, **107**, 789–798.

O'Bbien, P. J., Kröner, A., Jaeckel, P., Hegner, E., Zelazniewics, a. K[sc]rfyza, R. 1997. Petrological and isotopic studies on Palaeozoic high-pressure granulites, Góry Sowie, Polish Sudetes. *Journal of Petrology*, **38**, 433–456.

Oliver, G., Corfu, F. & Krogh, T. 1993. U–Pb ages from SW Poland: evidence for a Caledonian suture between Baltica and Gondwana. *Journal of the Geological Society, London*, **159**, 355–369.

Pallister, J. S., Stacey, J. S., Fisher, L. B. & Premo, W. R. 1987. Arabian shield ophiolites and late Proterozoic microplate accretion. *Geology*, **15**, 320–323.

Parry, M., Stípská, P., Schulmann, K., Hrouda, F., Jeîek, J. & Kröner, A. 1997. Tonalite sill emplacement at an oblique plate boundary: northeastern margin of the Bohemian Massif. *Tectonophysics*, **280**, 61–81.

Patchett, P. J. & Bridgwater, D. 1984. Origin of the continental crust of 1.9–1.7 Ga age defined by Nd isotopes in the Ketilidian terrain of South Greenland. *Contributions to Mineralogy and Petrology*, **87**, 311–318.

—— & Ruiz, J. 1987. Nd isotopic ages of crust formation and metamorphism in the Precambrian of eastern and southern Mexico. *Contributions to Mineralogy and Petrology*, **96**, 523–528.

Pbikryl, R., Schulmann, K. & Melka, R. 1996. Perpendicular fabrics in the Orlické hory orthogneisses (western part of the Orlica–Snieznik Dome, Bohemian Massif) due to high temperature E–W deformational event and the lower temperature N–S overprint. *Journal of the Czech Geological Society*, **41**, 156–166.

Priem, H. N. A., Kroonenberg, S. B., Boelrijk, N. A. I. M. & Hebeda, E. H. 1989. Rb–Sr and K–Ar evidence for the presence of a 1.6 Ga basement underlying the 1.2 Ga Garzon–Santa Marta granulite belt in the Colombian Andes. *Precambrian Research*, **42**, 315–324.

Reischmann, T. & Kröner, A. 1994: Late Proterozoic island arc volcanics from Gebeit, Red Sea Hills, NE Sudan. *Geologische Rundschau*, **83**, 547–563.

Restrepo-Pace, P. A., Ruiz, J., Gehrels, G. & Cosca, M. 1997. Geochronology and Nd isotopic data of Grenville-age rocks in the Colombian Andes: new constraints for late Proterozoic–Early Paleozoic palaeocontinental reconstructions of the Americas. *Earth and Planetary Science Letters*, **150**, 427–441.

Richard, P., Shimizu, N. & Allègre, C. J. 1976. $^{143}Nd/^{146}Nd$ a natural tracer: an application to oceanic basalts. *Earth and Planetary Science Letters*, **31**, 269–278.

Rocci, G., Bronner, G. & Deschamps, M. 1991. Crystalline basement of the West African Craton. *In*: Dallmeyer, R. D. & Léchorché, J. P. (eds) *The West African Orogens and Circum-Atlantic Correlatives*. Springer, Berling, 31–61.

Samson, S. D., McLelland, W. C., Patchett, P. J., Gehrels, G. E. & Anderson, R. G. 1989. Evidence from neodymium isotopes for mantle contributions to Phanerozoic crustal genesis in the Canadian Cordillera. *Nature*, **337**, 705–709.

Schulmann, K. & Gayer, R. 2000. A model for continental accretionary wedge developed by oblique collision: the NE Bohemian massif. *Journal of the Geological Society, London*, **157**, 401–416.

Stacey, J. S. & Stoeser, D. B. 1983. Distribution of oceanic and continental leads in the Arabian-Nubian shield. *Contributions to Mineralogy and Petrology*, **84**, 91–105.

Stein, M. & Goldstein, S. L. 1996. From plume head to continental lithosphere in the Arabian–Nubian shield. *Nature*, **382**, 773–778.

Štípská, P. & Schulmann, K. 1995. Inverted metamorphic zonation in a basement-derived nappe sequence, eastern margin of the Bohemian Massif. *Geological Journal*, **30**, 385-413.

——, ——, Kröner, A. & Jeîek, J. T. 2000. Thermomechanical role of Cambro-Ordovician paleorift during Variscan collision: NE margin of the Bohemian Massif. *Tectonophysics*, in press.

Teixeira, W., Tassinari, C. C. G., Cordani, U. G. & Kawashita, K. 1989. A review of the geochronology of the Amazonian craton: tectonic implications. *Precambrian Research*, **42**, 213–227.

Turniak, K., Mazur, S. & Wysoczaski, R. 1999. SHRIMP zircon geochronology and geochmistry of the Orlica–Snieznik gneisses (Sudetes, SW Poland) and implications for the evolution of the Variscides in East–Central Europe. Tectonophysics, in press.

Unrug, R., Haranczyk, C. & Chocyk-Jaminska, M. 1999. Easternmost Avalonian and Armorican terranes of central Europe and Caledonian–Variscan evolution of the polydeformed Krakow mobile belt: geological constraints. *Tectonophysics*, **302**, 133–157.

WASSERBURG, G. J., JACOBSEN, S. B., DEPAOLO, D. J., MCCULLOCH, M. T. & WEN, T. 1981. Precise determination of Sm/Nd ratios, Sm and Nd isotopic abundances in standard solutions. *Geochimica et Cosmochimica Acta*, **45**, 2311–2323.

WILLIAMS, I. S. & CLAESSON, S. 1987. Isotopic evidence for the Precambrian provenance and Caledonian metamorphism of high grade paragneisses from the Seve Nappes, Scandinavian Caledonides. II. Ion microprobe zircon U–Th–Pb. *Contributions to Mineralogy and Petrology*, **97**, 205–217.

WINDLEY, B. F. 1995. *The Evolving Continents*, 3rd edn. Wiley, Chichester.

ZIEGLER, P. A. 1986. Geodynamic model for the crustal consolidation of western and central Europe. *Tectonophysics*, **126**, 303–328.

Appendix 1: Sample localities, rock descriptions, and zircon ages

Lugian Domain

Jizerské Mountains

CS-3 and 4	Medium- to coarse-grained augne–gneiss from porphyritic Jizera Gneiss. Smeda River NW of Frydlant, Jizerské Hory. Zircon age: 506 Ma (5).
CS-5	Coarse-grained granite–gneiss, outcrop N of Nové Město, Jizerské Hory. Zircon age: 515 Ma (5).

Krkonoše Mountains

CS-2	Quartz–sericite schist (metadacite) of Velká Úpa Group. Roadcut NW of Zelezny Brod. No zircon age. Assumed depositional age *c.* 520 Ma.
CS-6	Strongly foliated (almost mylonitic) granitoid augen–gneiss, tectonically interlayered and folded with felsic gneiss (metadacite) of sample CS 7. Spindleruv Mlyn, N of Vrchlabi. Zircon age: 502 Ma (5).
CS-7	Massive felsic gneiss (metadacite?), interlayered and folded with granitic gneiss of CS 6. Village of Bedrichov, N of Vrchlabi. No zircon age. Assumed depositional age: 520 Ma.
CS-8	Coarse-grained, strongly foliated granitic augne–gneiss. Roadcut N of Vrchlabi. Zircon age: 503 Ma (5).
CS-9	Well-foliated, medium-grained granite–gneiss of GK-type (Opletal *et al.* 1980). Roadcut NW of Horni Marsov. Zircon age: 511 Ma (5).

Orlica–Snieznik Dome

CS-11	Two-mica–albite schist (felsic metavolcanic or metatuff of the *c.* 520 Ma Stronie Formation, Poland, or Nové Město Group, Czech Republic). Intruded by granite–gneisses of Orlica–Snieznik dome. No emplacement age. Presumed depositional age *c.* 520 Ma.
CS-12	Red, well-foliated granite–gneiss, locally augne–gneiss, of G-type. Roadcut in village of Zákouti. Zircon age: 507 Ma (5).
CS-14	Strongly foliated granitoid augne–gneiss of G-type (Opletal *et al.* 1980). Roadcut E of Jablonné nad Orlici. Zircon age: 505 Ma (5).
CS-20	Strongly foliated, coarse-grained granitic augne–gneiss. Forest SE of Cenkovice. Zircon age: 503 Ma (5).
CS-21	Strongly foliated, coarse-grained granitic augne–gneiss. Village of Hor. Hédec E of Králíky. Zircon age: 511 Ma (5).
CS-23	Amphibolite from interlayered amphibolite–metakeratophyre sequence of Nové Město Group. Forest NW of Nebeská Rybná. No zircon age. Assumed depositional age: 520 Ma.
CS-24	Very coarse augne–gneiss, almost porphyritic granite of G-type. Roadcut W of Bartosovice. Zircon age: 510 Ma (5).
CS-25	Strongly foliated coarse-grained granitic augne–gneiss. Small quarry in village of Mostovice (Poland). Zircon age: 510 Ma (5).
CS-26	Well-foliated metagreywacke of Stronie Group. Forest outcrop near Mloty, Poland. No emplacement age. Assumed depositional age *c.* 520 Ma.
CS-44a	Layered felsic granulite (metavolcanic rock?), Rychleby Mts, forest SW of Javorník. Granulite facies at 340 Ma (3).
CS-44b	Leucocratic, homogeneous, garnetiferous felsic granulite (restite?), Rychleby Mts, forest SW of Javorník 4 m S of CS 44a. Granulite facies at 340 Ma (3).
CS-45	Felsic granulite, similar to CS 44a, Rychleby Mts, forest road SW of Javorník. Granulite facies at 340 Ma (3).
CS-46	Schlieric, migmatitic granite–gneiss, interpreted as migmatitic Snieznik Gneiss. Rychleby Mts, SW of Javorník. Not dated, assumed to be identical to Snieznik Gneiss at 502 Ma (4).
CS-47	Layered felsic metavolcanic rock of dacitic composition from Stronie Formation. W of the Staré Město Belt, Rychleby Mts, *c.* 500 m W of the village of Vojtovice. Zircon age: 523 Ma (1).

Appendix 1 continued on next page

Appendix 1: Continued

Silesian Domain (Brunia)

Desná Dome

CS-22	Well-foliated granite–gneiss of Desná Dome. Bank of Desná River, E of village of Kouty nad Desnou. Zircon age: 507 Ma (1).
CS-37	Strongly foliated and isoclinally folded leucogranitic gneiss, *c.* 20 km SSE of Jesenik, cliff NW of Ludvikov cemetery. Northern termination of Desná Dome. Zircon age: 685 Ma (1).
CS-38	Porphyritic phase of leucocratic granite–gneiss cross-cutting the gneiss of sample CS-37. Same locality as CS-37. Zircon age: 517 Ma (1).
CS-48	Coarse- to medium-grained, almost pegmatitic granite cross-cutting the 502 Ma fabric of the Desná migmatitic gneiss. Forest between Medvudi and Tupu Hills, *c.* 1.5 km SSE of sample CS-22. Zircon age: 502 Ma (1).
CS-49a, b	Interlayered light grey (sample 49a) and dark grey (sample 49b) biotite schists. Large quarry near Zámcisko at hill Medvedí dul, SE of Barborka. (a) Plag–bt schist with strong mylonitic fabric, typical of the Desná Dome. Chemically similar to a trondhjemite and dacite. Interpreted as volcanic metagreywacke or metatuff. Zircon age: 598 Ma (1). (b) Plag–bt–musc schist, interpreted as metagreywacke with mylonitic fabric. Detrital zircon ages: 629–665 Ma (1).

Cervenohorské sedlo Belt

CS-35	Fine-grained ultramylonitic orthogneiss of granitic composition from rock sheet located between Keprník Nappe and Desná Dome. Collected near Jeseník–Sumperk Road, forest road *c.* 1 km S of Filipovice. Zircon age: 613 Ma (1).

Keprník Nappe

CS-34	Sheared granite–gneiss with ultramylonititic texture. Hill slope below railway line from Ramzova to Horni Lipova. Zircon age: 555 Ma (1).
CS-36	Porphyritic, well-foliated leucocratic Keprník Gneiss (mylonitic granite–gneiss). Forest N of Koutyn Desnou. Zircon age: 584 Ma (1).

Moravo-Silesian Transitional Zone

Velké Vrbno Unit

CS-39	Strongly foliated orthogneiss of tonalitic composition. Forest WNW of Horni Lipová. Zircon age: 574 Ma (1).

Staré Město Belt

CS 30a, c	Tonalitic gneiss (30a) and *in situ* melt patch (30c) in shear zone, quarry NW of Hanusovice. Zircon age: *c.* 503 Ma (1).
CS-31	Gt–sill-bearing granulite facies pelitic gneiss. Roadcut S of village of Îleb, NW of Hanusovice. Granulite facies at 504 Ma. Depositional age *c.* 520 Ma. (1).
CS-32	Tonalite from Variscan sill, roadcut W of Skorosice, Rychleby Mts Zircon age: 332 Ma (2)
CS-33	Coarse-grained alaskitic granite intruding mica schist. Hill Certovy kameny NE of Jeseník. Zircon age: 334 Ma (3).
CS-40	Fine-grained, light-coloured felsic rock (leptite) of leptite–amphibolite sequence. Quarry NW of Hanusovice. Not dated but equivalent to CS 50 Zircon age: 503 Ma (1).
CS-41	Anorthositic gneiss from a ductile-deformed layered gabbro unit (meta-ophiolite?). Forest NW of the village of Vlaské, *c.* 4 km W of Hanusovice. Zircon age: 504 Ma (1).
CS-42	Well-foliated tonalitic gneiss, collected from roadcut W of Vlaské. Zircon age: 345 Ma (3).
CS-50	Strongly folded and foliated leptite (rhyodacite) with melt patches from a leptyno-amphibolite sequence. Cliff in forest near Stríbrné Hill, SW of Zulová. Zircon age: 503 Ma (1).

(1) Kröner *et al.* (this volume); (2) Parry *et al.* (1997); (3) Kröner *et al.* (unpubl. data); (4) Turniak *et al.* (2000); (5) Kröner *et al.* (2000*a*).

Appendix 2: U–Pb and Pb–Pb ages of xenocrystic and detrital zircons in Meoproterozoic to Palaeozoic rocks from the NE margin of the Bohemian Massif

Samples	Magmatic or depositional age (Ma)	Xenocrystic and detrital zircon ages
Erzgebirge		
EZ5	484	587, 905, 1081
EZ	11	550 1099
EZ12	(520)	1019|†, 1074†, 1182†, 1628†, 1712†, 1740†, 1762†, 1829†, 2233†, 2418†
EZ 16	556	852, 953, 1453, 1677
EZ 19	(485)	823, 949, 1005, 1284, 1509, 1519, 1601, 2034
EZ 20	486	572, 726, 1021
EZ 21	567	788, 1215, 1328, 1466, 1516, 1752, 1984, 1025, 2602
EZ 23	458	564, 637, 936
CS 15	551	1487, 2464
CS 16	524	618, 700, 946, 1262
CS 17	553	723, 1501, 2066
CS 18	553	855, 1025, 1262, 1488, 1723
CS 19	550	745, 1059
DDR 16	550	1042, 1056, 1076, 1599, 2181
DDR 17	482	913, 1858
Góry Sowie, West Sudetes, Poland		
Pl 1	(520)	767†, 1618†
Pl 2	(520)	512†, 1255†, 1753†, 1855†
Pl 4	484	1125, 1174, 1325, 2246, 2368
Pl 6	486	1182, 1270, 1700, 1840
Pl 8	(520)	544†*, 680†*, 842†*, 909†*, 1023†*, 1059†*, 1138†*, 1736†*, 2155†
Pl 9	(490)	499†, 1401†, 2198†
Pl 14	(520)	2215†, 2394†, 2416†, 2620†
Pl 15	473	1267, 2537, 2675
Pl 16	487	1937
Pl 17	482	1937
Pl 20	334	1558
Pl 22	(520)	1080†, 20113†, 2055†
Pl 23	(520)	816†
Lusatian Nassif		
La 2	(600)	1395†, 2574†
La 4	585	2020
La 5	(600)	1136†, 1201†
La 7	564	1886
La 8	587	1282, 1785
La 9	584	1112, 1335, 1798, 1941, 1958, 1983
La 11	542	706
H 517	571	1600
DDR-2	(600)	1708†, 1881†, 2621†, 2932†
Lugian Domain		
Jizerské Mountains		
CS 3	506	846, 1706
CS 4	506	829, 842, 859, 916
CS 5	515	546, 642, 662, 728, 816, 1105
Krkonoše Mountains		
CS 6	502	814, 1040, 1260, 1848
CS 7	(520)	no data
CS 8	503	536, 547, 571, 683, 696, 882

Appendix 2 continued on next page

Appendix 2: Continued

Samples	Magmatic or depositional age (Ma)	Xenocrystic and detrital zircon ages
Orlica–Snieznik Dome		
CS 11	(520)	842†, 1497†
CS 12	507	1066
CS 20	503	692, 867, 1184
CS 22	507	604, 608, 612, 630, 722, 1019, 1106
CS 24	510	1372
CS 26	(520)	805†, 1574†, 2055†
CS 47	523	1753
Silesian Domain		
Desná Dome		
CS 22	507	604, 608, 612, 630, 722, 1019, 1106
CS 37	685	864
CS 38	517	594, 891, 911
CS 49b	(600)	629†, 642†, 665†
Keprník Nappe		
CS 34	555	721*, 782
CS 36	584	629*, 661, 746, 802* 1073*
Moravo-Silesian Transitional Zone		
Staré Město belt		
CS 30B	503	641*, 684*, 818*
CS 31	(520)	664†, 682†, 984†*

*Ages by the U–Pb method; those without asterisk are by Pb–Pb evaporation.
†Detrital zircon age; those without dagger are xenocryst ages.

From Cadomian subduction to Early Palaeozoic rifting: the evolution of Saxo-Thuringia at the margin of Gondwana in the light of single zircon geochronology and basin development (Central European Variscides, Germany)

U. LINNEMANN[1], M. GEHMLICH[2], M. TICHOMIROWA[2], B. BUSCHMANN[2], L. NASDALA[2], P. JONAS[2], H. LÜTZNER[3] & K. BOMBACH[2]

[1]*State Museum of Mineralogy and Geology Dresden, Research Centre, A.B. Meyer-Bau, Königsbrücker Landstrasse 159, Dresden, D-01109, Germany (e-mail: linnemann@snsd.de)*

[2]*Mining Academy Freiberg, Institute of Mineralogy, Werner-Bau, Brennhausgasse 14, Freiberg, D-09596, Germany*

[3]*Friedrich-Schiller-University Jena, Institute of Geosciences, Burgweg 11, Jena, D-07749, Germany*

Abstract: Saxo-Thuringia is classified as a tectonostratigraphic terrane belonging to the Armorican Terrane Collage (Cadomia). As a former part of the Avalonian–Cadomian Orogenic Belt, it became (after Cadomian orogenic events, rift-related Cambro-Ordovician geodynamic processes and a northward drift within Late Ordovician to Early Silurian times), during Late Devonian to Early Carboniferous continent–continent collision, a part of the Central European Variscides. By making use of single zircon geochronology, geochemistry and basin analysis, geological processes were reconstructed from latest Neoproterozoic to Ordovician time: (1) 660–540 Ma: subduction, back-arc sedimentation and tectonomagmatic activity in a Cadomian continental island-arc setting marginal to Gondwana; (2) 540 Ma: obduction and deformation of the island arc and marginal basins; (3) 540–530 Ma: widespread plutonism related to the obduction-related Cadomian heating event and crustal extension; (4) 530–500 Ma: transform margin regime connected with strike-slip generated formation of Early to Mid-Cambrian pull-apart basins; (5) 500–490 Ma: Late Cambrian uplift and formation of a chemical weathering crust; (6) 490–470 Ma: Ordovician rift setting with related sedimentation regime and intense igneous activity; (7) 440–435 Ma: division from Gondwana and start of northward drift. The West African and the Amazonian Cratons of Gondwana, as well as parts of Brittany, were singled out by a study of inherited and detrital zircons as potential source areas in the hinterland of Saxo-Thuringia.

Since the early definition of Kossmat (1927), and the work of von Gaertner (1944), the interpretation of the Saxo-Thuringian Zone of the Variscan mountain chain has considerably changed. Stratigraphic and geotectonic constraints suggest that the Saxo-Thuringian Zone is a tectonostratigraphic unit that had a peri-Gondwanan origin before a complex Variscan evolution. Many workers have interpreted the Saxo-Thuringian Zone as a part of Armorica *sensu* Van der Voo (1979).

Underestimated for a long time, this pre-Variscan evolution has to be located in the general framework of peri-Gondwanan evolution as represented by the distribution of Cadomian basement areas within the so-called Avalonian–Cadomian Orogenic Belt situated at the Gondwana margin (Nance & Murphy 1994, 1996). Characterized by widespread Neoproterozoic arc volcanism, calc-alkaline plutonism, large-scale volcanigenic turbidite deposition, often accompanied by mafic volcanism, and shallow-marine Cambro-Ordovician overstep sequences with 'Acado-Baltic' (i.e. peri-Gondwanan) faunas, the reconstructed Avalonian–Cadomian Orogenic Belt includes Western and Eastern Avalonia, Florida, Iberia, Carolina and Armorica. The last of these was named 'Cadomia' by Nance & Murphy (1994). As differential drift paths and individual block

From: FRANKE, W., HAAK, V., ONCKEN, O. & TANNER, D. (eds). *Orogenic Processes: Quantification and Modelling in the Variscan Belt.* Geological Society, London, Special Publications, **179**, 131–153. 0305-8719/00/$15.00

rotations were calculated, Bachtadse *et al.* (1998) and Franke (this volume) proposed the term 'Armorican Terrane Assemblage' for the different crustal units composing 'classical' Armorica *sensu* Van der Voo (1979). New palaeomagnetic data (Tait *et al.* 1995, 1997, this volume) show that Armorica was not a coherent microplate or terrane. Following the guidelines of terrane descriptions by Howell (1995), the term 'Armorican Terrane Collage' should be used for the group of peri-Gondwana-derived terranes with important common features ('Armorican affinities') such as the occurrence of Cadomian basement, peri-Gondwanan faunas, the tillites of the Sahara glaciation (Ashgill) and corresponding geomagnetic data (Linnemann *et al.* 1998).

It is the aim of this paper to present age data, aspects of basin development and geochemical data that are representative for the Neoproterozoic–Cadomian and early Palaeozoic plate tectonic evolution of the pre-Variscan units of the Saxo-Thuringian Zone. This tectonostratigraphic unit is interpreted as a peri-Gondwanan terrane (Saxo-Thuringian Terrane; shortened to Saxo-Thuringia).

The Avalonian–Cadomian Orogenic Belt

Nance *et al.* (1991) and Nance & Murphy (1994, 1996) gave the Avalonian–Cadomian Orogenic Belt completely new dimensions by large-scale reconstructions. When the Cadomian unconformity first was described in the last century in France (e.g. Bunel 1835), nobody could suspect that the Cadomian mountain chain was an orogen of thousands of kilometres extent (Fig. 1).

The palaeogeographical reconstruction of the Avalonian–Cadomian Orogenic Belt in Fig. 1 is based on the drawings of Nance & Murphy (1994, 1996). A number of important changes were made in the modified reconstruction based on aspects of local basin development and new geochronological data (Fig. 1).

The basement of the Urals was included in the Avalonian–Cadomian Orogenic Belt because of the occurrence of an unconformity dividing deformed Neoproterozoic sedimentary rocks from overlying Ordovician sandstones in at least three localities (Maslov *et al.* 1997). These unconformities were interpreted by the present authors as Cadomian unconformities. In addition, Glasmacher *et al.* (1999) obtained new geochronological data from the Urals related to Cadomian orogenic events.

The position of Iberia within the Avalonian–Cadomian Orogenic Belt is suspected to be very close to West African Craton, because the Ordovician overstep sequence includes the Ashgill diamictite of the Sahara glaciation (Robardet & Doré 1988). That requires a former position of Iberia within the Cadomian mountain chain at the margin of the West African Craton, which was the centre of the ice sheet during Ashgill time (Fig. 1).

New palaeomagnetic data and differences in the Early Palaeozoic basin development divide the classical Armorica of Van der Voo (1979) into at least three terranes, which have a few similarities ('Armorican affinities'), but also some differences. These crustal fragments are Armorica in its original sense (Brittany and Normandy, France), Saxo-Thuringia (Germany, this paper) and Perunica (Czech Republic) (Figs 1 and 2). Perunica was proposed by Havlíček *et al.* (1994) as a useful term for the low-grade Barrandian Basin with Neoproterozoic to Devonian rock suites tectonically overlying the high-grade rocks of the Moldanubian Zone (Bohemian Massif). Tait *et al.* (1995, 1997) and Kössler *et al.* (1996) published a number of palaeomagnetic data, which show block rotations up to 170° during the Late Silurian drift. This event is completely missing in Saxo-Thuringia. Apart from these palaeomagnetic data, large differences of the basin development concerning subsidence rates, the stratigraphic record and different timing of Variscan collision events (Linnemann *et al.* 1998; Franke this volume) require a differentiated pre-Variscan geotectonic development of Saxo-Thuringia

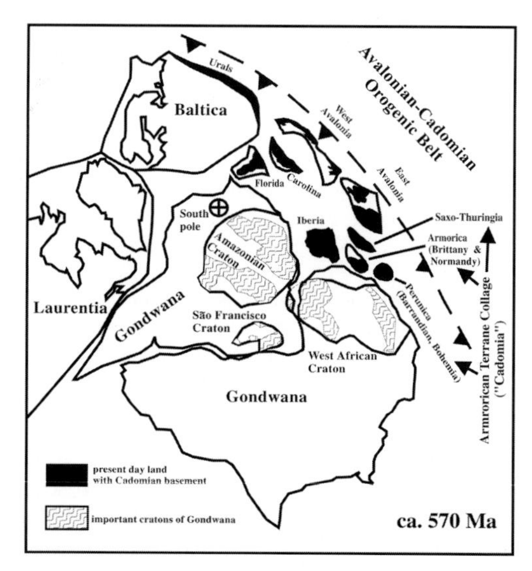

Fig. 1. Palaeogeography of the Avalonian–Cadomian Orogenic Belt and proposed distribution of peri-Gondwanan terranes (modified after Nance & Murphy (1994, 1996); Neoproterozoic continental reconstruction by Dalziel (1992)).

and Perunica during the Palaeozoic era. Also, differences of Saxo-Thuringia and Perunica in comparison with Normandy and Brittany (classical Armorica) could be not ignored. In Brittany and Normandy the first transgression took place during Arenig time. Saxo-Thuringia and Perunica were flooded much earlier during Tremadoc time. That constrains a position of Saxo-Thuringia and Perunica more to the external zones of the denuded and eroded Avalonian–Cadomian Orogenic Belt. These are only some of the reasons to split off 'Armorica' *sensu* Van der Voo (1979) into three relatively independent terranes. The early Palaeozoic overstep sequence could be used as a key to the Cadomian history of these fragments. The individual development during Cambro-Ordovician and Silurian time require varying positions of all three terranes within the Avalonian–Cadomian Orogenic Belt and resulting different Palaeozoic drift histories. On the other hand, some similarities (Armorican affinities) show clearly that the distance between Normandy–Brittany, Saxo-Thuringia and Perunica could be not too far during the Palaeozoic era. These Armorican affinities include the glaciomarine diamictite of the Sahara glaciation (Ashgill), the formation of oolitic iron ores during Ordovician time and peri-Gondwanan (Acado-Baltic) faunas.

All these arguments give a reason to present Normandy–Brittany (Armorica), Saxo-Thuringia and Perunica as tectonostratigraphic units of the so-called Armorican Terrane Collage derived from the Avalonian–Cadomian Orogenic Belt or peri-Gondwana (Fig. 1).

Saxo-Thuringia

Saxo-Thuringia is part of the Armorican Terrane Collage and is a fault-bounded crustal fragment with a distinct tectonostratigraphic evolution, preserved within the external zones of the Bohemian Massif (Linnemann *et al.* 1998) (Fig. 2a). As mentioned above, Saxo-Thuringia contain Neoproterozoic to Early Palaeozoic lithologies. Linnemann & Buschmann (1995) reinterpreted the Neoproterozoic units of Saxo-Thuringia as Cadomian basement and defined a Cadomian unconformity at several localities (Wurzelberg section, Schwarzburg Anticline; Hohe Dubrau section, Lausitz Anticline; boreholes Gumperda near Jena, and 5505/77 near Jena) (Fig. 2b).

When considering the distribution of low-grade Cadomian to Early Palaeozoic units in Saxo-Thuringia, five main outcrop areas with their corresponding Cadomian lithostratigraphic 'groups' are distinguished (Figs 2 and 3; Schwarzburg Anticline: Katzhütte and Frohnberg Groups; Northern Saxonian Anticline: Clanzschwitz Group; Elbe Zone: Weesenstein Group; Lausitz Anticline: Lausitz Group; Delitzsch–Torgau–Doberlug Syncline: Rothstein Group), which correspond to main structural units of this region. The Cadomian basement in general is dominated by greywacke turbidites, debris flows, quartzites, hydrothermal cherts and submarine volcanic intercalations. Large Cadomian plutons appear in nearly all zones. Following the lithostratigraphic standard cross-section and the local lithostratigraphic profiles (Fig. 3), two main periods can be distinguished after the formation of the Cadomian basement, the Cambrian period of denudation and formation of local, shallow-marine sediments. The Upper Cambrian sequence, in general, is not developed, as a result of uplift, combined with denudation and weathering processes (Linnemann & Buschmann 1995). Apart from these Cambrian occurrences, most parts of the Cadomian basement have a widespread Ordovician sedimentary succession starting in Tremadoc time, which is found in the Schwarzburg, Northern Saxonian and Lausitz Anticlines, as well as within the Elbe Zone. Characteristic magmatic events, during all periods, are represented by intrusion of granitoids and acidic volcanic rocks, and by corresponding pebbles in conglomerates. All of these were the subject of dating. Although this general geological evolution is valid for all five units, local differences appear and are described in detail below (Fig. 3).

In the Schwarzburg Anticline, Neoproterozoic and Lower Palaeozoic lithostratigraphic units were defined by von Gaertner (1944), who tried to compare these units with lithologies from the Barrandian Basin (Bohemia). Bankwitz & Bankwitz (1995) published a more detailed subdivision of Neoproterozoic lithologies, and the Lower Palaeozoic units received a new subdivision using sedimentological criteria (Lützner *et al.* 1986; Ellenberg *et al.* 1992). The sedimentary rocks of the Cadomian basement are represented by greywacke turbidites, debris flows, hydrothermal cherts and shales. Some mafic intrusions, dykes and sills are distributed in the lower part of the Cadomian section (Fig. 3). Acidic dykes and porphyroid intrusions within the Cadomian rock suite are most probably of Ordovician age. In some cases this is confirmed by a number of new datings, which are not given here. In the Schwarzburg Anticline, a hiatus between the Cadomian basement and its Palaeozoic cover was recognized by Heuse in Estrada *et al.* (1994), Linnemann & Buschmann (1995)

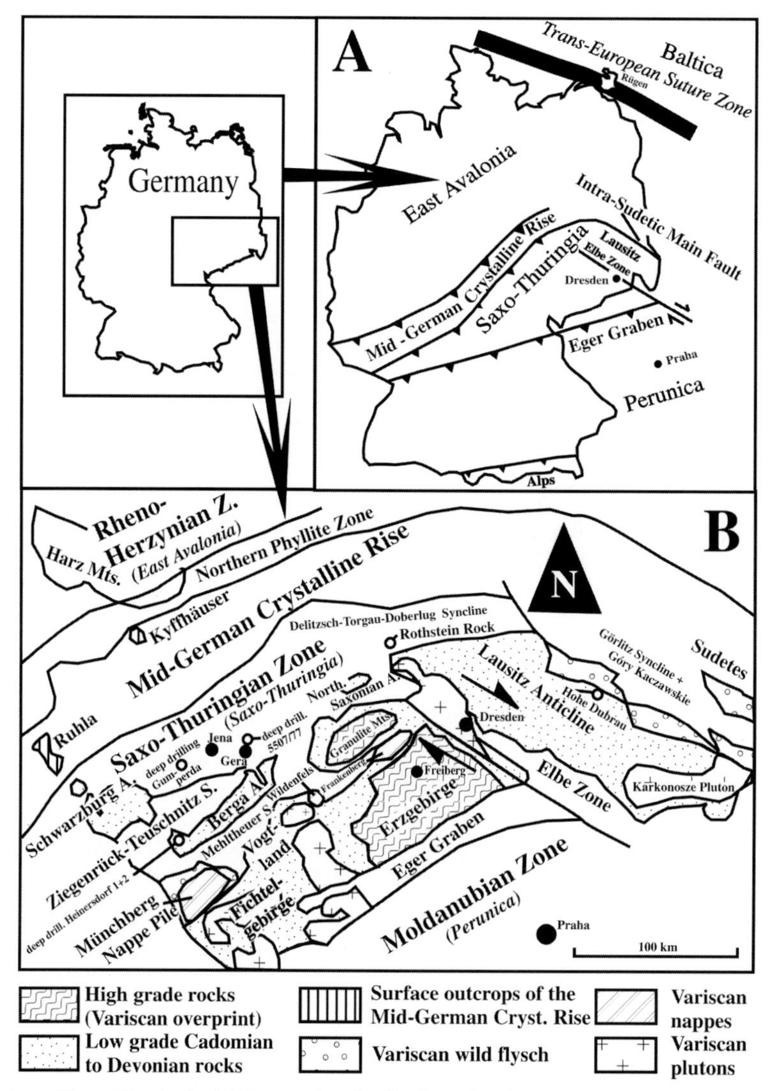

Fig. 2. Location of Saxo-Thuringia. (**a**) Present-day distribution of peri-Gondwanan terranes as well as important geological structures in Germany and bordering countries. (**b**) Tectonostratigraphic units of Saxo-Thuringia and adjoining areas as well as the position of some deep drillings and locations concerning this paper (note that the Mid-German Crystalline Rise does not belong to Saxo-Thuringia).

and Gehmlich *et al.* (1997). The gap comprises the entire Cambrian sequence (Fig. 3; see also Fig. 8, below). The earliest sedimentary cover above the Cadomian rock unit is the Goldisthal Group of Tremadoc age, which starts with the so-called 'Basal Quartzite' (Wurzelberg near Goldisthal), or as 'Conglomeratic Arkose', a conglomerate containing rhyolitic pebbles locally (Falk 1964). In the Blambach Valley, a rhyolite appears between the Cadomian basement and the Palaeozoic sediments (Bankwitz 1977). In general in the Schwarzburg Anticline, an Ordovician sediment pile of about 4000 m is exposed, containing siliciclastic shelf deposits such as quartzites, siltstones and shales, subdivided into the Goldisthal, Frauenbach, Phycodes and Gräfenthal Groups (see Fig. 9, below). Sequence stratigraphy and geotectonic setting were presented by Linnemann *et al.* (1997*b*) (see Fig. 9). Special intercalations are oolitic iron ores at the top of two lowstand systems tracts, a limestone layer up to 1 m thickness ('Kalkbank'), and a 250 m glaciomarine tillite that provides evidence of the Sahara glaciation

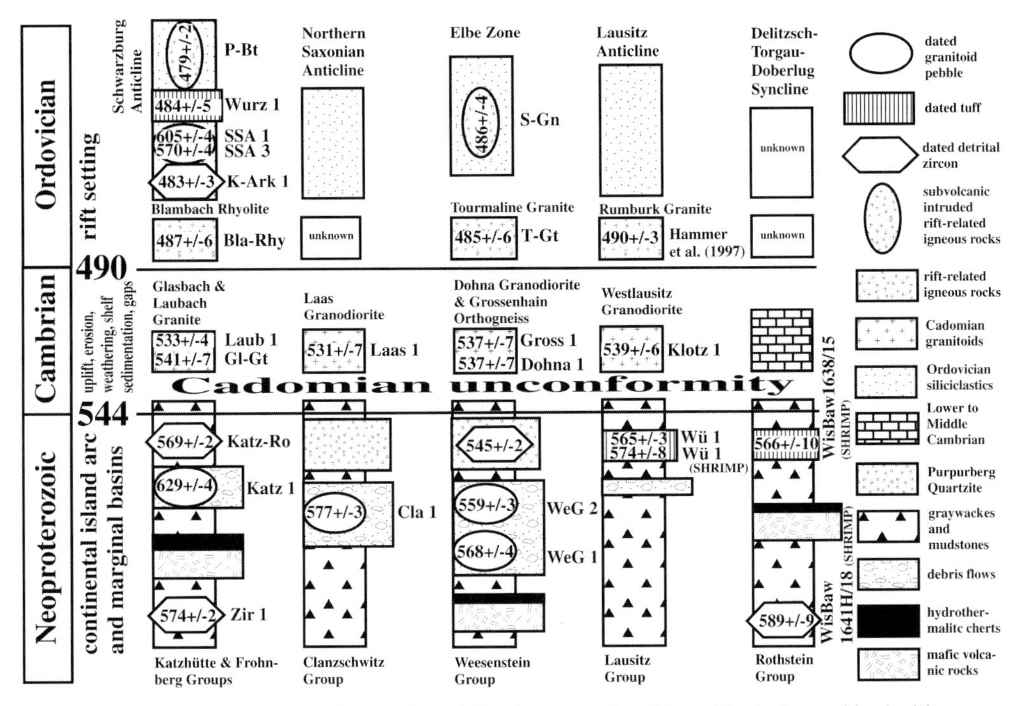

Fig. 3. Generalized lithosections of low-grade peri-Gondwanan units of Saxo-Thuringia combined with significant $^{207}Pb/^{206}Pb$ and SHRIMP zircon ages, as well as the general geotectonic setting. The full dataset is given in Tables 1 and 2. The sample indications in this figure relate to the samples listed in both tables. The U–Pb datings carried out using the ion microprobe are marked 'SHRIMP'.

during Ashgill time. At 2800 m above the Cadomian unconformity (Lauschenstein Member, Phycodes Group), one of the key fossils of the Upper Tremadoc, *Aranograptus* cf. *murrai* (Hall 1865), has been described (Heuse *et al.* 1994). This key fossil and the radiometric age of the Conglomeratic Arkose, the Basal Quartzite and the Blambach Rhyolite (Fig. 3; see also Fig. 9) require a Tremadoc unit of shelf deposits with a thickness of about 3000 m (see Fig. 9), which only can be explained by a rift setting.

In the neighbouring antiform, the Berga Anticline (Fig. 2b), Lower to Middle Cambrian sedimentary rocks with only local extension were discovered by the deep boreholes Heinersdorf 1 and 2 (Wucher, 1967) (Fig. 2b). The Cambrian sequence occurring in these boreholes was transgressed by the Upper Frauenbach Quartzite (of Tremadoc age) with a disconformity at its base (see Fig. 8). The Upper Cambrian sequence and the Goldisthal Group (of Tremadoc age) is missing (see Fig. 8).

In the Northern Saxonian Anticline (Figs 2b and 3), the Cadomian basement appears as the Leipzig and Clanzschwitz Groups. The Leipzig Group, mostly known from boreholes and from a small outcrop in the city of Leipzig–Zschocher, consists of greywacke turbidites, and one thin conglomerate horizon has been reported from a drill core (Schildau 1/63; Sehm 1976). These sediments are intruded by Cadomian granitoids in the Leipzig–Ellenberg Massif (borehole). The Cadomian unconformity and a Lower Ordovician overlap sequence were described from a borehole near Gera (5507/77; Linnemann & Buschmann 1995). The Clanzschwitz Group, fault-bounded to the Leipzig Group, is dominated by greywackes and debris flows containing a large amount of granitoid pebbles. The Collmberg Quartzite and its equivalent, the Hainichen–Otterwisch Quarzite, represents the Lower Ordovician cover of the Clanzschwitz Group.

The Elbe Zone (Figs 2 and 3), characterized as a Late Variscan dextral shear zone (Linnemann 1995), contains some relics of Cadomian basement and Palaeozoic rocks. The best-preserved Cadomian remnant, the Weesenstein Group, is composed of greywackes, debris flows with granitoid pebbles, mafic volcanites and the Purpurberg Quartzite, the last of these being a shallow-marine lowstand deposit (Linnemann 1991). Ordovician rocks, defined only by making use of lithostratigraphy, are distributed within the so-called 'Phyllite Unit' of the Elbe Zone (Pietzsch 1917; Kurze *et al.* 1992).

In the Lausitz Anticline (Figs 2 and 3), the Cadomian sedimentary sequence has been defined by Kemnitz & Budzinski (1994) as the Lausitz Group, consisting mainly of greywacke turbidites. Locally, ash layers from extrabasinal volcanic activity are preserved. These sediments were intruded by large granodioritic plutons, and transgressive siliciclastic deposits of Tremadoc age (Hohe Dubrau, Gross-Radisch) define the sedimentary cover. The area of the Monumentenberg (Hohe Dubrau) has been defined as the type locality for the Cadomian unconformity in Germany (Linnemann & Buschmann 1995).

The Delitzsch–Torgau–Doberlug Syncline (Figs 2 and 3), with the exception of the Rothstein Rock near Herzberg, is known only from boreholes. The Cadomian basement, represented by the Rothstein Group (Buschmann 1995) contains greywacke turbidites, synsedimentary melanges, hydrothermal cherts and mafic pillow lava flows; an assemblage that represents a Cadomian back-arc basin fragment (Buschmann *et al.* 1995). Slightly deformed during Cadomian orogenic processes, this series has an undeformed sedimentary cover of Lower to Middle Cambrian carbonates and siliciclastic rocks (e.g. Elicki 1997).

Method of single zircon dating

The geochronological Pb/Pb data were measured by M. Gehmlich at the Isotopic Laboratory of the Mining Academy Freiberg, with the exception of sample Katz-Ro, which was analysed by M. Tichomirowa. Three datings from the Delitzsch–Torgau–Doberlug Syncline were carried out by L. Nasdala using SHRIMP.

The Pb/Pb data were measured by single zircon evaporation (Kober 1986, 1987) on a FINIGAN MAT 262 mass spectrometer with a Philips 6665 ion counter. After reduction of common lead from cracks and inclusions and from discordant parts of the zircons at about 1450 °C, the remaining, highly crystallized parts of the zircons were heated for evaporation in one step from 1600 to 1620 °C. The $^{207}Pb/^{206}Pb$ ages were calculated from the corresponding measured ratios applying (1) correction of common lead (Stacey & Kramers 1975), (2) determination of the fractionation of the mass spectrometer by two zircon standards with 0.036 a.m.u. The quoted error of the mean ages is the standard deviation of the mean of single determinations.

Apparent ages were produced by iteration of the two equation systems of the decay chains of ^{238}U to ^{206}Pb and ^{235}U to ^{207}Pb, and the mean error 2σ of the corrected $^{207}Pb/^{206}Pb$ ratios was calculated by K. Bombach (Isotope Laboratory, Mining Academy Freiberg), using the errors of the measured ratios.

The obtained values were checked by determining the $^{207}Pb/^{206}Pb$ ages of standard zircons. Repeated analysis of two zircon standards yields mean ages in agreement with reported data. For zircon 91500 reported by Wiedenbeck *et al.* (1995) we obtained a value of 1065.4 ± 5.3 Ma ($n = 29$) in comparison with the given age in the cited paper of 1065.4 ± 0.4 Ma. The 2σ mean of all single measurements is about ± 2.0. For the zircon S-2-87 from the Wenham Monzonite separated from the Original Zartmann Zircon Standard of the Geological Survey of Canada with a reported age of 381.5 ± 4 Ma we obtained 379 ± 5.6 Ma ($n = 31$). The 2σ error of all single measurements again was about ± 2.0.

A very low statistical error of zircon evaporation ages can be achieved by using a large number of measured isotope ratios as well as a large number of single zircon grain measurements. In agreement with age and error calculations of the standard zircons, we separated the outliers of ages by using the Dixon test after Dean & Dixon (1951). The error of the mean age is calculated by the standard deviation.

For age estimation it is very important to obtain reproducible ages by measurements of several zircons from the same sample, because the $^{207}Pb/^{206}Pb$ ages produced are model ages with no information about the degree of discordance. Therefore zircons without cracks, inclusions or metamict areas were selected for the evaporation method. Zircons with older cores were selected by making use of cathodoluminescence microscopy (CL). From the investigated zircon populations, only long prismatic zircons were dated; these most probably represent the intrusion or effusion age of the igneous rock or the protolith.

Additional ages were determined by SHRIMP (Buschmann *et al.* 2000) on zircons from the Rothstein group (Figs 2 and 3), and from a Cadomian tuff (sample Wü 1, Wüsteberg, Lausitz Anticline, Figs 2 and 3). The latter give overlapping ages when compared with the evaporation data and are, consequently, a good proof for the reliability of the evaporation data.

Considering the age data, two main groups of age information have to be distinguished, those obtained from zircons sampled from former magmatic rocks (Table 1, Fig. 3) and those obtained from inherited or detrital zircons (Table 2, Fig. 5). A few typical zircons are shown in Fig. 4. When compared with the stratigraphic columns presented in the preceding

Table 1. *$^{207}Pb/^{206}Pb$ single zircon evaporation ages of Cadomian to Ordovician igneous rocks of Saxo-Thuringia*

Grain number/zircon type after Pupin (1980)	Mass scans	$^{207}Pb/^{206}Pb$	$^{204}Pb/^{206}Pb$	$^{207}Pb/^{206}Pb$ (corrected)	Age (Ma)
Sample P-Bt; porphyroid (Bärentiegel Porphyroid); Bärentiegel near Katzhütte; Schwarzburg Anticline; Frauenbach Group, Ordovician					
1/P1	49	0.0727643	0.0011255	0.056571	475 ± 8
2/S15	89	0.058518	0.0001406	0.056692	479 ± 2
3/S10	17	0.057562	0.00007335	0.056679	479 ± 6
4/S14	68	0.061889	0.00036765	0.0566936	480 ± 7
5/P5	85	0.0592346	0.0001876	0.056712	480 ± 5
Sample contains 0% rounded zircons				mean age:	479 ± 2
Sample Bla-Rhy; rhyolite (Blambach Rhyolite); Blambachtal near Sitzendorf; Schwarzburg Anticline; cropping out at the base of the Goldisthal Group; Ordovician					
1/S2	87	0.0603104	0.0002611	0.056693	480 ± 3
2/S12	18	0.0621537	0.00037855	0.056783	483 ± 27
3/P5	85	0.064869	0.0005711	0.056777	483 ± 3
4/S17	35	0.0586291	0.0001266	0.056866	486 ± 15
5/S2	86	0.06141	0.0003215	0.056965	490 ± 5
6/S1	70	0.0581485	0.00009105	0.057064	494 ± 8
7/S7	84	0.060315	0.00024315	0.057061	494 ± 4
Sample contains 10% rounded zircons				mean age:	487 ± 6
Sample Wurz 1; tuff (Wurzelberg Tuff); eastern flank of the Wurzelberg near the Frauenbach Valley; Goldisthal; Schwarzburg Anticline; Frauenbach Group; Ordovician					
1/P4	90	0.05783	0.0001214	0.056282	463 ± 3
2/P4	18	0.057333	0.0000624	0.056637	477 ± 2
3 to 4/S15	90	0.0568705	0.0000484	0.05668	479 ± 3
5/S20	90	0.05668	0.0000279	0.05678	483 ± 2
6 to 9/P5	90	0.05972	0.0002185	0.0567746	483 ± 5
10/P3	69	0.062863	0.000431	0.056831	485 ± 11
11/S14	18	0.0593211	0.0002069	0.056852	486 ± 8
12/P2	90	0.058275	0.00011	0.056894	487 ± 3
13 to 15/P4	90	0.0588266	0.0001426	0.0569862	491 ± 4
16 to 17/S20	54	0.062737	0.0003896	0.0573324	504 ± 12
Sample contains 5% rounded zircons				mean age:	484 ± 10
Sample T-Gt; tourmaline granite (Turmalingranit); Bahratal; Elbe Zone; Elbtalschiefergebirge; Ordovician					
1/S6	54	0.0789854	0.001565	0.0566202	477 ± 14
2/P2	90	0.062905	0.0004413	0.0567494	482 ± 4
3/S16	51	0.106503	0.003466	0.0565764	483 ± 13
4/S7	54	0.0965508	0.002774	0.0567765	483 ± 10
5/S6	54	0.0765816	0.001384	0.0568466	486 ± 5
6/S7	71	0.066018	0.0006435	0.056934	489 ± 9
7/S6	54	0.0985323	0.002888	0.0571464	497 ± 11
Sample contains 0% rounded zircons				mean age:	485 ± 6
Sample S-Gn; sericite gneiss (orthogneiss) (Serizitgneis); 500 m to the south of the castle in Blankenstein; Elbe Zone; Nossen–Wilsdruff–Schiefergebirge; Ordovician					
1/S6	88	0.0572213	0.0000492	0.056695	480 ± 3
2/S7	54	0.0596	0.0002181	0.056732	481 ± 8
3/S2	90	0.05714	0.000034	0.05684	485 ± 2
4/S6	66	0.057162	0.00003845	0.056834	485 ± 6
5/S7	88	0.0570218	0.0000213	0.056893	487 ± 2
6/S2	36	0.0576427	0.0000668	0.056902	488 ± 7
7/S2	86	0.0582087	0.00008695	0.056841	489 ± 2
8/S2	86	0.057146	0.0000275	0.056951	490 ± 2
Sample contains 0% rounded zircons				mean age:	486 ± 4

Table 1 continued over page

Table 1. *Continued*

Grain number/zircon type after Pupin (1980)	Mass scans	$^{207}Pb/^{206}Pb$	$^{204}Pb/^{206}Pb$	$^{207}Pb/^{206}Pb$ (corrected)	Age (Ma)
Sample Laub 1; granite (Laubach Granite); Laubachtal near Katzhütte, car park; Schwarzburg Anticline; Cadomian basement					
1 to 3/S2	90	0.059054	0.000106	0.057736	520 ± 3
4 to 6/S3	90	0.0593478	0.0001181	0.057856	524 ± 2
7 to 8/S3	90	0.062014	0.0002948	0.05798	529 ± 3
9/S2	90	0.0594844	0.0001128	0.05807	532 ± 2
10 to 11/S7	35	0.0587214	0.00004995	0.0581658	536 ± 8
12/S8	90	0.0595721	0.0000991	0.058355	543 ± 3
13 to 15/S6	90	0.0598844	0.0001149	0.0584416	546 ± 2
Sample contains 5% rounded zircons				mean age:	533 ± 10
Sample Gl-Gt; granite (Glasbach Granite); Steinberg Mtn near Mellenbach; Schwarzburg Anticline; Cadomian basement					
1/P1	85	0.060716	0.0001899	0.05815	535 ± 2
2/S5	88	0.05906	0.000071	0.058235	538 ± 2
3/S15	86	0.0597899	0.00012205	0.058284	540 ± 4
4/S10	86	0.059555	0.0000865	0.058544	550 ± 6
Sample contains 5% rounded zircons				mean age:	541 ± 7
Sample Laas 1; granodiorite (Laas Granodiorite); Dürrenberg, Laas; Northern Saxonian Anticline; Cadomian basement					
1/S2	82	0.067515	0.000685	0.057781	521 ± 2
2/S2	86	0.070071	0.000855	0.05795	528 ± 5
3/S6	69	0.0627047	0.000343	0.058104	534 ± 6
4/S7	87	0.0610844	0.0002169	0.05819	537 ± 4
5 to 6/S2	90	0.05934	0.00009	0.05818	537 ± 1
Sample contains 0% rounded zircons				mean age:	531 ± 7
Sample Dohna 1; granodiorite (Dohna Granodiorite); Gamighübel near Dresden–Torna; Elbe Zone; Elbtalschiefergebirge; Cadomian basement					
1/S2	18	0.061292	0.0002506	0.057793	522 ± 3
2/S2	51	0.0905496	0.0022487	0.058042	531 ± 10
3/S8	85	0.0686287	0.00074435	0.058129	535 ± 6
4/P1	71	0.0597537	0.00012275	0.05815	535 ± 4
5/S7	36	0.062498	0.00030165	0.05816	536 ± 3
6/S2	51	0.0906195	0.002253	0.0582074	537 ± 11
7/S3	33	0.064994	0.0004826	0.058235	539 ± 7
8/S2	17	0.058188	0.00002075	0.058207	539 ± 7
9/S2	43	0.0586186	0.0000410	0.058239	539 ± 2
10/S5	87	0.060648	0.0001785	0.058296	541 ± 3
11/S2	90	0.0650462	0.0004651	0.0585211	549 ± 4
Sample contains 0% rounded zircons				mean age:	537 ± 7
Sample Gross 1; orthogneiss (Grossenhain Orthogneiss); Grossenhain (Kupferberg); Elbe Zone; Cadomian basement					
1/S2	90	0.06231	0.00032	0.05797	529 ± 2
2/S1	87	0.05878	0.00006	0.05811	534 ± 1
3/S7	72	0.0634	0.00038	0.05812	534 ± 1
4/S7	90	0.06008	0.00014	0.05832	542 ± 1
5/S2	90	0.05944	0.00008	0.05846	547 ± 2
Sample contains 0% rounded zircons				mean age:	537 ± 7

Table 1 continued over page

Table 1. *Continued*

Grain number/zircon type after Pupin (1980)	Mass scans	$^{207}Pb/^{206}Pb$	$^{204}Pb/^{206}Pb$	$^{207}Pb/^{206}Pb$ (corrected)	Age (Ma)
Sample Klotz 1; granodiorite (Westlausitz Granodiorite); Dresden–Klotzsche in front of the State Museum of Mineralogy and Geology, Dresden, A.B. Meyer-Bau; Lausitz Anticline; Cadomian basement					
1 to 2/S10	54	0.05901	0.00009	0.05798	529 ± 2
3/S2	54	0.05877	0.00006	0.05817	536 ± 4
4/P1	90	0.05911	0.00008	0.05824	539 ± 1
5/P3	54	0.05871	0.00005	0.05825	539 ± 1
6/S5	90	0.05889	0.00006	0.05829	541 ± 1
7/P2	90	0.05892	0.00006	0.05833	542 ± 2
8/P1	84	0.06095	0.00018	0.05849	548 ± 3
Sample contains 0% rounded zircons				mean age:	539 ± 6
Sample Wü 1; tuff (Wüsteberg Tuff); Wüsteberg Mtn. near Kamenz; Lausitz Anticline; Lausitz Group; Cadomian basement					
1/D	71	0.0599493	0.000087	0.058869	562 ± 4
2/P5	88	0.0599892	0.0000873	0.058869	562 ± 4
3/P5	144	0.059626	0.00006098	0.058961	566 ± 4
4/D	144	0.0693135	0.0007292	0.0059048	569 ± 5
Sample contains 20% rounded zircons				mean age:	565 ± 3
Sample WeG 1; granite pebble from a debris flow; near the railway station, Weesenstein; Elbe Zone; Elbtalschiefergebirge; Weesenstein Group; Cadomian basement					
1/S15	88	0.065857	0.000518	0.058615	553 ± 4
2/P4	89	0.06071	0.0001501	0.05872	557 ± 3
3/P5	86	0.060775	0.0001424	0.058914	564 ± 3
4/S25	89	0.0592103	0.000028	0.05903	569 ± 3
5/S25	88	0.059655	0.000051	0.059093	570 ± 2
6/S23	33	0.109405	0.0034942	0.059092	570 ± 13
7/S25	90	0.059741	0.0000562	0.059121	571 ± 2
8/D	84	0.060325	0.0000946	0.059157	573 ± 2
9/D	87	0.059849	0.0000628	0.059185	573 ± 3
10/S23	88	0.072024	0.000904	0.059172	573 ± 4
11/P5	85	0.0596395	0.000039	0.059298	578 ± 2
Sample contains 0% rounded zircons				mean age:	568 ± 4
Sample WeG 2; granite pebble from a debris flow; near the railway station, Weesenstein; Elbe Zone; Elbtalschiefergebirge; Weesenstein Group; Cadomian basement					
1/S20	87	0.062643	0.000305	0.058422	546 ± 3
2/D	85	0.073375	0.0010396	0.058577	552 ± 3
3/P4	88	0.062625	0.0002857	0.058705	556 ± 2
4/S20	86	0.064942	0.0004294	0.058919	564 ± 4
5/S19	87	0.06569	0.0004834	0.058911	564 ± 3
6/P5	83	0.07295	0.0009762	0.059122	571 ± 5
Sample contains 0% rounded zircons				mean age:	559 ± 3
Sample Cla 1; granite pebble from a debris flow; Schlangnberg Mtn near Wellerswalde; Northern Saxonian Anticline, Clanzschwitz Group; Cadomian basement					
1/D	84	0.069311	0.000724	0.059059	570 ± 3
2/J5	86	0.068455	0.0006586	0.059066	570 ± 4
3/J5	85	0.0626223	0.0002572	0.059131	572 ± 4
4/P5	86	0.070315	0.000791	0.059145	572 ± 6
5/D	78	0.0701223	0.0007	0.059187	574 ± 3
6/D	89	0.062355	0.000227	0.059277	577 ± 1
7/P5	89	0.061449	0.0001584	0.059343	578 ± 4
8/J5	84	0.06447	0.0003652	0.059381	581 ± 2
9/J5	87	0.0601055	0.0000633	0.059388	581 ± 3
10/J5	87	0.061104	0.0001306	0.059448	583 ± 2
11/D	81	0.0631175	0.0002629	0.059428	583 ± 6
12/D	89	0.080692	0.001485	0.059462	584 ± 2
Sample contains 5% rounded zircons				mean age:	577 ± 3

Table 1 continued over page

Table 1. *Continued*

Grain number/zircon type after Pupin (1980)	Mass scans	$^{207}Pb/^{206}Pb$	$^{204}Pb/^{206}Pb$	$^{207}Pb/^{206}Pb$ (corrected)	Age (Ma)
Sample SSA 1; granite pebble; firing range near Scheibe-Alsbach; Schwarzburg Anticline; Frauenbach Group; Ordovician					
1/S13	71	0.0671986	0.0005088	0.05995	602 ± 10
2/S13	52	0.0655127	0.000384	0.060111	607 ± 12
Sample contains 0% rounded zircons				mean age:	605 ± 4
Sampla SSA 3; granite pebble; firing range near Scheibe-Alsbach; Schwarzburg Anticline; Frauenbach Group; Ordovician					
1/J5	87	0.066159	0.0005123	0.058947	565 ± 3
2/S25	84	0.0602929	0.0001051	0.0589785	566 ± 2
3/J4	89	0.061649	0.0001952	0.059045	569 ± 4
4/S24	89	0.0611967	0.0001627	0.0590506	569 ± 3
5/J4	70	0.066648	0.0005278	0.0592254	575 ± 12
6/J4	90	0.05958	0.0000377	0.059252	576 ± 2
Sample contains 5% rounded zircons				mean age:	570 ± 4
Sample Katz 1; conglomerate with granite and greywacke pebbles; Grendel Mtn near Sachsendorf; Schwarzburg Anticline; Frohnberg Group; Cadomian basement					
1/S12	88	0.06085	0.00004	0.06054	623 ± 3
2/S17	87	0.10021	0.0027539	0.060595	625 ± 9
3/S17	85	0.0613	0.0000413	0.060779	631 ± 4
4/S7	86	0.06282	0.0001571	0.060764	631 ± 2
5/S23	89	0.070255	0.0006732	0.060759	631 ± 4
6/S8	85	0.062924	0.0001561	0.060851	634 ± 5
7/S7	92	0.06932	0.0005582	0.061437	655 ± 3
Sample contains 60% rounded zircons				mean age:	629 ± 4

section, it becomes evident that distinct well-defined magmatic events can be recognized, and that a true Cadomian evolution can be distinguished from an Early Palaeozoic one.

Radiometric ages of intrusive and effusive igneous rocks

The database for the geochronological datings is shown in Table 1 (intrusion and effusion ages). The position of the samples within the lithostratigraphic units is shown in Fig. 3. Table 1 gives the locations of the samples. The shape of the individual zircons is described in table 1 after Pupin (1980).

Some granitoid pebbles could be separated from Cadomian debris flows (Fig. 3). Two pebbles from the Weesenstein Group were dated at 568 ± 4 Ma (WeG 1) and 559 ± 3 Ma (WeG 2). A pebble from the Clanzschwitz Group gave a mean age of 577 ± 3 Ma (Cla 1). From a matrix-supported granite-bearing conglomerate of the Frohnberg Group non-rounded magmatic zircons were separated having ages ranging from 623 ± 3 Ma to 631 ± 4 Ma (Katz 1). Detrital zircons of some greywacke samples gave similar ages. A sample of greywacke from the Frohnberg Group (Katz-Ro) contains non-rounded magmatic zircons with ages ranging from 569 ± 2 Ma to 596 ± 3 Ma. A detrital zircon from the Rothstein Group (WisBaw 1641H/14) was dated by SHRIMP with an age of 589 ± 9 Ma. All these data represents the most probable maximum ages of the Weesenstein, Clanzschwitz, Rothstein and Frohnberg Groups, at *c.* 560–580 Ma.

From the Lausitz and Rothstein Groups two tuffs were dated. The Wüsteberg Tuff (Wü 1) represents with a mean age of 565 ± 3 Ma (Pb/Pb) and 574 ± 8 Ma (SHRIMP) a sedimentation age for the Lausitz Group of around 570 Ma. A tuff from the Rothstein Group (WisBaw 1638/15) shows a similar age of 566 ± 10 Ma (SHRIMP), which relates to the sedimentation age of the Rothstein Group.

Within the lowermost part of the Katzhütte Group in the Schwarzburg Anticline a highly deformed fault-bounded unit exists. Von Gaertner (1944) used the term 'Kernzone' ('Core Zone') for these rocks. A sample of meta-sediments from the Zirkelstein (Zir 1) contains, apart from detrital Cadomian zircons ranging from 556 ± 4 Ma to 588 ± 3 Ma, one Ordovician zircon with an age of 481 ± 8 Ma. This young dating may be caused by a loss of lead. However, if the dating is correct, it gives reason to suspect that the Kernzone, which represents the deepest

Table 2. *$^{207}Pb/^{206}Pb$ single zircon evaporation ages of detrital and inherited zircons of Cadomian to Ordovician igneous and sedimentary rocks of Saxo-Thuringia*

Grain number/zircon type after Pupin (1980)	Mass scans	$^{207}Pb/^{206}Pb$	$^{204}Pb/^{206}Pb$	$^{207}Pb/^{206}Pb$ (corrected)	Age (Ma)
Sample Gp 28a; rhyolite pebble; deep drilling Gumperda near Jena; drillhole position: longitude 63771.7, magnitude 30407.0; sample from 1095 m below surface; Goldisthal Group; Ordovician; covered by Permian and Mesozoic strata of the Thuringian Basin					
1/rounded, red	90	0.11171	0.00004032	0.111568	1825 ± 3
All zircons of the rock sample: 20% rounded; 80% non-rounded					
Sample Gp 28b; sandstone; deep drilling Gumperda near Jena; drillhole position: longitude 63771.7, magnitude 30407.0; sample from 1095 m below surface; Goldisthal Group; Ordovician; covered by Permian and Mesozoic strata of the Thuringian Basin					
1 to 5/S2	90	0.0612179	0.0003434	0.0564709	471 ± 6
6 to 7/S21	71	0.070098	0.000949	0.056615	477 ± 4
8 to 10/S12	90	0.064524	0.0005426	0.056911	488 ± 2
All zircons of the rock sample: 50% rounded; 50% non-rounded					
Sample Bla-Rhy; rhyolite (Blambach Rhyolite); Blambachtal; Schwarzburg Anticline; Goldisthal Group; Ordovician					
8/P3	85	0.1483126	0.0003285	0.1445	2282 ± 2
All zircons of the rock sample: 10% rounded; 90% non-rounded					
Sample Wurz 1; tuff (Wurzelberg Tuff); Wurzelberg Mtn near Goldisthal; Schwarzburg Anticline; Frauenbach Group; Ordovician					
18/S15	90	0.0601155	0.0001494	0.0581752	536 ± 8
19/P4	36	0.1307	0.0005148	0.12437	2020 ± 10
All zircons of the rock sample: 5% rounded; 95% non-rounded					
Sample T-Gt; tourmaline granite (Turmalingranit); Bahratal; Elbe Zone; Elbtalschiefergebirge; Ordovician					
8/S7	90	0.060473	0.0001612	0.0583638	543 ± 5
9/S3	54	0.0992482	0.002828	0.058818	560 ± 5
10/S4	51	0.0755286	0.00165	0.0590069	567 ± 11
11/S7	90	0.0599272	0.00007564	0.0590523	569 ± 4
12/S8	52	0.1034141	0.00017205	0.10159	1651 ± 6
13/S13	87	0.11397	0.00066955	0.10525	1719 ± 9
All zircons of the rock sample: 0% rounded; 100% non-rounded					
Sample S-Gn; sericite gneiss (orthogneiss) (Serizitgneis); 500 m to the south of the castle in Blankenstein; Elbe Zone; Nossen–Wilsdruff–Schiefergebirge; Ordovician					
9/S7	89	0.06101	0.00006495	0.060271	613 ± 3
10/S12	18	0.10608	0.00003365	0.10608	1732 ± 4
11/S13	18	0.0124921	0.00034088	0.1206	1965 ± 16
12/S6	35	0.22886	0.00001935	0.22939	3048 ± 3
All zircons of the rock sample: 0% rounded; 100% non-rounded					
Sample K-Ark 1; rhyolite pebble-bearing conglomerate at the base of the Goldisthal Group with pebbles derived most probably from the Blambach Rhyolite (Conglomeratic Arkose, Konglomeratische Arkose); Birkenweg at Mellenbach; Schwarzburg Anticline; Goldisthal Group; Ordovician					
1 to 3/S6	90	0.064685	0.0005745	0.056608	476 ± 4
4 to 8/S6	90	0.0610634	0.0003152	0.056724	481 ± 3
9 to 11/S1	90	0.065167	0.0005955	0.056789	483 ± 3
12 to 13/S11	71	0.0627456	0.0004249	0.056828	485 ± 3
14/P3	54	0.059924	0.0001975	0.057206	499 ± 13
15 to 20/S2	60	0.0585925	0.000102	0.057332	504 ± 7
21/S3	85	0.06116	0.0002738	0.057414	507 ± 4
22 to 25/S1	90	0.0625967	0.0003688	0.057494	511 ± 2
26 to 28/S3	85	0.0600524	0.00016	0.057995	530 ± 9
29/S2	87	0.0989977	0.0028201	0.058359	543 ± 6
30 to 31/S7	90	0.059796	0.00009731	0.0586073	553 ± 5
32 to 33/S15	54	0.064974	0.000174	0.062714	699 ± 8
34 to 39/S7	90	0.0760204	0.0005694	0.06818	874 ± 2
All zircons of the rock sample: 85% rounded; 15% non-rounded					

Table 2 continued over page

Table 2. *Continued*

Grain number/zircon type after Pupin (1980)	Mass scans	$^{207}Pb/^{206}Pb$	$^{204}Pb/^{206}Pb$	$^{207}Pb/^{206}Pb$ (corrected)	Age (Ma)
Sample K-Ark 3; conglomerate at the base of the Goldisthal Group (Conglomeratic Arkose, Konglomeratische Arkose); Weisse Schwarza at the Viehberg Mtn near Katzhütte; Schwarzburg Anticline; Goldisthal Group; Ordovician					
1 to 2/S2	36	0.06657	0.00069	0.05679	483 ± 3
3 to 4/S16	90	0.06207	0.00037	0.05691	488 ± 1
5 to 6/S2	18	0.0609	0.0003	0.05683	485 ± 2
7 to 9/S1	90	0.05986	0.00021	0.05701	492 ± 1
10 to 12/S1	18	0.06206	0.00036	0.05704	493 ± 4
13/S1	18	0.06081	0.00022	0.05792	527 ± 17
14/S1	54	0.06038	0.00018	0.05794	527 ± 9
15 to 17/S20	90	0.10063	0.00066	0.09181	1464 ± 3
All zircons of the rock sample: 75% rounded; 25% non-rounded					
Sample K-Ark 4; quartzite at the base of the Goldisthal Group (Basal Quartzite, Basisquarzit of the Basis-Schichten); roadcut Wurzelberg Mtn near Goldisthal; Schwarzburg Anticline; Goldisthal Group; Ordovician					
1/S15	52	0.0675689	0.0006538	0.058355	543 ± 11
2/P4	53	0.059929	0.000123	0.058346	543 ± 3
3/P5	87	0.060969	0.0001789	0.058576	551 ± 3
4/S24	86	0.060239	0.0001089	0.058807	560 ± 3
5/S23	89	0.061391	0.0001805	0.058993	567 ± 2
6/P4	85	0.0611372	0.000144	0.059281	577 ± 3
All zircons of the rock sample: 80% rounded; 20% non-rounded					
Sample Laub 1; granite (Laubach Granite); Laubachtal near Katzhütte, car park; Schwarzburg Anticline; Cadomian basement					
16/S8	36	0.068608	0.0001095	0.0672929	847 ± 7
17/S11	90	0.104151	0.0000588	0.103727	1692 ± 8
All zircons of the rock sample: 5% rounded; 95% non-rounded					
Sample Gl-Gt; granite (Glasbach Granite); Steinberg Mtn near Mellenbach; Schwarzburg Anticline; Cadomian basement					
5/S10	32	0.060936	0.00008085	0.059952	602 ± 4
6/S12	86	0.12764	0.00007115	0.127201	2060 ± 4
7/S25	89	0.2351776	0.00005745	0.235366	3060 ± 8
All zircons of the rock sample: 5% rounded; 95% non-rounded					
Sample Laas 1; granodiorite (Laas Granodiorite); Laas; North Saxonian Anticline; Cadomian basement					
7/S7	52	0.0685392	0.000622	0.05986	598 ± 13
8/S7	84	0.069143	0.0001327	0.067509	853 ± 3
All zircons of the rock sample: 0% rounded; 100% non-rounded					
Sample Gross 1; orthogneiss (Grossenhain Orthogneiss); Grossenhain, Kupferberg; Elbe Zone; Cadomian basement					
6/S7	90	0.1283	0.00003	0.12835	2075 ± 2
All zircons of the rock sample: 0% rounded; 100% non-rounded					
Sample Klotz 1; granodiorite, (Westlausitz Granodiorite); Dresden–Klotzsche; in front of the State Museum of Mineralogy and Geology Dresden, A.B. Meyer-Bau; Lausitz Anticline; Cadomian basement					
9/S5	54	0.11774	0.00007	0.11728	1915 ± 14
All zircons of the rock sample: 0% rounded; 100% non-rounded					
Sample Pur 1; quartzite (Purpurberg Quartzite), Purpurberg Mtn near Friedrichswalde; Elbe Zone; Elbtalschiefergebirge; Cadomian basement					
1/S12	88	0.108332	0.003461	0.058148	538 ± 5
2/S7	87	0.073457	0.001045	0.058415	545 ± 2
3/S12	54	0.0944234	0.0025012	0.058444	546 ± 3
4/S20	87	0.091514	0.0022895	0.058549	550 ± 6
5/S8	87	0.0650122	0.0004398	0.058883	563 ± 5
6/S20	89	0.079304	0.0014088	0.059161	573 ± 3
7/S20	89	0.0758225	0.001104	0.0602	611 ± 6
8/S12	82	0.112077	0.003574	0.060676	628 ± 4

Table 2 continued over page

Table 2. *Continued*

Grain number/zircon type after Pupin (1980)	Mass scans	$^{207}Pb/^{206}Pb$	$^{204}Pb/^{206}Pb$	$^{207}Pb/^{206}Pb$ (corrected)	Age (Ma)
9/S6	90	0.073322	0.0008517	0.061244	649 ± 2
10/S13	88	0.075271	0.0009255	0.062179	680 ± 4
11/P3	88	0.08185	0.0001313	0.080044	1198 ± 5
12/P2	89	0.104515	0.0001517	0.10277	1675 ± 6
13/S25	89	0.106338	0.0001646	0.104462	1705 ± 22
14/S23	89	0.117978	0.0001348	0.116579	1904 ± 1
15/rounded, red	86	0.127476	0.0000229	0.12759	2065 ± 1
16/rounded, red	88	0.12741	0.00002055	0.12759	2065 ± 1
17/rounded, red	87	0.023045	0.000538	0.22466	3014 ± 1
All zircons of the rock sample: 90% rounded; 10% non-rounded					
Sample Wü 1; tuff (Wüsteberg Tuff); Wüsteberg Mtn near Kamenz; Lausitz Anticline; Lausitz Group; Cadomian basement					
5/P5	34	0.0600543	0.000038	0.059847	598 ± 7
6/S20	34	0.0744535	0.001002	0.059968	602 ± 13
7/P5	86	0.0605122	0.0001048	0.0605122	621 ± 4
8/P4	87	0.071936	0.0006897	0.062292	680 ± 9
9/S25	72	0.068214	0.0001632	0.0661272	810 ± 1
10 to 11/P4	36	0.09071	0.00018	0.08856	1395 ± 5
12/P4	88	0.116242	0.000238	0.11346	1855 ± 4
13/P5	88	0.116452	0.000087	0.1157	1891 ± 3
14/S15	71	0.131331	0.00019	0.1294	2090 ± 2
15/P3	90	0.133216	0.000163	0.13157	2119 ± 1
All zircons of the rock sample: 20% rounded; 80% non-rounded					
Sample WeG 2; granite pebble from a debris flow; railway station, Weesenstein; Elbe Zone; Elbtalschiefergebirge; Weesenstein Group; Cadomian basement					
7/S20	84	0.084138	0.001703	0.059752	595 ± 14
All zircons of the rock sample: 90% rounded; 10% non-rounded					
Sample Katz-Ro; greywacke; Rotseifenbach near Goldisthal; Schwarzburg Anticline; Frohnberg Group; Cadomian basement					
1/P3	90	0.059527	0.000047	0.059058	569 ± 2
2/P3	12	0.059838	0.000052	0.059295	578 ± 20
3/P4	90	0.064461	0.000367	0.059365	580 ± 6
4 to 6/P3	90	0.061296	0.00012	0.059783	596 ± 3
7/P4	17	0.086415	0.000063	0.085824	1334 ± 16
8/P3	90	0.118099	0.000173	0.116214	1899 ± 4
All zircons of the rock sample: 90% rounded; 10% non-rounded					
Sample Katz 1; conglomerate with granite and greywacke pebbles; Grendel Mtn near Sachsendorf; Schwarzburg Anticline; Frohnberg Group; Cadomian basement					
8/S15	92	0.06932	0.00056	0.061437	655 ± 3
9/S20	49	0.063383	0.00006	0.062738	699 ± 7
10/S20	17	0.06345	0.00006	0.062751	700 ± 13
11/P3	83	0.065625	0.00016	0.063551	727 ± 6
All zircons of the rock sample: 90% rounded; 10% non-rounded					
Sample Zir 1; meta-greywacke and intercalated quartzite (Kernzonenquarzit); Zirkelstein near Mellenbach; Schwarzburg Anticline; Katzhütte Group (Kernzone; tectonic melange of Cadomian basement and Early Palaeozoic rocks)					
1/S25	90	0.06305	0.00045	0.05673	481 ± 8
2/P5	90	0.06623	0.00054	0.05869	556 ± 4
3/D	90	0.06401	0.00038	0.05875	558 ± 3
4/P5	36	0.07113	0.00087	0.0588	560 ± 18
5/P4	85	0.06599	0.00049	0.05915	573 ± 3
6/D	90	0.0594	0.00003	0.05921	575 ± 2
7/D	90	0.06708	0.00055	0.05942	582 ± 3
8/S25	90	0.06091	0.00011	0.05951	586 ± 4
9/P4	90	0.06193	0.00018	0.05956	588 ± 3
All zircons of the rock sample: 95% rounded; 5% non-rounded					

Fig. 4. Scanning electron photomicrographs of typical zircons of significant igneous rocks of Saxo-Thuringia. (**a**) Cadomian zircon of a granitoid pebble of a island-arc-derived debris flow (sample Cla 1, *c.* 575 Ma, type D). (**b**) Cadomian zircon of the Dohna Granodiorite (sample Dohna 1, *c.* 540 Ma, type S 2). (**c**) Cadomian zircon of the Glasbach Granite (sample Gl-Gt, *c.* 540 Ma, type L 5). (**d**) Ordovician zircon of a conglomerate (classical German term: 'Konglomeratische Arkose') at the base of the Goldisthal Group (sample K-Ark 1, *c.* 490 Ma, type S 1). (**e**) Cadomian zircon of a quartzite (classical German term: 'Basisquarzit') at the base of the Goldisthal Group (sample K-Ark 4, *c.* 560 Ma, type S 25). (**f**) Ordovician zircon of the Bärentiegel Porphyroid (sample P-Bt, *c.* 480 Ma, type S 14). (**g**) Ordovician zircon of a particular sericite gneiss (classical German term: 'Serizitgneis' of the Elbe Zone) (sample S-Gn, *c.* 490 Ma, type S 12). Zircon types after Pupin (1980).

outcropping crustal level in the Schwarzburg Anticline, could be a fault-bounded tectonic melange containing Cadomian to Ordovician rocks. This would be require thrusting of the low-grade Cadomian to Palaeozoic rock units above the Kernzone during Variscan collision tectonics.

A very important set of detrital zircons was found in a sample of the Purpurberg Quartzite (Pur 1), which was characterized by Linnemann (1991) as a lowstand fan within Cadomian greywackes of the Weesenstein Group. Apart from the 'normal' Cadomian ages (550 ± 6 Ma to 680 ± 4 Ma) and several Meso- to Palaeoproterozoic and Archaean zircons (Table 3), three zircons (grains 1–3) with ages overlapping the Precambrian–Cambrian boundary were found (538 ± 5 Ma to 546 ± 3 Ma). The Precambrian–Cambrian boundary was fixed recently by several workers between 543 and 545 Ma (e.g. Bowring *et al.* 1993). Taking into account the 2σ error of the measurements of grains 1–3 (sample Pur 1) the sedimentation of the Purpurberg Quartzite was most probably an event during the earliest Cambrian time or represents the base of the Cambrian sequence. This means that the Precambrian–Cambrian boundary is hidden within the lower part of the Weesenstein Group and the sedimentation related to Cadomian processes continued into earlist Cambrian time, because the Purpurberg Quartzite is overlain by similar greywackes with a signature of a continental island arc, which also occurs in the lower part of the Weesenstein Group (Fig. 3).

The probable continuation of Cadomian sedimentation into Cambrian time is also confirmed by the radiometric ages of the Cadomian plutons, which have been intruded into Cadomian sedimentary rocks after Cadomian deformation (Fig. 3). The ages of detrital zircons within the Purpurberg Quartzite (lower time mark) and the intrusion of Cadomian plutons (upper time mark) constrain the deformation related to the Cadomian Orogeny to have occured during earliest Cambrian time at around 540 Ma.

The Glasbach Granite (Gl-Gt, 541 ± 7 Ma) and the Laubach Granite (Laub 1, 533 ± 10 Ma) were intruded into the Frohnberg and Katzhütte Groups of the Schwarzburg Anticline. The Laas Granodiorite (Laas 1) with an intrusive contact to the Clanzschwitz Group, was dated at 531 ± 7 Ma. The Weesenstein Group was intruded by the Dohna Granodiorite (Dohna 1) at 537 ± 7 Ma. The Grossenhain Orthogneis (Gross 1), cropping out within the Elbe Zone, was deformed during the Variscan dextral strike-slip along the Elbe Zone and is interpreted as an equivalent of the Dohna Granodiorite because of its similar intrusion age of 537 ± 7 Ma. The Westlausitz Granodiorite (Klotz 1) was intruded into the Lausitz Group. The age was determined at 539 ± 6 Ma (Fig. 3).

A post-Cadomian group of ages is represented by Tremadoc intrusive and effusive rocks as well as detrital zircons from the lowermost Ordovician sedimentary rocks. According to Davidek *et al.* (1998) and McKerrow & van Staal (this volume), the age of 490 Ma marks the base of the Tremadoc sequence.

With only two exceptions the Cadomian basement is transgressed by Tremadoc siliciclastic rocks. Only in the Delitzsch–Torgau–Doberlug Sycline is the Cadomian basement covered by Lower to Middle Cambrian siliciclastic rocks and carbonates (Fig. 3). In the deep boreholes Heinersdorf 1 and 2 (Berga Anticline, Fig. 2b, see Fig. 8) Lower to Middle Cambrian rocks were found underlying the Ordovician sequence (Wucher 1967) (Fig. 8). However, the relationship to the Cadomian basement is unknown, because it was not reached by the boreholes (Fig. 8).

The German Ordovician type section for the Thuringian facies is situated in the Schwarzburg Anticline. The age of the lowermost sediments has been in discussion for at least the last 50 years. The lowermost sedimentary rock is represented by a conglomerate called traditionally 'Konglomeratische Arkose' ('Conglomeratic Arkose')(Falk 1964), which belongs to the so-called 'Basis-Schichten' ('Basal Layer') of the Goldisthal Group. At some localities the conglomerate is replaced by a quartzite, the 'Basisquarzit' ('Basal Quartzite') (Fig. 9). At the locality from the Birkenweg in Mellenbach the Conglomeratic Arkose (sample K-Ark 1) contains rhyolite pebbles. Magmatic zircons from this sample were determined with Tremadoc ages ranging from 476 ± 4 Ma to 485 ± 3 Ma. In addition, another sample of the Conglomeratic Arkose (K-Ark 3) from the Weisse Schwarza Valley near Katzhütte, which does not have rhyolite pebbles, contains detrital zircons with ages from 483 ± 3 Ma to 493 ± 4 Ma. The ages of both samples clearly show that the shales and quartzites of the Goldisthal Group overlying the Conglomeratic Arkose must be younger than about 485 Ma. This leads to the conclusion that the largest part of the Cambrian sequence between the Cadomian basement and the Goldisthal Group is missing. A sample from the Basal Quartzite (K-Ark 4), the equivalent of the Conglomeratic Arkose unfortunately contains only detrital zircons with Cadomian ages (Table 2).

Table 3. *Average of trace element values of some Peri-Gondwanan sedimentary rocks of Saxo-Thuringia, which were used in the discrmination diagrams of Fig. 3 (values were measured by ICP-MS at ACTLABS, Ancaster/Ontario, Canada)*

Lithostratigraphy	Geological unit	System	Lithology	Number of samples	La (ppm)	Th (ppm)	Sc (ppm)	Zr (ppm)
Rothstein Group	Delitzsch–Torgau–Doberlug Syncline	Neoproterozoic (Cadomian basement)	greywacke	5	27	8	11	176
Frohnberg Group	Schwarzburg Anticline	Neoproterozoic (Cadomian basement)	greywacke	5	19	9	13	266
Clanzschwitz Group	Northern Saxonian Anticline	Neoproterozoic (Cadomian basement)	greywacke	5	35	8	7	232
Weesenstein Group	Elbe Zone	Neoproterozoic (Cadomian basement)	greywacke	5	41	15	15	241
Lausitz Group	Lausitz Anticline	Neoproterozoic (Cadomian basement)	greywacke	5	44	12	16	191
Lower and Upper Frauenbach Quartzite (Frauenbach Group)	Schwarzburg Anticline	Ordovician	quartzite	5	29	27	6	375
Dubrau Quartzite (Dubrau Fm)	Lausitz Anticline	Ordovician	quartzite	5	23	8	5	293
Collmberg Quartzite (Collmberg Fm)	Northern Saxonian Anticline	Ordovician	quartzite	5	34	11	7	336

An ash layer, the Wurzelberg Tuff (Wurz 1), distributed within the Frauenbach Group (Fig. 9), which overlies the Goldisthal Group, contains non-rounded zircons with an mean age of 484 ± 10 Ma. This age is interpreted as the sedimentation age of the Frauenbach Group.

The Blambach Rhyolite (Bla-Rhy) crops out in the Blambach Valley near Sitzendorf. The rhyolite is covered by the Conglomeratic Arkose. The Blambach Rhyolite was dated with an mean age of 487 ± 6 Ma. The contact between rhyolite and arcose is overprinted by a Variscan brittle-shear zone, but nevertheless interpreted as a transgressive one. If geologists accept the interpretation of this contact with the overlying Conglomeratic Arkose as the base of the Ordovician transgression, the age of the rhyolite is further argument for the age of the Goldisthal Group to be younger than *c.* 485 Ma.

A special dating was carried out with two small granite pebbles from a debris flow of the Frauenbach–Wechsellagerung (Frauenbach Group), from a locality at the firing range near Scheibe–Alsbach first described by Falk (1967). The first pebble was dated at 605 ± 4 Ma, the second one at 570 ± 4 Ma (samples SSA 1 and SSA 3, Table 1, Fig. 3). Both pebbles are obviously derived from the Cadomian basement. This clearly demonstrates that a part of the Cadomian island arc was preserved in the hinterland of Saxo-Thuringia and must have cropped out with a deep erosion level at the surface during Early Ordovician times.

A sample of sandstone (Gp 28b) from the deep borehole Gumperda was taken at 1095 m below surface, only a few metres above the transgressive contact of the Goldisthal Group over the underlying Cadomian basement. The detrital zircons gave ages between 471 ± 6 Ma and 488 ± 2 Ma. This also confirms the Tremadoc maximum age of the Goldisthal Group.

A porphyroid, the so-called Bärentiegel Porphyroid (P-Bt), was intruded in a sub-volcanic niveau into siliciclastic deposits of the Frauenbach Group with a mean age of 479 ± 2 Ma.

Other Ordovician igneous rocks are known from the Elbe Zone (Fig. 3). The Tourmaline Granite (T-Gt) shows an age of intrusion of 485 ± 6 Ma. The Sericite Gneiss (S-Gn), a meta-rhyolite, was dated at 486 ± 4 Ma. Both bodies occur within the 'Phyllitic Unit' of the Elbe Zone, which was interpreted by Kurze *et al.* (1992), based on lithostratigraphical criteria with respect to the Ordovician type section in the Thuringia section, as an Ordivician rock suite. The section, including the Tourmaline Granite and the Sericite Gneiss, is strongly deformed as a result of Variscan strike-slip movements along the Elbe Zone (Linnemann 1995).

In the eastern part of the Lausitz Anticline the Rumburk Granite was intruded into Cadomian igneous rocks. The post-Cadomian granite was dated by Hammer *et al.* (1997) to 490 ± 3 Ma (Pb/Pb).

Significance of ages from inherited and detrital zircons

The ages of exactly 100 inherited zircons from igneous rocks and detrital zircons from various sedimentary rocks were determined (for dataset, localities of sampling and zircon types, see Table 2). The ages reflect, apart from Cadomian and Lower Palaeozoic tectonomagmatic events, the history of the Gondwanan cratons in the hinterland of peri-Gondwana, such as the West African or the Amazonian Cratons (Fig. 1). On the one hand, the zircons of the Gondwanan cratons were transported as detritus into the marginal basins of the Avalonian–Cadomian Orogenic Belt. On the other hand, old African crust was recycled by magmatic activity during Cadomian orogenic processes.

Detrital zircons are an important tool to reconstruct the provenance of a peri-Gondwana-derived crustal fragment. The comparison of zircon ages of peri-Gondwanan terranes with the ages of the large cratons in the hinterland of the Avalonian–Cadomian Orogenic Belt was first used by Nance & Murphy (1994, 1996). The ages of the Amazonian Craton (Teixeira *et al.* 1989), of the West African Craton (Rocci *et al.* 1991) and of Brittany (Vidal *et al.* 1981; Egal *et al.* 1996) were used in Fig. 5 to relate the ages of Saxo-Thuringian zircons to one of the source areas. The groups of zircon ages are shown with intervals of 10 Ma (Fig. 5).

The first group is represented by pre-Rodinian events, originated before the formation of the Rodinia and Pannotia Supercontinents, as reconstructed by Dalziel (1997). The group contains zircons of cratonal provinces of the West African Craton (Leonian, Eburian) as well as of the Amazonian Craton (Trans-Amazonian, Rio Negro, Rodonian) (Fig. 5). In addition, the age of the Icartian event (Vidal *et al.* 1981), with zicon ages of around 2000 Ma, occurs within suite.

The formation of Rodinia (1200–1000 Ma) and Pannotia (<730 Ma) (Dalziel 1997; Unrug 1997) is represented in the next two groups. The age interval ranging from 670 to 730 Ma is not known from the West African Craton or Amazonian Craton, but more or less similar

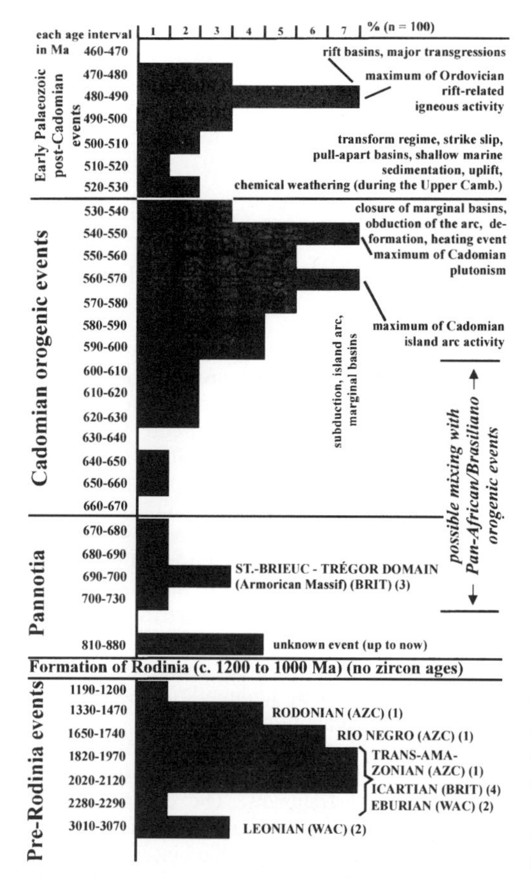

Fig. 5. Geotectonic events and significance of ages of inherited and detrital zircons on the basis of 100 measured zircon grains from various rocks of Saxo-Thuringia (dataset is given in Table 2). AZC, Amazonian Craton; WAC, West African Craton; BRIT, Brittany (Armorican Massif). References for geochronological provinces: (1) Teixeira *et al.* (1989); (2) Rocci *et al.* (1991); (3) Egal *et al.* (1996); (4) Vidal *et al.* (1981).

ages have been described by Egal *et al.* (1996) from the St Brieuc–Trégor Domain of the Armorican Massif. A 810–880 Ma interval of ages is unknown to the authors.

The largest group contains ages of 670 to 530 Ma, which reflect Cadomian orogenic events. The maximum of Cadomian island-arc activity is given by the peak at 560–570 Ma. The maximum of Cadomian plutonism was reached at *c.* 540 Ma.

The last part in the diagram is occupied by Early Palaeozoic ages, with a peak between 490 and 480 Ma related to the maximum of Ordovician rifting processes.

The most important information of these zircon ages is the mixing of zircons derived from the Amazonian as well as from the West African Cratons. Both Gondwanan cratons were obviously the source area for sedimentary input into marginal Cadomian basins preserved as remnants in Saxo-Thuringia. Most probably also parts of the Armorican Massif (Brittany) were eroded and redeposited in Saxo-Thuringian Cadomian basins. This requires a position of Saxo-Thuringia in the external zones of the Avalonian–Cadomian Orogenic Belt (Fig. 1).

Geochemical data

Some geochemical data of trace elements are represented in Table 3 and Fig. 6. Figure 6 shows the position of 25 samples of greywackes, all from low-grade overprinted Cadomian units of Saxo-Thuringia in two discrimination diagrams of Bhatia & Crook (1986). In all cases the data plot into the field for 'continental island-arc' signature and characterize the provenance and the geotectonic setting. For this reason the radiometric ages of the Cadomian rocks were interpreted as island arc related.

The geotectonic signature of the 15 quartzites taken from the Ordovician overstep sequence shown in Fig. 5 is completely different. These siliciclastic rocks show a passive margin provenance and are very mature. This important change in comparison with the signatures of the Cadomian sedimentary rocks will be discussed in the next section.

Interpretation of data: peri-Gondwanan orogenic processes preserved in Saxo-Thuringia

To reconstruct peri-Gondwanan orogenic processes in Saxo-Thuringia we have combined the geochronological dataset with geochemical data on the provenance of Cadomian and Lower Paleozoic sedimentary rocks and aspects of the basin development during Ordovician time.

The remnants of Cadomian basins preserved in Saxo-Thuringia are interpreted as fragments of back-arc basins. All sedimentary rocks (mostly greywackes) show a continental island-arc signature (Fig. 6). The input of a large amount of detrital zircons from the Gondwanan cratons requires a position of the Saxo-Thuringian marginal basins between continent and island arc (Fig. 7). The maximum activity of the arc terminated between 560 and 570 Ma (Fig. 5). The input of the arc into the back-arc basins is documented by granitoid pebbles with ages ranging between *c.* 577 and 560 Ma (Fig. 7). Two island-arc-derived ash layers with ages of

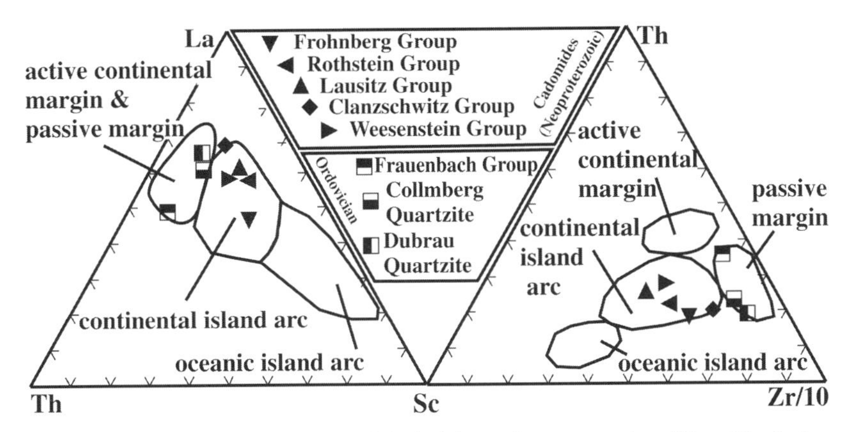

Fig. 6. Geotectonic significance of Cadomian and Ordovician sedimentary rocks of Saxo-Thuringia (discrimination diagrams by Bhatia & Crook (1986)). Each symbol represents an average value of five samples (geochemical data are listed in Table 3).

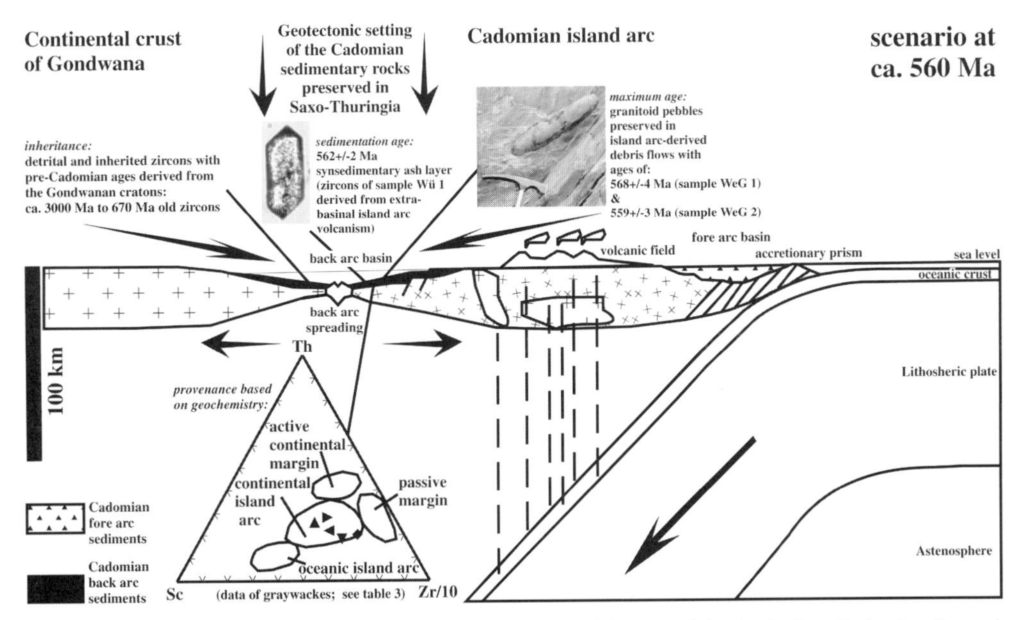

Fig. 7. Reconstruction of the geotectonic scenario at *c.* 560 Ma of the part of the Avalonian–Cadomian Orogenic Belt preserved in Saxo-Thuringia on the basis of the geochemistry of Cadomian sedimentary rocks, zircon ages of granitoid pebbles of an island-arc-derived debris flow and an ash layer (Wüsteberg Tuff) intercalated in Cadomian greywackes, and of detrital and inherited zircons showing pre-Cadomian ages.

562–574 Ma (Figs 3 and 7) reflect the extra-basinal volcanic activity of the arc and mark the sedimentation age of the back-arc filling. Detrital and inherited zircons (Fig. 5) suggest that the Amazonian as well as the West African Craton and most probably Brittany are potential source areas for the Cadomian sedimentary rocks of Saxo-Thuringia. The sedimentation within the back-arc basins was continued through the Precambrian–Cambrian boundary into earliest Cambrian time (see ages of the detrital zircons of the Purpurberg Quartzite; Fig. 3, Table 2).

The Cadomian arc was obducted towards the Gondwanan cratons at around 540 Ma. The back-arc basins were closed and deformed. Because of the Cadomian obduction and the resulting crustal growth, a heating event produced granitoids, which intruded into the orogen

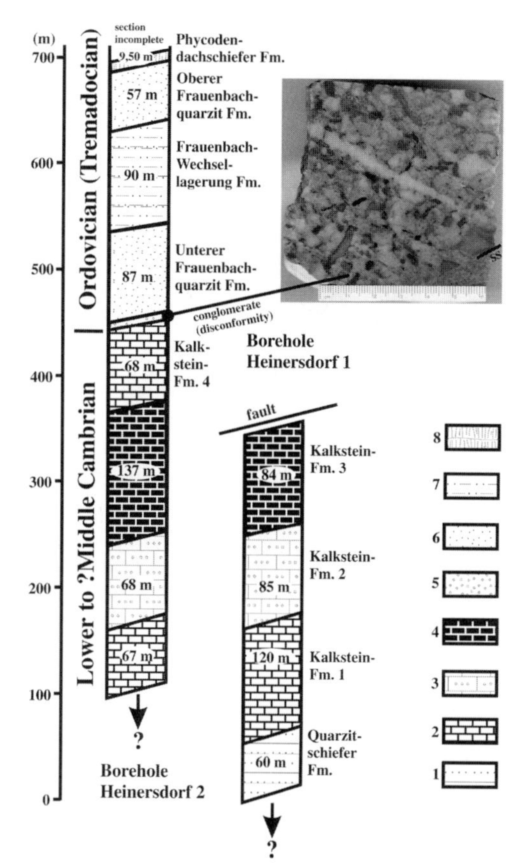

Fig. 8. Lithosections of Lower to Middle(?) Cambrian and Tremadoc sedimentary rocks of the boreholes Heinersdorf 1 and 2 (Berga Anticline). (Note the disconformity between Cambrian and Ordovician units and the gap in the Upper Cambrian sequence.) The sections of these boreholes were overprinted by Variscan greenschist facies metamorphism. 1, quartzite and phyllite; 2, intercalation of carbonate and phyllite (carbonate dominates); 3, intercalation of phyllite and carbonate (phyllite dominates); 4, black and dark grey sapropelitic carbonates and phyllites; 5, conglomerate (pebbles of quartz, quartzite, hornfels and chert); 6, quartzite; 7, intercalation of quartzite and phyllite; 8, phyllite.

at around 540–535 Ma (Fig. 3) as a result of post-orogenic extension.

During the Cambrian time, the margin to the oceanic plate after the model of Nance & Murphy (1994, 1996) probably became a transform margin. As a result of large-scale strike-slip movement within parts of the Cadomian Orogen, pull-apart basins formed and were transgressed by shallow-marine sediments of Early to Middle Cambrian age.

During the Upper Late Cambrian time, the Saxo-Thuringian part of the Avalonian–Cadomian Orogen was uplifted. In general, the Upper Cambrian sequence in Saxo-Thuringia is missing. Only a gap can be recognized (Fig. 8). A chemical weathering crust under humid conditions formed (this was the origin of kaolinite found in the region) (Linnemann & Buschmann 1995).

During the Tremadoc transgression the weathering crust was recycled. This process is represented by very mature siliciclastic rocks. An example is the conglomerate at the base of the Ordovician succession of the borehole Heinersdorf 2 (Fig. 8). The Tremadoc sandstones and conglomerates mostly consist of stable components (quartz, cherts, hornfelses). This maturity is also documented in the geochemical passive margin signatures of the Ordovician siliciclastic rocks, which are shown in Fig. 6.

The sedimentation regime of Ordovician time was dominated by a rift setting. The Tremadoc sequence alone consists of at least 3000 m of shallow-marine siliciclastic rocks containing four second-order sequences in terms of sequence stratigraphy (Fig. 9). The resulting enormous sedimentation rate only can be explained by a rift setting.

The igneous rocks showing Lower Ordovician ages (Figs 3 and 9) are related to the extension of the crust and the mobilization of magmas starting with an acidic first stage (490–470 Ma, Figs 3 and 9). During rifting, the composition of the magmatism became increasingly basic (Fig. 9).

Finally, the rifting process led to the separation of Saxo-Thuringia from Gondwana during Late Ordovician time (Fig. 9). After drifting in Silurian times, Saxo-Thuringia became incorporated into the Variscan Orogeny (Franke this volume) and became a part of the Central European Variscides.

The authors benefited from funding by the Deutsche Forschungsgemeinschaft (grants Lu 544/1 and 2, Li 521/6-1) and acknowledge incorporation into the DFG priority programme 'Orogenic processes—Quantification and Modelling in the Variscan Belt'. W. Franke (Giessen) is thanked for intense discussions and much support, and U. Kempe and J. Götze (Freiberg) are thanked for assistance with scanning electron microscopy and cathodoluminescence microscopy. Special thanks are dedicated to J. von Raumer (Fribourg) and F. Neubauer (Graz) for critical constructive reviews. D. Tanner is thanked for his corrections in the manuscript. Finally, many thanks are due to D. Nance (Athens, Ohio) and B. Murphy (Antigonish, Canada) for very interesting discussions during the International Conference on Cadomian Orogens in Badajoz (Spain, 1999).

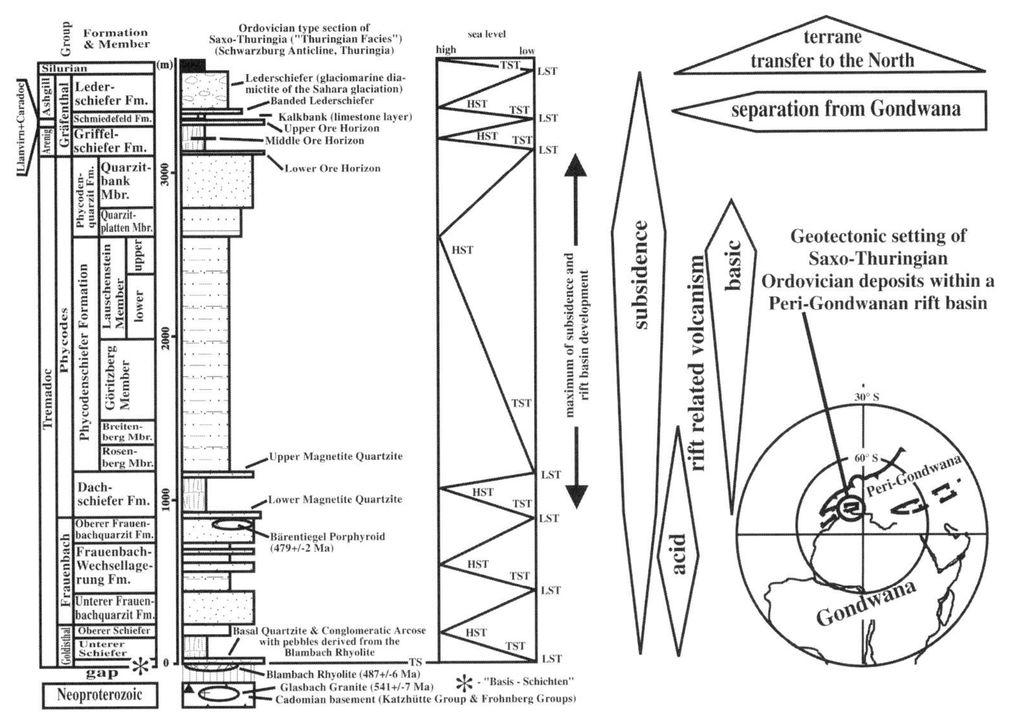

Fig. 9. Reconstruction of the geotectonic scenario during Ordovician time on the basis of basin development, sequence stratigraphy, volcanic activity and geochronology preserved in the Ordovician type section of Saxo-Thuringia in the Schwarzburg Anticline (TS, transgressive surface; TST, transgressive systems tract; HST, highstand systems tract; LST, lowstand systems tract) (palaeogeographical sketch after Noblet & Lefort (1990)).

References

Bachtadse, V., Schätz, M. R., Zwing, A., Tait, J. A. & Soffel, H. C. 1998. Paleogeography of the Paleozoic Terranes in the Variscan and Alpine foldbelts. *Terra Nostra—Schriften der Alfred-Wegener-Stiftung*, **98**(2), 12–15.

Bankwitz, P. 1977. Überblick über den Magmatismus und die Metamorphose im Altpaläozoikum Mittel- und Westeuropas. *Veröffentlichungen des Zentralinstituts für Physik der Erde*, **44**(1), 7–30.

——, P. & Bankwitz, E. 1995. Proterozoikum/Schwarburger Antiklinorium. *In*: Seidel, G. (ed.) *Geologie von Thüringen*. Schweitzerbart'sche, Stuttgart, 46–77.

Bhatia, M. R. & Crook, K. A. W. 1986. Trace element characteristics of greywackes and tectonic setting discrimination of sedimentary basins. *Contributions to Mineralogy and Petrology*, **92**, 181–193.

Bowring, S. A., Grotzinger, J. P., Isachsen, C. E., Knoll, A. H., Pelechaty, S. M. & Kolosov, P. 1993. Calibrating rates of early Cambrian evolution. *Science*, 261, 1293–1298.

Bunel, H. 1835. Observations sur les terrains intermédiaires du Calvados. *Mémoires de la Société Linnéenne de Normandie*, **5**, 99–100.

Buschmann, B. 1995. *Geotectonic facies analysis of the Rothstein Formation (Neoproterozoic, Saxothuringian Zone, Germany)*. PhD thesis, Bergakademie Freiberg, Freiberg.

——, Linnemann, U., Schneider, J. & Süss, T. 1995. Die cadomische Entwicklung im Untergrund der Torgau–Doberluger Synklinale. *Zeitschrift für Geologische Wissenschaften*, **23**(5/6), 729–749.

——, Nasdala, L., Jonas, P., Linnemann, U. & Gehmlich, M. 2000. Ages, sedimentology, provenance and paleogeographical affinity of late cadomian marginal basin fragments in the north-eastern part of the Saxo-Thuringian terrane (Germany): first SHRIMP U–Pb dating of tuffogenic and detrital zircons from Neoproterozoic sequences. *Neues Jahrbuch für Geologie und Paläontologie*, in press.

Chantraine, J., Auvray, B., Brun, J. P., Chauvel, J. J. & Rabu, D. 1994. Introduction. *In*: Keppie, J. D. (ed.) *Pre-Mesozoic Geology in France and Related Areas*. Springer, Berlin, 75–80.

Dalziel, I. W. D. 1992. On the organization of American plates in the Neoproterozoic and the breakout of Laurentia. *GSA Today*, **2**, 240–241.

—— 1997. Overview: Neoproterozoic–Paleozoic geography and tectonics: review, hypothesis, environmental speculation. *Geological Society of America Bulletin*, **109**, 16–42.

Davidek, K., Landing, E., Bowring, S. A., Westrop, S. R., Rushton, A. W., Fortey, R. A. & Adrain,

J. M. 1998. New uppermost Cambrian U–Pb date from Avalonian Wales and age of the Cambrian–Ordovician boundary. *Geological Magazine*, **135**(3), 305–309.

DEAN, R. B. & DIXON, W. J. 1951. Simplified statistics for small numbers of observations. *Analytical Chemistry*, **23**, 17–35.

EGAL, E., GUERROT, C., GOFF, E. L., THIÉBLEMONT & CHANTRAINE, J. 1996. The Cadomian orogeny revisited in northern Brittany (France). *In*: NANCE, R. D. & THOMPSON, M. D. (eds) *Avalonian and Related Peri-Gondwanan Terranes of the Circum-North Atlantic*. Geological Society of America, Special Papers, **304**, 281–318.

ELICKI, O. 1997. Biostratigraphic data of the German Cambrian—present state of knowledge. *Freiberger Forschungsheft, C*, **466**, 155–165.

ELLENBERG, J., FALK, F. & LÜTZNER, H. 1992. Siliciclastic shelf sedimentation of Early Paleozoic deposits in the Thüringer Schiefergebirge (Thuringian Slate Mountains). *In*: FALK, F. (ed.) *13th IAS Regional Meeting of Sedimentology (Guidebook)*. Friedrich-Schiller-Universität Jena, Institut für Geowissenschaften, Jena, 133–158.

ESTRADA, S., HEUSE, T. & SCHULZ, E. 1994. Zur regionalgeologischen Interpretation kambro-ordovizischer Schichten an der Nordwest-Flanke des Schwarzburger Antiklinoriums, Thüringen. *Zeitschrift für Geologische Wissenschaften*, **22**(5), 537–553.

FALK, F. 1964. Die Konglomeratische Arkose der Goldisthaler Schichten. *Abhandlungen der Deutschen Akademie der Wissenschaften zu Berlin*, **2**, 3–26.

FALK, F. 1967. Der Gerölltonschiefer von Scheibe—Eine grobklastische Sonderausbildung der Frauenbach-Serie (Tremadoc) des Schwarzburger Sattels in Thüringen. *Jahrbuch für Geologie*, **3**, 1–56.

FRANKE, W. 2000. The Mid-Euorpean Segment of the Variscides: Tectono-Stratigraphic Units, Terrane Boundaries and Plate Evolution. *This Volume*.

GEHMLICH, M., LINNEMANN, U., TICHOMIROWA, M., LÜTZNER, H. & BOMBACH, K. 1997. Datierung und Korrelation neoproterozoisch–frühpaläozoischer Profile des Schwarzburger Antiklinoriums und der Elbezone auf der Basis der Geochronologie von Einzelzirkonen. *Zeitschrift für Geologische Wissenschaften*, **25**(1–2), 191–201.

GLASMACHER, U. A., REYNOLDS, P., ALEKSEYEV, A. A., PUCHKOV, V. N., TAYLOR, K., GOROZHANIN, V. & WALTER, R. 1999. $^{40}Ar/^{39}Ar$ thermochronology west of the Main Uralian fault, southern Urals, Russia. *Geologische Rundschau*, **87**, 515–525.

HAMMER, J., BRÖCKER, M. & KRAUSS, M. 1997. Alter und geologische Signifikanz von Deformationszonen im östlichen Teil des Lausitzer Granitoidkomplexes. *Terra Nostra—Schriften der Alfred-Wegener-Stiftung*, **97**(5), 62–64.

HAVLIČEK, V., VANEK, J. & Fatka, O. 1994. Perunica microcontinent in the Ordovician (its position within the Mediterranean province, series division, benthic and pelagic associations). *Sbornik Geologickych Ved*, **46**, 23–56.

HEUSE, T., ERDTMANN, B. D. & KRAFT, P. 1994. Early Ordovician microfossils (acritarchs, chitinozoans) and graptolites from the Schwarzburg Anticline, Thuringia (Germany). *Veröffentlichungen des Naturhistorischen Museums Schleusingen*, **9**, 41–68.

HOWELL, D. G. 1995. *Principles of Terrane Analysis*. Chapman & Hall, London.

KEMNITZ, H. & BUDZINSKI, G. 1994. Die Grauwacken der Lausitz und ihre cadomische Prägung. *Abhandlungen des Staatlichen Museums für Mineralogie und Geologie zu Dresden*, **40**, 37–97.

KOBER, B. 1986. Whole-grain evaporation for $^{207}Pb/^{206}Pb$-age-investigation on single zircons using a double-filament thermal ion source. *Contributions to Mineralogy and Petrology*, **93**, 482–490.

—— 1987. Single-zircon evaporation combined with Pb^{+} emitter bedding for $^{207}Pb/^{206}Pb$-age investigation using thermal ion mass spectrometry and implications to zirconology. *Contributions to Mineralogy and Petrology*, **96**, 63–71.

KÖSSLER, P., TAIT, J., BACHTADSE, V., SOFFEL, H. C. & LINNEMANN, U. 1996. Paleomagnetic investigations of Lower Paleozoic rocks of the Thüringer Schiefergebirge. *Terra Nostra—Schriften der Alfred-Wegener-Stiftung*, **96**(2), 115–116.

KOSSMAT, F. 1927. Gliederung des varistischen Gebirgsbaues. *Abhandlungen des Sächsischen Geologischen Landesamts*, **1**, 1–39.

KURZE, M., LINNEMANN, U. & TRÖGER, K. A. 1992. Weesensteiner Gruppe und Altpaläozoikum in der Elbtalzone (Sachsen). *Geotektonische Forschungen*, **77**, 101–167.

LINNEMANN, U. 1991. Glazioeustatisch kontrollierte Sedimentationsprozesse im Oberen Proterozoikum der Elbezone (Weesensteiner Gruppe/Sachsen). *Zentralblatt für Geologie und Paläontologie, Teil I*, **12**, 2907–2934.

—— 1995. The Neoproterozoic terranes of Saxony (Germany). *Precambrian Research*, **73**, 235–250.

—— & BUSCHMANN, B. 1995. Die cadomische Diskordanz im Saxothuringikum (oberkambrisch–tremadocische overlap-Sequenzen). *Zeitschrift für Geologische Wissenschaften*, **23**(5–6), 707–727.

——, GEHMLICH, M., TICHOMIROWA, M. & BUSCHMANN, B. 1998. Introduction to the pre-symposium excursion (part I): the Peri-Gondwanan basement of the Saxo-Thuringian composite terrane. *In*: LINNEMANN, U., HEUSE, T., FATKA, O., KRAFT, P., BROCKE, R. & ERDTMANN, B.-D. (eds) *Pre-Variscan Terrane Analysis of Gondwanan Europe—Excursion Guides and Abstracts. Schriften des Staatlichen Museums für Mineralogie und Geologie zu Dresden*, **9**, 7–13.

——, ——, —— & LÜTZNER, H. 1997. Kalibrierung des saxothuringischen Ordoviziums mit Hilfe sequenzstratigraphischer und geochronologischer Methoden. *Terra Nostra—Schriften der Alfred-Wegener-Stiftung*, **97**(2), 130–131.

LÜTZNER, H., ELLENBERG, J. & FALK, F. 1986. Entwicklung der Sedimentationsrate und der Ablagerungsprozesse im Altpaläozoikum Thüringens. *Zeitschrift für geologische Wissenschaften*, **14**(1), 83–93.

MASLOV, A. V., ERDTMANN, B.-D., IVANOV, K. S., IVANOV, S. N. & KRUPENIN, M. T. 1997. The main tectonic events, depositional history, and the palaeogeography of the southern Urals during the Riphean–early Palaeozoic. *Tectonophysics*, **276**, 313–335.

MCKERROW, W. S. & VAN STAAL, C. R. 2000. The Palaeozoic time scale reviewed. *This volume*.

NANCE, R. D. & MURPHY, J. B. 1994. Contrasting basement isotopic signatures and the palinspastic restoration of peripheral orogens: example from the Neoproterozoic Avalonian–Cadomian belt. *Geology*, **22**, 617–620.

—— & —— 1996. Basement isotopic signatures and Neoproterozoic paleogeography of Avalonian–Cadomian and related terranes in the circum-North Atlantic. *In*: NANCE, R. D. & THOMPSON, M. D. (eds) *Avalonian and Related Peri-Gondwanan Terranes of the Circum-North Atlantic*. Geological Society of America, Special Papers, **304**, 333–346.

——, ——, STRACHAN, R. A., D'LEMOS, R. S. & TAYLOR, G. K. 1991. Late Proterozoic tectonostratigraphic evolution of the Avalonian and Cadomian terranes. *Precambrian Research*, **53**, 41–78.

NOBLET, C. & LEFORT, J. P. 1990. Sedimentological evidence for a limited separation between Armorica and Gondwana during the Early Ordovician. *Geology*, **18**(4), 303–306.

PIETZSCH, K. 1917. Das Elbtalschiefergebiet südwestlich von Pirna. *Zeitschrift der Deutschen Geologischen Gesellschaft*, **69**, 177–286.

PUPIN, J. P. 1980. Zircon and granite petrology. *Contributions for Mineralogy and Petrology*, **73**, 207–220.

ROBARDET, M. & DORÉ, F. 1988. The Late Ordovician diamictitic formations from southwestern Europe: north-Gondwana glaciomarine deposits. *Palaeogeography, Palaeoclimatology, Palaeoecology*, **66**, 19–31.

ROCCI, G., BRONNER, G. & DESCHAMPS, M. 1991. Crystalline basement of the West African Craton. *In*: DALLMEYER, R. D. & LECORCHE, J. P. (eds) *The West African Orogens and Circum-Atlantic Correlatives*. Springer, Berlin, 31–61.

SEHM, K. 1976. Lithologisch-petrofazielle und metallogenetische Untersuchung der Grauwacken-Pelit-Folge des Nordsächsischen Antiklinoriums. *Freiberger Forschungsheft*, **C311**, 8–135.

STACEY, J. S. & KRAMERS, J. D. 1975. Approximation of terrestrial lead isotope evolution by a two-stage model. *Earth and Planetary Science Letters*, **26**, 207–221.

TAIT, J. A., BACHTADSE, V., FRANKE, W. & SOFFEL, H. C. 1997. Geodynamic evolution of the European Variscan fold belt: paleomagnetic and geological constraints. *Geologische Rundschau*, **86**, 585–598.

——, —— & SOFFEL, H. 1995. Upper Ordovician palaeogeography of the Bohemian Massif: implications for Armorica. *Geophysical Journal International*, **122**, 211–218.

——, SCHÄTZ, M. BACHTADSE, V. & SOFFEL, H. 2000. Palaeomagnetism and Palaeotoic palaeogeography of Gondwana and European terranes. *This volume*.

TEIXEIRA, W., TASSINARI, C. C. G., CORDANI, U. G. & KAWASHITA, K. 1989. A review of the geochronology of the Amazonian Craton: tectonic implications. *Precambrian Research*, **42**, 213–227.

UNRUG, R. 1997. Rodinia to Gondwana: The geodynamic map of Gondwana Supercontinent Assembly. *GSA Today*, **7**, 1–6.

VAN DER VOO, R. 1979. Paleozoic assembly of Pangea: a new plate tectonic model for the Taconic, Caledonian and Hercynian orogenies. *EOS Transactions, American Geophysical Union*, **60**, 241.

VIDAL, P., AUVRAY, B., CHARLOT, R. & COGNÉ, J. 1981. Precadomian relicts in the Armorican Massif: their age and role in the evolution of the Western and Central Cadomian–Hercynian Belt. *Precambrian Research*, **14**, 1–20.

VON GAERTNER, H. R. 1944. Die Schichtgliederung der Phyllitgebiete in Thüringen und Nordbayern und ihre Einordnung in das stratigraphische Schema. *Jahrbuch der Reichsanstalt für Bodenforschung*, **62**, 54–80.

WIEDENBECK, M., CORFU, F., GRIFFIN, W. L. *et al.* 1995. Three natural zircon standards for U–Th–Pb, Lu–Hf trace element and REE analysis. *Geostandards Newsletter*, **19**, 1–23.

WUCHER, K. 1967. Ergebnisse der Kartierungsbohrungen Heinersdorf 1/60 und 2/62 (Thüringisches Schiefergebirge). *Jahrbuch für Geologie*, **1**, 297–323.

Review of geochemical variation in Lower Palaeozoic metabasites from the NE Bohemian Massif: intracratonic rifting and plume–ridge interaction

P. A. FLOYD[1], J. A. WINCHESTER[1], R. SESTON[1], R. KRYZA[2] & Q. G. CROWLEY[1]

[1]*Department of Earth Sciences, University of Keele, Keele ST5 5BG, UK (e-mail: p.a.floyd@esci.keele.ac.uk)*

[2]*Geological Institute, Wroclaw University, Wroclaw, Poland*

Abstract: During early Palaeozoic time the Cadomian basement of the northern margin of Gondwana underwent extensive rifting with the formation of various crustal blocks that eventually became separated by seaways. Attenuation of the continental lithosphere was accompanied by the emplacement of anatectic granites and extensive mafic-dominated bimodal magmatism, often featuring basalts with an ocean crust chemistry. Intrusive metabasites in deep crustal segments (associated with granitic orthogneisses) or extrusive submarine lavas at higher levels (associated with pelagic and carbonate basinal sediments) show a wide range of chemical characteristics dominated by variably enriched tholeiites. Most crustal blocks show the presence of three main chemical groups of metabasites: Low-Ti tholeiitic metabasalts, Main Series tholeiitic metabasalts and alkalic metabasalt series. They differ in the degree of incompatible element enrichment (depleted to highly enriched normalized patterns), in selected large ion lithophile (LIL) to high field strength element (HFSE) ratios, and abundances of HFSE and their ratios. Both the metatholeiite groups are characterized by a common enrichment of light REE–Th–Nb–Ta. High Th values (or Th/Ta ratios) and associated low ε_{Nd} values (especially in the Low-Ti tholeiitic metabasalts) reflect sediment contamination in the mantle source rather than at crustal levels, although this latter feature cannot be ruled out entirely. The range of chemical variation exhibited is a consequence of the melting of (a) a lithospheric source contaminated by a sediment component (which generated the Low-Ti tholeiites), and (b) a high-level asthenospheric mid-ocean ridge basalt (MORB)-type source that mixed with a plume component (which generated the range of enriched Main Series tholeiites and the alkali basalts). It is considered that a plume played an important role in the generation of both early granites and the enriched MORB-type compositions in the metabasites. Its significance for the initial fragmentation of Gondwana is unknown, but its presence may have facilitated deep continental crust melting and the fracturing into small crustal blocks. The early–mid-Jurassic plume-instigated break-up of the southern Gondwana supercontinent is considered to be a possible tectonic and chemical analogue for Early Palaeozoic Sudetic rifting and its magmatic products.

The Bohemian Massif of the central European Variscides is broadly interpreted as a collage of exotic terranes, separated by ductile shear zones generated by the collision between the Gondwana and Baltica continents during the Palaeozoic era (e.g. Franke 1989*a*; Matte *et al.* 1990; Oliver *et al.* 1993; Cymerman & Piasecki 1994; Cymerman *et al.* 1997). This sector of the Variscides records the development of various extensional sedimentary basins, partly floored by ocean crust, that opened and closed between late Precambrian and mid-Carboniferous times (Franke 1989*b*; Franke *et al.* 1995). Major episodes of continental rifting are recognized in Ordovician and Devonian time when the Cadomian crust at the leading edge of Gondwana underwent fragmentation, with the variable development of oceanic crust (Pin 1990). The preceding Early Ordovician phase of rifting was accompanied by the production of anatectic granites, bimodal mafic–felsic volcanic suites and ophiolite development (Perekalina 1981; Narebski 1993; Furnes *et al.* 1994). During the subsequent Variscan Orogeny these Lower Palaeozoic sequences and basement remnants were metamorphosed and deformed (Franke 1989*a*).

From: FRANKE, W., HAAK, V., ONCKEN, O. & TANNER, D. (eds). *Orogenic Processes: Quantification and Modelling in the Variscan Belt.* Geological Society, London, Special Publications, **179**, 155–174. 0305-8719/00/$15.00

This paper provides a geochemical overview of the metamorphosed basaltic products (metabasites) of the widespread late Cambrian–early Ordovician rift-related magmatism in the northeast of the Bohemian Massif. This sector includes part of what appears to be an 'ophiolitic belt' surrounding the Moldanubian of the Bohemian Massif (Finger & Steyrer 1995, fig. 10). The nature of well-documented metabasaltic amphibolites from the southeast of the Bohemian Massif (Austria), which are interpreted as representing the enriched MORB products of passive rifting in a back-arc environment (Misar *et al.* 1984; Fritz 1994; Finger & Steyer 1995) are also of relevance to our study. However, the interpretation of the tectonic environment or emplacement setting of the magmatism in the northeast Bohemian Massif, based on geochemical discrimination, has often been disputed. Some models, although identifying the presence of ocean floor, broadly favour a subduction-related setting (e.g. Narebski *et al.* 1986; Opletal *et al.* 1990; Oliver *et al.* 1993; Kröner & Hegner 1998), whereas others consider that all volcanism is rift related and no arcs can be identified in the Lower Palaeozoic sequences (e.g. Kryza & Kuszynski 1988; Narebski *et al.* 1988; Pin *et al.* 1988; Furnes *et al.* 1994; Kryza *et al.* 1995; Winchester *et al.* 1995, 1998; Floyd *et al.* 1996). Evidence presented in support of an arc environment largely derives from the chemical discrimination of early calc-alkali granitic material (e.g. Kröner & Hegner 1998) and the metabasites of the volcanic suites (e.g. Narebski *et al.* 1986). In the first case, the majority of the granites are continental crust partial melts and would thus inherit the 'arc-like' composition of their Cadomian protolith, and not necessarily reflect their emplacement environment. In the second case, *some* of the metabasites do have a low high field strength element (HFSE) characteristic that we believe has been mistaken for an arc signature. However, this chemical feature has to be explained within the context of all the other metabasite compositions, which are clearly not subduction-related. As indicated below, the full range of chemical compositions displayed are better explained in our model. For example, it has been suggested that the metabasites of the Sudetic crustal blocks show a geochemical commonality that could be interpreted as differential crustal contamination, together with mixing of mantle melts in a variably rifted continental margin (Floyd *et al.* 1997).

In the broader context of Lower Palaeozoic magmatism in the Variscides as a whole, bimodal volcanic suites are also found in other areas of the Bohemian Massif, as well as further afield in the Massif Central and the Iberian peninsula. Similar discussion concerning their eruptive environment (mainly rift v. arc) can be found in the literature (e.g. Pin 1990), although the Cambro-Ordovician magmatism is more probably related to rifted continental margins (e.g. Massif Central; Piboule & Briand 1985; Briand *et al.* 1995). Our study may therefore have a bearing on the development of Lower Palaeozoic magmatism across the entire Variscides.

Objectives

The objectives of the geochemical review are to (a) compare and identify the chemical characteristics of metabasites in different Sudetes crustal blocks, (b) determine the factors underlying the wide spectrum of metabasite compositions, and (c) provide a tectonic analogue that explains the range of metabasite compositions in a rift-related setting. However, we are aware of two major problems in this area, which with new data may modify the broad picture presented here: (a) the actual age of emplacement of different mafic units is still poorly constrained; apart from the concentration of magmatic activity in Ordovician time it may have extended over a long period from Cambrian to late Silurian time and changed in character with time; (b) the various crustal units experienced significant tectonic displacement, which largely obscures their original spatial relationships, such that their relationship one to the other is often contentious.

Sudetes crustal blocks

A variety of terranes or shear-bounded crustal blocks are recognized in the Sudetes of the Bohemian Massif (Fig. 1) all of which contain magmatic remnants of the Lower Ordovician rifting episode. In general, the magmatic suites are dominated by numerous metabasite bodies of both extrusive (massive and pillowed lavas) and shallow intrusive aspect (sills and dykes) and are often accompanied by minor felsic intrusive rocks. Details of the general geology and structural setting of the different blocks may be found in selected papers in Dallmeyer *et al.* (1995) and the references listed below that mainly relate to geochemical data.

Depending of the metamorphic grade and associated lithologies the crustal blocks investigated can be broadly divided into two tectonic groups: (a) high-level crustal rocks with recognizable lavas and associated carbonate and pelagic sediments, now metamorphosed in the greenschist to lower amphibolite facies (e.g. Kaczawa, Rudawy-Janowickie, Rychory

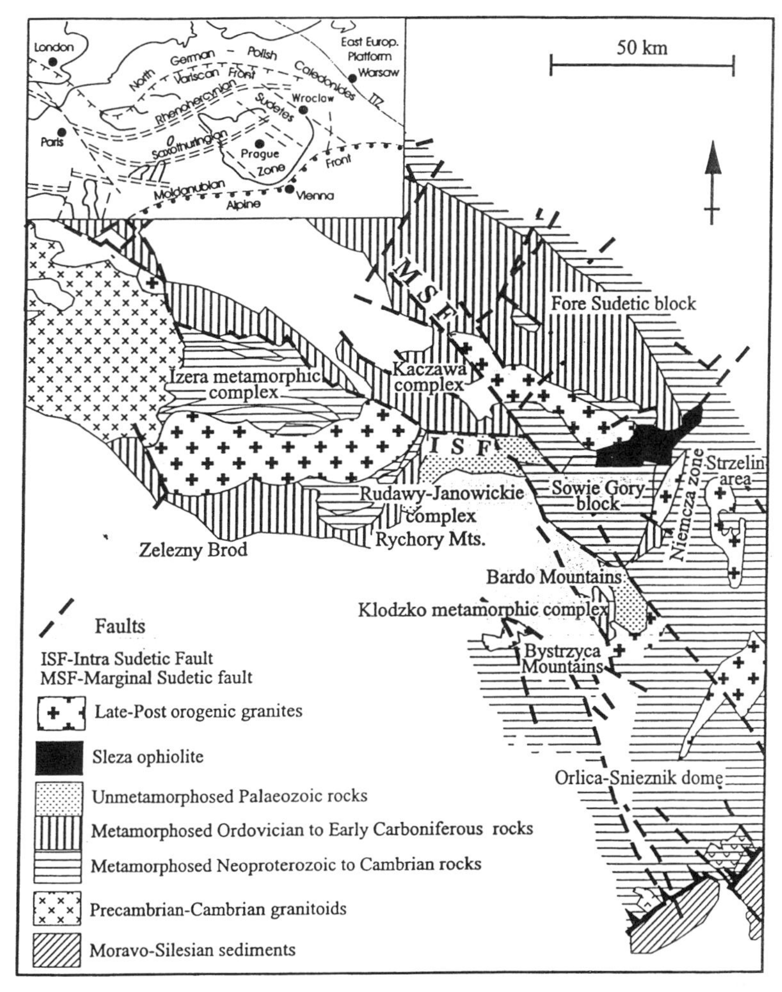

Fig. 1. Sketch map showing the location of the main crustal blocks mentioned in this paper. Inset indicates the position of the Sudeten Mountains in the NE of the Bohemian Massif and the tectonic zones of the Variscan orogen.

Mountains, Zelezny Brod; Fig. 1); (b) deeper-level gneissic crust with massive amphibolite bodies and associated granite–gneiss, paragneisses, mica schists and minor marbles, largely metamorphosed within the amphibolite to granulite facies (Izera, Orlica–Snieznik, Góry Sowie, Fore-Sudetic Block; Fig. 1). In simple terms, and bearing in mind the significant tectonic displacements, the broad geochemical correspondence between the metabasites in the high-level group and the deeper-level group might suggest an original cover and basement relationship with basaltic extrusive rocks in the former and their feeders in the latter. The temperature-dominated greenschist to amphibolite facies regional metamorphism represents a widespread Variscan tectonothermal overprint that has largely masked an earlier Variscan blueschist metamorphism recorded at some higher levels (Kryza *et al.* 1990; Patocka *et al.* 1996) and a remnant eclogite–granulite facies event locally preserved in fragments of the lower crust (Carswell & O'Brian 1993; Kryza *et al.* 1996; O'Brien *et al.* 1997).

Aspects of the geochemistry and petrology of the metabasites within the crustal blocks that form part of this review are outlined below.

High-level volcanic associations

The volcanic rocks of the Kaczawa complex are mostly pillowed and massive metabasalts accompanied by rhyodacites and trachytes (the latter provisionally dated at 511 ± 39 Ma, U–Pb zircon; Pin, pers. comm.). The volcanic rocks range in composition from alkaline to tholeiitic of MORB type. They are associated with mostly shallow-water metasedimentary rocks (volcaniclastic rocks, siliciclastic rocks and limestones) in the lower part, and pelagic shales and cherts in the upper part of the succession (Baranowski & Lorenc 1986; Furnes *et al.* 1989, 1994; Kryza 1993; Kryza *et al.* 1994; Muszynski 1994). From supposed temporal changes in the composition of the volcanic rocks, Furnes *et al.* (1994) recognized a progressively evolving rift from an ensialic stage to basin development underlain by MORB-type ocean crust. Interlayered metarhyodacitic lavas have isotopic compositions indicative of an origin by crustal melting. The metavolcanic rocks of the nearby Rudawy-Janowickie complex also exhibit a mafic–felsic association, although the metabasites are now largely amphibolites and greenschists. Relict textures and field relationships suggest they were mainly massive lavas, with some intrusive sheets and small diapirs. Combined chemical and structural data indicate that a number of stacked tectonic slices containing specific basaltic compositions (ranging from alkali basalts to enriched and depleted MORB) and separated by major thrusts can be recognized (Winchester *et al.* 1995; Kryza *et al.* 1995). The complex as a whole was interpreted as a former intracratonic rift partly floored by ocean crust and bounded to the west by an attenuated continental margin.

The Lower Palaeozoic successions of the Rychory Mountains and the Zelezny Brod complexes in the Czech Republic are considered to be equivalent to the Rudawy-Janowickie complex in Poland. Both of the former successions contain low-grade mafic–felsic volcanic assemblage including greenstones, metadiabases and aquagene tuffs associated with pelites, marbles and quartzites (Maluski & Patocka 1997; Patocka & Smulikowski 1997; Fajst *et al.* 1998; Kachlik & Patocka 1998). Geochemical and isotopic data for the Rychory Mountains metavolcanic rocks suggest a change from early, deep mantle melting, with the production of alkali basalts and enriched tholeiites, to shallow melting associated with extensional tectonics and the development of MORB-like rocks (Patocka *et al.* 1997).

In summary, three common features are exhibited by all the metavolcanic sequences: (a) the presence of a mafic–felsic volcanic association dominated by submarine metabasites and associated with both shallow-water and pelagic sediments (including carbonate); (b) the ubiquitous association of minor within-plate alkali basalts and dominant tholeiites, which exhibit a range of depleted to enriched incompatible element contents (depleted to enriched MORB types); (c) associated felsic volcanic rocks, which comprise both evolved lavas related to the metabasites by fractionation processes and other varieties generated by differential crustal melting and/or heavily contaminated by continental crust.

Deeper-level plutonic associations

The Izera metamorphic complex largely consists of variably deformed and mylonitized granitoids, orthogneisses and belts of meta-sedimentary mica schists. Metabasites are in minor proportion to the dominant acid plutonic rocks and appear as either intrusive sheets and dykes or isolated pod-like relicts within tectonically deformed granite (Nowak 1998; Seston 1999). Chemically the metabasites are predominantly alkaline in composition with minor depleted tholeiites.

The Orlica–Sniezník dome metamorphic complex comprises various orthogneisses of granitic parentage and paragneisses, together with metabasites exhibiting mainly amphibolite facies mineralogy, although some granulite facies relicts also occur. Reviews of the structural and lithostratigraphic complexity of the region have been given by Don *et al.* (1990), Urbanek *et al.* (1995) and Zelazniewicz *et al.* (1995). Previous geochemical data on the metabasites identified various tectonic environments, including island arcs and spreading centres (e.g. Wojciechowska 1986; Wojciechowska *et al.* 1989; Opletal *et al.* 1990), although Floyd *et al.* (1996) suggested the range of composition could best be accommodated in a rifted ensialic environment with variable crustal contamination.

The Góry Sowie block is a tectonically isolated gneissic massif with disputed geotectonic significance and tectonometamorphic history. The main lithologies consist of various paragneisses (some with retrogressed granulite facies lenses) and granitoid augen to flaser gneisses. High-grade amphibolite facies metabasites (some with relict magmatic textures) occur as small lens-like bodies throughout the gneisses. Textural evidence suggests that many of the arphibolites are retrogressed from an earlier granulite facies metamorphism (Kryza *et al.* 1996; Cymerman *et al.* 1997). Recent geochemical analyses of the metabasites have identified three metatholeiitic types with variably enriched characteristics, together with a minor group of alkalic metabasalts (Winchester *et al.* 1998). The overall chemical composition of the metabasites and their geological setting is consistent with emplacement in continental crust undergoing extension at a passive continental margin.

The Strzelin crystalline massif forms part of the Fore-Sudetic block and is mainly composed of various orthogneisses, micaceous paragneisses with calc-silicates, marbles and intercalated amphibolites (Oberc-Dziedzic 1995). Most amphibolites are interpreted as metadolerite sill-like bodies within the orthogneisses, although transitions into mica schists suggest some may have a volcaniclastic derivation (Szczepanski & Oberc-Dziedzic 1998). Geochemical data on amphibolites sampled from boreholes indicates that they represent a fractionated group of enriched within-plate tholeiites that were probably emplaced during the early rifting of continental crust (Szczepanski & Oberc-Dziedzic 1998).

In summary, a similar group of variably enriched tholeiitic and alkalic metabasites as found in the volcanic associations appears to be present, but associated with a much higher proportion of earlier orthogneisses of granitic parentage, as befits their deeper crustal level.

Sudetes geochemical database

The database on which this geochemical review is based comprises 575 full major oxide and trace element analyses (115 include rare earth element (REE) analyses) largely generated by the authors and obtained from the published literature (see references above). The majority of samples were analysed by X-ray fluorescence (XRF) spectrometry with REE determinations by inductively coupled plasma (ICP) and instrumental neutron activation analysis (INAA) methods. Analysis at Keele was performed on an ARL 8420 XRF spectrometer (Department of Earth Sciences) calibrated against both international and internal standards of appropriate composition. Details of methods, accuracy and precision have been given by Floyd & Castillo (1992). Representative chemical data for each of the separate crustal units can be found in the papers cited above.

Alteration effects

All the metabasite samples show varying degrees of mineralogical alteration (within the greenschist to granulite facies) and as such can be expected to have suffered selected element mobility, especially involving the large ion lithophile elements (LILE) (e.g. Thompson 1991). However, characteristic magmatic relationships are often maintained by those elements that are considered relatively immobile during metamorphism, such as high field strength elements (HFSE) and the REE (e.g. Pearce & Cann 1973; Smith & Smith 1976; Floyd & Winchester 1978). These elements and/or their ratios have been utilized to determine petrogenetic features and sources, especially as they appear to act isochemically during high-grade metamorphism (e.g. Tarney & Windley 1977; Floyd *et al.* 1996, for Orlica–Snieznik amphibolites). In some circumstances, such as the extensive carbonatization of metabasites, the REE and HFSE can be mobilized or their abundances diluted (e.g. Hynes 1980), although all carbonate-rich samples were excluded from this study. Although we recognize the limitations of using such trace elements, it is considered that they represent the closest approximation to magmatic compositions available for the metabasites.

Geochemical classification and general features

Viewed overall, the chemistry of the Sudetic metabasites displays a wide range of chemical variation, which can be broadly grouped into three main magma series: Low-Ti tholeiitic metabasalts, Main Series tholeiitic metabasalts, and meta-alkalic basalt series. Chemical discrimination of these series is based on the Nb/Y (Nb/Y = 0.7 distinguishes tholeiitic from alkaline basalts) and V/TiO_2 ratios (Ti/V = 20 distinguishes Low- from High-Ti or Main Series basalts). Each crustal unit contains varying proportions of each magma series (in general, there is a good sample coverage), with dominant tholeiitic compositions throughout (Fig. 2); the exception is the Izera metamorphic complex, which exhibits mainly alkaline metabasites. Chemical characteristics of the Low-Ti series broadly match those displayed by similar basalts elsewhere, especially the Low-Ti suite of Mesozoic Gondwanan basalts, which are characterized by low Zr/Y (<260), Ti/Zr (<70), V/Ti (<20) and light REE (e.g. Fodor 1987; Cox 1988; Erlank *et al.* 1988).

Chondrite-normalized REE patterns for the three magma series are also similar in each crustal unit (Fig. 3), with the metabasites displaying varying degrees of LREE enrichment. As seen in Fig. 4 the Low-Ti series are generally typified by relatively flat normalized patterns, with average $(Ce/Yb)_N = 1.7$, although some have very minor LREE depletions and enrichments. The Main Series metabasites are more enriched (average $(Ce/Yb)_N = 2.9$) and display variable degrees of enrichment, whereas the alkalic series shows the greatest LREE enrichment (average $(Ce/Yb)_N = 10.3$). Other distinctive chemical features

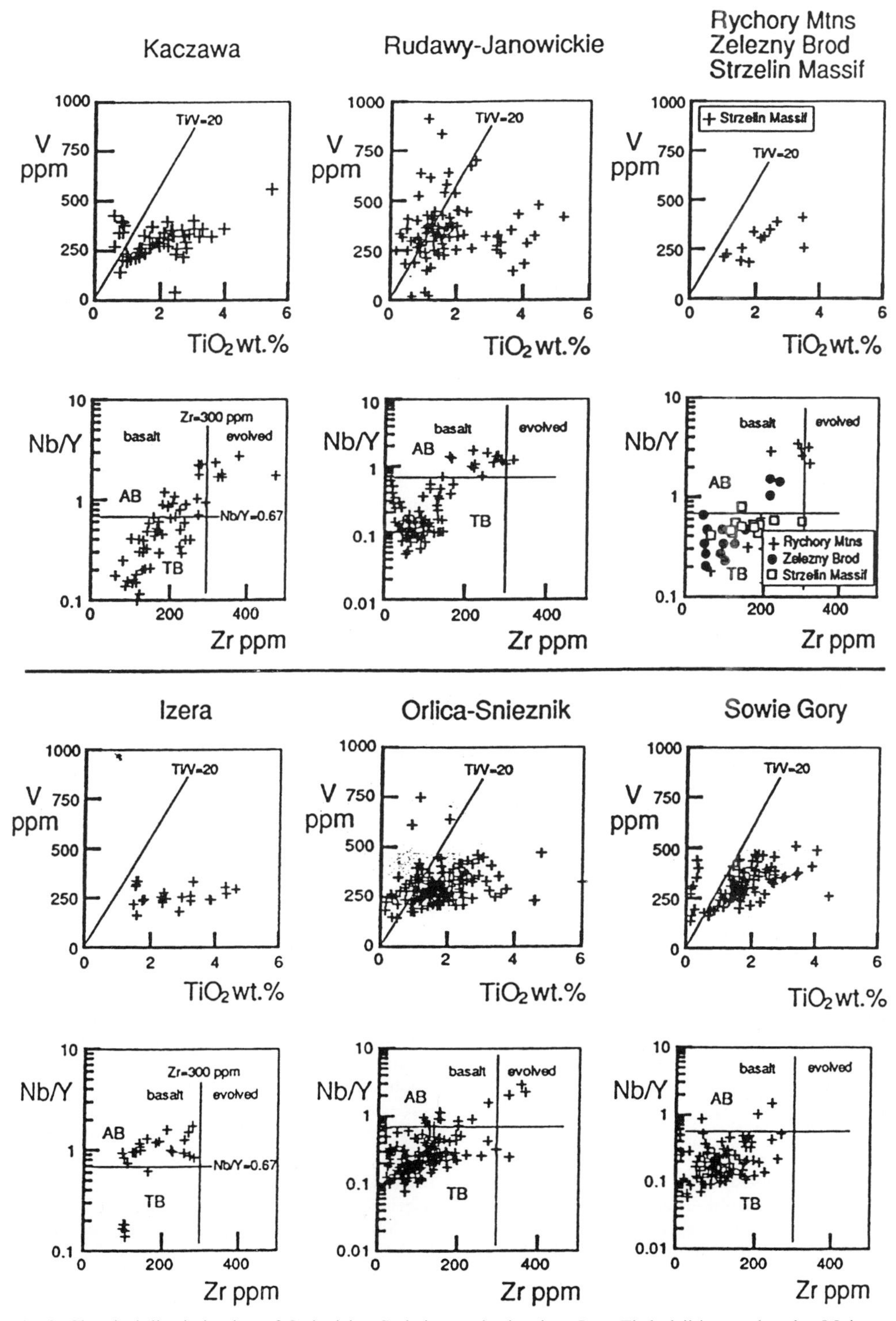

Fig. 2. Chemical discrimination of Ordovician Sudetic metabasites into Low-Ti tholeiitic metabasalts, Main Series tholeiitic metabasalts and alkalic metabasalts, on the basis of Ti/V and Nb/Y ratios (see text) in each crustal block: high-level volcanic associations (above line) and deeper crustal plutonic associations (below line).

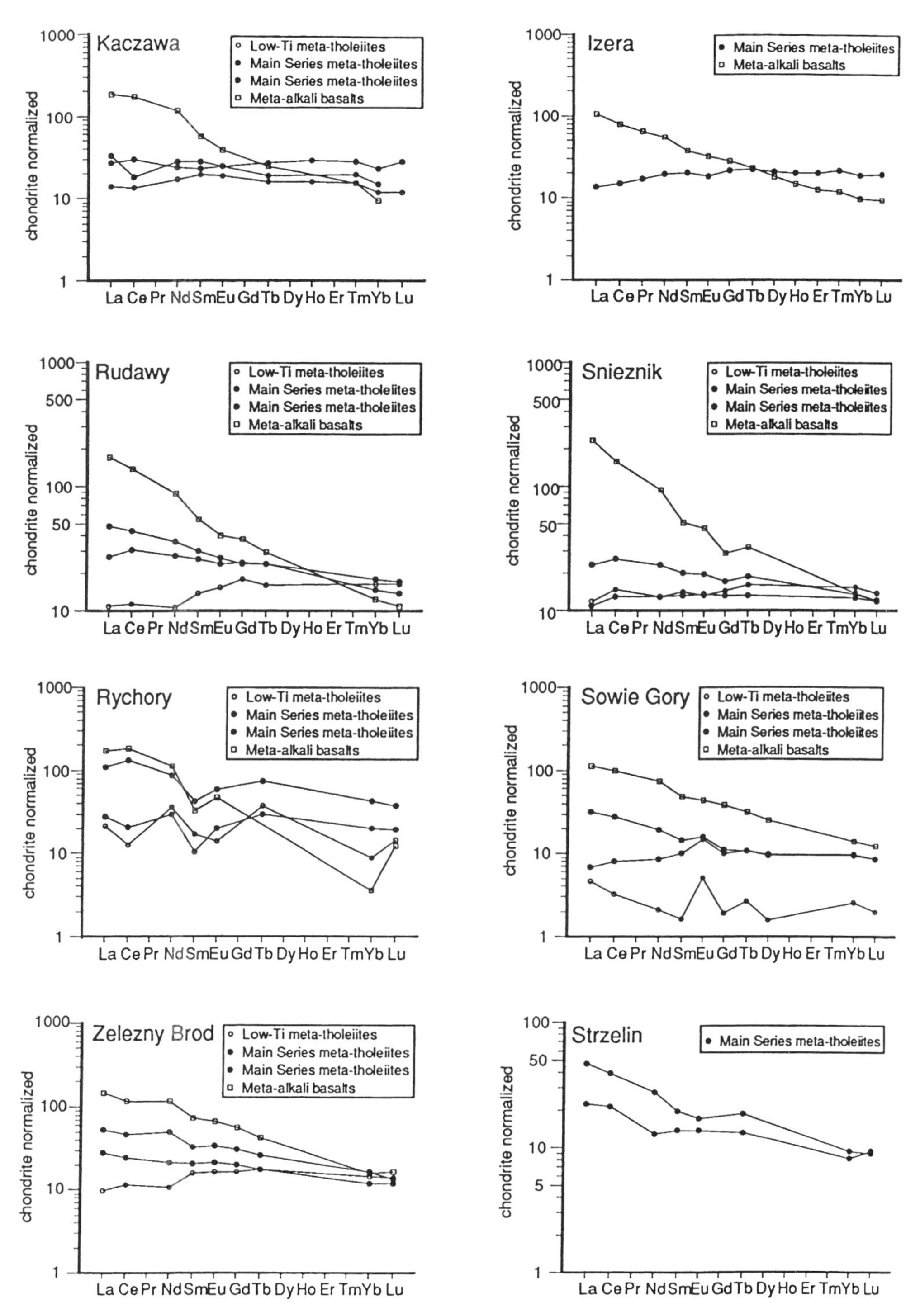

Fig. 3. Typical chondrite-normalized REE patterns for the three magmatic series (Low-Ti tholeiitic metabasalts, Main Series tholeiitic metabasalts and alkalic metabasalts) found in each of the crustal blocks.

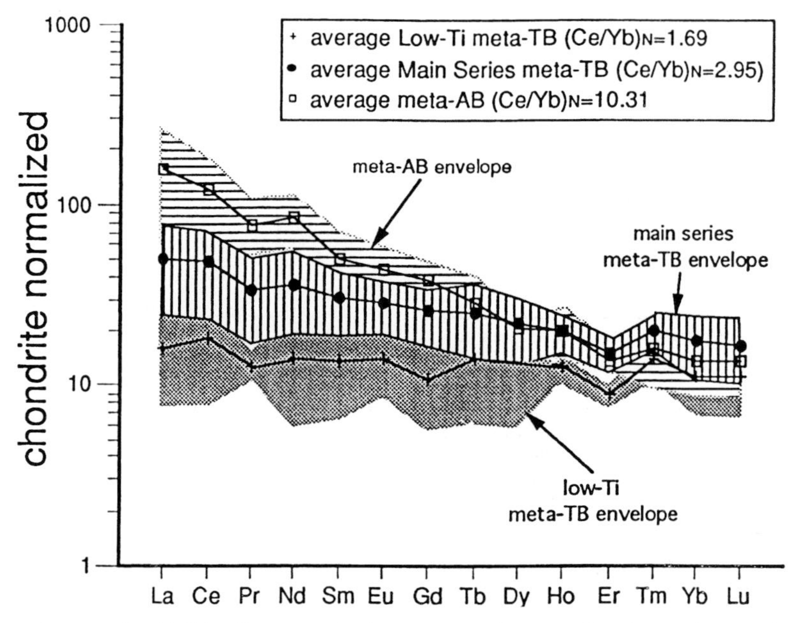

Fig. 4. Chondrite-normalized REE patterns showing the average and compositional range (mean values ± SD) for each of the three magma series: Low-Ti tholeiitic metabasalts, Main Series tholeiitic metabasalts and alkalic metabasalts. (Note the range of the LREE enrichments for the tholeiites (TB) and the alkali basalts (AB).)

of the three series in the volcanic suites are exhibited by representative MORB-normalized multi-element patterns (Fig. 5), although variation of LILE contents are partly a consequence of mobility during subsequent metamorphism. The Low-Ti series is characterized by generally flat HFSE patterns marginally depleted relative to N-MORB, whereas most Main Series metabasites show progressive enrichment patterns for the least incompatible elements (taking Th as a stable LILE). The steepest enrichment patterns are typical of the meta-alkalic basalts (Fig. 5). The negative K anomaly is common to all patterns and is probably a consequence of the removal of K during the development of amphibolite facies assemblages, although such anomalies can be a feature of continental basalts (Saunders *et al.* 1992). Of more significance is the mild enrichment of the LREE, Nb, Ta and Th shown by the tholeiitic metabasites relative to the adjacent HFSE. This appears to be a feature typical of many metabasites in all the crustal units and will be considered below.

Internal chemical variation

Each of the three series shows considerable internal variation, which is largely a consequence of the generation of different partial melts and their variable fractionation in high-level magma chambers. The generally low MgO, Ni and Cr contents of the least evolved metabasite compositions imply that few can be considered very primitive picritic or primary melts (Sato 1977; Frey *et al.* 1978). As seen in Fig. 6 relative to progressive fractionation indices, FeO*/MgO and Zr, the broad increase in incompatible TiO_2 coupled with the rapid decrease in compatible Cr is characteristic of magmatic fractionation trends; the latter being indicative of mafic ± spinal fractionation. However, the general lack of negative Eu anomalies among the REE (Fig. 3) implies that plagioclase fractionation was not significant or that magmas were relatively oxidized. Wide variation of some incompatible element ratios (e.g. Zr/Y, from 2 to 10), and in particular fractionation between LILE and HFSE, and LREE/HREE (heavy REE) cannot be explained by fractional crystallization alone. For example, normalized (Ce/Yb) ratios (Fig. 6c) display greater variation than would be produced by low-pressure fractionation of common silicates, although high-pressure eclogite fractionation could produce much of the range observed. However, differential partial melting of lherzolite mantle (30% to 1% range) can better accommodate much of the fractionation range shown by the REE (Fig. 6c). It is

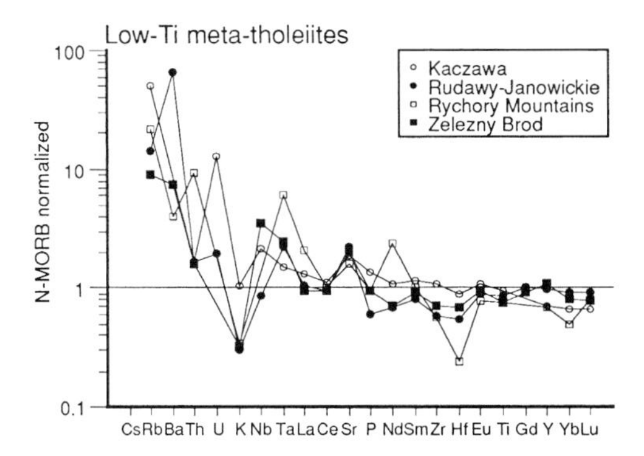

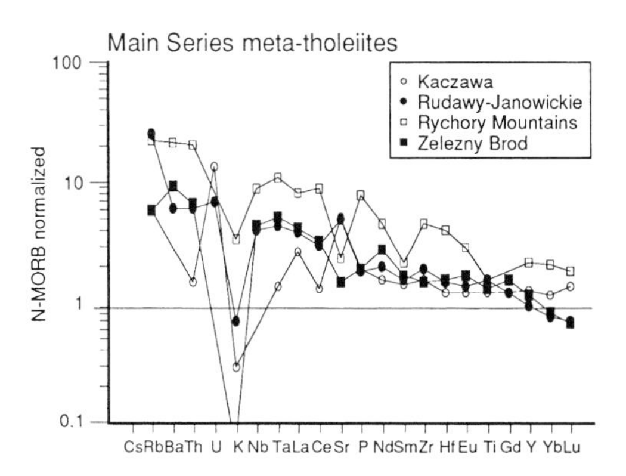

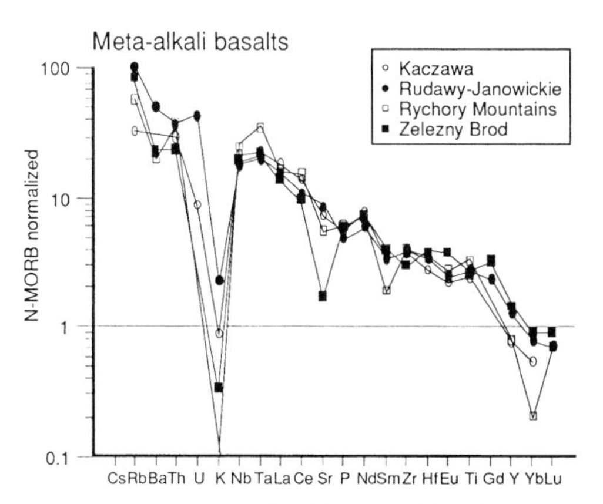

Fig. 5. N-MORB-normalized incompatible element patterns for the three magmatic series (Low-Ti tholeiitic metabasalts, Main Series tholeiitic metabasalts and alkalic metabasalts).

concluded that much of the internal broad scatter in (stable) incompatible element plots and lack of simple linear covariance, together with differential REE enrichment, is more likely to be the consequence of a number of partial melts each undergoing similar low-pressure fractionation. This model implies that during rifting the Sudetic continental crust was invaded by a number of separate partial melts that, on ponding at a high level, underwent low-pressure fractionation. This process allowed the development of zoned basaltic magma chambers that were subsequently tapped, producing sills and dykes in the lower crust and extrusive flows in rifted basins.

Crustal contamination

Open system fractionation for the generation of continental basalts that involve interaction between magma and crust is a widely recognized process (e.g. models by Cox & Hawkesworth 1985; Huppert & Sparks 1985: Devey & Cox 1987). Because the extensive Sudetic mafic magmatism is associated with rifted continental crust, contamination during ascent or in ponded magma chambers is a distinct possibility.

REE, Nb, Ta and Th enrichment (relative to HFSE) might be a geochemical consequence of a crustal contamination process. However, rather infrequent negative Nb–Ta anomalies limit the extent of contamination by the lower continental crust, which is characterized by a paucity of these elements relative to MORB (Weaver & Tarney 1984). Other element ratios (e.g. Nb/U, Ce/Pb, Ba/Nb, Ba/Th) that are often considered indicative of crustal contamination (Rehkamper & Hofmann 1997) cannot be considered here as they utilize mobile elements. We have, therefore, considered the observed differential enrichment in terms of two ratios: Ce/Zr and Th/Ta. To offer a possible explanation for the wide variation in Ce/Zr ratios (i.e. LREE enrichment), modelling of the assimilation with fractional crystallization process (AFC; DePaolo 1981) was undertaken. As seen in Fig. 6d, relative to closed system fractionation, the AFC process with variable contaminants (e.g. upper continental crust, Taylor & McLennan 1985; average pelagic sediment, Li 1991) and degrees of assimilation ($r = 0.2$–0.8) could account for some of the variation in Ce/Zr ratios, although the proportion of assimilated pelagic contaminant appears unrealistically high to achieve the highest Ce/Zr ratios and maintain a basic composition. On the other hand, Th is enriched in upper-crustal materials such that enhanced Th/Ta ratios greater than chondritic values (*c*. 2.1, Sun & McDonough 1989) might reflect contamination of mantle melts in continental settings. For example, trace element and supporting isotopic data for the rifted margin of the British North Atlantic Province show some Hebridean basalts selectively contaminated by continental crust and local sediments (Thompson *et al.* 1982, 1986). As illustrated in Fig. 7a,

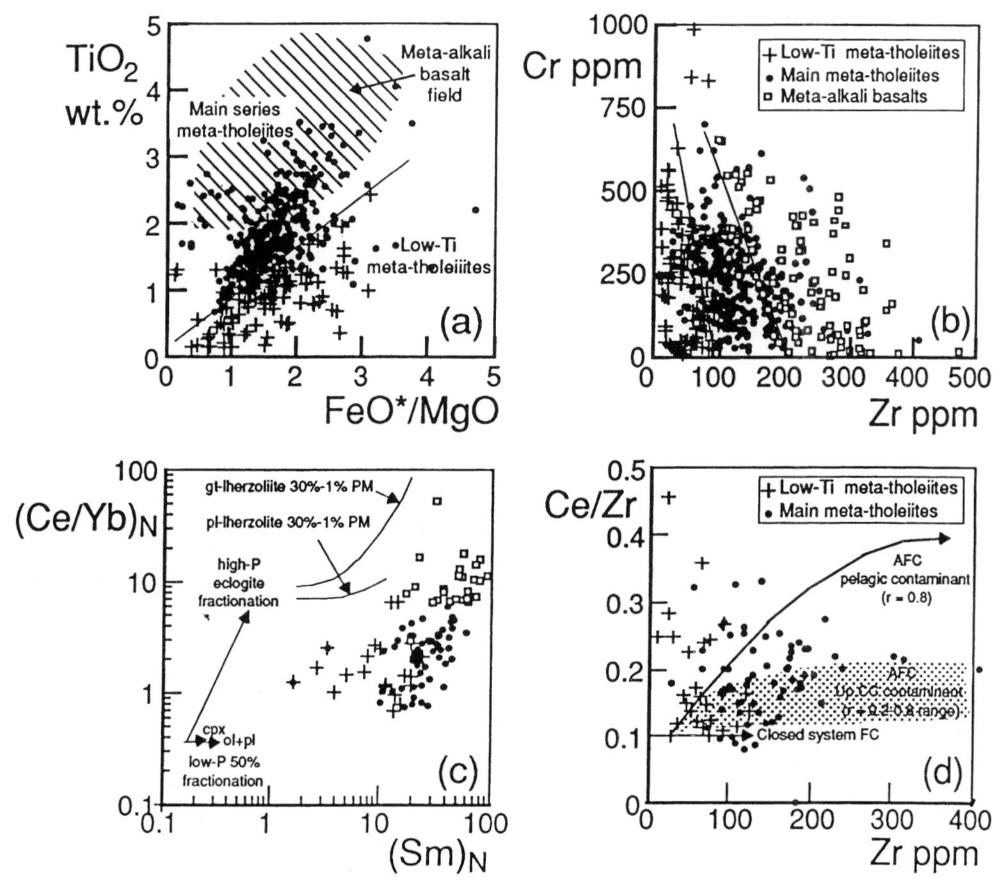

Fig. 6. Diagrams illustrating chemical variation within the Sudetic metabasite series as a result of different magmatic processes: (**a**, **b**) variable fractional crystallization (TiO_2–FeO*/MgO; Cr–Zr); (**c**) comparison of variable partial melting (PM) of lherzolite mantle sources and high-pressure eclogite fractionation (chondrite-normalized Ce/Yb–Sm); (**d**) AFC process with variable rates of crystallization to assimilation (*r*) and contaminants (average upper continental crust and average pelagic sediment).

distinct trend of increasing Th/Ta ratios reflects contamination by crustal materials relative to unaffected Icelandic basalts. A similar trend is shown by some Low-Ti and Main Series metabasites from the Sudetes; the former being influenced by a pelagic contaminant and the latter (possibly) by minor contamination by upper continental crustal material.

Current isotopic data do not provide very strong support for high-level crustal contamination, although a few metabasites with relatively low ε_{Nd} values (4–0 range, no negative values), which also exhibit minor negative Nb–Ta anomalies, are documented from the Kaczawa (Furnes *et al.* 1994) and Rychory Mountains (Patocka *et al.* 1997).

In general, it appears that high-level crustal contamination of the metabasites was probably very limited. Also, the greatest degree of contamination was apparently restricted to the least evolved or fractionated Low-Ti metabasites, a feature that is typical of some continental basalts and ascribed to source composition (e.g. Cox & Hawkesworth 1985). Similarly, the common enrichment factor (LREE, Nb, Ta and Th), which appears to be a characteristic feature of the Sudetic tholeiitic metabasites, is more likely to result from a source effect. For example, mantle melts that interact with ancient sub-continental lithosphere that has been enriched by subducted components can acquire a geochemical signature similar to that obtained by assimilating continental crust (e.g. Hawkesworth *et al.* 1983).

Sources

Apart from the chemical variation ascribed to magmatic processes, distinct normalized

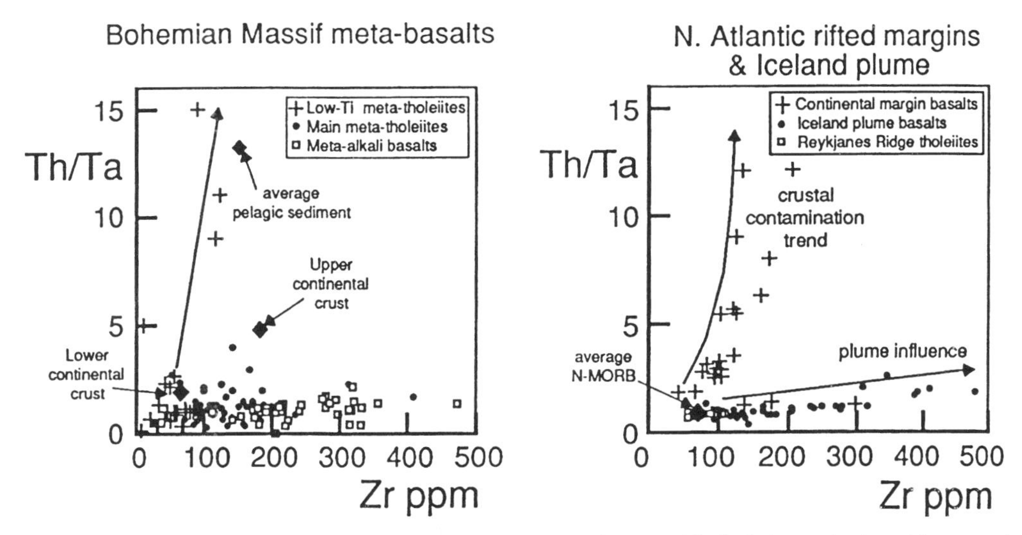

Fig. 7. Comparison of contamination trends (increasing Th/Ta ratios:arrow) in Sudetic metabasites with a crustal contamination trend exhibited by rift-related basalts from the North Atlantic continental margin (data from Thompson *et al.* (1982, 1986).

multi-element patterns and ratios of incompatible elements that are usually taken to reflect their source (e.g. involving stable elements: La/Nb, Zr/Nb, Ti/Zr, Ti/Y) suggest that the Sudetic metabasite series may be derived from a number of sources. Also, as suggested below, the characteristic enrichment feature is probably a consequence of mixing of melts derived from different sources, as well as minor source contamination.

General features

The three metabasite series show significant differences in the patterns and degree of enrichment exhibited by N-MORB normalized distributions (Fig. 5). For example, the strong progressive enrichment patterns of the meta-alkali series are typical of an enriched ocean-island basalt (OIB)-type source. In this context, OIB refers to a mantle source composition and does not imply a specific eruptive environment. However, this particular pattern is characteristic of many plume-related ocean islands, derived from an enriched end-member mantle of type II (EMII-type) with high Th/Nb and Th/La ratios (Weaver 1991). The low normalized values for the HFSE in the Low-Ti series (Fig. 5) suggest derivation from a strongly depleted asthenospheric MORB-like source. However, Low-Ti basalts are a characteristic feature of the Mesozoic flood basalts of Gondwanaland and a number of studies have suggested that they were derived by melting sub-continental lithosphere contaminated by LILE-enriched subducted sediment (e.g. Ellam & Cox 1989; Hergt *et al.* 1991). In terms of HFSE ratios (Fig. 8) many of the Sudeten Low-Ti metabasites overlap Gondwanan Low-Ti basalts and exhibit a distinctive trend towards low Ti/Zr values that could be generated by the involvement of sediment (e.g. Post-Archaean Sediment composite (PASC); Taylor & McLennan 1985). If, as indicated above, the enrichment feature might (in part) represent contamination by pelagic sediment, this was already in the source rather than being produced by continental crust interaction. In contrast, the Main Series metabasites have multi-element and REE patterns characteristic of a variably enriched asthenospheric source.

Mixing and plume interaction

The variable degrees of enrichment exhibited, especially by the Main Series metabasites, are considered the consequence of mixing between depleted and enriched asthenospheric sources. Figure 9 displays a mantle array for asthenospheric melts between a depleted N-MORB source and an enriched OIB-plume source. Although most of the variation for the Main Series and alkalic metabasites falls within the trend defined by the mantle array, the Low-Ti series exhibits minor vertical trends starting from low Ti/Y and Ta/Yb ratios (with increasing Zr/Y and Th/Yb) caused by the effects of sediment contamination. The fact that many of the Low-Ti

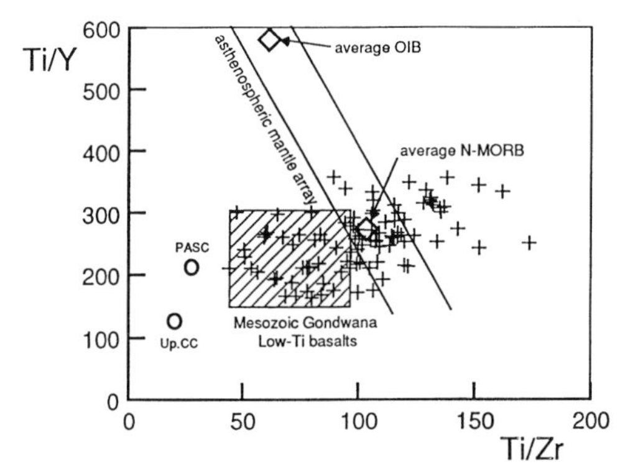

Fig. 8. Comparison of Sudetic Low-Ti tholeiitic metabasalts with Mesozoic analogues from southern Gondwana (data largely from Hergt *et al.* (1991) and Brewer *et al.* (1992). The decreasing Ti/Zr trend towards PASC away from N-MORB suggests that their source was contaminated by sediment. The clustering of the rest of the Low-Ti series around N-MORB implies that they were derived from a mantle at least as depleted as the MORB source. PASC (Post-Archaean Sediment composite) and Up.CC (upper continental crust average) from Taylor & McLennan (1985); average OIB and N-MORB from Sun & McDonough (1989).

series plot within the asthenospheric array suggests that the involvement of a sediment-contaminated lithosphere was probably limited (see also Fig. 8). Although some of the Sudetic metabasites have typical depleted N-MORB characteristics (e.g. Furnes *et al.* 1994; Kryza *et al.* 1985; Winchester *et al.* 1995), most show more enriched features similar to E-MORB. As illustrated in Fig. 9, it is considered that this results from mixing between plume-generated OIB-type melts and depleted MORB-type melts.

In this model the LREE–Nb–Ta–Th enrichment characteristic that is common to most tholeiitic metabasites is largely a feature of the variable influence of an enriched plume source. By analogy, however, the relatively high Th values and low Ti/Zr ratios in some Low-Ti series metabasites suggest minor sediment contamination of a lithospheric source.

Tectonic environment

There is a general consensus that Early Palaeozoic mafic magmatism, as represented by the metabasites of this review, accompanied the fragmentation of the northern margin of Gondwana by the development of intracontinental rifts. The magmatic episode was initiated by the development of (largely) anatectic granites that intruded the existing Cadomian continental basement. Subsequent mafic magmatism heralded the rifting of the continental crust with the extensive development of extrusive and intrusive tholeiites and minor alkalic basalts. Extension eventually reached the stage whereby ocean crust was developed as shown by the presence of depleted N-MORB compositions in a number of crustal units and the existence of nearby ophiolites (e.g. Sleza ophiolite, Fig. 1).

On the basis of metabasite compositions and the chemical discrimination of the tectonic

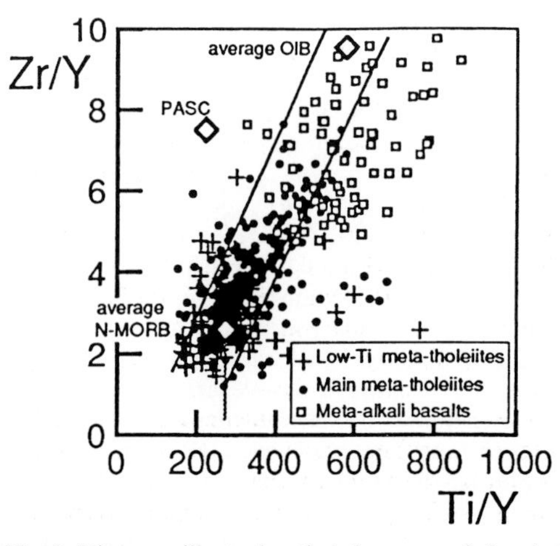

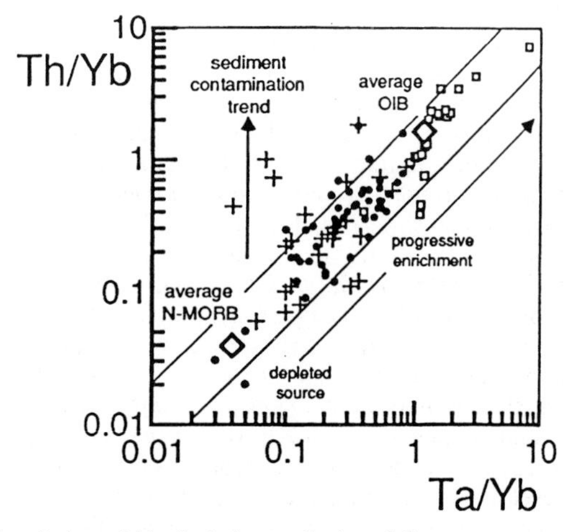

Fig. 9. Diagrams illustrating that the range of chemical variation within Sudetic metabasites defines an asthenospheric mantle array produced by mixing between a depleted N-MORB source and an enriched OIB (plume) source. (Note that metabasites pulled out of the array (vertical Zr/Y and Th/Yb trends), especially within the Low-Ti series, result from sediment contamination.) Average OIB and N-MORB from Sun & McDonough (1989).

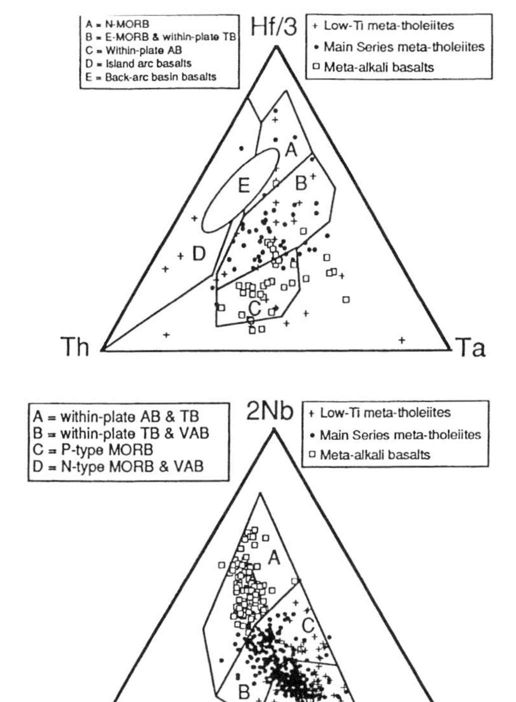

Fig. 10. Chemical tectonic environment discrimination diagrams illustrating the range of variation shown by the Sudetic metabasites, and the dominance of variably enriched MORB types. Hf–Ta–Th diagram after Wood (1980) with added back-arc field, Nb–Y–Zr diagram from Meschede (1986).

environment, a number of features are apparent (Fig. 10). The alkali metabasalts are typically intra-plate basalts and in view of their submarine setting probably represent sills or dykes and pillow lavas either associated with sea-floor seamounts or as accumulations in developing rift-related ensialic basins. Metatholeiite variation apparently exhibits a range of overlapping 'environments', although many (see Hf–Ta–Th plot, Fig. 10) are characterized as variably enriched MORBs (E- and P-types), as well as depleted MORB (N-type). Although, the E-MORB-designated metabasites have compositions that overlap with intra-plate tholeiites, they are closely associated in the field with oceanic sediments and true N-MORB. It is considered, therefore, that they are also representative of the development of incipient ocean crust, but have been derived from an enriched or contaminated source before extrusion at a speading centre. Enriched segments of ocean spreading centres are invariably found near ocean islands fed by active plumes; for example, on the Mid-Atlantic Ridge (MAR) adjacent to the Azores plume (Schilling 1975) and in segments of the southern MAR (Le Roex *et al.* 1987). Interaction between the active ridge and the nearby plume allows the dispersion of enriched asthenosphere into the depleted MORB source below the ridge (e.g. Vogt 1976; Schilling *et al.* 1985).

A possibly analogous situation for Sudetic Early Palaeozoic ocean crust development might be the splitting of southern Gondwana during early–mid-Jurassic time and the eventual development of an oceanic seaway between southern Africa and Antarctica (e.g. Elliot 1992; Storey 1995). At the present time the SW Indian Ridge and the American–Antarctic Ridge separate these two continental regions, both of which are characterized by variably enriched MORB whose composition has been influenced by the nearby Bouvet plume (Le Roex *et al.* 1983, 1985). Figure 11 illustrates the similar range of Y/Nb and Zr/Nb ratios of Sudetic metatholeiites with both modern ridges and the meta-alkali basalts with Bouvet OIB. The compositional range developed and the dominance of E-MORB in the SW Indian and the American–Antarctic Ridges is produced by mixing between a depleted N-MORB source and the enriched Bouvet plume source. By analogy, this implies that most Sudetic metabasites were formed in a similar ridge environment under the influence of nearby plumes.

One further consequence of the activity of mantle plumes is that nearby enriched ridge segments tend to be elevated above the normal ridge depth. For example, the progressive elevation of the Reykjanes Ridge from the MAR towards subaerial Iceland is also matched by a geochemical gradient generated by mixing the MORB source with the enriched Iceland plume (e.g. Schilling 1973). LREE fractionation (normalized La/Sm ratios) provides a measure of the enrichment factor, and the Zr/Nb ratio monitors the change from normal to enriched MORB with a cut-off value between the two of 20 (average MAR values for Zr/Nb are 31.3 and 15.9, respectively; Humphris *et al.* 1985). Both of these chemical variables are plotted in Fig. 12 and, by analogy with the Reykjanes Ridge chemical variation, indicate that the active ridge producing most of the Sudetic oceanic metabasites (both submarine lavas and deeper intrusive rocks) was shallower than normal segments. Klein & Langmuir (1987) demonstrated that there is a global correlation between average ridge depth and the major element chemistry (average Na_2O and FeO* values corrected for low-*P* fractionation) of axial basalts. Their approach was applied to submarine metabasites

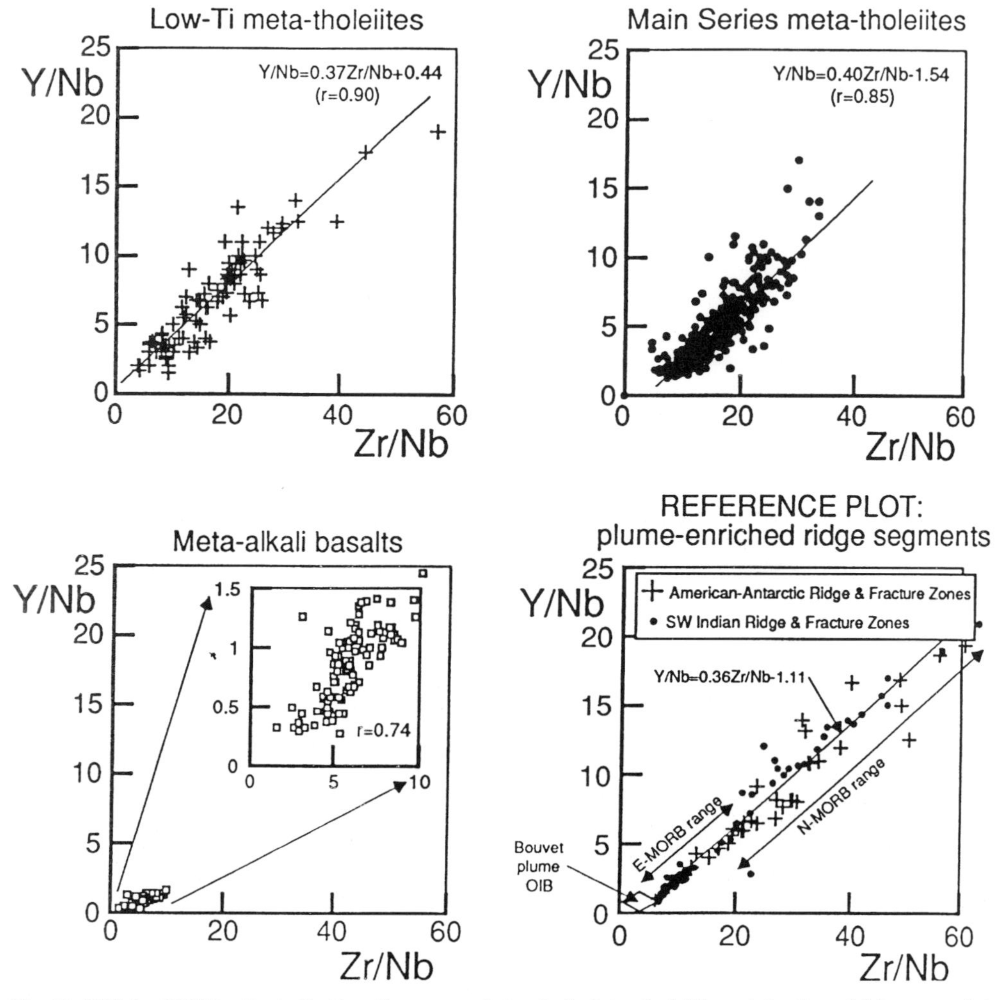

Fig. 11. Utilizing HFSE ratios indicative of source, variation in Sudeten tholeiitic metabasites exhibits a complete overlap with ridge segments variably enriched by the influence of a nearby plume. The Sudetic alkalic metabasalts have very low ratios similar to OIB of the Bouvet plume. Data for the American–Antarctic and SW Indian Ridges from Le Roex *et al.* (1983, 1985).

from four Sudetic crustal segments (Kaczawa, Rudawy-Janowickie, Rychory Mountains, Zelezny Brod) and averaged values for $Fe_{8.0}$ were calculated for each suite (Na_2O was not used because of its mobility during metamorphism) and an approximate axial depth was obtained. On the basis of Main Series metatholeiite compositions an average depth of 3.1 km (range 2.3–3.7 km) was obtained, with Zelezny Brod being significantly shallower (2.3 km) than the rest. The main distinction concerned the Low-Ti metatholeiites, which gave a deeper average axial depth of 4.6 km (range 3.6–5.0 km). If any reliance can be placed on these crude estimates, the axial depth for the Main Series metatholeiites was not significantly shallow like some elevated ridge segments (<2 km), but within the range (2.0–4.3 km) reported for the enriched American–Antarctic Ridge (Le Roex *et al.* 1985).

Discussion and conclusions

From the above geochemical review we identify three main aspects of the Sudetic metabasites: (a) although Early Palaeozoic magmatism was strictly bimodal, it became dominated by massive mafic volcanism, associated with rifting of continental crust and the eventual formation of oceanic lithosphere; (b) although high-level

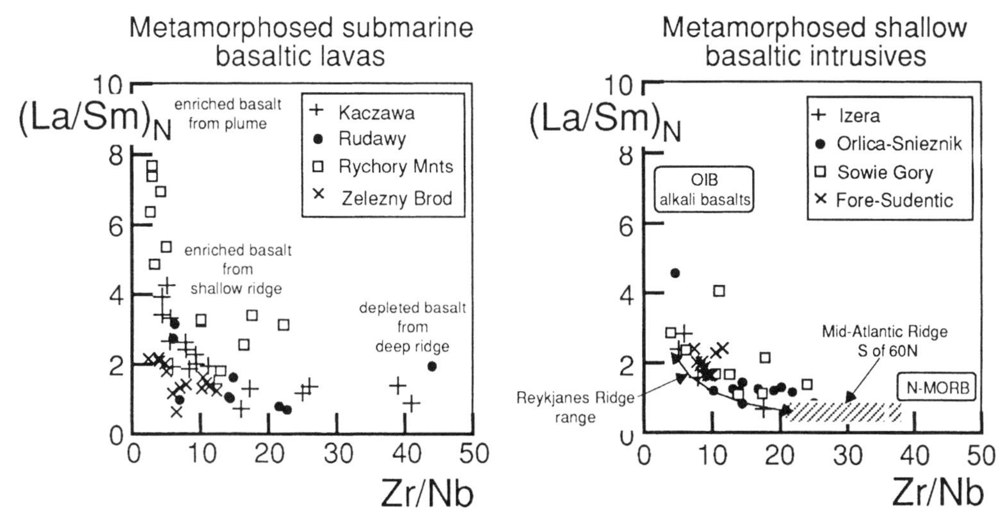

Fig. 12. Diagram showing the correlation between enrichment factors (La/Sm and Zr/Nb ratios) for ridge segments of variable axial depth ranging from depleted MORB (greatest depth) to enriched MORB (shallow depths) and plume-generated OIB. Many of the Sudetic metabasite data, as represented by both extrusive and intrusive suites, have enrichment factors similar to the relatively shallow Reykjanes Ridge. Data for Reykjanes Ridge largely from Schilling (1973).

crustal contamination was probably very limited, a lithospheric source already contaminated by pelagic sediments was available for the production of some Low-Ti metabasalts; (c) the chemical spectrum displayed by asthenospheric-derived basalts (Main Series metatholeiites and alkali metabasalts) is a consequence of mixing between a depleted MORB-like source and an enriched OIB-type plume source.

Plume involvement

Models for the generation of large magmatic provinces characterized by massive outpouring of basalt over a relatively short time interval (e.g. continental flood basalts) have generally appealed to the presence of a deep-seated mantle plume, either in an active (e.g. Duncan & Richards 1991) or passive role (e.g. White & McKenzie 1989). In the former case, impingement of the plume head against the lithosphere allows the penetration of plume-derived melts through a progressively domed and fractured lid, such that magmatism should precede continental rifting. In the latter process, by contrast, lithospheric stretching and thinning is initiated first, to be followed by melts from an uprising plume of asthenosphere that fills the space. In either model the progressive stretching of continental lithosphere may eventually lead to its failure, with the formation of new oceanic lithosphere between the rifted continental segments. Tectonic models for the fragmentation of northern Gondwana during early Palaeozoic time involve the extensive rifting of continental crust (e.g. Franke *et al.* 1995). In the light of the high volume of associated basaltic products, and the geochemical evidence for the development of enriched oceanic crust, it is considered that a plume or plumes played an important role during this rifting. Whether the plume played an active or passive role (i.e. did magmatism or rifting come first) is difficult to judge, although the development of early anatectic granites before mafic volcanism might suggest that the high heat flow necessary for their generation was the result of a plume head incubating below the sub-continental lithosphere before rifting (with associated ensialic basin information) occurred.

It seems unlikely that any hypothetical 'Sudetic plume' initiated or was the main driving force for continent margin rifting (e.g. see comments by Storey (1995) for Gondwana), in view of the progressive northward drift of microcontinental fragments (Avalonia, Armorica terranes) throughout early Palaeozoic time. However, it is possible that it assisted in the fragmentation of smaller continental blocks such as those now recognized in the Bohemian Massif.

Tectonic analogue

As introduced above, a possible tectonic analogue for Lower Palaeozoic fracturing of the northern Gondwana margin might be the disintegration of the Gondwana supercontinent during Jurassic time. In particular, the early–mid Jurassic period (*c.* 180–160 Ma) saw the separation of Africa + South America from Antarctica + Australia with the development of the Mozambique–proto-Weddell seaway (Storey 1995). Extensive continental flood basalts (CFBs) developed in southern Africa (Karoo province), Antarctica (Ferrar province) and Australia (Tasman province) at about this time; activity that might have been related to the Bouvet or Marion plumes. Although these basalts are continental in aspect they display a wide range of compositions, comprising both Low-Ti and High-Ti tholeiites that occupy different magmatic provinces (Cox 1988; Elliot 1992). The Ferrar province, in particular, exhibits Low-Ti basalts that are considered to have been derived from a sediment-contaminated sub-continental lithospheric mantle (Hergt *et al.* 1991; Brewer *et al.* 1992). The similarity in composition between these Jurassic basalts and the Sudetic Low-Ti metatholeiites (see above), together with the presence of a sediment contaminant, suggests that the latter may also represent lithospheric melts. Although the generation of CFB has been explained by the wholesale melting of metasomatized lithosphere (e.g. Hawkesworth *et al.* 1983), modelling suggests that the proportion of melt generated is relatively small, such that during plume–lithosphere interaction most melts are obtained from the asthenospheric source (Arndt & Christensen 1992; Saunders *et al.* 1992). The low HFSE signature of the Sudetic Low-Ti metabasites implies a source that was different from and more depleted than the enriched MORB-type source of the Main Series metabasites. Sub-continental lithosphere could represent this source and have undergone some thermal erosion during plume incubation and subsequently produced melts from this region.

With the further separation of fragments of Mesozoic Gondwana, Cretaceous and Tertiary ocean floor separated the continental fragments, with the eventual development of the present-day SW Indian Ridge and American–Antarctic Ridge. The sea-floor spreading stage saw the production of variably enriched ocean crust under the influence of the Bouvet plume and is considered analogous to the development of most of the submarine Sudetic metabasites. However, although ocean crust had developed in the Sudetic environment, continental separation was probably at a relatively elementary stage, as suggested by the high proportion of granitic material found in the deeper crustal levels.

Magmatic model

Rifting of the Gondwana margin was well advanced by early Ordovician time with the separation of Avalonia and then Armorica (Franke 1989*b*). This reflected extensive lithospheric thinning, followed by microcontinent migration and the development of seaways. The earliest magmatic activity is represented by the production of granites in late Cambrian–early Ordovician time (e.g. various orthogneisses from Izera, Snieznik, etc., Rumburk granite), which may be the first indication of a resident plume incubating under the sub-continental lithosphere of northern Gondwana. It is considered that most of the granitic material was produced by the melting of pre-existing Cadomian continental basement or at least had a strong crustal component (Klimas-August 1990; Oliver *et al.* 1993). Massive crustal anatexis would develop calc-alkali granites often with an inherited 'arc-like' signature typical of the lower continental crust (see Twist & Harmer 1987). The presence of a plume would provide additional heat for melting the lower continental crust as lithospheric extension alone was probably insufficient to sustain extensive melting from this source.

Continued rifting of the continental lithosphere with the eventual development of oceanic crust allowed the interaction of the uprising plume with both a sediment-contaminated lithosphere and a shallow-level asthenospheric MORB source. The volume and composition of the melts generated are directly related to the amount of lithospheric extension and the potential temperature of the asthenospheric mantle (McKenzie & Bickle 1988). Thus, during the early incubation stage, thermal erosion of the lithosphere by the plume produced a small volume of Low-Ti basaltic melts. This implies that the Sudetic Low-Ti metabasites are the earliest representatives of the voluminous mafic magmatism, although, because of the degree of deformation, unequivocal field proof is scarce. With further crustal attenuation, melting at shallow depths in the asthenosphere then produced the Main Series MOR-like basalts, whereas interaction between the plume and the MORB source allowed the development of variably enriched MORB compositions. In summary, the Sudetic metabasites document the change from magmatism featuring a minor lithospheric mantle component to one dominated by asthenospheric mantle melts, both of which were strongly influenced by the contribution of a plume.

Because of the small area represented by the Sudetic crustal segments (or the Bohemian

Massif as a whole) relative to (say) the fragmentation of Mesozoic southern Gondwana, no significant geochemical gradient or polarity can be recognized that might pinpoint the most likely location of the plume, and the distribution of normal versus enriched segments of the ocean ridge in relation to the plume centre. Comparison with more widely distributed contemporary rift-related volcanism across the Variscide terranes of Europe may provide this information.

Initial funding for part of this project grew out of the TEMPUS educational scheme (JEP 3656) and is now supported by EU TMR Network (Contract ERBFMRXCT97 0136) as a collaborative project 'Palaeozoic Amalgamation of Central Europe' (PACE). Many thanks are due to numerous colleagues in Germany, Poland and the Czech Republic for their expert advice and support in the field at various stages of the collaborative research.

References

ARNDT, N. T. & CHRISTENSEN, U. 1992. The role of lithospheric mantle in continental flood volcanism: thermal and geochemical constraints. *Journal of Geophysical Research*, **97**, 10967–10981.

BARANOWSKI, Z. & LORENC, S. 1986. A volcanic–carbonate association in the Göry Kaczawskie, Western Sudetes. *Geologische Rundschau*, **75**, 595–599.

——, HAYDUKIEWICZ, A., KRYZA, R., LORENC, S., MUSZYNSKI, A., SOLECKI, A. & URBANEK, Z. 1990. Outline of the geology of the Göry Kaczawskie (Sudetes, Poland). *Neues Jahrbuch für Geologie und Paläontologie, Abhandlungen*, **179**, 223–257.

BREWER, T. S., HERGT, J. M., HAWKESWORTH, C. J., REX, D. & STOREY, B. C. 1992. Coats Land dolerites and the generation of Antarctic continental flood basalts. *In*: STOREY, B. C., ALABASTER, T. & PANKHURST, R. J. (eds) *Magmatism and the Causes of Continental Break-up*. Geological Society, London, Special Publications, **68**, 185–208.

BRIAND, B., BOUCHARDON, J.-L., OUALI, H., PIBOULE, M. & CAPIEZ, P. 1995. Geochemistry of bimodal amphibolitic–felsic gneiss complexes from eastern Massif Central, France. *Geological Magazine*, **132**, 321–337.

CARSWELL, D. A. & O'BRIEN, P. J. 1993. Thermobarometry and geotectonic significance of high-pressure granulites: examples from the Moldanubian Zone of the Bohemian Massif in Lower Austria. *Journal of Petrology*, **34**, 427–459.

COX, K. G. 1988. The Karoo Province. *In*: MACDOUGALL, J. D. (ed.) *Continental Flood Basalts*. Kluwer, Dordrecht, 239–272.

—— & HAWKESWORTH, C. J. 1985. Geochemical stratigraphy of the Deccan Traps at Mahabaleshwar, Western Ghats, India, with implications for open system magmatic processes. *Journal of Petrology* **26**, 355–377.

CYMERMAN, Z. & PIASECKI, M. A. J. 1994. The terrane concept in the Sudetes, Bohemian massif. *Geological Quarterly*, **38**, 191–210.

——, —— & SESTON, R. 1997. Terranes and terrane boundaries in the Sudetes, northeast Bohemian Massif. *Geological Magazine*, **134**, 717–725.

DALLMEYER, R. D., FRANKE, W. & WEBER, K. (eds) 1995. *Pre-Permian Geology of Central and Eastern Europe*. Springer, Berlin.

DEPAOLO, D. J. 1981. Trace element and isotopic effects of combined wallrock assimilation and fractional crystallization. *Earth and Planetary Science Letters*, **53**, 189–202.

DEVEY, C. W. & COX, K. G. 1987. Relationships between crustal contamination and crystallization in continental flood basalt magmas with special reference to the Deccan Traps of the Western Ghats, India. *Earth and Planetary Science Letters*, **84**, 59–68.

DON, J., DUMICZ, M., WOJCEICHOWSKA, I. & ZELAZNIEWICZ, A. 1990. Lithology and tectonics of the Orlica–Snieznik dome, Sudetes—recent state of knowledge. *Neues Jahrbuch für Geologie und Paläontolgie, Abhandlungen*, **179**, 159–188.

DUNCAN, R. A. & RICHARDS, M. A. 1991. Hotspots, mantle plumes, flood basalts, and true polar wander. *Reviews of Geophysics*, **29**, 31–50.

ELLAM, R. M. & COX, K. G. 1989. A Proterozoic lithospheric source for Karoo magmatism: evidence from the Nuanetsi picrites. *Earth and Planetary Science Letters*, **92**, 207–218.

ELLIOT, D. H. 1992. Jurassic magmatism and tectonism associated with Gondwanaland break-up: an Antarctic perspective. *In*: STOREY, B. C., ALABASTER, T. & PANKHURST, R. J. (eds) *Magmatism and the Causes of Continental Break-up*. Geological Society, London, Special Publications, **68**, 165–184.

ERLANK, A. J., DUNCAN, A. R., MARSH, J. S. *et al.* 1988. A laterally extensive geochemical discontinuity in the subcontinental Gondwana lithosphere. *In*: *Geochemical Evolution of the Continental Crust*. Conference Abstracts, Procos de caldes, Brazil, 1–10.

FAJST, M., KACHLIK, V. & PATOCKA, F. 1998. Geochemistry and petrology of the Early Palaeozoic Zelezny Brod Volcanic Complex (W. Sudetes, Bohemian Massif): geodynamic interpretations. *GeoLines*, **6**, 14–15.

FINGER, F. & STEYRER, H. P. 1995. A tectonic model for the Eastern Variscides: indicators from the chemical study of amphibolites in the south-eastern Bohemian Massif. *Geologica Carpathica*, **46**, 137–150.

FLOYD, P. A. & CASTILLO, P. R. 1992. Geochemistry and petrogenesis of Jurassic ocean crust basalts, ODP Leg 129, Site 801. *In*: LARSON, R., LAUNCELOT, Y. *et al.* (eds). *Proceedings of Ocean Drinking Program, Scientific Results*, 129. College Station, TX, 361–388.

—— & WINCHESTER, J. A. 1978. Identificatin and discrimination of altered and metamorphosed volcanic rocks using immobile elements. *Chemical Geology*, **21**, 291–306.

——, ——, SESTON, R. 1997. Amphibolite geochemistry in the NE Bohemian Massif: the development of a fragmented continental margin. *Terra Nova*, **9**, *Abstract Supplement 1*, 145.

——, ——, CIESIELCZUK, J., LEWANDOWSKA, A., SZCZEPANSKI, J. & TURNIAK, K. 1996. Geochemistry of early Palaeozoic amphibolites from the Orlica–Snieznik dome, Bohemian massif: petrogenesis and palaeotectonic aspects. *Geologische Rundschau*, **85**, 225–238.

FODOR, R. V. 1987. Low- and high-TiO_2 flood basalts of southern Brazil: origin from picritic parentage and a common mantle source. *Earth and Planetary Science Letters*, **84**, 423–430.

FRANKE, W. 1989*a*. Tectonostratigraphic units in the Variscan belt of Central Europe. *In*: DALLMEYER, R. D. (ed.) *Terranes in the Circum-Atlantic Palaeozoic Orogens*. Geological Society of America, Special Paper, **230**, 67–90.

—— 1989*b*. Variscan plate tectonics in Central Europe—current ideas and open questions. *Tectonophysics*, **169**, 221–228.

——, DALLMEYER, R. D. & WEBER, K. 1995. Geodynamic evolution. *In*: DALLMEYER, R. D., FRANKE, W. & WEBER, K. (eds) *Pre-Permian Geology of Central and Eastern Europe*. Springer, Berlin, 579–593.

FREY, F. A., GREEN, D. H. & ROY, S. D. 1978. Integrated models of basalt petrogenesis: a study of quartz tholeiites to olivine melilitites from southeastern Australia utilizing geochemical and experimental petrological data. *Journal of Petrology*, **19**, 463–513.

FRITZ, H. 1994. The Raabs Serie, a Variscan ophiolite in the SE Bohemian Massif: a key for the tectonic interpretation. *Journal of the Czech Geological Society*, **39**, 32–33.

FURNES, H., KRYZA, R. & MUSZYNSKI, A. 1989. Geology and geochemistry of Early Palaeozoic volcanics of the Swierzawa Unit, Kaczawa Mts., W. Sudetes, Poland. *Neues Jahrbuch für Geologie und Paläontologie, Monatshefte*, **3**, 136–154.

——, ——, —— PIN, C. & GARMANN, L. B. 1994. Geochemical evidence for progressive, rift-related Early Palaeozoic volcanism in the eastern Sudetes. *Journal of the Geological Society, London*, **151**, 91–110.

HAWKESWORTH, C. J., ERLANK, A. J., MARSH, J. S., Menzies, M. A. & VAN CALSTEREN, P. 1983. Evolution of the continental lithosphere: evidence from volcanics and xenoliths in southern Africa. *In*: HAWKESWORTH, C. J. & NORRY, M. J. (eds) *Continental Basalts and Mantle Xenoliths*. Shiva, Nantwich, 111–138.

HERGT, J. M., PEATE, D. W. & HAWKESWORTH, C. J. 1991. The petrogenesis of Mesozoic Gondwana low-Ti flood basalts. *Earth and Planetary Science Letters*, **105**, 134–148.

HUMPHRIS, S. E., THOMPSON, G., SCHILLING, J. G. & KINGSLEY, R. H. 1985. Petrological and geochemical variations along the Mid-Atlantic Ridge between 46°S and 32°S; influence of the Tristan da Cunha mantle plume. *Geochimica et Cosmochimica Acta*, **49**, 1445–1464.

HUPPERT, H. & SPARKS, R. S. J. 1985. Cooling and contamination of mafic and ultramafic magmas during ascent through continental crust. *Earth and Planetary Science Letters*, **74**, 371–386.

HYNES, A. 1980. Carbonization and mobility of Ti, Y, and Zr in Ascot Formation metabasalts, S.E. Quebec. *Contributions to Mineralogy and Petrology*, **75**, 79–87.

KACHLIK, V. & PATOCKA, F. !998. Lithostratigraphy and tectonomagmatic evolution of the Zelezny Brod Crystalline Unit: some constraints for the palaeotectonic development of the W. Sudetes (NE Bohemian Massif). *GeoLines*, **6**, 34–35.

KLEIN, E. M. & LANGMUIR, C. H. 1987. Global correlations of ocean ridge basalt chemistry with axial depth and crustal thickness. *Journal of Geophysical Research*, **92**, 8089–8115.

KLIMAS-AUGUST, K. 1990. Genesis of gneisses and granites from the eastern part of the Izera metamorphic complex in the light of studies on zircons from selected geological profiles. *Geologia Sudetica*, **24**, 1–71.

KRÖNER, A. & HEGNER, E. 1998. Geochemistry, single zircon ages and Sm–Nd systematics, of granitoid rocks from the Göry Sowie (Owl Mts), Polish West Sudetes: evidence for early Palaeozoic arc-related plutonism. *Journal of the Geological Society, London*, **155**, 711–724.

KRYZA, R. 1993. Basic metavolcancic rocks of the central Kaczawa Mts. (Sudetes): a petrological study. *Prace Geologie–Mineralogie*, **39**.

——, MAZUR, S. & PIN, C. 1995. Leszczyniec meta-igneous complex in the eastern part of the Karkonosze–Izera Block, Western Sudetes: trace elements and Nd isotope study. *Neues Jahrbuch für Geologie und Paläontologie, Abhandlungen*, **170**, 59–74.

—— & MUSZYNSKI, A. 1988. Metamorphosed diabase sill–sediment complex from Wojcieszow: mineraological and geochemical evidence of Lower Palaeozoic early-stage rifting in the Western Sudetes, SW Poland. *Mineraologia Polonica*, **19**, 3–18.

——, ——, TURNIAK, K. & ZALASIEWICZ, J. A. 1994. A Lower Palaeozoic shallow-water sequence in the eastern European Variscides (SW Poland); provenance and depositional history. *Geologische Rundschau*, **83**, 5–19.

——, —— & VIELZEUF, D. 1990. Glaucophane-bearing assemblage overprinted by greenschist-facies metamorphism in the Variscan Kaczawa complex, Sudetes, Poland. *Journal of Metamorphic Geology*, **8**, 345–355.

——, PIN, C. & VIELZEUF, D. 1996. High-pressure granulites from the Sudetes (south-west Poland): evidence of crustal subduction and collisional thickening in the Variscan Belt. *Journal of Metamorphic Geology*, **14**, 531–546.

LE ROEX, A. P., DICK, H. J. B., ERLANK, A. J., REID, A. M., FREY, F. A. & HART, S. R. 1983. Geochemistry, mineralogy and petrogenesis of lavas erupted along the southwest Indian Ridge between the

Bouvet triple junction and 11 degrees east. *Journal of Petrology*, **24**, 267–318.

——, ——, GULEN, L., REID, A. M. & ERLANK, A. J. 1987. Local and regional heterogeneity in MORB from the Mid-Atlantic Ridge between 54.5°S and 51°S: evidence for geochemical enrichment. *Geochimica et Cosmochimica Acta*, **51**, 541–556.

——, ——, REID, A. M., FREY, F. A., ERLANK, A. J. & HART, S. R. 1985. Petrology and geochemistry of basalts from the American–Antarctic Ridge, Southern Ocean: implications for the westward influence of the Bouvet mantle plume. *Contributions to Mineraology and Petrology*, **90**, 367–380.

LI, Y. 1991. Distribution patterns of the elements in the ocean: a synthesis. *Geochimica et Cosmochimica Acta*, **55**, 3223–3240.

MCKENZIE, D. P. & BICKLE, M. J. 1988. The volume and composition of melt generated by extension of the lithosphere. *Journal of Petrology*, **29**, 625–679.

MALUSKI, H. & PATOCKA, f. 1997. Geochemistry and ^{40}Ar–^{39}Ar geochronology of the mafic metavolcanics from the Rychory Mts. complex (West Sudetes, Bohemian Massif): palaeotectonic significance. *Geological Magazine*, **134**, 703–716.

MATTE, P. H., MALUSKI, H., RAJLICH, P. & FRANKE, W. 1990. Terrane boundaries in the Bohemian massif: result of large-scale Variscan shearing. *Tectonophysics*, **177**, 151–170.

MESCHEDE, M. 1986. A method of discriminating between different types of mid-ocean ridge basalts and continental tholeiites with the Nb–Zr–Y diagram. *Chemical Geology*, **56**, 207–218.

MISAR, Z., JELINEK, E. & PACESOVA, M. 1984. The Letovice dismembered metaophiolites in the framework of the Saxo-Thuringian zone of the Bohemian massif. *Mineralia Slovaca*, **16**, 13–28.

MUSZYNSKI, A. 1994. *Acid metavolcanogenic rocks of the central Kaczawa Mts. (Sudetes): a petrological study*. Wydawnictwo Naukowe UAM, Poznan.

NAREBSKI, W. 1993. Lower to Upper Palaeozoic tectonomagmatic evolution of the NE part of the Bohemian Massif. *Zentralblatt für Geologie und Paläontologie*, **1**, 961–972.

——, DOSTAL, J. & DUPUY, C. 1986. Geochemical characteristics of Lower Palaeozoic spilite–keratophyre series in the western Sudetes (Poland): petrogenetic and tectonic implications. *Neues Jahrbuch für Mineralogie, Abhandlungen*, **155**, 243–258.

——, WOJCHIECHOWSKA, I. & DOSTAL, J. 1988. Initial rifting volcanics in the Klodzko Metamorphic complex (Polish Middle Sudetes), evidenced by geochemical data. *Bulletin of the Polish Academy of Sciences, Earth Sciences*, **36**, 261–269.

NOWAK, I. 1998. Geochemistry of metabasic dykes from the northern part of the Izera–Karkonosze Block, SW Poland. *GeoLines*, **6**, 48–49.

OBERC-DZIEDZIC, T. 1995. Problematyka badan serii metamorficznych Wzgorz Strzelineskich w swietle analizy materialu wiertniczego. *Acta Universitatis Wratislavensis, Prace Geologie–Mineralogie*, **5**, 77–103.

O'BRIEN, P. J., KRÖNER, A., JAECKEL, P., HEGNER, E., ZELANZNIEWICZ, A. & KRYZA, R. 1997. Petrological and isotopic studies on Palaeozoic high-pressure granulites, Gory Sowie Mts., Polish Sudetes. *Journal of Petrology*, **38**, 433–456.

OLIVER, G. J. L., CORFU, F. & KROUGH, T. E. 1993. U–Pb ages from Poland: evidence for a Caledonian suture zone between Baltica and Gondwana. *Journal of the Geological Society, London*, **150**, 355–369.

OPLETAL, M., JELINEK, E., PECINA, V., POSMOURTNY, K. & POUBOVA, E. 1990. Metavolcanites of the SE part of the Lugicum, their geochemistry and geotrectonic interpretation. *Sbornik Geologickych Ved, Geologie*, **45**, 37–64.

PATOCKA, F. & SMULIKOWSKI, W. 1997. Petrology and geochemistry of metabasic rocks of the Rychory Mts–Rudawy Janowickie Mts Complex (West Sudetes, NE Bohemian Massif). *Polskie Towarzystwo Mineralogiczne*, **9**, 146–150.

——, DOSTAL, J. & PIN, C. 1997. Early Palaeozoic intracontinental rifting in the central West Sudetes, Bohemian Massif: geochemical and Sr–Nd isotopic study on felsic–mafic metavolcanics of the Rychory Mts. complex. *Terra Nova, Abstract Supplement 1*, **9**, 144–145.

——, PIVEC, E. & OLIVERIOVA, D. 1996. Mineralogy and petrology of mafic blueschists from the Rychory Mts. crystalline complex (West Sudetes, Bohemian Massif). *Neues Jahrbuch für Mineralogie, Abhandlungen*, **170**, 313–330.

PEARCE, J. A. & CANN, J. R. 1973. Tectonic setting of basaltic volcanic rocks determined using trace element analyses. *Earth and Planetary Science Letters*, **19**, 290–300.

PEREKALINA, T. V. 1981. Variscan volcanism in central and western Europe. *Geologie en Mijnbouw*, **60**, 17–31.

PIBOULE, M. & BRIAND, B. 1985. Geochemistry of eclogites and associated rocks of the southeastern area of the French Massif central: origin of the protoliths. *Chemical Geology*, **50**, 189–199.

PIN, C. 1990. Variscan oceans: ages, origins and geodynamic implications inferred from geochemical and radiometric data. *Tectonophysics*, **177**, 215–227.

——, MAJEROWICZ, A. & WOJCIECHOWSKA, I. 1988. Upper Palaeozoic oceanic crust in the Polish Sudetes: Nd–Sr isotope and trace element evidence. *Lithos*, **21**, 195–209.

REHKAMPER, M. & HOFMANN, A. W. 1997. Recycled ocean crust and sediment in Indian Ocean MORB. *Earth and Planetary Science Letters*, **147**, 93–106.

SATO, H. 1977. Nickel content of basaltic magmas: identification of primary magmas and a measure of the degree of olivine fractionation. *Lithos*, **10**, 120–133.

SAUNDERS, A. D., STOREY, M., KENT, R. W. & NORRY, M. J. 1992. Consequences of plume–lithosphere interactions. *In*: STOREY, B. C., ALABASTER, T. & PANKHURST, R. J. (eds) *Magmatism and the Causes of Continental Break-up*. Geological Society, London, Special Publications, **68**, 41–60.

Seston, R. 1999. *Tectonic and geochemical studies in Palaeozoic rocks from part of the Polish Sudetes, south west of Wroclaw*. PhD thesis, University of Keele.

Schilling, J. G. 1973. Iceland mantle plume: geochemical study of Reykjanes Ridge. *Nature*, **242**, 565–571.

——, 1975. Azores mantle blob: rare earth evidence. *Earth and Planetary Science Letters*, **25**, 103–115.

——, Thompson, G., Kingsley, R. & Humphris, S. 1985. Hotspot-migrating ridge interaction in the South Atlantic. *Nature*, **313**, 187–191.

Smith, R. E. & Smith, S. E. 1976. Comments on the use of Ti, Zr, Y, Sr, K, P and Nb in classification of basaltic magmas. *Earth and Planetary Science Letters*, **32**, 114–120.

Storey, B. C. 1995. The role of mantle plumes in continental breakup: case histories from Gondwanaland. *Nature*, **377**, 301–304.

Sun, S. S. & McDonough, W. F. 1989. Chemical and isotopic systematics of oceanic basalts: implications for mantle composition and processes. *In*: Saunders, A. D. & Norry, M. J. (eds) *Magmatism in the Ocean Basins*. Geological Society, London, Special Publications, **42**, 313–345.

Szczepanski, J. & Oberc-Dziedzic, T. 1998. Geochemistry of amphibolites from the Strzelin crystalline massif, Fore-Sudetic block, SW Poland. *Neues Jahrbuch für Mineralogie, Abhandlungen*, **173**, 23–40.

Tarney, J. & Windley, B. F. 1977. Chemistry, thermal gradients and evolution of the lower continental crust. *Journal of the Geological Society, London*, **134**, 153–172.

Taylor, S. R. & McLennan, S. M. 1985. *The Continental Crust: its Composition and Evolution*. Blackwell, Oxford.

Thompson, G. 1991. Metamorphic and hydrothermal processes: basalt–seawater interactions. *In*: Floyd, P. A. (ed.) *Oceanic Basalts*. Blackie, Glasgow, 148–173.

Thompson, R. N., Dickin, A. P., Gibson, I. L. & Morrison, M. A. 1982. Elemental fingerprints of isotopic contamination of Hebridean Palaeocene mantle-derived magmas by Archaean sial. *Contributions to Mineralogy and Petrology*, **79**, 159–168.

——, Morrison, M. A., Dickin, A. P., Gibson, I. L. & Harmon, R. S. 1986. Two contrasting styles of interaction between basic magmas and continental crust in the British Tertiary Volcanic province. *Journal of Geophysical Research*, **91**, 5985–5997.

Twist, D. & Harmer, R. E. J. 1987. Geochemistry of contrasting siliceous magmatic suites in the Bushveld Complex: genetic aspects and implications for tectonic discrimination diagrams (South Africa). *Journal of Volcanology and Geothermal Research*, **32**, 83–98.

Urbanek, Z., Zelazniewicz, A., Kemnitz, H., Hermsdor, N. & Linnemann, U. 1995. Western Sudetes, VI.B Stratigraphy. *In*: Dallmeyer, R. D., Franke, W. & Weber, K. (eds) *Pre-Permian Geology of Central and Eastern Europe*. Springer, Berlin, 315–327.

Vogt, P. R. 1976. Plumes, subaxial pipe flow, and topography along the mid-ocean ridge. *Earth and Planetary Science Letter*, **29**, 309–325.

Weaver, B. L. 1991. The origin of ocean island basalt end-member compositions: trace element and isotopic constraints. *Earth and Planetary Science Letters*, **104**, 381–397.

—— & Tarney, J. 1984. Empirical approach to estimating the composition of the continental crust. *Nature*, **310**, 575–578.

White, R. & McKenzie, D. 1989. Magmatism at rift zones: the generation of volcanic continental margins and flood basalts. *Journal of Geophysical Research*, **94**, 7685–7729.

Winchester, J. A., Floyd, P. A., Awdankiewicz, M., Piasecki, M. A. J., Awdankiewicz, H., Gunia, P. & Gliwicz, T. 1998. Geochemistry and tectonic significance of metabasic suites in the Gory Sowie Block, SW Poland. *Journal of the Geological Society, London*, **155**, 155–164.

——, ——, Chocyk, M., Horbowy, K. & Kozdroj, W. 1995. Geochemistry and tectonic environment of Ordovician meta-igneous rocks in the Rudawy Janowickie complex, SW Poland. *Journal of the Geological Society, London*, **152**, 105–115.

Wojciechowska, I. 1986. Metabasites in the NW part of Snieznik metamorphic unit. *Geologische Rundschau*, **73**, 585–593.

——, Narebski, W., Dostal, J. & Pin, C. 1989. Palaeotectonic setting of metabasites of NW part of the Snieznik metamorphic unit: geological and preliminary geochemical evidences. *In*: Narebski, W. & Majerowicz, A. (eds) *Lower and Upper Palaeozoic Metabasites and Ophiolites of the Polish Sudetes*. Polish Academy of Science, Warsaw, 175–189.

Wood, D. A. 1980. The application of the Th–Hf–Ta diagram to problems of tectonomagmatic classification and to establishing the nature of crustal contamination of basaltic lavas of the British Tertiary volcanic province. *Earth and Planetary Science Letters*, **50**, 11–30.

Zelazniewicz, A., Kemnitz, H. & Hermsdorf, N. 1995. Western Sudetes, VI.C Structure. *In*: Dallmeyer, R. D., Franke, W. & Weber, K. (eds) *Pre-Permian Geology of Central and Eastern Europe*. Springer, Berlin, 315–327.

Chronological constraints on the pre-Variscan evolution of the northeastern margin of the Bohemian Massif, Czech Republic

A. KRÖNER[1], P. ŠTÍPSKÁ[2], K. SCHULMANN[2] & P. JAECKEL[1,3]

[1]*Institut für Geowissenschaften, Universität Mainz, D-55099 Mainz, Germany (e-mail: kroener@mail.uni-mainz.de)*

[2]*Department of Petrology, Charles University, Albertov 6, CS 12843, Praha 2, Czech Republic*

[3]*Max-Planck-Institut für Chemie, Postfach 3060, D-55020 Mainz, Germany*

Abstract: New single zircon ages enable us to provide an evolutionary scenario for the Neoproterozoic to Cambro-Ordovician tectonic history of part of the easternmost Sudetes along the northeastern margin of the Bohemian Massif. The easternmost crustal segment (Brunia) yields Neoproterozoic ages from both autochthonous and allochthonous Variscan units; these ages document a Cadomian (Pan-African) history that may be linked with the northern margin of Gondwana. A Cambro-Ordovician magmatic–thermal event in Brunia is represented by granitic to pegmatitic dykes intruding Neoproterozoic crust and by localized partial anatexis. Farther west a narrow zone of Cambro-Ordovician rifting is identified (Staré Městro belt), marked by gabbroic magmatism, bimodal volcanism and medium-pressure granulite facies metamorphism. The westernmost crustal domain (Orlica–Snieznik dome) is represented by Neoproterozoic crust intruded by Cambro-Ordovician plutons consisting of calk-alkaline granitoid rocks and affected by widespread Cambro–Ordovician anatexis. The geodynamic setting of the Neoproterozoic and Cambro-Ordovician domains is similar to that of the Western Sudetes, where both Cambro-Ordovician rifting and calc-alkaline magmatism were identified. We discuss the rifting mechanics in terms of sequential crustal thinning along the northern margin of Gondwana. The calc-alkaline magmatism, in conjunction with crustal rifting, is related to a back-arc geometry in front of a retreating south-dipping subduction zone during progressive closure of the Tornquist Ocean southeast of Avalonia.

The late Palaeozoic collisional history of the Variscan belt in central Europe is now reasonably well understood (Franke 1989, this volume; Matte *et al.* 1990), but there is still much discussion about the earlier, Neoproterozoic to early Palaeozoic, evolution of the various crustal segments constituting this belt. The Bohemian Massif is part of the axial zone of the Variscan belt and is composed of Mesoproterozoic to late Palaeozoic rock assemblages.

Three major zones have been recognized along the eastern margin of the Bohemian Massif (Schulmann & Gayer 2000) (see Fig. 1). They are, from bottom to top and east to west: (1) the undeformed lower plate Brunia, the Brunian domain; (2) a zone of intense Variscan deformation and metamorphism of the lower plate, the Moravo-Silesian domain; (3) the upper plate, consisting of middle- to upper-crustal rocks of the Moldanubian–Lugian domain. The Variscan structure of the eastern margin of the Bohemian Massif resulted from oblique collision between the combined Moldanubian–Lugian domain and the Neoproterozoic (Cadomian or Pan-African) Brunia domain (originally defined by Dudek (1980) as Brunovistulicum) to the east (see also Finger *et al.* this volume).

The Brunian domain, interpreted as a microcontinent (Matte *et al.* 1990), consists of Neoproterozoic high-grade metamorphic rocks intruded by numerous late Proterozoic granitoids (dated at *c.* 550 and 580 Ma) and overlain by a Devonian basin (see Chlupáč (1994) for review). The NE–SW-trending Moravo-Silesian domain is a belt of sheared and metamorphosed Brunia-derived rocks, piled up into a nappe sequence and consisting of three NE–SW-trending tectonic windows (Suess 1912). Variscan deformation and metamorphism have produced a tectonically inverted Barrovian metamorphic zonation ranging from a chlorite zone in the east to a kyanite–sillimanite zone in the west

From: FRANKE, W., HAAK, V., ONCKEN, O. & TANNER, D. (eds). *Orogenic Processes: Quantification and Modelling in the Variscan Belt.* Geological Society, London, Special Publications, **179**, 175–197. 0305-8719/00/$15.00

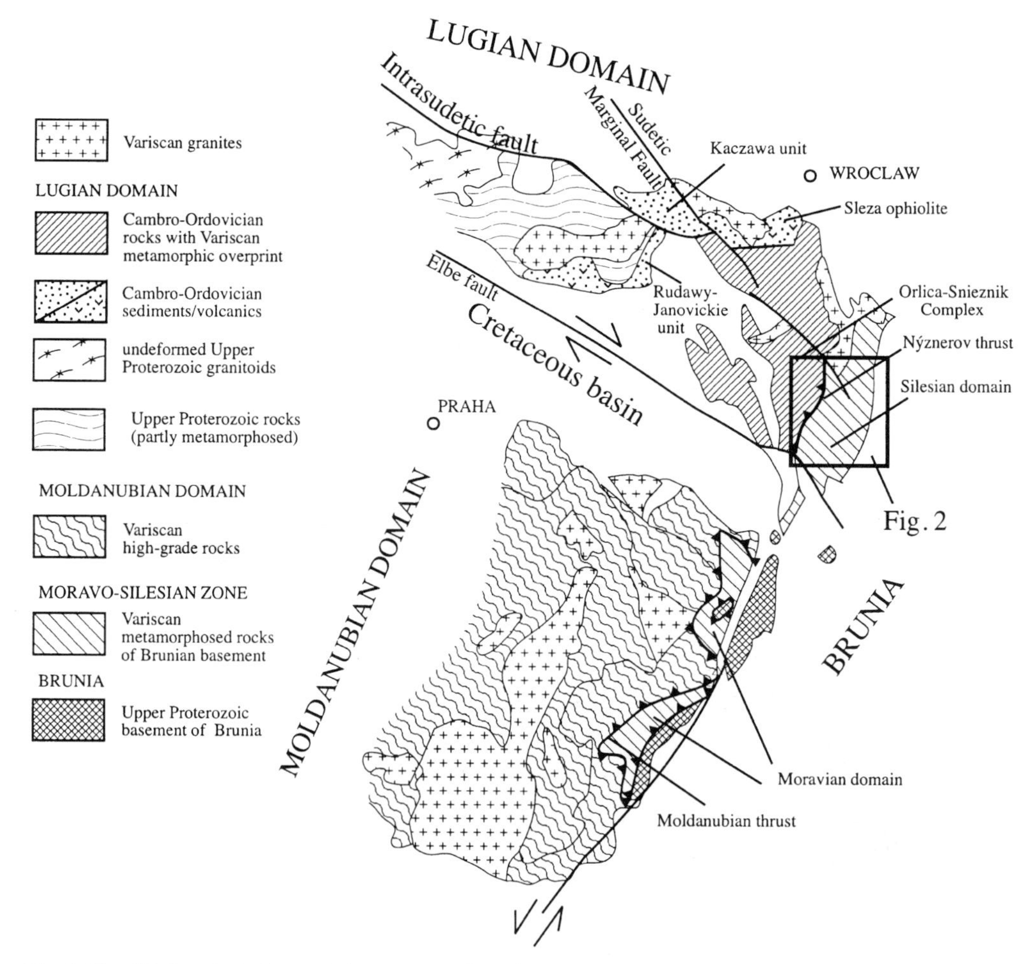

Fig. 1. Simplified geological map of the eastern and northeastern margin of the Bohemian Massif, showing major crustal units (after Matte *et al.* (1990) and Aleksandrowski *et al.* (1997)).

(Štípská & Schulmann 1995). The western Moldanubian domain is characterized by widespread Variscan anatexis, relics of high-pressure metamorphism and intrusion of large granitoid bodies. In the north, the Lugian domain consists mostly of medium-grade schists and high-grade rocks (migmatites and high-grade gneisses) as well as relics of high-pressure granulites and eclogites.

To be able to understand the geodynamic evolution of the contact zone between the Lugian and Brunia domains it is important to determine the composition, deformation, metamorphism and protolith age of the highly deformed rocks in this region. In this paper we present zircon age data from this critical tectonic boundary, in combination with whole-rock Nd mean crustal residence ages (Hegner & Kröner this volume), which document the pre-Variscan geological history of the Brunian and Lugian crustal segments.

Geology of the area studied

The geology of area studied is shown in the simplified map of Fig. 2, and a NW–SE cross-section is shown in Fig. 3. The Silesian domain has traditionally been subdivided into the low-grade eastern Desná dome and the allochthonous medium-grade western Keprník nappe. The precise western boundary of the Silesian domain is uncertain, and the rock assemblages west of the Keprník nappe are provisionally assigned to a transition zone between the Lugian and Silesian domains. From east to west this zone

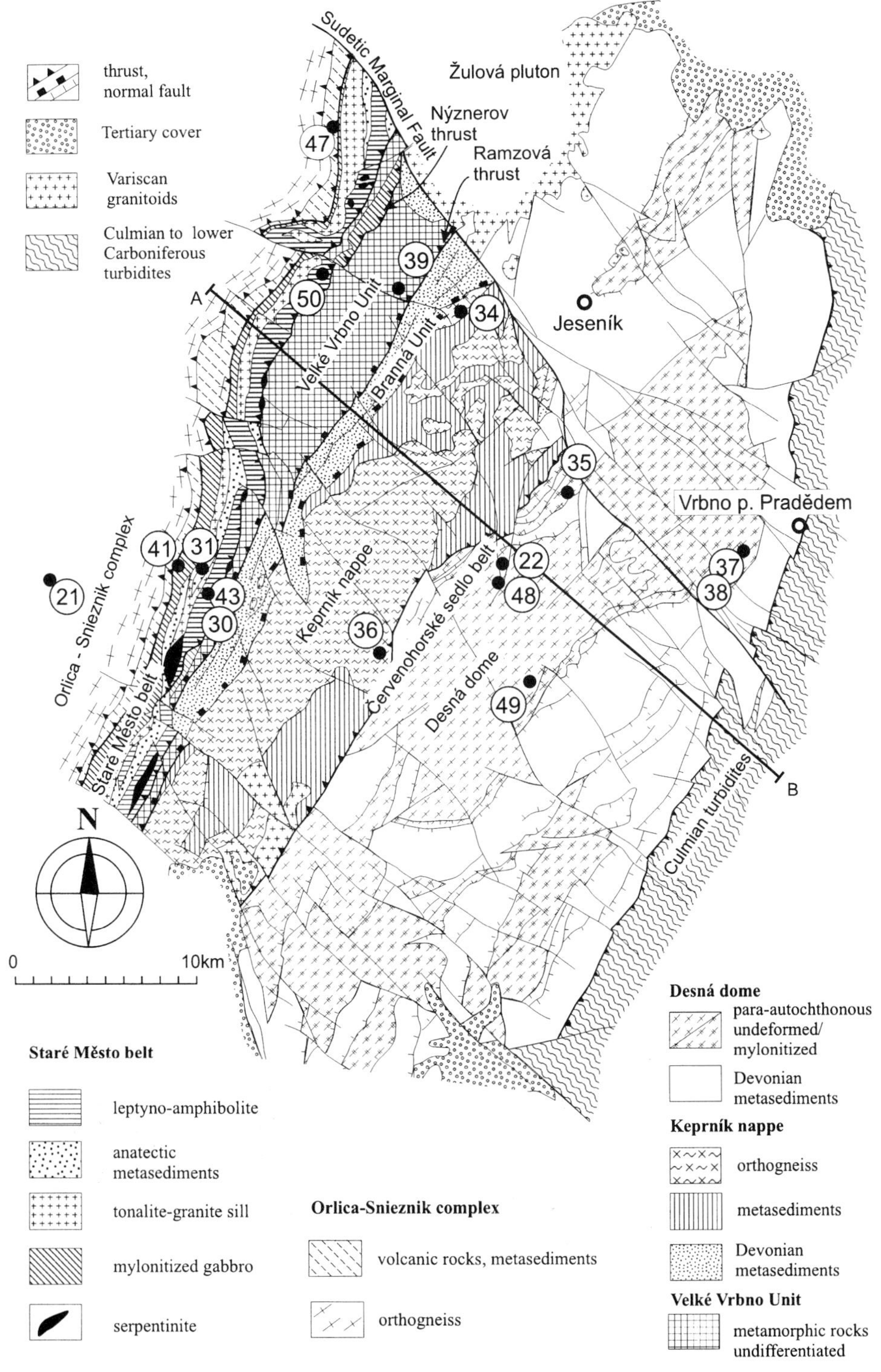

Fig. 2. Sketch map showing tectonic units of the Silesian and Lugian domains in the Jeseníky Mts, northeastern margin of the Bohemian Massif, eastern Czech Republic, and location of samples for which zircon ages were determined in this study (after Cháb, unpubl. map).

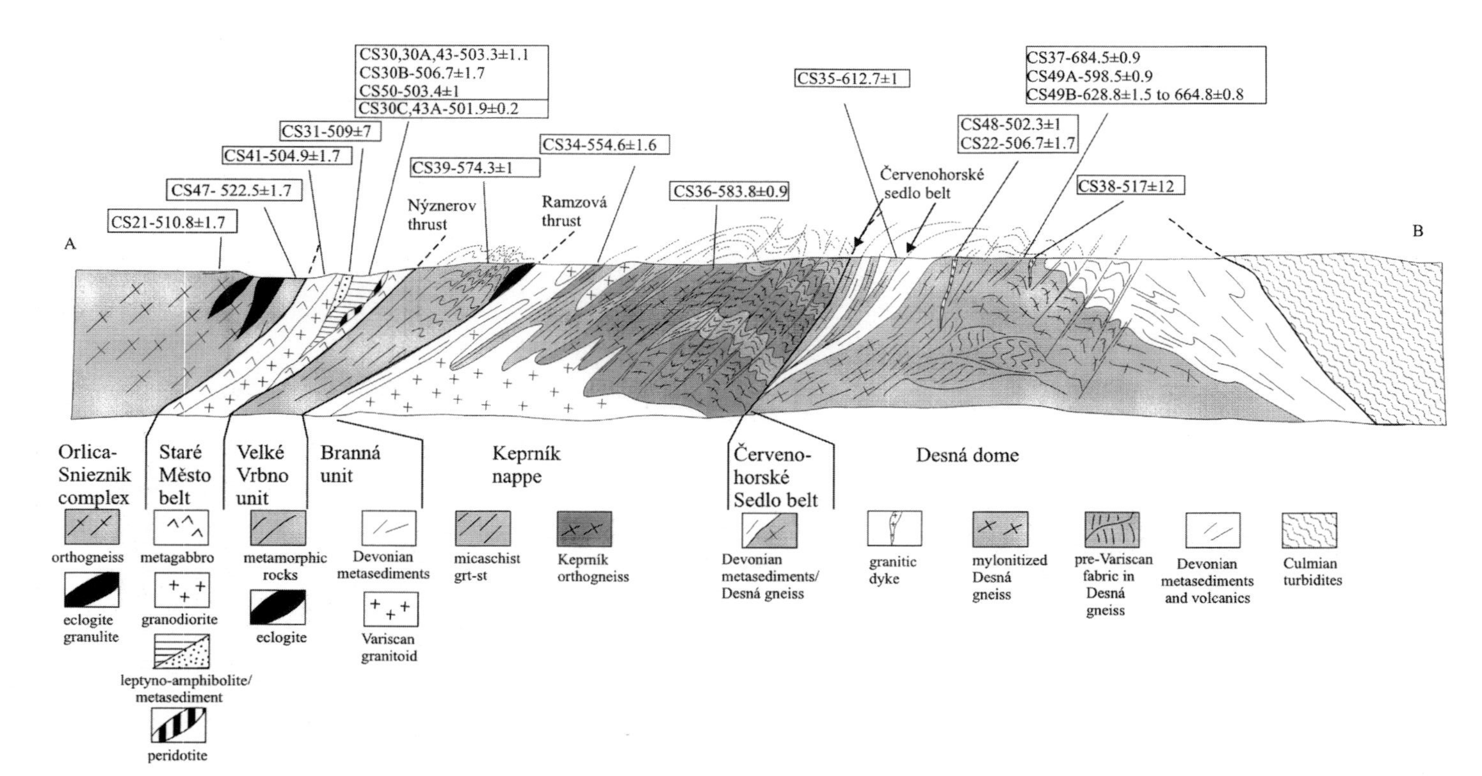

Fig. 3. Synoptic cross-section along line as indicated in Fig. 2 and showing various tectonic units, major tectonic boundaries and zircon ages reported in this paper. (For geological details, see Schulmann & Gayer (2000).)

consists of high-grade rocks of the Velké Vrbno unit and the Staré Město belt (Figs 2 and 3).

The Silesian domain has been affected by at least one pre-Variscan and two major Variscan tectono-metamorphic events. Pre-Variscan deformation (D1) and amphibolite facies metamorphism (M1) are locally preserved in crystalline complexes of the Desná dome and the Staré Město belt. The early Variscan deformation (D2) and metamorphism (M2) are related to major underthrusting of the entire Silesian domain below the Lugian domain (Schulman & Gayer 2000).

The parautochthonous core of the Desná dome is composed of medium-grade, biotite-rich layered gneisses and ophthalmitic to stromatitic migmatites. The cover sequence consists of quartzites of early Devonian (Pragian) age (Chlupáč 1994) as well as mica schists and metavolcanic rocks of mid-Devonian (Givetian) age (Fig. 3). The northwestern part of the cover is intruded by a Variscan calc-alkaline granite (the Žulová pluton, Fig. 2). The gneisses forming the core of the Desná dome display relics of pre-Variscan amphibolite facies metamorphism M1 (Fediuková *et al.* 1985), and these rocks were overprinted by Variscan greenschist to amphibolite facies deformation (Fig. 3).

The Keprník nappe is composed of a strongly deformed orthogneiss body tectonically intercalated with staurolite-bearing metapelites, calc-silicate rocks and quartzites. The Devonian sedimentary cover consists of low-grade polymictic conglomerates, quartzites, porphyroids, calc-silicate rocks and Devonian marbles. Variscan amphibolite facies deformation was later affected by strong D3 westward extension (Schulmann & Gayer 2000) (Fig. 3).

The Velké Vrbno unit (transitional zone) constitutes an anticlinal structure and is made up of sheets of felsic orthogneisses, metapelites, felsic and mafic metavolcanics, marbles and local occurrences of eclogites. All lithologies exhibit a strong amphibolite-facies metamorphic fabric.

Lower-crustal rocks of the Staré Město belt (transitional zone) define a narrow SW–NE trending linear domain (Fig. 2). The base is marked by a highly deformed sequence of leptyno-amphibolites associated with tonalitic gneisses and granulite facies metasediments that are intruded by a Variscan sill-like granodiorite body (Parry *et al.* 1997). In the hanging wall of the tonalite a layer of strongly sheared gabbro occurs. The earliest structures D1, and metamorphism M1, in rocks of the Staré Město belt are preserved in the leptyno-amphibolite sequence and associated metasediments. M1 reflects a transition from upper amphibolite to granulite facies conditions and exhibits temperatures of 800–900°C and pressures of 7–9 kbar (Parry *et al.* 1997; Štípská *et al.* 2000).

The eastern part of the Lugian domain consists of the Orlica–Snieznık dome, which is mainly built up of (1) granitic orthogneisses (Snieznık gneiss) and migmatites (Gieraltow gneiss) and (2) medium-grade staurolite schists with accompanying amphibolites, marbles and quartzites as well as low-grade felsic metavolcanic rocks, referred to in entirety as the Stronie Group. Locally there occur lensoid bodies of eclogite and intermediate to felsic high-pressure granulites (850°C, 19–22 kbar; Steltenpohl *et al.* 1993). The first fabric D1 is generally associated with high-temperature (HT) conditions and was affected by widespread D2 mylonitization under amphibolite facies conditions.

Previous geochronology

There are few precise ages for rocks in the area under discussion. Van Breeman *et al.* (1982) reported multigrain U–Pb zircon data for the Keprník gneiss and the Brno pluton, both part of the Brunia domain, and a Rb–Sr isochron age for the Snierznık gneiss of the Lugian domain. The Keprník gneiss sample revealed a complex zircon population of which six fractions are well aligned and yielded concordia intercept ages of $1422 + 191/-177$ and $546 + 6/-8$ Ma, respectively. The lower intercept age was interpreted to approximate the time of emplacement of the gneiss protolith, although considered by Van Breemen *et al.* as possibly being too low. The upper intercept age was ascribed to zircon inheritance from the source area of this crustally derived granite. Three discordant zircon fractions from a sample of Brno diorite (outside our study area) provided an upper concordia intercept age of 584 ± 5 Ma. Van Breemen *et al.* (1982) considered this to reflect the time of emplacement of the diorite. There are also K/Ar ages of uncertain significance ranging from 555 to 630 Ma on hornblende and biotite from the Brno pluton (Scharbert & Batík 1980).

Whole-rock samples of the Snieznık gneiss, sampled near the village of Žulová, yielded an Rb–Sr errorchron age of 487 ± 11 Ma (Van Breemen *et al.* 1982). Van Breemen *et al.* cautiously considered this to be the primary age of the gneiss protolith in view of the strong Variscan metamorphic overprint recorded in this area. Borkowska *et al.* (1990) reported an Rb–Sr whole-rock isochron age of 464 ± 18 Ma for Gieraltow migmatic gneisses and considered this to reflect the time of magmatic emplacement. Oliver *et al.* (1993) analysed three multigrain zircon fractions from a mylonitized Snieznik gneiss collected near the Polish–Czech border, and the ambiguous results indicate crystallization of the gneiss precursor at either $488 + 4/-7$ or 504 ± 3 Ma ago. A SHRIMP single zircon study of Turniak *et al.* (2000) yielded identical ages of *c.* 500 Ma for both the Snieznik and Gieraltow gneisses and demonstrated the presence of numerous inherited cores as old as 2.6 Ga in the zircons analysed. Brueckner *et al.* (1991) presented Variscan Sm–Nd mineral isochron ages between 327 and 352 Ma for eclogites occurring as tectonic lenses in the Snieznik and Gieraltow gneisses in Poland.

Scharbert & Batík (1980) reported a 14-point Rb–Sr isochron age of 551 ± 6 Ma for widely scattered samples from the Thaya (Dyje) granite in the Brunian domain in the Czech–Austrian border region (outside our study area) and interpreted this as reflecting the time of emplacement. Recent SHRIMP zircon data for the Bittesch gneiss within the Thaya Batholith yielded a concordant mean $^{206}Pb/^{238}U$ age of 586 ± 7 Ma, whereas inherited cores yielded ages as old as 1377 Ma (Friedl *et al.* 1998)

Twenty-eight muscovite, biotite and amphibole $^{40}Ar/^{39}Ar$ ages were reported by Maluski *et al.* (1995) for the Desná, Keprník and Snieznik domes in the present study area, and these document a strong Variscan overprint with cooling ages ranging between 340 and 300 Ma. Laser analyses on single muscovite grains from mylonitic gneisses along the eastern rim of the Desná dome also produced older ages between 380 and 520 Ma, with the cores of these grains showing the older ages whereas the rims were apparently reset during the Variscan event (Maluski *et al.* 1995). No information concerning the primary emplacement age of these rocks can be extracted from these data.

Finally, Fritz *et al.* (1996) reported Neoproterozoic $^{40}Ar/^{39}Ar$ hornblende and muscovite ages from rocks of the Brunian domain north of Brno (outside our study area). These range between 610 and 565 Ma were interpreted to reflect a strong Cadomian event.

Description of samples, geochemistry and ages

The samples studied here were analysed for major and trace elements by X-ray fluorescence (XRF). Single zircons or groups of two grains each were analysed isotopically by evaporation, vapour digestion or ion microprobe (SHRIMP II), and the analytical procedures are summarized in the Appendix. The major and trace element compositions of the dated samples, a brief characterization of the zircon morphology as well as the evaporation analyses and the vapour digestion and SHRIMP analytical data, respectively, can be obtained from the Society Library or the British Library Document Supply Centre, Boston Spa, Wetherby, West Yorkshire LS32 7BQ, UK, as Supplementary Publication No. SUP 18148 (7 pages) The Nd mean crustal residence ages quoted below are based on a two-stage model, and details have been given by Hegner & Kröner (this volume). We present and discuss our samples and isotopic data in three groups, namely those from the Silesian domain, the Transition zone and the Lugian domain.

Silesian domain (core of the Desná dome)

The first two samples come from a quarry near Zámčisko at the hill Medvědí důl (Figs 2 and 3), where we found interlayered light grey (sample CS 49A) and dark grey (sample CS 49B) biotite schists interpreted as metagreywacke and folded together with, and intruded by, granitoid orthogneisses (Kölbl 1929). Sample 49A is a typical Desná plagioclase–biotite schist marked by a strong mylonitic fabric. It is composed of plagioclase, quartz, biotite and epidote $\pm$ muscovite. The foliation is enhanced by elongation of quartz aggregates and by alignment of biotite. Chemically the dated sample resembles a trondhjemite or dacite and this, together with the zircon morphology, suggests that this rock is of igneous derivation, either a dacitic lava or a tuff. Sample 49B is chemically, mineralogically and texturally similar to sample 49A. It is composed of plagioclase, quartz, biotite, epidote and some muscovite. The microstructure is mylonitic, and the foliation is enhanced by alignment of dispersed biotite and by flattening of quartz aggregates.

The zircons of sample CS 49A are perfectly idiomorphic, unlike detrital grains in a sedimentary rock. Five grains were evaporated individually and produced a combined mean age of 598.4 ± 0.9 Ma (Fig. 4a) that we interpret to reflect either the time of igneous emplacement of the host rock if it is a tuff or the age of the source rock if it is a volcanic greywacke. By implication, this is also the approximate age of deposition of the supracrustal sequence to which samples CS 49A and -B belong. The host rock is most probably derived from melting of crust that, if it was of volcanic origin, probably reflects magmatic material, perhaps generated in an arc environment. An $\varepsilon_{Nd(t)}$ value of -1.6 and a mean crustal residence age of 1.2 Ga (Hegner & Kröner this volume) suggest that the sample is

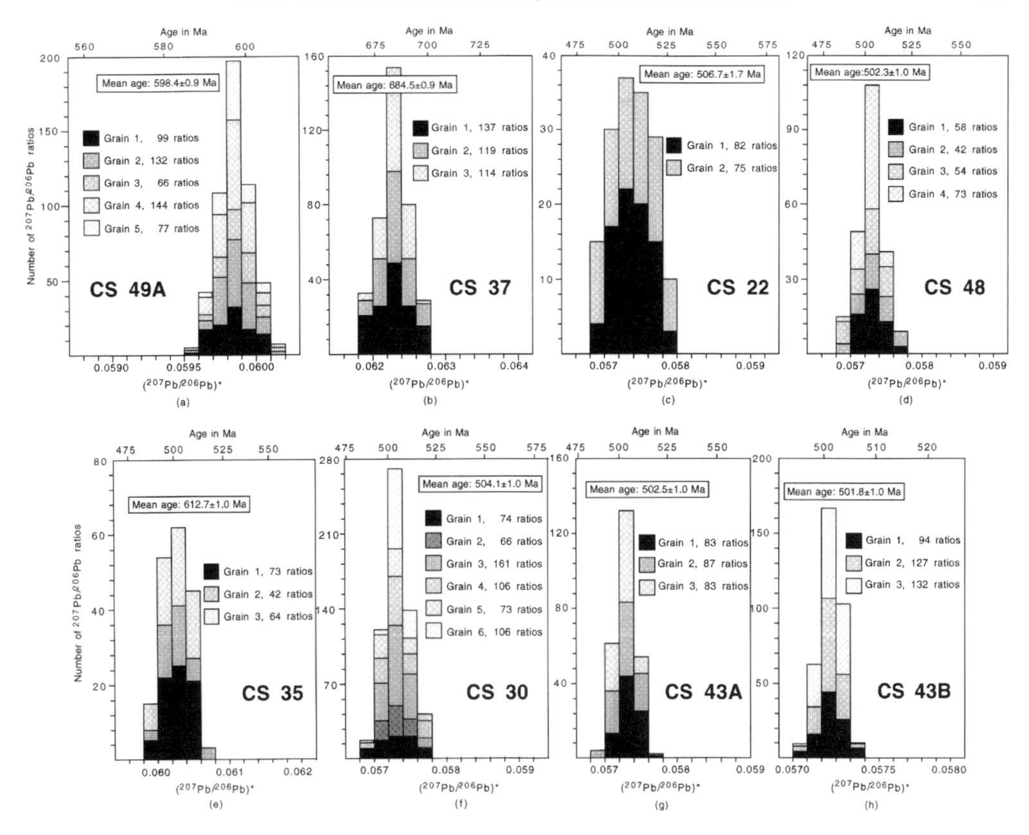

Fig. 4. Histograms showing distribution of radiogenic lead isotope ratios derived from evaporation of single zircons from Jeseníky Mts, northeastern Czech Republic. (**a**) Spectrum for five igneous zircons from metadacite or dacitic tuff sample CS 49A, quarry near Zámčisko. (**b**) Spectrum for three grains from leucocratic gneiss sample CS 37, Ludvíkov Cemetery, interpreted to reflect age of magmatic emplacement. (**c**) Spectrum for two grains from strongly foliated granite–gneiss sample CS 22, western Desná dome, east of Kouty nad Desnou, interpreted to reflect age of magmatic emplacement. (**d**) Spectrum for four grains from pegmatitic granite sample CS 48, forest SE of Kouty nad Desnou, Jeseníky Mts. (**e**) Spectrum for three grains from ultramylonitic orthogneiss sample CS 35, Filipovice. (**f**) Spectrum for six grains from tonalitic gneiss sample CS 30, Hanušovice quarry. (**g**) Spectrum for three grains from tonalitic gneiss sample CS 43A, Hanušovice quarry. (**h**) Spectrum for three grains from anatectic melt patch sample CS 43B, Hanušovice quarry. Analytical data are available as a Supplementary Publication (see p. 180).

derived from a source containing Grenvillian-age crustal material.

Metagreywacke sample CS 49B contains detrital zircons with morphologies that are typical of mechanical rounding during surface transport. Three grains were evaporated, and all have different ages ranging from 628.8 ± 1.5 to 664.8 ± 0.8 Ma, reflecting a Neoproterzoic source terrain. The $\varepsilon_{Nd(t)}$ value of -1.1 corresponds to a Nd mean crustal residence age of 1.2 Ga, identical to that of sample CS 49A (Hegner & Kröner this volume). This implies that the metagreywacke is derived from a source that contains a significant proportion of Grenvillian-age material, similar to the source of CS 49A.

Two other samples were collected from the northern termination of the Desná dome, which is composed of strongly mylonitic grandiorites and migratites (Novotný & Štelcl 1961; Fišera *et al.* 1986). The oldest rock dated in this region is a leucogranitic gneiss (sample CS 37, Figs 2 and 3) collected from a cliff NW of the Ludvíkov Cemetery some 20 km SSE of Jeseník, where it represents the main component of the Desná dome. The gneiss displays a strong foliation that is isoclinally folded with the development of a younger axial plane foliation. This younger foliation is folded together with a younger, 2 m wide dyke of porphyritic granite representing the youngest igneous phase of the pluton.

The dated sample is a mylonitized granite with relict clasts of plagioclase. It is composed of fine-grained albitic plagioclase, quartz, muscovite, relicts of biotite, chlorite and epidote. The microstructures as well as the mineralogy are consistent with very low grade conditions (chlorite zone) of deformation. Chemically the sample is a quartz monzonite.

The zircons have slightly rounded terminations, a feature common in metamorphic rocks and generally ascribed to metamorphic corrosion (for detailed discussion of this feature, see Kröner *et al.* (1994) and Kröner & Hegner (1998). Three grains were evaporated individually and yielded virtually identical $^{207}Pb/^{206}Pb$ isotopic ratios that combine to a mean age of 684.5 ± 0.9 Ma (Fig. 4b), which we interpret as reflecting the crystallization age of the gneiss protolith. One further grain with slightly more pronounced rounding at the ends yielded a significantly higher age of 864.0 ± 1.6 Ma; we interpret this grain as a xenocryst. As in the previous sample, the presence of xenocrysts suggests involvement of older crustal material in the generation of the host granitoid, but an $\varepsilon_{Nd(t)}$ value of +0.6 and a mean crustal residence age of 1.3 Ga (Hegner & Kröner this volume) make it likely that this crust has mixed with a considerable amount of juvenile material. We therefore suggest a continental-margin type magmatic arc for the generation of the Desná dome. The above granite emplacement age of *c.* 684 Ma is the oldest yet encountered anywhere along the eastern margin of the Bohemian Massif and clearly reflects the Cadomian (Pan-African)-age Brunia terrane outlined earlier.

The porphyritic dyke cross-cutting the gneiss has also been dated (sample CS 38) by both the vapour digestion and evaporation techniques. This rock is a porphyritic granite composed of K-feldspar, plagioclase, quartz, biotite and new phases of chlorite, muscovite and epidote. A weak foliation is underlined by microshear zones developed in the fine-grained matrix. Vapour digestion analysis of three very clear and idiomorphic zircon grains provided discordant results that can be fitted to a chord through the origin (mean square weighted deviates (MSWD) = 0.1) intersecting concordia at 517 ± 12 Ma (Fig. 5a). In view of the euhedral shape of the zircons we interpret this age as representing the time of crystallization of the porphyritic granite. Four further grains with slightly rounded terminations were evaporated and produced significantly older results between 594 and 911 Ma (Fig. 5a, inset); we interpret these grains as xenocrysts derived from the source region of the granite. The $\varepsilon_{Nd(t)}$ value of −4.3 and the Nd mean crustal residence age of 1.5 Ma (Hegner & Kröner this volume) support the conclusion that an appreciable amount of older crust was involved in the generation of the porphyritic granite. The emplacement age of 517 Ma is similar to early Palaeozoic ages elsewhere in the West Sudetes Mts (Oliver *et al.* 1993; Kröner *et al.* 1994, 1997; Kröner & Hegner 1998), and the significance of this will be discussed below.

Silisian domain (eastern margin of Desná dome)

The eastern part of the Desná dome is exposed in a broad belt east of the town of Šumperk (Figs 2 and 3), and two samples were dated from this area. The first is a massive, fine-grained ophthalmitic migmatite (sample CS 22) containing porphyroblasts of plagioclase and collected from the bank of the Desná River east of the village of Kouty nad Desnou (Figs 2 and 3). Chemically the rock is a granite. It has been described as a retrograded pearl gneiss (Fediuková *et al.* 1985) and consists of plagioclase, quartz, biotite and garnet (± chlorite and muscovite). Plagioclase, quartz and biotite constitute the matrix minerals and define the foliation. The rock is locally affected by microshears parallel to an early foliation.

There are two zircon generations in this rock, one idiomorphic and one with slightly rounded terminations. Two grains of the first variety produced a mean $^{207}Pb/^{206}Pb$ age of 506.7 ± 1.7 Ma (Fig. 4c), which we interpret as the time of emplacement of the gneiss protolith. Seven grains of the second variety yielded variable $^{207}Pb/^{206}Pb$ ages ranging between 604 and 1106 Ma, and we interpret these gains as xenocrysts from an older basement. However, it is unlikely that the Desná gneiss was derived from anatexis of this basement, as its $\varepsilon_{Nd(t)}$ value of −0.3 and the mean crustal residence age of 1.2 Ga (Hegner & Kröner this volume) suggest that a fair amount of relatively young material was involved in its generation. The most plausible explanation for the Nd isotopic systematics is mixing of an igneous source, possibly from an underplated juvenile reservoir, with older crust. Like sample CS 38, the emplacement age of 507 Ma is similar to Cambro-Ordovician ages for granitoids along the entire eastern margin of the Bohemian Massif in the West Sudetes Mts of Poland and the Czech Republic (Oliver *et al.* 1993; Kröner *et al.* 1994; Kröner & Hegner 1998).

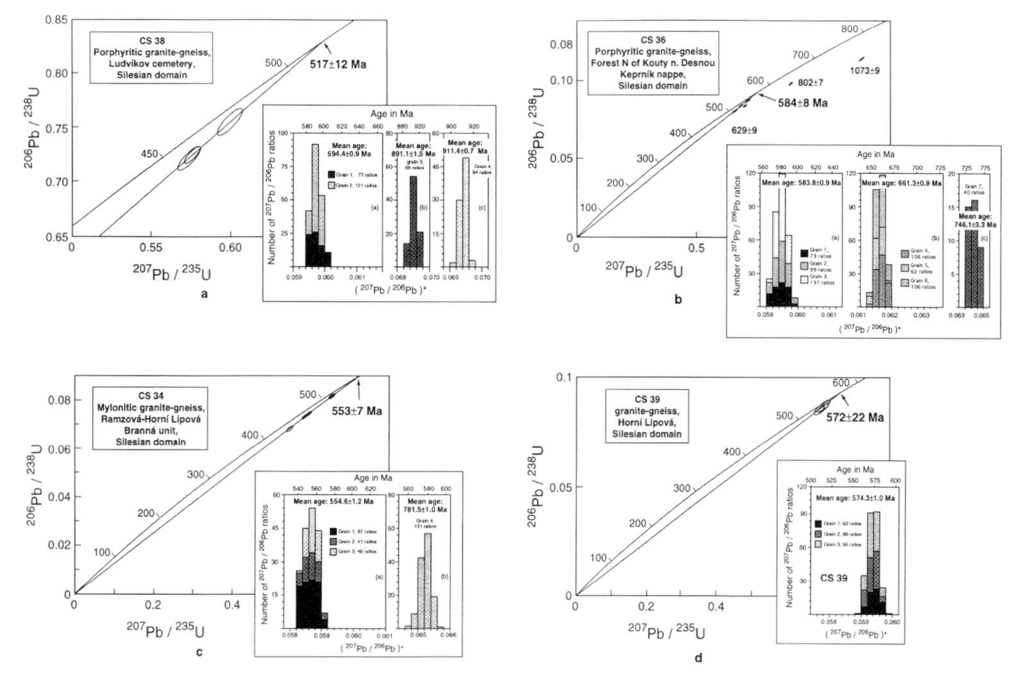

Fig. 5. Concordia diagrams showing analytical data for single zircons from rocks of the Jeseníky Mts, northeastern Czech Republic. (**a**) Porphyritic granite–gneiss sample CS 38. Inset shows histograms with distribution of radiogenic lead isotope ratios derived from evaporation of single zircons from same sample. (**b**) Keprník granite–gneiss sample CS 36, forest north of Kouty nad Desnou, Jeseníky Mts. Inset shows histograms with distribution of radiogenic lead isotope ratios derived from evaporation of single zircons from same sample. (**c**) Mylonitic orthogneiss sample CS 34, hill between Ramzová and Horní Lipová. Inset shows histograms with distribution of radiogenic lead isotopic ratios derived from evaporation of single zircons from same sample. (**d**) Granite–gneiss sample CS 39, Horní Lipová. Inset shows histogram with distribution of radiogenic lead isotope ratios derived from evaporation of single zircons from same sample. Analytical data are available as a Supplementary Publication (see p. 180).

The second sample (CS 48) is a whitish, coarse- to medium-grained, almost pegmatitic granite that cross-cuts the fabric in the Desná migmatitic gneiss and is itself slightly deformed. The sample was collected in the forest between the hills Medvědí and Tupý about 1.5 km SSE of sample CS 22 (Figs 2 and 3). Chemically this is a granite, and the sample is composed of K-feldspar, plagioclase, quartz and muscovite. It is affected by greenschist facies heterogeneous deformation and exhibits relict igneous textures. Four idiomorphic zircon grains were evaporated individually and yielded a combined mean $^{207}Pb/^{206}Pb$ age of 502.3 ± 1.0 Ma (Fig. 4d) that we interpret to reflect the time of magmatic emplacement. The $\varepsilon_{Nd(t)}$ value of -1.0 (Hegner & Kröner this volume) suggests that this rock is not entirely the result of remelting of old crust, but that some young material must have been involved in its generation. This sample is only marginally younger than our Desná gneiss sample CS 22, and this implies that the foliation seen in the Desná gneiss must have formed between 507 and 502 Ma ago, i.e. during an early Palaeozoic event and not during Variscan deformation as was previously thought (Fediuková *et al.* 1985).

Silesian domain (Červenohorské sedlo belt)

We dated only one sample from this domain, and this is a tightly folded, grey, fine-grained ultramylonitic granite (sample CS 35) that belongs to one of the sheets forming the strongly mylonitized Červenohorské sedlo belt, which occurs between the Keprník nappe and the Desná dome (Fig. 3). The sample was collected in a forest near the main Jeseník–Šumperk road about 1 km south of Filipovice (Figs 2 and 3). It is a mylonitic orthogneiss with rare relics of small porphyroclasts of K-feldspar (0.3 mm in size). The fine-grained matrix is composed of plagioclase, K-feldspar, quartz and biotite. The microstructure is consistent with medium-grade conditions (staurolite zone) of mylonitization.

The zircons are idiomorphic and appear completely unaffected by the intense mylonitization. Three grains were evaporated and provided a combined mean $^{207}Pb/^{206}Pb$ age of 612.7 ± 1.0 Ma (Fig. 4e) that we interpret as reflecting the time of the original granite emplacement. The $\varepsilon_{Nd(t)}$ value of 0.4 and the Nd mean crustal residence age of 1.2 Ga (Hegner & Kröner this volume) suggest that this rock is not derived from remelting of very old basement but that late Proterozoic material was involved in its generation.

Silesian domain (Keprník nappe)

We dated two samples from the pre-Variscan crystalline rocks of the Keprník nappe. The first is the porphyritic and well-foliated leucocratic Keprník gneiss (CS 36), which was collected in the forest north of Kouty nad Desnou (Figs 2 and 3). Chemically the gneiss is of granitic composition. Under the microscope the dated sample is a typical orthogneiss, composed of K-feldspar + plagioclase + quartz + muscovite. The foliation is defined by an alignment of biotite and elongation of quartz ribbons. The microstructures are characteristic of amphibolite facies conditions of deformation.

The zircons are slightly rounded at their terminations as a result of a metamorphic overprint. Six grains were analysed by vapour digestion, of which three are near-concordant and well aligned (MSWD = 0.06), providing a concordia intercept age of 584 ± 8 Ma (Fig. 5b). Two further grains have distinctly older $^{207}Pb/^{206}Pb$ minimum ages of 802 ± 7 and 1073 ± 9 Ma, and are interpreted as xenocrysts. The result for a further grain is slightly offset to the right of the discordia line, and the $^{207}Pb/^{206}Pb$ age is 629 ± 9 Ma (Fig. 5b). This grain is probably a mixture of old material and *c.* 584 Ma overgrowth and is not further considered here. Evaporation of three grains yielded a mean $^{207}Pb/^{206}Pb$ age of 583.8 ± 0.9 Ma, indistinguishable from the U–Pb age, whereas three further grains provided identical isotopic ratios that combine to a mean age of 661.3 ± 0.9 Ma. One further grain has an age of 746.1 ± 3.3 Ma (Fig. 5b, inset). We interpret all pre-584 Ma grains as xenocrysts, inherited from older crustal material, and suggest that there was a significant amount of 661 Ma crust in the source region of the gneiss, as suggested by the relative abundance of zircons of that age.

The two dating methods essentially provide the same picture, namely that crystallization of the gneiss precursor occurred at about 584 Ma, whereas the xenocrysts provide evidence of older crust dating back to 1073 Ma. The $\varepsilon_{Nd(t)}$ value is −5.8 and equivalent to a mean crustal residence age of 1.8 Ga (Hegner & Kröner this volume). These values clearly document a significant involvement of older crust in the generation of this rock. In fact, it is possible that the Keprník gneiss is entirely due to remelting of older curst. This is also supported by previous U–Pb zircon dating by Van Breemen *et al.* (1982), whose data for six strongly discordant multigrain fractions scatter about a discordia line with concordia intercept ages of $1422 + 191/-177$ Ma and $547 + 6/-8$ Ma, respectively. The disagreement between our crystallization age of 584 Ma and the lower intercept age of 547 Ma determined by Van Breemen *et al.* (1982) is not surprising, as the latter date is most probably the result of interpolation and Pb loss, and is therefore too low, as already noted by those workers. We conclude that the Keprník gneiss precursor was emplaced some 583 Ma ago and therefore belongs to the Cadomian-age Brunia terrane discussed above.

The second sample (CS 34) comes from the cover of the Keprník nappe, also known as the lower part of the Branná unit (Cháb 1986), and was collected on a steep hill below the railway line from Ramzová to Horní Lipová (Figs 2 and 3). This is a very fine-grained highly sheared pre-Devonian orthogneiss of granitic composition. Under the microscope it shows characteristic features of an ultramylonite composed of K-feldspar, plagioclase, quartz and biotite. The microstructural features are consistent with mylonitization of a granitic protolith under upper greenschist to lower amphibolite facies conditions.

The zircons are similar to those in sample CS 35 and were apparently not affected by mylonitization of the host rock. Four grains were analysed after vapour digestion and provided results that are between 5 and 25% discordant but can be fitted to a discordia line (MSWD = 0.44) with an upper concordia intercept at 553 ± 7 Ma (Fig. 5c). The same result was obtained by evaporation of three grains that yielded a mean $^{207}Pb/^{206}Pb$ age of 554.6 ± 1.2 Ma (Fig. 5c, inset). One grain had a significantly older $^{207}Pb/^{206}Pb$ age of 781.5 ± 1.0 Ma and is probably inherited from the source region of the original granite. Intracrustal melting is the most likely process by which the protolith of sample CS 34 was generated, and this is also supported by an $\varepsilon_{Nd(t)}$ value of −4.7, which corresponds to a mean crustal residence age of 1.6 Ga (Hegner & Kröner this volume). We consider that the age of *c.* 555 Ma approximates the time of emplacement of the original granite and note that the mylonitic deformation has not affected the isotopic systematics of the zircons. Therefore, the time of mylonite formation remains undated. The mylonitized granite, like the previous samples, belongs to the Cadomian-age Brunia terrane that we have now identified within all three tectonic units of the Silesian domain.

Transitional zone (Velké Vrbno unit)

We have dated only one sample from this unit, a strongly foliated orthogneiss (sample CS 39) collected in a forest WNW of Horní Lipová (Figs 2 and 3). Chemically the rock is classified as a tonalite, and consists of plagioclase, quartz, amphibole and chloritized biotite. The dated sample displays a banded structure marked by alternation of polycrystalline aggregates of subsequent grains of plagioclase with layers rich in amphibole and biotite. The microstructures are consistent with high-grade conditions of fabric acquisition.

The zircons have well-rounded terminations, and analysis of three grains after vapour digestion provided almost concordant results with a mean $^{207}Pb/^{206}Pb$ age of 570 ± 6 Ma. If the tightly grouped data points are fitted to a chord through the origin (MSWD = 0.07) the intercept age is identical, but less precise, at 572 ± 22 Ma (Fig. 5d). Three further grains were evaporated and yielded a combined mean $^{207}Pb/^{206}Pb$ age of 574.3 ± 1.0 Ma (Fig. 5d, inset). We consider the age of *c*. 574 Ma as most closely reflecting the time of intrusion of the granitic gneiss precursor. The $\varepsilon_{Nd(t)}$ value of 1.9 and the Nd model age of 1.1 Ga (Hegner & Kröner this volume) indicate a significant involvement of young or juvenile material in the generation of this rock. The age of 574 Ma is in the same range as most ages for the Silesian domain discussed above and suggests at least part of the Velké Vrbno unit to be derived from a Cadomian-age crustal unit.

Transitional zone (Staré Město belt)

The first six samples came from the leptyno-amphibolite sequence in a quarry just west of the village of Hanušovice (Figs 2 and 3) where the amphibolites are intercalated with tonalitic gneisses. The dominant rock type in this quarry is a strongly foliated tonalitic gneiss that locally contains thin layers of granodioritic gneiss. The tonalite is transected by numerous high-temperature ductile extensional shear zones in which *in situ* melt patches occur, indicating local anatectic mobilization of the tonalitic material during a high-grade metamorphic event. The tonalite is concordant with the planar fabric of the enclosing leptyno-amphibolite complex. It is essentially composed of quartz, plagioclase, biotite, amphibole and garnet. Elongate aggregates of amphibole and biotite separate the leucocratic layers consisting of coarse elongate plagioclase and well-annealed quartz that occurs in large aggregates or in the form of drops between the plagioclase grains. The microstructure corresponds to very high temperature conditions of fabric acquisition.

We have dated zircons from the tonalitic gneiss where it is unaffected by later anatexis (samples CS 30, 30A and 43A, taken from different locations in the quarry), from a granodioritic gneiss intercalation (CS 30B) and from the *in situ* melt patches (CS 30C, 43B). Three grains from sample CS 30 and six grains from sample CS 30A were analysed by vapour digestion, and the results are between 10 and 20% discordant, but can be fitted to a chord through the origin (MSWD = 0.38) that intersects concordia at 504 ± 4 Ma (Fig. 6a). Six further grains from sample CS 30 were evaporated individually and produced almost identical $^{207}Pb/^{206}Pb$ ratios that combine to a mean age of 504.1 ± 1.0 Ma (Fig. 4f). Evaporation of four grains from sample CS 30A yielded an identical age of 503.2 ± 1.0 Ma (Fig. 6a, inset), and a mean $^{207}Pb/^{206}Pb$ age of 502.5 ± 1.0 could be calculated from evaporation of three grains from sample CS 43A (Fig. 4g). In summary, both dating methods yielded identical results, and we adopt a weighted mean age of 503.3 ± 1.0 Ma for the intrusion of the tonalitic gneiss precursor.

Zircons from the more leucocratic gneiss variety represented by sample CS 30B are identical to those in the previous samples. Analysis of three grains by vapour digestion produced very consistent results that are about 15% discordant and, if fitted on a chord through the origin, intersect the concordia at 507 ± 19 Ma (diagram not shown). The mean $^{207}Pb/^{206}Pb$ age is 506.7 ± 1.7 Ma. We conclude from this that the leucocratic phase of the Hanušovice gneiss is identical in age to the tonalitic phase. One further grain yielded a much higher $^{207}Pb/^{206}Pb$ minimum age of 818 ± 16 Ma; this again is a xenocryst, probably derived from the source region of the gneiss or taken up during its ascent.

The combined weighted mean age for all zircons evaporated from the tonalitic and granodioritic gneiss samples is 503.3 ± 0.8 Ma. This is in the same range of Cambro-Ordovician granitoid emplacement ages as found in the Orlické Hory Mts to the west and in other parts of the Sudetes Mts (Oliver *et al.* 1993; Kröner *et al.*

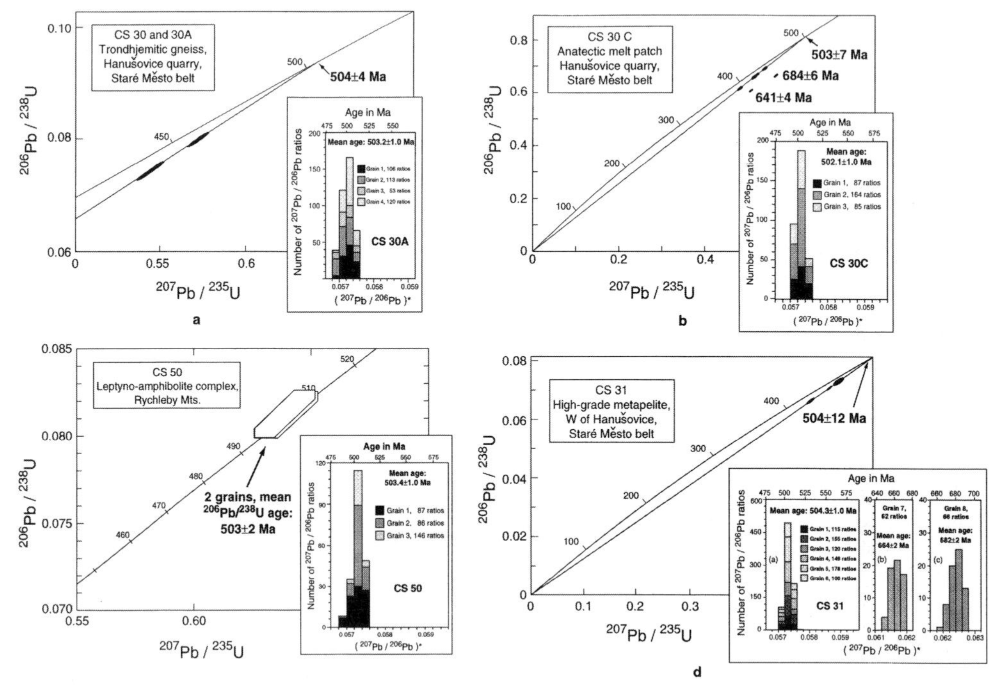

Fig. 6. Concordia diagrams showing analytical data from single zircons from rocks of the Jeseníky Mts, northeastern Czech Republic. (**a**) Eight grains from tonalitic gneiss samples CS 30 and 30A, Hanušovice quarry. Inset shows histogram with distribution of radiogenic lead isotope ratios derived from evaporation of four single zircons from sample CS 30A. (**b**) Six grains from anatectic melt patch sample CS 30C, Hanušovice quarry. Inset shows histogram with distribution of radiogenic lead isotope ratios derived from evaporation of three single zircons from same sample. (**c**) SHRIMP II analyses of two zircons from metarhyodacite sample CS 50, Skorošice, Rychleby Mts. Data boxes for each analysis are defined by standard errors in $^{207}Pb/^{235}U$, $^{206}Pb/^{238}U$ and $^{207}Pb/^{206}Pb$. Inset shows histogram with distribution of radiogenic lead isotope ratios derived from evaporation of three single zircons from same sample. (**d**) Three metamorphic zircons from metapelite sample CS 31, NE of Vlaské, Jeseníky Mts. Inset shows histograms with distribution of radiogenic lead isotope ratios derived from evaporation of single zircons from same sample. Analytical data are available as a Supplementary Publication (see p. 180).

1994). The Nd isotopic systematics of the two gneiss varieties are similar and preclude derivation of these rocks from anatectic melting of much older crust. Tonalite sample CS 30A has an $\varepsilon_{Nd(t)}$ value of +3.4 and a mean crustal residence age of 0.91 Ga, whereas granodioritic gneiss sample CS 30B has an $\varepsilon_{Nd(t)}$ value of 6.4 and an Nd model age of 0.67 Ga (Hegner & Kröner this volume). We suggest that the granitoid pluton from which the Hanušovice gneisses are derived formed from a mixture of juvenile material and minor late Proterozoic crust. Underplating of gabbroic magma at the base of the crust and interaction with the lower crust is a likely scenario for this process.

Anatectic melt patches of trondhjemitic composition within the tonalitic gneiss are coarse grained and, under the microscope, are composed of plagioclase (50–60 vol. %), weakly recrystallized quartz and rare (5 vol. %) biotite. Quartz is strongly annealed and forms isolated pockets in a framework of feldspar. Newly formed zircons in the anatectic melt samples CS 30C and 43B are exclusively short-prismatic and have sharp terminations. Six grains from sample CS 30C were analysed by vapour digestion and display a similar discordance pattern to the analyses discussed above. Four points are well aligned and define a chord intersecting concordia at 503 ± 7 Ma (Fig. 6b), whereas two grains are distinctly older with $^{207}Pb/^{206}Pb$ minimum ages of 641 ± 4 and 684 ± 6 Ma, respectively. Although no older cores could be seen under the optical microscope in these completely transparent grains, we suspect that the above ages reflect core-overgrowth relationships and signify the presence of older crustal material in the source region of the original tonalite from which the old grains must ultimately be derived. Three other idiomorphic grains from sample CS 30C were evaporated and yielded a combined mean age of 502.1 ± 1.0 Ma (Fig. 6b, inset), identical to, but more precise than, the U–Pb age. Zircons from sample CS 43B were only evaporated, and three grains produced a combined mean age of 501.8 ± 1.0 Ma(Fig. 4h). Again, this age is indistinguishable from the two previous determinations. The weighted mean evaporation age of the two melt samples CS 30C and 43B is 501.9 ± 0.6 Ma, slightly younger than the mean age of 503.3 ± 0.8 Ma for the gneissic samples. The $\varepsilon_{Nd(t)}$ value for sample CS 30C is 2.3 and the mean crustal residence age is 0.99 Ga, very similar to that of sample CS 30A (Hegner & Kröner this volume). This is not surprising, as the Nd isotopic systematics are normally not affected by intracrustal melting and differentiation.

The above data for the Hanušovice quarry can be interpreted in two ways. Either the fabric-forming event and subsequent anatectic melting occurred almost immediately after emplacement of the original tonalite, and in this case high-grade metamorphism and extensional deformation was associated with this event or, alternatively, *all* zircons dated from the melt patches are inherited, and the melting and deformational event remains undated. We give preference to the first interpretation, as we have already demonstrated a fabric-forming event at just above 500 Ma in the Desná dome of the Silesian autochthonous domain, and there is also evidence for a *c.* 500 Ma granulite-facies metamorphic event in the Hanušovice area as discussed below.

Another sample from the northern part of the leptyno-amphibolite sequence in the Rychleby Mts is from a felsic layer of this sequence, exposed on a cliff in the forest near Stříbrný hill and near the Polish border (sample CS 50, see Figs 2 and 3). The sequence is tightly isoclinally folded and is also affected by high-grade metamorphism as shown by local melt patches. Chemically the sample CS 50 is a rhyodacite, and the mineral composition is plagioclase, quartz and minor amphibole. The foliation is defined by flattened grains or aggregates of quartz and plagioclase, and by compositional banding marked by alternations of layers rich in amphibole with layers composed only of felsic minerals.

Two zircon grains of sample CS 50 were analysed on SHRIMP II and produced concordant results (Fig. 6c) with a mean $^{206}Pb/^{238}U$ age of 503 ± 2 Ma. Evaporation of three further grains yielded identical results, and the combined mean $^{207}Pb/^{206}Pb$ age is 503.4 ± 1.0 Ma (Fig. 6c, inset). We interpret this to reflect the time of emplacement of the original rhyodacite, and this is time-equivalent with the intrusion of the Hanušovice tonalitic gneiss precursor. This age underlines the regional significance of the *c.* 500 Ma event in the Staré Město belt in particular and in the Lugian domain in general.

The next sample (CS 41) is an anorthositic gneiss collected from a unit of layered gabbroic gneiss exposed in the forest west of the village of Vlaské and some 2 km NW of the Hanušovice quarry (Figs 2 and 3). The exposure from which this sample was taken clearly reveals relict magmatic layering in a ductilely deformed metagabbro, strongly reminiscent of layered gabbros in ophiolite suites. The rock is medium grained and dominantly composed of plagioclase enclosing irregular small grains of quartz.

Zircons in this sample were rare, small (60–80 μm in length) and displayed slight rounding at their ends. To obtain a strong enough signal for Pb isotopes in the mass spectrometer, three grains were evaporated together in two batches and yielded almost identical results that combine to a mean age of 504.9 ± 1.0 Ma (Fig. 7a). This is interpreted to reflect the crystallization age of the anorthosite and, by implication, of the layered gabbro unit. If the interpretation of this unit as part of a dismembered ophiolite suite is correct (Mísař *et al.* 1984), then this is also the formation age of the oceanic crust from which this ophiolitic fragment is derived. The $\varepsilon_{Nd(t)}$ value for the meta-anorthosite is 0.5, and the Nd model age is 1.1 Ga (Hegner & Kröner this volume). These values make it unlikely that the layered gabbro unit is derived from Nd-depleted oceanic crust of mid-ocean ridge basalt (MORB) composition. We rather suspect that the ophiolite suite formed in a marginal basin where the input of Nd-enriched detritus led to contamination of the upper-mantle source region, similar to processes envisaged for the generation of supra-subduction zone ophiolites (Pearce 1987; Zimmer *et al.* 1995).

The last sample (CS 31) is a pelitic, granulite facies, garnetiferous paragneiss occurring tectonically above the tonalitic gneiss and collected some 1.5 km NW of the Hanušovice quarry at a roadcut on a narrow road leading to the village of Žleb (Figs 2 and 3). Under the microscope the rock consists essentially of plagioclase, biotite, quartz, garnet and rutile in the matrix, and plagioclase forms porphyroblasts up to 0.5 cm in size.

There are two types of zircon in this sample, detrital and metamorphic. The latter type has been described from many granulite terrains and is believed to form at temperatures above 800°C near or at the peak of high-grade metamorphism

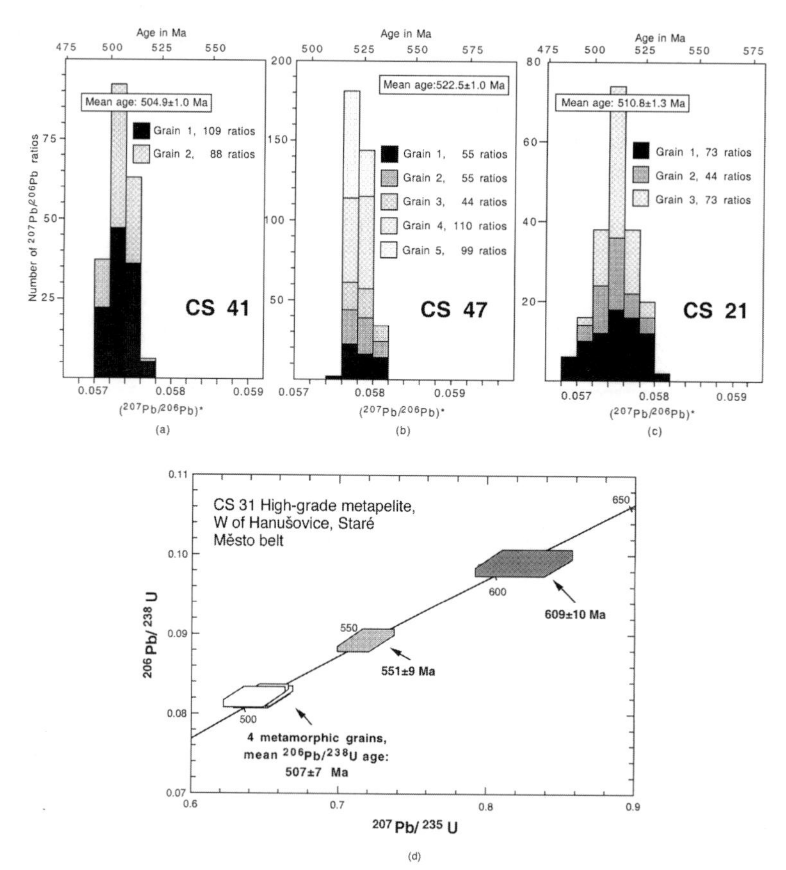

Fig. 7. Histograms (**a**–**c**) and concordia diagram (**d**) showing distribution of radiogenic lead isotope ratios for zircon analyses of samples from Jeseníky Mts, northeastern Czech Republic. (**a**) Spectrum for two fractions of three zircons each from anorthositic gneiss sample CS 41, west of Hanušovice. (**b**) Spectrum for five grains from felsic metadacite sample CS 47, Stronie Formation, Hraničná, Vojtovice, interpreted to reflect age of magmatic emplacement. (**c**) Spectrum for three grains from granitic augen-gneiss sample CS 21, Morava River, east of Králíky, Orlické Hory Mts. (**d**) SHRIMP II analyses of six metamorphic zircons from metapelite sample CS 31, NE of Vlaské, Jeseníky Mts. Data boxes as in Fig. 6c. Analytical data are available as a Supplementary Publication (see p. 180).

(see summary and references given by Kröner & Willner (1998). Cathodoluminescence photographs of these zircons reveal well-developed sector zoning, as is typical of metamorphic grains (Vavra 1990; Kröner & Willner 1998), and no older cores. Three metamorphic grains were analysed by vapour digestion and produced slightly discordant results that can be fitted on a chord through the origin (MSWD = 0.03) intersecting concordia at 504 ± 12 Ma (Fig. 6d). Six further grains were analysed by SHRIMP II and produced concordant results, of which four have a combined mean $^{206}Pb/^{238}U$ age of 509 ± 7 Ma (mean $^{207}Pb/^{206}Pb$ age is 502 ± 19 Ma), whereas the remaining grains have minimum ages of 551 ± 9 and 609 ± 10 Ma, respectively (Fig. 7d). The first four grains are clearly metamorphic, and their age is identical to that of the zircons analysed by vapour digestion, whereas the two additional grains are probably detrital (the difference between round, multifaceted grains and mechanically rounded detrital grains is almost impossible to see on mounted photographs for SHRIMP analysis). Six further multifaceted grains were evaporated and yielded virtually identical $^{207}Pb/^{206}Pb$ ratios that combine to a mean age of 504.3 ± 1.0 Ma (Fig. 6d, inset). Finally, we evaporated two of the detrital grains and obtained minimum ages of 663.5 ± 1.7 and 681.5 ± 1.5 Ma (Fig. 6d, inset). The detrital population analysed would suggest that the metapelite protolith was deposited after 551 Ma.

The $\varepsilon_{Nd(t)}$ value for sample CS 31 is 0.2, assuming depositional age of about 550 Ma, and the corresponding Nd mean crustal residence age is 1.2 Ga (Hegner & Kröner this volume). As in the case of metagreywacke sample CS 49B, the metapelite was not derived from erosion of ancient crust but from predominantly young (Neoproterozoic) crustal material, probably a Cadomian-age magmatic arc terrane. We conclude from the zircon and Nd data that the pelitic sediment from which sample CS 31 originated was deposited at some time after 551 Ma, probably as the erosion product of a Neoproterozoic magmatic arc terrane, and was metamorphosed to granulite-facies conditions at about 504 Ma.

Lugian domain

The first sample (CS 47) was collected just west of the Staré Město belt, in the Rychleby Mts about 500 m west of the village of Vojtovice (Figs 2 and 3). It is a well-layered felsic metavolcanic rocks, chemically a dacite, that is part of the sequence of metasediments. Under the microscope the rock consists essentially of plagioclase, quartz, K-feldspar and minor muscovite and biotite. The foliation is accentuated by elongation of quartz ribbons, plagioclase aggregates and by alignment of fine-grained muscovite and biotite. Five zircon grains were evaporated and produced a combined mean age of 522.5 ± 1.0 Ma (Fig. 7b) that we interpret to reflect the time of extrusion of the metadacite precursor. This age is identical to a single zircon evaporation age for a metadacite from the type locality of the Stronie formation near Romanov in the Orlica–Snieznik dome, Poland (Kröner & Don, unpubl. data), and confirms the correlation of the two units. One further zircon grain from sample CS 47 yielded a much higher $^{207}Pb/^{206}Pb$ age of 1752.5 ± 2.1 Ma and is most probably a xenocryst. The $\varepsilon_{Nd(t)}$ value of -5.0 and the mean crustal residence age of 1.6 Ga (Hegner & Kröner this volume) attest to the crustal origin of this metadacite whose source region is distinctly different from, and older than, the samples in the southern part of the Staré Město belt discussed above.

The Orlica–Snieznik dome is largely made up of granitoid gneisses (Fig. 2), and we report the results for only one sample from this rock type, as additional zircon ages and Nd isotopic characteristics will be published elsewhere (Kröner *et al.* 2000). Sample CS 21 is a well-foliated augen-gneiss of granitic composition, collected on the bank of the Morava River, east of Králíky (Figs 2 and 3). It is similar to many of the granitic gneisses exposed in the Orlické Hory Mts farther west and consists of plagioclase, quartz, biotite, muscovite and subordinate K-feldspar. The foliation is enhanced by the preferred orientation of biotite.

Three zircon grains were evaporated and yielded a combined mean $^{207}Pb/^{206}Pb$ age of 510.8 ± 1.3 Ma (Fig. 6c). The $\varepsilon_{Nd(t)}$ value is -3.9 and corresponds to a mean crustal residence age of 1.5 Ga (Hegner & Kröner this volume). These data suggest generation of the gneiss protolith by intracrustal melting, and the age and Nd isotopic characteristics of this granite gneiss are similar to those of several other granitoid gneisses from the Orlické Hory Mts farther west (Kröner *et al.* 2000), suggesting that this part of the Lugian domain is the result of a major Cambro-Ordovician magmatic event that is evident almost everywhere in the Czech and Polish Sudetes Mts along the eastern margin of the Bohemian Massif (Oliver *et al.* 1993; Kröner & Hegner 1998).

Discussion and conclusions

Our new zircon ages enable us to reconstruct the Neoproterozoic to Cambro-Ordovician evolution of part of the northeastern Bohemian Massif, in particular the contact zone between the Brunia and Lugian domains.

Silesian domain, westernmost part of Neoproterozoic Brunia

The myolitized gneisses of the major constituents of the Silesian domain (the Desná dome and the Keprník nappe) exhibit Neoproterozoic ages of emplacement. Moreover, the ultramylonitic rocks of thrust sheets located between the Desná dome and Keprník nappe (Červenohorské sedlo mylonitic belt) and in the footwall of the Devonian cover of the Keprník nappe (Branná Group) are also of Neoproterozoic age. We therefore interpret the Silesian domain as the northernmost part of the Brunia plate, which was underthrust below the Lugian plate, subsequently imbricated in the form of large crystalline nappes and then exhumed (Fig. 3). One of the major problems is the definition of the western limit of the Brunia plate, which is traditionally placed between the Devonian cover of the Keprník nappe and the Velké Vrbno unit. Our study has shown that at least part of the Velké Vrbno high-grade rocks are also derived from the Brunia plate. This, together with the presence of undoubtedly Variscan Barrovian metamorphism (kyanite schists) and medium-temperature (MT) eclogites, indicates that the Velké Vrbno unit represents the most deeply buried part of the Brunia microplate. Therefore, the Ramzová overthrust of Suess (1926) can no longer be considered as the easternmost limit of the Brunia plate; the boundary must be placed farther west, below the Staré Město belt.

Silesian domain; Cambro-Ordovician event

Our zircon ages document Cambro-Ordovician magmatic and metamorphic activity affecting the Brunia microplate. The thermal effect of this event was weak and is documented only by granitic (CS 38) and pegmatitic dykes (CS 48). Metamorphism locally reached the stage of partial melting, e.g. sample CS 22, but the precise PT conditions, as a result of the strong Variscan overprinting (Fediuková *et al.* 1985), cannot be determined. The Nd isotopic systematics for two of the above samples (CS 22 and CS 48) suggest participation of primitive material in the generation of these rocks. The Cambro-Ordovician dykes and metamorphism in the Brunia plate represent the easternmost extremity of this thermal even in central Europe.

Cambro-Ordovician event in the Staré Město belt

Felsic volcanic rocks of the leptyno-amphibolite sequence and associated tonalite–trondhjemite intrusive rocks at the base of the Staré Město belt yielded Cambro-Ordovician ages of emplacement (502–504 Ma). During high-temperature solidification and associated extensional deformation, these rocks have been metamorphosed under granulite facies conditions (8–10 kbar, 800–850°C, Štipská *et al.* 2000 that locally led to remelting. Zircons from melt patches in extensional lock-up shear bands as well as from an anorthositic layer in highly differentiated gabbro also yielded Cambro-Ordovician ages of emplacement. Finally, migmatitic, pelitic sediments, which are associated with these magmatic rocks and reached medium-pressure granulite facies metamorphism (7–9 kbar, 800–850°C, Štipská *et al.* 2000), contain metamorphic zircons, also of Cambro-Ordovician age.

We interpret the association of large volumes of gabbroic rocks (Fig. 2), highly differentiated dykes and bimodal volcanism (leptyno-amphibolite unit) as signifying an important rifting and underplating event at about 500 Ma ago. The geochemistry of these early Palaeozoic amphibolites in the Staré Město belt indicates MORB affinities, although the nature of crustal contamination suggests that the basaltic precursor was emplaced in a rifted ensialic environment (Floyd *et al.* 1996). Moreover, MP–HT metamorphism, also of Cambro-Ordovician age, and associated extensional tectonics suggest exceptionally high heat flow during thinning of the continental crust. Melting of crustal material was associated with this event, as documented by melt patches and the intrusion of granodioritic to tonalitic rocks that are probably products of melting of juvenile mafic (gabbroic?) material. The exceptionally hot geotherm, extensional tectonics and bimodal volcanism are all interpreted in terms of crustal thinning. The effect of gabbro underplating probably resulted in dehydration melting of lower mafic crust and the production of tonalitic magmas and associated granulite facies metamorphism. The geological environment and the chemistry of the Staré Město amphibolites (Floyd *et al.* 1996) as well as their MP–HT metamorphism indicate that crustal attenuation was insufficient to develop proper oceanic crust.

Cambro-Ordovician rifting is also documented in the Kaczawa Mountains (GK in Fig. 8) of the Polish West Sudetes Mts (Furnes *et al.* 1989), where diabase sills, pillow lavas, felsic volcanogenic calc-alkaline rocks and subsequent metabasalt flows of tholeiitic composition are attributed to this rifting event. Similar mafic volcanism (505 ± 5 Ma, Oliver *et al.* 1993) and gabbroic magmatism (494 ± 2 Ma, Oliver *et al.* 1993) was dated from the Rudawy–Janowicke region (Fig. 1) and is interpreted as ocean-floor magmatism and island-arc formation (Oliver *et al.* 1993). In addition, geochemical affinities of a mafic blueschist protolith to ocean-floor and ocean-island basalts in the Rýchory Mts (Maluski & Patočka 1997) suggest formation of oceanic crust in Ordovician time. However, no relics of Cambro-Ordovician granulite facies metamorphism were found in this region. In contrast to the Kaczawa Mts, the Rudawy–Janowicke region and southerly Rýchory Mts, where an oceanic stage had been reached, continental separation has not take place in the Staré Město belt.

Cambro-Ordovician event in the Orlica–Snieznỉk dome

Granitoid rocks in the Orlica–Snieznỉk dome have previously been dated between 487 and 505 Ma (Van Breemen *et al.* 1982; Liew & Hofmann 1988; Oliver *et al.* 1993). This age range is now extended to *c.* 522 Ma and suggests

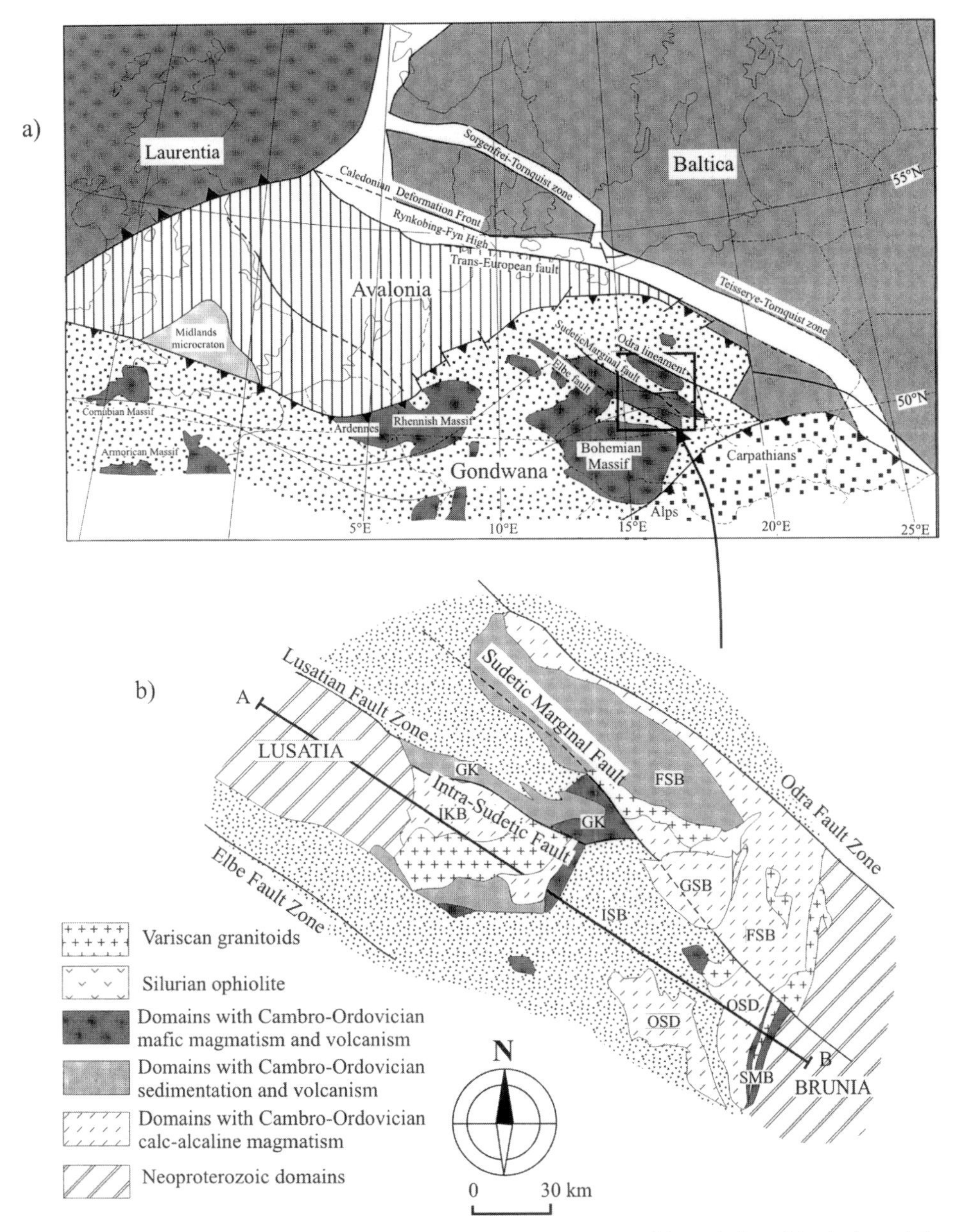

Fig. 8. (**a**) Position of western Sudetes with respect to recent geometry and boundaries of Avalonia and the Teisseyre–Tornquist zone (after Pharaoh *et al.* 1997). (**b**) Simplified geological map of the western Sudetes showing regional distribution of Neoproterozoic and Cambro-Ordovician domains (simplified after Żelaźniewicz 1997). FSB, Fore-Sudetic Block; GK, Gory Kaczawskie; GSB, Góry Sowie Block; IKB, Izera–Karkonosze Block; ISB, Intra-Sudetic Basin; OSD, Orlica–Sniezník dome; SMB, Staré Město belt.

that the dome constitutes a composite granitoid pluton. Nd isotopic characteristics of the dated samples (CS 21 and CS 47) suggest derivation from pre-existing continental crust (Hegner & Kröner this volume). The Orlica–Snieznik dome is a predominantly continental domain consisting of pre-Cadomian crust (Nd-model ages 1460–1660 Ma) and large volumes of Cambro-Ordovician granitoids of calc-alkaline composition similar to orogenic granite suites, which support a subduction-related origin (Kröner *et al.* 2000). This Cambro-Ordovician granitoid and gneissic complex is situated between two NE–SW-trending rifted domains (Kaczawa–Rudawy–Janowicke and Staré Město) that have been amalgamated during the Variscan Orogeny (Schulmann & Gayer 2000).

Cambro-Ordovician tectonics in the Sudetic region

The geometry and sedimentary sequences overlying Neoproterozoic and Cambro-Ordovician domains of the Sudetes Mts were discussed by Żelaźniewicz (1997). According to Cymerman *et al.* (1997), the lithostratigraphic terranes in the Sudetes exhibit a symmetrical distribution with a central region of Cambro-Ordovician basinal–oceanic rocks bordered by sialic domains with Cambro-Ordovician magmatic activity from the west (Izera and South Karkonosze region) and east (Orlica–Snieznik dome) (Fig. 8). This large domain of Cambro-Ordovician tectonothermal activity is bordered by Neoproterozoic Gondwana-derived terranes to the west (Lusatian domain) and east (Brunia domain). Our data demonstrate that, in addition to this simple scheme, Cambro-Ordovician crustal attenuation also occurred between the Orlica–Snieznik domain and the Brunia domain, but without the development of proper oceanic crust. Therefore, the major tectonic pattern of the Sudetes is probably represented by an alternation of rifted domains with sialic regions during Gondwana fragmentation at the Cambro-Ordovician boundary (Figs 8 and 9).

In modern coordinates, the fragmentation of northern Gondwana occurred in a NW–SE direction so that NE–SW-trending belts of rifted and sialic domains developed. These units were displaced along the NW–SE Intrasudetic fault system and are separated from the more southerly units by the Elbe zone and from more northerly units by the Odra lineament (Fig. 8). The age of these major NW–SE transcurrent fault systems is disputed (Aleksandrowski *et al.* 1997; Bula *et al.* 1997), but it is likely that they acted as transfer faults during the rifting period and were later reactivated.

Ordovician fragmentation and crustal thinning of the northern margin of Gondwana may have originated through several mechanisms. The Cambro-Ordovician rifted regions of the western Sudetes (e.g. the Rudawy–Janowicke and Staré Město domains) differ significantly from each other in terms of the degree of crustal thinning. The Staré Město belt experienced gabbroic underplating, granulite facies metamorphism and structures related to crustal extension. In contrast, the Rudawy–Janowicke domain exhibits extensive rifting leading to the development of oceanic crust (see Fig. 9).

The geometry of the Staré Město belt may therefore be interpreted in terms of rifting along the northern Gondwana passive margin with the formation of a narrow zone of crustal attenuation (several tens of kilometres), located several tens to a maximum of 100 km away from the major site of break-up (Rudawy–Janowicke terrane). Magmatic activity along the passive margin was relatively weak in the western part of Brunia, where only some granitic dykes were emplaced in extensional gashes (the Desná dome). Farther west, thinning of the crust in the Staré Město region was likely to be more pronounced, and the minimum crustal thickness was 24–28 km (equivalent to a pressure of 7–8 kbar).

Limited crustal thinning may have been responsible for a substantial increase in lithospheric strength (Bertotti *et al.* 1997), causing final break-up of the crust to have occurred in a different region. With analogy to the southern Alpine rifted margin, for example, we suggest that the Staré Město belt represents only a zone of local lithospheric thinning along a stable Neoproterozoic continental margin of Gondwana, whereas break-up occurred farther west in what is now the Rudawy–Janowicke domain.

However, early Palaeozoic rifting in the Rudawy–Janowicke and Staré Město zones seems difficult to reconcile with Cambro-Ordovician calc-alkaline magmatism in the Orlica–Snieznik dome, the Góry Sowie Mts and the Izera Mts, where a subduction environment has been postulated (Oliver *et al.* 1993; Kröner *et al.* 1994, 2000; Kröner & Hegner 1998). A geotectonic position similar to that of the present Indonesian Archipelago could be suggested, implying subduction of oceanic crust below continental domains associated with pulling and fragmentation of hanging-wall continental crust. This type of tectonic setting could explain both successive fragmentation of continental crust and the development of transfer fault systems, as well as crustal thinning associated with back-arc-related bimodal volcanism and subduction-related magmatism as postulated for the Lugian domain.

Analogous models of back-arc extension (Faccenna *et al.* 1996) use the concept of retreating subduction as a driving mechanism of successive break-up of the upper plate. In such models the zone of early thinning is progressively migrating away from the retreating subduction zone. Consequently, the zone of maximum

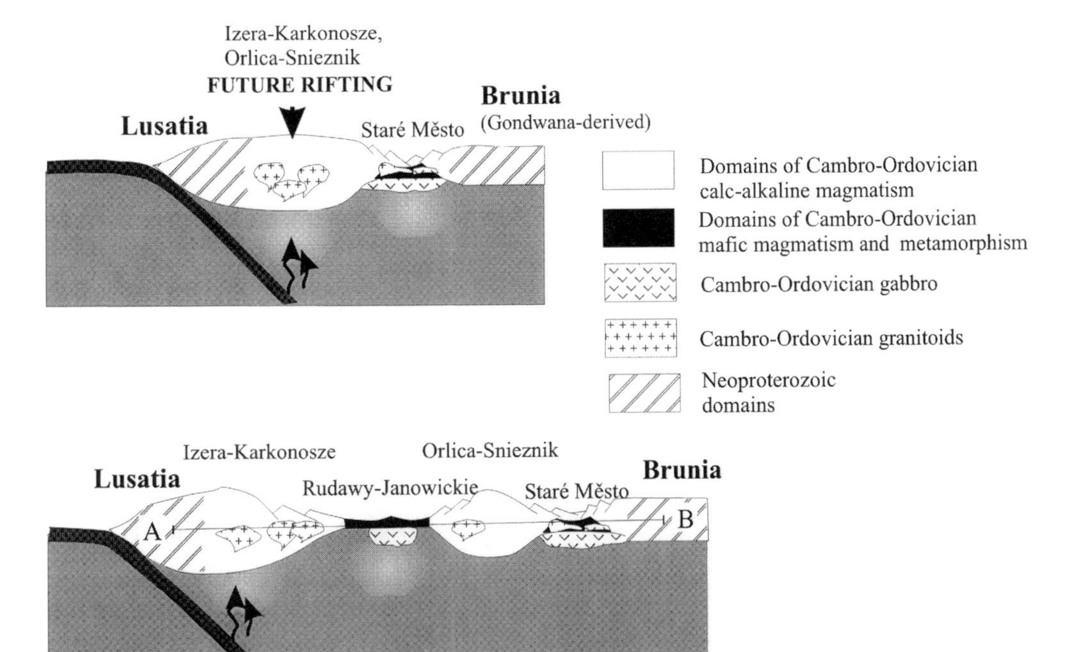

Fig. 9. Tectonic model of early Palaeozoic sequential rifting and related magmatism in the western Sudetes. The upper sketch shows early back-arc rifting in the Staré Město zone and supposed subduction-related calc-alkaline magmatism in sialic domains. Further trench retreat produced further rifting of softened sialic domains and development of oceanic crust west of the Staré Město rift.

crustal thinning is also migrating towards the trench. Therefore, the upper, extended, plate is made up of several rift zones separating large-scale boudins of continental crust.

A possible model for the development of a narrow intracontinetal rift zone in the eastern Bohemian Massif could be seen in the progressive closure of the Tornquist Ocean south of Avalonia (Tait *et al.* 1994, 1995; Torsvik *et al.* 1996), with a subduction zone dipping below the northern margin of Gondwana. This caused calc-alkaline magmatism, supra-subduction zone rifting and fragmentation on the overriding Gondwana plate (Fig. 9).

Relics of an early Palaeozoic rift, the Raabs sea, is preserved in a similar structural position as the Staré Město belt in the southern, Austrian, part of the eastern margin of the Bohemian Massif, as noted by Finger & Steyrer (1995). Those workers suggested that the Raabs sea was an immature intra-continental rift that opened through a process of continental back-arc rifting. In their model, the Staré Město belt, the Raabs unit and the Rudawy–Janowicke ophiolite represent relics of narrow oceans, which became curved after closure as a result of forceful indentation of the 'Moldanubian terrane' and clockwise rotational oroclinal bending around the eastern flank of the Bohemian Massif in early Carboniferous time (Tait *et al.* 1996). However, this model is not supported by sufficient data to rule out alternative opinions.

Financial support to A.K. by the Deutsche Forschungsgemeinschaft (DFG) within the Priority Programme 'Orogenic Processes' (Grant Kr 590/35) is gratefully acknowledged. A.K. also appreciates mass spectrometer analytical facilities in the Max-Planck-Institut für Chemie in Mainz. Some zircon analyses were carried out on the Sensitive High Resolution Ion Microprobe mass spectrometer (SHRIMP II) operated by a consortium consisting of Curtin University of Technology, the Geological Survey of Western Australia and the University of Western Australia with the support of the Australian Research Council. We appreciate the advice of A. A. Nemchin, A. Kennedy and D. Nelson during SHRIMP analysis and data reduction. This research was also funded by Grant 205/96/0270 of the Czech National Science Foundation to Pavla Štípská. We are grateful to Jan Cháb for making available his unpublished geological map of the Jeseníky Mts (Fig. 2) and for assistance during sample collection. F. Finger, V. Janoušek and G. Rogers considerably improved the manuscript through their comments. This is a contribution to EUROPROBE and the German Priority Programme 'Orogenic Processes'.

References

ALEKSANDROWSKI, P., KRYZA, R., MAZUR, S. & ZABA, J. 1997. Kinematic data on major Variscan strike-slip faults and shear zones in the Polish Sudetes, northeast Bohemian Massif. *Geological Magazine*, **134**, 727–739.

BERTOTTI, G., TER VOORDE, M., CLOETINGH, S. & PICOTTI, V. 1997. Thermomechanical evolution of the South Alpine rifted margin (North Italy): constraints on the strength of passive continental margins. *Earth and Planetary Science Letters*, **146**, 181–193.

BORKOWSKA, M., CHOUKROUNE, P., HAMEURT, J. & MARTINEAU, F. 1990. A geochemical investigation of the age, significance and structural evolution of the Caledonian–Variscan granite–gneisses of the Snieżnik metamorphic area, central Sudetes. *Geologica Sudetica*, **25**, 1–27.

BRUECKNER, H. K., MEDARIS, L. G., JR. & BAKUN-CZUBAROW, N. 1991. Nd and Sr age and isotope patterns from Variscan eclogites of the eastern Bohemian Massif. *Neues Jahrbuch für Mineralogie, Abhandlungen*, **163**, 169–196.

BULA, Z., JACHOWICZ, M. & ZABA, J. 1997. Principal characteristics of the Upper Silesian Block and Malopolska Block border zone (southern Poland). *Geological Magazine*, **134**, 669–677.

CHÁB, J. 1986. [Structure of the Moravo-Silesian branch of the European Upper Palaeozoic orogen (a working hypothesis)]. *Věstník Ústředního Ústavu Geologického*, **61**, 113–120 [in Czech].

CHLUPÁČ, I. 1994. Facies and biogeographic relationships in Devonian of the Bohemian Massif. *Courier Forschungsinstitut Senckenberg*, **169**, 299–317.

CYMERMANN, Z., PIASECKI, M. & SESTON, M. 1997. Terranes and terrane boundaries in the Sudetes, NE Bohemian Massif. *Geological Magazine*, **134**, 717–725.

DUDEK, A. 1980. The crystalline basement block of the Outer Carpathians in Moravia. *Rozpravy Československé Akademie Ved*, **90**, 1–85.

FACCENNA, C., DAVY, P., BRUN, J. P., FUNICIELLO, R., GIARDINI, D., MATTEI, M. & NALPAS, N. 1996. The dynamics of back-arc extension: an experimental approach to the opening of the Tyrrhenian Sea. *Geophysical Journal International*, **126**, 781–795.

FEDIUKOVÁ, E., FIŠERA, M., Cháb, J., Kopečný, V. & RYBKA, R. 1985. Garnets of the pre-Devonian rocks in the eastern part of the Hruby Jeseník Mts. (North Moravia, Czechoslovakia). *Acta Universitatis Carolinae Geologica*, **3**, 197–234.

FINGER, F. & STEYRER, H. P. 1995. A tectonic model for the eastern Variscides: indication from a chemical study of amphibolites in the south-eastern Bohemian Massif. *Geologica Carpathic*, **46**, 137–150.

——, HANŽL, P., PIN, C., VAN QUADT, A. & STEYRER, H. P. 2000. The Brunovistulian: Avalonian Precambrian sequence at the eastern end of the central European Variscides? *This volume*.

FIŠERA, M., SOUČEK, J. & NOVOTNÝ, P. 1986. [Blastomylonites of the Orlík nappes group, Hrubý Jesení Mts]. *Věstník Ústředního Ústavu Geologického*, **61**, 321–331 [in Czech].

FLOYD, P. A., WINCHESTER, J. A., CIESIELCZUK, J., LEWANDOWSKA, A., SZCZEPANSKI, J. & TURNIAK, K. 1996. Geochemistry of early Palaeozoic amphibolites from the Orlica–Snieznik dome, Bohemian Massif: petrogenesis and palaeotectonic aspects. *Geologische Rundschau*, **85**, 225–238.

FRANKE, W. 1989. Variscan plate tectonics in Central Europe—current ideas and open questions. *Tectonophysics*, **169**, 221–228.

—— 2000. The mid-European segment of the Variscides: tectono-stratigraphic units, terrane boundaries and plate evolution. *This volume*.

FRIEDL, G., MCNAUGHTON, N., FLETCHER, I. R. & FINGER, F. 1998. New SHRIMP-zircon ages for orthogneisses from the south-eastern part of the Bohemian Massif (Lower Austria). *Act Universitatis Carolinae—Geologica*, **42**, 251–252.

FRITZ, H., DALLYMEYER, R. D., NEUBAUER, F. & URBAN, M. 1996. Thick-skinned versus thin skinned thrusting: rheologically controlled thrust propagation in the Variscan collisional belt in the southeastern Bohemian Massif (Czech Republic–Austria). *Tectonics*, **15**, 1389–1413.

FURNES, H., KRYZA, R. & MUSZYNSKI, A. 1989. Geology and Geochemistry of Early Palaeozoic Volcanics of the Swierzawa Unit, Kaczawa Mts. W. Sudetes, Poland. *Nues Jahrbuch Für Geologie und Paläontologie, Montatshefte*, **3**, 136–154.

HEGNER, E. & KRÖNER, A. 2000. Review of ND isotopic data and xenocrystic and detrital zircon ages from the pre-Variscan basement in the eastern Bohemian Massif: speculations on palinspastic reconstructions. *This Volume*.

JAECKEL, P., KRÖNER, A., KAMO, S. L., BRANDL, G. & WENDT, J. I. 1997. Late Archaean to early Proterozoic granitoid magmatism and high-grade metamorphism in the central Limpopo belt, South Africa. *Journal of the Geological Society, London*, **154**, 25–44.

KOBER, B. 1986. Whole-grain evaporation for ^{207}Pb/^{206}Pb-age-investigations on single zircons using a double-filament thermal ion source. *Contributions to Mineralogy and Petrology*, **93**, 482–490.

—— 1987. Single-zircon evaporation combined with Pb^+ emitter-bedding for ^{207}Pb/^{206}Pb-age investigations using thermal ion mass spectrometry, and implications to zirconology. *Contributions to Mineralogy and Petrology*, **96**, 63–71.

KÖLBL, L. 1929. Die alpine Tektonik des Altvatergebirges. *Mitteilungen der Geologischen Gesellschaft*, **22**, 65–124.

KRÖNER, A. & HEGNER, E. 1998. Geochemistry, single zircon ages and Sm–Nd systematics of granitoid rocks from the Góry Sowie (Owl) Mts, Polish West Sudetes: evidence for early Palaeozoic arc-related plutonism. *Journal of the Geological Society, London*, **155**, 711–724.

—— & TODT, W. 1988. Single zircon dating constraining the maximum age of the Barberton greenstone belt, souther Africa. *Journal of Geophysical Research*, **93**, 15329–15337.

—— & WILLNER, A. P. 1998. Time of formation and peak of Variscan HP–HT metamorphism of quartz–feldspar rocks in the central Erzgebirge, Saxony, Germany. *Contributions Mineralogy and Petrology*, **132**, 1–20.

——, HEGNER, E. & JAECKEL, P. 1997. Cambrian to Ordovician granitoid orthogneisses in the Polish and Czech West Studies Mts and their geodynamic significance. *Terra Nostra*, **97**(11), 67–68.

——, JAECKEL, P., HEGNER, E. & OPLETAL, M. 2000. Single zircon ages and whole-rock Nd isotopic systematics of granitoid gneisses from the Czech Sudetes (Jizerské hory, Krkonoše and Orlice–Snieżník Dome). *Geologische Rundschau*, in press.

——, —— & OPLETAL, M. 1994. Pb–Pb and U–Pb zircon ages for orthogneisses from eastern Bohemia: further evidence for a major Cambro-Ordovician magmatic event. *Journal of the Czech Geological Society*, **39**, 61.

——, ——, REISCHMANN, T. & KRÖNER, U. 1998. Further evidence for an early Carboniferous ($\sim$340 Ma) age of high grade metamorphism in the Saxonian Granulite Complex. *Geologische Rundschau*, **86**, 751–766.

——, WINDLEY, B. F., JAECKEL, P., BREWER, T. S. & RAZAKAMANANA, T. 1999. New zircon ages and regional significance for the evolution of the Pan-African orogen in Madagascar. *Journal of the Geological Society, London*, **156**, 1125–1135.

LASKOWSKI, N. & KRÖNER, A. 1985. Geochemical characteristics of Archaean and late Proterozoic to Palaeozoic fine-grained sediments from southern Africa and significance for the evolution of the continental crust. *Geologische Rundschau*, **74**, 1–9.

LIEW, T. G. & HOFMANN, A. W. 1988. Precambrian crustal components, plutonic associations, plate environment of the Hercynian Fold Belt of Central Europe: indications from a Nd and Sr isotopic study. *Contributions to Mineralogy and Petrology*, **98**, 129–138.

LUDWIG, K. R. 1994. *ISOPLOT, a plotting and regression program for radiogenic-isotope data, version 2.75*. US Geological Survey, Open-File Report **91–45**.

MALUSKI, H. & PATOČKA, F. 1997. Geochemistry and ^{40}Ar–^{39}Ar geochronology of the mafic metavolcanic rocks from the Rychory Mountains complex (west Sudetes, Bohemian Massif): palaeotectonic significance. *Geological Magazine*, **134**, 703–716.

——, RAILICH, P. & SOUCEK, J. 1995. Pre-Variscan, Variscan and early Alpine thermo-tectonic history of the northeastern Bohemian Massif: a $^{40}Ar/^{39}Ar$ study. *Geologische Rundschau*, **84**, 345–358.

MATTE, Ph., MALUSKI, H., RAJLICH, P. & FRANKE, W. 1990. Terrane boundaries in the Bohemian Massif: results of large-scale Variscan shearing. *Tectonophysics*, **177**, 151–170.

NELSON, D. R. 1997. *Compilation of SHRIMP U–Pb zircon geochronology data, 1996*. Geological Survey of Western Australia, Recond. **1997/2**.

NOVOTNÝ, M. & ŠTELCL, J. 1961. *Petrograpy of the northeastern part of the Hrub Jeseník*. Státni Pedagogické Nakladatelství, Praha, Czech Republic.

O'CONNOR, J. T. 1965. *A classification for quartz-rich igneous rocks based on feldspar ratio*. US Geological Survey, Professional Papers, **525B**, 79–84.

OLIVER, G. J. H., CORFU, F. & KROUGH, T. E. 1993. U–Pb ages from SW Poland: evidence for a Caledonian suture zone between Baltica and Gondwana. *Journal of the Geological Society, London* **150**, 355–369.

PARRY, M., ŠTÍPSKÁ, P., SCHULMANN, K., HROUDA, F., JEŽEK, J. & KRÖNER, A. 1997. Tonalite sill emplacement at an oblique plate boundary: northeastern margin of the Bohemian Massif. *Tectonophysics*, **280**, 61–81.

PEARCE, J. A. 1987. An expert system for the tectonic characterization of ancient volcanic rocks. *Journal of Volcanology and Geothermal Research*, **32**, 51–65.

PHARAOH, T. C., ENGLAND, R. W., VERNIERS, J. & ŻELAŹNIEWICZ, A. 197. Introduction: geological and geophysical studies in the Trans-European suture zone. *Geological Magazine*, **5**, 585–590.

SCHARBERT, S. & BATÍK, S. 1980. The age of the Thaya (Dyje) pluton. *Verhandlungen der Geologischen Bundesanstalt*, **1980**, 325–331.

SCHULMANN, K. & GAYER, R. 2000. A model for an obliquely developed continental accretionary wedge: NE Bohemian massif. *Journal of the Geological Society, London*, **157**, 401–406.

STACEY, J. S. & KRAMERS, J. D. 1975. Approximation of terrestrial lead isotope evolution by a two-stage model. *Earth and Planetary Science Letters*, **26**, 207–221.

STELTENPOHL, M. G., CYMERMAN, Z., KROGH, T. E. & KUNK, M. J. 1993. Exhumation of eclogitized continental basement during Variscan lithospheric delamination and gravitational collapse, Sudety Mountains, Poland. *Geology*, **21**, 1111–1114.

ŠTÍPSKÁ, P. & SCHULMANN, K. 1995. Inverted metamorphic zonation in a basement-derived nappe sequence, eastern margin of the Bohemian Massif. *Geological Journal*, **30**, 385–413.

——, ——, THOMPSON, A. B. & JEŽEK, J. 2000. Thermomechanical role of a Cambro-Ordovician palaeorift during the Variscan collision: the NE margin of the Bohemian Massif. *Tectonophysics*, in press.

SUESS, F. E. 1912. Die moravischen Fenster und ihre Beziehung zum Grundgebirge des Hohen Gesenkes. *Denkschriften der kaiserlichen Akademie der Wissenschaften*, **88**, 541–631.

—— 1926. *Intrusionstektonik und Wandertektonik im variszischen Grundgebirge*. Borntraeger, Berlin.

TAIT, J., BACHTADSE, V. & SOFFEL, H. C. 1994. Silurian palaeogeography of Armorica: new palaeomagnetic data from central Bohemia. *Journal of Geophysical Research*, **99**, 2897–2907.

——, —— & —— 1995. Upper Ordovician palaeogeography of the Bohemian Massif: implications

for Armorica. *Geophysical Journal International*, **122**, 211–218.
——, —— & —— 1996. Eastern Variscan fold belt: Paleomagnetic evidence for oroclinical bending. *Geology*, **24**, 871–874.
TORSVIK, T. H., SMETHURST, M. A., MEERT, J. G. *et al.* 1996. Continental break-up and collision in the Neoproterozoic and Palaeozoic—a tale of Baltica and Laurentia. *Earth-Science Reviews*, **40**, 229–258.
TURNIAK, K., MAZUR, S. & WYSOCZAŃSKI, R. 2000. SHRIMP zircon geochronology and geochemistry of the Orlica–Snieznik gneisses (Sudetes, SW Poland) and implications for the evolution of the Variscides in East–Central Europe. *Geodynamica Acta*, in press.
VAN BREEMEN, O., AFTALION, M., BOWES, D. R., DUDEK, A., MÍSAŘ, Z., POVONDRA, P. & VRÁNA, S. 1982. Geochronological studies of the Bohemian Massif, Czechoslovakia, and their significance in the evolution of Central Europe. *Transactions of the Royal Society*, **73**, 89–108.
VAVRA, G. 1990. On the kinematics of zircon growth and its petrogenetic significance: a cathodoluminescence study. *Contributions to Mineralogy and Petrology*, **106**, 90–99.
YORK, D. 1969. Least squares fitting of a straight line with correlated errors. *Earth and Planetary Science Letters*, **5**, 320–324.
Żelaźniewicz, A. 1997. The Sudetes as a Palaeozoic orogen in central Europe. *Geological Magazine*, **134**, 691–702.
ZIMER, M., KRÖNER, A., JOCHUM, K. P., REISCHMANN, T. & TODT, W. 1995. The Gabal Gerf complex: a Precambrian N-MORB ophiolite in the Nubian Shield, NE Africa. *Chemical Geology*, **123**, 29–51.

Appendix

Analytical methods

Major oxides and trace elements were determined by XRF spectrometry on whole-rock fused glass discs and powder pellets, respectively, at the University of Mainz using a Philips PW 1404 XRF spectrometer and procedures as outlined by Laskowski & Kröner (1985). The results are presented as a Supplementary Publication (see p. 180). Heavy minerals were separated from whole-rock samples weighing about 5 kg each by standard procedures. Zircons were then handpicked for single grain analysis using the HF vapour digestion and evaporation techniques in Mainz and SHRIMP II in Perth.

Single zircon evaporation. Our laboratory procedures follow Kober (1986, 1987) with slight modifications (Kröner & Hegner 1988; Kröner & Todt 1988). Isotopic measurements were carried out on a Finnigan-MAT 261 mass spectrometer at the Max-Planck-Institute für Chemie in Mainz. Common lead was corrected, where necessary, using the model of Stacey & Kramers (1975).

Repeated analysis of single grains of an internal zircon standard during this study yielded a value of 0.126635 ± 0.000062 (2σ error of the population), and this error is considered the best estimate for the reproducibility of our evaporation data (Kröner *et al.* 1999). The calculated $^{207}Pb/^{206}Pb$ ratios and their 2σ (mean) errors are presented as a Supplementary Publication (see p. 180), together with a summary of the zircon morphology. The $^{207}Pb/^{206}Pb$ spectra are shown as histograms to permit visual assessment of the data distribution from which the ages are derived. As the evaporation technique provides only Pb isotopic ratios, all $^{207}Pb/^{206}Pb$ ages determined by this method are necessarily minimum ages. Kröner & Hegner (1998) discussed this problem and provided reliability criteria for evaporation analyses.

Vapour-digestion U–Pb analysis. Our laboratory procedures and instrumental conditions have been described by Jaeckel *et al.* (1997) and Kröner *et al.* (1999). Blank values were 3 pg for Pb and 10 pg for U. The Pb composition of the blank was $^{206}Pb/^{204}Pb = 17.78$, $^{207}Pb/^{204}Pb = 14.97$, $^{208}Pb/^{204}Pb = 36.57$. Isotopic measurements were performed on a Finnigan-MAT 261 mass spectrometer at the Max Planck-Institut für Chemie in Mainz. Common-Pb correction followed the two-stage model of Stacey & Kramers (1975). All errors are given at the $2\sigma_{ext.}$ level. The ages of 2σ errors of intercepts of the best-fit with concordia were determined using the regression calculation of York (1969). The analytical data are listed as a Supplementary Publication (see p. 180).

SHRIMP II procedure. Zircons were handpicked and mounted in epoxy resin together with chips of the Perth Consortium standard CZ3. The mount was then polished, cleaned, etched in 48% HF, photographed and then repolished as detailed by Kröner *et al.* (1998). Isotopic analyses were performed on the Perth Consortium SHRIMP II, and instrumental conditions have been summarized by Kröner *et al.* (1998). The 1σ error in the ratio $^{206}Pb/^{238}U$ during analysis of all standard zircons during this study was between 1.3 and 1.65%, and the primary beam intensity was between 2.2 and 2.8 nA. Sensitivity varied between 20 and 30 c.p.s. $ppm^{-1}Pb$. Raw data reduction followed the method described by Nelson (1997). Common-Pb corrections have been applied, assuming that common Pb is surface related and using the isotopic composition of Broken Hill. The analytical data are presented as a Supplementary Publication (see p. 180). Errors given on individual analyses are based on counting statistics and are at the 1σ level. Errors for pooled analyses are at 2σ or 95% confidence. The ages and 2σ errors of intercepts of the best-fit line with concordia were calculated using the method of Ludwig (1994).

Passive margin detachment during arc–continent collision (Central European Variscides)

O. ONCKEN[1], A. PLESCH[1], J. WEBER[1], W. RICKEN[2] & S. SCHRADER[2]
[1]*GeoForschungsZentrum Potsdam, Telegrafenberg, D-14473 Potsdam, Germany (e-mail: oncken@gfz-potsdam.de)*
[2]*University of Cologne, Institut für Geologie, Zülpicher Str. 49a, D-50674 Köln, Germany*

Abstract: In the Rhenish Massif and Ardennes, the Rheno-Hercynian fold and thrust belt of the Mid-European Variscides exposes a telescoped complete Late Palaeozoic passive margin, which was detached from the lower crust during Carboniferous collision with a continental arc (Mid-German Crystalline High on the leading edge of Armorica). Geometric analysis and isostatically corrected balanced cross-sections show that the basal detachment propagated from the oceanic realm into the passive margin by repeated ductile footwall failure during lithospheric flexure of the weak lower plate (effective elastic lithospheric thickness <4 km) under the load of the advancing upper plate. Early collision was associated with offscraping of the uppermost slope sediments in frontal accretion mode. Subsequent detachment of the remaining passive margin cover was initiated after subduction of the ocean–continent transition to about 6 kbar depth. Rocks and fabrics from the detachment show that, at this pressure, they crossed the brittle–plastic transition at the fossil 300–400 °C isotherm. Newly failed segments of the basal detachment propagated along this isotherm and branched off towards the foreland by progressively downstepping from the passive margin sediments into the basement of the downflexed lower-plate crust. This evolution coincided with cyclic changes in accretion mode of the lower-plate upper crust to the advancing orogenic wedge (repeated changes from basal to frontal accretion) as well as with a related stepwise propagation of a narrow foreland basin. Propagation of the detachment segments and the related imbricate fans in the lower plate, moreover, was controlled by the geometry of the basin structure by localizing branch lines and ramps along earlier growth faults.

As a rule, collisional orogeny starts with incorporation of a lower-plate passive continental margin into the evolving orogenic wedge that forms the upper-plate leading edge. Although foreland fold and thrust belts that developed from passive margin sequences have been extensively studied, the initial incorporation of passive margin material into a wedge is not well understood. Preservation and exposure of this transient stage of collisional orogeny seem to be extremely rare, as are active examples. At present, only few active margins are colliding with an incoming passive margin: the North Australian Margin with the Banda Arc, the North African margin with the Sicilian and Cretan Arcs, the China Margin with the Taiwan–Luzon Arc, etc. However, even high-resolution offshore reflection seismic data for the best-documented case, the Australian Margin–Banda Arc in the Timor area, do not image the initial continental accretion process (e.g. Karig *et al.* 1987; Richardson & Blundell 1996; Snyder *et al.* 1996). On Timor, Australian passive margin sediments are exposed following early margin accretion with frontal and basal accretion (Karig *et al.* 1987; Charlton *et al.* 1991), although the initiation of accretion has so far not been resolved from field data.

A deeper understanding of the initial stage of collisional orogeny requires preservation and exposure of a complete accreted passive margin, including the transition zone to the oceanic crust. A prime example of such a system is the Rheno-Hercynian fold and thrust belt on the northern flank of the European Variscides. As recently shown, it comprises a complete telescoped passive margin, with the sedimentary cover preserved in deeply eroded imbricates from the foreland basin in the north to the former shelf slope and ocean–continent transition in the south (Oncken *et al.* 1999). Accordingly, reconstruction of the detachment process of the sedimentary prism during collision and the factors controlling passive margin detachment seems feasible.

In general, the localization of detachments in different foreland thrust belts has been shown to be controlled by a number of factors, including (1) The basement–cover interface or weak sediments in a mechanically well-stratified sedimentary cover (e.g. Jamieson & Beaumont 1988;

From: FRANKE, W., HAAK, V., ONCKEN, O. & TANNER, D. (eds).
Orogenic Processes: Quantification and Modelling in the Variscan Belt.
Geological Society, London, Special Publications, **179**, 199–216. 0305-8719/00/$15.00

Hatcher & Hooker 1992); and (2) a rheological boundary such as the brittle–plastic transition (Ord & Hobbs 1989; Carminati & Siletto 1997). Additionally, the localization of ramps as well as the branching topology of the detachment and major thrusts have been shown to be controlled by lateral variations within the basin fill such as major facies changes or lithological contrasts across growth faults (Powell 1989; McClay & Buchanan 1992; Tavarnelli 1996).

Although a wealth of evidence has been collected on these aspects, no data are available to show how the basal detachment propagates from the approaching plate boundary into the passive margin and by which mode parts of the passive margin fail and are accreted to the upper-plate wedge. In the present paper, we attempt to explore in more detail the geometric evolution as well as the most important controlling factors of passive margin failure during early collision. A 3D geometric reconstruction of the detachment system and the former shelf basin from 2D lines (introducing a new section across the eastern Rhenish Massif; Fig. 1b), using additional constraints from flexural–isostatic modelling, geochronology, metamorphic petrology and synorogenic basin evolution form the basis of the analysis of initial passive margin accretion.

Geological framework

Carboniferous collision of the continental plate fragments building the European Variscides (e.g. Franke 1989; Matte 1991) generated several kinematically coupled orogenic wedges near the respective sutures. On the northern flank of the Variscides, the Rheno-Hercynian domain (exposed in the Harz Mts and the Ardennes–Rhenish Massif (Fig. 1a)) represents a foreland fold and thrust belt that developed from the sedimentary prism of a passive continental margin (Behrmann *et al.* 1991; Oncken *et al.* 1999). Passive margin evolution initiated with Early Devonian continental rifting that resulted in an intracontinental rift basin and a marginal plateau at the transition to the oceanic basin to the south. Opening of the oceanic basin started in late Early Devonian time.

A Devonian–Early Carboniferous siliciclastic basin fill of 3–12 km thickness was deposited on the rifted continental basement. In the north, the basement includes the foreland of the Caledonian Orogen to the north and is composed of thick Lower Palaeozoic sediments overlying an unknown crystalline basement (see André 1991). In the south, it is made up of metamorphic basement of Cadomian age that is partly overlain by Lower Palaeozoic island-arc type volcanic rocks (Franke & Oncken 1995; Klügel 1997; Oncken *et al.* 1999). In the eastern Rhenish Massif, the Giessen–Ostharz Nappe System comprises relics of an accretionary wedge with mid-ocean ridge basalt (MORB)-type volcanic rocks covered by Middle Devonian cherts and flysch of Late Devonian age (Engel & Franke 1983; Engel *et al.* 1983; Grösser & Dörr 1986). In the southernmost imbricates of the Phyllite Zone in the southwestern Rhenish Massif, similar rocks with intercalated pelagic sediments and megaslumps are neighboured to the north by imbricates with continental tholeiites. These imbricates were interpreted to trace the former ocean–continent transition (Oncken *et al.* 1999). The small ocean basin, which was the protolith of these rocks, was rooted between the Rheno-Hercynian and Saxo-Thuringian Zones and extended from southern England to east of the Harz Mts.

Closure of this oceanic basin started during earliest Late Devonian time with SE-directed subduction, and caused the formation of a flysch wedge and the evolution of a magmatic arc in the upper plate (Mid-German Crystalline High). In the early collision stage (*c.* 335–330 Ma, Viséan–Namurian boundary), the units of the southern Rhenish Massif (Phyllite Zone) were subducted and suffered a pressure-dominated metamorphism (Anderle *et al.* 1990; Klügel 1997). Accretion of the Rheno-Hercynian passive margin propagated from south to north through the basin from *c.* 325 to 300 Ma (Ahrendt *et al.* 1983) at an average rate of 14 mm a^{-1} (total shortening 52% or 180–200 km; Oncken *et al.* 1999). This phase is also recorded in a northwest-propagating sequence of narrow Lower Carboniferous turbidite basins that grade into a broad Upper Carboniferous molasse basin in the north (Engel & Franke 1983).

Structure of the Rhenish Massif and Ardennes

The following description summarizes the observations made from orogen-scale section balancing depicted in Figs 2 and 3. The westernmost Rhenish Massif (Ardennes) is characterized by a stack of thin-skinned imbricates involving Lower Devonian and younger sediments from the northern rift shoulder of the passive margin basin. A frontal thrust that roots in Lower and Upper Devonian shales above the basement–cover interface is exposed only west of the Rhine River: the Aachen Thrust (Meissner *et al.* 1981; Raoult & Meilliez 1987; DEKORP Research Group 1991; Fielitz 1992) and its western prolongation, the Faille du Midi (Fig. 3). In the east the basal detachment is located below the base of the folded basin fill and does not reach the surface (Fig. 2). The major anticlines at the southern margin of the Ardennes (see Fig. 1) expose Lower Palaeozoic sediments from the Pre-Devonian basement that were cut off from the northern rift shoulder by a footwall shortcut through a major inverted extensional growth fault. Across the resulting ramp, the upper-crustal detachment of the Aachen Thrust descends SE-wards to a mid-crustal décollement level at 15–17 km depth (Fig. 3).

In the central Rhenish Basin, characterized by a substantially increased thickness of marine Lower Devonian sediments (8–10 km), the thick-skinned detachment level is localized below the Devonian basin fill (Figs 2 and 3a). Major thrusts at regular spacing within this thick-skinned domain (Venn–Ebbe Thrust, Siegen Thrust, Boppard and Sackpfeife Thrust System, etc.) mark individual leading imbricate fans above narrow shear zones. Toward the south, the sediments from the marginal plateau and continental slope are again assembled into thin-skinned imbricate thrust sheets (Taunus Thrust, Phyllite Zone Thrust; Fig. 3a).

Throughout, the shear-zone rocks show a pervasive fabric (foliated cataclasites to mylonites), which is related kinematically to the northward-younging cleavage fabric in the hanging-wall and footwall rocks (Oncken 1989). In some cases, these thrusts were later reactivated (cataclastic zones crosscut earlier ductile shear-zone fabrics and structures, and offset peak metamorphic isograds, etc.; Oncken 1991). Also, balanced cross-sections by Oncken (1989), Klügel (1997) and Oncken *et al.* (1999), and a newly presented section in this paper through the eastern Rhenish Massif (Fig. 2), show that the more southward imbricate fans have been increasingly back-rotated and back-folded, affecting all above-mentioned features during the final growth stages of the wedge. Altogether, these observations are interpreted to show a simple in-sequence development of the imbricate fans with progressively later activation of basal thrusts toward the northwest and restricted reactivation of some of the more internal earlier thrusts.

The basal hanging walls of all thrusts expose syn-rift Siegenian to Lower Emsian rocks. South of the Siegen Thrust, the basal thrusts underlying the individual imbricate fans show an increasingly thinner Devonian sequence in their hanging walls: 14 km at the Siegen Thrust, 5 km at the Boppard Thrust, 3 km at the Taunus Thrust and 1–2 km at the Phyllite Zone thrust sheet. The stratigraphic section at the hanging-wall base remains the same (Siegenian to Lower Emsian units) with the exception of the Taunus unit, which, in parts, exposes the basement (metamorphic and volcanic, see above) under Gedinnian rocks. These thrusts therefore sole out above the basement–cover interface, at the imbricate fronts now exposed, and step down into the basement toward the hinterland. Section balancing clearly reveals that all root in a common basal detachment within the basement.

Two additional, mildly folded detachment generations appear at the eastern rim of the Rhenish Massif, as a result of an orogen-scale NE plunge of structures (Fig. 1). The upper one represents the base of the accretionary wedge (Giessen–Ostharz Nappes) and includes allochthonous flysch sediments along with some exotic rocks and oceanic fragments (Grösser & Dörr 1986), which trace the original suture zone. The other one below this suture, not well resolved structurally from the uppermost detachment, forms the roof thrust of the underlying thrust systems described above. Its hanging wall comprises thin slices (10–100 m) of Upper Devonian and Lower Carboniferous sediments from the continental slope and rise, and from parts of the Rheno-Hercynian marginal plateau to the south. These sediments were detached and accreted to the tip of the advancing accretionary wedge during initiation of continental subduction. The thin sequence has a very complex structural style. This is different from the earlier described deeper detachment segments and their associated hanging wall structures but similar to recent findings of equivalent settings where the uppermost, unconsolidated and overpressured sediments of a passive margin cover were accreted to the upper plate: on Timor, such sequences show melange-type structures, active mud volcanoes and complex faulting (Barber *et al.* 1986). In the Rhenish Massif, the subsequently formed basal detachment branched from the subduction zone into the passive margin and is independent of this early feature. It undercuts the entire basin fill in the southern and central parts of the basin, including the complex and deep rift structures, irrespective of the type of rocks encountered. Only in the northwest, where a discrete rift boundary had developed, does the detachment level rise across a ramp into the sedimentary cover.

A 3D model of the basal detachment fault based on available crustal-scale balanced cross-sections highlights its varied morphology between the western and eastern section of the belt (Fig. 4). The changes in depth and dip of the frontal parts of the basal detachment from the Faille du Midi (steep frontal splay) and the Aachen Thrust (gentle dips) to the east (gentle dips, blind) stand out. The linkage of eastern and western patches of some of the visualized detachment segments (basal detachment, Venn–Ebbe Detachment, Boppard Detachment) is strongly simplified and involves more complicated transfer structures than actually shown. The Venn–Ebbe System is the most extensive and features a strongly curving branch line. The Venn Thrust in the west is correlated with the Ebbe Thrust in the east rather than with the thrust north of it (see Fig. 1a). This correlation is suggested by the position of the branch lines of both Venn and Ebbe Thrust at the frontal edge of the blind duplex stack modelled in the balanced sections. This position indicates that both thrusts accommodated major parts of the substantial net shortening required by the probably laterally continuous duplex stack. The Siegen Thrust has a twisted shape, from generally more steeply dipping in the west to more gently dipping in the east. The Boppard–Sackpfeife Thrust System also shows a rather complex shape change in its western segment. The continuation to the east is not well constrained at surface, as a result of coverage with younger deposits, but is well identified from section balancing and restoration (see below). The most complex shape is the modelled Taunus Thrust. It must be continuous, but includes a steeply dipping western part and a backfolded and overturned central part. The Phyllite Zone Detachment and the slope sediment detachment are not included because of poor local preservation or exposure. The major frontal ramp of the basal detachment disappears gradually from the eastern end of the footwall cutoffs to the east. A western termination is less obvious.

In summary, the basal detachment mimics the rift geometry in parts, and therefore is clearly controlled by the crustal structure that resulted from Devonian extension. The three southernmost, first formed, segments develop in the sedimentary cover of the slope and rise and of the marginal plateau; the subsequent segments step below the basin fill.

Restored basin geometry

There are three domains in the palinspastic restorations across the entire Rhenish Basin (Figs 2c and 5). A thin, mostly post-rift Devonian platform in the NW is separated from the thick Rhenish basin fill to the south and east by a prominent rift margin fault with a cumulative downthrow of 6 km. Toward the east this margin fault evolves into a broader rift zone, which did not cause a major ramp during inversion. This margin developed during Early Devonian crustal stretching and separates nearly unextended crust to the northwest from strongly attenuated crust to the southeast (Raoult & Meilliez 1987; Behrmann *et al.* 1991; Oncken *et al.* 1999). The boundary fault system coincides with the southern termination of the seismically transparent middle and lower crust in the DEKORP 1A and DEKORP 2N sections (Franke *et al.* 1990; DEKORP Research Group 1991) attributed to the London–Brabant Massif in the subsurface. The variable geometry of this rift border fault played an important role during the synorogenic inversion of the western part of the Rhenish Basin by localizing the above-mentioned major oblique ramp and causing the basal detachment to shift from below the basin fill into the post-rift sequence deposited on the unrifted craton.

The restoration shows a more or less constant southward deepening of the central basin, which is more pronounced in the west. The southern margin of this central rift must be formed by a set of north-dipping normal faults that are responsible for the substantial decrease in thickness of syn-rift sediment from 14 to 5 km over a distance of <15 km. The restorations show the southern basin part (Hunsrück and Taunus Units) to be a marginal plateau typical of Atlantic-type passive margins. Schematic restoration of the exposed imbricates of the Phyllite Zone on the southern flank of the plateau aims only at a geologically admissible solution (see Oncken *et al.* (1999) for details). In the west, the stacked rocks were deposited on probably transitional, thinned crust with a thin sedimentary cover containing (a) the southern Early Devonian onlap region of the continental rift in the north, and (b) the northern slope of an Emsian to Late Devonian oceanic basin in the south. In the east, the Phyllite Zone comprises only the basement and sedimentary cover of the marginal plateau

Fig. 1. (a) Summary geological map. Major thrust systems are highlighted and numbered: 1, slope sediment detachment; 2, Phyllite Zone Detachment; 3, Taunus Thrust, 4: Boppard–Sackpfeife Thrust System; 5, Siegen Thrust; 6, Venn–Ebbe Thrust; 7, Faille de Midi–Aachen Thrust. 'Burial front' represents former tip of overthrust upper plate and is estimated from distribution of metamorphic grade. Circled numbers on right side denote syndeformational metamorphic ages by Ahrendt *et al.* (1983) and Klügel (1997). Inset: RH, Rheno-Hercynian; A, Ardennes; RM, Rhenish Massif; ST, Saxo-Thuringian; MO, Moldanubian. **(b)** Available balanced sections and seismic reflection profiles. References: ECORS, CO 83 and CO 99: Bois *et al.* (1994); Meilliez: Raoult & Meilliez (1987); Adams: Adams & Vandenberghe (1999); Hollmann and BGS-VL: Hollmann (1997); DEKORP 1A, 1B and 1C: DEKORP Research Group (1991); v. Winterfeld and Dittmar: Oncken *et al.* (1999); Klügel: Klügel (1997); Oncken: Oncken (1989); DEKORP 2N: Franke *et al.* (1990); Weber: Weber *et al.* (1997).

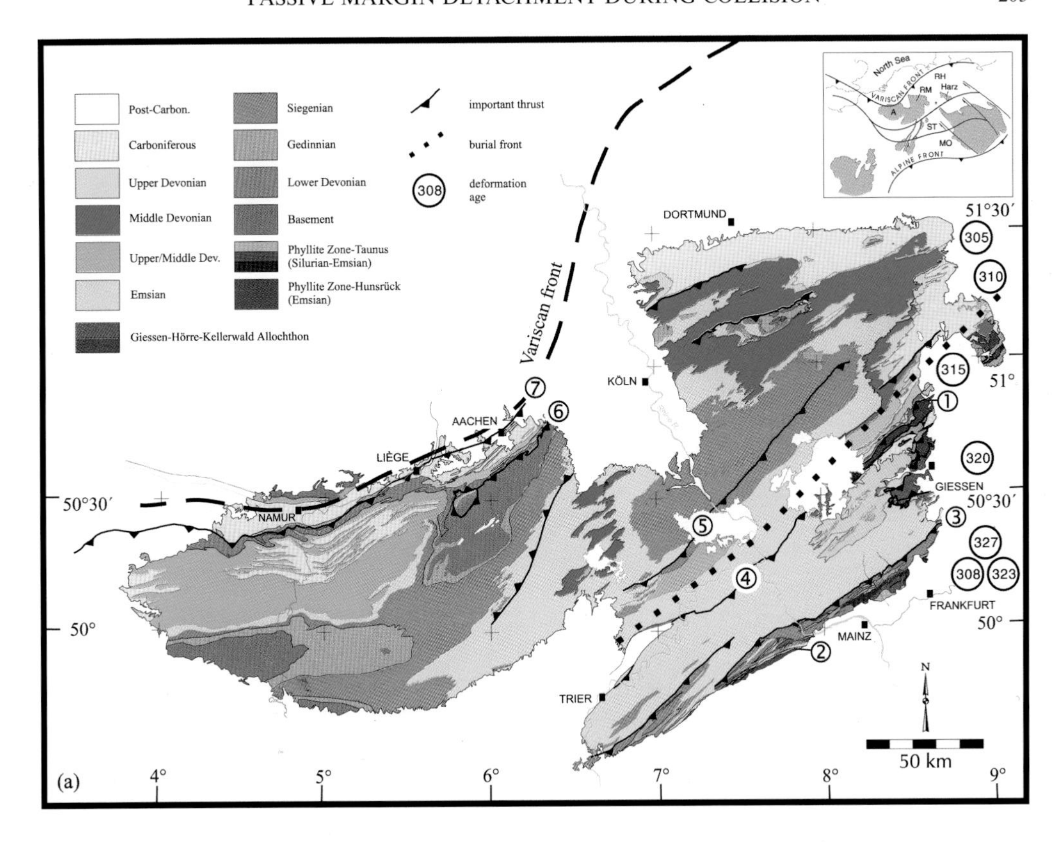
Post-Carbon.
Carboniferous
Upper Devonian
Middle Devonian
Upper/Middle Dev.
Emsian
Siegenian
Gedinnian
Lower Devonian
Basement
Phyllite Zone-Taunus (Silurian-Emsian)
Phyllite Zone-Hunsrück (Emsian)
Giessen-Hörre-Kellerwald Allochthon
important thrust
burial front
308 deformation age
North Sea
VARISCAN FRONT
RH
RM
Harz
A
ST
MO
ALPINE FRONT
Variscan front
DORTMUND
KÖLN
AACHEN
LIÈGE
NAMUR
GIESSEN
FRANKFURT
MAINZ
TRIER
305
310
315
320
327
308
323
51°30′
51°
50°30′
50°
N
50 km
(a)
4°
5°
6°
7°
8°
9°

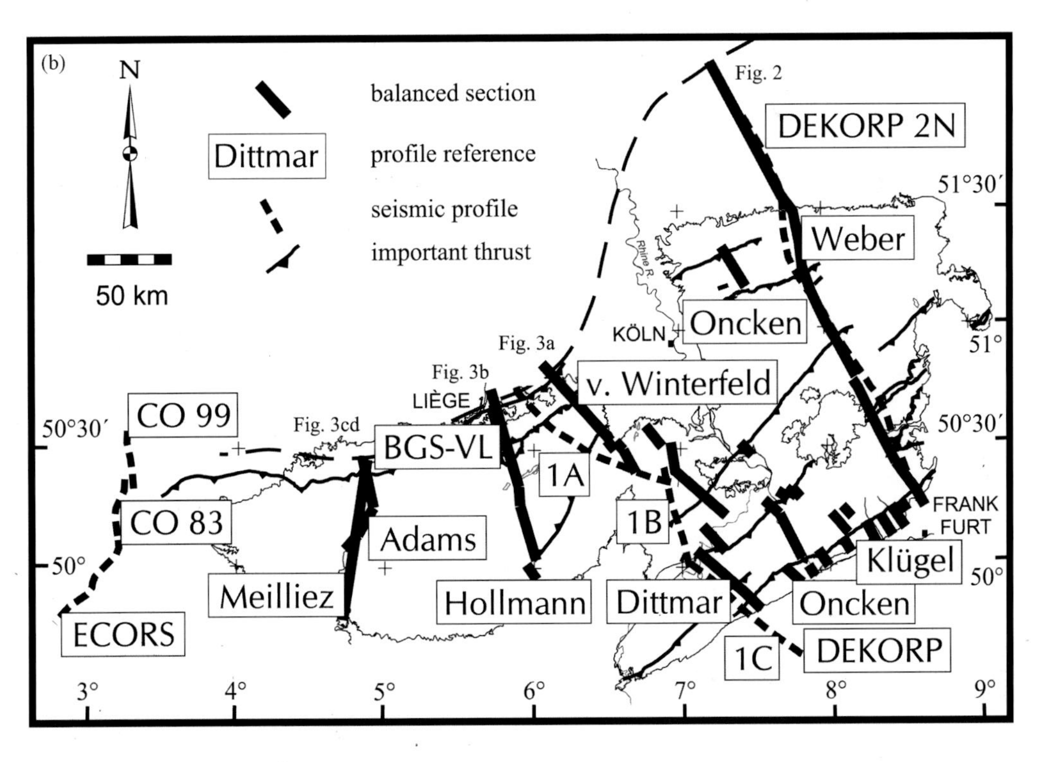
(b)
N
50 km
balanced section
Dittmar
profile reference
seismic profile
important thrust
Fig. 2
DEKORP 2N
Weber
Oncken
KÖLN
Rhine R.
Fig. 3a
Fig. 3b
LIÈGE
v. Winterfeld
Fig. 3cd
CO 99
BGS-VL
1A
1B
CO 83
Adams
Meilliez
Hollmann
Dittmar
1C
Oncken
Klügel
DEKORP
FRANK FURT
ECORS
51°30′
51°
50°30′
50°
3°
4°
5°
6°
7°
8°
9°

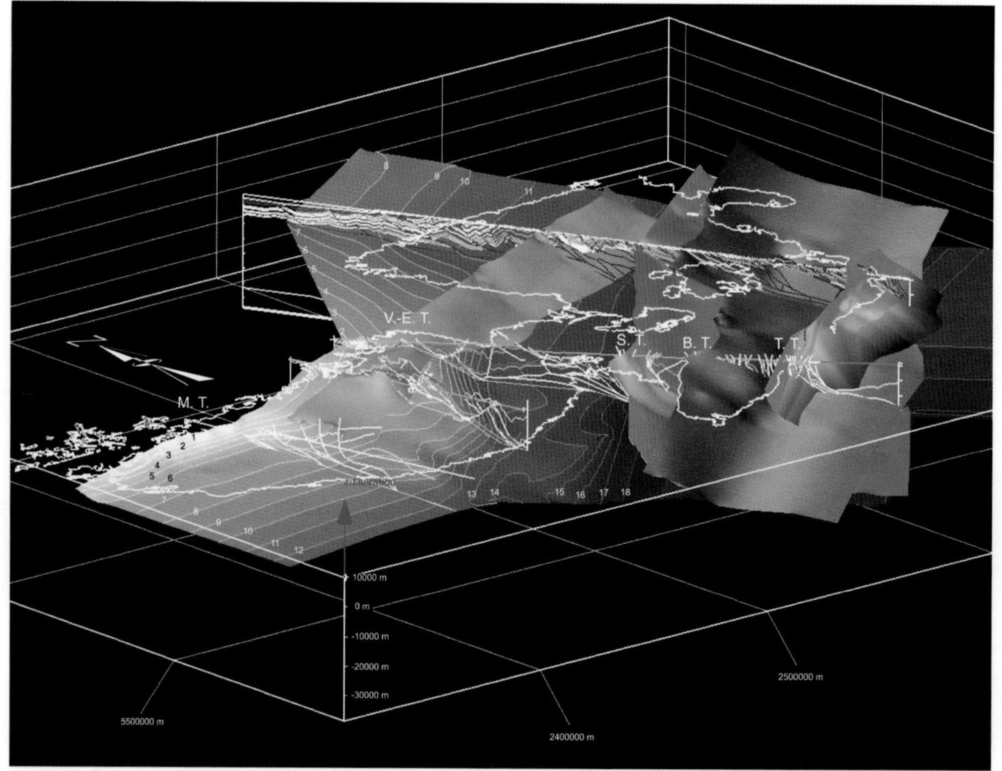

Fig. 4. The 3D detachment geometry of basal detachment and principal thrust faults that delimit the main imbricate–duplex systems. Colour shading shows variation in dip (gentle to steep) or depth (deep to shallow). T.T., Taunus Thrust (deep, blue; shallow, green); B.T., Boppard–Sackpfeife Thrust (gentle, orange; steep, red); S.T., Siegen Thrust (gentle, blue; steep, green); V.-E.T., Venn–Ebbe Thrust (gentle, blue; steep, brown); M.T., Faille du Midi Thrust. Basal detachment has depth shading (deep, dark brown; shallow, white) and depth contours (in km). Oblique 3D view from the southwest, showing the outline of the exposed Rhenish Massif and of the main balanced sections.

and shelf edge, indicating an eastward widening of the plateau.

In the restoration, the basal detachment trajectory is saucer shaped. This is also true for some of the basal thrusts of the individual leading imbricate fans (Figs 2c and 5a). Seven accretionary subsystems make up the wedge in the balanced and restored sections. The subsystems increase in size toward the north and are asymmetrically stacked at the southern rim of the restored basin. Each branch toward the north is located deeper within the passive margin (Figs 2c and 5a), with the exception of the northernmost system, which cuts across the northwestern rift shoulder. This succession results in a stratigraphic downward shift of the propagating basal detachment. The final and deepest detachment therefore undercuts the entire basin fill within the basement. This geometry and kinematic evolution is in contrast to standard foreland-propagating thrust systems, which generally show foreland-directed rise.

As a consequence, the in-sequence accretion of the thrust sheet systems first involved only thin units that included only the uppermost sediments from the slope and rise area and from the marginal plateau. Figures 2c, 5a and 6b show that this southernmost system, now immediately below the accretionary wedge, must have branched off near the base of the slope. This is also true for the Phyllite Zone Segment, comprising the remaining volcano-sedimentary slope

Fig. 2. (**a**) Balanced and (**c**) restored section through the eastern Rhenish Massif (Weber *et al.* 1997; for location see Fig. 1b; for details of the balancing technique, see Oncken *et al.* 1999). (**b**) Time-migrated seismic reflection line DEKORP 2N (see Franke *et al.* 1990).

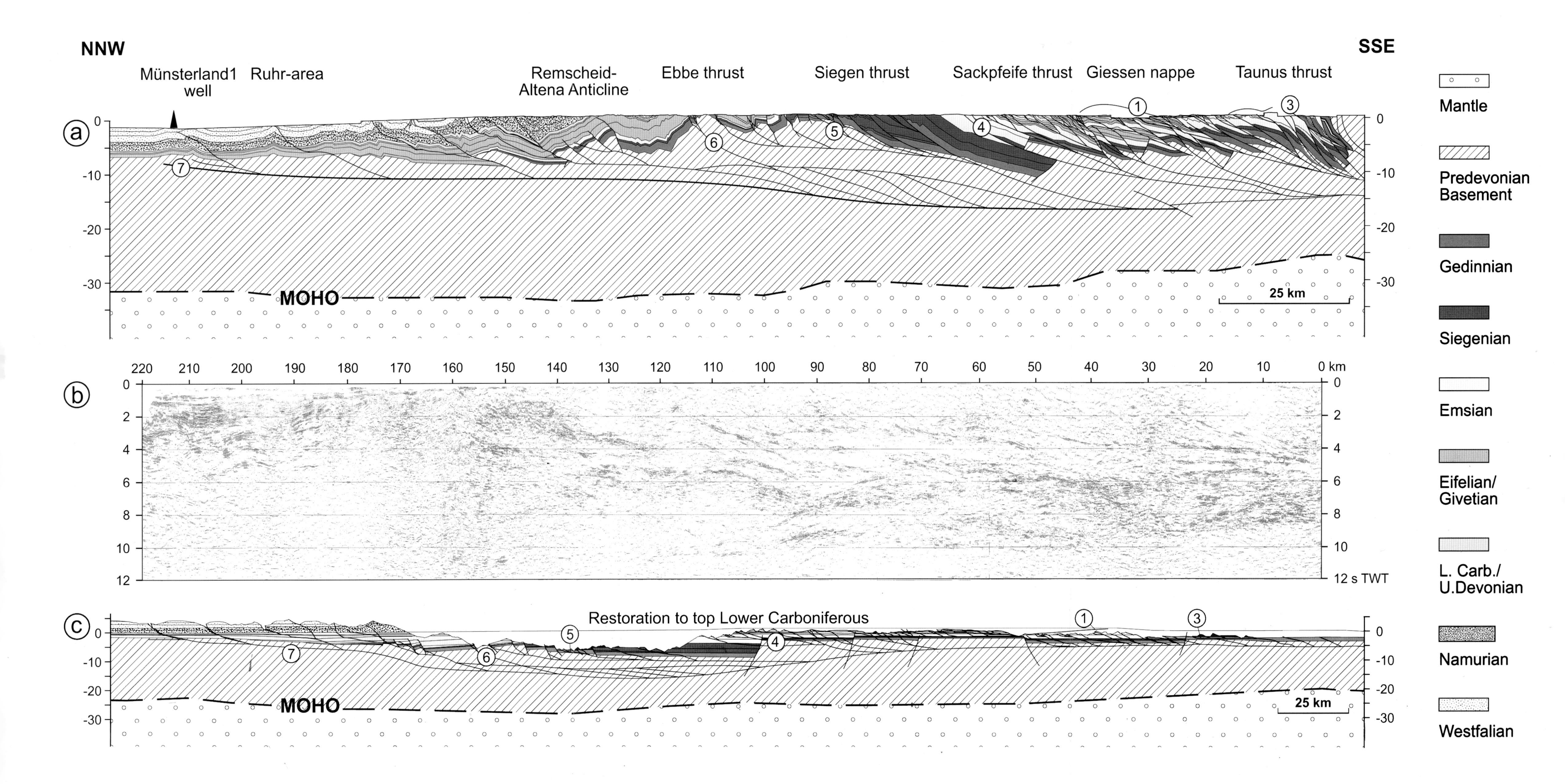
NNW
SSE
Münsterland1 well
Ruhr-area
Remscheid-Altena Anticline
Ebbe thrust
Siegen thrust
Sackpfeife thrust
Giessen nappe
Taunus thrust
a
b
c
MOHO
25 km
12 s TWT
0 km
Restoration to top Lower Carboniferous
Mantle
Predevonian Basement
Gedinnian
Siegenian
Emsian
Eifelian/ Givetian
L. Carb./ U.Devonian
Namurian
Westfalian

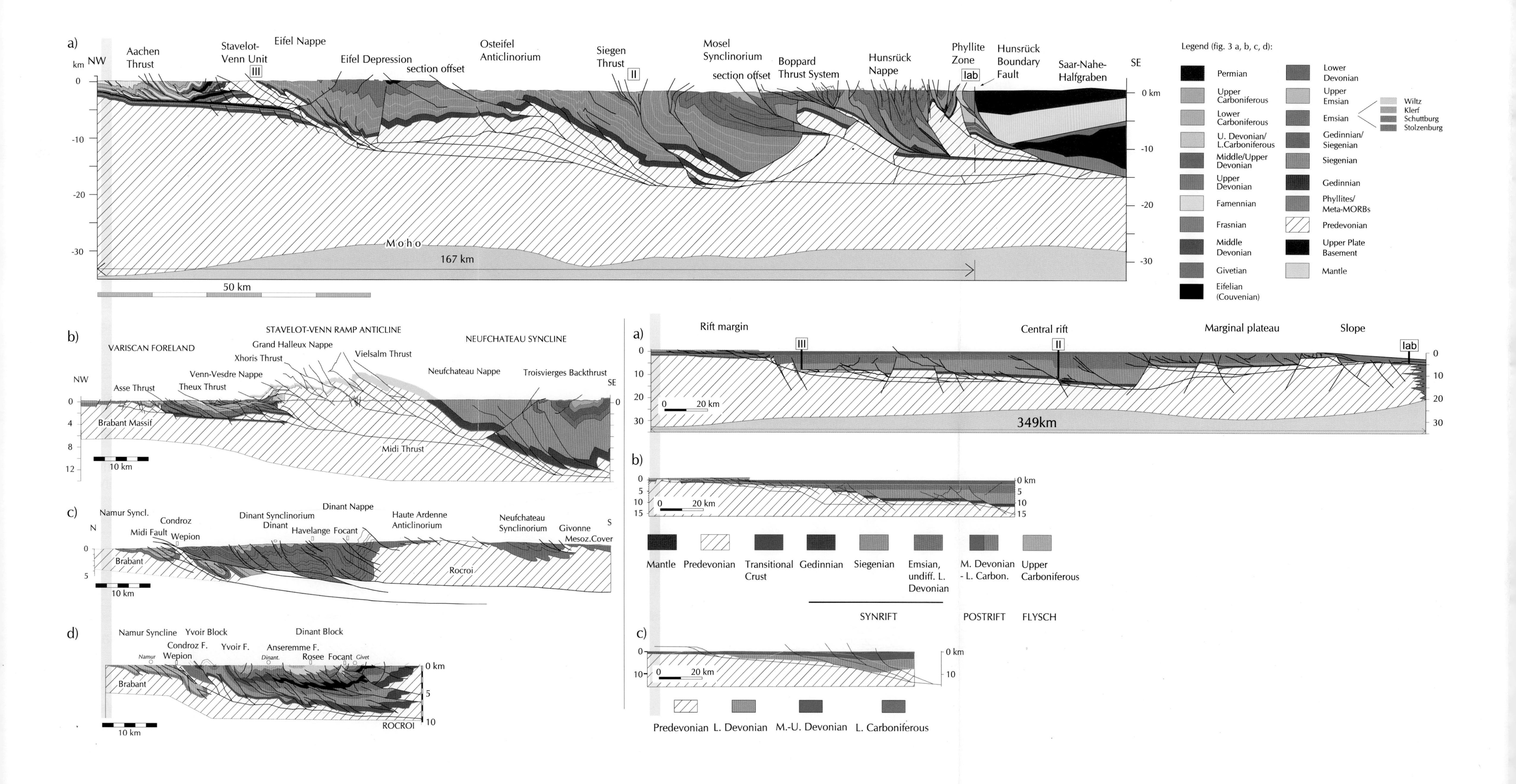
a)
NW
SE
Aachen Thrust
Stavelot-Venn Unit
III
Eifel Nappe
Eifel Depression
section offset
Osteifel Anticlinorium
Siegen Thrust
II
Mosel Synclinorium
section offset
Boppard Thrust System
Hunsrück Nappe
Phyllite Zone
Iab
Hunsrück Boundary Fault
Saar-Nahe-Halfgraben
Moho
167 km
50 km
Legend (fig. 3 a, b, c, d):
Permian
Upper Carboniferous
Lower Carboniferous
U. Devonian/ L.Carboniferous
Middle/Upper Devonian
Upper Devonian
Famennian
Frasnian
Middle Devonian
Givetian
Eifelian (Couvenian)
Lower Devonian
Upper Emsian
Emsian
Gedinnian/ Siegenian
Siegenian
Gedinnian
Phyllites/ Meta-MORBs
Predevonian
Upper Plate Basement
Mantle
Wiltz
Klerf
Schuttburg
Stolzenburg
b)
VARISCAN FORELAND
STAVELOT-VENN RAMP ANTICLINE
NEUFCHATEAU SYNCLINE
Grand Halleux Nappe
Xhoris Thrust
Vielsalm Thrust
Venn-Vesdre Nappe
Neufchateau Nappe
Troisvierges Backthrust
Asse Thrust
Theux Thrust
Brabant Massif
Midi Thrust
10 km
c)
Namur Syncl.
Condroz
Midi Fault
Wepion
Dinant Synclinorium
Dinant
Dinant Nappe
Havelange
Focant
Haute Ardenne Anticlinorium
Neufchateau Synclinorium
Givonne
Mesoz.Cover
N
S
Brabant
Rocroi
10 km
d)
Namur Syncline
Yvoir Block
Dinant Block
Condroz F.
Wepion
Yvoir F.
Anseremme F.
Namur
Dinant.
Rosee
Focant
Givet
Brabant
ROCROI
10 km
a)
Rift margin
Central rift
Marginal plateau
Slope
III
II
Iab
20 km
349km
b)
20 km
0 km
Mantle
Predevonian
Transitional Crust
Gedinnian
Siegenian
Emsian, undiff. L. Devonian
M. Devonian - L. Carbon.
Upper Carboniferous
SYNRIFT
POSTRIFT
FLYSCH
c)
20 km
0 km
Predevonian
L. Devonian
M.-U. Devonian
L. Carboniferous

cover, which forms the first segment of the basal detachment. This unit, however, started to detach only after subduction to about 6 kbar and heating above 300 °C (see below), indicating a switch from frontal to basal accretion. The Taunus Thrust branches off near the outboard boundary of the marginal plateau. The crustal horst forming the latter was then 'decapitated' and thrust with its cover onto the continental rift in two stages: the Boppard and Siegen Thrust units, both branching off at the major inboard boundaries of the marginal plateau and continental rift. However, the branch line of the Boppard–Sackpfeife System can be seen to be strongly curved: the offset of the eastern continuation to the north is clearly linked to a major transfer zone that separated segments of the marginal plateau and controlled the extensional style in the southern half of the rift basin. This transfer zone is also conspicuous from the map image of the Rhenish Massif showing the termination of major structures (Fig. 1). The Ebbe and Venn Thrusts branch off near the northern rift boundary. The localization of major ramps of these thrusts is again consistently controlled by growth faults (see frontal ramps of Taunus, Boppard and Venn Thrusts). In summary, the entire detachment architecture (branch lines and frontal ramps) is largely controlled by the previous basin architecture (Fig. 6).

As revealed by geometric analysis, detachment evolution was cyclic and switched from basal to frontal accretion (Plesch & Oncken 1999). Episodes of rapidly advancing wedge propagation followed basal accretion and were possibly related to initiation of a new deeper detachment within the lower-plate crust. This is corroborated by the stepwise migration of an intrabasinal high, trapping flysch sub-basins (Ricken *et al.* this volume).

Detachment rocks and fabrics

In the central and southern parts of the Rhenish Massif, the basal detachment is localized within lower Palaeozoic sediments in the north and in amphibolite facies rocks in the south. Apparently therefore, the localization of the detachment is independent of rock type and must be explained from the fabrics that originated near the detachment. The restoration in Fig. 5a shows that rocks from near the detachment are exposed in essentially four areas: the southern lower Palaeozoic core of the antiform that exposes the hanging-wall base of the Venn Thrust, the hanging-wall base of the Siegen Thrust (both in eroded back-limb positions), the imbricates of the Phyllite Zone, and the early-stage structures at the base of the Giessen Nappe (Fig. 2a and c).

In the first three areas, peak metamorphic conditions reached some 300 °C during pervasive deformation (Fig. 7). Arenitic sediments are mylonitic and show abundant evidence of early dislocation creep of quartz (Oncken 1989). In most cases, a change in the dominant deformation mechanism toward dissolution–precipitation creep is observed, especially in mica-rich rocks after substantial grain-size refinement by either cataclasis or dynamic recrystallization of quartz (Oncken 1989). Pressures were around 2 kbar at the two northern occurrences and 5–6 kbar at the southernmost (Massonne 1989; Holl 1995; Klügel 1997; Fig. 7). Moreover, from the Mosel Syncline to the Phyllite Zone (Fig. 3a), Holl (1995) demonstrated that pressures within the exposed Lower Devonian sediments increase continuously. In the Giessen Nappe Thrust, shear-zone fabrics and quartz deformation indicate peak conditions of only 200–300 °C (Grote 1983). Rocks from the northern thin-skinned flat, exposed in the imbricates of the Aachen Thrust and reached by boreholes farther west (Hollmann 1997), show only foliated cataclasites and synkinematic temperatures of 200–250 °C. Detachment localization here, as in the Giessen Nappe, was controlled by weak pelitic horizons in the Devonian sedimentary succession, after rising from the brittle–ductile boundary across the former rift flank in the NW part of the basin.

Klügel (1997) showed that deformation in the frontal part of the Phyllite Zone Detachment, indicating the start of detachment of the deeper parts of the sedimentary apron from the continental slope, initiated close to peak pressures, with a downdip stretching lineation and top to the NW shearing. Deformation outlived peak pressures to only a very minor degree before activation of the next segment.

Fig. 3. Simplified balanced sections through western Rhenish Massif and Ardennes. (**a**) Oncken *et al.* (1999); (**b**) Hollmann (1997); (**c**) Raoult & Meilliez (1987); (**d**) Adams & Vandenberghe (1999) (see Fig. 1b for locations). I, II, and III in (**a**) are positions of samples for thermobarometry shown in Fig. 7 (see text for sources and details).

Fig. 5. Simplified restored sections through western Rhenish Massif and Ardennes. (**a**) Oncken *et al.* (1999); (**b**) Hollmann (1997); (**c**) Adams & Vandenberghe (1999) (see Fig. 1 for locations; compare balanced sections in Fig. 3). Moho geometry according to modelling results by Heinen (1996).

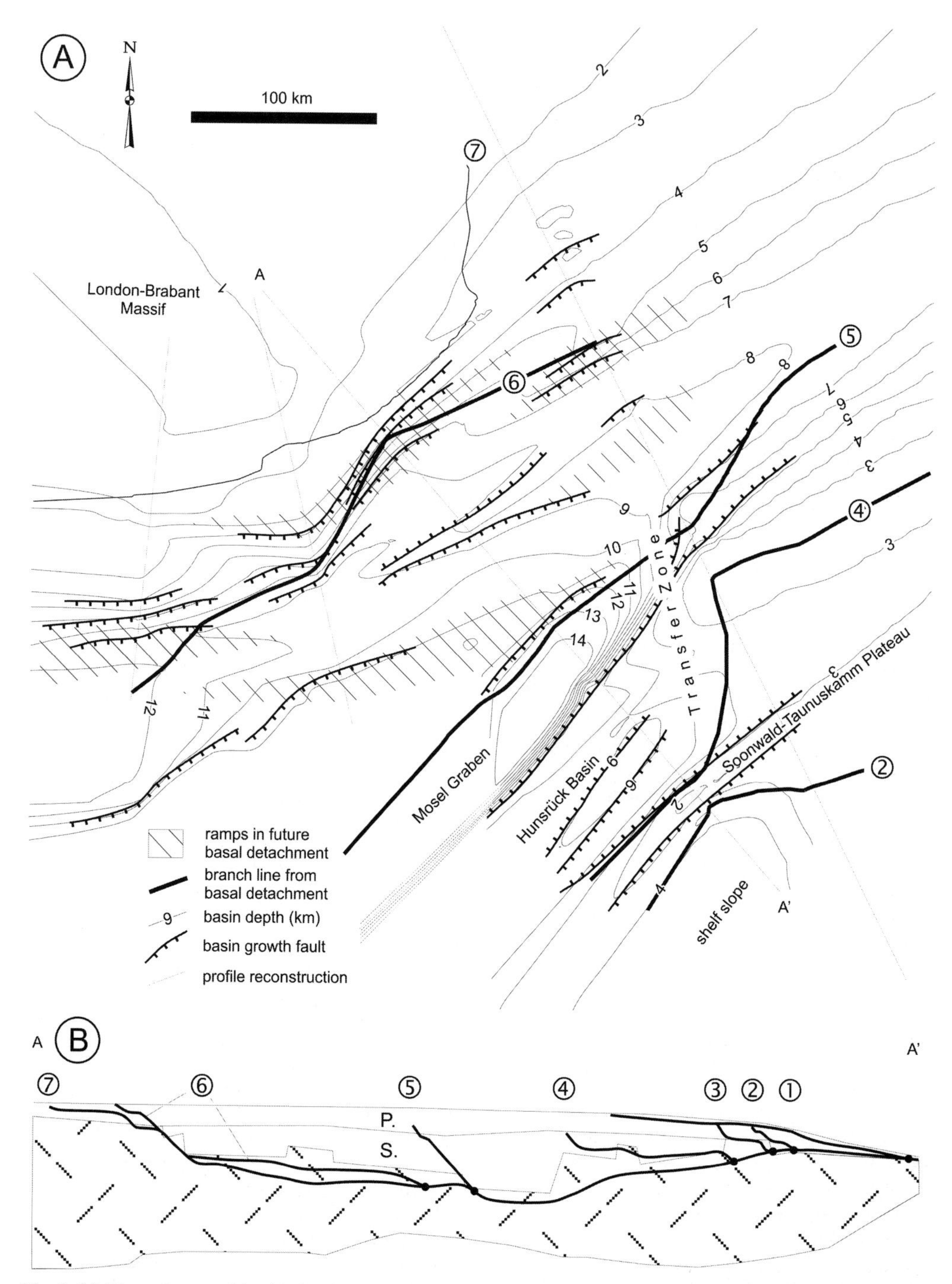

Fig. 6. (**a**) Map of restored basin showing growth fault traces, branch lines of major imbricate fans from basal detachment, and ramps in the basal detachment system. (**b**) Schematic 2D section (× 2 vertical exaggeration) showing basin and detachment branching geometry. Numbers of faults are same as in Fig. 1 and correspond to the activation sequence. S, synrift fill; P, postrift basin fill.

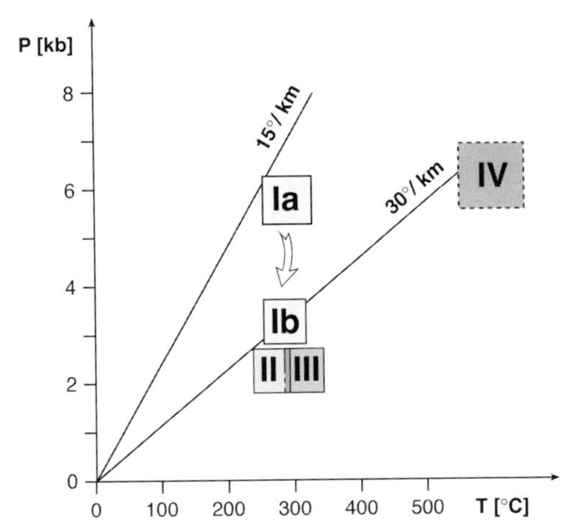

Fig. 7. *P–T* diagram for rocks from near basal detachment. I, Phyllite Zone Detachment (Ia, syn-detachment; Ib, late-orogenic; Klügel 1997); II, sample from Siegen Thrust branch line (Oncken 1989; Holl 1995); III, sample from base of Venn Thrust (Massonne 1989); IV, conservative extrapolation of late-stage *P–T* conditions near internal basal detachment (see text for details). Data from published mineral thermobarometry, rock deformation mechanisms, fluid inclusions and organic rank are as summarized by Oncken *et al.* (1999).

In combination, these observations have several consequences: (1) initiation of the basal detachment occurred when the subducting ocean–continent transition reached the isotherm around the brittle–plastic transition for quartz–mica-rich rocks; (2) foreland-directed propagation of the detachment tip into the lower plate followed this isotherm for a certain distance with more or less regular splays up through the sedimentary sequence either to a roof thrust (i.e. the base of the upper plate) or to the surface in the north; (3) during its initial propagation, the evolving detachment horizon dipped southeast under the upper plate.

From fabric analysis and thermobarometry (on phengites and amphiboles) along with isotopic dating (K/Ar on phengites), Klügel (1997) was able to identify two important stages of pervasive deformation and shearing in the Phyllite Zone rocks. An earlier one associated with stacking at some 5–6 kbar (*c.* 300 °C) at 328 Ma (Ia in Fig. 7) was succeeded by a transpressive stage at some 3 kbar (*c.* 300 °C) around 308 Ma (Ib in Fig. 7), i.e. indicating isothermal exhumation by some 3 kbar and a concomitant increase of the geothermal gradient during deformation. Analysis of the balanced sections and of the location of the deformation front at 308 Ma reveals that this was achieved by underplating during sequential activation of all later detachment segments. The isothermal uplift by 3 kbar from the active basal detachment toward the wedge surface therefore probably reflects substantial synkinematic heating of the internal part of the wedge and the basal detachment. This is a necessary consequence even if the advection of underplated material from the foreland had delayed heating of the wedge. Because of the latter effect, however, extrapolation of conditions from the present surface toward the basal detachment depth cannot be linear. A tentative solution with some 500 °C at some 8 kbar for this late stage in the south is shown by IV in Fig. 7.

Isostatic modelling of detachment propagation

Modelling of the geometric evolution under isostatic conditions relies on a set of different constraints in addition to those conventionally used in cross-section balancing. The distribution of synkinematic pressures (see above) can be explained by the known sediment thickness in the north (equivalent to 2 kbar above the base of the Siegenian sequence) and by the thickness of the overthrust accretionary wedge in the south (see below). The resulting flexural response of the lower-plate passive margin to loading by the upper plate is the key factor in producing the saucer-shaped detachment geometry in the restored sections. As the geochronological data reflect a continuous propagation of the collisional thrust wedge over 30 Ma, the above-mentioned pressures moreover record a time-transgressive evolution and may not be taken to constrain the detachment geometry at one specific time.

Reconstruction was performed using a combined balancing and isostatic modelling procedure. The modelling software 2D-MOVE (Midland Valley Company Ltd, Glasgow) is based on Egan's (1992) application of the principles of flexural isostasy as laid out by Turcotte & Schubert (1982). The cross-sectional shape of the applied load, its density, optional coverage with water, the hinterland-ward continuation of the load, and the flexural stiffness of the lower plate in terms of its effective elastic thickness control the amount of flexure. The software also allows for forward modelling of the advance of the upper-plate wedge by geometric deformation mechanisms. The large scale of the modelled deformation (50 km) requires a mechanism that averages over the deformed volume. Additional geological key observations that constrain the modelling process are as follows.

(1) Low effective elastic thickness (*c.* 4 km) of the lower-plate lithosphere, increasing to the north, as derived from basin modelling (Heinen 1996) for the Devonian period (expected error is less than 50%). Thermal relaxation after rifting until the onset of Late Carboniferous collision may have led to an increase of this value.

(2) Changes in accretion mode from basal to frontal as described above shifted the loading mass episodically to create new sub-basins (see Ricken *et al.* this volume). The number of individual sub-basins (three) roughly equals that of the imbricate fans in the southern massif. These basins evolved from below sea level during turbidite sedimentation to molasse-type fill in Namurian time indicating shallowing during the final stages of wedge propagation.

(3) Submergence of parts of the upper-plate fore-arc and accretionary wedge during convergence with a break during Namurian times (Saar Basin, Oncken 1998). Carbonate build-ups along a narrow shelf in the southeast indicate a partly subaerial wedge (Ricken *et al.* this volume). In conclusion, most of the propagating orogenic wedge is submerged and only some local subaerial topography is created during collision.

(4) Exhumation of a narrow belt of relatively high-pressure rocks at the collision boundary (Phyllite Zone, Klügel 1997) by about 3 kbar from 328 to 308 Ma (i.e. during Namurian time; barometric error is estimated at 1 kbar, time error at 5 Ma). Another 3 kbar were removed by early Permian time (290 Ma). The remaining Rheno-Hercynian units to the north show slower and less synconvergent exhumation. To the south a synconvergent basin with partly marine fill (Saar Basin, see above) separates this belt from the Mid-German Crystalline High (MGCH). The MGCH is largely composed of underplated Rheno-Hercynian rocks that were exhumed during Late Carboniferous time (Oncken 1997).

(5) A fore-arc width (i.e. arc–trench gap) of around 200 km (± 50 km) is estimated from barometric, seismic and balancing data (Oncken 1998). The wedge tip is defined by the northern limit of pressure-dominated metamorphism along the axis of the Mosel Syncline (Fig. 1); 50 km further south the wedge thickness must have been around 18 km at the onset of collision and accretion of the Phyllite Zone rocks from their barometry; this load was eroded by about 9 km during subsequent wedge propagation (see above). These data yield an average wedge taper of some 9° at the onset of collision with an error of $\pm 2°$ controlled by the input data.

An average density of the upper-plate wedge of 2.7 g cm^{-3} was calculated using the Nafe–Drake relationship (Nafe & Drake 1963) from the average *P*-wave velocity of 6.1 km s^{-1} along the European Geotraverse line (Aichroth *et al.* 1992). This is a typical value for intermediate-type rocks that agrees well with data from the Saar 1 borehole, which show the fore-arc to be composed of island-arc type magmatic basement with a thin sedimentary cover of mainly arc detritus and accretionary wedge relics (Giessen–Ostharz Nappes). Changes of the density of the added loads by less than 10% did not modify the results.

Modelling of the wedge deformation was performed by the incline shear mechanism (White *et al.* 1986), which divides the deformed block into parallel columns of a given orientation and preserves area. As the small preserved parts of the wedge do not constrain the overall shear angle, a neutral shear angle of 90° (vertical) was assumed. Also, test runs with other deformation algorithms (inclined shear at other shear angles, flexural slip or flow, fault-bend slip, etc.) did not yield any changes in wedge shape, with one minor exception (see below).

The restoration and inferences from thermal geometry, *P–T* estimates and rheology along with the listed geological boundary conditions suggest a single, best-fit model (Fig. 8). Although each of the geological constraints has an individual error, their number and their varied, independent nature has the potential to control the solution as well as to indicate inconsistencies between the geological constraints that prevent a solution. No inconsistency was observed and the modelling success indicates that the chosen strategy is able to yield an internally consistent solution accounting for all constraints. Moreover, modelling runs showed that lithospheric strength had an overwhelming influence as compared with the other constraints. Only an elastically weak lithosphere, in particular, is consistent with low emergence of the wedge surface above sea level and the initial deep burial of the margin slope, in addition to the observation of small turbidite basins. Variations in the wedge deformation mechanism only slightly shifted the topographic slope break that is transfered from the continental slope–plateau boundary into the upper plate and its surface (arrow in Fig. 8), and the weakness of the lithosphere ensures that this region remains the locus of most efficient uplift. Interestingly, this locus is related to the site of underplating and subsequent exhumation of the high-pressure belt after basal accretion of the Rheno-Hercynian cover. The modelled load geometry also has implications for the timing of accretion based on the following observations: The sequential detachment segments (saucer shaped in the restored sections) straighten out to a flat to

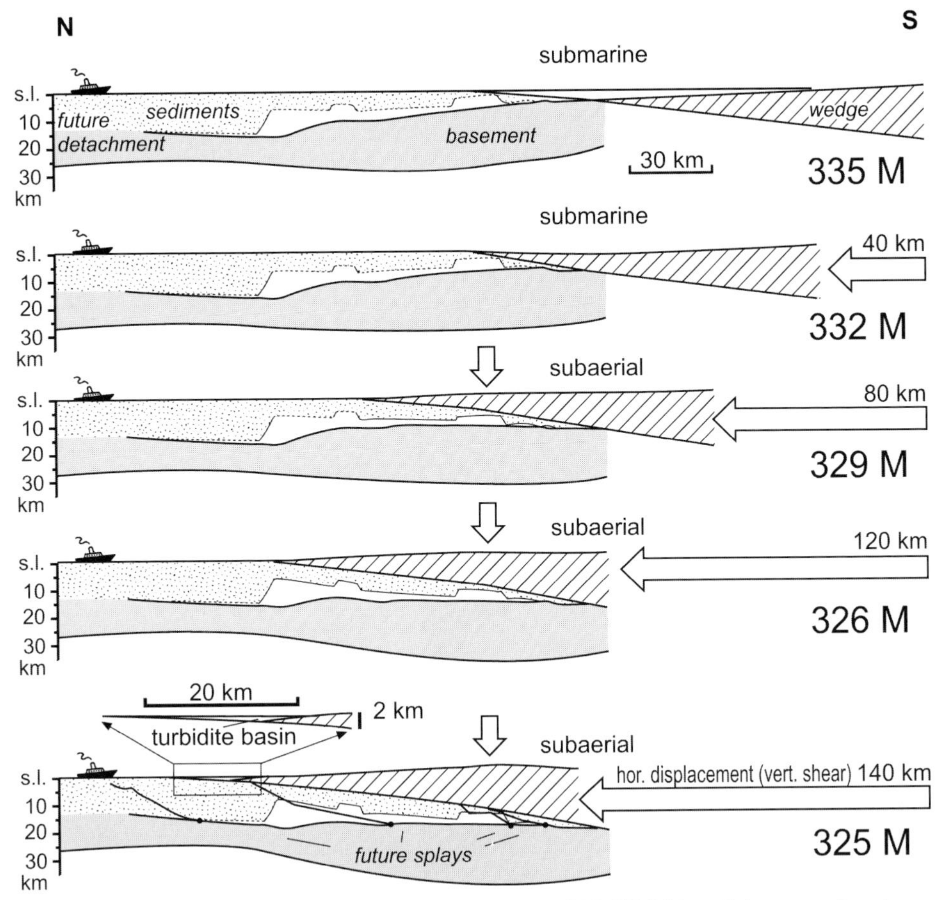

Fig. 8. Best-fit flexural isostatic model of early collision geometry (335–325 Ma) until the onset of passive margin failure and accretion. Arrow indicates site of preferential uplift above shelf-slope break (equivalent to later site of highest exhumation). Small turbidite basin and evolution from submerged to partly emerged state of the advancing wedge should be noted (see text for further detail).

weakly SE-dipping geometry. The approximate site of initiation of new detachment segments off the active detachment toward the foreland is constrained from the depth of the branch lines below the surface of the wedge. This value can be calculated from the observed barometric data above the branch lines at the exposed surface and their present-day depth in the balanced sections. The values are 6 kbar for the Phyllite Zone Thrust, 5 kbar for the Taunus Thrust and 4–5 kbar for the Boppard Thrust. Accordingly, the nucleation points of these detachments were located slightly south of the lower-plate bending zone (near wedge tip; see Fig. 9). As all were controlled by earlier basin structures, this observation indicates that nucleation occurred when a pronounced basin anisotropy with a related strength perturbation moved through the bending zone and reached the 300 °C isotherm. As a result of synkinematic heating of the wedge, this position is reached at decreasing pressures during wedge propagation.

The later detachments further north nucleated north of the bending zone on the flat-dipping detachment (Fig. 9). Higher crustal strength is here indicated by the substantial widening of the molasse basin that is succeeding the flysch stage. In the restored section this switch correlates with the initiation of accretion of the northern rift boundary and the deposition of sediments on the unrifted, stronger crust north of this boundary.

Discussion

Detachment localization

The above analysis has shown that detachment of the entire passive margin and its accretion to the

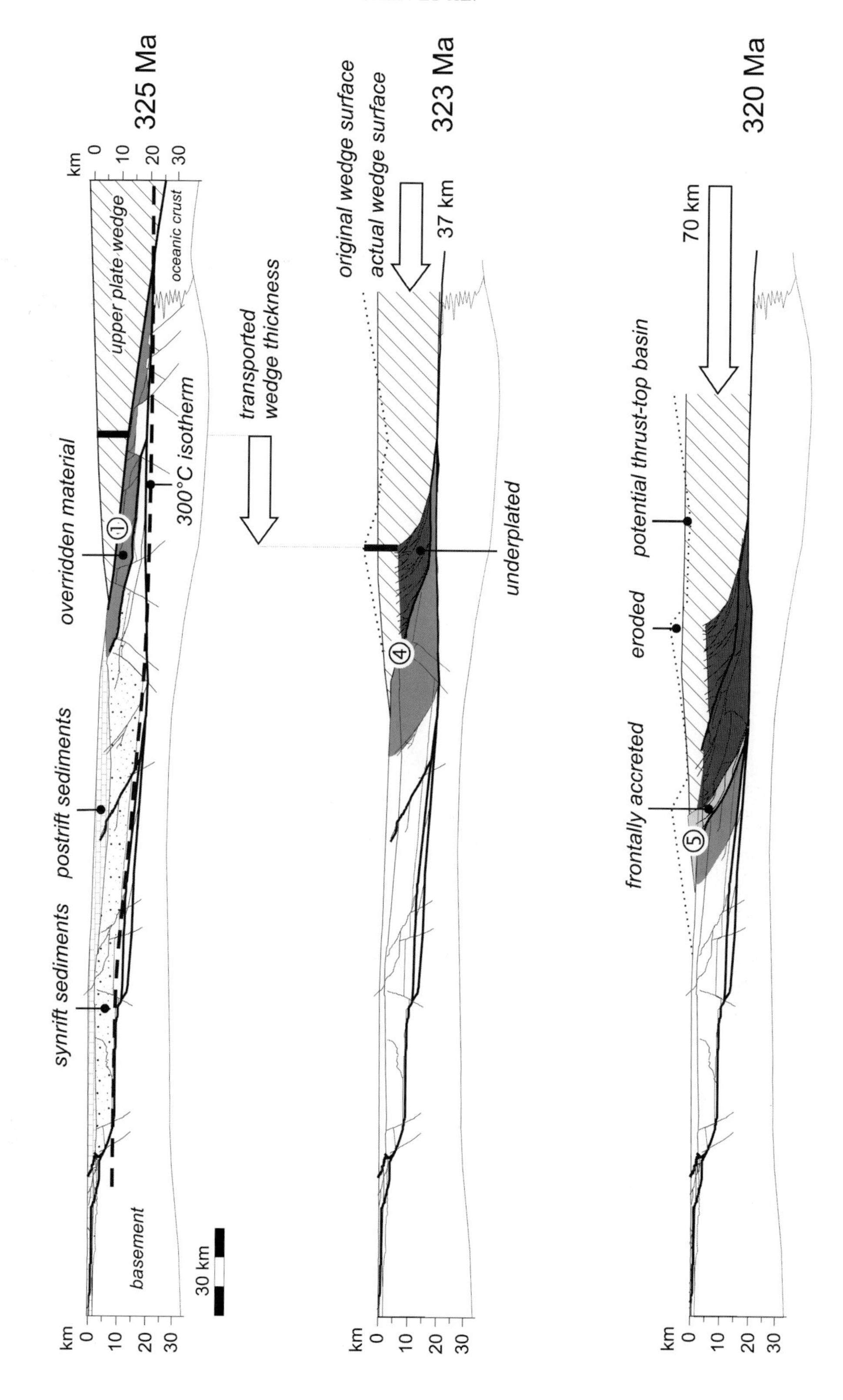
325 Ma
km
0
10
20
30
upper plate wedge
oceanic crust
overridden material
300°C isotherm
1
postrift sediments
synrift sediments
basement
30 km
transported wedge thickness
original wedge surface
actual wedge surface
37 km
323 Ma
underplated
4
320 Ma
70 km
potential thrust-top basin
eroded
frontally accreted
5

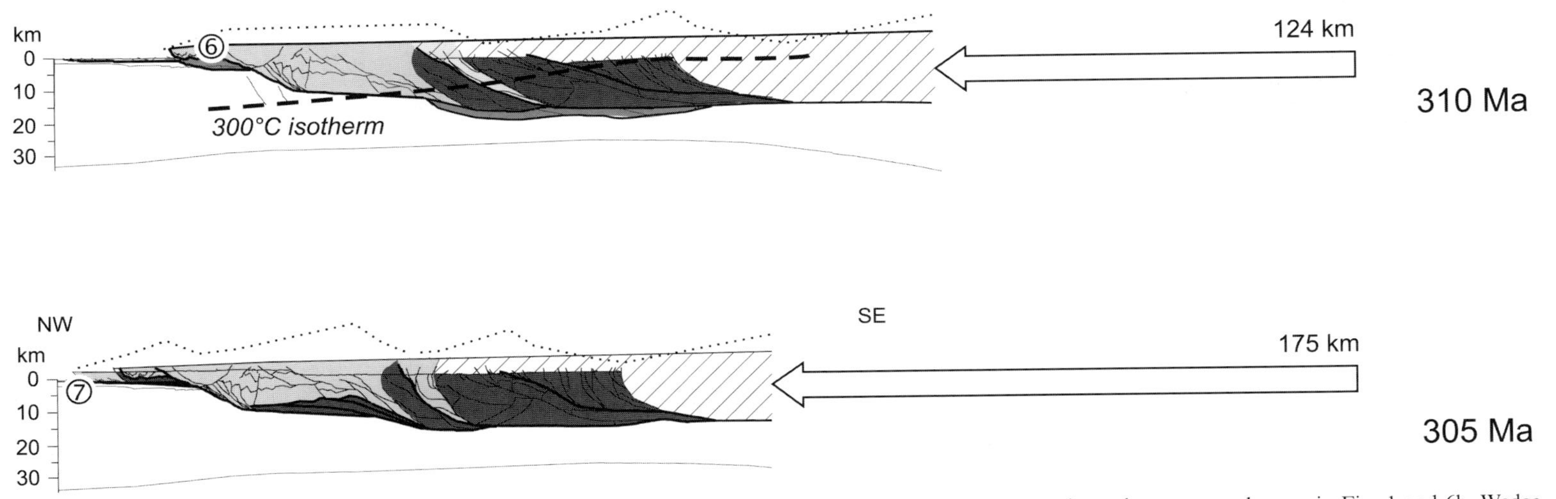

Fig. 9. Schematic sketch of detachment and thrust-belt growth (325–305 Ma). Activated detachment segments are shown by same numbers as in Figs 1 and 6b. Wedge surface is shown in equilibrated mode with average taper from barometry (see Plesch & Oncken 1999) here indicating uplifting (potentially eroded) and subsiding areas. Dark shading in duplexes indicates wedge growth by underplating; light shading corresponds to frontal accretion domains. Initial and late isotherm at brittle–plastic transition is shown from established P–T data (see text for details).

orogenic wedge progressed through a series of steps. A sequence of detachments forming individual imbricate fans is generated during early collision after offscraping of the uppermost sedimentary cover by the tip of the advancing upper-plate wedge. The sequential detachment segments increased in size and progressively stepped down into the crust of the southern Rhenish Massif. Their geometry in the restored sections is saucer shaped as a result of bending of a very weak lower-plate slab into the subduction zone. Its weakness (4 km of effective elastic thickness of the lithosphere) was inherited from previous crustal extension and related thermal weakening of the crust, as suggested by abundant volcanism in and at the margins of the central basin. Isostatic compensation of the upper-plate loading leads to mainly straight detachment segments branching off 'downwards' from the last active detachment toward the foreland when the future branch line has passed the bending zone and entered the brittle–plastic transition. Growth faults and horst blocks as well as extensional transfer zones systematically formed the sites of nucleation of all branch lines and frontal ramps in the detachment system. Similar findings have been reported from other fold and thrust belts (Powell 1989; McClay & Buchanan 1992), and suggest that even minor heterogeneities (e.g. growth faults within similar siliciclastic sediments) may cause a large enough stress or strength perturbation to nucleate or deflect thrust systems.

No localized extensional detachment can be established for the earlier stage of Early Devonian crustal extension. Integrated geometric, thermal and isostatic modelling by Heinen (1996) showed that the established basin-fill geometry and its thermal evolution (from maturation modelling of organic rank) can be accounted for only by straight normal fault systems of standard dip (50–70°) that root in a generally ductile lower crust at a deeper level than that used by the later contractional detachment. However, the rather shallow depth of the brittle–ductile transition during collision is probably a consequence of a not entirely re-equilibrated thermal state of the previously rifted basin as suggested from thermal modelling by Heinen (1996).

The imbricate splays within the hanging wall of each detachment segment do not seem to depend on the intrabasinal growth faults but rather show a relationship to sediment thickness. The length–thickness relationship of the internal imbricates (4:1 to 6:1 in the underplating-dominated area, 3:1 in the frontal accretion domain) seems to be a constant feature and therefore to be controlled by the mechanical properties of the basal detachment versus the hanging-wall sequence (see Platt 1988; Liu *et al.* 1992; Gutscher *et al.* 1996).

Whereas the uppermost detachments in the south are within the Devonian–Lower Carboniferous basin sequence, the final basal detachment is always located within the basement below the synrift sedimentary sequence. The basement–cover interface cannot be established as a relevant mechanical boundary; higher detachment segments occur above this boundary in pelitic sediments below quartzites or volcanic rocks in the Lower Devonian sequence and show dominant dissolution–precipitation creep. Detachment segments below the basement–cover interface have rheologies dominated by dissolution–precipitation and dislocation creep, and do not show any dependence on rock type.

Detachment strength

A change in the mechanical properties of the basal detachment is indicated by two important changes in wedge evolution: a substantial increase in wedge length and concomitant reduction in taper, as well as a switch from cyclic basal-to-frontal accretion to purely frontal accretion. From numerical studies and analogue experiments (Liu *et al.* 1992; Willett 1992; Gutscher *et al.* 1996), this has been suggested to reflect a substantial decrease in integral strength of the basal detachment for the Rheno-Hercynian thrust belt (Plesch & Oncken 1999).

Several other observations may help to explain this feature. Section balancing, exposures and xenoliths from young volcanoes, as well as analysis of the seismic reflection image of line DEKORP 1 by Oncken *et al.* (1999), have indicated that the metamorphic basement below the Devonian succession in the southern Rhenish Massif extends to about the area of the margin of the central basin. North of this boundary, a crystalline basement of unknown nature is covered by thick Lower Palaeozoic deposits in which the detachment developed. At this basin margin, therefore, the basal detachment crosses from gneissic basement into low-grade siliciclastic sediments.

The few exposures of the gneissic basement in the Taunus and Hunsrück Units show a variegated composition of amphibolites and mica-bearing quartz–feldspar rocks. In thin section, they exhibit brittle fractures and localized bands with dynamic recrystallization of quartz. Both features indicate a high-strength shear zone (for dislocation creep at around 300 °C; Paterson & Luan 1990) during this early stage of basement failure when the Taunuskamm–Soonwald Thrust was activated. The isothermal uplift observed in the adjoining Phyllite Zone rocks was shown to record substantial synkinematic heating of the internal wedge and the basal detachment above the value reported for feldspar and hornblende dislocation creep to occur (Voll 1976). Thus, synkinematic heating should have significantly contributed to mechanical softening of the crystalline basement in the south and to activation of the final basal detachment, which undercut the entire basin fill. A similar evolution is predicted from numerical modelling of small-scale orogenic wedges: Jamieson *et al.* (1998) have shown that detachment of the passive margin cover should initiate at the 300 °C brittle–plastic transition with a subsequent evolution to a deeper and warmer level (>500 °C). The main reason suggested is heating of the wedge system and concomitant weakening of its base near the upper–lower plate boundary through stacking of crustal material with a higher share of radioactive elements. Also, it may be expected in the case of the Rhenish Massif that frictional dissipation during the earlier stage of slip on a high-strength detachment has contributed to heating of the internal parts of the wedge.

Moreover, the described mechanism switch from cataclastic flow or from dislocation creep to dissolution–precipitation creep in the shear-zone rocks (Oncken 1989) is usually associated with a substantial drop in strength and shear localization (Rutter 1983; Groshong 1988; Schwarz & Stöckhert 1996; Stöckhert *et al.* 1999). A similar change of thermal conditions in the north as suggested for the south cannot be substantiated, as there was no substantial increase in overburden there.

Accordingly, it may be concluded that early accretion of the southern Massif occurred above a high-strength detachment (with high wedge taper and dominantly basal accretion), with a subsequent weakening during late accretion as a result of (a) thermal softening in the detachment zone (above feldspar yield strength) caused by synkinematic heating in the southern part of the wedge, and (b) a progressive increase of detachment length within mica-rich rocks that were controlled by another flow mechanism (with resulting low taper and dominantly frontal accretion). Initial localization of the basal detachment within the crustal pile must therefore have been caused by failure of the lower-plate crust at the brittle–plastic transition of quartz-dominated crust which, in a subduction zone setting, cut across the plate boundary at around the 300 °C isotherm. In summary, thermally activated mechanisms in a rheologically layered crust and subsequent localization of low-strength zones by a switch to grain-size sensitive deformation mechanisms and by synkinematic heating have controlled the depth level of passive margin detachment in the Rheno-Hercynian thrust belt.

Detachment propagation and sedimentation

The development of a number of imbricate fans each with individual architecture (Oncken *et al.* 1999) implies that detachment probably was not a stable process. The observed sequences from long underplated slabs to short frontally accreted imbricates have been suggested to reflect cyclic propagation (Plesch & Oncken 1999), a mode characteristic for high-strength basal detachments in analogue experiments (Gutscher *et al.* 1996). Supporting evidence on cyclicity comes from the analysis of turbidite basins (Ricken *et al.* this volume). Narrow, 40 km wide trench basins accumulated sedimentary sequences that can be linked to stepwise propagation and vertical growth of the thrust belt, probably as a result of cyclic underplating. The number of these basins (three) has been shown by Ricken *et al.* (this volume) to coincide temporally and spatially with some of the thrust systems. The age of the basins and the related respective location of the deformation front from geochronological data shows that the basins initiated in the trench position ahead of the frontal thrust. The lifetime of the basins (1.5 Ma) and the similar period of activity of the imbricate fans in the south is suggested to record alternation of slow wedge advance during thickening by underplating with periods of fast propagation during frontal accretion and wedge spreading. Initial coarsening followed by fining-upward filling sequences in the turbidite basins can be seen to corroborate this interpretation. Moreover, it tends to show that vertical motion in the hinterland was episodic (see above) with each return to dominantly basal accretion and wedge thickening.

The stepwise propagation of smaller detachment segments in the southern part of the downgoing continental plate caused repeated localized uplift above the site of basal accretion of the underplated units. The asymmetrically offset position of the detachment segments in the restored sections and their final stacking above each other in the south is a direct measure of the vertical material path component experienced by parts of the accreted material. Accordingly, the Phyllite Zone rocks were not only subducted deepest, they were also substantially uplifted and exhumed as a result of sequential stacking and underplating. A major consequence was shown to be an area of faster localized uplift in the advancing wedge, above the accreted Phyllite Zone, which is equivalent to the slope and rise material thrust over the shelf edge (Figs 8 and 9). Numerical studies have suggested that the inherited rifted margin and its strength have a very substantial role in the localization of uplift (Jamieson & Beaumont 1988). In the present case, the rift margin had this role during early collision until it was undercut by later detachment segments. Subsequent contraction of the shelf has displaced this area by some 150 km to the north.

A second such domain has been identified by Oncken (1997) under the site of the former magmatic arc (MGCH), where the first subducted Rheno-Hercynian basement rocks from under the original basal detachment have been detached under Barrovian metamorphic conditions and accreted to the upper plate, resulting in considerable uplift and erosion of parts of the upper plate. Basal accretion in both zones is roughly contemporaneous. The two zones are in part separated by remnant fore-arc crust that was not uplifted (Oncken 1998). The upper-plate material removed from both uplifting domains was feeding the foreland turbidite basin. Sequential failure of detachment segments with subsequent underplating of new material must have caused sequential stages of uplift and generation of subaerial topography suitable for erosional removal, as suggested from the stepwise propagation of the flysch basins and their filling sequence (Ricken *et al.* this volume). In conclusion, geometric analysis of balanced sections, isostatic modelling, fabric data from detachment rocks, and the propagation and fill history of turbidite basins all reflect the same linked process of stepwise wedge growth with cyclic frontal and basal accretion modes as a result of the propagation mode and the properties of the detachment.

Conclusions

Margin failure during arc–continent collision in the Variscides occurred above an initially high-strength detachment family. The detachment segments stepped down into the lower-plate crust with increasingly deeper parts of the continental outboard having been accreted in a cyclic mode to the base and tip of the upper-plate wedge. Detachment propagation from the oceanic realm into the lower-plate continental crust occurred when the ocean–continent transition crossed the brittle–plastic transition of quartz-bearing rocks. The detachment propagated by repeated ductile footwall failure along this boundary during large-scale flexural bending of the Rheno-Hercynian lower plate under the tectonic load of the advancing Saxo-Thuringian upper plate that moved as far north as the central graben. This early stage exhibits a cyclic wedge growth history, which is reflected in imbricate fan geometries, underplated and uplifted medium-pressure rocks, and stepwise turbidite basin propagation. Localization of the final basal detachment within the lower-plate crust was achieved by synkinematic thermal softening in gneissic feldspar-dominated basement under the internal orogenic wedge and by a mechanism switch from dislocation creep to dissolution–precipitation creep in clastic sediments near the brittle–ductile transition under the external parts of the wedge. Concomitant integral weakening of the basal detachment in this later stage is reflected by a change of wedge growth mode to dominantly frontal accretion, by a decrease in wedge taper, and by transition to a continuous foreland basin propagation. The localization of the downstepping branching topology toward the foreland and of the frontal ramps of the individual detachment generations is entirely controlled by the crustal architecture inherited from previous extension and basin formation at the passive margin: growth faults and horst blocks are consistently linked to later branch lines and frontal ramps formed during collision.

Research presented in this paper was supported by the Deutsche Forschungsgemeinschaft (Grant On 7/6), which is gratefully acknowledged. Furthermore, we wish to thank: H. Ahrendt, C. Büker, W. Franke, T. Kirnbauer, D. Meischner and R. Walter who contributed to the ideas expressed in this paper. The paper has especially benefited from constructive criticisms by S. Ellis, D. Tanner and an anonymous reviewer. Their support is gratefully acknowledged.

References

Adams, R. & Vandenberghe, N. 1999. The Meuse Valley section across the Condroz–Ardennes (Belgium) based on a predeformational sediment wedge. *Tectonophysics*, **304**, 179–195.

Ahrendt, H., Clauer, N., Hunziker, J. C. & Weber, K. 1983. Migration of folding and metamorphism in the Rheinisches Schiefergebirge deduced from K–Ar and Rb–Sr age determinations. *In*: Martin, H. & Eder, F. W. (eds) *Intracontinental Fold Belts. Case Studies in the Variscan Belt of Europe and the Damara Belt in Namibia*. Springer, Berlin, 323–338.

Aichroth, B., Prodehl, C. & Thybo, H. 1992. Crustal structure along the central segment of the EGT from seismic-refraction studies. *Tectonophysics*, **207**, 43–64.

ANDERLE, H.-J., MASSONNE, H.-J., MEISL, S., ONCKEN, O. & WEBER, K. 1990. Southern Taunus Mountains. *In*: *Field Guide 'Mid German Crystalline Rise and Rheinisches Schiefergebirge' (International Conference on Palaeozoic Orogens in Central Europe)*. Göttingen-Gießen, 125–148.

ANDRÉ, L. 1991. The concealed crystalline basement in Belgium and the 'Brabantia' microplate concept: constraints from the Caledonian magmatic and sedimentary rocks. *Annales de la Societe Géologique de Belgique*, **114**, 117–140.

BARBER, A. J., TJOKROSAPOETRO, S. & CHARLTON, T. R. 1986. Mud volcanoes, shale diapirs, wrench faults, and melanges in accretionary complexes, eastern Indonesia. *AAPG Bulletin*, **70**, 1729–1741.

BEHRMANN, J., DROZDZEWSKI, G., HEINRICHS, T., HUCH, M. & ONCKEN, O. 1991. Crustal-scale balanced cross sections through the Variscan fold belt, Germany: the central EGT-segment. *Tectonophysics*, **196**, 1–21.

BOIS, C., CAZES, M., CHOUKROUNE, P. *et al.* 1994. Seismic reflection images of the Pre-Mesozoic crust in France and adjacent areas. *In*: KEPPIE, J. D. (ed.) *Pre-Mesozoic Geology in France and Related Areas*. Springer, Berlin, 3–48.

CARMINATI, E. & SILETTO, G. B. 1997. The effects of brittle–plastic transitions in basement-involved foreland belts: the Central Southern Alps case (N Italy). *Tectonophysics*, **280**, 107–123.

CHARLTON, T. R., BARBER, A. J. & BARKHAM, S. T. 1991. The structural evolution of the Timor complex, eastern Indonesia. *Journal of Structural Geology*, **13**, 489–500.

DEKORP RESEARCH GROUP 1991. Results of the DEKORP 1 (BELCORP–DEKORP) deep seismic reflection studies in the western part of the Rhenish Massif. *Geophysical Journal International*, **106**, 203–227.

EGAN, S. 1992. The flexural response of the lithosphere to extensional tectonics. *Tectonophysics*, **202**, 291–308

ENGEL, W. & FRANKE, W. 1983. Flysch-sedimentation: its relations to tectonism in the European Variscides. *In*: MARTIN, H. & EDER, F. W. (eds) *Intracontinental Fold Belts. Case Studies in the Variscan Belt of Europe and the Damara Belt in Namibia*. Springer, Berlin, 289–321.

——, ——, GROTE, C., WEBER, K., AHRENDT, H. & EDER, F. W. 1983. Nappe tectonics in the southeastern part of the Rheinisches Schiefergebirge. *In*: MARTIN, H. & EDER, F. W. (eds) *Intracontinental Fold Belts. Case Studies in the Variscan Belt of Europe and the Damara Belt in Namibia*, Springer, Berlin, 267–287.

FIELITZ, W. 1992. Variscan transpressive inversion in the northwestern central Rheno-Hercynian belt of western Germany. *Journal of Structural Geology*, **14**, 547–563.

FRANKE, W. 1989. Tectonostratigraphic units in the Variscan belt of central Europe. *In*: DALLMEYER, R. D. (ed.) *Terranes in the Circum-Atlantic Palaeozoic Orogens*. Geological Society of America, Special Papers, **230**, 67–90.

—— & ONCKEN, O. 1995. Zur prädevonischen Geschichte des Rhenohercynischen Beckens. *Nova Acta Leopoldina*, **NF 71, 29**, 53–72.

——, BORTFELD, R. K., BRIX, M. *et al.* 1990. Crustal structure of the Rhenish Massif: results of deep seismic reflection lines DEKORP 2 North and 2 North-Q. *Geologische Rundschau*, **79**, 523–566.

GROSHONG, R. H. J. 1988. Low temperature deformation mechanisms and their interpretation. *Geological Society of America Bulletin*, **100**, 1329–1360.

GRÖSSER, J. & DÖRR, W. 1986. MOR-Typ-Basalte im östlichen Rheinischen Schiefergebirge. *Neues Jahrbuch für Geologie und Paläontologie, Monatshefte*, **12**, 705–712.

GROTE, C. 1983. *Strukturelle Untersuchungen im südwestlichen Verbreitungsgebiet der Gießener Grauwacke (Blatt 5516 Weilmünster)*. Dipl. thesis, Göttingen. 1–112.

GUTSCHER, M. A., KUKOWSKI, N., MALAVIEILLE, J. & LALLEMAND, S. 1996. Cyclical behavior of thrust wedges; insights from high basal friction sandbox experiments. *Geology*, **24**(2), 135–138.

HATCHER, R. D. & HOOKER, R. J. 1992. Evolution of crystalline thrust sheets in the internal parts of mountain chains. *In*: MCCLAY, K. R. (ed.) *Thrust Tectonics*. Chapman & Hall, London, 217–233.

HEINEN, V. 1996. Simulation der präorogenen devonisch-unterkarbonischen Beckenentwicklung und Krustenstruktur im Linksrheinischen Schiefergebirge. *Aachener Geowissenschaftliche Beiträge*, **15**, 1–161.

HOLL, H. G. 1995. Die Siliziklastika des Unterdevon im Rheinischen Trog (Rheinisches Schiefergebirge): Detritus-Eintrag und *P,T*-Geschichte. *Bonner Geowissenschaftliche Schriften*, **18**, 1–163.

HOLLMANNN, E. G. 1997. Der variszische Vorlandüberschiebungsgürtel der Ostbelgischen Ardennen—Ein bilanziertes Modell. *Aachener Geowissenschaftliche Beiträge*, **25**, 1–235.

JAMIESON, R. A. & BEAUMONT, C. 1988. Orogeny and metamorphism: a model for deformation and pressure–temperature time paths with applications to the central and southern Appalachians. *Tectonics*, **7**, 417–445.

——, ——, FULLSACK, P. & LEE, B. 1998. Barrovian regional metamorphism; where's the heat? *In*: TRELOAR, P. J. & O'BRIEN, P. J. (eds) *What Drives Metamorphism and Metamorphic Relations ?* Geological Society, London, Special Publications, **138**, 23–51.

KARIG, D. E., BARBER, A. J., CHARLTON, T. R., KLEMPERER, S. & HUSSONG, D. M. 1987. Nature and distribution of deformation across the Banda Arc–Australian collision zone at Timor. *Geological Society of America Bulletin*, **98**, 18–32.

KLÜGEL, T. 1997. Geometrie und Kinematik einer varistischen Plattengrenze—der Südrand des Rhenoherzynikums im Taunus. *Geologische Abhandlungen Hessen*, **101**, 1–215.

LIU, H., MCCLAY, K. R. & POWELL, D. 1992. Physical models of thrust wedges. *In*: MCCLAY, K. R. (ed.)

Thrust Tectonics. Chapman & Hall, London, 71–81.

MASSONNE, H. J. 1989. Metamorphoseentwicklung des Rhenoherzynikums am Beispiel eines Meta-Granodiorites aus dem Venn–Stavelot-Massiv. *Abstracts, 5th Rundgespräch zur Geodynamik des Europäischen Varistikums, Braunschweig, 16–18 November 1989*.

MATTE, P. 1991. Accretionary history and crustal evolution of the Variscan belt in Westem Europe. *Tectonophysics*, **196**, 309–337.

MCCLAY, K. R. & BUCHANAN, P. G. 1992. Thrust faults in inverted extensional basins. *In*: MCCLAY, K. R. (ed.) *Thrust Tectonics*. Chapman & Hall, 419–434.

MEISSNER, R., BARTELSEN, H. & MURAWSKI, H. 1981. Thin-skinned tectonics in the northern Rhenish Massif. *Nature*, **290**, 399–401.

NAFE, J. E. & DRAKE, C. L. 1963. Physical properties of marine sediments. *In*: HILL, M. N. (ed.) *The Sea 3*. Interscience, New York, 784–815.

ONCKEN, O. 1989. Geometrie, Deformationsmechanismen und Kinematik großer Störungzonen der hohen Kruste (Beispiel Rheinische Schiefergebirge). *Geotektonische Forschungen*, **73**, 1–215.

—— 1991. Aspects of the structural and paleogeothermal evolution of the Rhenish Massif. *Annales de la Societe Géologique de Belgique*, **113**, 139–159.

—— 1997. Transformation of a magmatic arc and an orogenic root during oblique collision and its consequences for the evolution of the European Variscides (Mid-German Crystalline Rise). *Geologische Rundschau*, **86**, 2–20.

—— 1998. Evidence for subduction–erosion in a fossil collisional belt—Mid European Variscides (Germany). *Geology*, **26**, 1075–1078.

——, VON WINTERFELD, C. & DITTMAR, U. 1999. Accretion and inversion of a rifted passive margin—the Late Palaeozoic Rhenohercyniam fold and thrust belt. *Tectonics*, **18**(1), 75–91.

ORD, A. & HOBBS, B. E. 1989. The strength of the continental crust, detachment zones and the development of plastic instabilities. *Tectonophysics*, **158**, 269–289.

PATERSON, M. S. & LUAN, F. C. 1990. Quartzite rheology under geological conditions. *In*: KNIPE, R. J. & RUTTER, E. H. (eds) *Deformation Mechanisms, Rheology, and Tectonics*. Geological Society, London, Special Publications, **54**, 299–307.

PLATT, J. P. 1988. The mechanics of frontal imbrication: a first-order analysis. *Geologische Rundschau*, **77**, 577–590.

PLESCH, A. & ONCKEN, O. 1999. Orogenic wedge growth during collision—constraints on mechanics of a fossil wedge from its kinematic record (Rheno-Hercynian fold-and-thrust belt, Central Europe). *Tectonophysics*, **304**, 117–139.

POWELL, C. 1989. Structural controls on Palaeozoic basin evolution and inversion in Southwest Wales. *Journal of the Geological Society, London*, **146**, 439–446.

RAOULT, J. F. & MEILLIEZ, F. 1987. The Variscan Front and the Midi Fault between the Channel and the Meuse River. *Journal of Structural Geology*, **9**, 473–479.

RICHARDSON, A. N. & BLUNDELL, D. J. 1996. Continental collision in the Banda arc. *In*: HALL, R. & BLUNDELL, D. (eds) *Tectonic Evolution of Southeast Asia*. Geological Society, London, Special Publications, **106**, 47–60.

RICKEN, W., SCHRADER, S., ONCKEN, O. & PLESCH, A. 2000. Turbidite basin and mass dynamics related to orogenic wedge growth: the Rheno-Hercynian case. *This volume*.

RUTTER, E. H. 1983. Pressure solution in nature, theory, and experiment. *Journal of the Geological Society, London*, **140**, 725–740.

SCHWARZ, S. & STÖCKHERT, B. 1996. Pressure solution in siliciclastic HP–LT-metamorphic rocks—constraints on the state of stress in deep levels of accretionary complexes. *Tectonophysics*, **255**, 203–209.

SNYDER, D. B, PRASETYO, H., BLUNDELL, D. J., PIGRAM, C. J., BARBER, A. J., RICHARDSON, A. & TJOKOSAPROETRO, S. 1996. A dual doubly vergent orogen in the Banda Arc continent–arc collision zone as observed on deep seismic reflection profiles. *Tectonics*, **5**, 34–53.

STÖCKHERT, B., WACHMANN, M., KÜSTER, M. & BIMMERMANN, S. 1999. Low effective viscosity during high pressure metamorphism due to dissolution–precipitation creep: a record of HP–LT metamorphic carbonates and siliciclastic rocks from Crete. *Tectonophysics*, **303**, 299–319.

TAVARNELLI, E. 1996. The effects of pre-existing normal faults on thrust ramp development: an example from the northern Apennines, Italy. *Geologische Rundschau*, **85**, 363–371.

TURCOTTE, D. L. & SCHUBERT, G. 1982. *Geodynamic Applications of Continuum Physics to Geological Problems*, Wiley, New York.

VOLL, G. 1976. Recrystallisation of quartz, biotite, and feldspars from Erstfeld to the Leventina nappe, Swiss Alps, and its geological significance. *Schweizerische Mineralogische und Petrographische Mitteilungen*, **56**, 641–647.

WEBER, J., ONCKEN, O. & WALTER, R. 1997. Inversion eines gedehnten passiven Kontinentalrandes—EDV-gestützte Profilbilanzierung im rechtsrheinischen Schiefergebirge. *Terra Nostra*, **97**, 193–195.

WHITE, N. J., JACKSON, J. A. & MCKENZIE, D. P. 1986. The relationship between the geometry of normal faults and that of sedimentary layers in their hanging walls. *Journal of Structural Geology*, **8**, 897–909.

WILLET, S. D. 1992. Dynamic and kinematic growth and change of a Coulomb wedge. *In*: MCCLAY, K. R. (ed.) *Thrust Tectonics*. Chapman & Hall, London, 19–31.

Deformation, metamorphism and exhumation: quantitative models for a continental collision zone in the Variscides

M. SEYFERTH & A. HENK

Institut für Geologie, Universität Würzburg, Pleicherwall 1, D-97070 Würzburg, Germany (e-mail: michael.seyferth@mail.uni-wuerzburg.de)

Abstract: Two-dimensional thermo-mechanical finite-element models are used to gain a quantitative insight into the complex strain partitioning in continental collision zones. If models with Moho temperatures of 700–900 °C, as is indicated by petrological data, are simulated, frequently used flow laws for the lower crust cannot reproduce significant crustal thickening. Instead, decoupling between crust and mantle occurs, resulting in the widening of a diffuse deformation zone. To reproduce observed petrological data and orogen geometries, a stronger lower crust, with viscosities between 10^{21} and 10^{23} Pa s, is required. Models are applied specifically to a collision zone from the Variscan Orogen of Central Europe to understand the tectonometamorphic history, strain partitioning within the collision zone, as well as the rapid synconvergent exhumation of metamorphic complexes. Model predictions agree with the observed distribution of peak metamorphic conditions and show systematic variations of contemporaneous pressure–temperature (P–T) paths across the collision zone.

Rocks in continental collision zones may experience burial, intense deformation and metamorphism. However, the collision of continental crust not only brings rocks to greater depths and temperatures; it also provides a very effective mechanism to bring them back to the surface. In general, two processes can contribute to the exhumation of rocks in orogenic settings: surface erosion and tectonic denudation by normal faulting and ductile thinning (England & Molnar 1990; Jamieson 1991). Both exhumation processes are related to the topography generated by the continent–continent collision. The topographic relief of an orogen not only enhances exhumation by erosion of the overburden; additionally, local gravitational instabilities induced by the high topography and overthickened crust can result in exhumation by synconvergent extension (Dewey 1988; Burg *et al.* 1994). The latter mechanism implies a pronounced strain partitioning within the colliding continental crust, as its lower parts are still under compression whereas its uppermost part may already be extending. The concept of strain partitioning is used to describe how strain varies both laterally and vertically in the modelled 2D section. Numerical models have proven to be a particularly valuable tool in gaining a quantitative understanding of the fundamental processes controlling the evolution and structure of orogenic belts (e.g. Willett *et al.* 1993; Beaumont *et al.* 1994). In particular, they provide information on how strain is distributed within a collision zone and help to assess to what extent the deformation commonly observed in upper- and middle-crustal rocks can be used as an indicator for the tectonic processes in the deeper crust and upper mantle.

In this study we use finite-element (FE) techniques to obtain quantitative insight into the temporal and spatial distribution of deformation and metamorphism, as well as the mechanisms for rapid synconvergent exhumation in areas of colliding, hot continental crust. We compare the model predictions with field data from a well-documented collision zone located at the northern flank of the Late Palaeozoic Variscan Orogen in Central Europe. Available datasets relevant to the orogenic evolution of this area include petrological and geochronological data (e.g. Okrusch 1995), seismic sections (Meissner & Bortfeld 1990) and convergence estimates (Oncken 1998). Although fossil orogens usually lack some information such as palaeotopography and erosion rates, they can provide insights into deeper crustal levels and the processes of their exhumation to the surface. In addition, they can provide information on the complete orogenic history rather than being restricted to early stages of orogenesis observable in active orogens. Thus, our modelling results may have general implications for the understanding of strain partitioning,

From: FRANKE, W., HAAK, V., ONCKEN, O. & TANNER, D. (eds).
Orogenic Processes: Quantification and Modelling in the Variscan Belt.
Geological Society, London, Special Publications, **179**, 217–230. 0305-8719/00/$15.00

metamorphism and exhumation processes in continental collision zones characterized by elevated crustal temperatures.

Modelling concept

Methodically, this study builds on the work of Willett *et al.* (1993) and Beaumont & Quinlan (1994), among others, who used 2D FE models to calculate crustal deformation within compressional orogens and compared model predictions with natural examples. The numerical simulations describe collision of two continental fragments by applying displacements to material points at the base of one crustal block, whereas the base of the other is fixed horizontally (Fig. 1). The point at which the discontinuity in the basal boundary condition occurs is termed S and represents the locus of asymmetric detachment and underthrusting of the mantle lithosphere. As a result of convergence, an orogen develops showing a marked asymmetry in general topographic relief, as well as in internal structure and strain distribution. Typically, the collision zone is characterized by a broad zone of thrusts and folds verging toward the subducting plate, the so-called pro-side, whereas a narrower zone of deformation verging towards the stationary plate forms on the retro-side. Deformation also follows a typical temporal sequence starting with an initial stage of block uplift between two conjugate deformation zones termed step-up shears, which ultimately leads to a pro-wedge with minimum critical taper and a retro-wedge comprising two segments with a maximum and minimum critical taper (Fig. 1; terminology after Willett *et al.* (1993)). This basic deformation pattern can be modified by external processes, i.e. asymmetric erosion, as well as thermal weakening of the thickened crust, which leads to plateau formation and gravitational collapse (Willett *et al.* 1993). Beaumont & Quinlan (1994), Beaumont *et al.* (1994) and Jamieson *et al.* (1998) gave comprehensive discussions of the modelling concept and variations in input factors, e.g. temperature gradient, rheological stratification, locus of the discontinuity in the prescribed displacement field and erosional denudation. It should be noted that the models presented in those papers assume very low to moderate crustal temperatures, i.e. Moho temperatures between 360 and 600 °C. Temperatures in the Variscides were significantly higher, which is fundamental in understanding the evolution of this orogen. Thus, before applying the model concept outlined above to the specific continental collision zone of the Variscan Orogen, we will briefly discuss the special features of our numerical modelling approach and some general implications of high crustal temperatures and weak rheologies on the resulting orogen geometry and strain distribution.

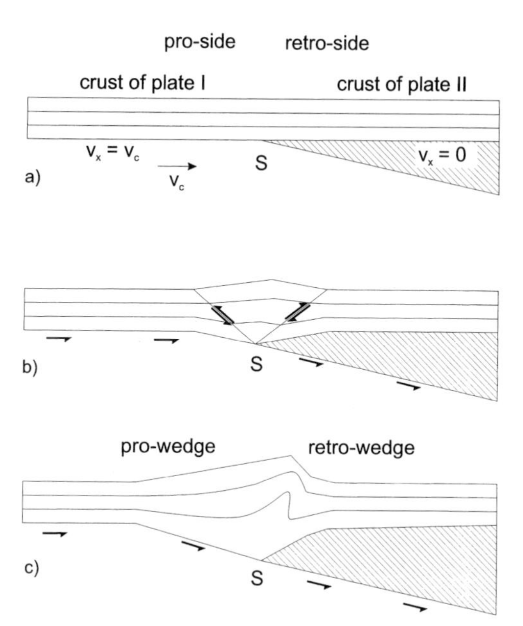

Fig. 1. Modelling concept (simplified after Willett *et al.* (1993)). (**a**) The initial model set-up describes collision of two continental fragments by applying displacement to the base of one crustal block, whereas the base of the other is fixed horizontally. The point at which the discontinuity in the basal boundary condition occurs is termed *S*, and represents the locus of asymmetrical detachment and underthrusting of the mantle lithosphere. (**b**) Deformation starts with block uplift between two conjugate deformation zones. (**c**) Continuing convergence leads to formation of a pro-wedge with minimum critical taper and a retro-wedge comprising two segments with maximum and minimum critical taper.

Numerical modelling approach

The numerical simulations are based on a 2D plane-strain FE approach using the ANSYS® (ANSYS Inc., Houston, TX, USA) software package. A comprehensive description of the capabilities of ANSYS® for the numerical simulation of geodynamic processes has been given by Henk (1998). The model in this study describes a crustal-scale section oriented perpendicular to the strike of the evolving orogen. The crust is divided into an upper and a lower part, each characterized by specific thermal and mechanical material properties (Table 1). The thermal calculations consider, among other

Table 1. *Thermal and mechanical material parameters used in this study; whenever possible, material parameters specific the Variscan crust in Central Europe are used*

Variable	Symbol	Value	Reference
Initial model geometry			
Thickness upper crust (m)	h_{uc}	15×10^3	
Thickness lower crust (m)	h_{lc}	15×10^3	
Model width (m)	w_{ini}	1200×10^3	
Rate of convergence (m a^{-1})	v_c	15×10^{-3}	
Mechanical material properties			
Poisson's ratio	ν	0.25	
Upper crust			
Density at 273 K (kg m^{-3})	ρ	2800	
Young's modulus (Pa)	E	0.5×10^{11}	
Strain rate coefficient (Pa^{-n} s^{-1})	a_0	3.16×10^{-26}	Hansen & Carter (1983)
Activation constant (K)	Q/R	22432	Hansen & Carter (1983)
Power-law stress exponent	n	3.3	Hansen & Carter (1983)
Bulk strain rate (s^{-1})	ϵ'	1×10^{-14}	
Lower crust			
Density at 273 K (kg m^{-3})	ρ	3000	
Young's Modulus (Pa)	E	0.8×10^{11}	
Strain rate coefficient (Pa^{-n} s^{-1})	a_0	5.05×10^{-28}	Mackwell *et al.* (1998)
Activation constant (K)	Q/R	58693	Mackwell *et al.* (1998)
Power law stress exponent	n	4.7	Mackwell *et al.* (1998)
Bulk strain rate (s^{-1})	ϵ'	1×10^{-14}	
Thermal material properties			
Surface temperature (K)	T_s	273	
Basal heat flow (W m^{-2})	q	0.04	
Upper crust			
Thermal conductivity (W m^{-1} K^{-1})	k	3–1.3	(1), Zoth & Hänel (1988)
Specific heat (m^2 s^{-2} K^{-1})	C	1.3×10^3	
Radiogenic heat production (W m^{-3})	H	2.3×10^{-6}	(2), Cermák (1995)
Lower crust			
Thermal conductivity (W m^{-1} K^{-1})	k	2.5–1.7	(1), Zoth & Hänel (1988)
Specific heat (m^2 s^{-2} K^{-1})	C	1.3×10^3	
Radiogenic heat production (W m^{-3})	H	0.52×10^{-6}	(2), Cermák (1995)

(1) Temperature-dependent; (2) corrected for 300 Ma.

factors, temperature-dependent thermal conductivities and radiogenic heat production. Thermal boundary conditions of the model are a constant heat flow applied to the base of the crust and a constant surface temperature of 0 °C. Lateral variations in the basal heat flow that may result from mantle subduction (Jamieson *et al.* 1998) and the preceding oceanic subduction stage are not considered. For the mechanical calculations the irreversible deformation in the brittle domain is described by an elastic–perfectly plastic material law with a pressure-dependent yield strength approximating Byerlee's relationship (Byerlee 1978) and assuming hydrostatic fluid pressure conditions. Deformation in the ductile domain can be described either by temperature-dependent viscous flow or by temperature- and strain-rate-dependent creep laws (see below). The top of the model is a free surface and isostatic forces act at its base. Convergence is modelled by a displacement boundary condition applied to the basal nodes left of a point S. This model concept implies that convergence is driven by the mantle lithosphere until it detaches at the mantle subduction point S. As the crustal rheology is strongly temperature dependent and substantial advective heat transport occurs, the thermal and mechanical calculations have to be coupled via temperatures and displacements (see Henk (1998) for further details on the modelling techniques).

High crustal temperatures and/or convergence in excess of 100 km can result in severe distortion of the basal elements and often cause numerical problems. To overcome these limitations a remeshing algorithm was developed as an add-on to ANSYS®. With this technique the deformed FE grid can be repeatedly replaced by a new one that is capable of experiencing further deformation. The new grid consists of the same number of elements and covers the geometry of the previously deformed model. It also keeps track of the material domains of the upper and lower crust. New nodes are arranged in vertical columns, thus approaching nearly rectangular element shapes.

The modelling results calculated before remeshing (temperature, strain) are interpolated to the new FE grid. To keep track of the evolution experienced by the initial grid points, a tracking grid of marker points is used. This grid describes the particle paths in space and time and can be directly compared with observed *P–T–t* paths. As four-node elements are used, the relative position of marker points can be described by three parameters, comprising the number of the element currently containing the marker point and the marker point's position relative to opposite element faces. Vice versa, the absolute marker point position is updated using the position of the corresponding edge nodes.

This technique not only permits large as well as localized deformation, but also allows changes in the domain geometry by adding or removing parts of the model. Thus, it is possible to describe sedimentation and erosion at the upper surface of the model as well as processes such as tectonic erosion and delamination at the base of the crust.

Impact of temperature and rheology on the orogen geometry and strain distribution

Crustal temperature and rheology are the two most important factors controlling the deformation within an evolving orogen, as well as its overall shape. Both factors are coupled, as the rheology in the ductile domain is strongly temperature dependent. High temperatures and the corresponding low integrated strength of the crust may impose important limitations on the development of significant crustal thickness and topography within continental collision zones. This relationship between rheology and orogen geometry can be illustrated by two numerical experiments that differ only in the assumed initial temperature field (Fig. 2). Both models apply the power-law creep parameters for dry anorthosite (Shelton & Tullis 1981), commonly used in geodynamic studies to represent the lower crust. After 100 km of convergence, the 'cold' model, i.e. with initial Moho temperatures of 500 °C, shows a localized crustal thickness increase by up to 55%. The strain distribution depicts the typical bivergent pattern rooting at point S with additional subhorizontal strain zones between upper and lower crust (Fig. 2a; see also Beaumont & Quinlan (1994, figs 2 and 11)). The comparatively high strength of the lower crust and the resulting good coupling between crust and mantle are responsible for the strain localization and crustal thickening in the vicinity of S. In contrast, the same model scenario but with an initial Moho temperature of 700 °C shows no such strain localization but only a wide zone of diffuse deformation (Fig. 2b; see also Beaumont & Quinlan (1994, fig. 6)). The crust is thickened only by 25%, indicating that crustal convergence is compensated by widening of the deformation zone rather than crustal thickening. Any asymmetry in orogen geometry and strain distribution is only very weakly developed. The reason for this is the low strength, particularly of the lower crust, so that the imposed movement of the mantle is only weakly coupled to the deformation within the crust. Consequently, a Moho-parallel high-strain zone develops in the lowermost crust spreading out towards the pro- and retro-side.

From these modelling results it appears very difficult to thicken hot continental crust to generate sufficient topography that would encourage substantial synconvergent exhumation by erosion and tectonic denudation. Obviously, such a conclusion would contradict field observations from several orogens, including the Variscan Orogen of Central Europe, which show focused deformation and crustal thickening although petrological data point to lower-crustal temperatures of 800 °C and more. Also, numerical models assuming a wide zone of weak, hot model crust, embedded between stronger and colder crust ('vice models'; see Ellis *et al.* (1998)) result in diffuse deformation and only minor crustal thickening. This inconsistency could be solved by recent experimental studies indicating that the effective strength of rocks representative of the lower crust may be much higher than previously thought (Mackwell *et al.* 1998; Vauchez *et al.* 1998). Inappropriate sample preparation may have changed the rheologies of supposed dry rocks, which were actually softened by water weakening, breakdown of hydrous minerals and/or dehydration melting (Mackwell *et al.* 1998). The revised flow laws would imply that the lower crust is much stronger and may be as stiff as the upper mantle. Consequently, a good mechanical coupling between lower crust and mantle would exist at the Moho, allowing localized deformation and

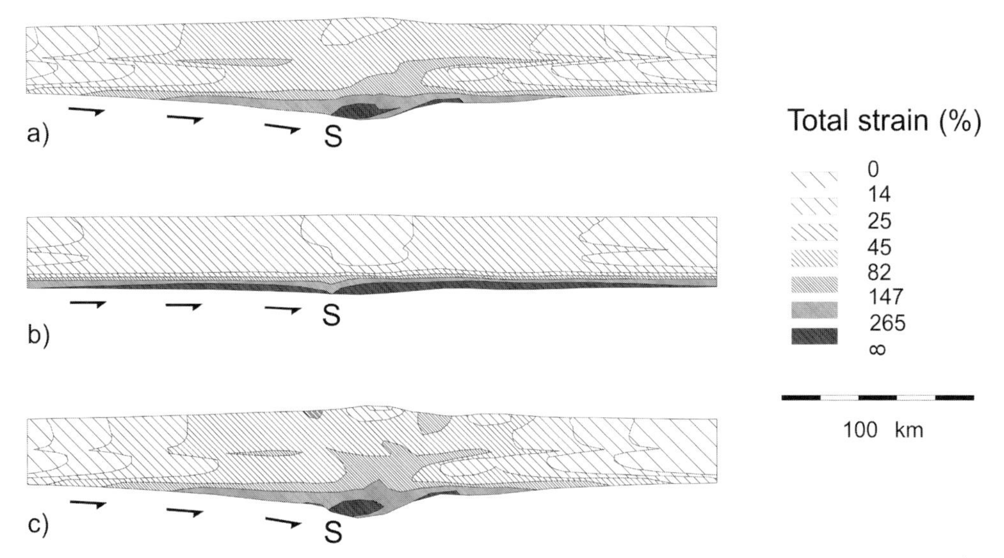

Fig. 2. Deformed FE models contoured for total von Mises strain after 100 km of convergence at 20 mm a^{-1}. (**a**) 'Cold' crust with an initial Moho temperature of 500 °C. For the lower crust the anorthosite flow law of Shelton & Tullis (1981) is assumed. (**b**) 'Hot' crust with an initial Moho temperature of 700 °C. For the lower crust the anorthosite flow law of Shelton & Tullis (1981) is assumed. (**c**) Same as (**b**) but for the lower crust the dry diabase flow law of Mackwell *et al.* (1998) is used.

crustal thickening. Additionally, the discrepancy between laboratory measurements and large-scale behaviour of the lower crust may in part be due to sample size and extrapolation of strain rates over several orders of magnitude (Kohlstedt *et al.* 1995).

The present study utilizes the flow laws for dry Westerly granite (Hansen & Carter 1983) and dry Maryland diabase (Mackwell *et al.* 1998) to represent upper crust and lower crust, respectively (Fig. 3). However, even these dry rheologies exhibit a strong strain-rate dependence and as a result of the imposed displacement boundary condition a high-strain zone develops at the base of the crust. Although these effects of strain softening and decoupling between crust and mantle, respectively, are less pronounced than for the wet rheologies, they still impede localized crustal thickening. If this strain-rate dependence of the power-law rheology is removed by assuming an average strain rate of 10^{-14} s^{-1}, effective viscosities between 10^{21} and 10^{23} Pa s, depending on temperature, result for the lower crust (Fig. 3). If such a temperature-dependent viscous rheology is applied to the model scenario outlined above, the resulting numerical model shows a pronounced thickness increase and localized strain accumulation in spite of high crustal temperatures (Fig. 2c).

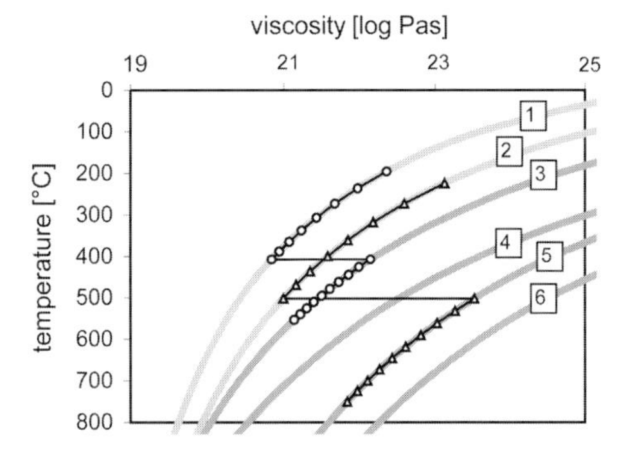

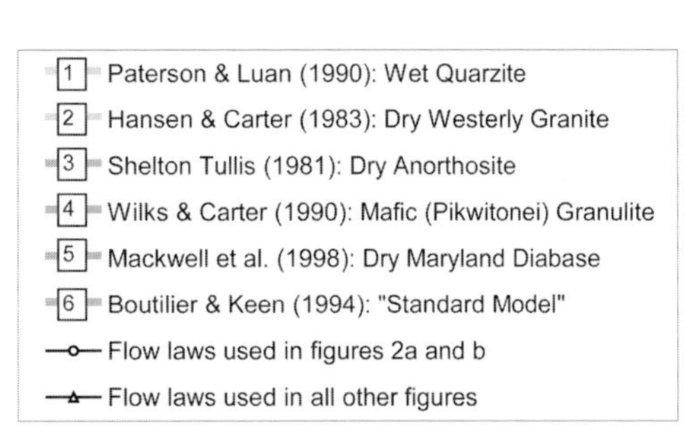

Fig. 3. Flow laws based on power-law creep parameters of upper- and lower-crustal rocks (for $\dot{\epsilon} = 10^{-14}$ s^{-1}).

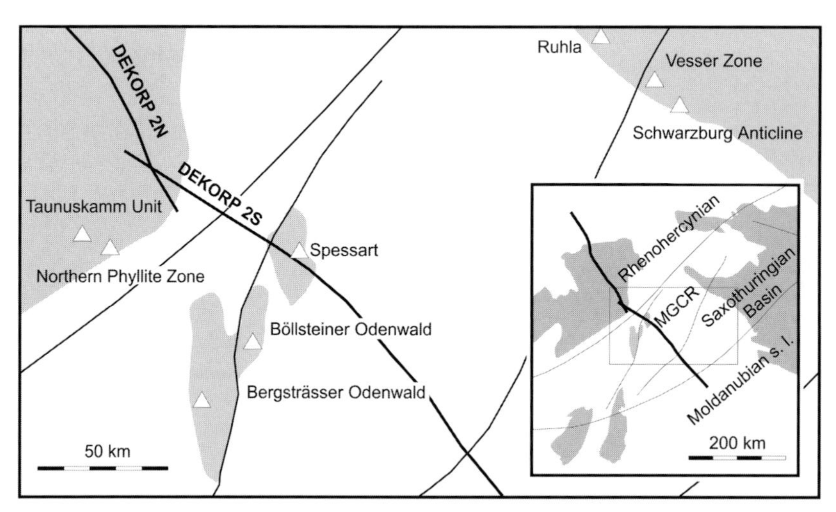

Fig. 4. Sketch map showing Variscan massifs in Central Europe and the location of the study area. Bold lines indicate deep seismic reflection profiles used for comparison with the results of the numerical simulations.

Case study: strain partitioning and rapid synconvergent exhumation at the boundary between internal and external zone of the Variscan Orogen

The Variscan Orogen in Central Europe (see Fig. 4) was formed by the sequential collision of microplates, derived from the northern margin of Gondwana, with Laurentia and Baltica during the Carboniferous period (see Franke (1992), Walter (1992), Oncken (1997) and this volume for comprehensive reviews of the Variscan evolution and regional geology). Typical features of the final collision stage are the widespread and almost synchronous high-temperature–low-pressure (HT–LP) metamorphism and granite magmatism as well as the rapid exhumation of metamorphic complexes. This case study concentrates on a cross-section through the northern flank of the Late Palaeozoic Variscan Orogen and aims for a quantitative analysis of the collision between a passive margin (Rheno-Hercynian Zone) in the north and a former magmatic arc situated on a fragment of continental crust (Mid-German Crystalline Rise and Saxo-Thuringian Zone) in the south. This collision zone also represents the boundary between the internal zone of the orogen and the external fold and thrust belt. This area was chosen because various datasets are available to constrain a numerical modelling approach, e.g., estimates for convergence velocity, crustal shortening (Oncken 1998), petrological and geochronological data (for comprehensive review see Okrusch (1995)). Additionally, deep seismic profiles (Meissner & Bortfeld 1990) can be used to infer the internal structure of the collision zone and the location of high-strain zones within the crust.

Geological constraints for numerical model

The collision zone between the Rheno-Hercynian and Saxo-Thuringian Zone traces the former site of the Lizard–Giessen–Harz Ocean, a relatively small oceanic basin presumably only a few hundred kilometres wide (Franke 1992; Oncken 1997). Closure of this basin from late Mid-Devonian time onwards resulted in the formation of a magmatic arc in the northern Saxo-Thuringian Zone, the so-called Mid-German Crystalline Rise (MGCR). Subduction of oceanic crust was essentially completed by the end of Devonian time. Continent–continent collision and final closure of the basin continued throughout Early Carboniferous time and probably accelerated from Late Viséan time onwards. Continental collision between the Rheno-Hercynian and Saxo-Thuringian Zones was contemporaneous with rapid exhumation of metamorphic rocks, as is indicated by thermal–kinematic modelling of various *P–T–t* data sets from the MGCR (Henk 1995). At *c.* 325 Ma, the MGCR was thrust above the southernmost Rheno-Hercynian Zone and subsequently the deformation front continued to migrate further northwestward. The youngest sediments that were affected by compression are located in the Ruhr Basin area and are Westphalian D (about 305 Ma) in age (Ziegler 1990).

Within the Rheno-Hercynian Zone, the rocks at present exposed at the Earth's surface are unmetamorphosed near the orogenic front, but peak metamorphic pressure conditions gradually increase towards the south and reach 6 kbar in the southernmost part (Northern Phyllite Zone; Anderle *et al.* 1990). Crustal shortening across this orogenic wedge was achieved mainly by thrusting and amounted to 42% (Dittmar *et al.* 1994), i.e. 116 km assuming a present width of 160 km. North-verging thrusts separate the Rheno-Hercynian Zone from the crystalline core of the Variscides. The MGCR represents a roughly triangular block bounded by outward-verging thrusts (Oncken 1998). Rocks at the surface show amphibolite-facies metamorphic grade with peak pressures of 6–8 kbar at 600–650 °C (Willner *et al.* 1991; Okrusch 1995). Immediately south of the MGCR the northern Saxo-Thuringian Zone consists of a small south-facing fold and thrust belt, which has been interpreted as the retro wedge of the collision zone (Schäfer 1997; Oncken 1998). Shortening within this wedge was less than 40% and was achieved by folding and ductile shortening rather than thrusting. Peak metamorphic pressures increase from close to zero near the southern tip of the wedge to 4–5 kbar at the northern rim (Kemnitz 1995; Schäfer 1997). Compared with the pro-side, the pressure gradient on the retro-side is much steeper.

In summary, the general structure of the collision zone between the Rheno-Hercynian and Saxo-Thuringian Zones shows several features that are typical for small, bivergent orogens. The total amount of shortening across the Rheno-Hercynian pro-wedge, the pop-up structure of the MGCR and the Saxo-Thuringian retro-wedge was estimated by Oncken (1998) to be at least 250 km.

Numerical model of the study area

The numerical model comprises an initially 1200 km long section, which represents a traverse through the Rheno-Hercynian and Saxo-Thuringian tectonometamorphic units at the onset of continental collision, i.e. after the closure of the Lizard–Giessen–Harz Ocean. The model is oriented perpendicular to the strike of the Variscan units and the large model width is necessary to avoid disturbance of the modelling results by side-wall effects. Initially, a uniform Moho depth of 26 km is assumed, averaging the crustal thicknesses estimated by Oncken *et al.* (this volume) for the Rheno-Hercynian passive margin, the magmatic arc of the evolving MGCR and the adjacent Saxo-Thuringian Basin. Altogether, the numerical model consists of 1960 elements.

The rheology of the upper and the lower crust is approximated by flow laws for granite and dry diabase. The assumption of dry conditions in the lower crust in the vicinity of the modelled traverse may be justified by its granulitic composition documented by xenoliths found in Tertiary volcanic rocks (Blundell *et al.* 1992).

The convergent plate boundary between the Rheno-Hercynian passive margin and the Saxo-Thuringian Terrane is indicated by the point S at the base of the model crust, which is initially located at a distance of 700 km from the left model margin. The mechanical boundary conditions reflect southeast-directed mantle subduction and underthrusting of the left-hand plate, i.e. the Rheno-Hercynian Zone. The convergence velocity is assumed to be 15 mm a^{-1} over a time span of 17 Ma, corresponding to a total amount of shortening of about 250 km, as was estimated by Oncken (1998) for the minimum amount of convergence.

The influence of surface processes on the deformation and exhumation pattern is taken into account by applying moderate erosion rates to the model surface. As no specific information on erosion rates exists for the Variscides, some general values published by Summerfield & Hulton (1994) have been used to establish multilinear functions describing erosion rate v. surface elevation. However, there are geological data from the study area that provide information about the relative distribution of erosion: the metamorphic profile shows more pronounced exhumation at the northwestern flank of the orogen, and syncollisional flysch sediments were transported almost entirely towards the northwestern foreland (Oncken *et al.* this volume). Both facts imply an asymmetric erosion, which was focused on the northwestern flank of the orogen, i.e. towards the pro-side. Therefore, initial model runs using a uniform erosion rate have been modified by introducing a watershed at the maximum model surface elevation and different erosion rates on either side of it. Erosion rates on the pro-side were assumed to be 10 times larger than on the retro-side.

The deformation of the initial crustal section in response to the prescribed displacement boundary condition was studied using the dynamic FE approach outlined above. The time step for the mechanical and thermal calculations is 100 ka, whereas remeshing of the deformed FE grid is carried out every 1 Ma.

Results of the numerical simulation

Modelling results are illustrated using the tracking grid, which represents the cumulative

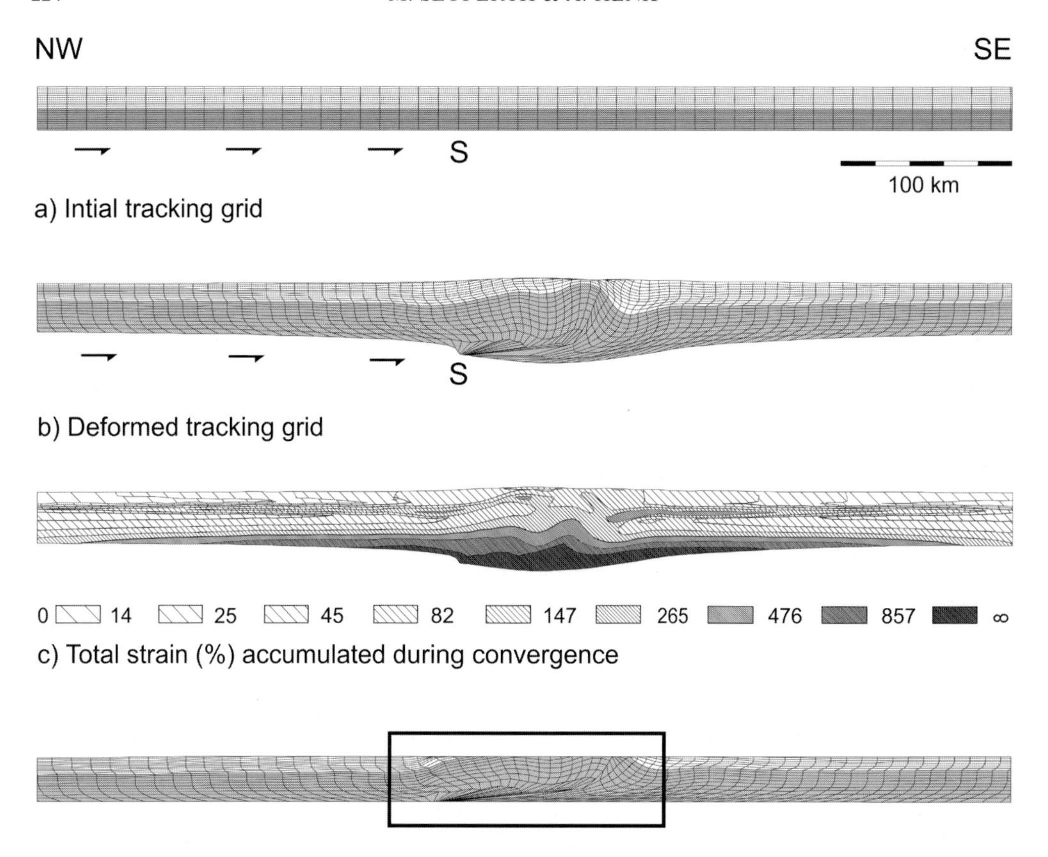

Fig. 5. (**a**) Initial tracking grid (only central part is shown). (**b**) Deformed tracking grid after the end of convergence. Shading shows the compositional layering within the crust. (**c**) Total von Mises strain accumulated during the convergence stage. (**d**) Deformed FE grid geometrically corrected for postconvergent extension and erosion to allow comparison with the present-day crustal configuration. Box indicates central part of the model shown in detail in Fig. 6.

displacement of the initial grid nodes. The deformed tracking grid at the end of the convergence stage, i.e. after 17 Ma and 250 km of convergence, is shown in Fig. 5b. The deformed grid of the upper plot shows a dome-like zone of uplifted lower-crustal rocks, whose most prominent part is related to the retro-zone of the orogen. There is also a notable difference in the dip angles of the top of the lower crust: a more gently dipping pro-side and a steeply dipping retro-side.

This asymmetry is also clearly visible on finite strain plots (Fig. 5c). Strain accumulated during the orogeny is mainly focused on the material boundaries at the Moho (i.e. the model base) and between the upper and the lower crust. Both boundaries act as major zones of subhorizontal ductile shearing, but, in contrast to the low-viscosity model depicted in Fig. 2a, coupling between the upper mantle and the model crust is still effective enough to thicken the crust by more than 50%. Because of the large amount of convergence, as well as the influence of surface erosion, the bivergent strain pattern is less prominent than in the initial stages (e.g. in Fig. 2a).

Comparison with seismic reflection profiles

The deformed FE grid shown in Fig. 5b represents the situation after continental collision ceased. Immediately afterwards, the thickened crust of the Variscan Orogen was extended by a combination of plate boundary stresses and gravitational forces (Henk 1997). This postconvergent crustal re-equilibration occurred within a very short time span of about 20 Ma and led to the rather uniform Moho depth of

30 km that still characterizes most of present-day Central Europe (Blundell *et al.* 1992). Thus, when comparing the modelling results with present-day datasets, such as deep seismic reflection profiles, the postcollisional evolution has to be taken into account. In a first approach, this is done by a two-stage transformation process that simulates geometrically postorogenic extension and erosion until a uniform 30 km thick crust is obtained. Future work will focus on fully dynamic FE models covering the whole orogenic evolution by combining the model presented in this paper with models describing postorogenic extension (e.g. Henk 1997).

During the first stage of geometrical transformation, the position of the tracking grid nodes is recalculated assuming constant element areas and a 40% pure shear horizontal extension in the central thickened part of the orogen, decreasing to 0% across two 50 km wide transition zones at the flanks of the orogen. The assumed amount of extension corresponds to estimates derived from subsidence analyses of the Saar–Nahe Basin located immediately west of the study area (Henk 1993). Extension already causes a significant decrease in crustal thickness, but to achieve the uniform thickness observed, the remnants of the orogen have to be removed by postconvergent erosion (Fig. 5d). Erosion partly occurred simultaneously with extension in reality, but is calculated in a separate second step. During this stage, erosion of all material exceeding the reference crustal thickness of 30 km is assumed.

Figure 6a shows the central part of the tracking grid transformed to the present-day situation, whereas Fig. 6b depicts the total von Mises strain accumulated during the convergence stage. Modelling results are compared with a line drawing of deep seismic reflection profiles DEKORP 2N and 2S (Fig. 6c; see Fig. 4 for location), which cross the study area. Both the numerical model and the seismic section show an anomaly within the crustal structure a little more than 100 km in width. In the modelled section, it consists of a dome-shaped uplifted lower crust, forming a plateau of deeply exhumed rocks at the Earth's surface.

The comparison of surface deformation, sandbox experiments and FE models implies that zones of high cumulative strain in continuum models relate to discrete tectonic structures in nature. Therefore, seismic reflectivity, which is partly created by discrete faults and shear zones, is assumed to correlate with high-strain zones within total deformation plots of continuum FE models (Beaumont *et al.* 1994). The position of the domed, outward-verging high-strain areas both at the pro- and at the retro-side (Fig. 6b) of the structure resembles the pattern of major reflectors (Fig. 6c) defining the central uplifted structure of the MGCR.

Comparison with petrological data

Another interesting dataset are the peak metamorphic pressure and temperature conditions achieved during the Variscan Orogeny. Petrological data published by Anderle *et al.* (1990) for the southernmost Rheno-Hercynian Zone (pro-wedge), by Willner *et al.* (1991) and Okrusch (1995) for the Odenwald and Spessart Mountains (central pop-up), respectively, and by Schäfer (1997) for the Vesser Zone and the Schwarzburg Anticline (retro-wedge) are used for comparison with the numerical simulation. The maximum pressures and temperatures recorded by the rocks now at the surface were horizontally projected into a NW–SE-trending profile parallel to the DEKORP 2N and 2S traverse (Fig. 4).

Comparison between observed peak metamorphic conditions and calculated maximum pressures and temperatures is shown in Fig. 7. As we want to reproduce P–T data obtained from rocks at the present-day surface, the modelled curves are based on syncollisional peak metamorphic conditions recorded by marker points, which reached the model surface after postcollisional extension and erosion. Field as well as model data show a bell-shaped pressure and temperature profile with plateaux at 6–7 kbar and 600–650 °C. The overall shape is rather symmetrical and shows only minor differences in the pressure and temperature decrease between pro- and retro-side. However, it should be kept in mind that the uncertainties in thermal parameters and petrological calibration data also have an influence on the rheological model, as crustal strength is strongly temperature dependent.

Some uncertainties exist concerning the age of the peak metamorphic pressure in the Böllsteiner and Bergsträsser Odenwald (indicated by question marks in Fig. 7a). According to discordant U–Pb age data on zircon of 380 Ma (Todt 1979; Willner *et al.* 1991), pressures of 8–10 kbar were already achieved before continental collision, i.e. during the preceding subduction stage. In the temperature curve (Fig. 7b), a major difference between model predictions and observed data from the TKU and NPZ is found. Real peak temperatures are 150–200 °C lower than suggested by the numerical model. This difference is probably related to rapid burial and continuous cooling of these units by underthrusting of cooler rocks, as was shown by Henk

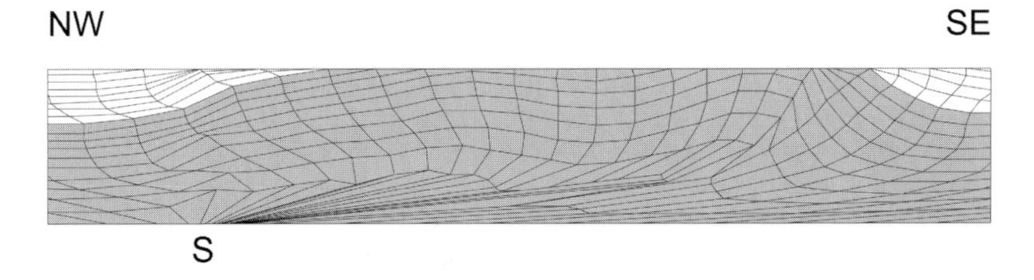

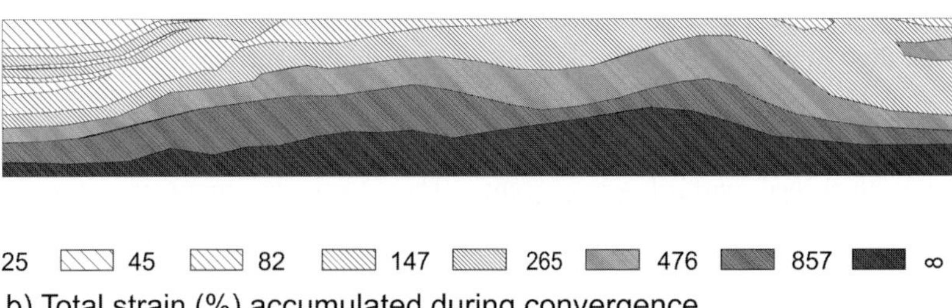

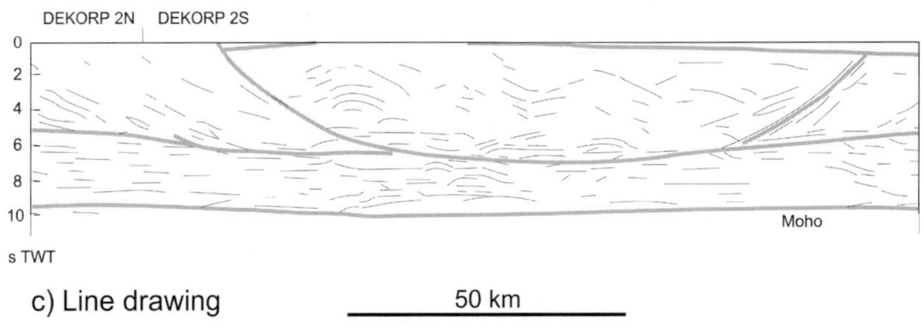

Fig. 6. (**a**) Deformed FE grid (detail of Fig. 5d). (**b**) Total strain accumulated during the convergence stage. (**c**) Line drawing of deep seismic reflection profiles DEKORP 2N and 2S (after Oncken (1998); see Fig. 4 for

(1995). This process is related to discrete fault planes and cannot be reproduced by the continuum approach used in the present study.

The numerical model can also be used to predict complete syncollisional *P–T* paths. Comparison of the contemporaneous *P–T* evolution of various particle points now located at the Earth's surface shows systematic variations across the model orogen. Figure 8 documents six *P–T* paths from the pro-side (paths 1–3) and retro-side (paths 4–6) including the time maximum metamorphic conditions were reached after the onset of convergence. The starting point of each path correlates with the conditions in a thermally equilibrated crust as prescribed by the initial model set-up. In reality, *P–T* paths will also be influenced by the thermal effects of the precollisional evolution, i.e. subduction of oceanic crust.

Only the three pro-side and the most external retro-side paths document an initial prograde evolution as a result of continental collision and burial of rocks, respectively. Decompression on the pro-side starts systematically later with increasing distance from the central block: the most external path shows a longer and more pronounced burial stage, whereas towards the centre of the orogen a short phase of burial is followed by rapid and accelerated exhumation. Consequently, all paths in the pro-wedge area have a clockwise shape with a rather tight loop. The maximum syncollisional exhumation in these parts of the model amounts to about 10 km (path 3). The far retro-side path again resembles pro-side paths with an initial phase of rapid burial, a prolonged period near maximum pressure conditions and some subsequent

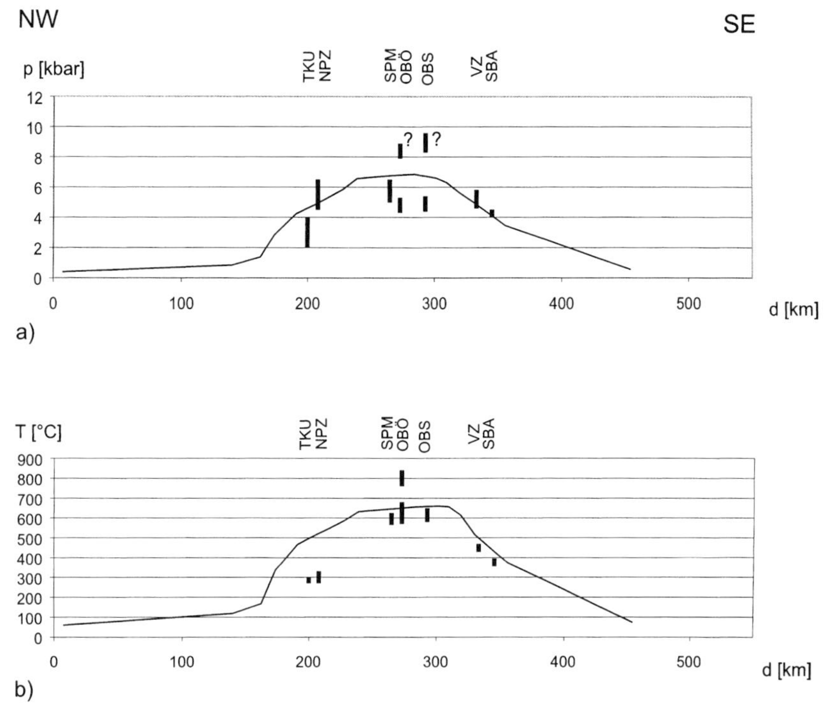

Fig. 7. Comparison of modelled peak metamorphic pressures (**a**) and temperatures (**b**) now at surface with observed petrological data along a NW–SE-trending profile across the collision zone. TKU, Taunuskamm Unit; NPZ, Northern Phyllite Zone; SPM, Spessart Mountains; OBÖ, Böllsteiner Odenwald; OBS, Bergsträsser Odenwald; VZ, Vesser Zone; SBA, Schwarzburg Anticline. Data after Anderle *et al.* (1990), Willner *et al.* (1991), Okrusch (1995) and Schäfer (1997).

exhumation. The sample path from the internal part of the retro wedge (path 4) also exhibits significant exhumation of up to 10 km during convergence, but starts with an anticlockwise evolution, i.e. some cooling during initial burial (path 4). The sense of curvature, also in path 5, changes to clockwise after about 10 Ma, contemporaneously with a distinct acceleration in exhumation velocity.

Although a direct reproduction of the observed metamorphic (P–T–t) paths in the study area is not within the scope of this work, there is some good agreement with the model predictions. Very rapid uplift and cooling (as documented by paths 3 and 4) is indicated for the Spessart Mountains by K–Ar dates on hornblendes, muscovites and biotites ranging between 318 and 324 Ma (Okrusch 1995). Counterclockwise stretches (paths 4 and 5) show striking similarities with published P–T paths from the Böllsteiner Odenwald (Willner *et al.* 1991) and for the Ruhla Formation of the Ruhla Crystalline Complex (Zeh *et al.* 2000).

Conclusions

The general thermo-mechanical models of orogenic evolution presented in the first half of this paper illustrate how crucial the rheology of the lower crust is for crustal thickening and strain localization. If orogens with Moho temperatures in excess of 700 °C are simulated using flow laws commonly applied in geodynamic studies, i.e. anorthosite (Shelton & Tullis 1981) to represent the bulk rheology of the lower crust, models fail to produce significant crustal thickening and strain localization within the crust. Instead, a basal shear zone develops, decoupling deformation in the crust from the mantle, which results in widening of the deformation zone rather than crustal thickening. With respect to petrological data of lower-crustal rocks from orogenic settings, which often point to Moho temperatures in the range of 700–900 °C, this modelling result may indicate that some of the frequently used flow laws are actually too weak. To reproduce observed petrological data and

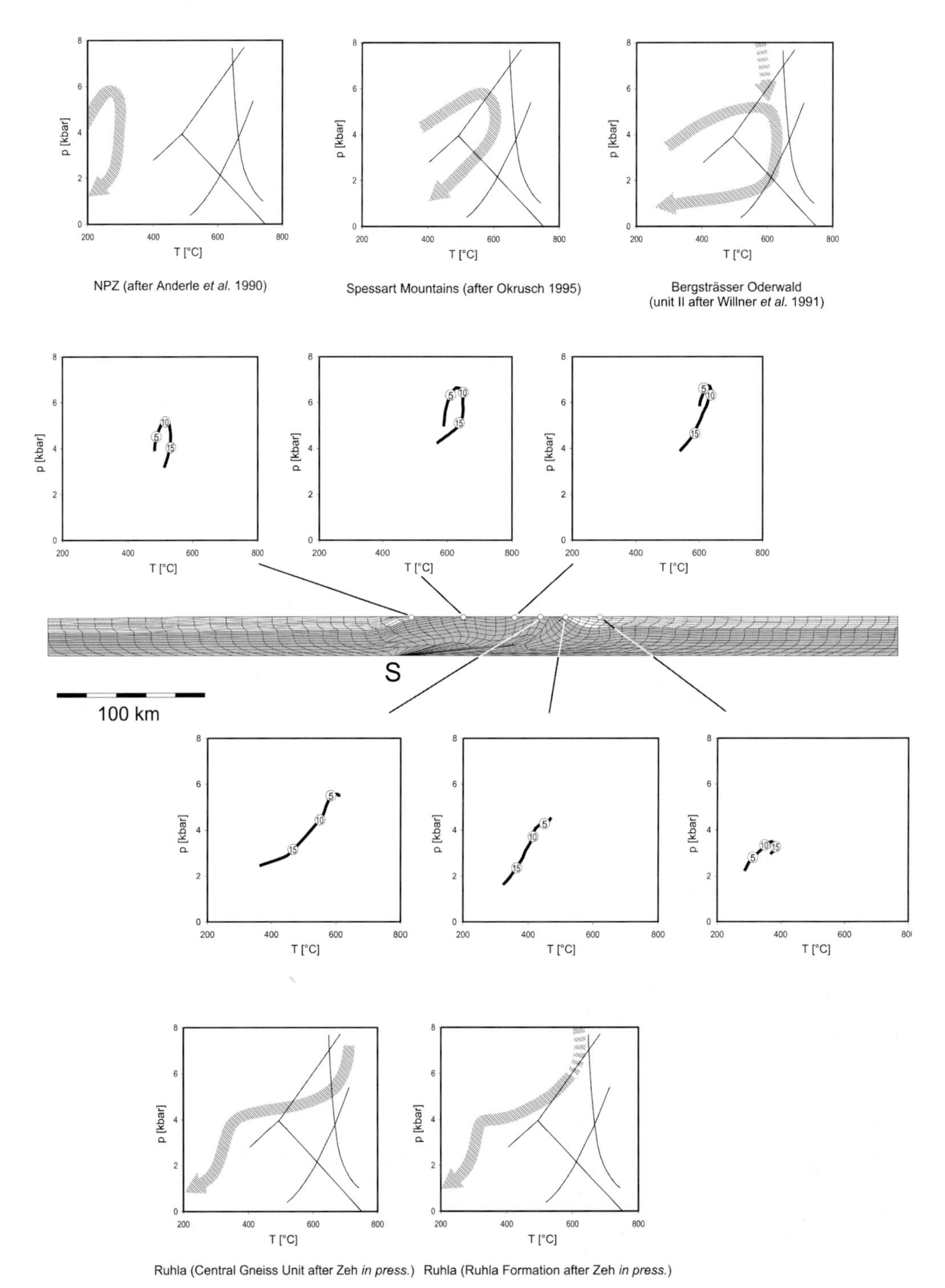

Fig. 8. Variation of contemporaneous *P–T* paths for selected points now at the Earth's surface across the modelled collision zone. Circled numbers indicate the time after the onset of continental collision in Ma. For comparison, schematic *P–T* paths from the MGCR are given at the top and at the bottom.

observed orogen geometries a stronger lower crust, with viscosities between 10^{21} and 10^{23} Pa s, is required. This can be achieved, for example, by using a dry diabase rheology (Mackwell *et al.* 1998) to represent the mechanical behaviour of the lower crust.

Application of the numerical model to the collision between two tectonometamorphic units of the Variscan Orogen, the Rheno-Hercynian and the Saxo-Thuringian Zones, yields quantitative insights into strain partitioning, peak metamorphism and the rapid synconvergent exhumation of metamorphic complexes forming the MGCR.

Observed seismic (DEKORP 2N and 2S) and metamorphic profiles across the MGCR are in good agreement with the presented modelling results, i.e. the orogen's width and geometry after post-convergent extension and erosion. As a result of erosion focused on the pro-side (NW), both peak metamorphic temperatures and pressures of rocks now at the surface show a rather symmetrical distribution. The profiles are characterized by a nearly 100 km wide plateau of pressures exceeding 6 kbar and temperatures in the range of 600–700 °C. Metamorphic temperatures determined for the outcrop areas in distal pro-side position (Taunuskamm Unit, Northern Phyllite Zone) are distinctly lower than model predictions, which can be explained by cooling of these units by underthrusting of colder rocks. The remaining model predictions agree well with the spatial distribution of peak metamorphic pressure and temperature conditions.

Additionally, numerical simulation results can be used to show systematic variations of contemporaneous P–T paths across the collision zone. Although metamorphic data derived from the MGCR locally show a more complex evolution and there are still uncertainties concerning the exact age of metamorphism, the shape and distribution of model-predicted paths are in first-order agreement with their observed petrological equivalents.

This study benefited greatly from discussions with S. Ellis, O. Oncken, E. Stein and A. Zeh. C. Beaumont and S. Willett are thanked for their helpful comments on an earlier version of the manuscript. Financial support by the Deutsche Forschungsgemeinschaft as part of the programme 'Orogene Prozesse—ihre Simulation und Quantifizierung am Beispiel der Varisciden' is gratefully acknowledged.

References

ANDERLE, H.-J., MASSONNE, H.-J., MEISL, S., ONCKEN, O. & WEBER, K. 1990. Southern Taunus Mountains. *In*: FRANKE, W. (ed.) *International Conference on Paleozoic Orogens in Central Europe 1990, Field Guide 'Mid German Crystalline Rise and Rheinisches Schiefergebirge'*. University of Giessen, Giessen, 125–148.

BEAUMONT, C. & QUINLAN, G. 1994. A geodynamic framework for interpreting crustal-scale seismic refectivity patterns in compressional orogens. *Geophysical Journal Interational*, **116**, 754–783.

——, FULLSACK, P. & HAMILTON, J. 1994. Styles of crustal deformation in compressional orogens caused by subduction of the underlying lithosphere. *Tectonophysics*, **232**, 119–132.

BLUNDELL, D., FREEMAN, R. & MUELLER, S. (eds) 1992. *A Continent Revealed—the European Geotraverse*. Cambridge University Press, Cambridge.

BURG, J.-P., VAN DEN DRIESSCHE, J. & BRUN, J.-P. 1994. Syn- to post-thickening extension in the Variscan Belt of Western Europe: modes and structural consequences. *Géologie de la France*, **3**, 33–51.

BYERLEE, J.D. 1978. Friction of rocks. *Pure and Applied Geophysics*, **116**, 615–626.

CERMÁK, V. 1995. A geothermal model of the central segment of the European Geotraverse. *Tectonophysics*, **244**, 51–55.

DEWEY, J. F. 1988. Extensional collapse of orogens. *Tectonics*, **7**, 1123–1139.

DITTMAR, D., MEYER, W., ONCKEN, O., SCHIEVENBUSCH, T., WALTER, R. & VON WINTERFELD, C. 1994. Strain partitioning across a fold and thrust belt: the Rhenish Massif, Mid-European Variscides. *Journal of Structural Geology*, **16**, 1335–1352.

ELLIS, S., BEAUMONT, C., JAMIESON, R. A. & QUINLAN, G. 1998. Continental collision including a weak zone: the vice model and its application to the Newfoundland Appalachians. *Canadian Journal of Earth Science*, **35**(11), 1323–1346.

ENGLAND, P. C. & MOLNAR, P. 1990. Surface uplift, uplift of rocks and exhumation of rocks. *Geology*, **18**, 1173–1177.

FRANKE, W. 1992. Phanerozoic structures and events in Central Europe. *In*: BLUNDELL, D., FREEMAN, R. & MUELLER, S. (eds) *A Continent Revealed—the European Geotraverse*. Cambridge University Press, Cambridge, 164–180.

HANSEN, F. D. & CARTER, N. L. 1983. Semibrittle creep of dry and wet Westerly Granite at 1000 MPa. *Proceedings—Symposium on Rock Mechanics*, **24**, 429–447.

HENK, A. 1993. Late orogenic basin evolution in the Variscan internides: the Saar–Nahe Basin, southwest Germany. *Tectonophysics*, **223**, 273–290.

—— 1995. Late Variscan exhumation histories of the southern Rhenohercynian Zone and western Mid-German Crystalline Rise: results from thermal modelling. *Geologische Rundschau*, **84**, 578–590.

—— 1997. Gravitational orogenic collapse vs. plate-boundary stresses: a numerical modelling approach to the Permo-Carboniferous evolution of Central Europe. *Geologische Rundschau*, **86**, 39–55.

—— 1998. Thermomechanische Modellrechnungen zur postkonvergenten Krustenreequilibrierung in den Varisciden. *Geotektonische Forschungen*, **90**, 1–124.

JAMIESON, R. A. 1991. *P–T–t* paths of collisional orogens. *Geologische Rundschau*, **80**, 321–332.

——, BEAUMONT, C., FULLSACK, P. & LEE, B. 1998. Barrovian regional metamorphism: where's the heat? *In*: TRELOAR, P. J. & O'BRIAN, P. J. (eds) *What Drives Metamorphism and Metamorphic Reactions?* Geological Society, London, Special Publications, **138**, 23–51.

KEMNITZ, H. 1995. Phyllitic islands in very low grade surroundings—nappes in Thuringia? A microanalytical approach on metamorphic phyllosilicates. *Journal of Czech Geological Society*, **40**(3), 20–21.

KOHLSTEDT, D. L., EVANS, B. & MACKWELL, S. J. 1995. Strength of the lithosphere: constraints imposed by laboratory experiments. *Journal of Geophysical Research*, **100**, 17587–17602.

MACKWELL, S. J., ZIMMERMAN, M. E. & KOHLSTEDT, D. L. 1998. High-temperature deformation of dry diabase with application to tectonics on Venus. *Journal of Geophysical Research*, **103**, 975–984.

MEISSNER, R. B[sc]ORTFELD, R. K. 1990. *Dekorp-Atlas*. Springer, Berlin.

OKRUSCH, M. 1995. Mid-German Crystalline High—Metamorphic evolution. *In*: DALLMEYER, R. D., FRANKE, W. & WEBER, K. (eds) *Pre-Permian Geology of Central and Eastern Europe*. Springer, Berlin, 201–213.

ONCKEN, O. 1997. Transformation of a magmatic arc and an orogenic root during oblique collision and its consequences for the evolution of the European Variscides (Mid-German Crystalline Rise). *Geologische Rundschau*, **86**, 2–20.

—— 1998. orogenic mass transfer and reflection seismic patterns—evidence from DEKORP sections across the European Variscides (central Germany). *Tectonophysics*, **286**, 47–61.

——, PLESCH, A., WEBER, J., RICKEN, W. & SCHRADER, S. 2000. Passive margin detachment during arc–continent collision (Central European Variscides). *This volume*.

SCHÄFER, F. 1997. Krustenbilanzierung eines variscischen Retrokeils im Saxothuringikum. Scientific Technical Report Geoforschungszentrum Potsdam, **97/16**.

SHELTON, G. & TULLIS, J. 1981. Experimental flow laws for crustal rocks. *EOS Transactions, American Geophysical Union*, **62**, 396.

SUMMERFIELD, M. A. & HULTON N. J. 1994. Natural controls of fluvial denudation rates in major world drainage basins. *Journal of Geophysical Research*, **99**, 13871–13883.

TODT, W. 1979. U–Pb-Datierungen an Zirkonen des kristallinen Odenwaldes. *Fortschritte der Mineralogie*, **57**(Beiheft 1), 153–154.

VAUCHEZ, A., TOMMASI, A. & BARUOL, G. 1998. Rheological heterogeneity, mechanical anisotropy and deformation of the continental lithosphere. *Tectonophysics*, **296**, 61–86.

WALTER, R. 1992. *Geologie von Mitteleuropa. Schweizerbart'sche, Stuttgart.*

WILLETT, S., BEAUMONT, C. & FULLSACK, P. 1993. Mechanical model for the tectonics of doubly vergent compressional orogens. *Geology*, **21**, 371–374.

WILLNER, A. P., MASSONNE, H.-J. & KROHE, A. 1991. Tectonothermal evolution of a part of a Variscan magmatic arc: the Odenwald in the Mid-German Crystalline Rise. *Geologische Rundschau*, **80**, 369–389.

ZEH, A., COSCA, M. A., BRAETZ, H., OKRUSCH, M. & TICHOMIROWA, M. 2000. Simultaneous horst–basin formation and magmatism during Late Variscan transtension: evidence from $^{40}Ar/^{39}Ar$ and $^{207}Pb/^{206}Pb$ geochronology in the Ruhla Crystalline Complex. *Geologische Rundschau*, in press.

ZIEGLER, P. A. 1990. *Geological Atlas of Western and Central Europe*. Shell Internationale Petroleum Maatschappij, The Hague.

ZOTH, G. & HÄNEL, R. 1988. Appendix. *In*: HÄNEL, R., RYBACH, L. & STEGENA, L. (eds) *Handbook of Terrestrial Heatflow Density Determination*. Kluwer, Dordrecht, 449–466.

Heat flow evolution, subsidence and erosion in the Rheno-Hercynian orogenic wedge of central Europe

R. LITTKE[1], C. BÜKER[1], M. HERTLE[1], H. KARG[1,4], V. STROETMANN-HEINEN[2] & O. ONCKEN[3]

[1]*Institute of Geology and Geochemistry of Petroleum and Coal, Aachen University of Technology, Lochnerstr. 4-20, D-52056 Aachen, Germany (e-mail: littke@lek.rwth-aachen.de)*

[2]*Institute of Mineralogy and Economic Geology, Aachen University of Technology, Wuellnerstr. 2, D-52056 Aachen, Germany*

[3]*GeoForschungsZentrum Potsdam, Telegrafenberg, D-14473 Potsdam, Germany*

[4]*Present address: Wintershall AG, Friedrich-Ebert Str. 160, D-34112 Kassel, Germany*

Abstract: Numerical, thermal and rheological modelling techniques are applied to unravel the basin-forming processes and the heat flow and burial history of the Rheno-Hercynian fold belt (Rhenish Massif), the adjacent Subvariscan foreland (Ruhr Basin), and the intramontane Saar–Nahe Basin. Thermal history and crustal architecture in the study areas were affected mainly by the Variscan Orogeny during late Palaeozoic times. Calibration of the simulated thermal histories is primarily based on vitrinite reflectance and fission-track data. Mechanical modelling reveals average β values of 1.7, reaching a maximum of 2.4 in the central basin (Mosel Graben) and at the transition to the Giessen Ocean to the south during Early Devonian rifting. This stage was associated with tholeitic magmatism and an elevated heat flow of up to 110 mW m^{-2}, preserved in weakly overprinted syn-rift sediments. Average basal heat flow during maximum burial at the end of the Carboniferous period (i.e. the end of crustal shortening) was between 50 and 70 mW m^{-2} with a slight decrease from the Subvariscan foreland basins towards the Rheno-Hercynian in the south. The values suggest average crustal thicknesses of between 32 and 36 km during late Carboniferous time. For the Saar–Nahe Basin, values between 50 and 75 mW/m^2 represent the thermal regime in the upper crust during the late Stephanian and early Permian time. Estimated eroded thicknesses of Palaeozoic sediments vary between 2500 m in the northern and central Ruhr Basin and more than 6000 m in the Osteifel and the Siegen Anticline within the Rheno-Hercynian, and between 1800 and 3600 m in the Saar–Nahe Basin. Fission-track data provide evidence for significant reheating during the Mesozoic era within the entire study area. This phase of heating, probably linked to North Atlantic rifting, coincides with post-Variscan ore formation and with major tectono-magmatic events in central Europe.

Tectonic processes in the Earth's crust are mainly a result of the temperature field in the Earth's interior and strongly govern the temperature field in sedimentary basins located in the upper crust (Allen & Allen 1990; Yalcin *et al.* 1997). Studies of palaeotemperature fields, therefore, have the potential to reveal important information on the tectonic processes and variables related to the tectonic evolution of a crustal unit. In this paper, palaeotemperature indicators such as vitrinite reflectance, fission-track and fluid-inclusion data were used to unravel the thermal evolution of the northern Rhenish Massif and the Ruhr Basin, situated in the Rheno-Hercynian foreland belt of the Variscides, and the Saar–Nahe Basin, which is one of several Late Carboniferous interior basins inside the Variscan belt. The application of numerical basin-modelling techniques facilitated a quantitative interpretation of the reconstructed thermal history, resulting from fundamental orogenic processes such as basin formation, uplift and erosion, and magmatism. In particular, information was obtained on the heat flow evolution and the amount of erosion related to the Variscan Orogeny. Figure 1 provides an overview of the areas studied.

The basic geological data for the Ruhr Basin, the Lower Rhein Embayment and the northern

From: Franke, W., Haak, V., Oncken, O. & Tanner, D. (eds).
Orogenic Processes: Quantification and Modelling in the Variscan Belt.
Geological Society, London, Special Publications, **179**, 231–255. 0305-8719/00/$15.00

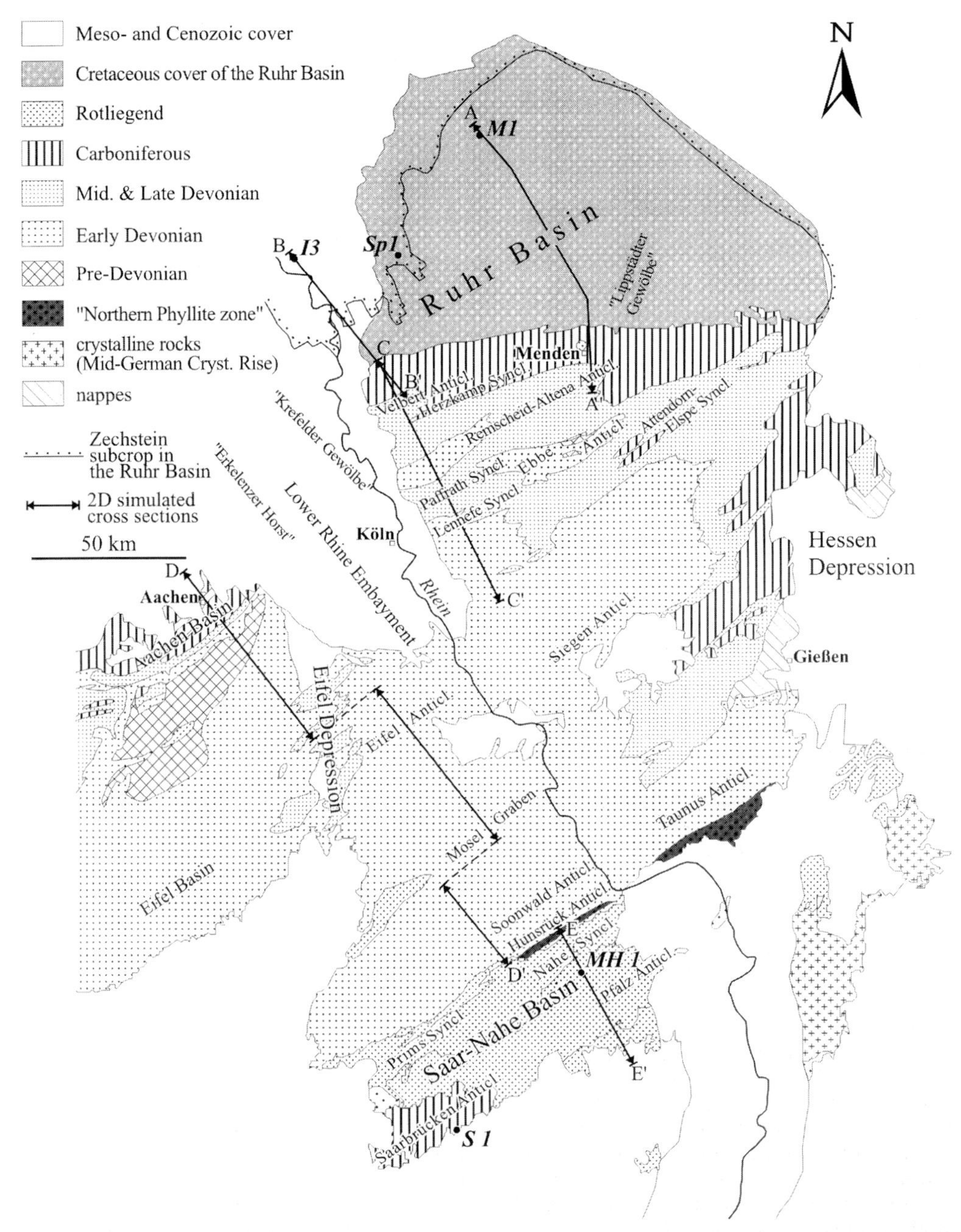

Fig. 1. Geological map of the studied areas, including locations of 1D modelled key wells and 2D modelled cross-sections.

Rhenish Massif east of the river Rhein were obtained from the PhD theses of Büker (1996) and Karg (1998), and for the Saar–Nahe Basin from Hertle & Littke (2000). Further details on the reconstruction of the crustal evolution and the heat flow in the Rhenish Massif west of the river Rhein have been given by Heinen (1996). The major objective of this paper is to present a comprehensive overview of the main results of the above-cited theses, in which a variety of geochemical and modelling techniques were applied to investigate the study area's geological and thermal evolution. Although a similar approach has been used for several oil- and gas-bearing sedimentary basins in the past decade, this study is one of the first to examine in detail the pre-, syn- and post-orogenic temperature evolution of an ancient orogenic wedge.

Study areas and their geological evolution

Rhenish Massif

The Rhenish Massif is part of the Rheno-Hercynian zone, which evolved from a rifted passive margin to an orogenic belt following collision with an upper plate continental fragment, the Mid-German Crystalline Rise (Fig. 1). Collision resulted in the formation of a doubly vergent orogenic wedge system with the Rheno-Hercynian fold belt on the northern flank, the Mid-German Crystalline Rise forming the axial zone of the collisional system, and parts of the Saxo-Thuringian fold-belt as the southern flank (Schäfer *et al.* this volume). At the surface, the Rheno-Hercynian is limited to the south by the Hunsrück–Taunus Boundary Fault, a steep wrench and normal fault, which has been repeatedly active since the early Late Carboniferous, that inverted the former suture zone.

The Rhenish Massif is segmented by the north–south-trending Eifel Depression (Fig. 1), where Cenozoic and Mesozoic sediments of, for example, the Lower Rhein Embayment rest unconformably on the deformed Palaeozoic basement rocks. Towards the east, the Rhenish Massif plunges beneath the Permo-Mesozoic units of the Hessen Depression (Fig. 1).

Recently published palinspastic restorations across the Rhenish Basin (west of the river Rhein; Oncken *et al.* 1999) show basin segmentation into several domains. The main basin segments are (from north to south) the stable platform of the Anglo-Brabant Massif ('Old Red Shelf'), the Eifel Basin, the Mosel Graben, the Hunsrück–Soonwald Plateau and the Phyllite Zone at the southern margin of the Rheno-Hercynian fold and thrust belt. A 1 km thick shallow-marine to littoral sequence of mostly post-rift Devonian platform sediments on the unrifted Anglo-Brabant Massif is separated from up to 15 km thickness of Rhenish basinal sequence to the southeast by a prominent rift margin fault with a cumulative downthrow of 6 km. This margin developed during Early Lower Devonian crustal stretching and separated unextended crust to the northwest from strongly attenuated crust to the southeast (von Winterfeld & Walter 1993; Oncken *et al.* 1999). The restoration shows a southward deepening of the Eifel–Mosel Basin, which is more pronounced in the west. In contrast, the southern margin of the Mosel Graben was shown to be formed by a set of north-dipping normal faults, with a substantial decrease in thickness of syn-rift sediments from 14 to 5 km. The southern basin (Taunus and Soonwald units) is a marginal plateau, typical of Atlantic-type passive margins, showing a decreasing sediment thickness toward the oceanic basin. The southernmost unit of the Rheno-Hercynian Zone, the Phyllite Zone, has been suggested to represent the former slope and rise (Oncken *et al.* 1999). The predominance of mid-ocean ridge basalt (MORB)-type meta-volcanic rocks in the southernmost imbricate of this zone indicates the transition to the Giessen Ocean. The decrease of clastic input after Early Emsian time suggests that this narrow ocean ('Lizard–Giessen–Ostharz Ocean') was opened during the peak of Early Emsian crustal extension (e.g. Oncken *et al.* 1999).

Throughout the Rhenish Basin, sedimentation commenced during Gedinnian time (Fig. 2) with continental clastic deposits of varying thickness (a few metres to 1 km) deposited on continental basement of variable composition (Oncken *et al.* 1999). During Siegenian time, a thick (0.3–7 km) continental to shallow-marine succession was deposited. The onlap limits of this basin were slightly north of the present-day outcrop of the Aachen–Midi Detachment (von Winterfeld & Walter 1993) and slightly south of the present-day Hunsrück Border Fault (Fig. 1). The development of extensional growth faults began in Siegenian time; their activity reached a maximum by the end of the Early Emsian time. Sediments deposited after the end of this initial rifting stage consist of shallow-marine clastic deposits and carbonates, intercalated with some bimodal volcanic rocks concentrated near the major rift boundaries (peak of magmatism at the northern rim of the basin is dated at $373 + 8/-9$ M, (Goffette *et al.* 1991) and 381 ± 16 Ma (Kramm & Buhl 1985)). Rifting in the western Rhenish Massif largely terminated by the end of Early Devonian time and was followed, after a minor hiatus, by shallow-marine to pelagic clastic deposits and carbonates of the post-rift stage (Eifelian to Viséan time).

East of the river Rhein, a post-rift sequence developed after Early Devonian time. Both a shallow-marine shelf and a deeper marine basin evolved as a result of locally different subsidence rates. On the shelf, clastic deposits and shallow-marine carbonates were deposited and accompanied by the growth of reefs (Rhenish facies). In the deeper parts of the basin, turbiditic limestones and shales predominate (Hercynian facies). The lower Lower Carboniferous pelagic

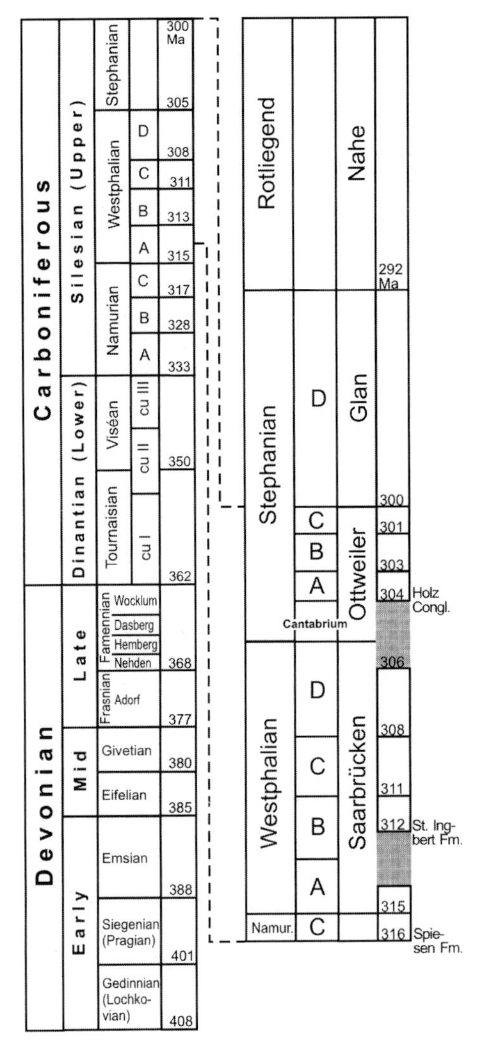

Fig. 2. Generalized stratigraphic column for the northern Rhenish Massif and the Ruhr Basin (left) and for the Saar–Nahe Basin (right).

shales are covered by fine- to coarse-grained, mostly turbiditic clastic rocks deposited in a foreland-basin setting (Kulm facies) which grades into a carbonate platform in the north-west (Kohlenkalk facies). In the eastern Rhenish Massif, northward migration of the flysch basin, beginning in the south during Early Viséan time, reflects the prograding front of the orogenic belt (Engel *et al.* 1983; Ricken *et al.* this volume). This development, however, is not preserved west of the river Rhein. Toward the north and west, the flysch deposits were gradually replaced during Namurian time by marine to continental molasse-type sediments, which persisted until Late Westphalian time.

The deformation style in the Rhenish Massif is generally characterized by large NW-verging anticlines and synclines replaced by imbricate fans further south (see Oncken *et al.* this volume). Deformation and crustal shortening during the Variscan Orogeny generally increased to the south and resulted in detachment of the Palaeozoic rocks from the deeper crust. The age of deformation and metamorphism was estimated by Ahrendt *et al.* (1983) on the basis of K–Ar ages of metamorphic white mica in volcanic rocks. The data indicate a northward propagation of the metamorphic front between 328 and 305 Ma, that is, from Namurian to Westphalian D time (Hess & Lippolt 1986; Fig. 2).

Subvariscan basins

During and after northwestward propagation of the Rheno-Hercynian fold and thrust belt, a large system of foreland molasse basins evolved north of the orogenic front. The largest, the Ruhr Basin, is situated at the northern margin of the Rheno-Hercynian belt, east of the exposed Anglo-Brabant Massif (Fig. 1), and is part of the external fold and thrust belt. The basin today has a length of about 150 km in a SW–NE direction and a width of 80 km across strike (e.g. Drozdzewski 1993). The southeastern border is eroded and is supposed to have been originally situated in the area of the Remscheid or Ebbe Anticline (Fig. 1; Drozdzewski 1993). Northwards, the Palaeozoic rocks of the Ruhr Basin dip under a mostly Upper Cretaceous cover. Maximum Cretaceous thicknesses of almost 2400 m were bored north of the Münsterland 1 well ('M1'), on the northern margin of the basin (Fig. 1). Only in the westernmost part are Permian, Triassic and Jurassic rocks preserved.

The Upper Carboniferous strata of the Ruhr Basin, conformably deposited on the earlier shelf sequence, reflect the molasse stage of the Variscan Orogen in a typical foreland basin setting. The sedimentary environment developed from marine to shallow marine and finally to paralic (Drozdzewski 1993). Coal formation (Littke 1987; Littke & ten Haven 1989) began in Namurian C time and continued at least until Westphalian C time. Maximum subsidence occurred along the northern rim of the Rheno-Hercynian belt. Regionally, the Namurian and Westphalian strata thin from about 5000 m in the south to less than 3000 m in the north.

The Ruhr coal district has been affected by northwest-propagating folding and thrusting at the end of Rheno-Hercynian wedge growth. The main folds and thrusts strike parallel to the basin axis and die out towards the northwest. The tectonic structure of the orogenic front in the Ruhr Basin differs from that exposed west of the river Rhein (Fig. 1). In the Eifel–Ardennes area, the Palaeozoic sequences were overthrust along the Aachen–Midi Thrust system on the Upper Carboniferous cover of the Brabant Massif (Fig. 1). Eastwards, thrusting at the surface decreases and changes into a broad fold belt involving detachment folds and buried imbricate thrusts above a blind detachment (see Oncken *et al.* this volume).

The Permo-Carboniferous Saar–Nahe Basin

The Saar–Nahe Basin, which shows a typical half-graben geometry, is an intramontane late orogenic sedimentary basin that developed on the axial zone of the propagating orogenic wedge system (Fig. 1). The South Hunsrück Fault, separating the Rheno-Hercynian and the Saxo-Thuringian zone of the Variscides, controlled basin subsidence during Late Carboniferous and Permian time (Henk 1993). The Saar–Nahe Basin has a SW to NE lateral extent of about 300 km and is about 100 km wide. Most of the Permo-Carboniferous sedimentary rocks are unconformably covered by Mesozoic strata, leaving only 100 km × 40 km of Palaeozoic outcrop. In places, at least 10 km of fluvial, lacustrine and alluvial fan sediments with intercalated coal seams accumulated from latest Namurian C–Westphalian A to Permian times (Schäfer & Korsch 1998). This evolutionary phase is coincident with the period of postcollisional contraction of the neighbouring Rheno-Hercynian foreland fold and thrust belt, the lower crust of which was underthrust by some 180 km under the extending Saar Basin (Oncken *et al.* 1999).

The Saar 1 well ('S1' in Fig. 1), which is more than 5000 m deep, reached the basement underneath the sedimentary sequence; this consists of a 403 ± 24 Ma granite that is part of the Mid-German Crystalline Rise (Sommermann 1993). Sedimentation commenced with some 1000 m of Middle Devonian to Lower Carboniferous marine carbonates and siliciclastic rocks covering this basement. Following a hiatus in Namurian time, sedimentation was resumed in latest Namurian C–Westphalian A time with the deposition of a basal conglomerate (Spiesen Formation; Fig. 2). During Permo-Carboniferous time, the depocentre shifted successively from SW to NE (Korsch & Schäfer 1995). In the Upper Carboniferous sequence, fluvial, deltaic and lacustrine sandstones and siltstones with intercalated coals predominate, whereas alluvial and playa-lake sediments characterize the Permian sequence. At the beginning of Permian time, intense bimodal volcanism took place.

In the area of the Saarbrücken Anticline (Fig. 1), sedimentation was interrupted by two erosional events. The first event occurred in Westphalian B time, recorded at the base of the St Ingbert Formation (Fig. 2). The second and more important event, at the Westphalian–Stephanian boundary, is indicated by the Holz Conglomerate above a regional unconformity (Fig. 2), where it is estimated that up to 1500 m of the sedimentary sequence is missing (Engel 1985). Within the synclines, no unconformity can be seen in the seismic reflection data, but a hiatus cannot be excluded. Generally, maximum burial was reached during Stephanian and Permian times, but it was reached earlier in the southwest and in the anticlinal structures and later in the northeast and in the synclines. The time of maximum burial was followed by uplift and erosion. During late Permian (Zechstein) or early Triassic (Buntsandstein) time, sedimentation recommenced, continuing into Jurassic time. The Cretaceous period is characterized by erosion. In the area of the Saarbrücken Anticline most of the Mesozoic rocks were eroded. During Eocene–Oligocene time, sedimentation took place only in the eastern part of the basin. This depositional phase was followed by moderate erosion, which has continued until recent times.

Analytical methods and numerical modelling

To unravel heat flow evolution, subsidence and erosion in the study areas, different numerical modelling techniques were used. Temperature history modelling was mainly calibrated by vitrinite reflectance and fission-track data.

Vitrinite reflectance

Vitrinite reflectance measurements were made on dispersed organic matter and coals using established procedures (Taylor *et al.* 1998). Random reflectance instead of maximum reflectance was measured because the data were compared with the random vitrinite reflectance calculated with the EASY%Ro algorithm of Sweeney & Burnham (1990). Vitrinite reflectance values are expressed as R_r (random reflectance) and represent the mean values measured in oil immersion. These values are equivalent to R_o and R_m used in earlier literature, but differ from R_{max} or R_{min} (see Taylor *et al.* 1998). Other geochemical maturity characteristics (Radke *et al.* 1997) and fluid inclusion homogenization temperatures were measured as well, but are not referred to in the following text.

Fission-track analysis

Fission-track analysis was performed in collaboration with the 'London Fission Track Research Group' (University College, London), on 40 apatite and seven zircon samples. Fission-track ages were determined using the external detector method (Gleadow 1981). In each grain, the numbers of spontaneous and induced tracks in the corresponding location within the mica external detector were recorded. Fission-track ages were calculated by replacing track counts by track densities, taking into account the total area in which counting was performed. Further details of the procedure have been given by, for example, Hurford & Green (1982), Naeser (1993), Büker (1996) and Karg (1998).

Heat flow and subsidence history modelling ('basin modeling')

The basic concept of numerical basin modelling is to obtain the present state of a sedimentary succession using a forward modelling approach. Based on the input data, the following processes are combined during the simulations: (1) deposition of decompacted sediments, starting with the oldest defined layer; (2) sediment compaction, that is, the decrease of porosity and permeability and the expulsion of the pore fill in response to overburden load and the evolving pore pressure; (3) the temporal and spatial evolution of the subsurface temperature field as a result of conductive and convective (forced convection) heat flow and changing thermal rock properties.

The concept and necessary input and calibration data were described by Welte & Yalcin (1988), Hermanrud (1993), Leischner *et al.* (1993), Poelchau *et al.* (1997) and Yalcin *et al.* (1997). Basin modelling results may not necessarily be true, but they should be consistent with the geological dataset, in particular with the temperature or maturity parameters. Calculation of vitrinite reflectance was performed according to the procedure of Sweeney & Burnham (1990). It should be noted that their EASY%Ro algorithm is defined only for vitrinite reflectance values between 0.2 and 4.6% R_r and, therefore, does not cover the entire maturity range observed in the study area.

Simulations were performed using the 1D PDI/PC and the 2D PETROMOD software of IES GmbH and the 1D BASINMOD software of Platte River Associates. Differences between the results calculated by these approaches are small as long as the assigned petrophysical parameters are identical. Predefined parameters may differ considerably, but we used mainly self-defined parameters reflecting the petrophysical inventory of the respective stratigraphic units. Simulations were carried out for about 100 wells in the Ruhr Basin and in the Saar–Nahe Basin.

In the Rhenish Massif, deep boreholes do not exist and reflectance values were measured on surface samples only. As the reflectance isolines run parallel to bedding in most of the northern Rhenish Massif (i.e. the maturation pattern is pre-orogenic), the pseudo-well concept could be applied. According to this concept, reflectance values are projected into the centres of synclines to build a pseudo-well (Oncken 1984, 1987, 1990). More than 30 pseudo-wells were constructed for different parts of the northern and western Rhenish Massif. The concept of pre-orogenic maturation does not apply for all parts of this mountain belt. For example, it does not apply for the eastern part of the transition zone between the Rhenish Massif and the Ruhr Basin (Karg 1998).

In addition, four 2D simulations were carried out along seismic profiles or previously published geological cross-sections (lines AA′, BB′, CC′ and EE′ in Fig. 1). Input and calibration data for the various wells, pseudo-wells and 2D sections have been given by Büker (1996), Heinen (1996), Karg (1998) and Hertle & Littke (2000).

Mechanical basin modelling

The evolution of heat flow through time and its spatial distribution is a key variable for thermal modelling. It was constrained along the westernmost section through the Rhenish massif (line DD′ in Fig. 1) by applying an independent mechanical basin modelling approach using the software STRETCH of Quantitative Basin Analysis Associates (QBAA) which is based on the flexural cantilever model as described by Kusznir & Ziegler (1992). This technique quantitatively relates the thermal evolution to the elastic lithospheric properties and the geometry of lithosphere deformation during extension. Input parameters and detailed simulation results have been given by Heinen (1996).

Modelling results

Syn-rift evolution

Heat flow and subsidence history. Heat flow was modelled for the period from Early Gedinnian to Late Viséan time (Fig. 2) using the 1D PDI/PC and the 2D PETROMOD software. The results show a clear heat flow maximum at the end of Emsian time for all basin segments with the exception of the Hunsrück Basin. However, in the latter part a calibration of the modelled heat flow evolution during Devonian time was not possible because of the metamorphic overprint of the thermal indicators during the Variscan Orogeny.

In the Anglo-Brabant Shelf, heat flow reached a maximum of 80 mW m^{-2} at the end of Emsian time after increasing slowly from Gedinnian times. This moderate heat flow value corresponds to the tectonic position as a 'rift-shoulder' of the Old Red Shelf during rifting. The Eifel Basin and the Mosel Graben (Fig. 1) show higher heat flow values at the end of Emsian time with 90 and 110 mW m^{-2} on average with an estimated error of $\pm$ 10 mW m^{-2}. In the Mosel Graben area, the highest sedimentation rates, therefore, correlate with the maximum heat flow values.

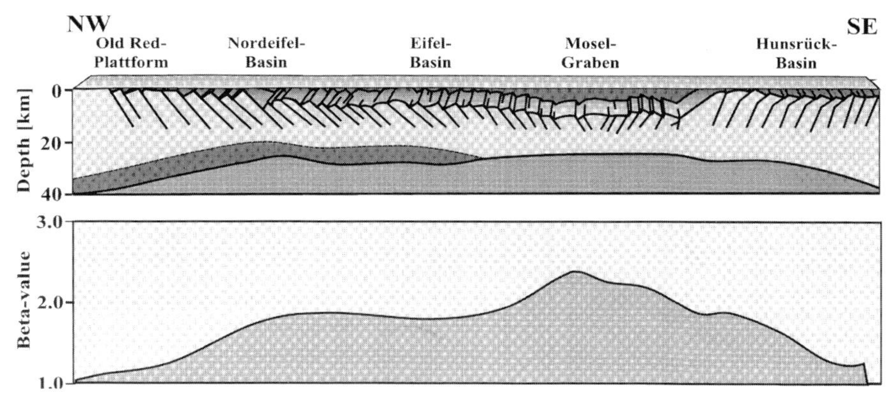

Fig. 3. Crustal structure (top) and stretching factor (bottom) for the syn-rift phase (latest Emsian time; see Fig. 2) along cross-section DD′ (see Fig. 1 for location). The dark grey shaded area in the northwestern part of the model marks unresolvable variations in crustal thickness. The length of the model is about 160 km.

Simulation of subsidence curves with concomitant decompaction of the sediments for the internal parts of the basin (Eifel Basin and Mosel Graben) shows a clear decrease of subsidence at the end of Emsian time. On the Anglo-Brabant Shelf, and the Eifel and Hunsrück Basins (Figs. 1 and 3), a hiatus is documented for this time period (Solle 1970). This hiatus can be interpreted as the uplift of the basins as a result of the maximum thermal expansion of the lithosphere during increased syn-rift heat flow. This result also coincides with the peak of magmatism. Subsequent thermal relaxation of the heated lithosphere during the post-rift phase was associated with low subsidence rates.

Crustal evolution. A best-fit model of the crustal evolution in the western Rhenish Massif, which reproduces the above observations, was achieved with the STRETCH software with a model showing distributed symmetric crustal attenuation with an average β value of 1.7 (2.4 in the rift axis) beginning with an initial crustal thickness of 35–40 km. The results are presented in Fig. 3, which shows the model of the western Rheno-Hercynian crust at the end of the synrift phase (Late Emsian time). Brittle extension is restricted to the upper 15 km. Below the brittle–ductile transition zone, only ductile deformation is assumed (for details, see Kusznir & Ziegler (1992)). The initial Moho depth of 40 km rose to *c.* 27 km (estimated error ± 1 km) in the Mosel Graben and 30 km in the Eifel Basin and Hunsrück–Soonwald Plateau.

The calculated β values are symmetrical with respect to the basin geometry. The highest β values occur in the Eifel Basin and Mosel Graben, whereas the marginal zones of the Rheno-Hercynian Basin experienced only moderate stretching. To calibrate the results of thermomechanical modelling, the heat flow anomalies calculated by the rise of the lithosphere–asthenosphere geotherm were compared with those from the basin modelling approach. The calculated heat flow anomalies during syn-rift stretching show the highest values at the end of Emsian time in the Mosel Graben with up to 40 mW m^{-2} heat flow anomaly. Assuming an average thermal heat flow of 60 mW m^{-2}, these values agree well with the results of the thermal basin modelling.

Late Variscan and Post-Variscan evolution

One-dimensional basin modelling

The Subvariscan Ruhr Basin. Isselburg 3 is the key well for the northwestern part of the Ruhr Basin ('I3' in Fig. 1; see also Littke *et al.* 1994). Underneath a thick Tertiary (232 m), Jurassic (66 m), Triassic (790 m) and Zechstein (229 m) cover, the well reached Upper Westphalian A units and penetrated the Carboniferous sequence over a depth of more than 3000 m down to the Devonian sequence (Fig. 4a). In contrast to the greater part of the Ruhr Basin, no Cretaceous deposits are present in this area. Excellent vitrinite reflectance data were available as calibration data (Teichmüller 1971), also including information from apatite fission-track analyses and Rock-Eval pyrolysis. The Kupferschiefer at the Zechstein base in this well and in the entire region is of much lower maturity (0.57% R_r; e.g. Teichmüller 1971) than the unconformably underlying Upper Carboniferous rocks (1.1% R_r). This indicates that the coalification of the Carboniferous and Devonian rocks is of pre-Zechstein age. This finding was confirmed by the

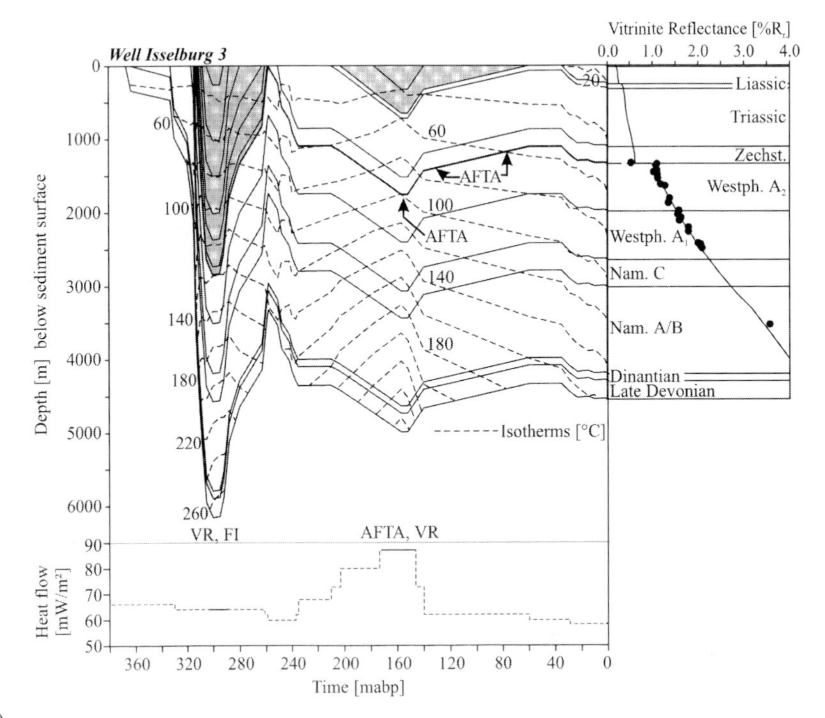
Well Isselburg 3
Vitrinite Reflectance [%R_r]
0.0 1.0 2.0 3.0 4.0
Liassic
Triassic
Zechst.
Westph. A_2
Westph. A_1
Nam. C
Nam. A/B
Dinantian
Late Devonian
AFTA
Isotherms [°C]
Depth [m] below sediment surface
VR, FI
AFTA, VR
Heat flow [mW/m²]
Time [mabp]

Fig. 4(a).

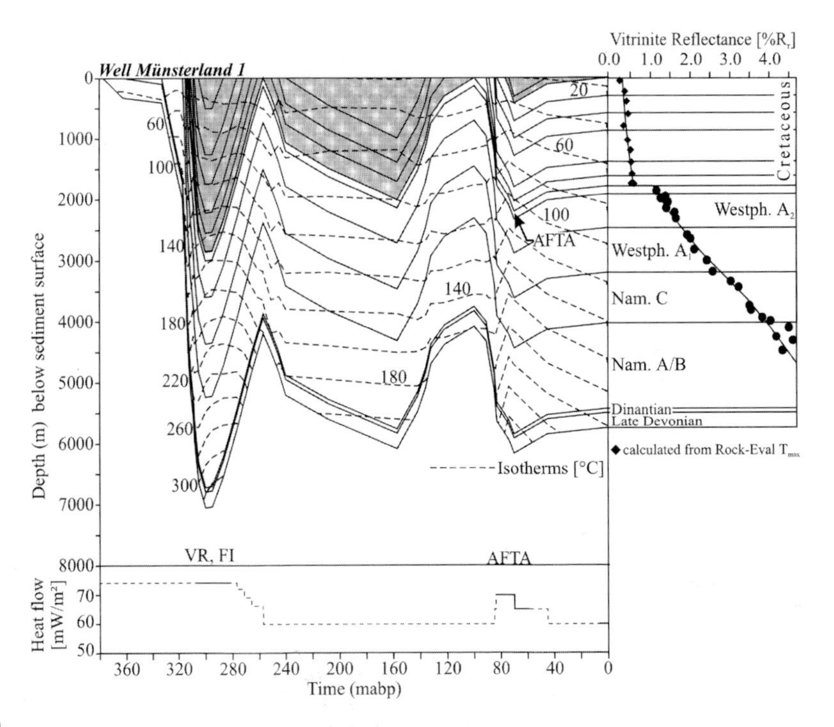
Well Münsterland 1
Vitrinite Reflectance [%R_r]
0.0 1.0 2.0 3.0 4.0
Cretaceous
Westph. A_2
Westph. A_1
Nam. C
Nam. A/B
Dinantian
Late Devonian
◆ calculated from Rock-Eval T_{max}
AFTA
Isotherms [°C]
Depth (m) below sediment surface
VR, FI
AFTA
Heat flow [mW/m²]
Time (mabp)

Fig. 4(b).

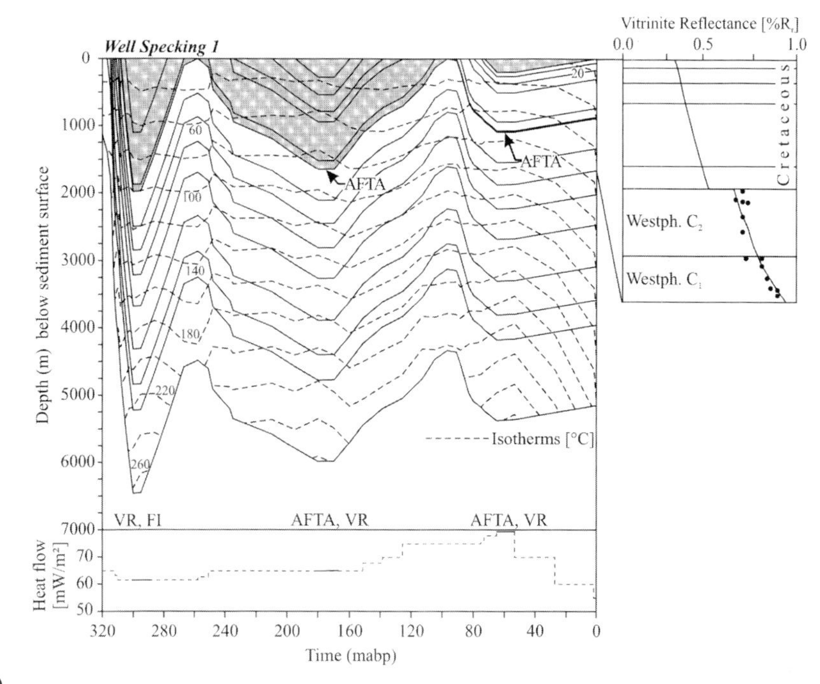

Fig. 4(c).

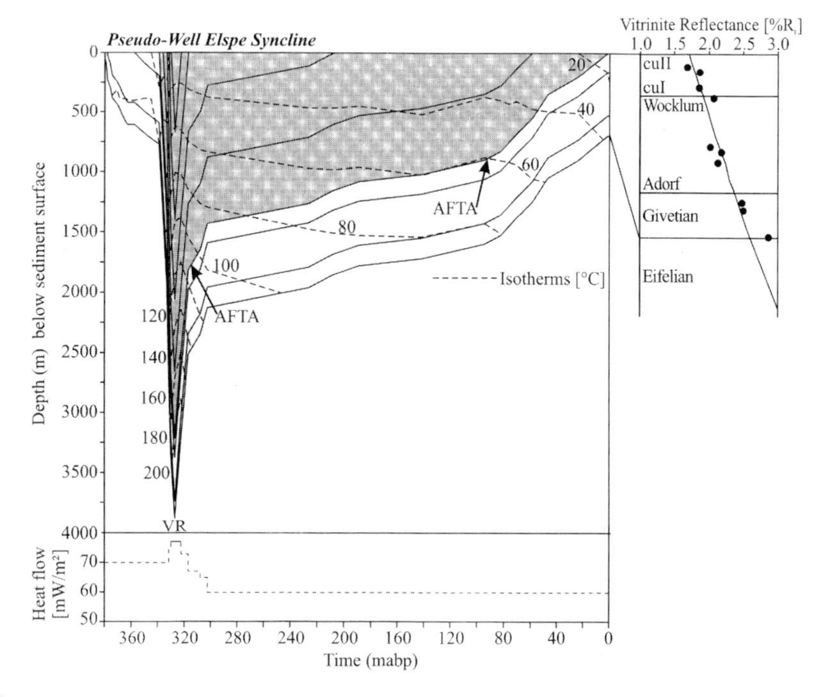

Fig. 4(d).

Fig. 4. Burial and heat flow history and calibration data for the Isselburg 3 (**a**), Münsterland 1 (**b**), and Specking 1 (**c**) wells and a pseudo-well located in the Elspe Syncline (d). (See Fig. 1 for well locations.)

Rock-Eval pyrolysis, which revealed T_{max} values of 430 and 440 °C for the Kupferschiefer (Type II kerogen) and the top of the Carboniferous sequence (Type III kerogen), respectively. Vitrinite reflectance of the pre-Permian sequence increases regularly from 1.09 (1348 m) to 2.08% R_r (2597 m). A value of 3.56% R_r was measured for the Devonian rocks at 3654 m depth (Fig. 4a). Fission-track analyses indicate that the uppermost Carboniferous rocks reached maximum temperatures, between 140 and 160 °C (derived from vitrinite reflectance modelling), soon after deposition; that is, during late Carboniferous or Early Permian time. Hence, the pre-Zechstein coalification of the Carboniferous and older section was confirmed. The Carboniferous surface was reheated to around 100 °C during Jurassic time, cooled to around 70 °C until Early Cretaceous time, and then further to the present-day temperatures of around 50 °C (Fig. 4a).

Numerical modelling revealed that the Palaeozoic vitrinite reflectance pattern in well Isselburg 3 can be explained only by assuming a total of about 3000 m of eroded Upper Carboniferous strata and a heat flow at times of maximum burial of about 65 mW m^{-2}. Deposition and erosion of this overburden occurred between Upper Westphalian A and Zechstein time.

The Münsterland 1 ('M1' in Fig. 1) borehole served as a key well for the northern part of the area studied (Fig. 4b). Beneath the 1800 m thick Late Cretaceous (Albian to Campanian) cover, the lower Westphalian B sequence was drilled. Middle Devonian rocks were reached at a final depth of nearly 6000 m. Teichmüller *et al.* (1979) published a wealth of coalification data for the Palaeozoic section encountered by the well that show a regular increase from 1.14 to more than 6% R_r with stratigraphic age (Fig. 4b). T_{max} values from Rock-Eval pyrolysis increase from 423 to 433 °C between the top and the base of the Cretaceous sequence. For the uppermost Carboniferous rocks, much higher T_{max} values of around 455 °C were determined. This would suggest a pre-Cretaceous maturation of the Palaeozoic rocks. Fluid inclusion analysis revealed maximum homogenization temperatures of 174, 193 and 198 °C for Westphalian A samples at 2090, 2476 and 2665 m depth, respectively. These values are in good agreement with the modelled coalification temperatures (Fig. 4b). Apatite fission-track analyses on two Westphalian samples (1850 and 2090 m) revealed that they cooled to temperatures below 110–120 °C only 60–70 Ma ago. Hence, we could not extract information on the Mesozoic temperature history from the data.

The construction of a numerical model of this well was based on the assumption that the Palaeozoic sequence in this region reached maximum temperatures, as recorded by the degree of coalification, in pre-Permian times, as is valid for the southern part of the Ruhr Basin. Hence, the coalification pattern of the Palaeozoic section is best explained assuming an eroded overburden of around 2700 m and a heat flow at the time of maximum burial of around 75 mW m^{-2} (Fig. 4b).

The Specking 1 well is located within a major syncline in the central Ruhr Basin ('Sp1' in Fig. 1) and drilled the youngest preserved Carboniferous strata in the Ruhr Basin, Westphalian C units. According to Fiebig & Groscurth (1984), only 100 m of the Upper Westphalian C sequence are missing in this well. Nevertheless, the Westphalian C sequence is of a relatively high thermal maturity, as indicated by vitrinite reflectance values which range between 0.65 and 0.90% R_r (Schwarzkopf & Schoell 1985; Fig. 4c). Nearby, the overlying Zechstein sequence is of remarkably low maturity (around 0.3% R_r). These circumstances allowed the most precise stratigraphic classification of the eroded overburden, as deposition *and* erosion must have occurred between the latest Westphalian C and Zechstein time. The simulations revealed that the thermal maturity of the Westphalian C rocks can be reached only if about 1800 m of eroded overburden and a heat flow of around 60 mW m^{-2} at times of maximum burial (Fig. 4c) are assumed.

The same procedure as described above was used in more than 50 wells all over the Ruhr Basin. General trends of the late Carboniferous burial and heat flow evolution are shown in Fig. 5. The simulations show that heat flow values did not exceed 68 mW m^{-2}, except for the northeastern part close to the Münsterland 1 well. The overall average is about 58 mW m^{-2}, which agrees with the tectonic setting at a folded and imbricated passive continental margin. The amount of eroded overburden varies between 1500 m in the central synclines of the basin and about 4000 m at the southern margin, with a continuous southward increase. The large amounts of eroded post-Westphalian C sediments (1000–2000 m) confirm earlier results from Littke *et al.* (1994) and Büker *et al* (1995) and indicate that sedimentation did not end before Stephanian times.

The Northern Rhenish Massif. The results achieved with the 1D basin-modelling programs PDI/PC and BASINMOD for the northern Rhenish Massif are summarized in Fig. 5. The modelled heat flow during maximum burial scatters

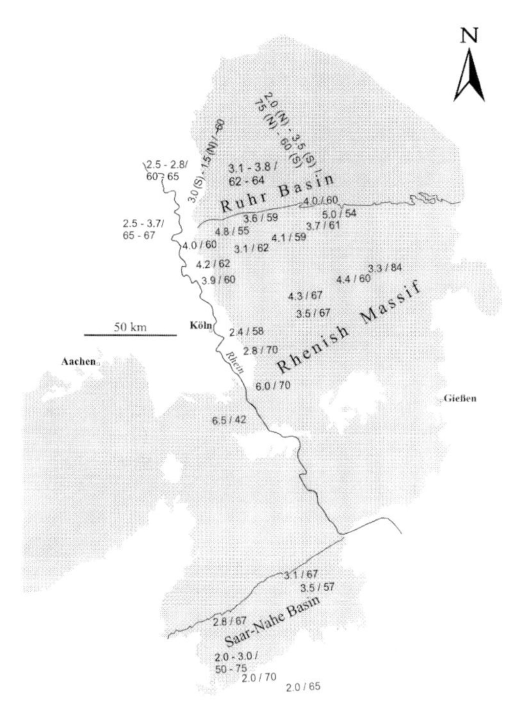

Fig. 5. Overview of thicknesses of eroded Palaeozoic overburden (km) (first of each pair of values) and heat flow values (mW m^{-2}) during maximum burial for the areas studied.

between 54 and 70 mW m^{-2} with an average of less than 60 mW m^{-2}. This indicates a slight decrease from the Ruhr Basin towards the Rhenish Massif. Major positive heat flow anomalies were not observed. The highest heat flow of 67–70 mW m^{-2} was calculated for the eastern and western rims of the Attendorn–Elspe Syncline (Fig. 4d) as well as for the northern limb of the Siegen Anticline (Fig. 1). However, it has to be pointed out that calibration of the heat flow by vitrinite reflectance data is less unequivocal for the pseudo-wells in the Rhenish Massif than for the real wells in the Ruhr and Saar–Nahe basins.

The estimated eroded thicknesses depend extensively on the structural position of the pseudo-wells, that is, whether they are located in synclines or anticlines. The highest values of eroded Palaeozoic sediments amount to about 6000 m in the Siegen Anticline and in the Eifel Anticline. In the Attendorn–Elspe Syncline, 3300–4400 m thickness of sediment was removed. At the northern rim of the Rheno-Hercynian, the amount of eroded sediment thickness ranges from 3100 m (Herzkamp Syncline) to 5500 m (Remscheid–Altena Anticline). Anomalously little erosion is calculated for the area east of Köln, that is, the Paffrath Syncline and the Lennefe Syncline (Fig. 1). Vitrinite reflectance values for Middle and Upper Devonian shales range between 0.7 and 1.5% R_r, which is very low compared with adjacent areas. A good fit between measured and calculated reflectance values is achieved only by assuming erosion of a 2400 m thick upper Palaeozoic sequence in the Paffrath and a 2800 m thick one in the Lennefe Syncline. The low thermal maturity of the Devonian and older rocks in this area has already been discussed in the literature (e.g. Wolf & Braun 1994).

The Permo-Carboniferous Saar–Nahe Basin. The general coalification pattern of the Saar–Nahe Basin (see also Hertle & Littke (2000)) indicates that maturation was syn-kinematic (Teichmüller *et al.* 1983), in contrast to the mainly pre-kinematic maturation in the areas discussed above. Therefore, the calibration of simulated temperature histories is completely based on information from wells and does not include pseudo-wells as in the case of the northern Rhenish Massif.

The two key wells for understanding the evolution of the Saar–Nahe Basin are the 5800 m deep Saar 1 well and the 1700 m deep Meisenheim 1 well ('S1' and 'MH1' in Fig. 1; see Fig. 8a and b, below). To obtain a good fit between measured and calculated maturity data, an additional sediment thickness of 2000 m and a heat flow of 70 mW m^{-2} for the Saar 1 well, and 3600 m and 60 mW m^{-2} for the Meisenheim 1 well had to be included in the model. The sedimentation lasted for about 40 Ma in the Saar–Nahe Basin and was followed by the erosion of the now-missing section until sedimentation recommenced in Late Permian (Zechstein) time. Apatite fission-track data provided evidence for a reheating of the Palaeozoic sediments during Jurassic and early Cretaceous time. This is explained by a moderate burial below several hundred metres (up to 1400 m) of lower Mesozoic sediments and elevated heat flow reaching a maximum of *c.* 80 mW m^{-2} during Jurassic–Cretaceous time. High heat flow during Early Jurassic time has been reported from several areas in central Europe (Leischner *et al.* 1993; Vercoutere & van den Haute 1993; Petmecky *et al.* 1999). This phenomenon can be tentatively interpreted as related to the opening of the North Atlantic Ocean.

Later, during Jurassic and Cretaceous time, uplift and erosion with decreasing heat flow values followed. This scenario, however, does not exclude small-scale sedimentation or hydrothermal activity during this period, but such

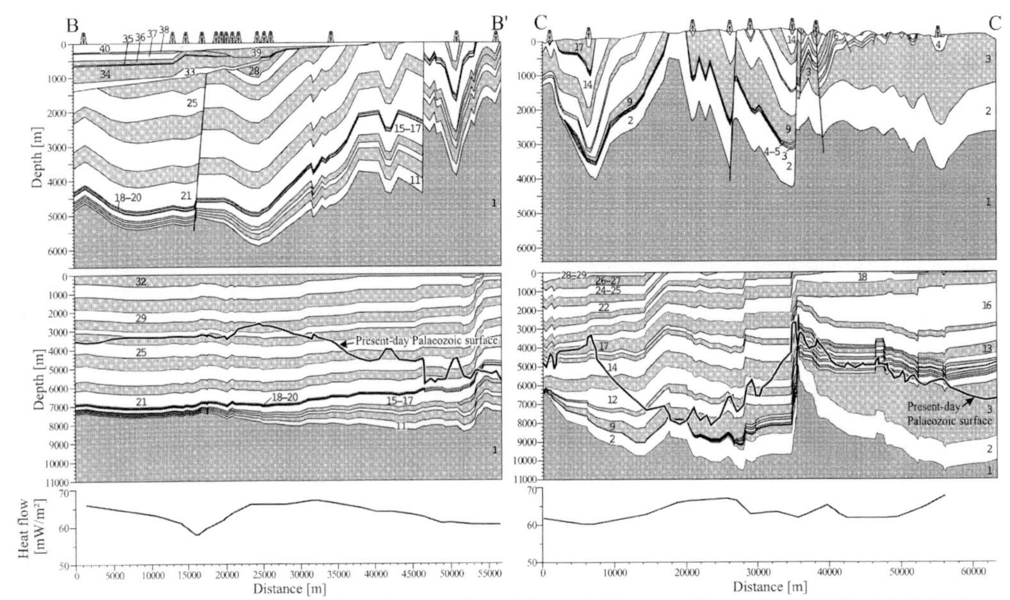

1: "Basement"; 2: Gedinnian; 3-5: Siegenian; 6-9: Emsian; 10-11: Eifelian; 12-13: Givetian; 14: Frasnian; 15-17: Fammen.; 18: Tournais.; 19: Viséan; 20: Nam. A; 21-22: Nam. B; 23: Nam. C; 24-25: Westph. A; 26-27: Westph. B; 28-29: Westph. C; 30: Westph. D; 31-32: Steph.; 33: Zechstein; 34-36: Early Trias.; 37: Mid. Trias; 38: Juras.; 39: Cret.; 40: Tertiary

Fig. 6. Top: finite-element models of the cross-sections along the river Rhein (see Fig. 1 for locations). Well symbols mark the locations of calibration wells and pseudo-wells. The models have been constructed based on interpreted seismic reflection, and outcrop and borehole data. Bottom: possible basin configuration and heat flow at times of maximum burial along the above cross-sections. Calibration of the models using palaeo-temperature indicators (see text for details) has resulted in the reconstruction of thicknesses of eroded Palaeozoic overburden (layers above the present-day Palaeozoic surface) and the heat flow distribution. A stratigraphic classification of the eroded overburden based on geological considerations resulted in the basin configuration shown. (Note that the cross-sections are not restored, and the basin configuration and heat flow pattern are not neccesarily valid for the same time along the cross-section.

phenomena have not been detected in the available temperature-sensitive data. Present-day bottom-hole temperatures can be explained by a heat flow of 55 to 63 mW m^{-2}. The calculation of heat flow included new data on thermal conductivities on 18 samples from the Saar–Nahe Basin as well as earlier published thermal conductivity data (Hückel & Kappelmeyer 1966).

Two-dimensional basin modelling

Western Ruhr Basin and Rhenish Massif. Two-dimensional basin-modelling was carried out along one cross-section through the western part of the Ruhr Basin and one through the Rhenish Massif to the south (lines BB′ and CC′ in Figs 1 and 6, respectively). The finite-element mesh of the numerical model was created by subdividing the cross-sections into 102 (Ruhr Basin) and 150 (Rhenish Massif) vertical grid lines and 53 depositional, non-depositional and erosional events. The Palaeozoic geology along the northern section is dominated by a thick sequence of folded Namurian and Westphalian sedimentary rocks. The underlying sequence has been drilled only at a few sites and consists mainly of lower Carboniferous shales and carbonates and Upper Devonian clastic deposits. Fifteen wells along the section, each penetrating a considerable part of the Upper Carboniferous sequence, provided the basic data for the 2D modelling. In the southern part, that is, the Rhenish Massif, vitrinite reflectance data for the calibration of the model were derived from eight pseudo-wells.

Modelling revealed that reflectance isolines (isomaturation, isocoalification, isorank) within the Palaeozoic sequence are more or less parallel to bedding and thus confirm that maximum burial and heating occurred before folding and deformation. In more detail, there is a slight increase of vitrinite reflectance from north to south for all stratigraphic units. On the basis of detailed rank analysis of coal seams, Juch (1991) has already referred to this pattern, which can be explained by a deeper burial of the sediments in the south.

For the base of the Bochum Beds (Westphalian A; see Fig. 2) vitrinite reflectance increases from 1.40% R_r in the Isselburg 3 well to 1.90% R_r in the Wattenscheid Anticline in the central part of the section, pointing to maximum temperatures increasing from about 140 to 160 °C. The modelled heat flow at the end of Carboniferous time and the estimated amount of eroded overburden leading to a best fit between measured and calculated vitrinite reflectance data (Fig. 6) are in agreement with the results from 1D simulations. Maximum burial occurred during Stephanian time in most of the northern section (Fig. 6). In the south, however, there is much more uncertainty with respect to the end of sedimentation and onset of uplift and erosion. In the models a Westphalian age was considered reasonable, taking into account the northward propagation of the Variscan deformation front (see Drozdzewski 1993).

The maximum thickness of eroded Palaeozoic sediments was calculated for the centre of the Remscheid Anticline (7800 m), whereas on the crest of the Velbert Anticline only about 5000 m were removed. The very low thicknesses calculated further to the south for the Paffrath Syncline can be attributed to a ramp-bounded position of this area above a major intracrustal flat detachment (Oncken *et al.* 1999, this volume), causing locally restricted uplift and erosion during imbrication in contrast to the neighbouring thrust-bounded units. Towards the Siegen Anticline in the southernmost part of the section, an increase of the thickness of eroded Palaeozoic sediments to about 6000 m has been modelled.

The age of maximum burial and, hence, maximum temperatures decrease from south to north in our models. K–Ar ages (Ahrendt *et al.* 1983) indicated a timing that ranges from earliest Namurian time in the south (Siegen Anticline) to late Westphalian time in the north (Velbert Anticline). Zircon fission-track data from the centre of the Remscheid Anticline indicate that cooling started at about 312 Ma (Westphalian A time; Karg 1998). These results all suggest a northward-propagating deformation front during the Variscan Orogeny.

Eastern Ruhr Basin and Rhenish Massif. The seismic reflection line DEKORP 2-N crossed nearly the entire Rheno-Hercynian zone (Franke *et al.* 1990). The northern 90 km of the geological interpretation of the seismic line (line AA′ in Fig. 1) were transformed into a finite-element model (Fig. 7) using additional published and unpublished geological information (e.g. Drozdzewski 1993; Oncken *et al.* this volume). The input data forming the numerical model (e.g. strata thicknesses, depositional ages and lithologies) were stored in a finite-element grid formed by 177 grid points along the section and 52 depositional, non-depositional and erosional events (Fig. 7). More details on the construction of the finite-element model have been given by Büker *et al.* (1995). The model presented here makes use of 20 calibration wells (Fig. 7), the southernmost three of which, situated in the northern Rhenish Massif, are pseudo-wells as described above. The vitrinite reflectance data for some of the wells have been calculated from volatile matter (data from Juch (1991)), and thus, carry some uncertainty.

For the large synclines, greater amounts of eroded overburden above a distinct reference horizon (e.g. the Namurian C base), have been calculated than for the anticlines (Fig. 7). Stratigraphic classification of the eroded overburden (see below) suggests that sedimentation either lasted longer in the present-day synclines (i.e. that they formed during sedimentation) or that deposition rates were higher. Assuming a longer-lasting sedimentation results in the basin configuration shown in Fig. 7. The calculated heat flow at the base of the defined sequence decreases regularly from around 75 mW m^{-2} in the north to values of around 60 mW m^{-2} in the south. This decrease can be explained either by crustal thickening and shortening in the south as a result of the advancing orogenic wedge or by rifting in the North German Basin in early Permian times (e.g. Ziegler 1990). Sensitivity analyses, however, have shown that temperature modelling using the assumption of an eroded overburden with decreasing thermal conductivities towards the north (i.e. a northward facies change from sandy to shaly lithologies), would require higher heat flow values in the south and lower ones in the north. With this reasonable geological scenario, the heat flow would decrease southwards only half as much as that mentioned above.

The Permo-Carboniferous Saar–Nahe Basin. A 40 km long NW–SE-southeast trending cross-section through the eastern part of the Saar–Nahe Basin (line EE′ in Fig. 1; burial and temperature histories in Fig. 8) is shown in Fig. 9, which illustrates the half-graben geometry. The finite-element model of the basin consists of 87 vertical grid lines and 53 events. Along the section, data from three wells and eight outcrop samples were used to calibrate the model. The modelling results indicate that isocoalification lines cut across bedding and thus reveal syn-kinematic coalification in the basin. A good

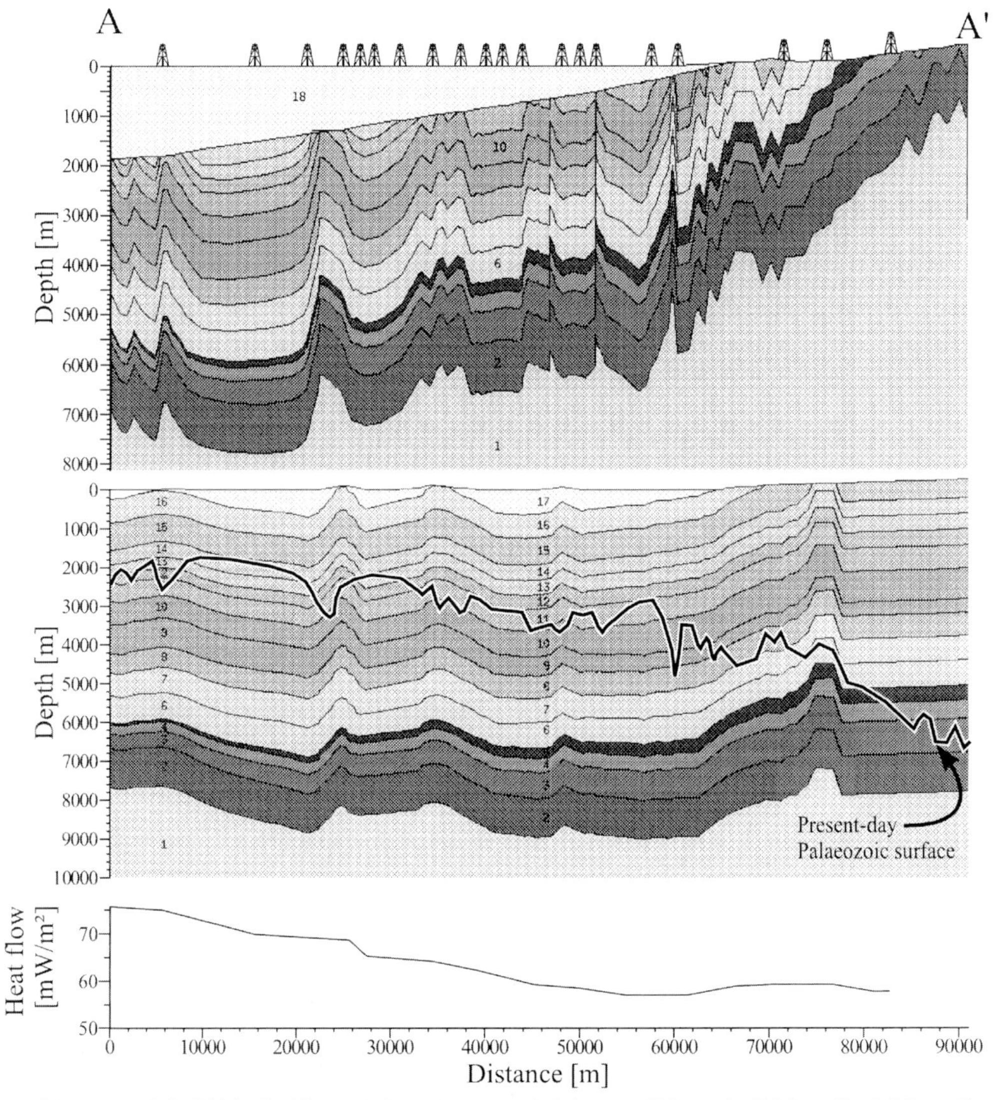

1: "Basement"; 2: Eifel.; 3: Givet.; 4: Late Devon..; 5: Dinant.; 6 Nam. A; 7: Nam. B; 8: Nam. C; 9-10: Westph. A; 11-12: Westph. B; 13-14: Westph. C; 15: Westph. D; 16-17: Steph.; 18: Cretac..

Fig. 7. Top: the finite-element model of the eastern cross-section through the Ruhr Basin and northern Rhenish massif (see Fig. 1 for location). Well symbols mark the locations of calibration wells and pseudo-wells. The models have been constructed based on interpreted seismic reflection, and outcrop and borehole data (see Franke *et al.* (1990) for details). Bottom: possible basin configuration and heat flow at times of maximum burial along the above cross-sections (see Fig. 6 for more information).

fit between measured and calculated data was obtained with palaeo-heat flow values between 67 (Nahe Syncline; well B1 in Fig. 9) and 57 mW m^{-2} (Pfalz Anticline; wells B2 and B3 in Fig. 9) as well as eroded sediment thicknesses from 3100 to 3600 m. Accordingly, 3700 m of Permian (Rotliegend) sediments must have been deposited in the Nahe Syncline (600 m are still preserved), whereas the thickness of the Permian (Rotliegend) sequence at the NW flank of the Pfalz Anticline must have been about 2900 m. The sedimentation pattern reflects the synsedimentary growth of the syncline and anticline structures.

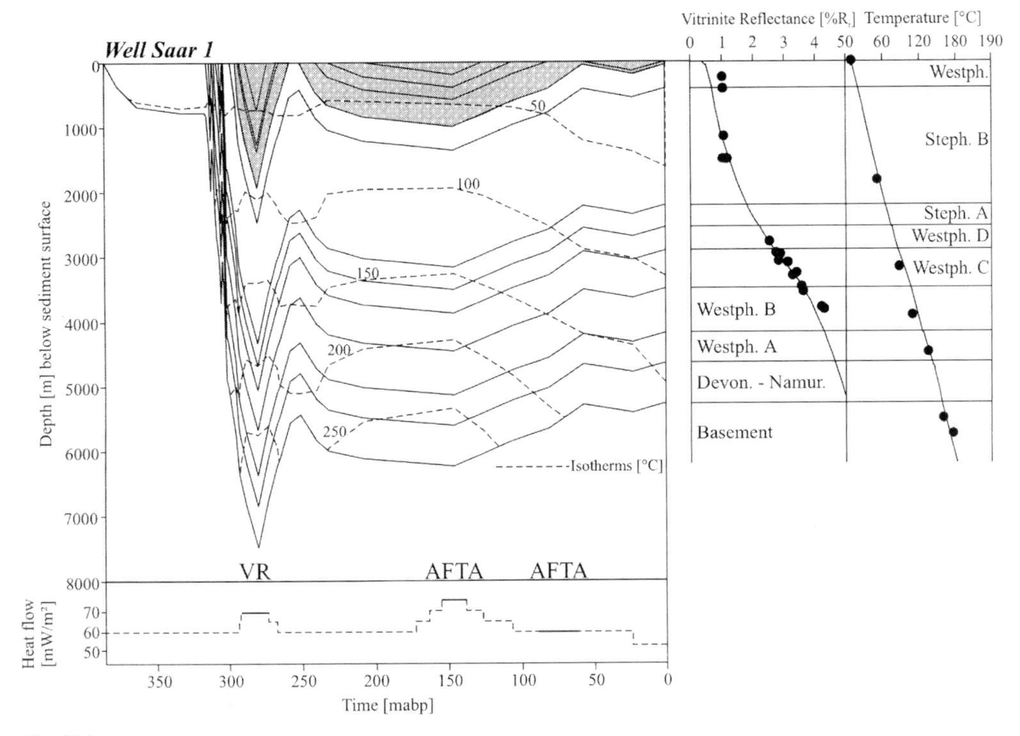

Fig. 8(a).

The asymmetry of the Saar–Nahe Basin is related to its origin as an internal pull-apart basin at the border between the Rheno-Hercynian and Saxo-Thuringian Zones of the Variscides (e.g. Korsch & Schäfer 1995; see other explanation by Henk (1993)). The depocentre shifted from southwest to northeast. This greatly influenced the timing of maximum burial and the age of the eroded Palaeozoic strata. In the southwest, sedimentation ceased and erosion started much earlier than in the northeast. There are also differences between synclines and anticlines. Table 1 gives an overview of the total amount of eroded Palaeozoic sediments and their stratigraphic position for different areas of the Saar–Nahe Basin including the heat flow during time of maximum burial. The data were calculated for individual wells and provide information on the very complex differential burial and temperature history of the basin.

Discussion

Palaeogeographical constraints

Burial and temperature histories for different parts of the Variscan Orogen (see Fig. 1) were calculated on the basis of maturity and palaeotemperature information derived mainly from vitrinite reflectance and apatite fission-track measurements. The reliability of the simulation results depends clearly on the reliability of the input and calibration data. Even if these are undisputable, as in most of the Ruhr Basin, there are still alternative geological conceptual models that fit the observed data. Therefore, numerous scenarios were tested to verify or falsify the respective geological conceptual models. These alternative scenarios have been published elsewhere (Littke *et al.* 1994; Büker *et al.* 1995; Karg 1998; Hertle & Littke 2000), so that only few important alternatives are treated here.

One such scenario, previously published for the Ruhr Basin, assumes that deposition during Late Carboniferous time ceased in Westphalian C time, which led earlier workers to the conclusion that heat flow at the time of maximum burial was extremely high, that is, of the order of 120 mW m^{-2} or more (Buntebarth *et al.* 1982*a*). These scenarios, however, would produce a strong vitrinite reflectance gradient, and this was not observed in any of the boreholes. The consequence of this is that a large amount (several hundred metres to 3 km) of now eroded Palaeozoic overburden must have existed which

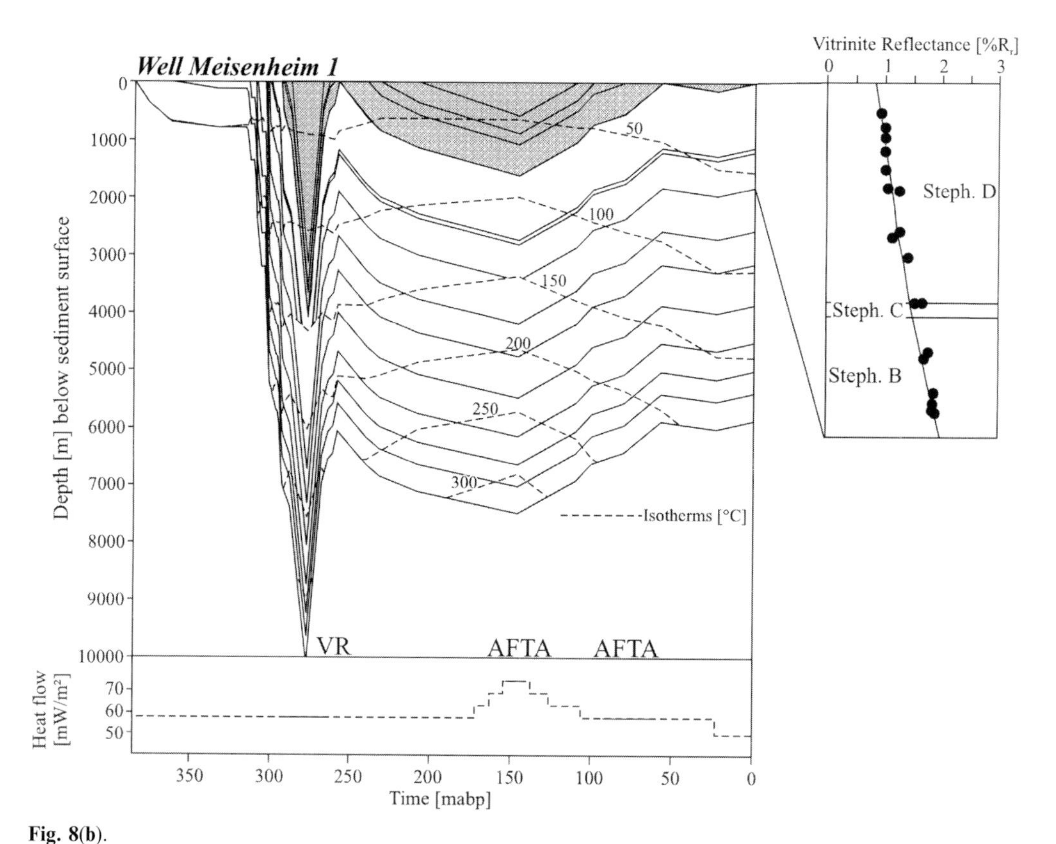

Fig. 8(b).

Fig. 8. Burial and heat flow histories and calibration data for the Saar 1 (**a**) and Meisenheim 1 (**b**) wells. (See Fig. 1 for well locations.)

was completely eroded before Late Permian (Zechstein) time in the western part of the Ruhr Basin and before the late Early Cretaceous in the eastern and southern part. Permian sediments unconformably overlie the Rhenish Massif in various places; for example, Zechstein deposits at the eastern margin and along the Lower Rhein Embayment and earlier Permian (Rotliegend) deposits in the Malmedy Graben (Belgium) and at the southern and northern rim of the Rhenish Massif (Fig. 1). Therefore, it is concluded that much of the erosion of the Palaeozoic rocks occurred before Zechstein time.

As the thicknesses of the Westphalian A, B and C units are well known in the Ruhr Basin, a part of the eroded overburden there has to be attributed to the Westphalian D and the Stephanian sequence (see discussion of the sedimentation rate evolution by Littke *et al.* (1994)). This fact modifies earlier assumptions on the palaeogeography of Stephanian sequence (e.g. Ziegler 1990) in which its distribution is restricted to northern Germany.

Assigning calculated thicknesses of eroded Palaeozoic overburden to different stratigraphic units becomes increasingly difficult with deeper erosion of the Rheno-Hercynian mountain belt. In this case, knowledge of sedimentation rates in adjacent areas (e.g. Drozdzewski 1993) and sedimentation rate evolution in young sedimentary basins in general (Allen & Allen 1990; Cloetingh & Banda 1992) has to be applied. Furthermore, tectonic stacking and complete subsequent erosion of the nappes has to be considered (see discussion by Littke *et al.* (1994)). With these uncertainties, the simulation results suggest that Stephanian beds of considerable thickness were deposited in the area of the present Ruhr Basin, but did not extend much further to the south. The former presence of Westphalian sediments has to be assumed for the entire northern Rhenish Massif, with calculated thicknesses of up to more than 4 km. Only for the Ebbe Anticline (Fig. 1) was an earlier end of sedimentation constructed in Namurian time. Whereas the thickness of eroded Palaeozoic

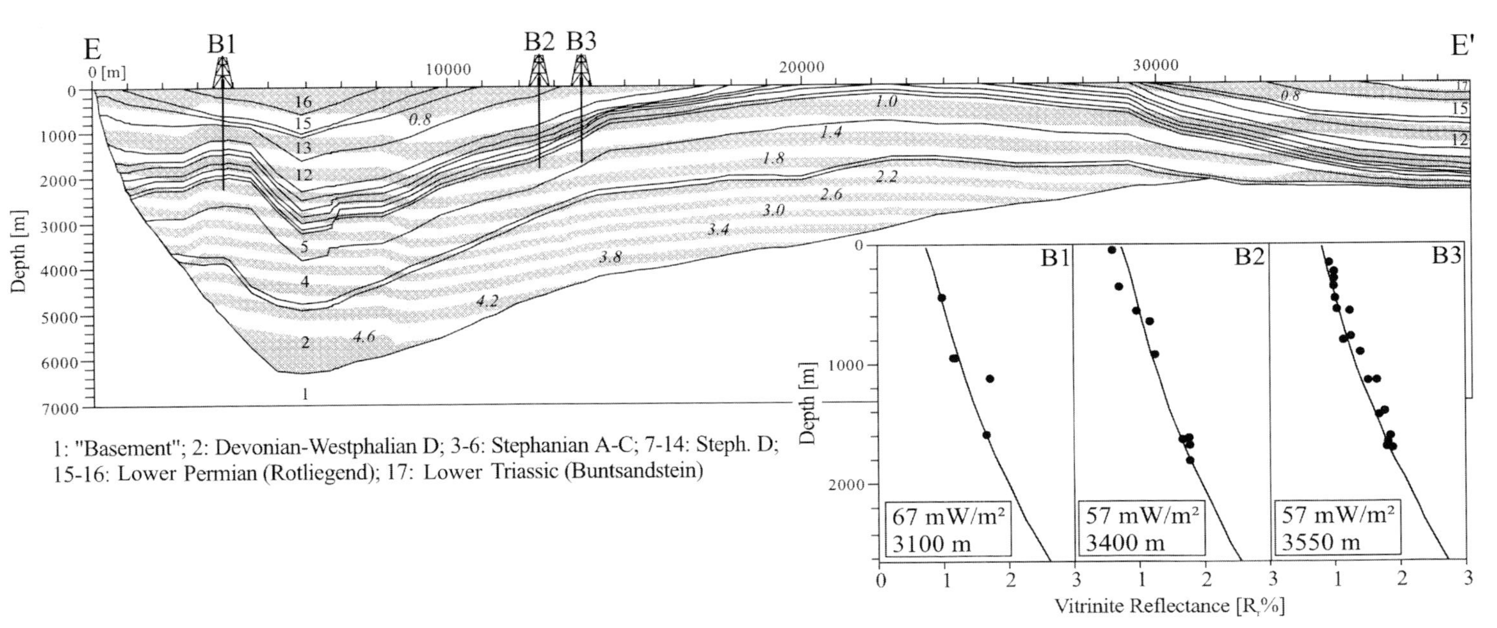

Fig. 9. Result of a 2D simulation of a cross-section through the eastern part of the Saar–Nahe Basin (see Fig. 1 for location). The finite-element model of the cross-section and the calculated present-day vitrinite reflectance pattern (% R_r; grey shadings) are shown. The model has been calibrated by comparing calculated (lines) and measured (dots) vitrinite reflectance data from three wells (B1–B3) and outcrop samples. Also shown is the heat flow at times of maximum burial and the amount of eroded Palaeozoic overburden calculated for wells B1–B3.

Table 1. *Heat flow during times of maximum burial, and amount of eroded Palaeozoic thicknesses as well as their stratigraphic classification for different areas in the Saar–Nahe Basin (for locations see Figs 1 and 9)*

Well	Heat flow (mW m^{-2})	Total eroded thickness (m)	Stratigraphic classification of eroded thicknesses (m)						
			Westph. C	Westph. D	Steph. A	Steph. B	St. C	Steph. D	Rotlieg
Saarbrücken Anticline, East									
ZIEH	75	1830		390	340	1100			
WEMN	48	2600				1020	80	1420	80
HIRZ	46	3040				540	80	1420	1000
Saarbrücken Anticline, Centre									
HOHL	60	2560		550	470	1450	90		
QUNO	62	2460		450	470	1450	90		
FRÖH	60	2540			260	1450	80	750	
EIVO	60	2300				720	80	1420	80
PETE	60	2110						1260	850
PRIM	57	2110						1310	800
Saarbrücken Anticline, West									
EMME	53	3130			200	1450	80	1400	
FÜOS	50	2940		760	630	1450	80	20	
FÜVÖ	50	2950		760	630	1450	80	30	
LAOS	53	3170	80	1080	630	1380			
NESA	55	2690			310	1450	80	850	
FRIE	56	2750				1370	80	1300	
Prims Syncline									
STWE	67	2800							2800
Nahe Syncline									
B1	67	3100							3100
Pfalz Anticline									
B2	57	3400						500	2900
MH1	57	3550						650	2900
Zweibrücken Syncline									
SAAR 1	70	2000						800	1200
BEXB	75	1800							1800
LAND	65	2000							2000
Lorraine									
GIRO	60	1700	1700						

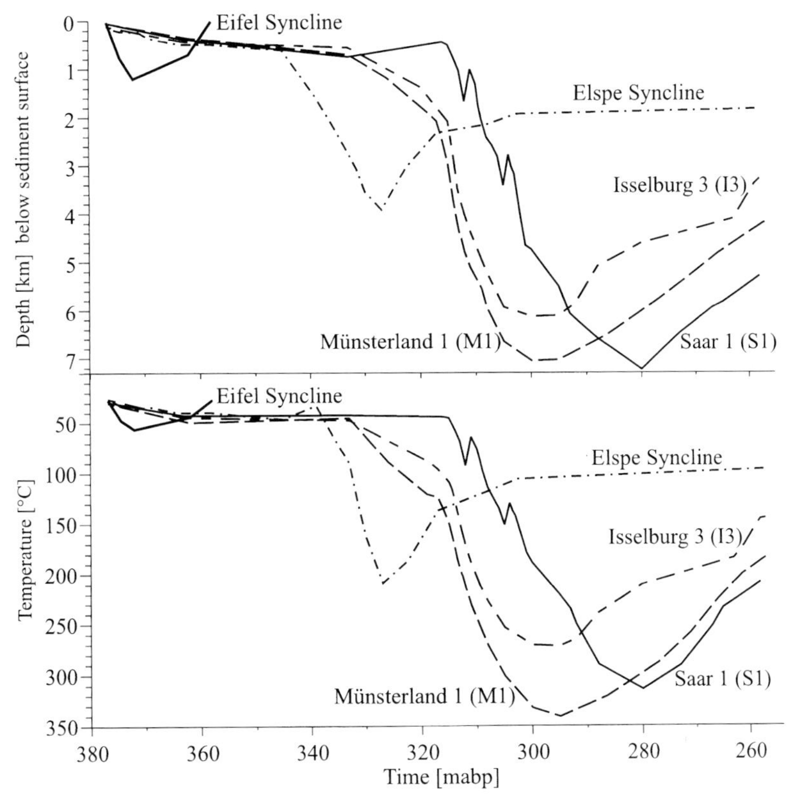

Fig. 10. Comparison of simulated burial (top) and temperature (bottom) histories for the base of the Late Devonian sequence at different locations in the studied area (see Fig. 1 for locations).

(i.e. Devonian, Lower and Upper Carboniferous) sediment clearly increases southward from the external Ruhr Basin towards the Rhenish Massif, the amount of deposited Upper Carboniferous sediments decreases in this direction.

Heat flow evolution

A major result of this study is the insight into heat flow evolution and in particular the heat flow during maximum burial before the onset of orogenic contraction. The timing of the temperature peak, for which heat flow was based on vitrinite reflectance data, is well constrained for areas in which immature, post-orogenic sediments overlie the pre- and synorogenic sedimentary rocks. This is the case in the western Ruhr Basin, where immature Zechstein deposits overlie more mature Upper Carboniferous rocks. In other areas, the timing is less obvious, but, where available, fission-track data support a temperature peak during the Late Carboniferous or early Permian time for all investigated areas. Figure 10 shows reconstructed burial and temperature histories for the base of the Namurian C sequence at different locations in the studied area and Fig. 11 the temperature history for the Gedinnian sediments along the western cross-section through the northern Rhenish Massif east of the river Rhein. The figures imply that uplift related to folding and imbrication started earlier in the south than in the north, as also indicated by K–Ar ages (Ahrendt *et al.* 1983). There are no indications for higher temperatures during Mesozoic or Cenozoic time other than those attained before or during Late Carboniferous orogeny. Three stages can be identified in our data:

(1) The syn-rift stage is preserved in the syn-rift sediments of the central parts of the western Rhenish Massif and characterized by a relatively high, late Early Devonian heat flow of 80–120 mW m^{-2}. These values record basin formation with associated lithospheric thinning ($\beta = 1.7$) and magmatism.

(2) The arrival of the deformation front in the northern Rhenish Basin occurred during Westphalian time and is coincident with maximum

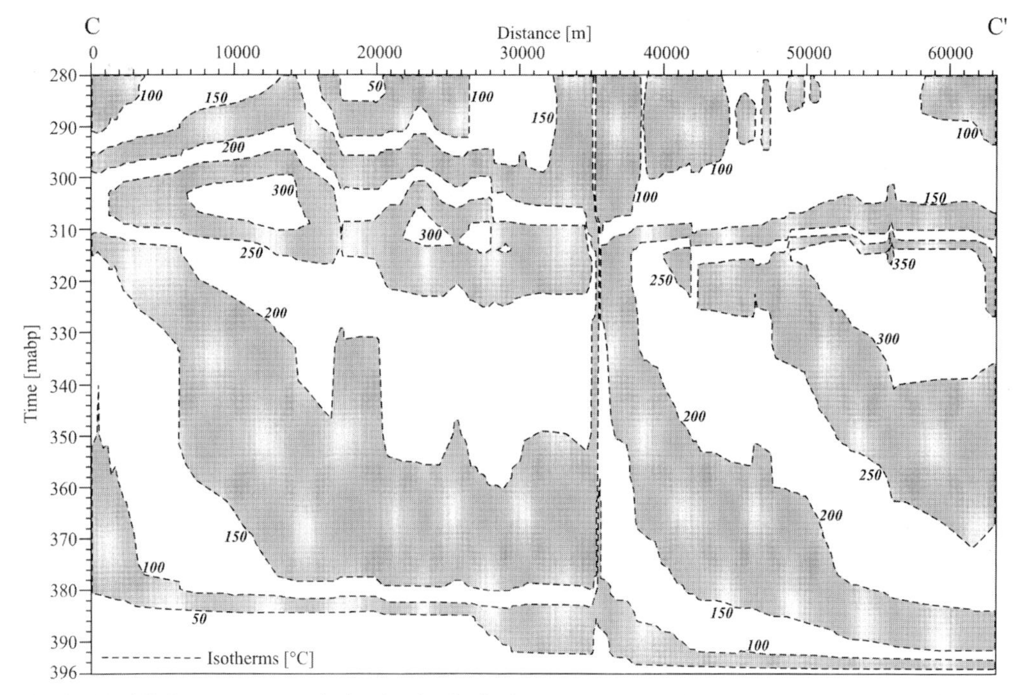

Fig. 11. Modelled temperature evolution for the Gedinnian sequence (see Fig. 2) along cross-section CC′ east of the river Rhein (for locations see Figs 1 and 6).

burial and heating in the post-rift and syn-orogenic sediments. This stage is characterized by heat flow values of some 50–60 mW m^{-2} corresponding to a crust of about 35 km thickness (Bodri & Bodri 1985). Later, during Stephanian time, the advancing deformation front reached the Ruhr Basin, where heat flow values between 55 and 75 mW m^{-2} were typical. The slightly higher values might suggest slightly lower crustal thicknesses, but can also be due to such features as higher heat conductivities of the now eroded overburden, that is, higher sandstone/shale ratios.

(3) The syn-kinematic stage is preserved in the maturation pattern of the Saar–Nahe Basin and reflects the temperatures during Stephanian or Early Permian time. Basically, identical heat flow values are recorded as for Stephanian time in the Ruhr Basin.

Heat transfer occurs either by conduction or by convection. The influence of the latter process is known to be important in some geological settings (Clauser & Villinger 1990). However, the limited lateral continuity of aquifers in the study area does not seem to support a major effect of convection on heat transfer before and during deformation. In addition, it has to be kept in mind that the sedimentary rocks investigated were at depths of 2–7 km during maximum burial; that is, below the highly porous and more permeable uppermost crust, where groundwater movement is most effective. Furthermore, large, small-scale variations in calculated heat flow densities, which would be explainable by convective influence, are not observed. The Saar–Nahe Basin may be an exception here, because small-scale variations are more pronounced than in the Ruhr Basin and northern Rhenish Massif. Also, larger aquifer systems such as widespread and thick sandstone sequences did exist here.

The greatest portion of the heat distributed in sedimentary systems is known to be derived from the deeper crust or mantle. Heat production inside the sedimentary basins is negligible for the Devonian sequence of the Rhenish Massif because of its lithology. Higher heat production can be expected for the shaly Carboniferous sequences rich in organic matter (see Yalcin *et al.* (1997) for a review). A more substantial contribution to the energy budget can be derived from frictional dissipation of deformation in orogenic wedges (Barr & Dahlen 1991) and from simple stacking of continental material (Jamieson *et al.* 1998). Both factors can be seen to have influenced the thermal evolution of the more southerly Rhenish Massif. One-dimensional numerical modelling of P–T–t paths from the southern massif by Henk (1995) has revealed an increase in heat flow from Late Viséan to Westphalian time, reaching values of some 85 mW m^{-2}. The axial zone of the orogenic wedge further south (the Mid-German Crystalline Rise) was shown to have similar to even higher heat flow values during this time. In agreement with this result, abundant granitic magmatism in Late Carboniferous time in this area suggests substantial heating of the deeper parts of the orogenic wedge. The mechanical analysis by Plesch & Oncken (2000) and Oncken *et al.* (this volume), moreover, indicates that the base of the southern, internal part of the orogenic wedge was heated from some 300 °C to more than 500 °C during Carboniferous wedge propagation. This is in contrast to the northern external wedge forming the northern Rhenish Massif and Ruhr Basin, which does not show any record of substantial syn-orogenic heat flow changes. These combined observations were interpreted to reflect two major aspects: (1) heating by advective material flow and uplift with increasing synkinematic exhumation to the south (from about 6 km at the Siegen Anticline to 18 km in the Phyllite Zone, Oncken *et al.* (this volume)); (2) moreover, an initially strong detachment in the south that weakened during wedge growth probably contributed to the energy budget through frictional dissipation in the internal parts of the wedge. This, along with tectonic stacking (Oncken 1990), destroyed the pre-orogenic thermal record in the sediments. In the adjoining axial zone, stacking of crustal, heat-producing material had an important role in generating heat and promoting crustal melting. This is also reflected in the thermal record of the sediments in the Saar–Nahe Basin that formed on this axial zone.

Possible magmatic intrusions and hydrothermal systems

Several magmatic intrusions have been suggested for the northern study area. These include the regions of the 'Lippstädter Gewölbe' (Lippstadt High), the 'Krefelder Gewölbe' (Krefeld High; Buntebarth *et al.* 1982*b*) and the 'Erkelenzer Horst' (Erkelenz High; for locations see Fig. 1). However, the vitrinite reflectance values in the areas of Lippstadt and Krefeld do not indicate any magmatic influence. Only in the area of Erkelenz do high coalification gradients point towards a high heat flow at times of maximum temperatures of about 110–120 mW m^{-2}, possibly coupled with magmatic activities. Even here, however, the vitrinite reflectance gradients and the necessary heat flow are not spectacularly high, if compared, for instance, with the Styrian Basin, which was affected by volcanism (Yalcin *et al.* 1997). In the Saar–Nahe Basin, Rotliegend volcanism is a well-known phenomenon which occurred at *c.* 290 Ma (e.g. Lippolt *et al.* 1989). Of course, the basin models presented here cannot exclude hydrothermal activities, but they can definitely exclude that such systems influenced the general coalification pattern.

Post-orogenic temperature history

Apatite and zircon-fission track analysis has shown that the investigated Palaeozoic rocks have been subjected to elevated temperatures during Triassic and Jurassic time and, in the Ruhr Basin, again during Tertiary time (Figs 4 and 8). The present-day Palaeozoic surface in the Ruhr and Saar–Nahe Basin and in the northernmost Rhenish Massif reached temperatures up to 100 °C and in the western Rhenish Massif (Remscheid–Altena Anticline and Paffrath Syncline; Fig. 1) even above 120 °C. Although it is impossible to extract clear Mesozoic burial and heat flow histories from the fission-track data, the simulations revealed that the high temperatures were not related to elevated heat flow values while the samples remained at shallow burial depths. Such a scenario would have resulted in very high vitrinite reflectance gradients, overprinting the well-calibrated trends established for Late Carboniferous times. Accordingly, a reburial of the Palaeozoic sequence under up to 1500 m of Upper Permian, Triassic and Jurassic rocks in combination with elevated although not very high heat flow values had to be assumed to satisfy both calibration datasets (vitrinite reflectance and fission-track data; Fig. 4a–c). For the Ruhr Basin, sedimentation in latest Cretaceous and early Tertiary times with subsequent erosion of up to 800 m is established (Fig. 4c; Büker 1996). In all cases, heat flow values at times of maximum Mesozoic and Cenozoic temperatures had to be kept below 90 mW m^{-2} to preserve the Palaeozoic vitrinite reflectance pattern. As these are 'minimum erosion–maximum heat flow' scenarios, it would nevertheless be possible to assume lower heat flow values with the consequence that the respective amount of eroded overburden would be even higher. However, from various information on high Mesozoic temperatures in adjacent areas in northern Germany (Leischner *et al.* 1993; Vercoutere & van den Haute 1993; Petmecky *et al.* 1999), we suggest that central and northern Europe were in an above-average heat flow regime during latest Triassic–Early Jurassic time, probably related to the initial opening of the North Atlantic Ocean.

The reconstructed amounts of eroded overburden contradict earlier assumptions of a long period of non-deposition in these parts of the study area during Mesozoic and latest Cretaceous–early Tertiary time (e.g. Murawski *et al.* 1983; Ziegler 1990).

Conclusions

Numerical basin modelling combined with petrographic studies is a powerful research instrument to study the evolution of external, non-metamorphic parts of orogenic fold belts. Up to now, this technique has been mainly applied to Alpine complexes such as the Cordilleran Foreland Basin in Western Canada (Osadetz *et al.* 1992) or the external parts of the Swiss Alps (Schegg & Leu 1998), but even older fold belts can be successfully studied. In the case of the Rheno-Hercynian orogenic wedge of central Europe, this approach revealed new information on the uplift and erosion history in the course of orogeny, as well as on heat flow history and post-orogenic processes. These results fit well into the context of age dating (K–Ar) in the Rheno-Hercynian mountain belt (Ahrendt *et al.* 1983). Together with other studies on metamorphic and more internal parts of the orogenic wedges (e.g. Franke & Stein this volume; Henk *et al.* this volume), they provide the fundamental basis for a more holistic and quantitative understanding of orogenic evolution.

The thermal evolution and late erosional removal of sedimentary rocks in an orogenic wedge system has been reconstructed from analysis of the thermal record of pre- to post-orogenic sediments deposited in different parts of the Rheno-Hercynian fold and thrust belt, from the almost undeformed foreland to the axial collision zone with the Saxo-Thuringian upper plate. The temporal evolution of heat flow patterns shows an early stage in (late Early Devonian time) of high heat flow (*c.* 110 mW m^{-2}) restricted to a central rift area within the former Rheno-Hercynian passive margin. This feature can be linked to crustal stretching (β1.7) and it decayed over time in the post-rift stage with thermal contraction and isostatic subsidence of the crust. The unrifted northern basin margin had an average heat flow (50–75 mW m^{-2}) and did not experience this anomaly and also does not show distinct thermal changes from basin evolution to orogenic deformation. The internal wedge, which developed from the external shelf and ocean–continent transition, shows evidence of syn-kinematic heating to higher than average values (75–85 mW m^{-2}), reflected in some preserved syn-orogenic sediments as well as in *P–T–t* paths and the occurrence of late-stage crustal melting.

Although the thick syn-orogenic sediments deposited in the foreland molasse basin reflect substantial syn-orogenic erosion of the internal wedge during its propagation, additional post-orogenic erosional removal has affected the entire fold and thrust belt as well as the foreland molasse basin between Westphalian and Zechstein time by values of at least 2 km and partly more than 6 km. In the Carboniferous Ruhr Basin a southward increase of late orogenic uplift and erosion is evident. The thickness of eroded Palaeozoic sediments depends on the structural position, that is, whether they occur in synclines or in anticlines. According to these results, sedimentation lasted longer than previously assumed for this area: until Stephanian time in the Ruhr Basin and until Westphalian time in the northern Rhenish Massif. The coalification pattern in the Saar–Nahe Basin differs from that in the Ruhr Basin reflecting a predominantly syn-orogenic maturation. Eroded overburden and heat flow values during maximum burial are, however, within a similar range.

Apatite fission-track data provide evidence for a reheating of the Palaeozoic sediments during Jurassic and early Cretaceous time throughout the study area. This is explained by moderate burial under several hundred metres (up to 1400 m in the Saar–Nahe Basin) of lower Mesozoic sediments and elevated heat flow of up to 80 mW m^{-2} during Jurassic–Cretaceous time. High heat flow values during Early Jurassic time have been reported from several areas in central Europe and are tentatively interpreted to be related to the opening of the North Atlantic Ocean.

The work presented here was mainly financed by the Deutsche Forschungsgemeinschaft (DFG) in the framework of the programme 'Simulation orogener Prozesse am Beispiel der Varisziden'. This support is gratefully acknowledged. Furthermore, we thank the coal mining companies Ruhrkohle AG and Saarberg-werke AG, the Ruhrverband Essen, the CONOCO CBM-Division Essen and the GLA Krefeld, which provided a wealth of information. Reviews of an earlier version of this manuscript by T. McCann, S. Egan and D. Tanner improved the quality of the text and are gratefully acknowledged.

References

AHRENDT, H., CLAUER, N., HUNZIKER, J. C. & WEBER, K. 1983. Migration of folding and metamorphism in the Rheinisches Schiefergebirge deduced from K–Ar and Rb–Sr age determinations. *In*:

MARTIN, H. & EDER, F. W. (eds) *Intracontinental Fold Belts: Case Studies in the Variscan Belt of Europe and the Damara Belt in Namibia*. Springer, New York, 323–338.

ALLEN, P. A. & ALLEN, J. R. 1990. *Basin Analysis—Principles and Applications*. Blackwell, Oxford.

BARR, T. D. & DAHLEN, F. A. 1991. Brittle frictional mountain building; 2, Thermal structure and heat budget. *Journal of Geophysical Research*, **94**, 3923–3947.

BODRI, L. & BODRI, B. 1985. On the correlation between heat flow and crustal thickness. *Tectonophysics*, **120**, 69–81.

BÜKER, C. 1996. Absenkungs-, Erosions- und Wärmeflußgeschichte des Ruhr-Beckens und des Rechtsrheinischen Schiefergebirges. *Berichte des Forschungszentrums Jülich*, **3319**.

——, LITTKE, R. & WELTE, D. H. 1995. 2D-modelling of the thermal evolution of Carboniferous and Devonian sedimentary rocks of the eastern Ruhr basin and northern Rhenish Massif, Germany. *Zeitschrift der Deutschen Geologischen Gesellschaft*, **146**, 321–339.

BUNTEBARTH, G., KOPPE, I. & TEICHMÜLLER, M. 1982*a*. Palaeogeothermics in the Ruhr basin. *In*: CERMAK, V. & HAENEL, R. (eds) *Geothermics and Geothermal Energy*. Schweizerbart, Stuttgart, 45–55.

——, ——, & ——, 1982*b*. Das permokarbonische Intrusiv von Krefeld und seine Einwirkung auf die Karbon-Kohlen am linken Niederrhein. *Fortschritte in der Geologie von Rheinland und Westfalen*, **30**, 31–45.

CLAUSER, C. & VILLINGER, H. 1990. Analysis of conductive and convective heat transfer in a sedimentary basin, demonstration for the Rhine Graben. *Geophysical Journal International*, **100**, 393–414.

CLOETINGH, S. & BANDA, E. 1992. Europe's lithosphere—physical properties. Mechanical structure. *In*: BLUNDELL, D., FREEMAN, R. & MÜLLER, S. (eds) *A Continent Revealed: the European Geotraverse*. Cambridge University Press, Cambridge, 80–91.

DROZDZEWSKI, G. 1993. The Ruhr coal basin (Germany): structural evolution of an autochthonous foreland basin. *International Journal of Coal Geology*, **23**, 231–250.

ENGEL, H. 1985. Zur Tektogenese des Saarbrücker Hauptsattels und der Südlichen Randüberschiebung. *In*: DROZDZEWSKI, G., ENGEL, H., WOLF, R. & WREDE, V. (eds) *Beiträge zur Tiefentektonik westdeutscher Steinkohlenlagerstätten*. Geologisches Landesamt Nordrhein-Westfalen, Krefeld, 217–235.

ENGEL, W., FRANKE, W., GROTE, C., WEBER, K., AHRENDT, H. & EDER, F. W. 1983. Nappe tectonics of the southeastern part of the Rheinisches Schiefergebirge. *In*: MARTIN, H. & EDER, F. W. (eds) *Intracontinental Fold Belts*. Springer, Heidelberg, 267–287.

FIEBIG, H. & GROSCURTH, J. 1984. Das Westfal C im nördlichen Ruhrgebiet. *Fortschritte in der Geologie von Rheinland und Westfalen*, **32**, 257–267.

FRANKE, W. & STEIN, E. 2000. Exhumation of high-grade rocks in the Saxo-Thuringian Belt: geological constraints and geodynamic concepts. *This volume*.

——, BORTFELD, R., BRIX, M. *et al.* 1990. Crustal structure of the Rhenish Massif: results of deep seismic reflection lines DEKORP 2-North and 2-North-Q. *Geologische Rundschau*, **79**(3), 523–566.

GLEADOW, A. J. W. 1981. Fission track dating methods; what are the real alternatives? *Nuclear Tracks*, **5**, 3–14.

GOFFETTE, O., LIEGEOIS, J. P. & ANDRE, L. 1991. Âge U–Pb sur zircon dévonien moyen a supérieur du magmatisme bimodal du massif de Rocroi (Ardenne, France); implications géodynamiques. *Comptes Rendus de l'Académie des Sciences, Série 2*, **312**(10), 1155–1161.

HEINEN, V. 1996. *Simulation der präorogenen devonisch-unterkarbonischen Beckenentwicklung und Krustenstruktur im Linksrheinischen Schiefergebirge*. Aachener Geowissenschaftliche Beiträge, **15**.

HENK, A. 1993. Late orogenic evolution in the Variscan Internides: the Saar–Nahe Basin, Southwest Germany. *Tectonophysics*, **223**, 273–290.

—— 1995. Late-Variscan exhumation histories of the southern Rhenohercynian Zone and western Mid-German Crystalline Rise—results from thermal modelling. *Geologische Rundschau*, **84**, 578–590.

—— & VON BLANCKENBURG, F., FINGER, F., SCHALTEGGER, U. & ZULAUF, G. 2000. Syn-convergent high-temperature metamorphism and magmatism in the Variscides; a discussion of potential heat sources. *This volume*.

HERMANRUD, C. 1993. Basin modelling techniques—an overview *In*: DORÉ, A. G. *et al.* (eds) *Basin Modelling: Advances and Applications*. NPF Special Publications, **3**, 1–34.

HERTLE, M. & LITTKE, R. 2000. Coalification pattern and thermal modelling of the Permo-Carboniferous Saar Basin (SW-Germany). *International Journal of Coal Geology*, **42**, 273–296.

HESS, J. C. & LIPPOLT, H. J. 1986. $^{40}Ar/^{39}Ar$ ages of tonstein and tuff sanidines: new calibration points for the improvement of the Upper Carboniferous time scale. *Chemical Geology (Isotope Geoscience Section)*, **59**, 143–154.

HÜCKEL, B. & KAPPELMEYER, O. 1966. Geothermische Untersuchungen im Saarkarbon. *Zeitschrift der Deutschen Geologischen Gesellschaft*, **117**, 280–311.

HURFORD, A. J. & GREEN, P. F. 1982. A user's guide to fission track dating calibration. *Earth and Planetary Science Letters*, **59**, 343–354.

JAMIESON, R. A., BEAUMONT, C., FULLSACK, P. & Lee, B. 1998. Barrovian regional metamorphism; where's the heat? *In*: TRELOAR, P. J. & O'BRIEN, P. J. (eds) *What Drives Metamorphism and Metamorphic Reactions?* Geological Society, London, Special Publications, **138**, 23–51.

JUCH, D. 1991. Das Inkohlungsbild des Ruhrkarbons—Ergebnisse einer Übersichtsauswertung. *Glückauf-Forschungshefte*, **52**, 37–47.

KARG, H. 1998. *Numerische Simulation der thermischen Geschichte, Subsidenz und Erosion des westlichen*

Rechtsrheinischen Schiefergebirges, des Ruhrbeckens und des Paläozoikums der Niederrheinischen Bucht. Berichte des Forschungszentrums Jülich, **3618**.

KORSCH, R. J. & SCHÄFER, A. 1995. The Permo-Carboniferous Saar–Nahe Basin, south-west Germany and north-east France: basin formation and deformation in a strike-slip regime. *Geologische Rundschau*, **84**, 293–318.

KRAMM, U. & BUHL, D. 1985. U–Pb zircon dating of the Hill Tonalite, Venn–Stavelot Massif, Ardennes. *Neues Jahrbuch für Geologie und Paläontologie, Abhandlungen*, **171**, 329–337.

KUSZNIR, N. J. & ZIEGLER, P. A. 1992. The mechanics of continental extension and sedimentary basin formation; a simple-shear/pure-shear flexural cantilever model. *Tectonophysics*, **215**, 117–131.

LEISCHNER, K., WELTE, D. H. & LITTKE, R. 1993. Fluid inclusions and organic maturity parameters as calibration tools in basin modelling. *In*: DORÉ, A. G. *et al.* (eds) *Basin Modelling: Advances and Applications*. NPF Special Publications, **3**, 161–172.

LIPPOLT, H. J., HESS, J. C., RACZEK, I. & VENZLAFF, V. 1989. Isotopic evidence for the stratigraphic position of the Saar–Nahe Rotliegende volcanism. II. Rb–Sr investigations. *Neues Jahrbuch für Geologie und Paläontologie, Abhandlungen*, **9**, 539–552.

LITTKE, R. 1987. Petrology and genesis of Upper Carboniferous seams from the Ruhr region, West Germany. *International Journal of Coal Geology*, **7**, 147–184.

—— & TEN HAVEN, H. L. 1989. Palaeoecologic trends and petroleum potential of Upper Carboniferous coal seams of western Germany as revealed by their petrographic and organic geochemical characteristics. *International Journal of Coal Geology*, **13**, 529–574.

——, BÜKER, C., LÜCKGE, A., SACHSENHOFER, R. F. & WELTE, D. H. 1994. A new evaluation of palaeo-heatflows and eroded thicknesses for the Carboniferous Ruhr basin, western Germany. *International Journal of Coal Geology*, **26**, 155–183.

MURAWSKI, H., ALBERS, H. J., BENDER, P. *et al.* 1983. Regional tectonic setting and geological structure of the Rhenish Massif. *In*: FUCHS, K., VON GEHLEN, K., MÄLZER, H., MURAWSKI, H. & SEMMEL, A. (eds) *Plateau Uplift*. Springer, Berlin, 381–403.

NAESER, N. D. 1993. Apatite fission-track analysis in sedimentary basins—a critical appraisal. *In*: DORÉ A. G. *et al.* (eds) *Basin Modelling: Advances and Applications*. NPF Special Publications, **3**, 147–160.

ONCKEN, O. 1984. Zusammenhänge in der Strukturgenese des Rheinischen Schiefergebirges. *Geologische Rundschau*, **73**(2), 619–649.

—— 1987. Heat flow and kinematics of the Rhenish Basin. *In*: VOGELS, A., MILLER, H. & GREILING, R. (eds) *The Rhenish Massif*. Earth Evolution Series. Vieweg, Braunschweig, 63–78.

—— 1990. Aspects of the structural and paleogeothermal evolution of the Rhenish Massif. *Annales de la Société Géologique de Belgique*, **T113**(2), 139–159.

——, PLESCH, A., WEBER, J., RICKEN, W. & SCHRADER, S. 2000. Passive margin detachment during arc–continent collision (Central European Variscides). *This volume*.

——, VON WINTERFELD, C. & DITTMAR, U. 1999. Accretion of a rifted passive margin: the late Paleozoic Rhenohercynian fold and thrust belt (Middle European Variscides). *Tectonics*, **18**(1), 75–91.

OSADETZ, K. G., JONES, F. W., MAJOROWICZ, J. A., PEARSON, D. E. & STASIUK, L. D. 1992. Thermal history of the Cordilleran Foreland Basin in Western Canada: a review. *In*: MACQUEEN, R. W. & LECKIE, D. A. (eds) *Foreland Basins and Fold Belts*. American Association of Petroleum Geologists, Memoirs, **55**, 259–278.

PETMECKY, S., MEIER, L., REISER, H. & LITTKE, R. 1999. High thermal maturity in the Lower Saxony Basin: intrusion or deep burial? *Tectonophysics*, **304**, 317–344.

PLESCH, A. & ONCKEN, O. 2000. Orogenic wedge growth during collision—constraints on mechanics of a fossil wedge from its kinematic record (Rhenohercynian FTB, Central Europe). *Tectonophysics*, **309**, 117–139.

POELCHAU, H. S., BAKER, D. R., HANTSCHEL, T., HORSFIELD, B. & WYGRALA, B. 1997. Basin simulation and the design of the conceptual basin model. *In*: WELTE, D. H., HORSFIELD, B. & BAKER, D. R. (eds) *Petroleum and Basin Evolution*. Springer, Berlin, 3–70.

RADKE, M., HORSFIELD, B., LITTKE, R. & RULLKÖTTER, J. 1997. Maturation and petroleum generation. *In*: WELTE, D. H., HORSFIELD, B. & BAKER, D. R. (eds) *Petroleum and Basin Evolution*. Springer, Berlin, 171–229.

RICKEN, W., SCHRADER, S., ONCKEN, O. & PLESCH, A. 2000. Turbidite basin and mass dynamics related to orogenic wedge growth; the Rheno-Hercynian case. *This volume*.

SCHÄFER, A. & KORSCH, R. J. 1998. Formation and sediment fill of the Saar–Nahe Basin (Permo-Carboniferous, Germany). *Zeitschrift der Deutschen Geologischen Gesellschaft*, **149**, 233–269.

——, ONCKEN, O., KENITZ, H. & ROMER, R. L. 2000. Upper-plate deformation; a case study from the German Variscides (Saxo-Thuringian Zone). *This volume*.

SCHEGG, R. & LEU, W. 1998. Analysis of erosion events and palaeogeothermal gradients in the North Alpine Foreland Basin of Switzerland. *In*: DÜPPENBECKER, S. J. & ILIFFE, J. E. (eds) *Basin Modelling: Practice and Progress*. Geological Society, London, Special Publications, **141**, 137–155.

SCHWARZKOPF, TH. & SCHOELL, M. 1985. Die Variation der C- und H-Isotopenverhältnisse in Kohlen und deren Abhängigkeit von Maceralzusammensetzung und Inkohlungsgrad. *Fortschritte in der*

Geologie von Rheinland und Westfalen, **33**, 161–168.

SOLLE, G. 1970. Die Hunsrück-Insel im oberen Unterdevon. *Notizblätter des Hessischen Landesamt für Bodenforschung*, **98**, 50–80.

SOMMERMANN, A. 1993. Zirkonalter aus dem Granit der Bohrung Saar 1. *Beihefte zum European Journal of Mineralogy*, **5**(1), 145.

SWEENEY, J. J. & BURNHAM, A. K. 1990. Evaluation of a simple model of vitrinite reflectance based on chemical kinetics. *AAPG Bulletin*, **74**, 1559–1570.

TAYLOR, G. H., TEICHMÜLLER, M., DAVIS, A., DIESSEL, C. F. K., LITTKE, R. & ROBERT, P. 1998. *Organic Petrology*. Bornträger, Berlin.

TEICHMÜLLER, M. 1971. Das Inkohlungsprofil des flözführenden Oberkarbons der Bohrung Isselburg 3 nordwestlich von Wesel. *Geologische Mitteilungen*, **11**, 181–184.

——, TEICHMÜLLER, R. & LORENZ, V. 1983. Inkohlung und Inkohlungsgradienten im Permokarbon der Saar–Nahe-Senke. *Zeitschrift der deutschen geologischen Gesellschaft*, **134**, 153–210.

——, —— & WEBER, K. 1979. Inkohlung und Illit-Kristallinität–Vergleichende Untersuchungen im Mesozoikum und Paläozoikum von Westfalen. *Fortschritte in der Geologie von Rheinland und Westfalen*, **27**, 201–276.

VERCOUTERE, C. & VAN DEN HAUTE, P. 1993. Post-Paleozoic cooling uplift of the Brabant Massif as revealed by apatite fission track analysis. *Geological Magazine*, **130**(5), 639–646.

VON WINTERFELD, C. & WALTER, R. 1993. Die variszische Deformationsfront des nordwestlichen Rheinischen Schiefergebirges—Ein bilanziertes geologisches Tiefenprofil über die Nordeifel. *Neues Jahrbuch für Geologie und Paläontologie, Monatshefte*, **1993**, 305–320.

WELTE, D. H. & YALCIN, M. N. 1988. Basin modelling—A new comprehensive method in petroleum geology. *Advances in Organic Geochemistry*, **13**(1–3), 141–151.

WOLF, M. & BRAUN, A. 1994. Eine neue Inkohlungskarte des nördlichen Rechtsrheinischen Schiefergebirges. *Neues Jahrbuch für Geologie und Paläontologie, Monatshefte*, **1994**, 449–475.

YALCIN, N. M., LITTKE, R. & SACHSENHOFER, R. F. 1997. Thermal history of sedimentary basins. *In*: WELTE, D. H., HORSFIELD, B. & BAKER, D. R. (eds) *Petroleum and Basin Evolution*. Springer, Berlin, 71–167.

ZIEGLER, P. A. 1990. *Geological Atlas of Western and Central Europe*. Shell Internationale Petroleum Maatschappij, Den Haag.

Turbidite basin and mass dynamics related to orogenic wedge growth; the Rheno-Hercynian case

W. RICKEN[1], S. SCHRADER[1], O. ONCKEN[2] & A. PLESCH[2]

[1]*Department of Geology, University of Cologne, Zuelpicher Str. 49a, D-50674 Köln, Germany*

[2]*GeoForschungsZentrum Potsdam, D-14473 Potsdam, Germany*

Abstract: Investigation of the Rheno-Hercynian Turbidite Basin suggests an interrelation between orogenic wedge migration and changes of the basin architecture, facies types and progradation rates of turbidite systems. The orogenic wedge was built dominantly by the Mid-German Crystalline Rise and accreted older parts of the Rheno-Hercynian Turbidite Basin. Turbidite sequences seem to have developed in stepwise prograding sub-basins. They show systematic trends of widely correlative (tens of kilometres) large-scale cycles that show an upward transition from coarse-grained clastics to highly diluted mud turbidites with highstand shedding of carbonates. These observations suggest the existence of accommodation cycles on the assumed shelf along the Mid-German Crystalline Rise. Exposure of the shelf above sea level was associated with bypass and transport of coarse-grained clastics into the basin, whereas flooding is thought to have caused storage of clastic deposits on the shelf and the simultaneous deposition of mud turbidites in the basin. Accommodation cycles, estimated to reflect an average duration of 10^5–10^6 a, were probably related to cyclic changes of underthrusting in the internal parts of the orogenic wedge with concomitant uplift. The lifetime and onset of new sub-basins, of the order of a few million years, may show stages of imbricate fan formation. A pattern of major change in sedimentation rates, basin geometry and subsidence style of the Rheno-Hercynian Turbidite Basin and the Sub-Variscan Molasse Basin was interpreted to reflect the different change in crustal strength. This occurred because the hinge line between previously rifted and unrifted crust was overrun during wedge progradation. Numerical mass balance that compares exhumed and intra-basinal masses of the Rheno-Hercynian Turbidite Basin, including the proportional fill of the Sub-Variscan Molasse Basin and the Saar–Nahe Basin, indicates a significant lack of sediment mass (i.e. more than one-third of the exhumed masses). In this estimate, syn- to post-depositional truncation was included. As extension in the internal parts of the orogenic wedge has only a minor role, the above deficit is probably related to subduction of sediment during the early stages of convergence.

A significant amount of the mass in orogenic belts around the world is contained in fore-arc and foreland basins. Investigation of their sediment fill allows a detailed monitoring of the processes associated with the growth of orogenic belts, including crustal loading, uplift and thrusting (e.g. Allen & Homewood 1986; Dorobek & Ross 1995).

In this paper, we present a new interpretation of the subdivided basin fill, using several new concepts in turbidite sedimentology (Mutti 1992; Pickering *et al.* 1995), including the role of mud turbidites (Piper & Stow 1991) and the application of sequence stratigraphy of clastic and carbonate sediments (Eberli 1991; Vail *et al.* 1991; Schlager *et al.* 1994; Kolla *et al.* 1995). This allows the recognition of accommodation cycles on the shelf of the former upper plate. In addition, mass balancing is applied (Wetzel 1993; Einsele *et al.* 1996) based on numerical inversion of erosion and deposition, in which the basin fill is backstripped and becomes redistributed on the former catchment areas (Hay 1995; Einsele & Hinderer 1998), and thus yields an estimate of denudation rates. These, in turn, can be linked to and cross-checked with the exhumation history, derived from thermochronological data. Most of the published mass balances concern the post-collisional stage of young orogenic belts (Ibbeken & Schleyer 1991; Wetzel 1993; Einsele *et al.* 1996), and very few studies have attempted to assess the mass-flux and its controlling parameters in ancient orogenic systems from the precollisional to the postorogenic stage.

From: FRANKE, W., HAAK, V., ONCKEN, O. & TANNER, D. (eds).
Orogenic Processes: Quantification and Modelling in the Variscan Belt.
Geological Society, London, Special Publications, **179**, 257–280. 0305-8719/00/$15.00

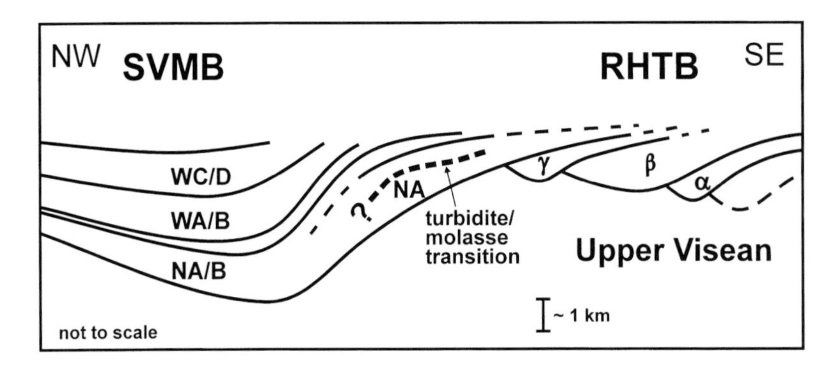

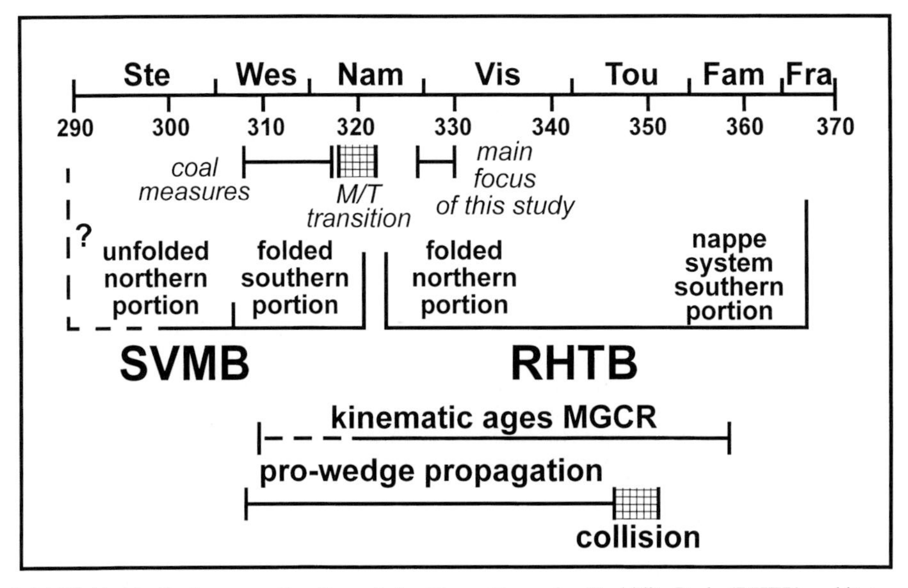

Fig. 1. (**a**) Highly idealized cross section through the Rheno-Hercynian Turbidite Basin (RHTB) and its transition into the Sub-Variscan Molasse Basin (SVMB). The RHTB shows progradation with internal sub-basins, whereas the SVMB displays a layer-cake filling pattern. Stratigraphic symbols used for the RHTB (i.e. α, β, and γ) are the *goniatites* stage subdivision as given in Fig. 2, whereas the SVMB is indicated by the symbols NA to NC for the Namurian stages A to C, and WA to WD for the Westphalian stages A to D. (**b**) Major events in the history of the two stages of the investigated basins, the Rheno-Hercynian Turbidite Basin (RHTB) and the Sub-Variscan Molasse Basin (SVMB). Bold type shows major tectonic events and the tectonic conditions of the basin, and italics give depositional stages in basin fill.

An ideally suited system for such a study is the Mid-European Variscides, which, in spite of their incomplete exposure, have been investigated recently by a series of quantitative studies related to the thermobarometric history and the kinematic processes of wedge formation (e.g. Franke *et al.* 1990; Henk 1995; Oncken *et al.* 1999; Plesch & Oncken 1999) that form a sound basis for mass-flux reconstruction and basin dynamics. Observations in north central Europe show two major stages of synorogenic evolution of the same foreland basin during the Variscan Orogeny: an older turbidite stage, the Rheno-Hercynian Turbidite Basin (RHTB), which shows a transition from early trench–fore-arc to foreland basin (but represents foreland conditions during the main phase of fill from Late Viséan to Early Namurian time), and a younger molasse stage, the Sub-Variscan Molasse Basin (SVMB, Fig. 1). Both basin stages were dominantly filled from sediment sources to the south, and together monitor a nearly continuous 60 Ma record (from Late Devonian to Late Westphalian time) of convergent processes in central Europe that were related to subduction and collision. The early history of the basin stage dominantly documents the denudation history, and therefore, uplift rates and thrusting of the prograding active margin orogenic wedge, the Mid-German Crystalline Rise (MGCR). In contrast, the younger basin stage, the SVMB, resulted from crustal loading of the lower plate, at the final stage of orogenic wedge progradation. Both basin stages are characterized by a typical asymmetric architecture. However, only the early type in basin evolution, the RHTB, is associated with a bulge-like structure.

In this paper we demonstrate the influence of orogenic growth on the filling pattern of the RHTB, reconstructing (1) the architecture and history of basin fill, and the time-transgressive turbidite progradation front as a monitor of the sedimentary expression of oceanic convergence, collision and early processes of fold belt formation; (2) the major controls of the stratigraphic record; from these the accommodation cycles for the assumed former northern shelf along the MGCR will be derived and the importance of tectonic processes versus that of sea-level variations discussed; (3) the mass transfer from the eroded MGCR to the two stages of foreland basin; we determine whether the masses eroded from the exhumed upper plate and the related basin fill balance or not.

The aim of this paper is to show the close interrelation between tectonic processes of an ancient prograding orogenic wedge with basin-forming processes. In comparison with younger Alpine or recent counterparts in well-studied foreland basins (e.g. Ricci Lucchi 1986; Zoetemeijer 1993), the Variscan Belt allows a much lower resolution. Nevertheless, this paper may propose a strategy for investigating wedge-controlled foreland basins that formed in an early stage after collision in ancient fold belts.

Geological frame and stratigraphy

Kossmat's (1927) traditional subdivision of the Mid-European Variscides into a number of zones has become reinterpreted in terms of microplates (see Franke this volume). The MGCR (Scholtz 1930; Brinkmann 1948), a zone of crystalline basement, was underthrust by the Rheno-Hercynian Zone, a former passive continental margin (Franke & Oncken 1990). The MGCR can be considered as the core of an orogenic wedge, which showed significant uplift, up to a depth equivalent to 9 kbar, and northwestward progradation of a pro-wedge system (Oncken 1997). This scenario is associated with the formation of the RHTB, a fore-arc to foreland basin that spans a time interval of roughly 45 Ma, from Late Devonian to the Mid–Namurian time. The asymmetric RHTB (Fig. 1a) was dominantly filled by turbidite wedges that show progradation from SE to NW. The depositional record of these turbidites dominantly monitors the uplift and thrust history of the upper-plate wedge, the MGCR and accreted Rheno-Hercynian material, and therefore, documents the early plate tectonic phases of subduction to the beginning of collision. During the period maximum sediment supply, from Early Tournaisian to Early Namurian time, the RHTB already represented foreland basin conditions.

At the northwestern margin, the RHTB shows a transition to the second type of foreland basin evolution, the SVMB (Fig. 1a), which represents the molasse stage of the foreland basin as defined by Ricci Lucchi (1986). It covers a time span from Namurian to Late Westphalian time, continues into the Stephanian Basin, and shows a record of about 30 Ma. Coal measures in the Ruhr area occur from Late Namurian to Late Westphalian time. The southern part of the SVMB includes the deformation front. However, most of this wide and asymmetric basin is undeformed (Drozdzewski & Wrede 1994). Deposition is characterized by shallow deltaic sequences and coal measures because abundant clastic supply outpaced subsidence (Fig. 1b). Unlike the RHTB, where the basin fill dominantly reflects the thrust history at the MGCR (Schrader 1999), the depositional architecture of the SVMB was also sea level controlled (Süss 1996).

As mentioned above, turbidites were entirely derived from the MGCR (Engel & Franke 1983; Engel *et al.* 1983*a*, *b*), thus monitoring sensitively uplift processes of the upper plate, related to arc formation, collision and pro-wedge progradation (Franke & Oncken 1990; Franke 1992; Oncken 1997). Tectonically, two zones of the RHTB exist, the southern one (i.e the Giessen–Harz Nappes; Franke & Oncken 1995) is represented by a nappe system, whereas the northern one is characterized by a fold and thrust belt (Oncken *et al.* 1999). The northern portion of the RHTB is the focus of this study (Schrader 1999), but significant conclusions can be drawn when the entire basin is considered.

The time frame

The stratigraphic frame of the investigated Late Viséan interval is based on the goniatite zonation introduced by Schmidt (1925). He subdivided the *goniatites* stage (cd III), which is the focus of this study, into three zones, the *crenistria* zone (cd IIIα), the *striatus* zone (cd IIIβ) and the *granosus zone* (cd IIIγ). Later, these zones were further subdivided into 11 subzones by various

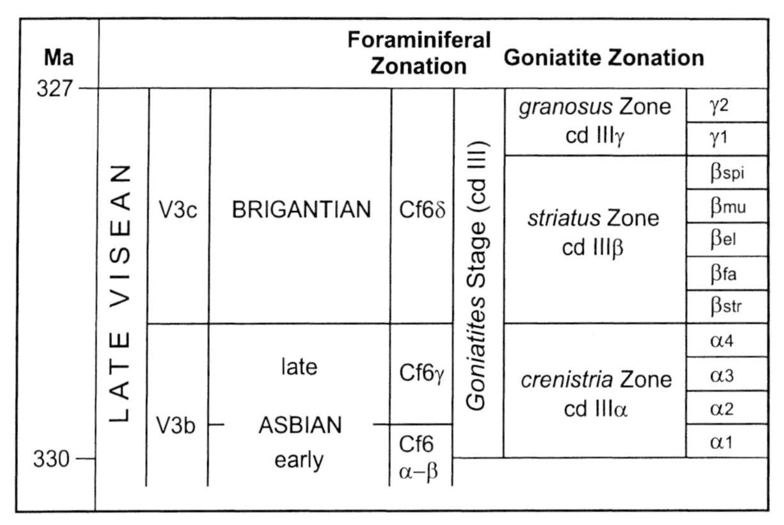

Fig. 2. Late Viséan goniatite stratigraphy of various researchers for Central Europe and its correlation with the stratigraphy of the Asbian and Brigantian sequence as referred to in the text. The stratigraphic frame used here is that of the Dinantian Stratigraphy Group (1971).

researchers, including Kulick (1960) and the Dinantian Stratigraphy Group (1971) (Fig. 2). Compared with the stages of the British Isles, the *goniatites* stage spans the time interval from Early Asbian to Late Brigantian time, based on investigations by Conil & Paproth (1983), Riley (1993), Jones & Sommerville (1996) and Korn (1996).

The absolute duration of this interval can be assessed, when compared with the time scales of Roberts *et al.* (1995) and Gradstein & Ogg (1996). On the basis of age dating of zircons (Claoué-Long *et al.* 1995), an estimated duration of this interval from Early Asbian to Late Brigantian time is of the order of 3–5 Ma. Here a mean duration of about 4 ± 1 Ma is used (Roberts, pers. comm.). However, as the duration of the goniatite stage (cd III) is not entirely equivalent to the total Asbian and Brigantian period, we assume an even smaller duration for this interval of about 3 Ma. Under the assumption that each goniatite subzone represented an identical amount of time, an average duration of such a zone was 0.3 Ma. These assumptions allow monitoring of sediment and basin fill processes on a high-resolution scale.

The Rheno-Hercynian Turbidite Basin (RHTB)

The Rheno-Hercynian Turbidite Basin (RHTB), located in the central Rheno-Hercynian Zone in central Europe, is one of the larger synorogenic sediment basins in the Variscan Belt (Fig. 3a). Its turbidite deposits are exposed in two major areas, the eastern Rhenish Massif and the Harz Mountains, and some smaller areas between them (i.e. the Werra and Flechtingen Hills). The central parts of the basin, where clastic sediments were supplied from the MGCR, are buried by Mesozoic to Cenozoic sediments. The turbidites show a preserved thickness of more than 2 km; especially the eastern portion of the basin, the Harz Mountains, is characterized by thick turbidite accumulations. The turbidites were assembled in two evolutionary steps. An earlier step is related to trench–fore-arc sedimentation in an oceanic realm (i.e. remnants of the Giessen Ocean) during pre-collisional convergence. These turbidites were accreted into an accretionary wedge from Frasnian time to the onset of collision in Tournasian time (Engel & Franke 1983) and subsequently thrust onto the passive margin of the lower plate (i.e. the Rheno-Hercynian Zone). Minor remnants of this stage are preserved in the Giessen Harz Nappes. Turbidite sedimentation continued with deposition on the subsiding continental shelf, which was loaded by overthrusting of the prograding orogenic wedge in a foreland basin setting (Franke this volume; Oncken *et al.* this volume).

The total width of this second-step turbidite basin, from SE to NW, was roughly 300 km, and along strike a length of at least 370 km is assumed. This estimate includes the eastern Rhenish Massif in the SW to at least the Oder–Elbe Fault Zone (Franke 1992) in the NE. However, its continuation in the latter area is

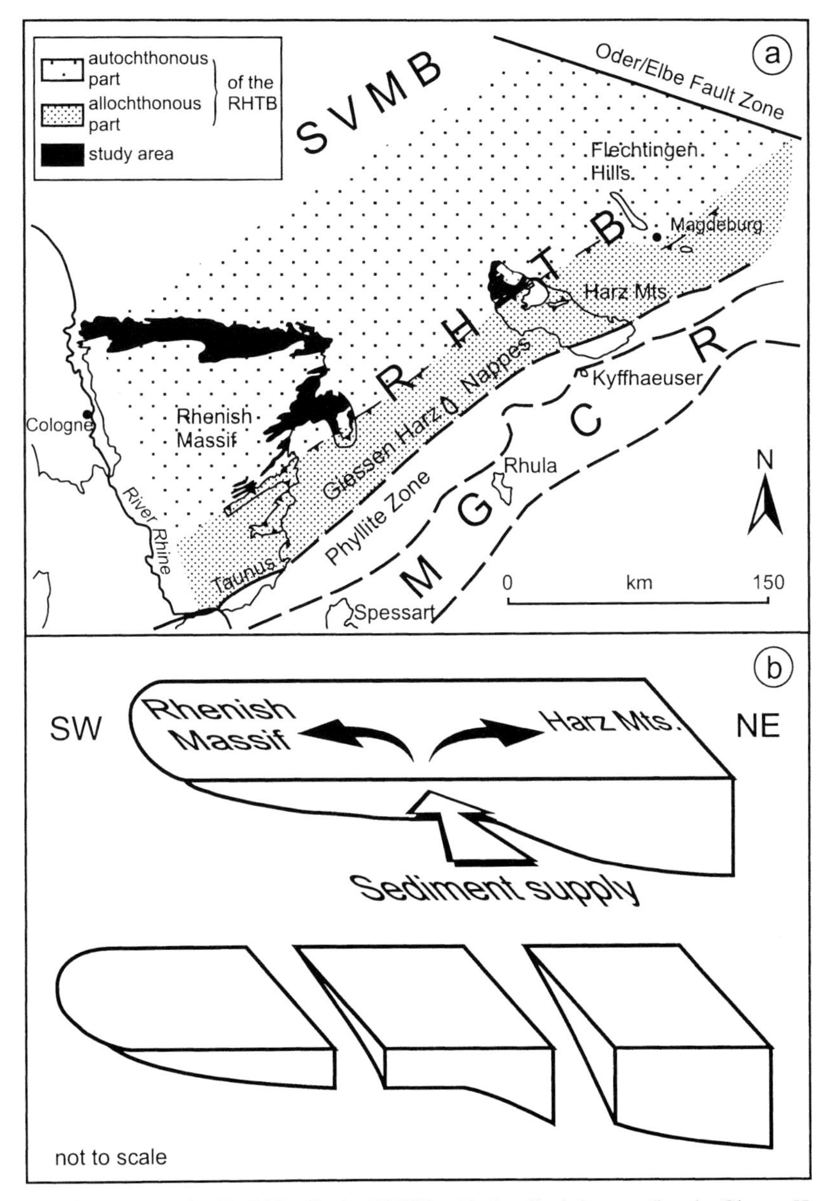

Fig. 3. (**a**) The Rheno-Hercynian Turbidite Basin (RHTB) with the allochthonous (i.e. the Giessen Harz Nappes) and the autochthonous portion. In the northwest, the transition towards the Sub-Variscan Molasse Basin (SVMB) is shown. (**b**) Highly schematized diagram of the Rheno-Hercynian Turbidite Basin (RHTB) during Late Viséan time, viewed from the Mid-German Crystalline Rise (in the foreground, not shown) in the direction of dominant sediment transport (see arrows). The position of the assumed submarine canyon (main arrow) is indicated.

under debate. In the northeastern Rhenish Massif a dominant pattern of distal onlap of turbidite systems is observed. Our studies cover the northern portion of the RHTB (Schrader 1999).

Palaeocurrent pattern and grain-size distribution suggest that two major fan systems developed (Fig. 3b). The fan system in the Rhenish Massif is characterized by transport dominantly to SW, whereas similar deposits in the Harz show an opposite transport direction. This observation, as reported by, for example Kuenen & Sanders (1956), Kulick (1960), Plessmann (1961), Sadler (1983) and Schrader (1999), resulted in a general model with two elongated turbidite fans with opposite orientation. This pattern suggests the existence of a major clastic source. Its sediments were transported through an assumed submarine canyon located somewhere between these fan systems in the area of the Mesozoic Hessian Basin (Engel & Franke 1983). The fact that this system with two opposite fan lobes was stable for a relatively long time period of about 40 Ma (i.e. from Late Devonian to Mid-Namurian time) suggests that the angle of obliqueness of collision between the two plates was not large enough to generate a discernible offset of the clastic source. These results are in agreement with the observation that the highest uplift rates, equivalent to 6–9 kbar, occur in the central part of the MGCR (Oncken 1997; Oncken *et al.* this volume), where the position of the entrance of the feeder channel is assumed from turbidite fan geometry.

Stepwise progradation

On the basis of goniatite stratigraphy, a relatively high time resolution of on average 0.3 Ma per subzone can be assumed for the studied Upper Viséan turbidite systems. This allows the monitoring of the pattern of progradation of turbidite prisms in detail. Although average progradation rates of 1.5 cm a^{-1} are observed (in the palinspastically restored basin from Late Devonian to the Mid-Namurian time), detailed investigation yields a more differentiated view that seems to show intervals of dominantly aggradation (with minor progradation) and sudden shifts in onlap of turbidite sedimentation.

These observations (Schrader 1999) are based on (1) the internal pattern of sub-basin fill and its sequence stratigraphic interpretation; (2) the thickness and cutting relationship of turbidite units; (3) the location of the depocentre within individual depositional prisms.

For the Late Viséan time interval, which comprises 11 goniatite subzones (see Fig. 2), the formation of three elongated sub-basins from SE to NW, the Wildungen Sub-basin, the Waldeck Sub-basin, and the Korbach Sub-basin, can be established in a section perpendicular to strike along the eastern Rhenish Massif. Relatively good evidence is observed only for the transition from the Waldeck to the Korbach Sub-basin (Fig. 4a), whereas the detailed sediment pattern and final evidence for the existence of the underlying Wildungen Sub-basin remain uncertain, as only the basin margin with a reduced sedimentary thickness is observed. Inferred sub-basins are thought to represent asymmetric prisms of turbidite fill, although palaeocurrent pattern indicated an inflection into the former sub-basin axis (i.e. mainly parallel to strike).

The Wildungen Sub-basin, characterized by deposits of the cd IIIα4, is thought to have been located further to the southeast. The following Waldeck Sub-basin (Fig. 4b) shows dominantly aggradation of up to 1000 m of turbidite sequences, from the cd IIIβstr to the cd IIIβspi subzones. Within this sub-basin, a minor northwestward shift of the depocentre (here taken as the zones of maximum turbidite thickness) by a few kilometres is observed with the onset of the upper cd IIIβ, which covers the cd IIIβmu and the cd IIIβspi subzones (for stratigraphic symbols, see Fig. 2). At the boundary between the cd IIIβspi and cd IIIγ1 subzones, northwestward movement by more than 15 km is observed with the onset of the Korbach Sub-basin. Internally, this sub-basin shows aggradation of the depositions of cd IIIγ1 and cd IIIγ2 subzones.

Each elongated sub-basin seems to have been bounded by a minor topographic high at its northwestern end, as observed for the Waldeck Sub-basin (Fig. 4b). These assumed minor highs are characterized by onlapping time lines, and, finally, turbidite overflow. A reduction of the stratal thicknesses can be observed from the sub-basin centres to the intra-basinal highs. In the sub-basin centre thick turbiditic successions were deposited, whereas pelitic sediments dominate the area of the intra-basinal highs. These assumed highs were located in areas where the seismic reflection line of DEKORP-2N indicates the position of well-known major thrust faults, which developed with the onset of thrust belt formation (Franke *et al.* 1990). These are the Eisemroth Thrust for the transition between the Wildungen and the Waldeck Sub-basins, the Siegen Thrust between the Waldeck and Korbach Sub-basins and the Ebbe Thrust at the northwestern end of the Korbach Sub-basin (Fig. 5). With the help of a balanced section for the eastern Rhenish Massif (see Oncken *et al.* this volume), the original widths of the individual sub-basins were established, at more than 70, 85 and 80 km for the Wildungen, Waldeck and Korbach Sub-basins, respectively. The association of the assumed intra-basinal highs with the modelled trajectories of major fault and thrust zones may suggest a common origin of both structures. The time for generation of new sub-basins is, within the biostratigraphic resolution

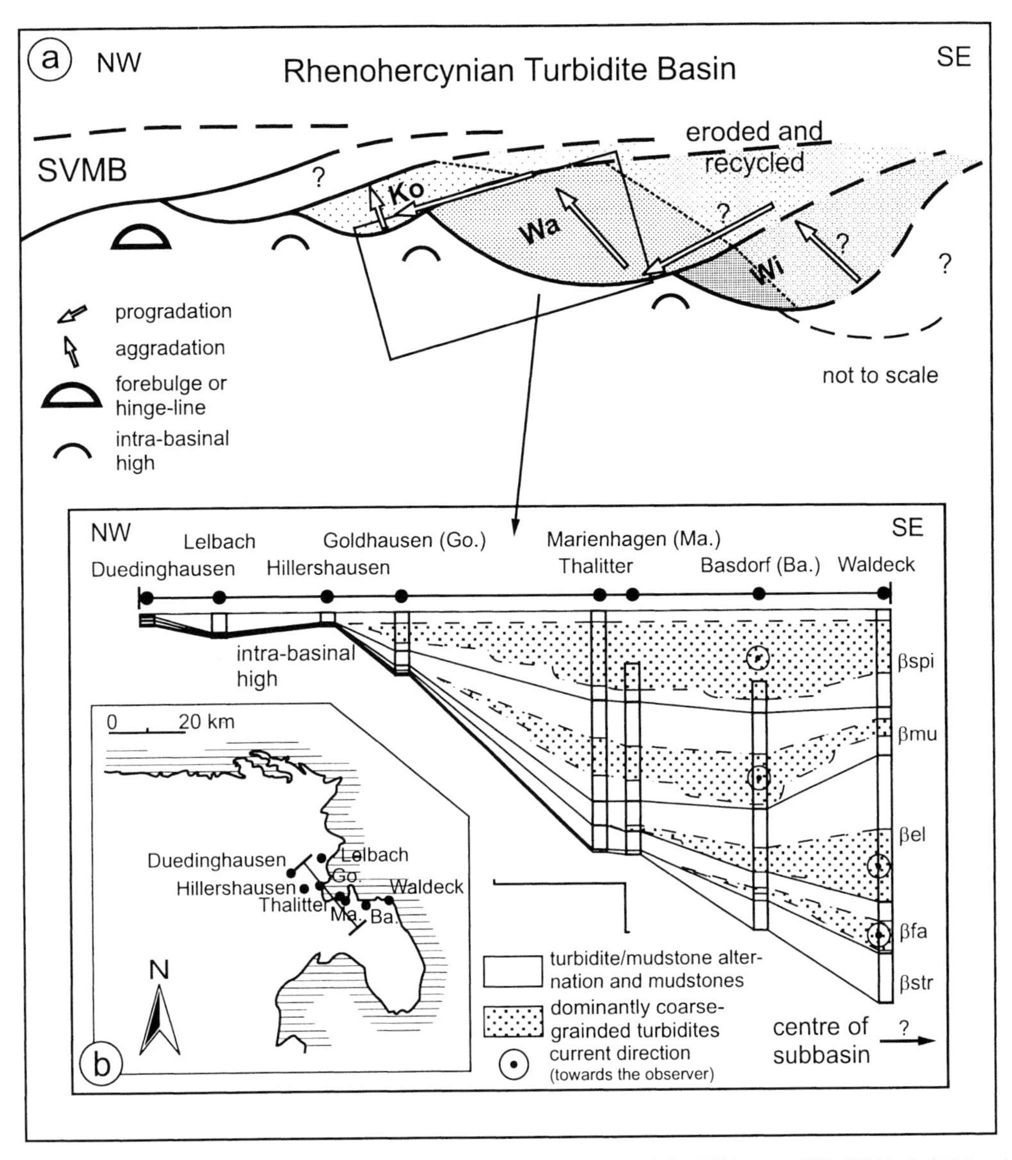

Fig. 4. (**a**) Conceptional interpretation of the turbidite filling pattern of the Wildungen (Wi), Waldeck (Wa) and Korbach Sub-basins (Ko), which are thought to show periods of dominantly aggradation (oblique-oriented arrows) that alternate with those of sudden shifts (sideward-oriented arrows). The various levels of erosion (dashed line) should be noted; this makes the evidence for the Wildungen Sub-basin (Wi) more difficult to interpret. Assumed position of submarine highs is indicated. (**b**) Inset gives the detailed architecture of the Waldeck Sub-basin with correlation of coarse- and fine-grained systems; unfolded (after Schrader 1999). The transition to mud turbidites during onlap on the Rheno-Hercynian shelf should be noted. Palaeocurrents are directed towards the observer (circle with dot).

used here, of the order of 0.5–1.4 Ma (Roberts *et al.* 1995; see Fig. 2).

Two possible explanations for this relationship can be discussed using the various approaches related to timing and other constraints. However, the quality of data is not good enough to choose either.

(1) The time lag of about 20 Ma between turbidite sedimentation and the activation of the underlying thrust faults from mica

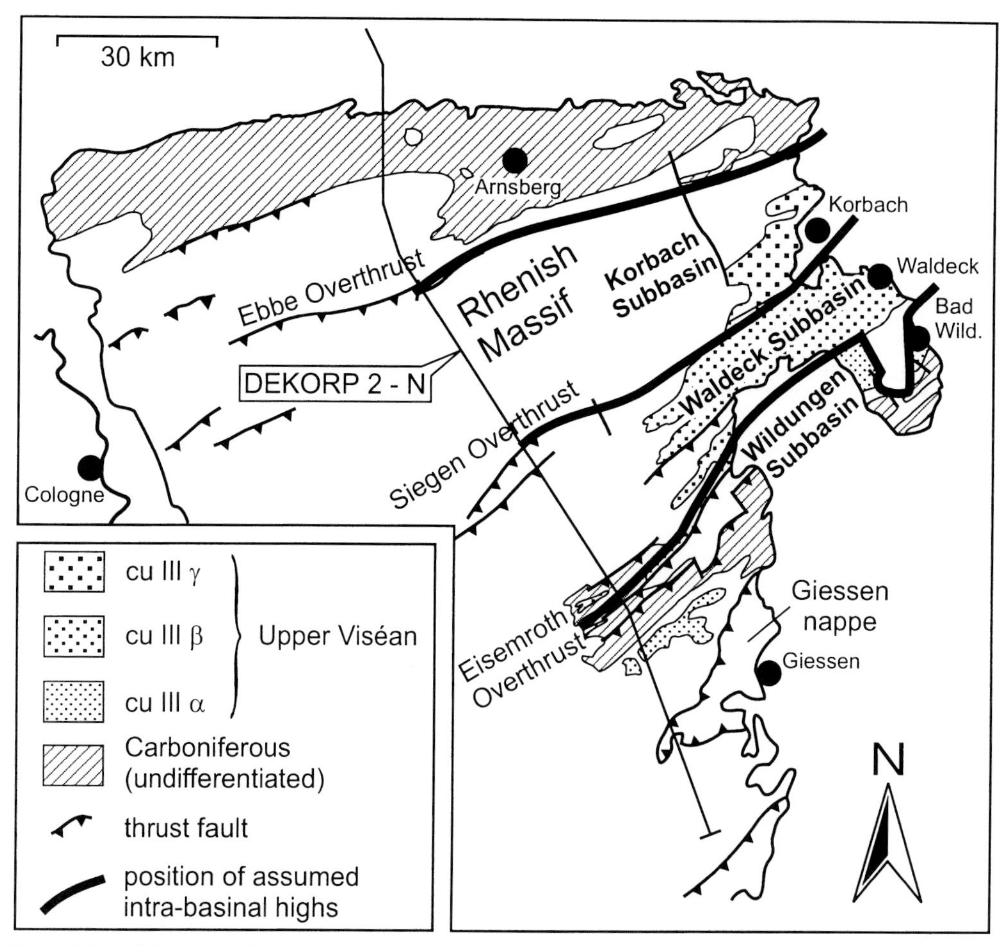

Fig. 5. Map of the eastern Rhenish Massif with major thrust systems. The diagram shows the coincidence of the position of smaller sub-basin highs (bold lines) with thrust systems (after Schrader 1999). Areas of sub-basins are indicated (i.e. the Wildungen, Waldeck and Korbach Sub-basins). Also shown is the DEKORP 2-N seismic line.

geochronology can be related to various reasons including the relatively large time interval of synorogenic mica growth and mixing with detrital ages. As age dating mainly involved the smallest and latest grown mica grain size <2 µm (Ahrendt *et al.* 1983), cleavage ages may be thought to date the end of deformation. The spread of ages from different mica fractions within a single area may encompass more than 10 Ma, with the ages decreasing with mica size, supporting this view (Klügel 1997; see compilation given by Plesch & Oncken (1999)). Unless this fact is due to other issues (sample selection, preparation, detrital components, etc.) this may be taken to indicate a broader active accretion zone of more than 100 km width. Therefore, the deformation tip of the propagating wedge may well be close to the youngest, active turbidite basin. Interestingly, the major fault systems found to correlate with the assumed position of the major pattern of sub-basin highs (best documented for the Waldeck Sub-basin and its transition to the Korbach Sub-basin) include only the fault systems that were related to mainly frontal accretion of Rheno-Hercynian shelf sediments to the advancing orogenic wedge (Plesch & Oncken 1999). In addition, there is a rough overlap between the lifetime of these fault systems (3–4 Ma) and the duration of the basin fill, which, depending on time scale used, is 0.5–1.4 Ma. It is obvious that the duration of basin fill can be only roughly estimated, based on the time scale used here (see Fig. 2). The lifetime of imbricate systems is estimated from the average wedge propagation rate derived from geochronological data, i.e. 1.5 cm a^{-1}, and the spacing of the thrust systems in the restored southern basin, i.e. 40–50 km. The smaller highs, observed between the various turbidite basins (termed 'sub-basins', with respect to the entire RHTB), may thus have been related to relief formed by the growing fault systems during frontal accretion. Under these conditions, these basins would be classified as on-thrust fan, piggy-back-like turbidite basins. The imbricate systems exposed further south (Taunus Mountains; see Oncken *et al.* this volume), which are not related to turbidite basins, were entirely active during wedge growth by underplating.

(2) Alternatively, however, the active deformation zone was narrower with the tip south of the active turbidite sub-basin. This would be more in accord with the isostatic modelling results of Oncken *et al.* (this volume). Loading of a mechanically weak lithosphere will create a minor basin (depth a few 100 m; width 30–60 km) in front of the wedge tip. Also, accreted material south of the thrust front will invariably be uplifted and recycled, thereby preventing the formation of thrust-top basins. Thus, wedge propagation involving the substratum of the turbidites by uplifting newly accreted former passive-margin material in a stepwise manner and depressing the foreland by loading may alternatively be suggested to have controlled the active turbidite basin location (see Oncken *et al.* this volume).

Changing accommodation space at the MGCR

As will be shown later by mass balance calculation, the MGCR was a mountain range of several kilometres altitude at the time of turbidite deposition and it was subject to erosion during rapid uplift and thrust sheet underplating. Erosion was dominantly subaerial, as indicated by probably river-derived conglomerates that occur in many turbidite channel fills. The existence of a narrow shelf is suggested by shallow-water carbonate turbidites that document the existence of carbonate build-ups. Additionally, abundant plant fragments contained in many of the turbidite deposits imply the occurrence of deltaic coastal plain environments with coastal swamps (Fig. 6).

Relative sea level at the MGCR shelf was an important factor controlling the accommodation space. In sequence stratigraphy, accommodation space is the vertical space available for deposition of sediments in shallow waters between the sea floor and the surface of the sea (Posamentier *et al.* 1988). Theoretically, the accommodation space is controlled by tectonic processes and eustatic sea-level variations. Detailed facies analysis of the turbidite succession of the cd IIIα4 to cd IIIβstr (Schrader 1999) shows that phases of decreasing accommodation space during uplift in the shelf area and increased erosion on the MGCR are major controlling factors for coarse-grained turbidite sedimentation in the RHTB.

Observations on the sedimentary record by Schrader (1999) indicated a hierarchy pattern of turbidite cycles in the RHTB of three orders of magnitude. As demonstrated by Schrader, on the basis of various techniques including sequential facies distribution and numerical frequence analysis, the first two cycles, small-scale and medium-scale, reflect autocyclic processes, including migration of channels, lobes and other structures on the ancient fan surface. Only the large-scale cycles, which recorded an estimated time span of the order of 10^5–10^6 a (see Fig. 2), were interpreted as dominantly documenting accommodation cycles on the assumed former shelf of the MGCR. This interpretation by Schrader was based on two observations: first, the systematically upward change in turbidite lithology from coarse-grained clastic deposits to highly diluted clastic deposits with carbonate turbidites; and second, the basinwide correlation of these cycles.

(1) Related to the first of these observations, the lower portion of an idealized large-scale sequence is characterized by siliciclastic turbidite deposits, indicating previous active uplift at the MGCR and no or strongly reduced accommodation space on the shelf of the MGCR (Schrader 1999) (Fig. 6). These coarse-grained turbidites show an upward transition into very fine grained mud turbidites with intercalated calcareous turbidites. This transition indicates a significant decrease of clastic supply into the RHTB, suggesting that the shelf was an area of active sedimentation. Periods characterized by flooding are thought to be associated with shelfal subsidence and decreasing uplift in the MGCR. In addition, environmental conditions on the flooded shelf were favourable for the carbonate factory, reflected by highstand shedding (Schlager *et al.* 1994) of carbonates and their export by turbidity currents (Fig. 6). The existence of shallow-water carbonates in areas of favourable conditions on the assumed former shelf of the MGCR is further supported by the well-known carbonates of 'Schreufa', reflecting resedimented carbonate clasts contained in the spirale conglomerate (cd IIIβspi) (Paproth 1953).

(2) Related to the second of Schrader's (1999) observations, the large-scale sequences can be correlated over at least 50 km, especially within

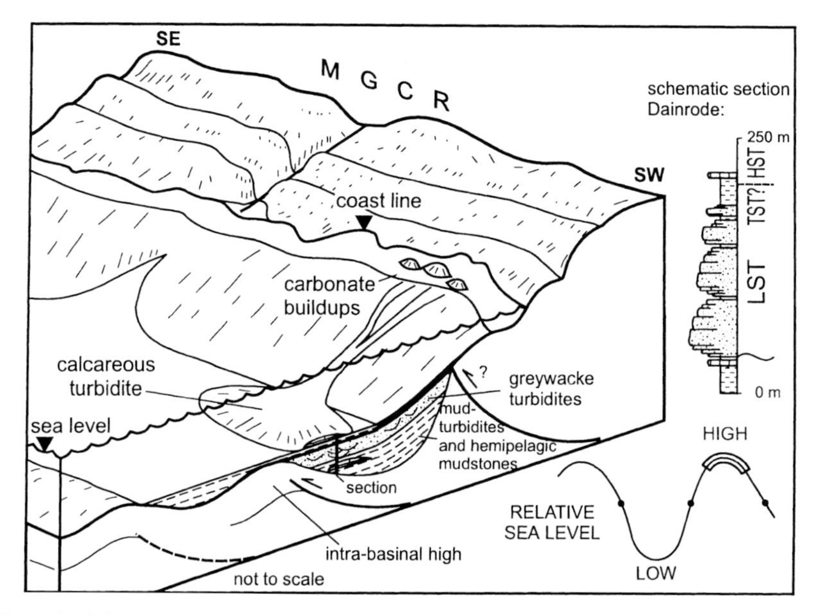

Fig. 6. Schematized diagram of a major turbidite sequence associated with assumed accommodation space changes on the shelf along the Mid-German Crystalline Rise (MGCR), here based on the Dainrode section, i.e. the Wildungen sub-basin (after Schrader 1999).

the Waldeck Sub-basin at the eastern margin of the Rhenish Massif. Although correlation into the Harz area was not established, because of poorer stratigraphic resolution, the principal order of the three cycles can also be found. Correlation in the Waldeck Sub-basin is based on lithostratigraphic observation and is further supported by the well-constrained time frame in this region (Fig. 7). The widely correlative large-scale cycles, with systematic change in lithology as described above, suggest a dominantly allocyclic control.

A composite, basinwide stacking pattern of accommodation cycles was established for the shelf along the MGCR (Schrader 1999). Compared with an integrative coastal onlap curve of relatively stable shelfs and passive continental margins in the USA, Europe and Russia (Ross & Ross 1985, 1987, 1988; Herbig, 1998), a higher-order cyclicity of major turbidite sequences is observed, although the onset of new sub-basins is roughly coincident with periods of basinward shifts of the coastal onlap curve. This may indicate a combination of tectonic and eustatic factors influencing many second- to third-order coastal onlap curves (Fig. 8). Observed length of turbidite large-scale cycles is of the order of 10^5–10^6 a and compares well with periods of 10^4–10^7 a, which are typical for underthrusting and other tectonic processes (Miall 1997). Thus, the accommodation curve established here for the former shelf along the MGCR indicates dominantly tectonic processes and seems to reflect changing thrust wedge activity (Schrader 1999).

Subsidence pattern and turbidite progradation rates

With the onset of turbidite sedimentation the pattern of subsidence in the RHTB changed significantly. The pre-turbidite sedimentation is associated with low subsidence rates, characteristic for thermal relaxation of the evolved Rheno-Hercynian passive margin (Buchholz & Wachendorf 1993), whereas the turbidite system shows significantly higher subsidence rates (i.e. by a factor of 10). Using the high-resolution goniatite stage stratigraphy by the Dinantian Sratigraphy Group (1971) (see Fig. 2), and an average value of 0.3 Ma per goniatite subzone as discussed above, two phases of subsidence can be distinguished during turbidite sedimentation (Fig. 9): a first subsidence phase with low rates and a second phase with high rates. The first phase, which pre-dates the onset of significant turbidite deposition, is characterized by incipient turbidite overflow and distal turbidites. In contrast, the second phase is associated with the typical turbidite sequences described here.

Tracing the subsidence pattern in direction of turbidite prism fill from SW to the NE, two major patterns are observed: (1) a time-transgressive

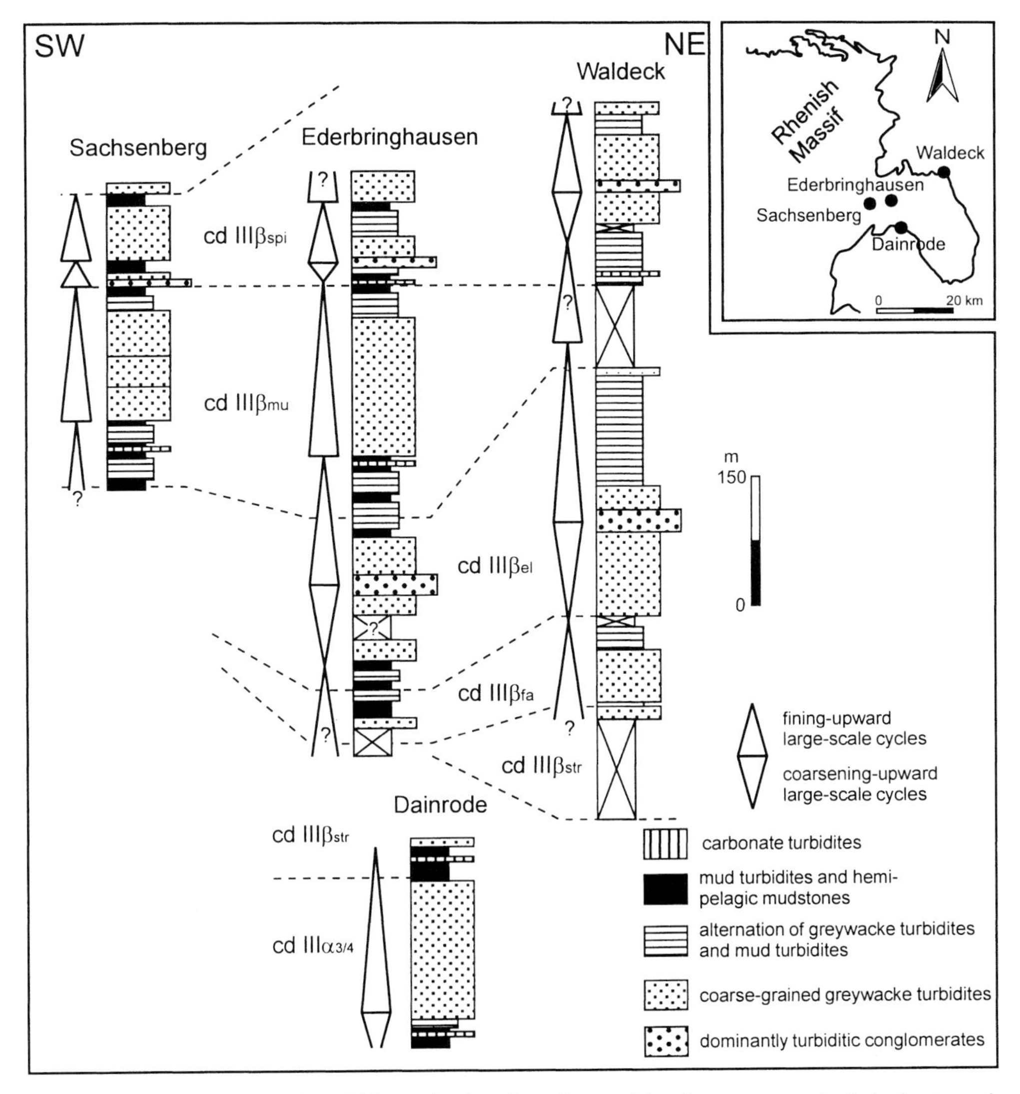

Fig. 7. Correlation of large-scale turbidite cycles that show the transition from coarse-grained clastics to mud turbidites with intercalated carbonate turbidites at the eastern Rhenish Massif (after Schrader 1999). Large-scale cycles were interpreted to indicate dominantly accommodation cycles at the former shelf of the Mid-German Crystalline Rise (MGCR). Stratigraphic symbols for the Late Viséan sequence are given in Fig. 2.

later onset of enhanced turbidite sedimentation can be observed from SE to NW, reflecting the general northwestward migration of turbidite prism and thus of the orogenic wedge (Fig. 9); (2) the thickness of each turbidite unit decreases northwestward, indicating mainly the filling processes of the RHTB. This suggests the existence of an asymmetric foreland basin. The results reported here were supported by the gradual change towards prodelta deposits during the turbidite–molasse transition on approaching the SVMB in the NW (see Fig. 1a).

Migration of turbidite prisms towards the NW is recognized by successively younger ages of the onset of turbidite sedimentation during cd III time, with younger ages from about 329 to 327 Ma for a cross-section of 52 km (restored), documenting migration rates of 3.4–6.3 cm a^{-1} in the restored basin section, based on the stratigraphy and time assessment per subzone used here, as given before (see Fig. 2). This rate, from the base of the cd IIIα4 subzone to the beginning of Namurian C time at the turbidite–molasse transition, is significantly higher than

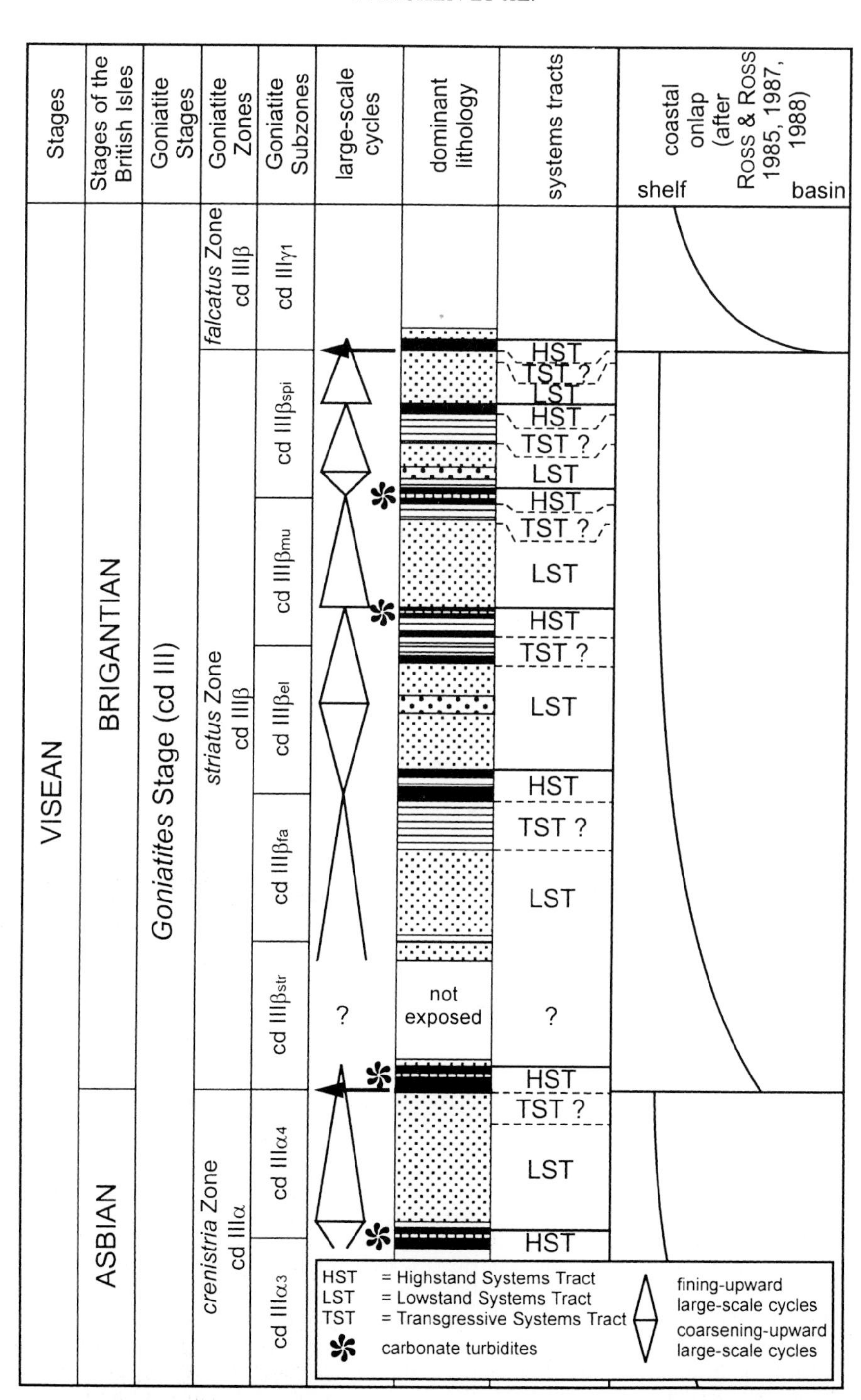

Fig. 8. Sequence stratigraphic interpretation of the eastern Rheno-Hercynian Turbidite Basin showing higher-frequency pattern of accommodation cycles compared with the Ross & Ross coastal onlap curve, suggesting underplating pulses (i.e. base of LST) at the Mid-German Crystalline Rise (after Schrader 1999). Position of carbonate turbidites is indicated (*). Stippled, coarse-grained turbidites; horizontal lines, medium-grained turbidites; black, mud turbidites.

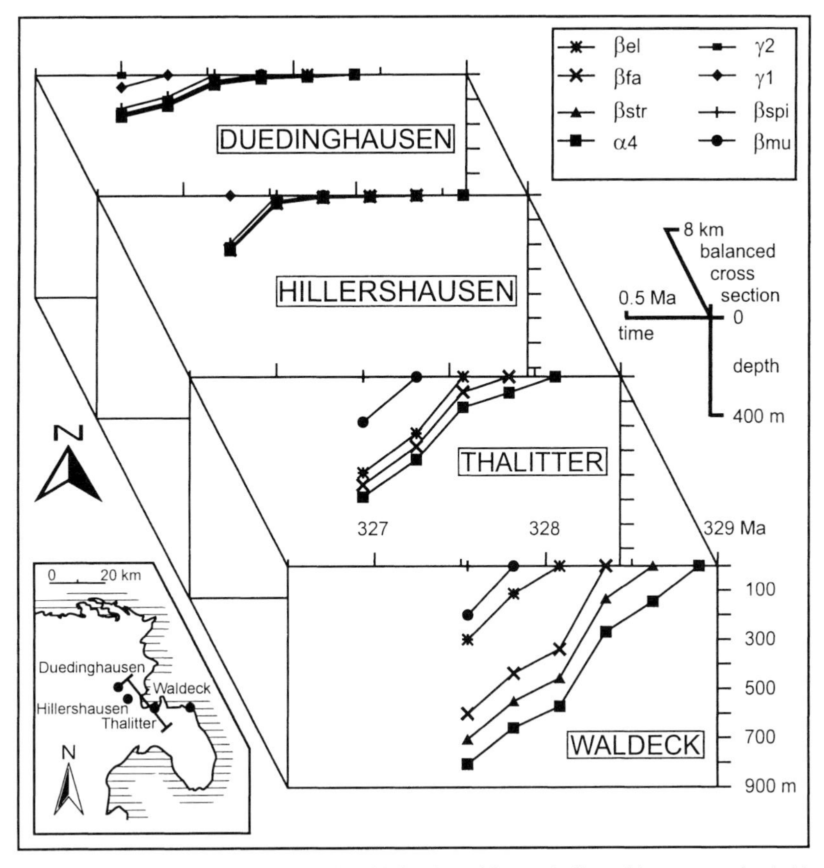

Fig. 9. Time and spatial transgressive onlap of turbidite deposition as indicated by a general subsidence pattern for a suite of sections along the eastern margin of the Rhenish Massif (after Schrader 1999). Subsidence includes sediment compaction; not shown are effects of changing water depths.

the average Viséan to Early Namurian progradation rate (1.5 cm a^{-1} in the restored basin). The above given interval is estimated to span from 329 to 318 Ma (Roberts *et al.* 1995; Burger *et al.* 1997). Accordingly, wedge propagation must have fluctuated substantially between stillstand and rapid propagation for at least the above growth period. On the basis of geometric analysis, Plesch & Oncken (1999) have recently suggested cyclic growth of the orogenic wedge, which corroborates the above observation. This growth type was suggested to be largely controlled by successive activation of individual imbricate thrust systems and repeated associated changes from basal to frontal accretion (Oncken *et al.* this volume). Alternatively, for a short time period turbidite progradation rates and that of the orogenic wedge were decoupled because turbidite migration rates may reflect the wedge volume above sea level available for erosion and also depend on the relief of the basin floor.

Changing sedimentation rates for the RHTB and the SVMB

Sedimentation rates of individual intervals of the two investigated basins, the RHTB and the SVMB, suggest a close interrelation between tectonic forcing, subsidence rate and sediment supply. A compilation of sedimentation rates (Fig. 10), calculated with different time estimates, was plotted against the time scale of Roberts *et al.* (1995). Sedimentation rates are based on known or best estimates of the stratigraphy of preserved sediments and were derived with significant ranges of error, using the upper Viséan time frame as given in Fig. 2 and assuming a factor of two in duration of each interval in which these rates were calculated. For the SVMB, sedimentation rates of Drozdzewski & Wrede (1994), based on a comparable time scale, were used.

Although sedimentation rates in the early nappe system of the RHTB have to be somewhat

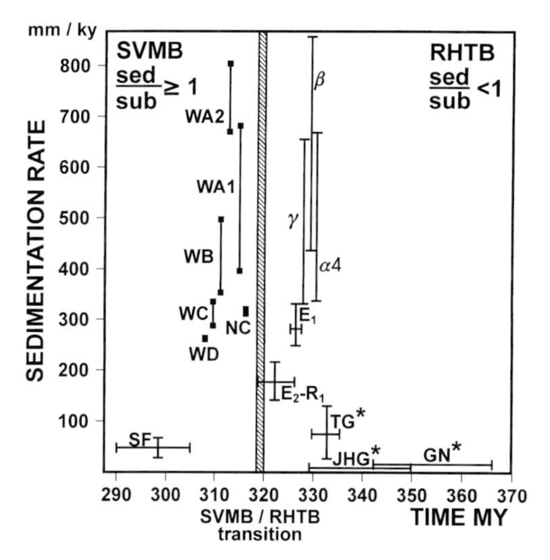

Fig. 10. Average sedimentation rates for the two investigated basins, the Rheno-Hercynian Turbidite Basin (RHTB) and the Sub-Variscan Molasse Basin (SVMB). Squares are data from Drozdzewski & Wrede (1994). SF, Stephanian, WA to WD, Westphalian A to D; NC, Namurian C; E_2–R_1, Arnsbergian to Alportian; E_1, Pendleian; $\alpha 4$, β, γ, Late Viséan Goniatite stage (cd III); TG, Tanner Greywacke; JHG, Jesberg–Hundshäuser Greywacke; GN, Giessen Nappe. Time scale from Roberts *et al.* (1995). Starred symbols indicate sedimentation rates that are reduced by reduction of stratigraphic thickness related to nappe processes.

enhanced because of their tectonic dissection, the general pattern of sedimentation rate is characterized by two maxima, related to both the RHTB and the SVMB, and a minimum at the transition between these two stages in basin evolution. The different pattern of sedimentation rates that dominantly represent subsidence rates is interpreted to indicate different crustal properties of the Rheno-Hercynian Zone. Also, the geometry and type of basin changed along with the change in subsidence pattern from a marine type with narrow sub-basins in the RHTB to the development of a broad, continental type in the SVMB. In restored balanced sections (see Oncken *et al.* this volume), the position of this change correlates with the transition from previously rifted crust in the southeast (stretching factor $\beta = 1.7$–2.4 in the central basin) to nearly unrifted crust in the northwest ($\beta = 1.0$–1.5; Littke *et al.* this volume). As thermomechanical modelling of basin evolution by Littke *et al.* (this volume) has clearly shown that this change in crustal reaction was associated with the effective elastic strength of the lithosphere, we suggest that an inherited strength increase towards the northwest is responsible for the change in basin style. The boundary zone between the two domains constituted a stationary hinge line in the RHTB, which was overrun by the propagating orogenic wedge during Namurian time (see also Oncken *et al.* this volume). Concomitant uplift of the wedge was shown to be related to an increase in erosion and substantial internal out-of-sequence shortening (Plesch & Oncken 1999; Oncken *et al.* this volume).

The very high sedimentation rates in the RHTB around 328 Ma are associated with underthrusting and related erosion of the Rheno-Hercynian continental shelf under the MGCR and initiation of underplating and uplift (Fig. 10; see Oncken *et al.* this volume). The second maximum in sedimentation rate is related to the SVMB and shows the highest values at around 310–315 Ma. The models of Plesch & Oncken (1999) and Oncken *et al.* (this volume) show that around this time (*c.* 310 Ma) substantial internal thickening of the orogenic wedge occurred, because it prograded into stronger, unstretched crust across the northern rift boundary of the Rheno-Hercynian Basin. Consequently, thickening of the wedge resulted in higher crustal loading, which was compensated by increased erosion and sedimentation during the formation of the SVMB. The transition between the RHTB and the SVMB is shown by relatively low sedimentation rates, which seem to reflect the former position of weakened crust at the hinge line between the stretched and unstretched crust.

The two stages in foreland basin evolution, the RHTB and the SVMB, show a different pattern of subsidence as indicated by the transition from turbidites to deltaic deposition. The RHTB shows a higher subsidence and was an underfilled basin, whereas the SVMB was moderately overfilled, as indicated by the occurrence of continental sequences. The predominance of deltaic systems and coal cycles with systematically occurring marine transgressions (Süss 1996) suggests that subsidence was mainly compensated by the sediment fill (Fig. 10). During the entire history of the RHTB, turbidite fans developed and the basin floor remained in deeper-water conditions (several hundred metres or more). Towards the turbidite–molasse transition, during deposition of the upper Arnsberg Beds (i.e. the Namurian A sequence; Kraft 1992), the first indicators of pro-delta systems are observed (see Fig. 1a). This suggests that the RHTB was finally filled up during this late stage of turbidite sedimentation.

Mass balance between exhumation and sediment fill

Mass balance calculation is a major tool to estimate the importance of different processes active within an orogenic wedge, including tectonic exhumation, the degree of erosion in the hinterland and denudation rates (Wetzel 1993; Einsele *et al.* 1996). The basic idea is numerical inversion of erosion, sediment transport, and deposition from the sediment source, the MGCR, into the related sedimentary basins. A critical point is whether erosion and sedimentation of such a system were in balance or not. Related to the complex processes of collision, wedge growth and the formation of various syn- to post-deformational basin fills, various factors are included, namely: (1) size and volume of the restored RHTB, including the estimated sediment proportion contained in the allochthonous Giessen Harzs Nappes system; (2) geometry and thermobarometric history of the MGCR from Mid-Devonian to Late Carboniferous time; (3) the proportion of erosion of the MGCR during the molasse stage from Mid-Namurian to Late Westphalian time, contained as a proportional sediment mass in the SVMB, and the total volume of a smaller wedge-top molasse basin, the Saar–Nahe Basin (SNB, Henk 1991; Korsch & Schäfer 1995); (4) estimates of the subducted sediment volume (S) and the amount of syn- to post-depositional erosion at the top of the basin fill (i. e. truncation, T); (5) additional factors, influencing the calculations made here, such as density and areal factors, and assumptions on the degree of chemical denudation in the source area.

For simplicity, several factors that influence the mass balance calculation only slightly were not considered. They include the sediment fill of various Stephanian basins, of both the smaller intra-orogenic basins and the larger Stephanian basin in northern Europe. Intra-orogenic molasse basins were not considered, as their volumes are so much smaller than those of the other basins that they may be ignored. This includes, for example, the Delitzsch Basin. Also, the internal recycling of turbidite deposits was not incorporated as the material involved remained within the basin itself. Retro-arc sedimentation into the southern Saxo-Thuringian Basin was insignificant, as this basin was also mostly filled from southwestern (i.e. Moldanubian) sources (Franke 1984*a*). Thus, exhumation of the MGCR was mainly compensated by turbidite deposition into the RHTB; and, when this basin was filled and underwent deformation, sedimentation continued into the SVMB.

A further factor, the amount of tectonic extension to total exhumation, is difficult to estimate from thermobarometric data alone. Structural data from the southern Rheno-Hercynian belt (Oncken *et al.* 1999), the MGCR (Oncken 1997) and the SNB (Henk 1991) show extension of about 5–10% perpendicular to strike, *c.* 60% parallel to strike, and <30% perpendicular to strike, respectively. This deformation was shown to have been acquired during continuing regional convergence and does not represent post-orogenic extension. The large value of orogen-parallel extension in the MGCR during initial collision (between 340 and 325 Ma) was entirely compensated by transpressive shortening, and was shown by Oncken (1997) to have had no impact on the thickness of the overlying units, and must therefore not be considered in the calculation. Most of the exhumation in the MGCR and the adjoining southern Rhenish Massif was accomplished during this transpressional stage. The above-cited minor degrees of extension of the upper crust (5–10%, <30%) were achieved in the final stages of exhumation in this core area of the collisional system (after some 315 Ma), when the currently exposed rocks had almost reached the present surface. The net effect of this process is local redistribution of mass over a larger area equivalent to a minor reduction of the barometric values (well below their methodological error of *c.* 1 kbar). As a result of these joint observations, exhumation must have been largely driven by erosion. At most, the cited increase of exhumed area in map view during the final extensional stage may lead to a slight overestimate in former volume overlying the affected areas. Considering the relatively large area of the system studied here (i.e. the catchment and basin system of the mid-European Variscides) and the extension figures given above, the overestimated erosion volume may reach a value of 5–15%. As this value is probably below the detection value, it is not considered in the mass balance, but discussed for its consequences on the final values.

All given numbers are best geological estimates; however, their values represent only broad orders of magnitude. Also, the various thermobarometric data and ages represent mean values and contain the typical errors. For clarity, this calculation was performed for two basic scenarios of wedge progradation and associated basin formation, the turbidite and the molasse stages, as given by scenarios I and II, respectively, in Fig. 11.

Estimates of uplift rates of the MGCR are based on thermobarometry, age dating (e.g. Lippolt *et al.* 1984; Nasir *et al.* 1991; Hirschmann

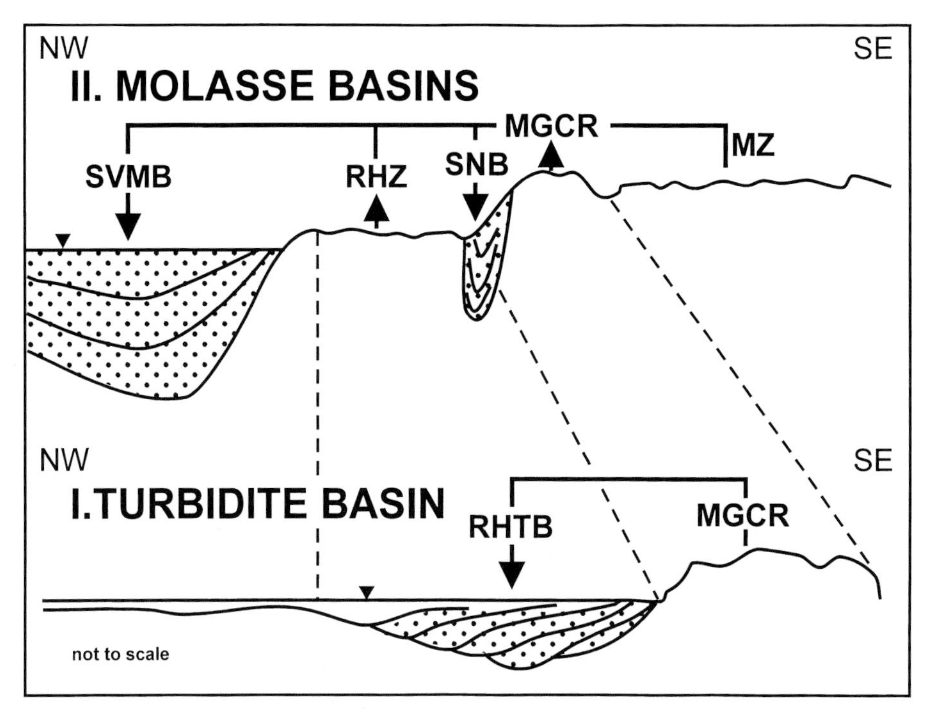

Fig. 11. Volumes involved during mass balance calculation. Scenarios I and II: turbidite and molasse stages. Arrows: solid sediment paths. RHTB, Rheno-Hercynian Turbidite Basin; MGCR, Mid-German Crystalline Rise; SVMB, Sub-Variscan Molasse Basin; Saar–Nahe Basin; RHZ, Rheno-Hercynian Zone; MZ, Moldanubian Zone.

1995; Okrusch 1995), and numerical simulations by Henk (1995). The latest compilation by Oncken (1997) is the database of our study. Unroofing since Viséan time is related to continental subduction of Rheno-Hercynian crust below the MGCR, underplating of part of this material and uplift. In the models of Henk (1995) and Oncken (1997), maximum unroofing in the central zone of the MGCR is estimated to be as much as 15–25 km (see Franke this volume).

Upper-plate exhumation affected the MGCR, represented by a small zone of metamorphic basement on average 64 km wide and some 670 km long, between the Elbe and the Bray Fault. From Oncken's (1997) map, giving the metamorphic pressure data of outcropping rocks, an average exhumation of 3.5 kbar is derived, corresponding to 13 km depth. The total exhumation volume (TEV_{MGCR}) that was available for possible erosion is estimated as *c.* 540×10^3 km^3. It may, however, include an undetected portion of tectonic unroofing, although evidence for synconvergent extensional deformation outside the MGCR is very minor (see above; Oncken 1997). In addition, unroofing may contain a fraction of post-Stephanian uplift, which is small enough to be neglected, because the Stephanian basin is thought to be dominantly filled by recycling and truncation of the SVMB and not due to significant erosion on the MGCR.

On the basis of the following arguments, the area of the MGCR is assumed to stretch from the Bray Fault to the Elbe Line, and is well documented by various drill-holes (see compilation by Oncken (1997)): (1) during continuing convergence, the MGCR underwent significant extension of *c.* 60% parallel to strike that may explain the length of this zone; (2) various workers agree that no major basin can be found related to the MGCR except the three basins investigated in this study, the RHTB, the SVMB, and the SNB, which received their detritus from the MGCR (e.g. Deneke 1977); (3) for the two contrarily developed elongated fan systems of the RHTB, with an assumed submarine canyon somewhere underneath the later Hessian Basin, a dominantly river-derived point source can be assumed, suggesting that major river systems and related sediment transport developed parallel to the strike (Schrader 1999); (4) the only small continuation of the MGCR east of the Elbe Line by forming a thinned, narrow belt, as suggested by Franke (this volume) will create a comparably small additional area subjected to possible erosion and sediment supply, and will thus be ignored. Also, barometric data suggest that this part has not been significantly exhumed (Oncken 1997).

The Saar–Nahe Basin (SNB) represents a relatively small on-arc molasse basin that developed above an earlier upper-plate fore-arc basin (Henk 1991, 1993; Korsch & Schäfer 1995; Oncken 1997). Unlike the molasse stage (scenario II, Fig. 11), sediment volumes, dominantly carbonates, can be neglected for scenario I. The Phyllite Zone, located north of the MGCR, represents the underplated Rheno-Hercynian slope facies (Oncken *et al.* 1999). Its synkinematic pressure decrease (from 6 to 3 kbar; Klügel 1997) records erosion of the advancing upper-plate orogenic wedge by a depth equivalent to *c.* 3 kbar between 325 and 310 Ma. The depth equivalent to the final 3 kbar was eroded by early Permian time. This material, as well as the eroded parts of the underlying Rheno-Hercynian Zone and Phyllite Zone were the main contribution to the fill of the SVMB.

A comprehensive mass balance calculation can be performed only if the exhumation history during the formation of the RHTB and the SVMB is considered. Henk (1995) showed that uplift in the MGCR occurred from 330 to 310 Ma, which includes the major filling intervals of both the RHTB and the SVMB. By 310 Ma, several basins in the south, including the Saxo-Thuringian Basin and the oceanic part of the RHTB, were already closed, and the entire area of the central European Variscides (Fig. 11, II) was subject to erosion (Franke 1989; Franke & Oncken 1995). As the catchment area of the SVMB is much larger than the small zone along the MGCR, and the various Variscan zones contributed different amounts to erosion, only part of the sediment fill of the SVMB was used for mass balance calculation. This proportional volume (PV_{SVMB}) is thought to represent the sediment contribution solely from the MGCR. The SNB is not considered as a proportional fill, because it received sediment entirely from the MGCR during wedge growth.

The major step in balancing masses, therefore, uses the average total exhumation volume of the MGCR (TEV_{MGCR}) that is balanced against the intra-basin sediments, including the total cumulative sediment volume (TCV_{RHTB}) of the RHTB (i.e. from the Frasnian to Lower Namurian sequence), the proportional volume of sediment contained in the SVMB (PV_{SVMB}) and the sediment volume of the SNB (V_{SNB}; i.e. from the Namurian to the top of the Westphalian sequence; Fig. 11). In addition, one has to consider the lacking intra-basin volumes, the amounts of pre- to post-Stephanian truncation (T) and precollisional (i.e. pre-Tournasian) subduction below the MGCR (S). Equation 1 can be considered as a first approximation, because intra-basinal mass volumes have to be transformed to denudation volumes, depending on various factors explained below:

$$TEV_{MGCR} \approx TCV_{RHTB} + PV_{SVMB} + V_{SNB} + T + Sz \qquad (1)$$

The total cumulative sediment volume in the RHTB (TCV_{RHTB}) includes the cumulative sediment volume of the largely preserved basin fill and the remaining Giessen–Harz Nappes volume that was accumulated during the oceanic subduction stage. Schrader (1999) showed that the preserved portion of the RHTB has an asymmetric geometry with the highest sediment thickness in the SE. Various models of this basin (i.e. with scenarios of different extension to the NE) give a cumulative sediment volume of *c.* 93×10^3 km^3, whereas for the intra-nappe system a volume of 37×10^3 km^3 is estimated. These assumptions are based on restored sections of 55% and 35% shortening for the allochthonous and autochthonous systems, respectively (Oncken 1997). Turbidites of the allochthonous system were mainly recycled and contributed to the sediment fill of the autochthonous portion of the RHTB. For the allochthonous system only the present thickness of turbidites, which escaped erosion, was utilized. Thus, truncation of the RHTB is not included, as this can be considered either as internal recycling or the contribution to the various molasse basins. However, the truncation of the SVMB is, because it is the final step that is considered here. This latter truncation event covers the mass transfer (i.e. basically recycling) into the Stephanian basin.

The volume of denudation of the hinterland depends on the relationship between the clastic and the dissolved load of river systems contributing to the basin fill. When this denudation volume (DV) is derived by using only the porous clastic basin fill, one has to consider the equivalent amount of chemical denudation (CD) that occurred in the catchment area and was transported to the world ocean. The denudation volume ($DV_{MGCR,I}$) of the MGCR during the formation of the RHTB was (see scenario I in Fig. 11)

$$DV_{MGCR,I} = DF_{RHTB} \times (1 + CD) \times TCV_{RHTB} \qquad (2)$$

where DF_{RHTB} is the density factor. The total amount of erosion at the MGCR was higher, as it was composed of both clastic sediments and solutes. When using only the volume of clastic deposits contained in the basin fill to derive the degree of erosion in the hinterland, this value becomes too low, as chemical dissolution is not included. An assumed value of 20% ($CD = 0.2$) for chemical denudation of the total river load in a tropical mountain range can be considered as a conservative estimate (Einsele 2000; Einsele & Hinderer 1998). This value has been consistently employed for the catchments of the other basins.

The density conversion factor (DF) for the two groups of basins is used to minimize the effect of pore space and of the lower density of the intra-basinal sediments than of the eroded crystalline rocks at the MGCR:

$$DF = \frac{\rho_s - (\rho_s \times 0.01\phi)}{\rho_c} \quad (3)$$

where ϕ is the sediment porosity, and ρ_s and ρ_c are the densities for sediment grains and exhumed crystalline rocks, respectively. We chose $\rho_s = 2.65$ g cm^{-3} and $\rho_c = 2.78$ g cm^{-3} (Henk 1995), and a final porosity of 5% and 10% is assumed for the RHTB and the two molasse basins, the SVMB and the SNB, representing typical values of late diagenetic sandstones (Houseknecht 1987).

As mentioned above, the proportion of sediment derived from the MGCR and contained in the SVMB is an important factor influencing the mass balance calculations. Thus, volumes described in equation (1) ($PV_{SVMB} + V_{SNB}$) have to be converted to derive the denudation volume at the MGCR for the molasse stadium ($DV_{MGCR,II}$) as given in scenario II in Fig. 11. Equivalent to the expression in equation (2), one has to use the factors of density ($DF_{SVMB,SNB}$) and of chemical denudation (CD):

$$DV_{MGCR,II} = DF_{SVMB,SNB} \times (1 + CD) \times [PV_{SVMB} + V_{SNB} + T_{SNB}] \quad (4)$$

PV_{SVMB} is the proportional volume of the SVMB, which depends on the total volume (V_{SVMB}) including truncation (T_{SVMB}), and a proportional factor (PF_{MGCR}) of sediment yield that shows the contribution only from the MGCR [$PV_{SVMB} = PF_{MGCR} \times (V_{SVMB} + T_{SVMB})$]. The proportional factor (PF_{MGCR}) is a combination of two different effects. (1) The first effect is the areal distribution of the MGCR, which comprised only a fraction of the entire mid-European catchment area. The latter (taken as 100%) is estimated to stretch from the turbidite–molasse transition, as defined in Fig. 1a, to the middle Moldanubian Zone, that is, crudely the Averno-Vosgian Zone. Within this mid-European catchment area, the area of the MGCR is only 24%, yielding an areal distribution of 0.24. (2) The second effort is the different exhumation volumes, which seem to dominantly reflect surface erosion per Variscan zone after 330 Ma, that is, during formation of the SVMB. These show an average exhumation for the MGCR of 1.5 kbar or 5.4 km, whereas the Moldanubian Zone, with an areal distribution of 49%, yields 4 kbar (i.e. 14.4 km) of exhumation. The lowest amount of erosion, with a mean of 4 km, can be estimated for the Rheno-Hercynian thrust wedge, which shows an areal distribution of 27%. These numbers are based on age-calibrated thermobarometric data compiled by Oncken (1997) and burial temperatures given by Oncken (1984) and Littke *et al.* (this volume). According to the values given above the proportional factor of sediment contribution to the SVMB from the MGCR is $PF_{MGCR} = 0.14$.

The estimated sediment volume of the SVMB (V_{SVMB}) is roughly 873×10^3 km^3 for a section along the Rheno-Hercynian Front to the Elbe Fault and includes the southern Namur Basin up to the latitude of Brussels (based on isopach maps by Drozdzewski & Wrede (1994)). For the Ruhr area and the southern rim of the Rhenish Massif, a truncation volume of 40×10^3 km^3 (Littke *et al.* 1994, this volume) can be roughly estimated. This total volume is necessary to calculate the proportional fill (PV_{SVMB}), which gives a volume of 128×10^3 km^3. Additional sediment is contained in the Saar–Nahe Basin (SNB; Fig. 11, II), which is estimated to contain 22×10^3 km^3 and a truncation value of 12×10^3 km^3. The SVMB was corrected for a southern zone of 50 km, where an average shortening of 27% is assumed (Drozdzewski & Wrede 1994), whereas the SNB shows no relevant shortening (Korsch & Schäfer 1995). As mentioned above, smaller on-wedge molasse basins with volumes significantly lower than the SNB were ignored.

The mass balance calculation was performed by rewriting equation (1) according to the various definitions given in equations (2)–(4), resulting in the following equation:

$$TEV_{MGCR} = DF_{RHTB} \times (1 + CD) \times (TCV_{RHTB} + S_{RHTB}) + DF_{SVMB,SNB} \times (1 + CD) \times \{[PF_{MGCR} \times (V_{SVMB} + T_{SVMB})] + V_{SNB} + T_{SNB}\} \quad (5)$$

In equation (5), TEV_{MGCR} is the total exhumed sediment volume along the MGCR, DF are the density factors for the RHTB and the two molasse basins (SVMB, SNB) as given in equation (3), and CD is chemical dissolution; TCV_{RHTB} shows the total cumulative volume of the sediment fill of the RHTB, whereas V_{SVMB} and V_{SNB} are the volumes of the SVMB and the SNB. T is truncation of previously deposited sediments for the SVMB and the SNB, and S_{RHTB} is the loss of subducted sediment volume below the MGCR. PF_{MGCR} is the proportional factor of the sediment contribution solely from the MGCR.

Equation (5) can be considered as a general equation, in which each intra-basinal volume, either present or lacking, is related to the effects of density changes and chemical dissolution. When this mass balance calculation is performed by only using the correct (i.e. those found at present) sediment volumes contained in the restored basins, the derived masses are significantly out of balance compared with the exhumation volumes. A mass deficit by a maximum factor of 0.6 for the intra-basin sediments is indicated, whereas an equilibrated balance would give a factor of one.

The total exhumed volume of the MGCR shows a maximum possible erosion volume of 540×10^3 km^3, if the exhumed volume is equivalent to that of erosion. The calculated denudation volume ($DV_{MGCR,I}$) for the RHTB, including the effects of density change and chemical dissolution, is estimated at 141×10^3 km^3, whereas that for the MGCR ($DV_{MGCR,II}$) for the two molasse basins, including the numbers for truncation, is estimated at 167×10^3 km^3. The lacking intra-basinal sediment mass, expressed as denudation volume, can be roughly estimated at 230×10^3 km^3.

The significant deficit of sediment mass by a factor of 0.6 may be related, among other controlling factors given below, to two major processes: overestimate of eroded sediment mass as a result of an underestimate of extensional tectonics, and loss of sediment mass caused by subduction.

If the imbalance of 0.6 was to be entirely explained by late-stage extension, this would require an average of 40% extension over the entire exhumed area that contributed to the basin fill. This, clearly, is not supported by structural observations that indicate maximum values of only 10–30% in parts of the core area of the collisional system (see above). From the figures cited above, a maximum of some 5–15% of apparent eroded volume distributed over the entire area of the Rheno-Hercynian and the MGCR may be due to late-stage extensional unroofing. Subtraction of this figure from the above mass deficit has only a minor impact by slightly reducing the deficit.

Only when the other variable of equation (5), subduction of sediment mass (S), is taken into account, does the imbalance between exhumed masses and intra-basinal sediments disappear. Reconstruction of the pre-collisional geometry of the western MGCR by Oncken (1998), using criteria established for the present circum-Pacific margins, including the geometry and evolution of subduction zones and fore-arc areas, has identified substantial subduction erosion before collision. The removed masses would also have included sediments deposited in the trench. A conservative estimate of sediments lost by subduction yields a minimum of 450 km^2 in cross-section. Extrapolation to the entire southern length of the RHTB (i.e. 370 km) gives a subducted volume of some 170×10^3 km^3. However, subduction values related to trench migration estimated from geometric balancing may well have been higher by a factor of 2–3 (e.g. up to 1100 km^2; see figs 3 and 4 of Oncken 1998). This number is expected to have included sediments derived from the slope to the oceanic plate as well as the fore-arc basement. These values are typical of the rates of mass loss at some of the recent circum-Pacific margins for similar time spans (von Huene & Scholl 1991).

Undetected and previously deposited sediment masses, mainly at the margins of the basin fill (i.e. truncation), may further contribute to lacking volumes. For the RHTB, Oncken (1984) has estimated an erosional uplift of the NE Rhenish Massif, the area of the onlap of the RHTB, as between 2 and 4 km. Most of this previous cover is likely to be related to Carboniferous sedimentation and erosion (also see Littke *et al.* this volume). This truncation, however, is considered as 'internal recycling', as the sediment was successively transported either into other parts of the RHTB or into the SVMB. Other truncation values, established for the various molasse basins, were already included as discussed above (equation (5)).

Unknown palaeogeographical patterns may also be responsible for an underestimate of sediment mass, mainly as a result of the unknown prolongation of the considered basins to the east (i.e. east of the Elbe Fault; see above) and for the RHTB undetected retro-arc sedimentation. Another process that may have affected the mass balance is undetected strike-slip movements that moved significant mass volumes into or out of the parts of the Variscides analysed here. Because of the size of the studied system this will require displacements in excess of some 100 km, which are highly unlikely to have occurred, for several reasons, as follows. A quantitative estimate from Oncken (1997) yields values well below those given above. His results were based on 3D crustal balancing, considering also the sequence of sinistral and dextral motions at the boundary between the MGCR and the Rheno-Hercynian Zone. Also, the coincidence described here of the area of peak exhumation and related denudation in the upper plate, the MGCR, with the position of the feeder channel does not support finite strike-slip displacement beyond a few tens of kilometres. The position of the feeder channel was assumed to be at the centre of the two identified turbidite lobes of the lower plate, the RHTB.

The preserved fill of the Saxo-Thuringian Basin was derived from Moldanubian sources (Franke 1984*a*, *b*) with the exception of a smaller, Early Carboniferous molasse basin at the southern margin of the MGCR (i.e. Basin of Delitzsch), which received additional sediments from the north (Franke, pers. comm.). In addition, Schäfer *et al.* (1997) have shown that the northwestern Saxo–Thuringian fold belt, kinematically linked to the Rheno-Hercynian orogenic wedge, was eroded by a depth equivalent to 1–2 kbar from above the preserved Upper Viséan turbidites. This may well have been material from the MGCR.

On the basis of the intra-basin sediment masses derived here, the minimum depth of denudation at the MGCR can be quantified for the periods represented by scenarios I and II (Fig. 11), by relating the calculated denudation volume (equations (2) and (4)) to the area of the MGCR. Accordingly, the minimum denudation depth of the MGCR was between 3.3 and 3.9 km for scenarios I and II, respectively. When these values are transformed to average denudation rates, this correlates with either 70 mm ka^{-1} (Mid-Frasnian time to the turbidite–molasse transition, *c.* 49 Ma) or 300 mm ka^{-1} (this transition to Late Westphalian, *c.* 13 Ma). The relatively low denudation rate for scenario I changes to 180 mm ka^{-1} if the lacking volume (equivalent to 5.4 km of erosion), most probably related to subduction, is included.

Denudation rates are relatively low when compared with present-day mountain rivers, which have average denudation rates between 40 and 500 mm ka^{-1} depending on climate zone (Einsele 2000). The observed low rates may indicate a subtropical climate, whereas true tropical conditions would suggest some 200–500 mm ka^{-1} (Einsele 2000). In addition, the derived average denudation rates are somewhat lower than the uplift rates (decreasing from 3.2 mm a^{-1} in Namurian time to 0.2 mm a^{-1} in Latest Carboniferous time) as modelled by Henk (1995). This may indicate that tectonic uplift was not entirely compensated by erosion during the beginning of this process, suggesting that the MGCR was a true mountain belt.

This scenario is supported by various arguments given above and includes a simple isostatic Airy compensation, which is highly idealized in an active orogenic wedge. The calculation was based on an assumed *c.* 27 km thickness of the Rheno-Hercynian crust, whereas that of the MGCR was assumed to have, during early convergence and collision, *c.* 40 km, to compensate for the 13 km of exhumation that on the average occurred at the MGCR. Such a scenario yields an average height difference between the floor of the RHTB and the MGCR of 2.2 km. This number will decline to about 1.4 km altitude difference, when 9 km of Devonian sediments were assumed to have underlain the RHTB. These calculations were based on using various rock densities given in equation (3) and a mantle density of 3.33 g cm^{-3}. Flexural isostatic modelling from Oncken *et al.* (this volume) of the advancing upper-plate orogenic wedge over the Rheno-Hercynian Zone even yields a value of some 3 km difference in altitude. The numbers derived here support the arguments given above that the MGCR was a true mountain belt during the formation of the RHTB.

Discussion and conclusion

During its geodynamic history, the RHTB changed from a pre-collisional trench–fore-arc basin to a post-collisional foreland basin. The major phase of post-collisional turbidite sedimentation was in Late Viséan time (cd IIIα to cd IIIγ), representing the early foreland basin stage. The catchment area of the RHTB was located in the MGCR and the internal subaerial part of the orogenic wedge to the north of it. The latter is thought to be partly covered by the northern shelf zone of the MGCR. This shelf zone can be characterized by sensitive interrelation of tectonically driven processes and various factors of the depositional regime, influencing the factors of shelfal storage and the style of turbidite deposition. Sub-basins that suggest stepwise migration are dominantly associated with several major turbidite sequences, which allow the interpretation of changing accommodation space on the assumed shelf along the MGCR.

The major conclusions are as follows

(1) Although the restored turbidite progradation rate is on the average 1.5 cm a^{-1} for the

interval from Viséan to Namurian time, the studied Late Viséan basin architecture shows a relatively rapid northwestward progradation at a higher rate of up to 6.3 cm a^{-1}. These rates are based on the stratigraphic frame and the time scale by Roberts *et al.* (1995). The assumed pattern of stepwise progradation is associated with distinct thickness variations along turbidite wedges. Sub-basins are thought to be associated with smaller intrabasinal highs, characterized by onlap, and finally turbidite overflow. The position of these highs shows a coincidence with some major imbricate thrusts, suggesting coupling between foreland accretion of the prograding wedge and the turbidite basin architecture.

(2) The large-scale turbidite sequences that can be widely correlated show systematic trends from coarse-grained turbidites to mud turbidites with calcareous turbidites. These sequences document changing accommodation space on the assumed shelf along the MGCR. During exposure of the shelf, turbidites were transported into the RHTB, whereas during flooding, accommodation space was created and coarse-grained clastics remained on the shelf, whereas dominantly mud turbidites were deposited in the RHTB. At the same time, carbonate build-ups on the flooded shelf contributed to highstand shedding of carbonate turbidites. When compared with other coastal onlap curves of passive margins, a higher order of cyclicity (10^5–10^6 a) is visible, which is dominated by tectonic processes for at least the Late Viséan stage. This observation may be linked to cyclic thrust-sheet underplating under the MGCR.

(3) A compilation of sedimentation rates allows a geodynamic interpretation of the transition from the RHTB to the SVMB. The RHTB shows a sediment undersupply and a higher subsidence rate, whereas the SVMB was partly oversupplied or shows a balance between subsidence and clastic fill. After collision, sedimentation rates rose significantly, with the highest values during Late Viséan time, whereas the transition zone between the RHTB and the SVMB is characterized by low sedimentation rates. This trend is associated with an important strength boundary between unrifted crust (in the north) and rifted crust (in the south) that was inherited from Devonian crustal attenuation. Apart from the change in sedimentation (or subsidence) rates, propagation of the orogenic wedge across this crustal hinge is also reflected by a change in synorogenic basin geometry (e.g. decreasing depth), as well as by a phase of internal wedge thickening and uplift. This effect is well observed in the higher sedimentation rates of the SVMB that portray the higher erosion rates of the uplifting wedge.

Transfer of sediment mass in the RHTB was investigated by balancing the exhumed mass at the MGCR against the various related intrabasin sediment masses. In spite of the application of comprehensive techniques, a significant lack of sediment mass is observed. However, inclusion of other factors, such as recently suggested sediment subduction and minor degrees of extensional exhumation, as described below, yields a more equilibrated system.

The major conclusions, related to mass transfer, are as follows.

(1) As upper-plate material was exhumed and eroded from the Mid-Frasnian to molasse stage during Westphalian time, MGCR-derived sediment masses were contained in the two foreland basin stages, the RHTB and SVMB, and the top-wedge SNB. During the molasse stage, however, central Europe was the sediment source of the SVMB with different erosion rates per Variscan zone, thus only some of the fill of this basin is related to erosion at the MGCR.

(2) Including all potential contributing factors, a considerable deficit between exhumed and deposited masses is derived. The mass deficit ratio between deposited and exhumed masses is about 0.6 or around 230×10^3 km^3 (expressed as denudation volumes) for both the sediment fill of the RHTB and the proportional fill of the SVMB including the volume of the SNB. For both molasse basins, the effects of truncation were considered, whereas that of the RHTB was understood as internal recycling. Apart from principal errors, which may be due to other controlling factors (e.g. other palaeogeographical conditions), a more equilibrated balance is obtained when subduction of eroded material below the MGCR before collision is assumed. A conservative estimate yields a subducted slab of former sediment for the southern length of the RHTB that is approximately in the same order of magnitude as the lacking intrabasinal masses. This suggests that subduction of sediment masses derived from the upper plate under the MGCR was relevant.

(3) Most of the exhumation of the MGCR and the adjoining southern Rhenish Massif was accomplished during conditions of plate convergence. The minor late-stage extension of the upper crust was achieved in the final stages of exhumation in the core area of the collisional system (after some 315 Ma), when the now-exposed rocks had almost reached the present surface. A maximum of some 5–15% of apparent eroded volume distributed over the entire area of the Rheno-Hercynian zone and the MGCR can

be attributed to late-stage extensional unroofing. Subtraction of this figure from the above mass deficit has only a minor impact by slightly reducing the deficit. Therefore, exhumation must have been largely compensated by erosion.

(4) Denudation rates were overall relatively low and thus suggest subtropical climate conditions. They change to reasonable but still low values for the interval represented by the RHTB, when the lack of sediment mass, most probably related to subduction, is included. Isostatic Airy backstripping and independant flexural forward modelling shows an altitude difference of 1.4 km and some 3 km between the top of the advancing orogenic wedge and the floor of the RHTB, suggesting that during this stage of wedge formation, the MGCR can be considered as a true mountain belt.

Our work has benefited from critical dicussion with E. Deneke (NLfB Hannover), G. Einsele (University of Tübingen), W. Franke (University of Giessen), D. Meischner (University of Göttingen) and J. Weber (RWTH Aachen). In addition, we acknowledge the critical review and helpful suggestions on the text by D. Bernoulli (ETH Zürich), W. Franke, J. L. Mansy, (University of Lille) and D. Tanner (Universities of Freiburg and Göttingen). The late M. Horn (HLfB Wiesbaden), who died so early, spent encouraging days with us in the field. We benefited from his profound knowledge and personal greatness. Research of this paper was funded by the German Science Foundation, which is gratefully acknowledged.

References

AHRENDT, H., CLAUER, N., HUNZIKER, J. C. & WEBER, K. 1983. Migration of folding and metamorphism in the Rheinische Schiefergebirge deduced from K–Ar and Rb–Sr age determinations. *In*: MARTIN, H. & EDER, W. F. (eds) *Intracontinental Fold Belts*. Springer, Heidelberg, 323–338.

ALLEN, P. A. & HOMEWOOD, P. (eds) 1986. *Foreland Basins*. International Association of Sedimentologists, Special Publications, **8**.

BRINKMANN, R. 1948. Die Mitteldeutsche Schwelle. *Geologische Rundschau*, **36**, 56–66.

BUCHHOLZ, P. & WACHENDORF, H. 1993. Abschätzung der mittleren Sedimentations- und Subsidenzraten im Devon und Karbon des Oberharzes. *Zeitschrift der Deutschen Geologischen Gesellschaft*, **144**, 159–172.

BURGER, K., HESS, J. C. & LIPPOLT, H. J. 1997. Tephrochronologie mit Kaolin-Kohlentonsteinen: Mittel zur Korrelation paralischer und limnischer Ablagerungen des Oberkarbons. *Geologisches Jahrbuch*, **147**.

CLAOUÉ-LONG, J. C., COMPSTON, W., ROBERTS, J. & FANNING, C. M. 1995. Two Carboniferous ages: a comparision of SHRIMP zircon dating with conventional zircon ages and $^{40}Ar/^{39}Ar$ analysis. *In*: BERGGREN, W. A., KENT, D. V. & AUBRY, M.-P. (eds) *Geochronology, Time Scales and Global Stratigraphic Correlation*. SEPM, Special Publications, **54**, 3–21.

CONIL, R. & PAPROTH, E. 1983. Foraminifers from the uppermost Devonian and the Dinantian of the Rhenish Massif (Federal Republic of Germany). *Paläontologische Zeitschrift*, **57**, 27–38.

DENEKE, E. 1977. Die Petrographie der Kulm-Grauwacken des Edergebietes (NE Rheinisches Schiefergebirge). *Geologisches Jahrbuch Hessen*, **105**, 75–97.

DINANTIAN STRATIGRAPHY GROUP 1971. II. Unterkarbon (Dinantium). *Fortschritte in der Geologie um Rheinland und Westfalen*, **19**, 5–18.

DOROBEK, S. L. & ROSS, G. M. 1995. *Stratigraphic Evolution of Foreland Basins*. SEPM Special Publications, **52**.

DROZDZEWSKI, G. & WREDE, V. 1994. Faltung und Bruchtektonik—Analyse der Tektonik im Subvariscikum. *Fortschritte in der Geologie von Rheinland und Westfalen*, **38**, 7–187.

EBERLI, G. P. 1991. Calcareous turbidites and their relationship to sea-level fluctuations and tectonism. *In*: EINSELE, G., RICKEN, W. & SEILACHER, A. (eds) *Cycles and Events in Stratigraphy*. Springer, Heidelberg, 340–359.

EINSELE, G. 2000. *Sedimentary Basins—Evolution, Facies, and Sediment Budget*, 2nd Edn. Springer, Heidelberg.

—— & HINDERER, M. 1998. Quantifying denudation and sediment-accumulation systems (open and closed lakes): basic concepts and first results. *Palaeogeography, Palaeoclimatology, Palaeoecology*, **140**, 7–21.

——, RATSCHBACHER, L. & WETZEL, A. 1996. The Himalaya–Bengal Fan denudation–accumulation system during the past 20 Ma. *Journal of Geology*, **104**, 163–184.

ENGEL, W. & FRANKE, W. 1983. Flysch-sedimentation: its relations to tectonism in the European Vaiscides. *In*: MARTIN, H. & EDER, F. W. (eds) *Intracontinental Fold Belts*. Springer, Heidelberg, 289–321.

——, FLEHMING, W. & FRANKE, W. 1983*a*. The mineral composition of Rheno-Hercynian flysch sediments and its tectonic significance. *In*: MARTIN, H. & EDER, F. W. (eds) *Intracontinental Fold Belts*. Springer, Heidelberg, 289–321.

—— & FRANKE, W. & LANGENSTRASSEN, F. 1983*b*. Palaeozoic sedimentation in the northern branch of the Mid-European Variscides—essay of an interpretation. *In*: MARTIN, H. & EDER, F. W. (eds) *Intracontinental Fold Belts*. Springer, Heidelberg, 9–41.

FRANKE, W. 1984*a*. Late events in the tectonic history of the Saxothuringian Zone. *In*: HUTTON, D. W. H. & SANDERSON, D. J. (eds) *Variscan Tectonics of the North Atlantic Region*. Blackwell, Oxford, 33–45.

—— 1984*b*. Variscischer Deckenbau im Raume der Münchberger Gneismasse, abgeleitet aus der Fazies, Deformation und Metamorphose im umgebenden Paläozoikum. *Geotektonische Forschungen*, **68**, 1–253.

—— 1989. Variscan plate tectonics in central Europe—current ideas and open questions. *Tectonophysics*, **169**, 221–228.

—— 1992. Phanerozoic structures and events in central Europe. *In*: BLUNDELL, D., FREEMAN, R. & MUELLER, S. (eds) *A Continent Revealed. The European Geotraverse*. European Science Foundation. Cambridge University Press, Cambridge. 164–180.

—— 2000. The mid-European segment of the Variscides: techtono-stratigraphic units, terrane boundaries and plate evolution. *This volume*.

—— & ONCKEN, O. 1990. Geodynamic evolution of the North-Central Variscides—a comic strip. *In*: FREEMAN, R., GIESE, P. & MUELLER, S. (eds) *The European Geotraverse: Integrative Studies. European Science Foundation, Strasbourg, 187–194.*

—— & ——1995. Zur prädevonischen Geschichte des Rhenohercynischen Beckens. *Nova Acta Leopoldina NF*, **71**(291), 53–72.

——, BORTFELD, R. K., BRIX, M. *et al.* 1990. Crustal structure of the Rhenish Massif: results of deep seismic reflection lines DEKORP 2-North and 2-North-Q. *Geologische Rundschau*, **79**, 523–566.

GRADSTEIN, F. M. & OGG, J. 1996. A Phanerozoic time scale. *Episodes*, **19**(1–2), 3–5.

HAY, W. W. 1995. Tectonics and climate. *Geologische Rundschau*, **85**, 409–437.

HENK, A. 1991. Structure of the Saar–Nahe-Basin (SW-Germany) from DEKORP profiles 1-C and 9-N. *American Geophysical Union Geodynamic Series*, **22**, 91–95.

—— 1993. Late orogenic evolution in the Variscan Internides: the Saar–Nahe Basin, southwest Germany. *Tectonophysics*, **223**, 273–290.

—— 1995. Late Variscan exhumation histories of the southern Rheno-Hercynian Zone and western Mid-German Crystalline Rise. *Geologische Rundschau*, **84**, 578–590.

HERBIG, H.-G. 1998. The late Asbian transgression in the central European Culm basins (Late Viséan, cd IIIα). *Zeitschrift der Deutschen Geologischen Gesellschaft*, **149**(1), 39–58.

HIRSCHMANN, G. 1995. Mid-German Crystalline High—lithological characteristics. *In*: DALLMEYER, R. D., FRANKE, W. & WEBER, K. (eds) *Pre-Permian Geology of Central and Eastern Europe*. Springer, Heidelberg, 155–163.

HOUSEKNECHT, D. 1987. Assessing the relative importance of compaction processes and cementation to reduction of porosity in sandstones. *AAPG Bulletin*, **71**, 633–642.

IBBECKEN, H. & SCHLEYER, R. 1991. *Source and Sediment. A Case Study of Provenance and Mass Balance at an Active Plate Margin (Calabria, Southern Italy)*. Springer, Heidelberg.

JONES, G. L. & SOMERVILLE, I. D. 1996. Irish Dinantian biostratigraphy: practical applications. *In*: STROGEN, P., SOMERVILLE, I. D. & JONES, G. L. (eds) *Recent* Advances in Lower Carboniferous Geology. Geological Society, London, Special Publications, **107**, 371–385.

KLÜGEL, T. 1997. *Geometrie und Kinematik einer variscischen Plattengrenze—der Südrand des Rhenoherzynikums im Taunus*. Geologische Abhandlung Hessen, **101**.

KOLLA, V., POSAMENTIER, H. W. & EICHENSEER, H. 1995. Stranded parasequences and the forced regressive wedge systems tract: deposition during base-level fall—discussion. *Sedimentary Geology*, **95**, 139–145.

KORN, D. 1996. Revision of the Rhenish Late Viséan goniatite stratigraphy. *Annales de la Société Géologique de Belgique*, **117**(1), 129–136.

KORSCH, R. J. & SCHÄFER, A. 1995. The Permo-Carboniferous Saar–Nahe Basin, south-west Germany and north-east France: basin formation and deformation in a strike-slip regime. *Geologische Rundschau*, **84**, 293–318.

KOSSMAT, F. 1927. Gliederung des varistischen Gebirgsbaues. *Abhandlung Sächsische Geologische Landes-Anstalt*, **1**, 1–39.

KRAFT, T. 1992. *Faziesentwicklung vom flözleeren zum flözführenden Oberkarbon (Namur B–C) im südlichen Ruhrgebiet*. DGMK-Berichte.

KUENEN, P. H. & SANDERS, J. E. 1956. Sedimentation phenomena in Kulm and Flözleeres greywackes, Sauerland and Oberharz, Germany. *American Journal of Science*, **254**, 649–671.

KULICK, J. 1960. Zur Stratigraphie und Palaeogeographie der Kulm-Sedimente im Eder-Gebiet des nordöstlichen Rheinischen Schiefergebirges. *Fortschritte in der Geologie von Rheinland und Westfalen*, **3**, 243–288.

LIPPOLT, H. J., HESS, J. C. & BURGER, K. 1984. Isotopische Alter von pyroklastischen Sanidinen aus Kaolin-Kohlentonsteinen als Korrelationsmarken für das mitteleuropäische Oberkabon. *Fortschritte in der Geologie von Rheinland und Westfalen*, **32**, 119–150.

LITTKE, R., BÜKER, C., HERTLE, M., KARG, H., STROETMANN-HEINEN, V. & ONCKEN, O. 2000. Heat flow evolution, subsistence and erosion in the Rheno-Heraynian orogenic wedge of central Europe. *This volume*.

——, ——, LÜCKGE, A., SACHSENHOFER, R. F. & WELTE, D. H. 1994. A new evaluation of paleo-heatflows and eroded thickness for the Carboniferous Ruhr basin, western Germany. *International Journal of Coal Geology*, **26**, 155–183.

MIALL, A. D. 1997. *The Geology of Stratigraphic Sequences*. Springer, Heidelberg.

MUTTI, E. 1992. *Turbidite Sandstones*. Agip, Parma.

NASIR, S., OKRUSCH, M., KREUZER, H., LENZ, H. & HÖHENDORF, A. 1991. Geochronology of the Spessart Crystalline Complex, Mid-German Crystalline Rise. *Mineralogy and Petrology*, **44**, 39–55.

OKRUSCH, M. 1995. Mid German Crystalline High—metamorphic evolution. *In*: DALLMEYER, R. D., FRANKE, W., WEBER, K. (eds) *Tectonographic Evolution of the Central and East European Orogens*. Springer, Heidelberg.

ONCKEN, O. 1984. Zusammenhänge in der Strukturgenese des Rheinischen Schiefergebirges. *Geologische Rundschau*, **73**, 619–649.

—— 1997. Transformation of a magmatic arc and an orogenic root during oblique collision and it's

consequences for the evolution of the European Variscides (Mid-German Crystalline Rise). *Geologische Rundschau*, **86**, 2–20.
—— 1998. Evidence for precollisional subduction erosion in ancient collisional belts: The case of the Mid-European Variscides. *Geology*, **26**, 1075–1078.
——, Plesch, A., Weber, J., Ricken, W. & Schrader, S. X. 2000. Passive margin detachment during arc–continent collision (Central European Variscides). *This volume*.
——, von Winterfeld, C. & Dittmar, U. 1999. Accretion of a rifted passive margin: The Late Paleozoic Rheno-Hercynian fold and thrust belt (Middle European Variscides). *Tectonics*, **18**, 75–91.
Paproth, E. 1953. Eine Kohlenkalkfauna aus dem Kulmkonglomerat von Frankenberg an der Eder. *Paläontologische Zeitschrift*, **27**(3/4), 169–207.
Pickering, K. T., Clark, J. D., Smith, R. D. A., Hiscott, R. N., Ricci Lucchi, F. & Kenyon, N. H. 1995. Architectural element analysis of turbidite systems, and selected topical problems for sand-prone deep-water systems. *In*: Pickering, K. T., Hiscott, R. N., Kenyon, N. H., Ricci Lucchi, F. & Smith, R. D. A. (eds) *Atlas of Deep Water Environments: Architectural Style in Turbidite Systems*. Chapman & Hall, London, 1–10.
Piper, D. J. W. & Stow, D. A. V. 1991. Fine-grained turbidites. *In*: Einsele, G., Ricken, W. & Seilacher, A. (eds) *Cycles and Events in Stratigraphy*. Springer, Heidelberg, 361–376.
Plesch, A. & Oncken, O. 1999. Orogenic wedge growth during collision—constraints on mechanics of a fossil wedge from its kinematic record (Rheno-Hercynian FTB, Central Europe). *Tectonophysics*, **309**, 117–139.
Plessmann, W. 1961. Strömungsmarken in klastischen Sedimenten und ihre geologische Auswertung. *Geologisches Jahrbuch*, **78**, 503–566.
Posamentier, H. W., Jervey, M. T. & Vail, P. R. 1988. Eustatic controls on clastic deposition I—conceptual framework. *In*: Wilgus, C. K., Hastings, B. S., Kendall, C. G. St C., Posamentier, H. W., Ross, C. A. & Van Wagoner, J. C. (eds) *Sea-level Changes—an Integrated Approach*. Society of Economic Paleontologists and Mineralogists, Special Publications, **42**, 107–124.
Ricci Lucchi, F. 1986. The Oligocene to recent foreland basins of the northern Apennines. *In*: Allen, P. A. & Homewood, P. (eds) *Foreland Basins*. International Association of Sedimentologists Special Publications, **8**, 105–139.
Riley, N. J. 1993. Dinantian (Lower Carboniferous) biostratigraphy and chronostratigraphy in the British Isles. *Journal of the Geological Society London*, **150**, 427–446.
Roberts, J., Claoué-Long, J. C. & Jones, P. J. 1995. Australian early Carboniferous time. *In*: Berggren, W. A., Kent, D. V. & Aubry, M.-P. (eds) *Geochronology, Time Scales and Global Stratigraphic Correlation*. Society of Economic Paleontologists and Mineralogist Special Publications, **54**, 23–40.
Ross, C. A. & Ross, J. R. P. 1985. Late Paleozoic depositional sequences are synchronous and worldwide. *Geology*, **13**, 194–197.
—— & —— 1987. Late Paleozoic sea levels and depositional sequences. *Cushman Foundation for Foraminiferal Research, Special Publications*, **24**, 137–149.
—— & —— 1988. Late Paleozoic transgressive–regressive deposition. *In*: Wilgus, C. K., Hastings, B. S., Kendall, C. G. St C., Posamentier, H. W., Ross, C. A. & Van Wagoner, J. C. (eds) *Sea-level Changes—an Integrated Approach*. Society of Economic Paleontologists and Mineralogists Special Publications, **42**, 227–247.
Sadler, P. M. 1983. Depositional models for the Carboniferous flysch of the eastern Rheinisches Schiefergebirge. *In*: Martin, H. & Eder, W. F. (eds) *Intracontinental Fold Belts*. Springer, Heidelberg, 125–143.
Schäfer, J., Neuroth, H., Ahrendt, H., Dörr, W. & Franke, W. 1997. Accretion and exhumation at a Variscan active margin, recorded in the Saxothuringian flysch. *Geologische Rundschau*, **86**, 599–611.
Schlager, W., Reijmer, J. J. G. & Droxler, A. 1994. Highstand shedding of carbonate platforms. *Journal of Sedimentary Research*, **64**, 270–281.
Scholtz, R. 1930. Das Variscische Bewegungsbild entwickelt aus der inneren Tektonik eines Profiles von der Böhmischen Masse bis zum Massiv von Brabant. *Fortschritte Geologie Paläontologie*, **8**(25), 235–316.
Schmidt, H. 1925. Die carbonischen Goniatiten Deutschlands. *Jahrbuch Preußische* Geologische Landes-Anstalt, **45**, 489–609.
Schrader, S. (1999). *Das Rhenohercynische Turbiditbecken (Spätes Viseum)—Faziesentwicklung und Beckenanalyse*. PhD thesis, University of Cologne.
Süss, M. P. 1996. *Sedimentologie und Tektonik des Ruhr-Beckens: Sequenzstratigraphische Interpretation und Modellierung eines Vorlandbeckens der Varisciden*. Bonner Geowissenschaftliche Schriften, **20**.
Vail, P. R., Audemard, F., Bowman, S. A., Eisner, P. N. & Perez-Cruz, C. 1991. The stratigraphic signatures of tectonics, eustasy and sedimentology—an overview. *In*: Einsele, G., Ricken, W. & Seilacher, A. (eds). *Cycles and Events in Stratigraphy*. Springer, Heidelberg, 617–659.
von Huene, R. & Scholl, D. W. 1991. Observations at convergent margins concerning sediment subduction, subduction erosion, and the growth of continental crust. *Reviews of Geophysics*, **29**, 279–316.
Wetzel, A. 1993. The transfer of river load to deep-sea fans: a quantitative approach. *AAPG Bulletin*, **77**, 1679–1692.
Zoetemeijer, R. 1993: *Tectonic modelling of foreland basins—thin skinned thrusting, syntectonic sedimentation and lithospheric flexure*. PhD thesis: Free University of Amsterdam.

Upper-plate deformation during collisional orogeny: a case study from the German Variscides (Saxo-Thuringian Zone)

FRAUKE SCHÄFER[1], ONNO ONCKEN[2], HELGA KEMNITZ[2] & ROLF L. ROMER[2]

[1]*Bundesanstalt für Geowissenschaften und Rohstoffe, Stilleweg 2, D-30655, Hannover, Germany*

[2]*GeoForschungsZentrum Potsdam, Telegrafenberg, D-14473 Potsdam, Germany (e-mail: heke@gfz-potsdam.de)*

Abstract: A doubly vergent orogenic wedge system within the Central European Variscides developed during Carboniferous collision of two continental fragments, the northwestern edge of the Saxo-Thuringian upper plate and the Rheno-Hercynian passive margin in the lower plate. The resulting thrust system in the upper plate above the SE-dipping subduction zone retains the memory of the mode of deformation partitioning and material flow pattern in its internal architecture, its kinematic, metamorphic and geochronological record, and its reflection seismic image. New data indicate a stepwise SE-ward progradation of the NW Saxo-Thuringian fold belt with two stages of shortening between about 340 and 335 and between 320 and 310 Ma above a NW-dipping basal detachment. The NW Saxo-Thuringian fold belt is reinterpreted as a retro-wedge that was kinematically coupled to the Rheno-Hercynian pro-wedge and subduction system. The two steps in retro-wedge growth are linked to (a) the onset of collision with the Rheno-Hercynian margin causing upper-plate uplift and (b) a widespread late-orogenic stage of wedge thickening. The retro-wedge accumulated mostly diffuse shortening of >100 km versus the shortening by imbrication of 180–200 km in the Rheno-Hercynian lower plate. Material advection and orogenic architecture were strongly affected by asymmetric erosional removal towards the lower-plate foreland and by transient mechanical properties of the wedge system.

Collision of continental plates usually involves the formation of doubly vergent orogens with thrust wedges on both sides of the former plate boundary. This has been shown through field studies in numerous young collisional belts such as the Alps and the Pyrenees, and corroborated by numerical and analogue modelling (Willett *et al.* 1993; Beaumont *et al.* 1996; Jamieson *et al.* 1998). The bulk of compressional strain in such a setting is normally accumulated in extensive fold-and-thrust belts on the lower plate (pro-wedge). The upper plate may, however, absorb some of the deformation in a smaller, steeper fold-and-thrust belt with opposite polarity to the main tectonic transport direction (retro-wedge). Whereas a doubly-vergent orogen may readily be identified by topographical and tectonic features in active belts, evidence for fossil doubly vergent orogenic wedges that were kinematically coupled is still scarce. Also, it is generally unclear how the rock record may help to constrain the factors that control the evolution and style of retro-wedges.

The Variscan Orogen of Central Europe allows such an analysis, as it consists of a number of minor deeply eroded continental fragments that collided in Late Devonian to Early Carboniferous times (Franke (this volume) and references therein). Predominating NW-vergent structural patterns indicate that collision and tectonic transport occurred in a generally northwestward direction above SE-dipping subduction zones. In the NW, the Rheno-Hercynian shelf, which evolved on Avalonia (Fig. 1), was overthrust by the Mid-German Crystalline Zone (MGCZ). The MGCZ acted as the active continental margin, comprising remnants of a magmatic arc (Willner *et al.* 1991; Flöttmann & Oncken 1992). To the SE, the MGCZ is followed by the continental Saxo-Thuringian basin, which in turn is overridden by the Teplá–Barrandian terrane in the SE.

In contrast to the extensively analysed Rheno-Hercynian thrust belt (Plesch & Oncken (1999) and references therein), relevant details of the deformation geometry and evolution of the

From: FRANKE, W., HAAK, V., ONCKEN, O. & TANNER, D. (eds). *Orogenic Processes: Quantification and Modelling in the Variscan Belt.* Geological Society, London, Special Publications, **179**, 281–302. 0305-8719/00/$15.00

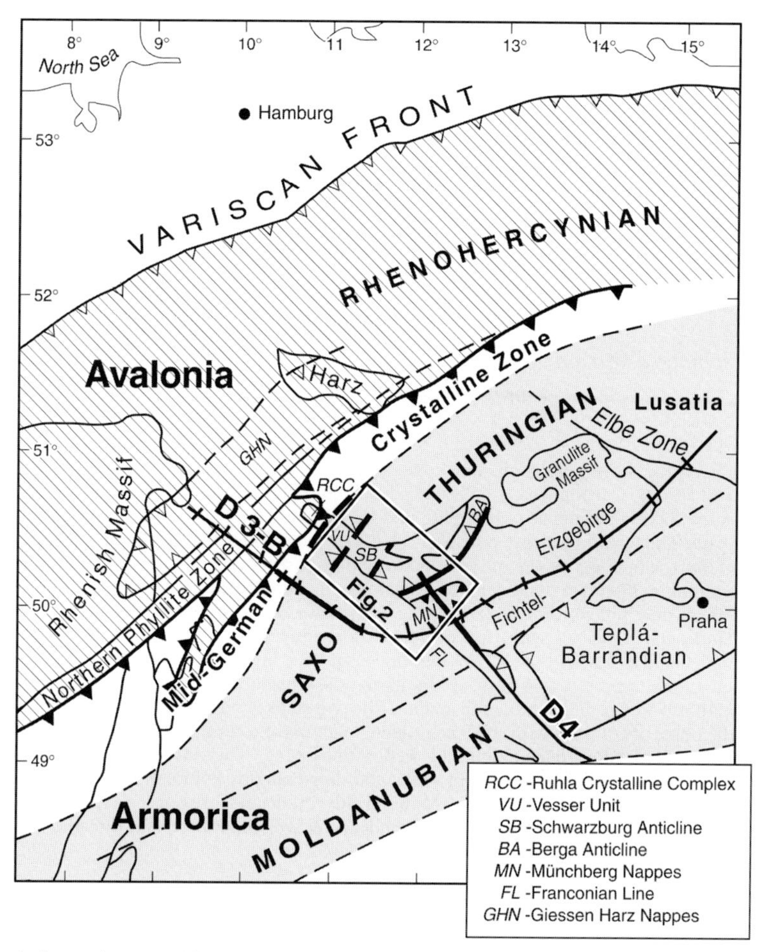

Fig. 1. Geological overview map of the Central European Variscides, showing the study area (inserted frame), major fault zones and structural units, and the position of the reflection seismic profiles (D 3-B, DEKORP 3B/MVE West; D4, DEKORP 4). Thick sections of seismic lines were projected along-strike on a common section (Fig. 4).

Saxo-Thuringian plate, which belonged to Armorica, are still under debate (Franke *et al.* 1995; Ahrendt *et al.* 1996; Oncken 1997; Oncken *et al.* 1999). The leading, northwestern part of the Saxo-Thuringian plate differs from its southeastern trailing edge in intensity, style and age of tectonometamorphic overprint, its tectonic vergency (SE rather than NW), and seismic signal patterns. It is therefore suspected to be part of a doubly vergent orogenic system formed during collision with the Rheno-Hercynian plate to the NW, rather than belonging to the fold-and-thrust belt formed ahead of the Saxo-Thuringian–Teplá–Barrandian collision zone in the SE (Fig. 1; Schäfer 1997).

An understanding of the conditions controlling the partitioning of collisional deformation into the upper plate from the structural and metamorphic architecture of the resulting orogen requires a complete reconstruction of material displacement during collision. For the Variscides, this can be achieved in a Central German section only through the largely exhumed and exposed leading part of the Saxo-Thuringian upper plate (Fig. 1). This area is also covered by crustal reflection seismic studies, providing information about its geometry at depth. Integrating new tectonic, metamorphic and geochronological data from this leading, NW part of the Saxo-Thuringian upper plate with available data from the Rheno-Hercynian zone and MGCZ, the present paper is an attempt to identify the main factors that controlled the evolution of this fossil, doubly vergent orogenic wedge system.

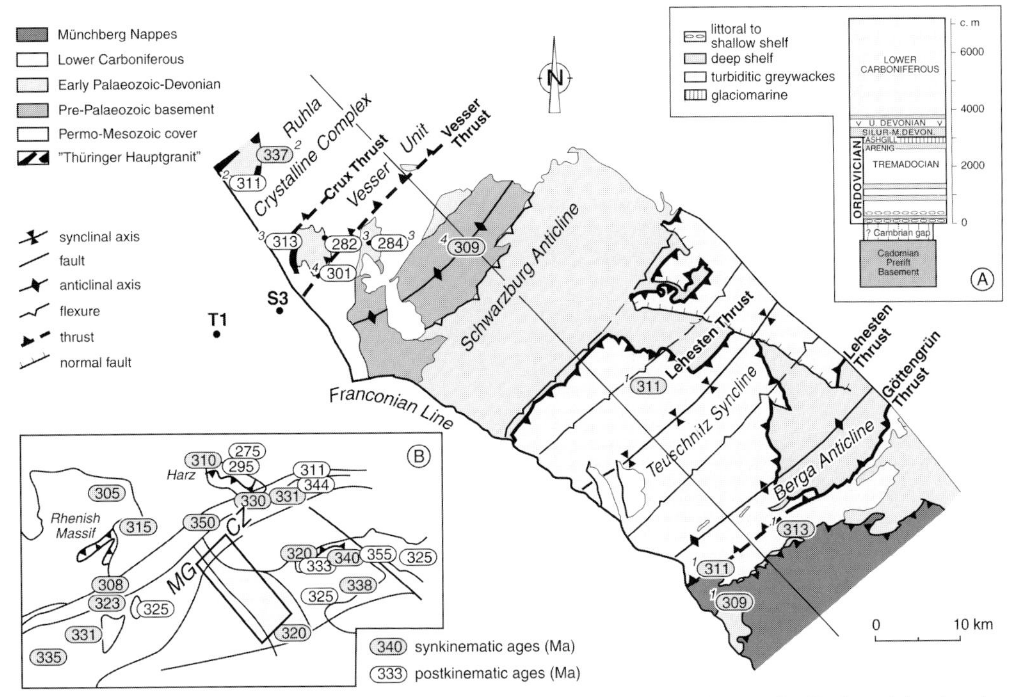

Fig. 2. Main tectonic units of western Saxo-Thuringia and associated age data. Sources for K–Ar and Ar–Ar mica age data, and Pb–Pb and U–Pb zircon age data (superscript numbers in figure): 1, Ahrendt *et al.* (1983, 1986); 2, Zeh *et al.* (2000); 3, Schäfer *et al.* (1997), Siebel (1999) and this paper; 4, Kemnitz *et al.* (1999). Straight line indicates location of cross-section (see Fig. 4c). T1 and S3 are wells drilling RCC rocks under Mesozoic cover (Wunderlich 1995). Inset figures: (**a**) Simplified stratigraphic reference section of the Saxo-Thuringian basin, derived from the SE flank of the Schwarzburg Anticline; (**b**) syn- and post-kinematic age data from the Variscan units surrounding the study area.

Geological development of NW Saxo-Thuringia

General geological framework

The study area covers a section across the former arc and back-arc area of the Rheno-Hercynian–Saxo-Thuringian collisional system east of the Franconian Line, a mainly Upper Cretaceous reverse fault (Fig. 1). The following tectonometamorphic units are covered from NW to SE: the southern Ruhla Crystalline Complex (RCC) representing the central MGCZ in the study area, the Vesser Unit, and the Saxo-Thuringian basin proper.

The present boundary between the Rheno-Hercynian lower plate and the Saxo-Thuringian upper plate is widely identical with the MGCZ (Fig. 1). The RCC is exposed as a composite complex of amphibolite facies units of presumed early Palaeozoic age and magmatic intrusions. Most of these rocks probably represent accreted and exhumed slabs of Rheno-Hercynian lower-plate crust (Oncken 1997), but the RCC also includes remnants of a Lower Carboniferous arc in the uppermost thrust sheets with calc-alkaline granites intruded in amphibolite facies rocks (Zeh 1996).

To the SE (Figs 1 and 2), the RCC is adjoined by a narrow belt of low-grade, strongly deformed Lower Palaeozoic volcani-sedimentary rift sequences (Bankwitz *et al.* 1994) that represent a remnant of thinned continental rift or ocean–continent transition from the Saxo-Thuringian shelf (Kemnitz *et al.* (2000) and references therein). It is exposed around the small locality of Vesser, and is therefore labelled here as the Vesser Unit.

In the Saxo-Thuringian basin proper, Variscan deformation affected a complete Palaeozoic sequence resting on metasedimentary Cadomian basement. Palaeozoic sedimentation started with typical Armorican Cambro-Ordovician shelf sequences (Lützner *et al.* 1986; Noblet & Lefort 1990; Linnemann & Buschmann 1995) and developed into neritic and pelagic Devonian deposits

(Fig. 2a). Volcanism related to a renewed pulse of NW–SE-directed crustal extension (Franke 1993) culminated in early Late Devonian times. Uppermost Devonian–Lower Carboniferous turbidites indicate basin deepening during the approach of the Teplá–Barrandian plate margin from the SE. The collision with this plate resulted in shortening and a NW directed tectonic transport of the Saxo-Thuringian basin sediments (Franke 1984; Stein 1988; Behrmann *et al.* 1991). Within this NW-vergent structural belt, the intensity of Variscan deformation and metamorphism decreases from upper greenschist grade at the suture with the Teplá–Barrandian zone to very low grade north of the Münchberg Nappes. The deformation front is located on the southern flank of the Berga Anticline (Figs 1 and 2). It is important to distinguish this NW-vergent structural belt at the trailing edge of the Saxo-Thuringian plate from the SE-vergent structural belt dealt with in this paper.

Structural features of NW Saxo-Thuringia

The SE-vergent structural belt of the Saxo-Thuringian plate involves, from NW to SE, the Vesser Unit, the Schwarzburg Anticline, the Teuschnitz Syncline and the Berga Anticline (Fig. 2). Each unit presents its own structural style (Fig. 3a), with the overall intensity of deformation and metamorphism decreasing towards the SE. In the study area, the deformation front of the SE-vergent structural belt roughly coincides with the Göttengrün Thrust on the southeastern flank of the Berga Anticline.

Deformation in the northernmost unit, the Vesser Unit, is shear-zone dominated along steeply NW-dipping mylonitic zones (Fig. 3a). Shear sense indicators show a general top-to-the-south transport direction with a minor sinistral component. Mylonite foliation parallels a pervasive, steeply NW-dipping main foliation that, in places, refolds an earlier cleavage. This older cleavage and bedding planes are superimposed with different intensity and often entirely recrystallized. Bedding (SS in Fig. 3a) tends to parallel the main foliation but in cases of obliquity, stratigraphic younging is towards the NW. This is corroborated from age dating of the volcanic rocks (Kemnitz *et al.* 2000) and the straightforward lithological succession that precludes both major internal folding and tectonic repetition. Several granites intruded the Early Palaeozoic volcano-sedimentary sequences of the Vesser Unit after deformation.

The boundaries of this unit towards the RCC in the north and the Schwarzburg Anticline in the south are not exposed, but can be inferred from a substantial change in metamorphic grade and structural style: greenschist facies Vesser Unit v. amphibolite facies RCC and low-grade to very low-grade facies of the Schwarzburg Anticline. The absence of extensional features, and the orientation of the partly mylonitic main foliation, suggest that major SE-directed thrusts border the Vesser Unit: the Crux Thrust (Schäfer 1997; see references) at the top of the Vesser Unit in the north, and the Vesser Thrust at its base to the south (Fig. 3a). This interpretation is corroborated by several wells (Wunderlich (1995) and references therein) in the along-strike continuation of the Vesser Unit to the west, that drilled RCC-type metamorphic rocks below the Mesozoic cover west of the Franconian Line (Fig. 2). Considering the 3 km vertical offset along the Franconian Line, these wells project above the exposed Vesser Unit and therefore indicate a NW dip of the intervening Crux Thrust.

The Vesser Thrust displaces the Vesser Unit over the exposed footwall ramp that forms part of the NW flank of the Schwarzburg Anticline. This antiformal unit can be subdivided into a Neoproterozoic basement, a greywacke–shale sequence with intercalated granite sills and intrusions outcropping in the core of the anticline, and its Palaeozoic syn- to post-rift cover sequence on the flanks. The Schwarzburg Anticline spans a half-wavelength of about 30 km with an average interlimb angle of 120–150°, and is slightly SE-vergent and asymmetric with a kink on its SE flank (Fig. 3a).

Rocks of the Neoproterozoic core show complex deformational patterns (Bankwitz & Bankwitz 1995) as a result of both Cadomian and Variscan deformation. The oldest (possibly Cadomian) deformation increment is characterized by a strong tectonic layering transposing bedding and sedimentary structures. In contrast, the syn- to post-rift cover sequences on both flanks of the Schwarzburg Anticline have well-preserved bedding and sedimentary structures (Fig. 3a). The first deformation fabric that can undoubtedly be related to the Variscan Orogeny is a steeply NW-dipping, pervasive foliation, characterized by syn-kinematic white mica growth. It continues into the post-Cadomian cover rocks, where it is the first foliation, without change of orientation or style. On the northwestern flank of the Schwarzburg Anticline, in the footwall of the Vesser Thrust, it parallels the main foliation of the Vesser Unit, showing stretching lineations with the same steep NE dip as in the Vesser Unit. The folds associated with the main foliation show mostly tight interlimb angles that decrease with distance to the core of the anticline. Additionally, there is a

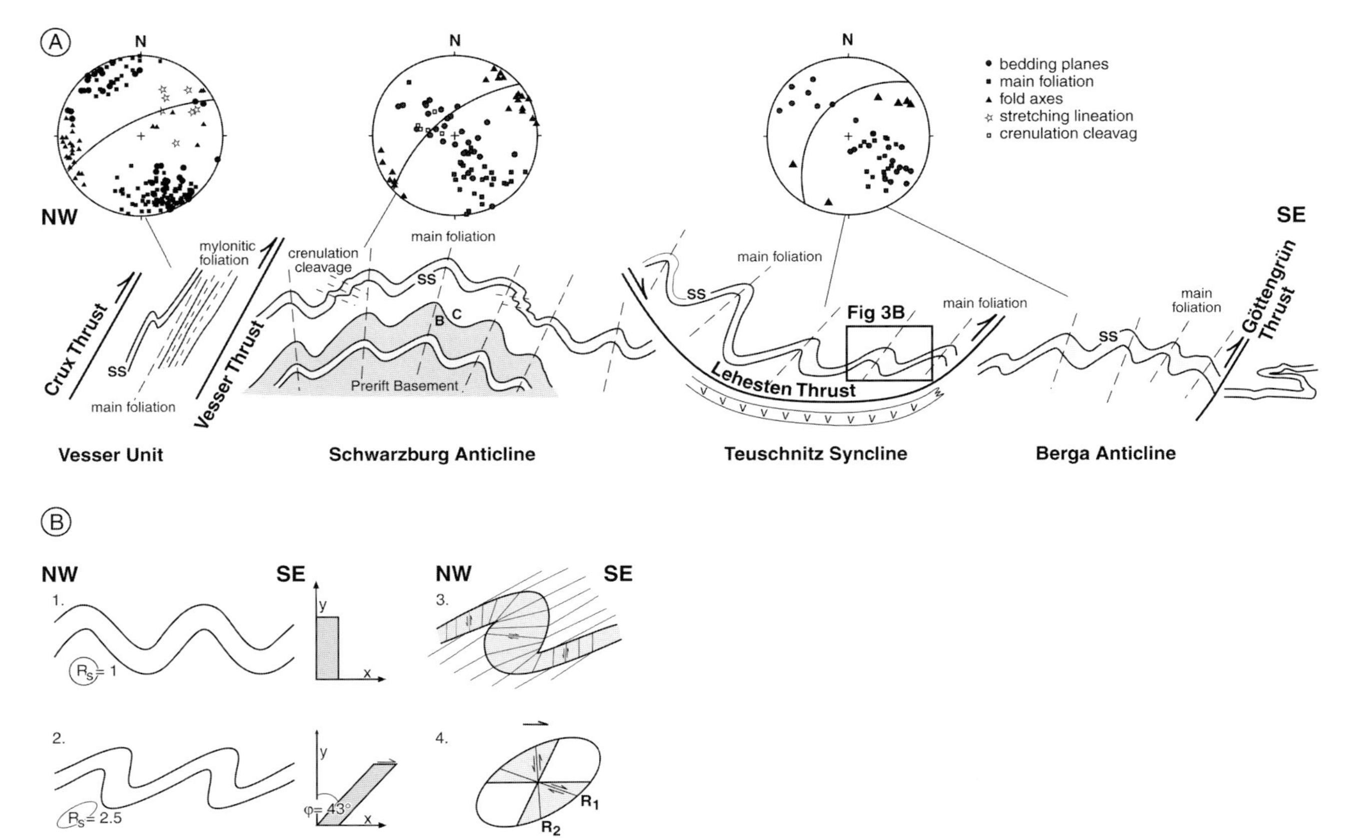

Fig. 3. (**a**) Section sketch of main tectonic units showing fabric relation as observed in the field. (See text for discussion.) (**b**) Geometry and genetic development of asymmetric folds in the Teuschnitz Syncline: (1) initial fold geometry before shear strain application; (2) fold geometry at $R_s = 2.5$ and shear angle $\phi = 43°$, corresponding to the geometries observed in the field; (3) fold sketch drawn at outcrop, with cleavage refraction through a competent layer (grey) and shear sense along cleavage planes; (4) simple shear strain ellipse corresponding to the shear strain deducted in (2). Shortening sector is shaded grey, stretching sector is unshaded; R_1, orientation of Riedel 1 shear planes; R_2, orientation of Riedel 2 shear planes. It should be noted that the SE-dipping limbs of the original fold in (1) fall into the shortening sector of the strain ellipse, the NW-dipping limbs fall into the stretching sector of the strain ellipse, resulting in the geometries depicted in (2) and (3). The cleavage planes in the competent layer act as Riedel 1 and Riedel 2 planes of the strain ellipse. Modified from Schäfer (1997).

younger crenulation cleavage across the central parts of the Schwarzburg Anticline, characterized by nonpenetrative pressure solution seams and a gentle dip towards the SE. It is present in both Neoproterozoic and Palaeozoic rocks but quickly dies out towards the southeastern flank of the anticline.

Towards the SE of the Schwarzburg Anticline, new grown mica disappears, and the only cleavage is characterized by progressively fewer penetrative pressure solution seams, dipping at varying angles to the NW (Fig. 3a). This area marks a transition from low-grade to unmetamorphized units. The related fold structures are open and slightly SE-verging, with a subhorizontal to slightly SE-dipping fold envelope.

The boundary of the Schwarzburg Anticline structural unit to the SE is linked to a detachment above the Devonian metabasalts, the Lehesten Thrust (Fig. 2). It is followed by the Teuschnitz Syncline, which forms another large-scale feature, with a half-wavelength of 20–25 km, moderately dipping flanks and a broad, flat bottom. The Lehesten Thrust, where exposed on the NW flank of the Teuschnitz Syncline, has a downdip, top-to-the-SE sense of shear, equivalent to a normal fault displacement. On the SE flank, the fault rises to the surface again, preserving its top-to-the-SE sense of shear, which is up-dip on this flank of the syncline. The Lehesten Thrust remains in a flat position relative to stratigraphy throughout the Teuschnitz Syncline. On the SE side of the syncline, the Upper Devonian sequence is characterized by thick (up to several 100 m) metabasalts that resisted folding at scales smaller than a few kilometres. This is probably the cause for the localization of the Lehesten detachment and the differential deformation between the lower-level Schwarzburg Anticline and the upper-level Teuschnitz Syncline.

The uppermost Devonian and Lower Carboniferous turbidites overlying the Lehesten detachment fault have been deformed into tight, upright to strongly SE-facing folds (Fig. 3a). The spaced transverse pressure solution cleavage clearly post-dates folding, as it is oblique to the trace of small-scale fold axial planes in both map and section view. It is largely parallel to the NW-dipping limbs of small-scale folds and overprints their symmetry by thinning and stretching as well as thickening and shortening the SE-dipping limbs (Schwan 1958). From shear box experiments, a best-fit reproduction of this geometry was achieved with a subhorizontal, SE-directed, penetrative simple shear overprint of formerly open, upright buckle folds with a flat-lying fold envelope (Schäfer 1997). In the strain environment of this deformation increment, the NW-dipping limbs are located in the extensional sector of the strain ellipse, whereas the steep SE-dipping limbs occupy the shortening sector (Fig. 3b). On the flank between the Teuschnitz Syncline and Berga Anticline, semi-ductile normal faults developed from the cleavage planes, suppressing the NW limbs of minor folds completely. This is interpreted to be the result of a late rotation of the fold envelope and fold axial planes on the flanks of the Teuschnitz Syncline, so that the previously developed cleavage planes can now act as Riedel 2 planes of the strain ellipse (Fig. 3b).

Altogether, the deformation history in the Teuschnitz Syncline can be summarized in three increments: first, upright buckle folds were formed; second, these buckle folds were overprinted by penetrative, top-to-the-SE simple shear; third, they were rotated into their present-day position within the large-scale Teuschnitz Syncline, which is the youngest feature relative to small-scale folding and cleavage.

To the SE, the Berga Anticline in the hanging wall of the Göttengrün Thrust marks the southernmost large-scale feature of the SE-vergent Saxo-Thuringian fold belt. The anticline forms an upright, open fold, with an interlimb angle of about 120° and a half-wavelength of 10–15 km (Fig. 3a). It represents a large-scale fault bend fold above a fault ramp in the subsurface. The Göttengrün Thrust itself is out of sequence with respect to the formation of the Berga Anticline, branching off from the top of this hidden ramp at a later stage (Schäfer 1997). This interpretation is based on the angular relationship between the Berga Anticline and the Göttengrün Thrust, whose dip (50–60°) is too steep to have caused the gently dipping flanks (about 30°) of the Berga Anticline (Jamison 1992). Farther east, SE-directed thrusts are also observed south of the Göttengrün Thrust, suggesting a partly blind tip of SE-directed deformation in this part of the Saxo-Thuringian zone.

The Göttengrün Thrust has a top-to-the-SE displacement, increasing along-strike from at most 1 km in the SW to about 3 km in the NE. The Berga Anticline is superseded by smaller-scale, SE-facing folds with open to tight interlimb angles and an axial-planar pressure solution cleavage. On the SE flank of the Berga Anticline, the SE-vergent structures refold earlier NW-vergent structures related to the Münchberg Nappe emplacement, which was caused by the collision between the Teplá–Barrandian and the Saxo-Thuringian (Franke 1984; Stein 1988; Behrmann *et al.* 1991).

Seismic data and balanced section

Two reflection seismic lines cross the area under consideration: DEKORP 3B/MVE-90 West and DEKORP 4 (Fig. 1). Parts of the migrated reflection seismic profiles and the interpretation (Schäfer 1997) are shown in Fig. 4a and b. On the NW margin of the section, the Vesser Unit is underlain by thick bundles of subhorizontal reflections (km 10–35, 2–8 s two-way travel time (TWT)). These are offset along two partly transparent, SE-dipping boundaries, indicating that they were thrust on each other. The lower of these fault structures also crosscuts the boundary to the MGCZ and its internal antiformal reflections (km 12, 3.5 s TWT), which suggests later backthrusting over the basement thrust stack. To the SE, the reflectors below the Vesser Unit are cut by a NW-dipping structural boundary that reaches into the units of the adjacent Schwarzburg Anticline (km 20–40, 7–2 s TWT). Farther to the SE, a similar band rises towards the SE, meeting a strong synclinal to flat-lying reflector (km 65–90, 1–2 s TWT) that has been interpreted as the top of the Devonian basalts of the central Saxo-Thuringian basin from its link with surface exposure (DEKORP Research Group 1988). An antiformal structure underneath this Devonian marker (km 80–105, 1.5–2.5 s TWT) has been related to the Saxonian Granulites by Franke (1993) and shows internal stacking related to a gently NW-dipping band of reflections on its NW flank. SE of this feature, the seismic section is dominated by flat-lying to SE-dipping structures interpreted to be related to the NW-directed emplacement of the Münchberg Nappes (km 95–115, 1–2 s TWT) and Fichtelgebirge antiform (km 115–135) (Franke 1993; DEKORP Research Group 1994).

On the basis of the reflectivity patterns of the seismic sections and the above-described structural field record, a simplified balanced cross-section (Fig. 4c) has been constructed through the SE vergent domain of the Saxo-Thuringian belt (see Schäfer (1997) for details). Only the architecture of the large-scale structures was constructed using the fold envelope of minor folds within the sequences at outcrop level. The reflectivity patterns reproduce the gross structural geometry at depth and show the imaged stratigraphic units with their deformed thickness (by internal folding and ductile strain). Accordingly, restoration uses an area balancing approach assuming plane-strain conditions. The plane-strain assumption is supported by along-strike continuity of large folds and thrusts of at least several 10 km to more than 100 km, and the absence of oblique strike-slip faults and of other features suggesting major mass transfer through the section plane.

Restoration of the section (Fig. 4d–f) was carried out in four steps: (a) line length balancing of thrust bodies and large-scale folds; (b) line length restoration of secondary folds at outcrop scale; (c) correction of line length and thickness changes from penetrative strain; (d) area balancing of the basement units. The total shortening of the section was then calculated by adding the shortening amounts of (a), (b) and (c).

(a) *Large-scale folds and thrust bodies.* Flexural unfolding of the large-scale anti- and synclines yields a shortening of only 2.5%. Pin lines were placed in the foreland SE of the Göttengrün Thrust, the crest of the Schwarzburg Anticline and the trailing edge of the Vesser Unit. As the hanging-wall cut-off on the Vesser Unit is completely eroded, displacement can only be estimated from conservative construction as well as from the difference in barometry between hanging wall and footwall (2–3 kbar) and from the fault dip, yielding some 12 km. From the same argument, the order of magnitude for the Crux Thrust is a minimum of 25 km. Together, thrusting and large-scale folding account for a minimum shortening of about 15% (excluding the Crux Fault) in the section plane.

(b) *Secondary folds at outcrop scale.* A much larger amount of shortening is contained in the ubiquitous secondary folds at outcrop scale (wavelengths are several metres to tens of metres). This has been accounted for through simple line length measurements of folded beds in exposed sections (sinuous length minus total length of fold). In the core zone of the Schwarzburg Anticline, shortening amounts to 60–70%, whereas the Palaeozoic rocks on the flanks have been shortened by an average of only 25%. The amount of shortening through folding increases again towards the Teuschnitz Syncline and Berga Anticline, where it reaches an average of 55% (for details see Pfeiffer (1970), Bankwitz & Bankwitz (1988, 1995) and Schäfer (1997)).

(c) *Penetrative strain.* In addition to pure buckle folding and large-scale thrusting, shortening has occurred through penetrative strain. Assuming plane strain, this penetrative deformation can be expressed as line length shortening through

$$(1 + e)^{-2} = 1/R_{ac} \cos^2 \alpha + R_{ac} \sin^2 \alpha$$

where e is the line length shortening, R_{ac} the ellipticity of the strain ellipsoid in the section plane, and α the angle between bedding and

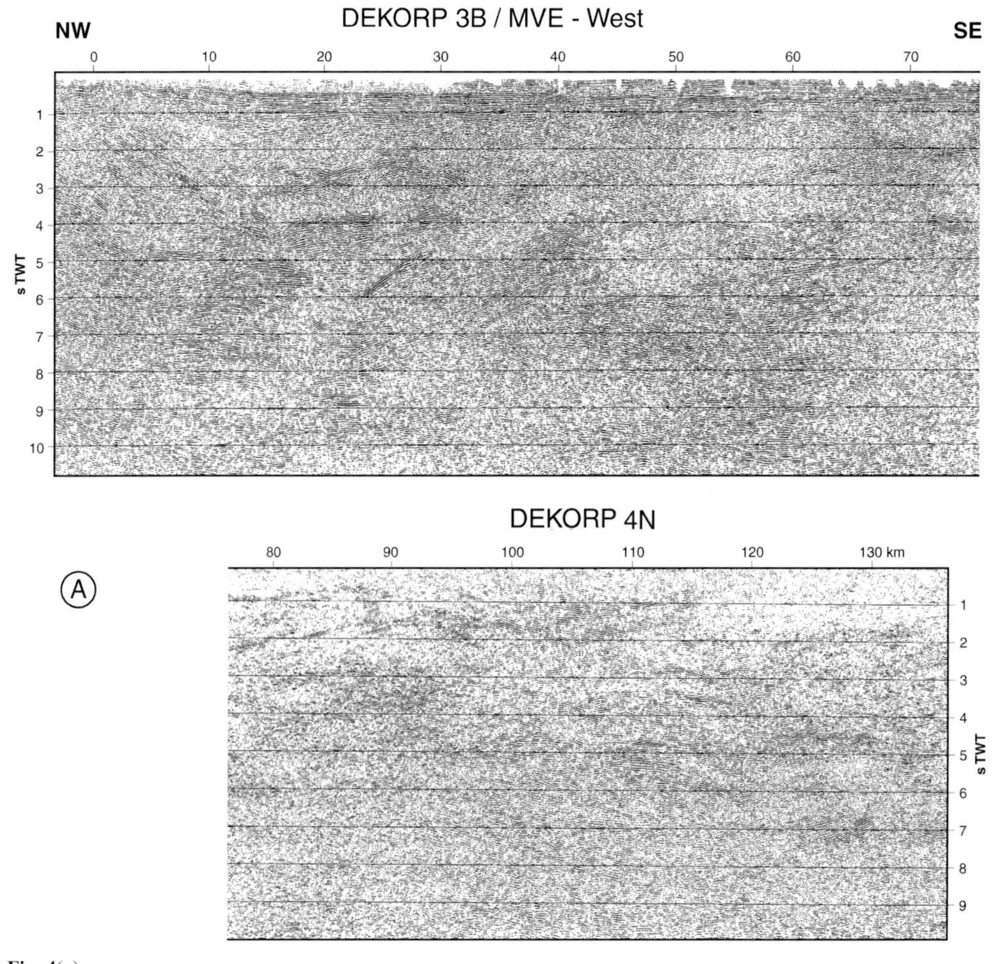

Fig. 4(a).

cleavage (Ramsay & Huber 1987). Strain on the SE flank of the Schwarzburg Anticline is in the range of $R_{ac} = 2.1–2.9$ (Ellenberg 1964; Bankwitz & Bankwitz 1988) and yields an average shortening of 29%.

In the Teuschnitz Syncline, the penetrative strain has been derived from shear box experiments, and corroborated with the geometric method of Lisle (1992). It is in the range of $R_{ac} = 2.5–2.9$. However, because of the asymmetry of the fold limbs in relation to the cleavage, the intersection angle α is different on the NW- and SE-dipping limbs. Furthermore, the NW limbs are substantially longer than the SE limbs. On the basis of an average length ratio of 2:1 between NW and SE limbs, line lengths are found to have actually been stretched during the shearing increment by 20–22%. This is due to the orientation of pre-existing folds relative to the shear strain ellipsoid.

Combining the shortening amounts from (b) and (c) for the Early Palaeozoic sequences of the Schwarzburg Anticline on the one hand and the Lower Carboniferous turbidites of the Teuschnitz Syncline on the other, both areas yield a total shortening of 46–47%. Together with the shortening from large-scale folding and thrusting found in (a), total shortening of the SE-vergent Saxo-Thuringian belt amounts to a minimum of 50%, or a palaeo-basin width of at least 160 km (Schäfer 1997). Uncertainties mainly include the amount of displacement and internal deformation of the RCC and the Vesser Unit. The Neoproterozoic basement has been restored by area-conserving inclined shear. It has been considered as pinned to the Palaeozoic cover in

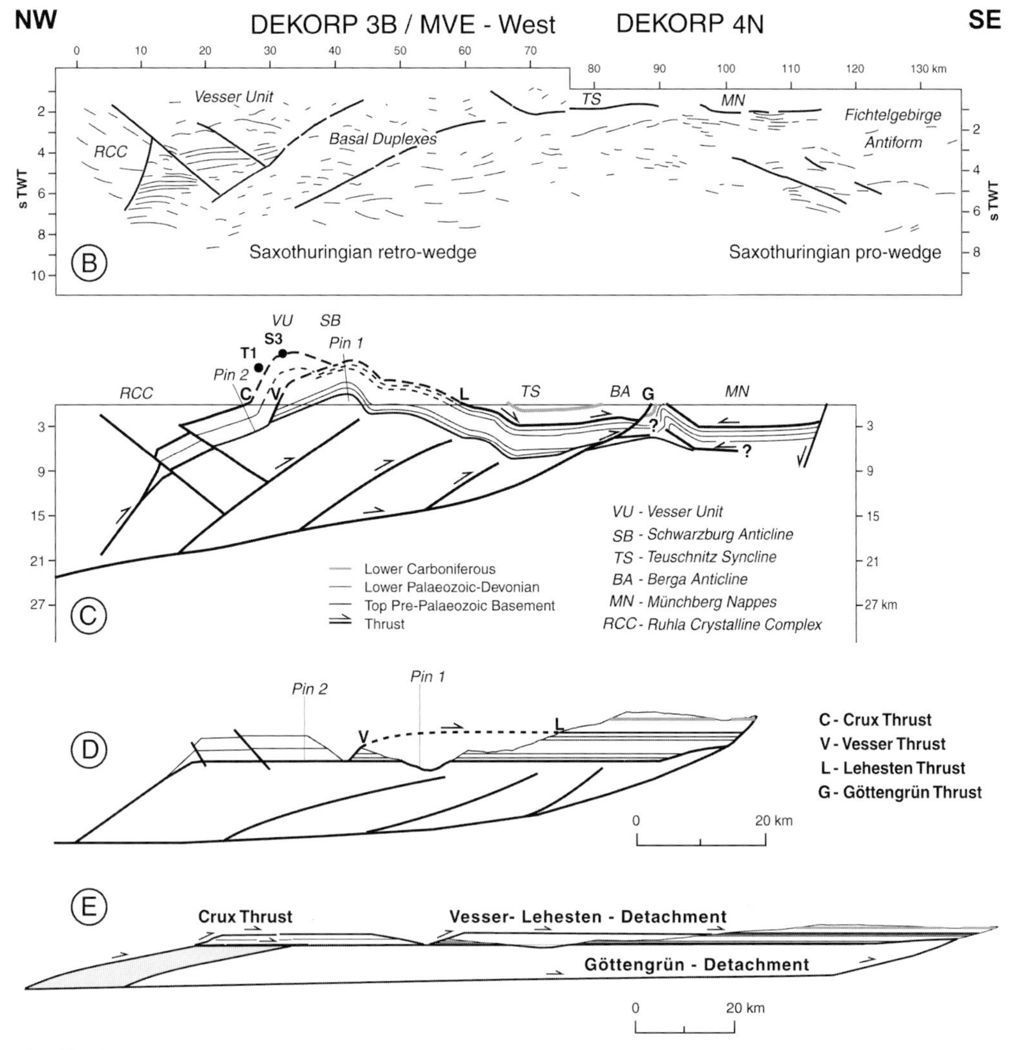

Fig. 4(b–e).

Fig. 4. (**a**) Migrated reflection seismic profiles DEKORP 4N and DEKORP 3B/MVE West. Line DEKORP 4N has 1 s TWT of Permo-Mesozoic sediments covering the Palaeozoic fold belt west of the Franconian Line (see Fig. 1). (**b**) Line drawing of lines in (**a**) and interpretation mounted at km 76 with 1 s TWT vertical offset of line DEKORP 3B, as required from the 3 km throw across the Franconian Line. Fine lines, reflectors; bold lines, reflection features interpreted as faults. (**c**) Balanced cross-section showing the anatomy of large-scale structures of the Saxo-Thuringian retro-wedge. (**d**) Line-length restoration of large-scale structures. It should be noted that the pre-Palaeozoic basement has been pinned to the Palaeozoic cover in the core of the Schwarzburg Anticline, as no slip along the boundary can be observed at outcrop. (**e**) Area-balanced restoration removing fold and ductile strain; light shaded area underneath the northern end of Vesser Unit represents missing basement (see text for explanation).

the core of the Schwarzburg Anticline, where no shearing along the contact can be detected at outcrop (Fig. 4c and d). Although the section is balanced, basement restoration also shows that part of the Vesser Unit lacks basement in the subsurface, which, from the restoration procedure, would be equivalent to a minimum amount. This observation can only be accounted for if this part of the basement was accreted to and uplifted with the hanging wall of the Crux Thrust (i.e. the MGCZ) during contemporaneous activity of the Crux and Vesser Thrusts.

Interpretation

The section in Figs 3a and 4c highlights the general structure of the SE-vergent belt. A narrow zone in the NW can be defined as an 'internal belt', controlled by steep imbricates of greenschist facies rocks (Vesser Unit), and limited to the south by the Vesser Thrust. Together with the Crux Thrust this internal belt accumulated more than 50 km of thrust-related shortening. A broad zone to the SE, reaching the deformation front, can be defined as an 'external belt' and is largely controlled by folding, with thrusts only at the tip, rear and top of the belt (total shortening *c.* 47 km). The rearward thrust, the Vesser Thrust, juxtaposes the internal fold belt with greenschist facies rocks against the external fold belt with mainly anchizone grade rocks. The Vesser Thrust is tentatively interpreted to flatten out in the exposed Lehesten Thrust at the base of the strongly folded Lower Carboniferous turbidites overlying the nearly undeformed Upper Devonian metabasalts in the Teuschnitz Syncline (Fig. 4c). This upper thrust system has been refolded by the later activation of several crustal scale duplexes underneath, resulting in the large-scale anti- and synclines seen today (Schwarzburg Anticline, Teuschnitz Syncline). This interpretation is in accordance with the observed structural evolution in the Lower Carboniferous turbidites of the Teuschnitz Syncline: (1) an early stage of detachment buckle folding on the Vesser–Lehesten Thrust with decreasing displacement towards the SE; (2) a stage with pervasive thickening of the wedge and shearing of previously formed buckle folds; (3) a late stage of formation of a new, deeper basal detachment and accretion with reorientation of fabric elements during anti- and syncline formation. The duplex thrusts in the subsurface do not show major displacement compared with their size, indicating that the external wedge seems to have shortened and thickened mostly by non-localized shear. The formation of a basal thrust and activity on localized thrusts cannot be established before this latest increment. The basal detachment of the entire system rises from nearly 25 km depth in the NW towards the SE, reaching the surface in the Göttengrün Thrust at the southern flank of the Berga Anticline. Backthrusts linked to the basal detachment fault system cut across the entire internal belt and its hanging wall, which are therefore identified as an earlier feature.

Geochronological data

The northern section of NW Saxo-Thuringia has been intruded by a late- to post-kinematic granitic suite, the so-called Thuringian Granite. Its older, northern parts in the RCC, which intruded amphibolite facies rocks (Brotterode Formation, Zeh 1996), show weak cataclasis and some greenschist facies to very low grade deformation features. The southernmost branch intruded the Lower Palaeozoic Vesser rocks and lacks deformation features. Three magmatic rocks, two from the vicinity of the Vesser Unit and one from the NW flank of the Schwarzburg Anticline (Fig. 2), belonging to this post-collisional stage, have been dated by the U–Pb zircon method (Table 1). These ages are related to different phases of post-kinematic Variscan magmatism that constrain the final stages of the Variscan structural and metamorphic event.

A calc-alkaline amphibole–biotite granite (Teuschelsbrunnen, west of Vesser village), shows a weak alignment of mafic to intermediate enclaves that may be related to magma emplacement. This granite represents a part of the 'Thüringer Hauptgranit', which includes a variety of granitoids and diorites.

There are several different populations of zircon, including long-prismatic crystals as well as short-prismatic and rounded ones. Only needle-shaped, colourless to slightly brownish tinted crystals were selected for analyses. Three zircon samples are concordant, whereas a fourth sample is slightly discordant. The weighted mean $^{206}Pb/^{238}U$ age of the concordant zircon samples is 313.0 ± 1.0 Ma (Fig. 5a). This age is interpreted as the emplacement age of the granite. Using this age to constrain the discordia through the discordant sample yields an upper intercept age at 823 ± 148 Ma, which, however, has no strict age meaning. Instead, it represents a minimum age for the oldest source that contributed inherited zircon and a maximum age for the youngest source.

A biotite- and hornblende-bearing granite (Schmiedefeld locality) was emplaced into Lower Permian sediments and the volcanisedimentary sequences of the Vesser Units. It is dominated by deep red porphyritic orthoclase phenocrysts. Zircon from this granite forms stubby tetragonal prisms with simple pyramid faces. Zircon is commonly highly fractured and has various inclusions of opaque minerals, melt and apatite. Although care was taken to analyse only slightly fractured and inclusion-poor crystals, all zircon samples are characterized by a low $^{206}Pb/^{204}Pb$ ratio. This unradiogenic lead signature is mainly caused by a large contribution of common lead (2.1–4.2 ppm). The four zircon samples scatter about a discordia with intercepts at 282 ± 11 and 1697 ± 983 Ma (2σ; mean square weighted deviation

Table 1. *U–Pb analytical data of zircon from the post-Variscan intrusions in the Vesser area in the Saxo-Thuringian Zone of the Variscan Orogen, Germany*

		Concentration (ppm)		$\frac{^{206}Pb}{^{204}Pb}$	Radiogenic Pb (at%)‡			Atomic ratios‡			Apparent age (Ma)§		
Sample*	Weight (mg)	U	Pb_{tot}	Measured ratios†	^{206}Pb	^{207}Pb	^{208}Pb	$\frac{^{206}Pb}{^{238}U}$	$\frac{^{207}Pb}{^{235}U}$	$\frac{^{207}Pb}{^{206}Pb}$	$\frac{^{206}Pb}{^{238}U}$	$\frac{^{207}Pb}{^{235}U}$	$\frac{^{207}Pb}{^{206}Pb}$
VE9603													
1	0.308	1003	57.4	661.3	84.68	4.52	10.80	0.05152	0.37916	0.05338	324	326	345
2	0.358	723	38.0	1714	84.19	4.43	11.38	0.04973	0.36086	0.05263	313	313	313
3	1.493	959	49.6	1749	85.60	4.50	9.90	0.04968	0.36048	0.05263	313	313	313
4	1.533	899	46.9	1677	85.37	4.49	10.14	0.04983	0.36156	0.05262	314	313	312
VE9061													
5	0.682	191	12.1	255.6	57.58	3.92	20.51	0.04522	0.32333	0.05186	285	284	279
6	0.228	183	14.0	140.3	75.90	4.16	19.94	0.04673	0.36311	0.05481	294	307	404
7	1.055	172	11.4	218.3	74.82	3.94	21.24	0.04505	0.32712	0.05267	284	287	315
8	1.154	165	11.6	198.8	74.42	4.07	21.51	0.04691	0.35402	0.05474	296	308	401
SB9602													
9	0.287	3582	174	3370	81.75	4.31	13.94	0.04549	0.33077	0.05274	287	290	317
10	0.241	3655	178	3626	81.40	4.28	14.32	0.04529	0.32868	0.05263	286	289	313
11	0.303	3233	160	1672	81.67	4.27	14.06	0.04529	0.32669	0.05232	286	287	299
12	0.711	3201	155	2828	81.57	4.27	14.16	0.32528	0.32528	0.05234	284	286	300
13	1.128	3445	167	3700	81.53	4.27	14.20	0.04528	0.32683	0.05235	286	287	301

*Zircon concentrates were obtained using standard mineral separation techniques and concentrated by hand under the binocular microscope. All samples were dissolved with 52% HF in Parr autoclaves at 220°C for 4 days dried and converted into chloride form using 6 N HCl at 220°C in the autoclave. Ion-exchange chromatography after taking aliquots by weight and tracer addition as described by Krogh (1973). Measured on single Re filaments using a silica-gel emitter and H_3PO_4 (Gerstenberger & Haase 1997) at 1200–1260°C on a VG 54-30 (Faraday collectors and Daly probe) and Finnigan MAT262 (Faraday collectors and ion counting) multicollector mass spectrometer, respectively.

†Lead isotope ratios corrected for fractionation with 0.1% per a.m.u.

‡Lead corrected for fractionation, blank and initial lead $^{206}Pb/^{204}Pb = 18.3$, $^{207}Pb/^{204}Pb = 15.63$, $^{208}Pb/^{204}Pb = 38.4$).

During the measurement period total blanks were less than 30 pg for lead and less than 1 pg for uranium for samples analysed with a ^{208}Pb–^{235}U mixed tracer.

§Apparent ages were calculated using the constants recommended by IUGS (Steiger & Jäger 1977).

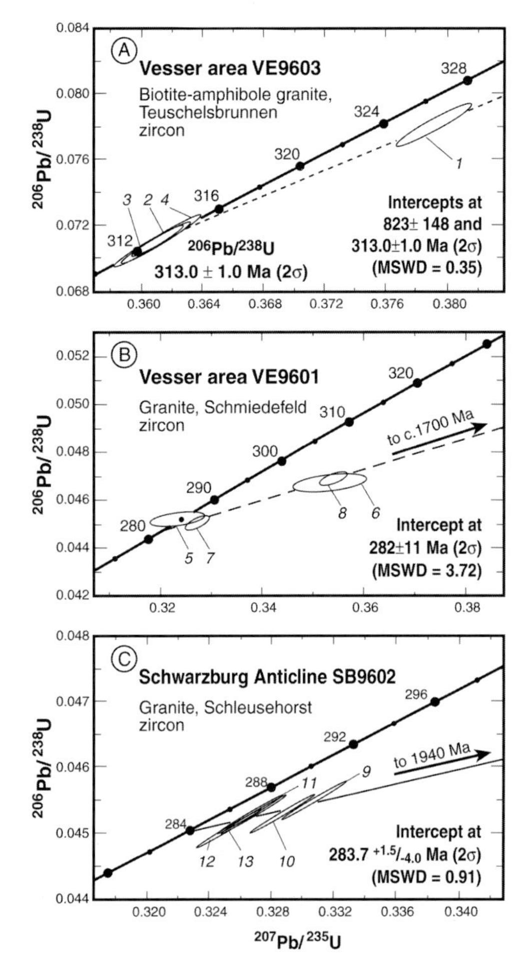

Fig. 5. Concordia diagram for several post-kinematic Variscan granites. (**a**) Alkaline, biotite–amphibole granite intruded into the Vesser Unit (Teuschelsbrunnen locality); (**b**) porphyritic hornblende–biotite-bearing granite intruded into the Vesser Unit (Schmiedefeld locality); (**c**) granite from the NW flank of the Schwarzburg Anticline (Schleusehorst locality).

(MSWD) = 3.72; Fig. 5b). The large uncertainty of the intercepts is due to the small spread of the data and the excessive scatter. Sample 5 (Table 1) included only grains free of inclusions and fractures, and therefore might represent the least disturbed sample. It appears concordant and yields a $^{206}Pb/^{238}U$ age of 285 ± 3 Ma. It is unclear whether the scatter of the data is related to a combination of various inherited components or a later thermal overprint, as recorded in Mesozoic K–Ar illite ages (Franzke *et al.* 1996).

A medium-grained leucogranite with biotite and muscovite (Lichter Gabelskopf, Schleusehorst) intruded into Lower Palaeozoic sequences on the NW flank of the Schwarzburg Anticline. This granite is characterized by abundant late veinlets and voids filled with purple fluorite. Zircon was separated from a sample free of these secondary features. The analysed zircon is distinguished by its high contents of uranium (3200–3650 ppm; Table 1). All five zircon samples are slightly discordant. A discordia fitted through the five zircon samples yields a lower intercept at $283.7 + 1.5/-4.0$ Ma (2σ; MSWD = 0.91; Fig. 5c). This intercept is interpreted as the time of granite emplacement. The inherited zircon is dominantly of Palaeoproterozoic provenance signature.

The sampled granitoids are related to late members of the so-called 'Thüringer Hauptgranit' (Fig. 2), which is composed of numerous intrusions. We interpret the older age of about 313 Ma as one of the earliest intrusion pulses, which followed the end of deformation in this area, as well as at the southern limit of the RCC (311 Ma; Zeh *et al.* 2000). Earlier granitic magmatism farther north in the RCC area (337 Ma; Zeh *et al.* 2000) shows minor low-temperature deformation following intrusion into partly cooled and exhumed country rock. Their intrusion ages can therefore be interpreted as a minimum age for peak deformation and metamorphism in the MGCZ. Ar–Ar data from white mica in the gneissic country rock show cooling through its closure temperature at this time (Zeh *et al.* 2000). Accordingly, activation of the Vesser Thrust and deformation of its hanging-wall units during concomitant peak metamorphism, i.e. peak activity of the internal fold belt, occurred before 313 Ma. Most probably it can be linked to the stage of underthrusting under the MGCZ during uplift and cooling of the latter, i.e. at around 337 Ma. Deformation of the internal NW Saxo-Thuringian thrust belt would then be bracketed between 337 and 313 Ma with a bias towards the earlier age.

Metamorphic evolution is further constrained by K–Ar ages of metamorphic white mica from metasediments on the size fractions <2 μm, 2–1 μm, <1 μm and <0.2 μm (Kemnitz *et al.* 1999). From microfabric observation, the mineral fractions <2 and 2–1 μm contribute to the main foliation. Considering the greenschist to very low grade conditions of the Saxo-Thuringian belt, they thus represent minimum values for deformation. For the Vesser units, a K–Ar cooling age of 301 ± 7 Ma has been obtained, which, according to the undeformed granite of 313 Ma and the observed late mica generation from microfabric studies (see below), should record post-kinematic mica growth during cooling. An age of

309 ± 8 Ma was found for the Schwarzburg Anticline; this is identical to the ages in the Teuschnitz Syncline and Berga Anticline determined by Ahrendt *et al.* (1986). Because of lower peak metamorphism in these areas these values are interpreted to yield minimum deformation ages. In all areas, the fractions with Permian ages are thought to be associated with a minor overprint, which occurred during widespread post-orogenic magmatism (280–290 Ma). Ubiquitous Triassic and Jurassic ages, especially of the small size fractions (<0.2 μm), reflect secondary alterations, which seem to be related to localized thermal overprint (Ahrendt *et al.* 1986; Wemmer 1991; Franzke *et al.* 1996).

On a more regional scale (Fig. 2b), K–Ar ages of metamorphic and detrital white mica (Ahrendt *et al.* 1983, 1986; Schäfer *et al.* 1997) suggest that convergence and collision of the Saxo-Thuringian plate with the Teplá–Barrandian active margin occurred between 370 and 320 Ma. NE and SW of the Münchberg Nappes, K–Ar ages of the <2 μm fraction around 325 Ma were interpreted as maximum ages for the regional metamorphic and deformation event (Ahrendt *et al.* 1986). The earliest undeformed granites intruded this part of the thrust belt at around 325–320 Ma (see compilation by Siebel (1999)). In the frontal, northwestern part of the Saxo-Thuringian Zone, on the other hand, metamorphism and deformation occurred later: both the K–Ar data presented here and crosscutting relationships with the NW-vergent deformation of the Münchberg Nappes indicate that the external part of the fold belt above the Göttengrün Thrust system was activated after 320 Ma and was terminated at around 310 Ma. There is no evidence for later shortening. The entire SE-facing thrust belt thus shows two evolutionary stages, each with a stepwise SE propagation of deformation and accretion during late Viséan to Westphalian time (between 337 and 313 Ma and at *c.* 310 Ma) above two related detachment systems (Vesser–Lehesten Thrusts and Göttengrün Thrust).

Data on metamorphism

Metamorphic studies exist for parts of the low to very low grade external NW Saxo-Thuringian fold-and-thrust belt. In the Berga Anticline and Teuschnitz Syncline, illite crystallinity (Franke 1984; Kunert 1999) indicates average metamorphic temperatures of the anchizone–epizone transition during deformation (between 280 and 320°C; see also Brand (1980)). For the core of the Schwarzburg Anticline, greenschist facies metamorphism with minimum temperatures of about 350°C was deduced from the mineral assemblage hornblende + chlorite + zoisite in metabasites (Hirschmann 1959). This is in agreement with the appearance of metamorphic biotite. For the Vesser Unit, greenschist facies conditions are indicated by garnet-bearing metapelites. Deformed basaltic and gabbroic rocks of the Vesser Unit record metamorphic changes from the magmatic magnesio-hornblende to actinolitic amphibole and the formation of albite, although the majority of coexisting feldspars indicate a preserved magmatic composition and thus do not correspond to the equilibrium composition of greenschist facies (see Schäfer (1997) for details).

New P–T data from mineral associations and microfabrics

In the Vesser Unit, metapelites are dominated by the mineral assemblage white mica + chlorite + quartz + albite ± garnet ± biotite ± ilmenite ± epidote ± apatite ± hematite. Analytical methods and microprobe data for garnet and white mica can be obtained from the Society Library or the British Library Document Supply Centre, Boston Spa, Wetherby, West Yorkshire LS23 7BQ, UK as Supplementary Publication No. SUP 18752 (5 pages). Metamorphic white mica is present in two phengite generations that mainly differ by their growth orientation with respect to the fabric generations: the first (I) forming an older, now folded cleavage S1, the second (II) paralleling the penetrative main foliation S2 and the locally developed mylonitic foliation. Their average Si contents increase from generation I to II (Si_I 6.23–6.27, Si_{II} 6.24–6.52 p.f.u.), indicating an increase in pressure and/or decrease in temperature. A retrograde third generation (III) of small, undeformed, unoriented white mica grains overgrows both cleavage generations.

Chlorite, which also occurs in three generations, shows the same temperature trend (decrease) from generation I to II and further to III, as indicated by increasing Si contents and decreasing *X*Fe ratios. Likewise, compositional changes in garnet point to a decrease in temperature towards the last deformational increment. Garnet composition is spessartine–almandine. A chemically homogeneous garnet I with a Mn:Ca ratio $\geqslant 1$ grew syn-kinematically to the formation of S1, which appears (only in pure metapelitic horizons) as a crenulation cleavage that is differentiated into cleavage lamellae and microlithons. S1 is deflected around garnet grains and strain shadows have been developed. White mica I is preserved in the microlithons surrounding this garnet I. In most observed cases S2 occurs as the penetrative main foliation in metapelites, where a garnet rim phase II around garnet I has been formed with a Mn:Ca ratio decreasing to <1. The garnet rim is syn-kinematic with respect to the slightly deflected S2 foliation that is defined by white mica II. Internal, mildly rotated quartz inclusion trails grade into the external S2 foliation at the garnet rims.

Garnets are partly pseudomorphosed by chlorite II and III. Chlorite II traces S2 but also overgrows the garnets. It is therefore concluded that chlorite II is not equilibrated with the garnet rim phase but represents more retrograde conditions. Biotite forms an important component only in some garnet-free metapelitic rocks. It usually is restricted to very fine-grained flakes smaller than 10 μm in length on the penetrative cleavage S2.

The garnet–phengite thermometer (Hynes & Forest 1988) has been applied to the garnet bearing metapelites of the Vesser Unit. The thermometer has been derived for the special case of Mn-rich garnets in low- to medium-pressure and low-temperature rocks. With a high MnO:FeO ratio in bulk rock geochemistry, garnet formation can be forced at the cost of biotite and chlorite at temperatures clearly below the stability limit of almandine (Hynes & Forest 1988). In the Vesser Units, the mineral composition (i.e. contents of celadonite in white mica, presence of garnet, presence of ilmenite instead of magnetite or hematite) is similar to that of the rocks for which the thermometer has been calibrated.

The thermometer proved to be highly pressure independent as calculated temperature changes do not exceed 25°C for a pressure estimate in the interval between 1 and 10 kbar. Metamorphic temperatures were calculated applying the ideal-mixing and the non-ideal mixing calculation models for garnet (Hynes & Forest 1988). The ideal-mixing model yields temperatures of 408–448°C for garnet I and muscovite I, and 380–409°C for garnet II and muscovite II (Fig. 6). The non-ideal mixing model yields temperatures that are about 45–60°C higher. As the growth of biotite is largely suppressed in the garnet-bearing association, the ideal-mixing model seems to yield the more realistic temperature estimate (Table 2).

In recrystallized sections of shear zones, where re-equilibration of all mineral phases has occurred, syn-S2 mylonitic fabrics show transitions from σ- to δ-clasts and a non-penetrative development of shear bands. These fabric features and recrystallization of quartz as well as incipient recrystallization of feldspar

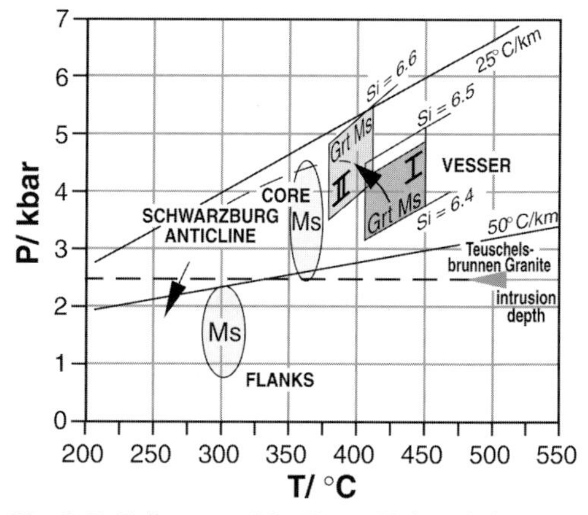

Fig. 6. *P–T* diagram of the Vesser Unit and the Schwarzburg Anticline. Lower and upper temperature stability limits have been calculated from garnet–phengite thermometry in metapelites (Hynes & Forest 1988), and the pressure limits are based on the Si isopleths for phengite after Massonne (1991). The granite intrusion depth is based on the Al contents in hornblende, after Anderson & Smith (1995). Grt, garnet; Ms, muscovite.

porphyroclasts in metavolcanic rocks indicate formation temperatures in the range of 350–450°C. In contrast to the metapelites, phengite-bearing metapsammitic rocks contain some detrital potassium feldspar, minor chlorite and biotite, i.e. the critical assemblage required for phengite barometry (Massonne & Schreyer 1987; Massonne 1991). Detrital muscovite grains can be clearly distinguished from metamorphic white mica by grain size, shape and chemical composition ($Si_{Det} \leqslant 6.2$ p.f.u.). The average Si contents in both phengite phases is significantly higher; Si_{I} ranges from 6.4 to 6.48 p.f.u. and Si_{II} from 6.46 to 6.6 p.f.u. Phengite I during S1 formation grew at temperatures of up to 450°C at pressures of about 3.2–4.8 kbar; for the second fabric generation pressure estimates at temperatures of up to 410°C are at 3.5–5.5 kbar (Fig. 6).

The post-kinematic granite that intruded the Vesser Unit at about 313 Ma (Fig. 5a) contains the critical assemblage plagioclase + quartz + orthoclase + biotite + chlorite + hornblende + muscovite + epidote + ilmenite. The average Al_{tot} of 1.3 p.f.u. in hornblende allows a semi-quantitative estimate of intrusion depth (Anderson & Smith 1995) of $\leqslant$2.5 kbar at 750°C. This is supported by evidence from the early stages of the 'Thüringer Hauptgranit' in the RCC, which were found to intrude at $\leqslant$2 kbar in a largely cooled basement (Zeh *et al.* 2000).

Table 2. *Formation temperatures of homogeneous garnets (grt I) and growth-zoned garnets (grt II) from metapelitic rocks of the Vesser Unit, NW part of Saxo-Thuringia*

	Formation temperature (°C)									
Sample V4	grt I									
Analyses:	5	7	8	9	10					
wm I1.1	399	408	468	415						
	444	453	511	461	460					
*VE9518**	core (grt I)					rim (grt II)				
Analyses:	C1.8	C1.13	C1.16	C1.18	C1.19	C1.3	C1.26	C1.27	C1.28	C2.8
wm II C3	400	402	418	427	403					
	449	456	470	481	459					
wm II C9						380	394	389	399	394
						424	434	431	441	435
VE9516	core (grt I)					rim (grt II)				
Analyses:	A1.2					A1.1	A1.3	A1.4	A1.5	
wm II A1.2	410									
	460									
wm II A1.1						409	404	407	409	
						446	443	447	453	
VE9522†	grt I									
Analyses:	B1.1	B1.2	B1.3	B1/4	B1.5					
wm I B2.1	444	439	448	445	444					
	499	493	505	502	503					
Analyses:	B2.1	B2.2	B2.3	B2./4	B2.5	B2.6				
wm I B2.1	454	439	444	444	438	448				
	510	491	501	501	491	501				

Calculation was performed with the garnet–phengite thermometer, using (first row for each entry) the ideal-mixing model and (second row) the model after Hoinkes 1986 (see Hynes & Forrest (1988)).

*In samples VE9516 and VE9518, which are mylonitic metasediments bearing growth-zoned garnets, only white mica paralleling S2 (wm II) could be recognized. Calculated temperature of the garnet cores, therefore, possibly is too low, although it is somewhat higher than for the garnet rims.

†This metasiltstone sample, which contains both phengite I and II besides diagnostic detrital white mica, gave consistent temperatures for two neighbouring homogeneous garnet grains (grt I).

For the Autunian granite (around 282 Ma) in the Vesser Unit (Fig. 5b) a subvolcanic intrusion depth can be derived from its porphyric fabric and its $Al_{tot} < 1$ p.f.u. in hornblende.

In the Schwarzburg Anticline and its flanks, new metamorphic pressure data were obtained for the Glasbach granite (538 Ma, Gehmlich *et al.* 1997), a sill-like intrusion in Neoproterozoic sediments. During Variscan Orogeny it was deformed together with its host rock (Bankwitz & Bankwitz 1995). Metamorphic phengite forming the foliation within the sheared granite is therefore considered syn-kinematic to Variscan deformation. The limiting assemblage of the granite is given by phengite + quartz + K-feldspar + biotite, whereas that of its host rock (Neoproterozoic metagreywacke) is quartz + K-feldspar + albite + phengite + chlorite ± biotite ± calcite + titanite. Quartz recrystallization is widely distinctive. These observations constrain peak temperatures to *c.* 350–375°C. With an average Si content of 6.49 p.f.u. in the phengites of the granite, the estimated maximum pressures are in the range of 3.2–3.8 kbar (Schäfer 1997). Similar pressure estimates are recorded in the host rock, where metamorphic phengites (Si ranging from 6.39 to 6.6 p.f.u.) yield maximum pressures of 2.5–4.5 kbar (Fig. 6).

Farther to the SE, peak metamorphism reached very low grade conditions only. Chlorite porphyroblasts that grew epitactically on detrital muscovite are the only indicator of metamorphism. Despite the limitations of the method, chlorite thermometry (Cathelineau 1988) has been applied to the Ordovician metasediments of the flanks of the anticline. The resulting temperatures in the range of 275–315°C, although only interpretable to give approximate estimates, however, can be taken as an indicator of an even temperature distribution within the phyllitic Ordovician slates (Schäfer 1997).

Interpretation

The peak metamorphic conditions of the internal fold belt exceeded those experienced by the external units to the south, although the character of deformation and metamorphism in both areas is similar. A temperature maximum of 450°C at an estimated maximum pressure of 3.2–4.5 kbar was followed by a second syn-kinematic stage with temperatures of up to 410°C and 3.5–5.5 kbar, which was related to the formation of ductile shear zones. This syn-kinematic *P–T* sequence, which obviously differs from paths elsewhere in the Saxo-Thuringian by its temperature decrease at increasing or constant pressure (Fig. 6), would imply deformation during burial and cooling in response to overriding cold crustal sequences. This finding probably links the metamorphic evolution of the internal belt to overriding by the RCC along the Crux Thrust to the north.

Higher metamorphic conditions are known from the Ruhla Crystalline Complex (RCC), the transitional unit between the magmatic arc (MGCZ) to the north and the Saxo-Thuringian basin dealt with here. The units of the RCC have all undergone amphibolite facies metamorphism, but show a very complex uplift history (Zeh 1996). This evolution was largely concluded by 337 Ma, upon the intrusion of a late-kinematic granite, which was found to intrude at $\leqslant 2$ kbar in a largely cooled basement (Zeh *et al.* 2000). Minor uplift and block faulting continued through Late Carboniferous time. The granite ranks among the early stages of the so-called 'Thüringer Hauptgranit' suite, which also forms intrusive contacts with the Lower Palaeozoic sequence of the Vesser Unit. Here, the granite of calc-alkaline character is undeformed and marks one of the earliest post-kinematic intrusions at about 313 Ma to a depth of possibly more than 6 km. The final granite stage in the Vesser Unit (282 Ma) suggests exhumation to subvolcanic conditions. Thus, these post-kinematic magmatic pulses record exhumation and cooling of the internal belt during continuing shortening and metamorphism in the external belt through Late Carboniferous time. Towards the southeast, the *P–T* conditions drop below 400°C at 4–4.5 kbar recorded in the core of the Schwarzburg Anticline. In the SE part of the fold-and-thrust system (Teuschnitz Syncline, Berga Anticline), metamorphic temperatures are below 300°C at pressures of 1–1.5 kbar, as implied from the absence of quartz recrystallization and estimated from the metamorphic field gradient (Fig. 6). The SE-directed deformation at the northern Saxo-Thuringian margin can therefore be related to a decrease in intensity of deformation and metamorphism, translating to a shallower burial and exhumation towards the SE.

Discussion

The evidence presented shows that the NW part of the Saxo-Thuringian Zone underwent a different evolution from its southeastern counterpart, which was related to collision with the Teplá–Barrandian plate fragment to the south. The fold-and-thrust belt resulting from the latter tapers out slightly north of the Göttengrün Thrust, and is crosscut from the north by a later deformation with SE-directed tectonic transport. On the basis of mainly timing and geometric constraints, we suggest that this later stage was linked to the absorption of deformation in the Saxo-Thuringian upper plate upon collision with the Rheno-Hercynian lower plate to the north. Structural data, section balancing, and isotopic dating from the Rheno-Hercynian belt (Ahrendt *et al.* 1983; Oncken *et al.* 1999), the former passive margin of Avalonia, document a southeastward-directed subduction of the Rheno-Hercynian shelf and the oceanic basin to the south. Build-up of an oceanic accretionary wedge (remnants in the Giessen–Harz Nappes, Fig. 1) was dated between 375 and 340 Ma in the Harz area (Ahrendt *et al.* 1996). South of the Rhenish Massif (Fig. 1), shelf subduction started around 335 Ma (Figs 2c and 7). Basal accretion of the passive margin sediments to the base of the advancing upper plate during early collision was related to major uplift, metamorphism and cooling of the MGCZ site at around 337 Ma in the east (Zeh *et al.* 2000), the area under consideration (Figs 2c and 7) and slightly later in the west (325 Ma; Nasir *et al.* 1991; Anthes 1998, see also Oncken 1997). Thrust belt growth on the lower plate proceeded by continuous foreland accretion until 300 Ma at a continuous rate of 1.5 cm a^{-1} (Plesch & Oncken 1999) above a SE-dipping basal detachment—the former subduction zone (see Oncken *et al.* this volume). A stage of major internal reactivation and thickening in large parts of the wedge occurred at around 308–312 Ma. This stage was related to a subcritical state of the Rheno-Hercynian wedge with major wedge readjustment as a result of changes in the foreland mechanics (Plesch & Oncken 1999). This entire evolution is directly linked to upper-plate deformation as shown in the results presented for the NW Saxo-Thuringian fold-and-thrust belt (Fig. 7). In our model, the basal detachment under the latter branches antithetically from the subduction zone and ascends to the SE, forming the detachment

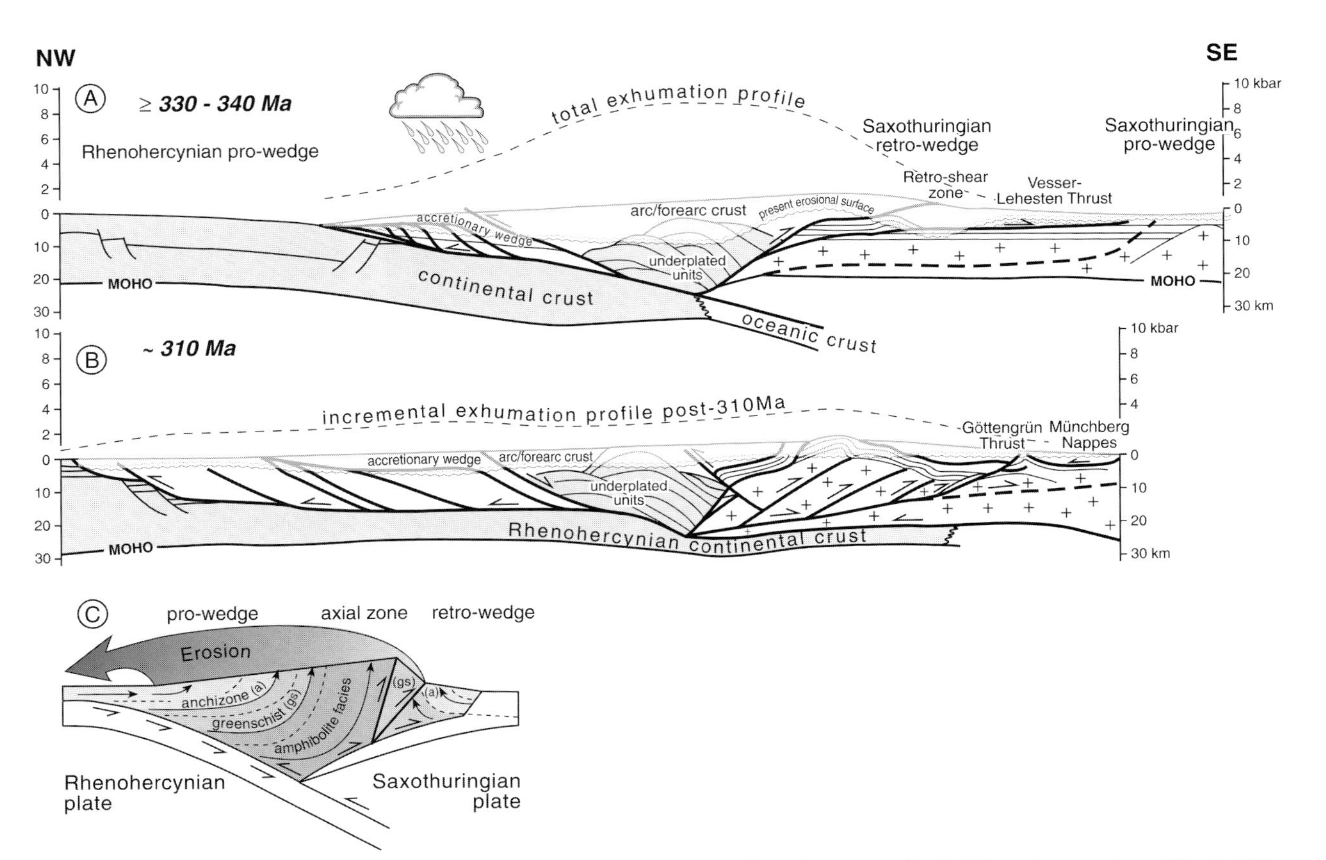

Fig. 7. Schematic section across the Rheno-Hercynian–Saxo-Thuringian doubly vergent orogenic wedge system. Dashed lines, barometric profiles; bold lines, faults; wavy lines, erosion depth; grey lines, present erosional surface (double wavy lines) and inferred former topographic surface (single line). Crosses indicate Cadomian basement of Armorica; Rheno-Hercynian pre-Devonian basement rocks are shaded. (**a**) Early stage of collision at *c.* 340–330 Ma. It should be noted that increased erosion on the pro-wedge side results in a steeper gradient than predicted from numerical modelling (e.g. Willett *et al.* 1993). (**b**) Late stage of collision at *c.* 310 Ma. The increase of wedge lengths and related decrease in taper angles and exhumation of this late increment should be noted; this indicates a change of the erosional pattern and detachment strength. (**c**) Conceptual model of doubly vergent orogens showing material flow paths resulting from asymmetric erosional patterns inferred for the Rheno-Hercynian–Saxo-Thuringian (modified from Willett *et al.* (1993)).

of an independent, SE-vergent fold-and-thrust belt. The Rheno-Hercynian, the MGCZ and the NW part of the Saxo-Thuringian thus form a kinematically coupled system of a pro-wedge, an axial zone and a retro-wedge, operating during the Carboniferous collision of Avalonia with the NW part of Armorica. On the basis of the data presented, evolution and style of the retro-wedge are mainly governed by three aspects: the details of the collision history, the mechanical properties controlling deformation partitioning and the pattern of erosional mass transfer.

The retro-wedge developed at the onset of collision with formation of a narrow internal thrust stack, here described as the retro-shear belt, and a broad later fold belt. On the basis of crosscutting faults and the uplift history of the MGCZ in the hanging wall of the Crux Thrust, supported also by the anticlockwise shape of the P–T paths of the footwall, the earlier retro-shear belt accumulated most deformation between *c.* 340 and 335 Ma. This retro-shear belt includes the southern parts of the RCC and the internal belt of the retro-wedge (Crux and Vesser Thrusts with intervening mylonites; see also conspicuous NW-dipping reflectors in seismic line DEKORP 3B/MVE West, Fig. 3). It nucleated at the downdip end of the site of basal underplating in the subduction zone and accommodated underplating-driven uplift of the leading edge of the upper plate. In numerical and analogue modelling, as well as in well-studied young orogenic belts such as the Alps, this location consistently is the nucleation point of retro-shears (e.g. Beaumont *et al.* 1996; Gutscher *et al.* 1996). This system was inactive during the final stages and replaced by the shallower dipping detachment, surfacing in the Göttengrün Thrust. Activation of this external part was coeval with the stage of major readjustment of the Rheno-Hercynian thrust belt at about 310 Ma (see above). Retro-wedge evolution is thus intimately linked to specific stages during collision: initial continental subduction with underplating below and uplift of parts of the upper plate, and to a mechanically controlled change during foreland accretion in the lower plate.

The Rheno-Hercynian pro-wedge forms a classical foreland imbricate thrust belt with predominantly localized deformation along thrusts (Oncken *et al.* 1999) and total shortening of 180–200 km. In contrast, the Saxo-Thuringian retro-wedge shows a different mode of deformation partitioning. Most localized shortening (>50 km) occurred at the internal retro-shear belt, and some 50 km of diffuse shortening were accumulated subsequently in the much wider external belt. The earlier localized thrusting in the internal part seems to be strongly restrained by the occurrence of thick mafic layers in the sedimentary sequence as well as by the proximity to the zone that accommodated uplift of the MGCZ during underplating. The younger external belt mostly involved the sediments below the Devonian basalts, which apparently lack a major mechanical anisotropy. Also, slow deformation at high average temperatures may have enhanced distributed rather than localized deformation.

Mechanical properties can also be estimated qualitatively from analysis of the wedge taper (angle between topographic slope and detachment). A first-order estimate of the original thrust belt geometry and topographic slope was derived by adding the respective peak pressure data to the basal detachment in the balanced cross-sections, as these data will generally record the depth of accretion into the orogenic wedge. The resulting taper of the short retro-wedge developed from 16° in the early collisional stage to 30° in the late stage (Fig. 7) and shows the high angles typical of retro-wedges (Willett *et al.* 1993). A similar reduction in taper (from 10° to 6°) for the same period was found for the Rheno-Hercynian pro-wedge (Plesch & Oncken 1999). From mechanical analysis of critically tapered orogenic wedges this is an indication of relative weakening of the basal detachments with respect to their hanging walls through time (e.g. Huiqi *et al.* 1992; Willett 1992). Probably, this feature is related to radiogenic heating during crustal thickening, which is also suggested by the observed increase of crustal melts in the late stage.

A related feature of the Saxo-Thuringian retro-wedge is its high transient thermal gradient, despite small and probably slow uplift. This suggests a thermally disturbed retro-wedge system, whose internal part went through cooling during deformation and burial. Syn-kinematic gradients in the internal Rheno-Hercynian pro-wedge increased from 15°C km^{-1} to <30°C km^{-1} during deformation, which is typical for flow paths in accretionary wedges. In the internal Saxo-Thuringian retro-wedge, gradients decrease syn-kinematically from 45°C km^{-1} to 25°C km^{-1}. In the external belt, they were around 50°C km^{-1}. This suggests a 'hot' basin in the back-arc area of the precollisional upper-plate arc system, which was underthrust as well as overthrust by cooler material from the north, i.e. by Rheno-Hercynian basement rocks with a divergent flow path during collisional indentation (see Fig. 7a and b). Consequently, a thermally weakened upper plate was affected during collision and retro-wedge formation in addition to its weakening through time. This aspect is expected to have had a profound influence on the deformation style and the large amount of upper-plate shortening.

Further conclusions on retro-wedge controls can be drawn from the distribution of barometric data, which show that erosional removal during this evolution was not symmetric. A substantially larger volume of material has been removed from the pro-wedge side in comparison with the retro-wedge area. This is also reflected in the large syn-orogenic flysch and molasse volume deposited on the lower plate north of the axial zone with so far no eroded material identified in the Saxo-Thuringian basin (Schäfer *et al.* 1997; Ricken *et al.* this volume). In consequence, runoff must have been mainly to the north, and the catchment area of the erosional system must have crossed the axial zone and parts of the retro-wedge, thus preventing major material transport to the SE. Also, the geometry of the exhumation profile across the system, with a broad symmetric peak above the axial zone and the internal pro-wedge, indicates that material flow and, therefore, strain as well as exhumation rates were concentrated in this part of the system (see Fig. 7). Accordingly, orogenic material path patterns in the Rheno-Hercynian–Saxo-Thuringian system with strong advection in the axial zone above the retro-shear belt are closely linked to, if not driven by, the asymmetric removal of this material to the north. These findings corroborate theoretical predictions (e.g. numerical modelling by Willett *et al.* (1993), Beaumont *et al.* (1996) and Jamieson *et al.* (1998)) that were made for the case of asymmetric erosional mass removal focused on the pro-wedge side of a bivergent kinematically coupled orogenic wedge.

Conclusions

The Saxo-Thuringian fold belt has been affected by two widely independent processes originating from two different active subduction zones. The regionally important domain of SE-vergent structures in the NW differs in its orogenic development from the SE part. It acted as a thrust-linked retro-wedge behind the active margin in an upper-plate position with respect to the Rheno-Hercynian lower plate as the pro-wedge. Collisional deformation led to an imbricate style in the Rheno-Hercynian lower plate (nearly 200 km shortening), whereas the Saxo-Thuringian upper plate absorbed a minimum of 100 km of contraction, about half of it localized in a narrow retro-shear belt south of the MGCZ. Three factors of particular importance for Saxo-Thuringian retro-wedge evolution can be identified from the rock record, as follows.

The geochronological data of the metamorphic and magmatic stages allow the identification of a stepwise progradation of the retro-wedge deformation front, and confirm its link to the main stages of collision. The Saxo-Thuringian retro-wedge initiated with a NW-dipping retro-shear belt branching from the SE-dipping subduction zone underneath the MGCZ at the onset of continental collision (about 340 Ma) during underplating and uplift of parts of the active margin. After a period of quiescence, the retro-wedge propagated into the Saxo-Thuringian basin (about 310 Ma), probably driven by major internal readjustment of the Rheno-Hercynian pro-wedge at this time.

These two stages are associated with thermal weakening and, in deeper parts, with crustal melting at the collision boundary zone and retro-wedge area as observed from an increase of late- to post-orogenic granitic magmatism. A related decrease in taper of both wedges indicates a relative weakening of the basal detachments through time. Last, not least, the apparently 'hot state' of the upper plate versus the lower plate suggests that deformation style and magnitude were influenced by reduced upper-plate strength.

Orogenic particle flow geometry was deflected north of the retro-shear belt encompassing the axial zone and the internal pro-wedge. As virtually all erosional debris of the coupled wedge system is found in syn-orogenic basins in the north, asymmetric erosional removal towards the pro-wedge side in the north must have controlled the mass transfer mode in the Rheno-Hercynian–Saxo-Thuringian wedge system.

Research was funded by GeoForschungsZentrum Potsdam and by Deutsche Forschungsgemeinschaft (DFG) as part of the priority programme 'Orogenic Processes'. We wish to thank our co-operation partners W. Franke, H. Ahrendt† and K. Wemmer for refreshing discussions. We further thank D. Tanner and the two reviewers, J. Behrmann and an anonymous reviewer, for constructive comments that helped to improve the presentation of this paper. Special thanks go to L. Franz and J. Rötzler, who gave valuable advice and support in all issues of metamorphism and thermobarometry. Thanks are also due to W. Seifert and C. Brauer, for guiding through microprobe and RDA analyses, respectively. Thin sections were prepared by B. Schikowski, and the mineral separation for U–Pb zircon geochronology was carried out by M. Boche and J. Müller. M. Dziggel helped to draw the figures.

References

AHRENDT, H., CLAUER, N., FRANKE, W., HANSEN, B. T. & TEUFEL, S. 1986. Geochronologie. Kontinentales Tiefbohrprogramm der Bundesrepublik Deutschland (KTB)—Ergebnisse der Vorerkundungsarbeiten Lokation Oberpfalz. *2. KTB-Kolloquium 19–21.9.1986, Seeheim/Odenwald*, 32–38.

——, ——, HUNZIKER, J. C. & WEBER, K. 1983. Migration of folding and metamorphism in the Rheinische Schiefergebirge deduced from K–Ar and Rb-Sr age determinations. *In*: MARTIN, H. & EDER, F. W. (eds) *Intracontinental Fold Belts: Case Studies in the Variscan Belt of Europe and the Damara Belt in Namibia*. Springer, Berlin, 323–338.

——, FRANZKE, H.-J., MARHEINE, D., SCHWAB, M. & WEMMER, K. 1996. Zum Alter der Metamorphose in der Wippraer Zone/Harz—Ergebnisse von K/Ar-Altersdatierungen an schwachmetamorphen Sedimenten. *Zeitschrift der Deutschen Geologischen Gesellschaft*, **147**(1), 39–56.

ANDERSON, L. J. & SMITH, D. R. 1995. The effects of temperature and fO_2 on the Al-in-hornblende barometer. *American Mineralogist*, **80**, 549–559.

ANTHES, G. 1998. *Geodynamische Entwicklung der Mitteldeutschen Kristallinschwelle: Geochronologie und Isotopengeochemie*. PhD thesis, Universität Mainz.

BANKWITZ, P. & BANKWITZ, E. 1988. Intensität der Deformation in der saxothuringischen Zone, einschließlich der Lausitz. *Zeitschrift für Geologische Wissenschaften*, **16**(5), 373–392.

—— & —— 1995. Proterozoikum/Schwarzburger Antiklinorium. *In*: SEIDEL, G. (ed.) *Geologie von Thüringen*. Schweizerbart, Stuttgart, 46–77.

——, ——, KRAMER W. & PIN, C. 1994. Early Paleozoic bimodal volcanism in the Vesser area, Thuringian Forest, eastern Germany. *Zentralblatt für Geologie und Paläontologie, Teil I*, **1992**(9–10), 1113–1132.

BEAUMONT, C., ELLIS, S., HAMILTON, J. & FULLSACK, P. 1996. Mechanical model for subduction–collision tectonics of Alpine-type compressional orogens. *Geology*, **24**(8), 675–678.

BEHRMANN, J., DROZDZEWSKI, G., HEINRICHS, T., HUCH, M., MEYER, W. & ONCKEN, O. 1991. Crustal-scale balanced cross sections through the Variscan fold belt, Germany: the central EGT-segment. *Tectonophysics*, **196**, 1–21.

BRAND, R. 1980. Die niedriggradige Metamorphose einer Diabas-Assoziation im Gebiet von Berg/Frankenwald. *Neues Jahrbuch für Mineralogie Abhandlungen*, **139**(1), 82–101.

CATHELINEAU, M. 1988. Cation site occupancy in chlorites and illites as a function of temperature. *Clay Minerals*, **23**, 471–485.

DEKORP RESEARCH GROUP 1988. Results of the DEKORP 4/KTB Oberpfalz deep seismic reflection investigations. *Journal of Geophysics*, **62**, 69–101.

—— 1994. The deep reflection seismic profiles DEKORP 3/MVE-90. *Zeitschrift für Geologische Wissenschaften*, **22**(6), 623–825.

ELLENBERG, J. 1964. Beziehungen zwischen Ooid-Deformation in den ordovizischen Eisenerzen und der Tektonik an der SE-Flanke des Schwarzburger Sattels (Thüringen). *Geologie*, **13**, 168–197.

FLÖTTMANN, T. & ONCKEN, O. 1992. Constraints on the evolution of the Mid-German Crystalline Rise—a study of outcrops west of the river Rhine. *Geologische Rundschau*, **81**, 515–543.

FRANKE, W. 1984. Variszischer Deckenbau im Raume der Münchberger Gneismasse—abgeleitet aus der Fazies, Deformation und Metamorphose im umgebenden Paläozoikum. *Geotektonische Forschungen*, **68**, 1–253.

—— 1993. The Saxonian Granulites—a metamorphic core complex? *Geologische Rundschau*, **82**, 505–515.

—— 2000. The mid-European segment of the Variscides: tectono-stratigraphic units, terrane boundaries and plate evolution. *This volume*.

——, RAUCHE, H. & STEIN, E. 1995. Saxothuringian Basin: autochthon and nonmetamorphic nappe units: structure. *In*: DALLMEYER, R. D., FRANKE, W. & WEBER, K. (eds) *Pre-Permian Geology of Central and Eastern Europe*. Springer, Heidelberg, 235–248.

FRANZKE, H. J., AHRENDT, H., KURZ, S. & WEMMER, K. 1996. K–Ar-Datierungen von Illiten aus Kataklasiten der Floßbergstörung im südöstlichen Thüringer Wald und ihre geologische Interpretation. *Zeitschrift für Geologische Wissenschaften*, **24**(3–4), 441–456.

GEHMLICH, M., LINNEMANN, U., TICHOMIROWA, M., LÜTZNER, H. & BOMBACH, K. 1997. Datierung und Korrelation neoproterozoisch-frühpaläozoischer Profile des Schwarzburger Antiklinoriums und der Elbezone auf der Basis der Geochronologie von Einzelzirkonen. *Zeitschrift für Geologische Wissenschaften*, **25**, 191–201.

GERSTENBERGER, H. & HAASE, G. 1997. A highly effective emitter substance for mass spectrometric Pb isotope ratio determinations. *Chemical Geology*, **136**, 309–312

GUTSCHER, M. A., KUKOWSKI, N., MALAVIEILLE, J. & LALLEMAND, S. 1996. Cyclical behavior of thrust wedges; insights from high basal friction sandbox experiments. *Geology*, **24**(2), 135–138.

HIRSCHMANN, G. 1959. Beitrag zur Kenntnis von prävariskischem Magmatismus und Metamorphose im SW-Abschnitt des Schwarzburger Sattels (Thüringen). *Geologie*, **8**, 523–534.

HOINKES, G. 1986. Effects of grossular-content in garnet on the partitioning of Fe and Mg between garnet and biotite; an empirical investigation on staurolite-zone samples from the Austroalpine Scheeberg Complex. *Contributions to Mineralogy and Petrology*, **92**, 393–399

HUIQI, L., MCCLAY, K. R. & POWELL, D. 1992. Physical models of thrust wedges, *In*: MCCLAY, K. R. (ed.)

Thrust Tectonics. Chapman & Hall, London, 71–81

HYNES, A. & FOREST, R. C. 1988. Empirical garnet–phengite geothermometry in low-grade metapelites, Selwyn Range (Canadian Rockies). *Journal of Metamorphic Geology*, **6**, 297–309.

JAMIESON, R. A., BEAUMONT, C., FULLSACK, P. & LEE, B. 1998. Barrovian regional metamorphism: where's the heat? *In*: TRELOAR, P. J. & O'BRIEN, P. J. (eds) *What Drives Metamorphism and Metamorphic Reactions?* Geological Society, London, Special Publications, **138**, 23–51.

JAMISON, W. R. 1992. Stress controls on fold thrust style. *In*: MCCLAY, K. R. (ed.) *Thrust Tectonics*. Chapman & Hall, London, 155–164.

KEMNITZ, H., ROMER, R. L. & ONCKEN, O. 1999. Gondwana breakup and the northern margin of the Saxothuringian belt (Variscides of Central Europe). *International Journal of Earth Sciences*, **89**, in press.

——, SCHÄFER, F., ROMER, R. L. & ROSNER, M. 1999. Die geodynamische Geschichte eines variszischen Retrokeils unter besonderer Berücksichtigung des Vesser-Gebietes. *Beiträge zur Geologie von Thüringen, Neue Folge*, **6**, 41–91.

KROGH, T. E. 1973. A low-contamination method for hydrothermal decomposition of zircon and extraction of U and Pb for isotopic age determinations. *Geochimica et Cosmochimica Acta*, **37**, 485–494.

KUNERT, V. 1999. *Die Frankenwälder Querzone: Entwicklung einer thermischen Anomalie im Saxothuringikum*. PhD thesis, Universität Giessen.

LINNEMANN, U. & BUSCHMANN, B. 1995. Die cadomische Diskordanz im Saxothuringikum (oberkambrisch–tremadocische overlap-Sequenzen). *Zeitschrift für Geologische Wissenschaften*, **23**(5–6), 707–727.

LISLE, R. J. 1992. Strain estimation from flattened buckle folds. *Journal of Structural Geology*, **14**(3), 369–371.

LÜTZNER, H., ELLENBERG, J. & FALK, F. 1986. Entwicklung der Sedimentationsrate und der Ablagerungsprozesse im Altpaläozoikum Thüringens. *Zeitschrift für Geologische Wissenschaften*, **14**(1), 83–93.

MASSONNE, H. J. 1991. *High-pressure, low-temperature metamorphism of pelitic and other protoliths based on experiments in the system K_2O–MgO–Al_2O_3–SiO_2–H_2O*. Habilitation thesis, Universität Bochum.

—— & SCHREYER, W. 1987. Phengite geobarometry based on the limiting assemblage with K-feldspar, phlogopite, and quartz. *Contributions to Mineralogy and Petrology*, **96**, 212–224.

NASIR, S., OKRUSCH, M., KREUZER, H., LENZ, H. & HÖHNDORF, A. 1991. Geochronology of the Spessart Crystalline Complex, Mid-German Crystalline Rise. *Mineralogy and Petrology*, **44**, 39–55.

NOBLET, C. & LEFORT, J. P. 1990. Sedimentological evidence for a limited separation between Armorica and Gondwana during the Early Ordovician. *Geology*, **18**, 303–306.

ONCKEN, O. 1997. Transformation of a magmatic arc and an orogenic root during oblique collision and its consequences for the evolution of the European Variscides (Mid-German Crystalline Rise). *Geologische Rundschau*, **86**, 2–20.

——, PLESCH, A., WEBER, J., RICKEN, W. & SCHRADER, S. 2000. Passive margin detachment during arc–continent collision (Central European Variscides). *This volume*.

——, VON WINTERFELD, C. & DITTMAR, U. 1999. Accretion of a rifted passive margin; the late Paleozoic Rhenohercynian fold and thrust belt (middle European Variscides). *Tectonics*, **18**, 75–91.

PFEIFFER, H. 1970. Zum inneren Bau des Ostthüringischen Kulm-Synklinoriums. *Jahrbuch Geologie, Mitteilungen des Zentralen Geologischen Instituts Berlin*, **5–6**, 165–173.

PLESCH, A. & ONCKEN, O. 1999. Orogenic wedge growth during collision; constraints on mechanics of a fossil wedge from its kinematic record (Rhenohercynian FTB, Central Europe). *In*: SINTUBIN, M., VANDYCKE, S. & CAMELBEEK, T. (eds) *Palaeozoic to Recent tectonics in the NW European Variscan front Zone. Tectonophysics*, 117–139.

RAMSAY, J. G. & HUBER, M. I. 1987. *The Techniques of Modern Structural Geology. 2. Folds and Fractures*. Academic Press, New York.

RICKEN, W., SCHRADER, S., ONCKEN, O. & PLESCH, A. 2000 Turbidite basin and mass dynamics related to orogenic wedge growth; the Rheno-Hercynian case. *This volume*.

SCHÄFER, F. 1997. *Krustenbilanzierung eines variscischen Retrokeils im Saxothuringikum*. Scientific Technical Report, GFZ Potsdam, **16**.

SCHÄFER, J., NEUROTH, H., AHRENDT, H., DÖRR, W. & FRANKE, W. 1997. Accretion and exhumation at a Variscan active margin, recorded in the Saxothuringian flysch. *Geologische Rundschau*, **86**, 599–611.

SCHWAN, W. 1958. Untervorschiebungen und Aufbruchsfalten. *Neues Jahrbuch für Geologie und Paläontologie*, **8–9**, 356–377.

SIEBEL, W. 1999. Variszischer spät- bis postkollisionaler Plutonismus in Deutschland: Regionale Verbreitung, Stoffbestand und Altersstellung. *Zeitschrift für Geologische Wissenschaften*, **26**(5–6), 1–30.

STEIGER, R. H. & JÄGER, E. 1977. Convention on the use of decay constants in geo- and cosmochronology. *Earth and Planetary Science Letters*, **36**, 359–362.

STEIN, E. 1988. Die strukturgeologische Entwicklung im Übergangsbereich Saxothuringikum/Moldanubikum in NE-Bayern. *Geologica Bavarica*, **92**, 5–131.

WEMMER, K. 1991. *K/Ar-Altersdatierungsmöglichkeiten für retrograde Deformationsprozesse im spröden und duktilen Bereich—Beispiele aus der KTB-Vorbohrung (Oberpfalz) und dem Bereich der Insubrischen Linie (N-Italien)*. Göttinger Arbeiten zur Geologie und Paläontologie, **51**.

WILLETT, S. D. 1992. Dynamic and kinematic growth and change of a Coulomb wedge. *In*: MCCLAY, K. R. (ed.) *Thrust Tectonics*. Chapman & Hall, London, 19–31.

WILLETT, S., BEAUMONT, C. & FULLSACK, P. 1993. Mechanical model for the tectonics of doubly vergent compressional orogens. *Geology*, **21**, 371–374.

WILLNER, A., MASSONNE, H. J. & KROHE, A. 1991. Thermal evolution of a Variscan magmatic arc: the Odenwald (FR Germany) as an example. *Geologische Rundschau*, **80**, 369–390.

WUNDERLICH, J. 1995. Mitteldeutsche Kristallinzone. *In*: SEIDEL, G. (ed.) *Geologie von Thüringen*. Schweizerbart, Stuttgart, 22–46.

ZEH, A. 1996. Die Druck–Temperatur–Deformationsentwicklung des Ruhlaer Kristallins (Mitteldeutsche Kristallinzone). *Geotektonische Forschungen*, **86**, 1–212.

—— & COSCA, M. A., BRATZ, H., OKRUSH, M. & TICHOMIROVA, M. 2000. Simultaneous horst–basin formation and magmatism during Late Variscan transtension: evidence from $^{40}Ar/^{39}Ar$ and $^{207}Pb/^{206}Pb$ geochronology in the Ruhla Crystalline Complex. *International Journal of Earth Sciences*, **89**, 52–71.

Geophysical constraints on exhumation mechanisms of high-pressure rocks: the Saxo-Thuringian case between the Franconian Line and Elbe Zone

C. M. KRAWCZYK[1], E. STEIN[2], S. CHOI[3], G. OETTINGER[1], K. SCHUSTER[1,5], H.-J. GÖTZE[3], V. HAAK[1], O. ONCKEN[1], C. PRODEHL[4] & A. SCHULZE[1]

[1]*GeoForschungsZentrum Potsdam, Telegrafenberg, D-14473 Potsdam, Germany (e-mail: lotte@gfz-potsdam.de)*

[2]*Institut für Mineralogie, TU Darmstadt, Schnittspahnstr. 9, D-64287 Darmstadt, Germany*

[3]*Institut für Geologie, Geophysik und Geoinformatik, FU Berlin, Malteserstr. 74-100, D-12249 Berlin, Germany*

[4]*Geophysikalisches Institut, Universität Karlsruhe, Hertzstr. 16, D-76187 Karlsruhe, Germany*

[5]*Present address: Bundesanstalt für Geowissenschaften und Rohstoffe, D-30631 Hannover, Germany*

Abstract: Major bodies of high-pressure (HP) rocks in the Saxo-Thuringian Belt in East Germany (Saxonian Granulite Massif, Erzgebirge) are investigated using a variety of geophysical methods (seismic reflection and refraction survey, magnetotelluric studies, gravity modelling). The Saxonian Granulite Massif and the Erzgebirge are not a continuous feature, as can be seen from discontinuous reflections, offset of upper-crustal seismic refraction velocity layers, and crustal resistivity increasing towards the Erzgebirge. Their juxtaposition during the evolution of two Variscan-age thrust wedges may have controlled this geometry. The earlier thrust wedge emplaced the supracrustal Erzgebirge HP nappes from the southeast to the northwest onto the Saxo-Thuringian Basin, whereas the later one propagated southwards and uplifted the Saxo-Thuringian granulites from deeper levels. To the southwest, the granulites are observed at shallow depth as far as the Franconian Line; to the southeast they extend down to the Moho, or they continue at mid-crustal levels. The granulites beneath the Saxo-Thuringian Belt can only have originated in one of two subduction zones: either through 'subduction erosion' and subsequent underplating of parts of the Saxo-Thuringian Plate from the north, or by intracrustal plug flow of overheated material from the southeast.

The exhumation of deeply buried crust during collisional orogenesis is among the hotly debated geodynamic problems. Within this context, the high-pressure (HP) and ultra-high-pressure (UHP) rocks in the Saxo-Thuringian part of the Variscan Orogen are ideally suited for analyses and modelling of fundamental principles of syncollisional exhumation. This unit shows an almost complete sedimentary and P–T–t record from before to after the collisional evolution and is well imaged by seismic reflection and other geophysical observations (Fig. 1).

Unroofing of the HP granulites in the Variscides has been interpreted in diverse ways. Whereas early interpretations favoured compressional mechanisms, Weber & Behr (1983) interpreted the Saxonian Granulite Massif (SGM) as an 'isolated diapir' that originated from deep crustal levels. Similarly, Franke (1993) and Reinhardt & Kleemann (1994) used a metamorphic core complex model to explain the exhumation of the SGM by a sequence of processes in Palaeozoic time. Although these earlier models are now maintained only for the final exhumation stage, the provenance and early emplacement of the granulites still is an open question: the exhumed SGM and related rocks were recently suggested to be derived through subduction erosion from the overriding Saxo-Thuringian Plate with subsequent underplating (Oncken 1998), or by return flow from the Teplá–Barrandian subduction zone to the SE

From: FRANKE, W., HAAK, V., ONCKEN, O. & TANNER, D. (eds).
Orogenic Processes: Quantification and Modelling in the Variscan Belt.
Geological Society, London, Special Publications, **179**, 303–322. 0305-8719/00/$15.00

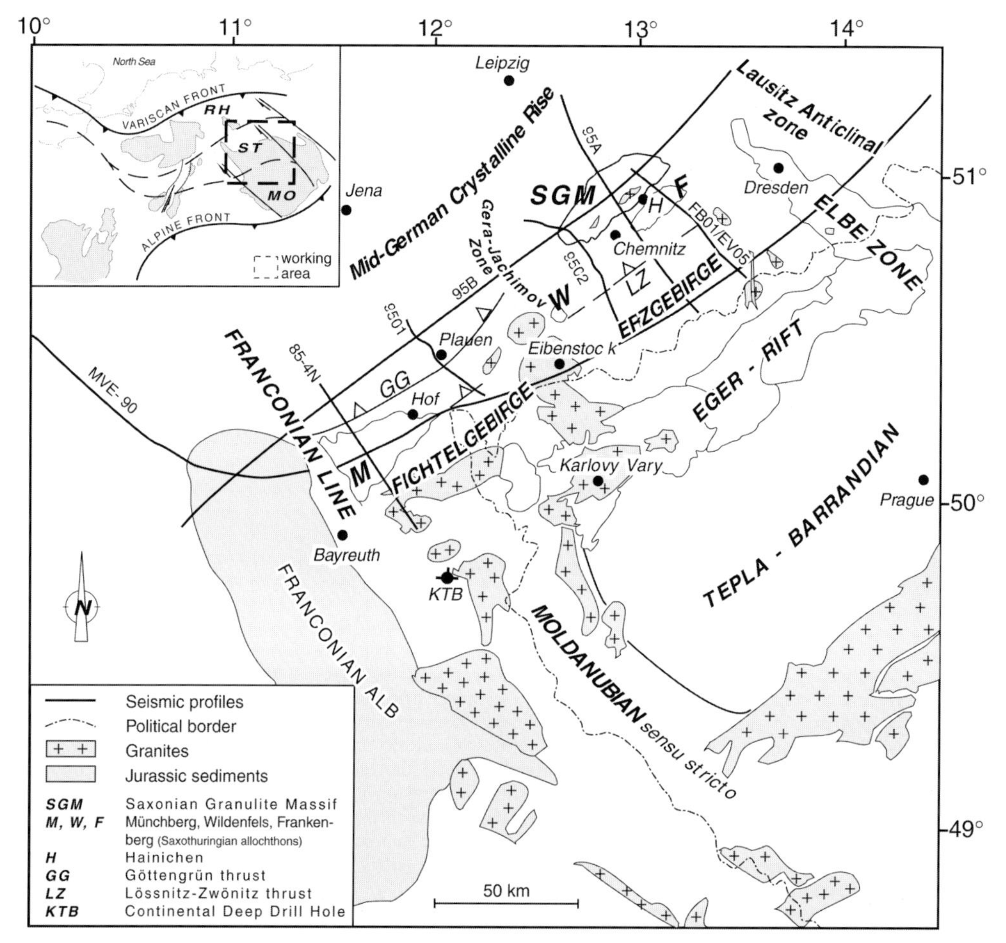

Fig. 1. Map of the investigated area with location of geophysical profiles. RH, Rheno-Hercynian Unit, ST, Saxo-Thuringian Unit, MO, Moldanubian Unit. (For more detailed information, see geological profiles above seismic reflection lines (Figs 2 and 3), and enclosure at back for overview.)

(DEKORP & Orogenic Processes Working Groups 1999).

The internal structure of the Saxo-Thuringian Belt has been investigated by several seismic projects (Fig. 1): by the NW–SE-trending DEKORP 85-4N line (DEKORP Research Group 1988), and by the NE–SW-trending DEKORP MVE-90 line along the southern margin of the Saxo-Thuringian Belt (DEKORP Research Group 1994). The crustal structure can be divided into four seismic layers comprising a transparent upper crust (except for the bowl shape of the Münchberg Gneiss Massif), mid-crustal, slightly dipping reflector lenses, a transparent middle crust, and finally a reflective lower crust (DEKORP Research Group 1994). The overall Moho depth of ± 30 km decreases to 27 km below the Erzgebirge, coincident with a marked gravity low (Plaumann 1986). Lower-crustal velocities in the Saxo-Thuringian Belt are higher ($V_p = 6.8$–7.0 km s^{-1}) than in the Rheno-Hercynian or Moldanubian Belts ($V_p = 6.5$–6.8 km s^{-1}; Ansorge *et al.* 1992). Recent geophysical observations reveal a possible continuation of the Saxonian Granulite Massif antiform structure to the SW for *c.* 150 km (DEKORP & Orogenic Processes Working Groups 1999). The current debate focuses on the succession of events and the geodynamic setting of the processes that led to the formation and unroofing of the HP–HT granulites.

In this paper, we will summarize various types of geophysical information and data, with the focus on reconciling possible geophysical models with geological ideas concerning the evolution of

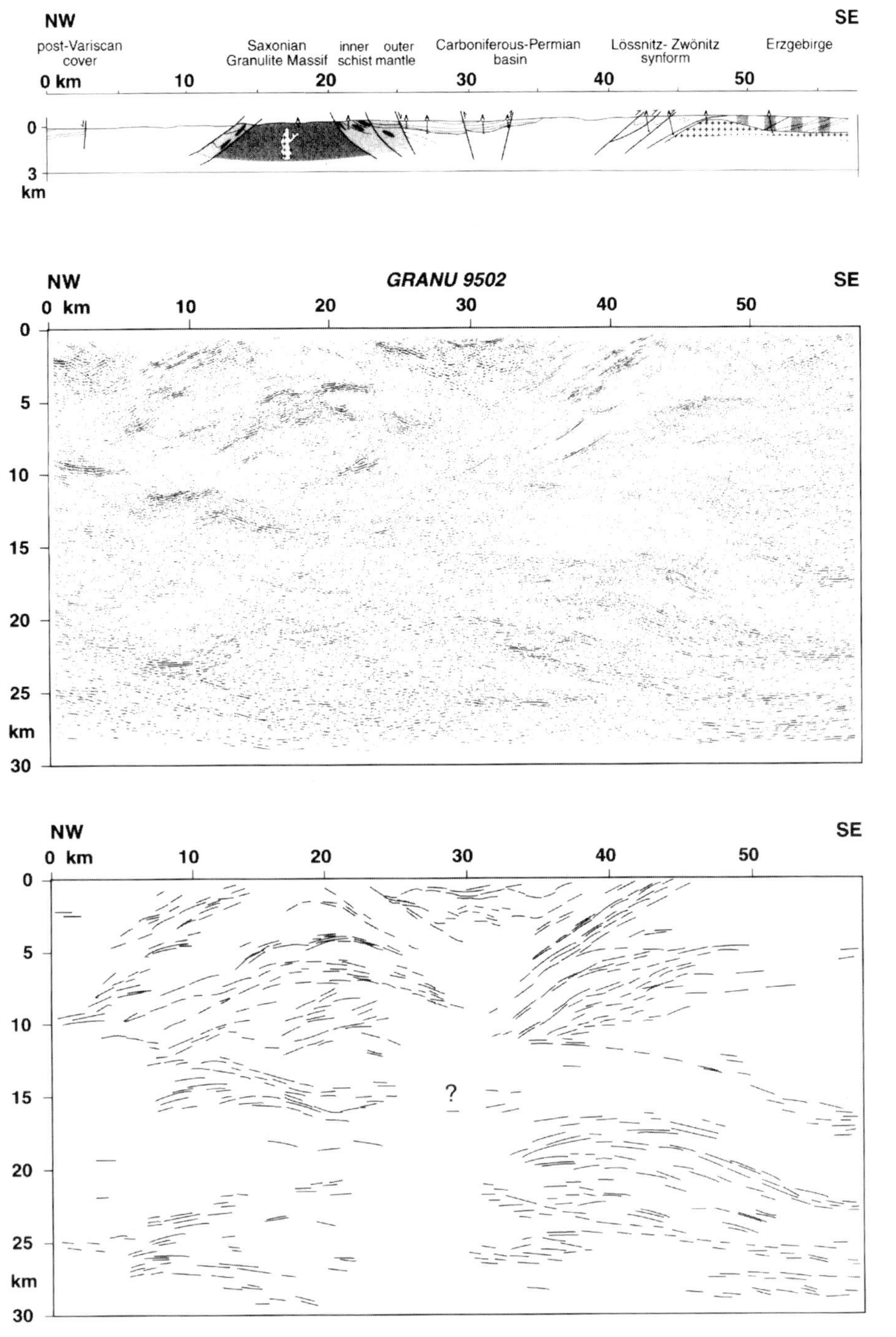

Fig. 2. Depth-migrated seismic section of profile GRANU 9502, with line drawing below. The schist mantle of the exposed granulites can be traced to at least 5 km depth. The granulites themselves are seismically more transparent in the uppermost crust with two strong domal reflective bands below. A Permo-Carboniferous sedimentary basin is outlined (km 25–35), and the Lössnitz–Zwönitz Zone is imaged by strong NW-dipping reflections at the northern flank of the Erzgebirge.

the area through time and space, and to thus differentiate between geodynamic models for the evolution of the Saxo-Thuringian Belt (see also Franke & Stein this volume).

Geological setting of the Saxo-Thuringian Zone and the Granulite Massif

Within the Variscan collisional belt, the Saxo-Thuringian Unit is situated between the Rheno-Hercynian foreland unit to the north and the Teplá–Barrandian Unit to the south (e.g. Franke 1989; Matte 1991; see enclosure at back). These units form continental plate fragments that were separated by minor oceanic basins. In this plate assemblage, the Rheno-Hercynian Belt evolved as a passive margin that was subsequently subducted southward under the Saxo-Thuringian Belt; the latter thus was in an upper-plate position with respect to the Rheno-Hercynian Belt. However, the Saxo-Thuringian Belt was also subducted southward under the Teplá–Barrandian Unit. Both subduction zones were active in Late Devonian to Carboniferous times, forming a 'tandem subduction zone'. At its northern margin, the Saxo-Thuringian Unit shows remnants of a magmatic arc–active margin on continental crust, which developed a SE-ward extending retro-wedge as described by Schäfer *et al.* (this volume). To the SE, a fold-and-thrust belt shows crustal thickening at the margin to the Teplá–Barrandian Unit. In conclusion, the Saxo-Thuringian Unit is in a back-arc position at its northern plate margin, and in a foreland setting with respect to its southern margin. The overall convergence and collision of these plate fragments started in early Late Devonian time and culminated in late Early Carboniferous time, with a shortening of the Rheno-Hercynian and Saxo-Thuringian sedimentary basins of *c.* 50% (Behrmann *et al.* 1991; Schäfer *et al.* this volume).

The *c.* 100–180 km wide Saxo-Thuringian Belt shows geological and geophysical characteristics that are difficult to reconcile. Unmetamorphosed to low-grade marine Palaeozoic sedimentary rocks are in immediate contact with highly metamorphosed basement rocks. Also, there is a prominent influence of extensional features that accommodated continuous marine sedimentation through Carboniferous convergence, and the contemporaneous rise of granulite-facies HP rocks below these sediments (Franke 1993).

The SGM exhibits an isothermal decompressional P–T history (Rötzler 1992), with granulite metamorphism at *c.* 340 Ma (von Quadt 1993; Vavra & Reinhardt 1999), subsequent cooling at *c.* 333 Ma (Werner & Reich 1997), and an extremely condensed metamorphic succession in its roof (P_{max} granulite 20 kbar; P contact metamorphism of schist mantle 2–3 kbar).

Towards the north and south, the Granulite Massif is bounded by younger, late- or post-orogenic basins, partly containing upper Viséan (Hainichen Molasse) to Rotliegend sediments. These are either characterized by normal and strike-slip faulting or onlap structures, and the base of the basin is slightly folded. Analyses of synkinematic illites (fraction <2 μm) in the Frankenwald area date the most recent shortening at *c.* = 310 Ma (Ahrendt *et al.* 1996). As interpreted by Schäfer *et al.* (this volume) these data record SE-directed shortening and stacking that also involved the cooled granulites and their cover.

The Erzgebirge, in contrast, was overthrust on the Palaeozoic sediments from the southeast before post-orogenic granite intrusion at *c.* 320 Ma. Its internal structure reveals a partly inverted stack of thin UHP to medium-pressure (MP) nappes that were assembled before collision and overthrusting from *c.* 360 ± 7 Ma to late cooling at 330 Ma (Schmädicke *et al.* 1995; Werner & Lippolt this volume). It thus records a similar (albeit somewhat later) evolution to that recorded in the Münchberg–Frankenberg Nappes to the north and west (Franke 1984). These are considered to contain Teplá–Barrandian remnants and accreted Saxo-Thuringian basin sediments overthrust on the shelf in Lower Carboniferous time. From geochemical arguments, Mingram (1996) suggested that the Erzgebirge Nappes contain only subducted and exhumed Saxo-Thuringian rocks.

Geophysical database

Seismic, magnetotelluric and gravimetric modelling studies were performed to further investigate the internal structure of the SGM, the extensional shear zone separating the granulite core from the surrounding schist mantle, and the nature and extent of lower-crustal material. The geophysical database presented here was mainly acquired and compiled from 1995 onwards. We aim here to show the confidence levels of the various datasets and to discuss geometric boundary conditions for further interpretation and thermo-mechanical modelling of the exhumation history of HP rocks in the back-arc and fore-arc of a convergent regime (see Henk this volume).

Seismic refraction database

The seismic refraction experiment in May 1995 studied the deeper structure of the Saxo-Thuringian Terrane to obtain an overall velocity model as a key to material properties, for example, for gravity modelling (see below). First results and models of the lines GRANU 95A and 95B have been published by Enderle *et al.* (1998), supplementing studies of Schulze & Bormann (1990) and the MVE-90 experiment (DEKORP Research Group 1994). Ray-tracing techniques for the wide-angle P-wave seismic data were applied yielding *c.* 1–2 km depth error. The error of the finite difference first arrival modelling for the P_g and P_s phases amounts to ± 0.1 km s^{-1} in V_p between 0 and 5 km depth, equivalent to *c.* ± 100 m in depth. The main features of the refraction velocity model are: (1) *P*-wave velocities are less than 6.0 km s^{-1} in the upper 5 km of the exposed SGM; (2) a high-velocity layer ($V_p = 6.3$–6.4 km s^{-1}) occurs at 5 km depth below the SGM, and its surface deepens to 8 km depth below the Erzgebirge; (3) *P*-wave velocities in the middle crust vary laterally between 6.3 and 6.6 km s^{-1} between the exposed SGM and the area NE of the Franconian Line; (4) below 22 and 23 km depth, velocities of 7.0 km s^{-1} define a prominent high-velocity lower crust; (5) the crust–mantle boundary at 31 km depth does not show any pronounced undulation.

Seismic reflection database

The part of the Saxo-Thuringian Belt under consideration has been seismically investigated by several DEKORP campaigns (see Fig. 1 for location): the latest survey, GRANU '95 (DEKORP & Orogenic Processes Working Groups 1999), will be discussed here in more detail. Several reflection profiles covered the Saxo-Thuringian Belt in a NW–SE trend: line GRANU 9502 across the SGM, paralleled to the SW by profiles GRANU 9501 and DEKORP '85-4N (DEKORP Research Group 1988), and by FB01/EV05 further to the NE (see reprocessed data of DEKORP & Orogenic Processes Working Groups (1999)). Perpendicular to all these profiles, the MVE-90 line runs NE–SW along the southern margin of the Saxo-Thuringian Belt (DEKORP Research Group 1994). Thus, this seismic grid provides spatial control, and could help delimit regional structures.

In addition to the standard seismic processing of profiles GRANU 9501 and 9502 (see DEKORP & Orogenic Processes Working Groups 1999), the final seismic processing of the stacked datasets has here been adapted to handle dipping reflections by the post-stack plane-wave decomposition technique. This method allows the enhancement of relevant, laterally coherent events by local slant stacks over a limited aperture with semblance weighting of individual dip components. Only coherent events actually existent in the data can be enhanced by this technique. The dip-enhanced data were then depth migrated using a Kirchhoff algorithm. The seismic reflection velocity model was calibrated and blended with the velocities derived from refraction modelling. For the reflection line, an average velocity function of the root-mean-square velocity model was calculated and then weighted by percentage, determined from comparison of refraction and reflection data. This weighted function was finally blended with the complete velocity field of the reflection line. This procedure was performed for both profiles and has two advantages. First, the reflection velocity model is kept as a background model by the weighting, and is not simply replaced by the refraction model. Second, strong variations in the overall velocity field are smoothed by blending.

The resulting depth sections and line drawings are shown in Figs 2 and 3. Profile GRANU 9502 shows that the schist mantle of the granulitic core can be followed from the surface to at least 5 km depth (Fig. 2: km 5–14, km 21–30), whereas the granulite itself is seismically transparent in the upper 4 km. Two strong, dome-shaped reflection bands are imaged below, with the apex at 4 and 5 km depth (km 17, km 21), coincident with an increase of refraction velocity from <6.0 to 6.4 km s^{-1}. SE of the granulites, the Carboniferous sedimentary basin is outlined (km 23–36), deepening from the surface to *c.* 3.5 km depth. Further to the SE, the basin is bound by strong, NW-dipping reflections (km 36–46) cutting down from the northern flank of the Erzgebirge to 10 km depth. These reflections correlate with an important shear zone (Lössnitz–Zwönitz Zone) composed of alternating graphitic schists and metabasalts, which have been drilled down to 1.5 km depth and are known from mining (Hösel *et al.* (1997) and references therein). The SE end of profile 9502 images a weaker set of subhorizontal reflections at 5 km depth (km 47–57), soling towards the NW into a reflective band at *c.* 12 km depth. Imaging in the transparent zone in the upper and middle crust that coincides with the transition from the SGM to the Erzgebirge (Fig. 2; km 28–35) cannot be further improved, although seismic data processing had been focused in this area to handle and image

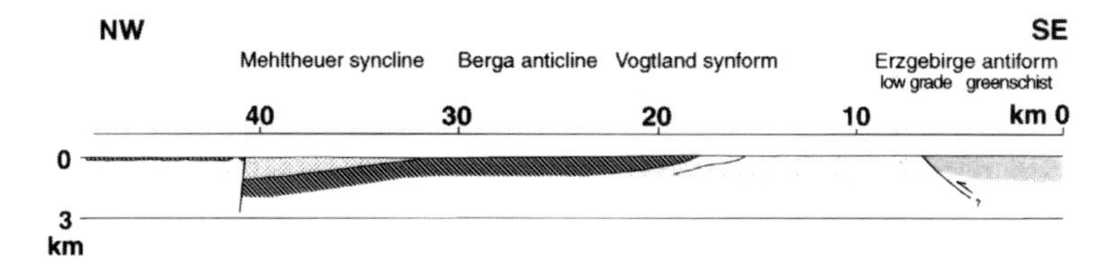

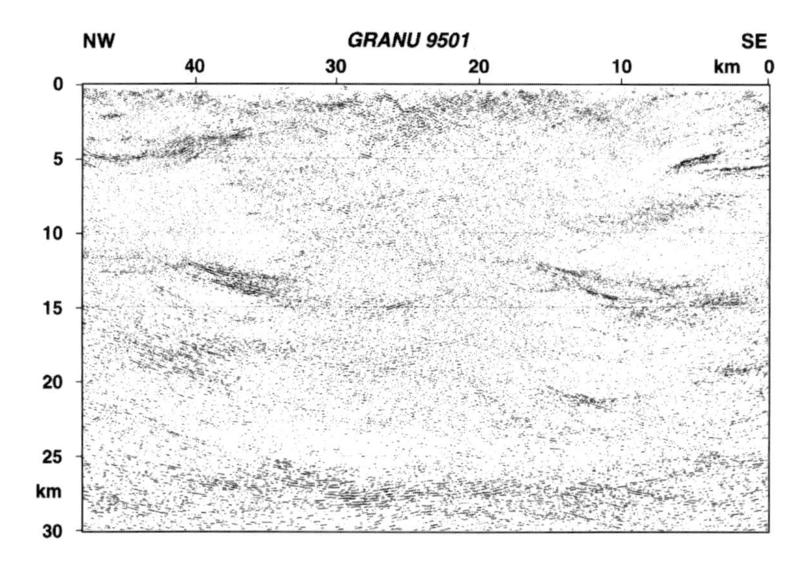

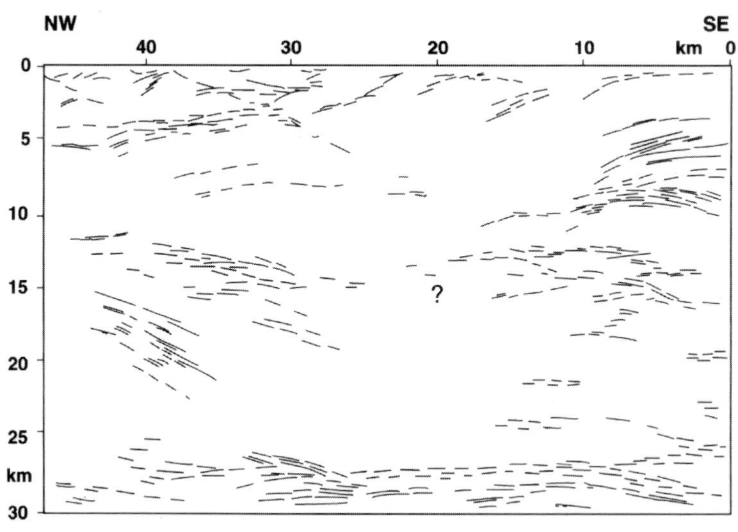

Fig. 3. Depth-migrated seismic section of profile GRANU 9501, with line drawing below. The reflections at 5 km depth in the NW half of the profile correlate with the high velocities (V_p = 6.3–6.4 km s^{-1}) observed on refraction line 95B. In the central part of the profile, seismic imaging is hampered by the influence of nearby granites.

steep dips (e.g. dip enhancement, migration in a broad frequency range with large aperture and up to 65° dip). As there was no obvious lack of energy penetration, the transparency might be attributed to a steepening of the reflectors imaged above. The middle crust shows weaker reflectivity in the NW half of the profile (km 0–20, 15–23 km depth), whereas reflective structures are present below the Erzgebirge (km 30–57, 17–27 km depth).

The overall seismic appearance of profile GRANU 9501 (Fig. 3) is more transparent than that of 9502, especially in the central part of the profile where the influence of granites hampers seismic imaging (see Fig. 1 and enclosure for location). In the upper crust, the SE part of the profile shows again NW-ward dipping reflections between 5 and 6 km depth (Fig. 3, km 0–7). These elements are delimited by subhorizontal reflections at *c.* 8 km depth (km 3–12), and are also confirmed by reflections recorded on profile MVE-90 (DEKORP Research Group 1994). Another very pronounced structure in the upper crust lies at the NW end of the profile (km 33–47) sloping down to the NW from 3 km to 5 km depth. This reflection band is located in the SW continuation of the granulite dome, along-strike of its longitudinal axis, and correlates with the high-velocity layer ($V_p = 6.4$ km s^{-1}) observed by the seismic refraction experiment (Enderle *et al.* 1998). Whereas the central segment of profile 9501 is transparent (km 16–32), the upper middle crust NW and SE of it images strong reflective bands between 12 and 15 km depth (km 40–32, km 16-0). Strong and uniform, mainly horizontal reflectivity is observed in the lower crust between 25 and 30 km depth (Fig. 3).

MT database

The magnetotelluric (MT) survey was carried out at 17 locations (site distance 4–10 km; sampling rate 20 Hz; period range 0.1–1000 s; 10–15 days recording) along the seismic refraction line GRANU 95A. Additionally, audiomagnetotelluric (AMT) measurements were carried out at each location and at another 26 locations close to the schist mantle of the Granulite Massif (location distance 500 m). Transfer functions in a broad frequency range from 8 kHz to 0.125 mHz were derived. The data quality is good in the northern and middle part of the profile, but poor in the Erzgebirge. The decomposition of the impedance tensor shows that the regional structure is 2D, and the electromagnetic strike-angle corresponds to the Saxo-Thuringian structural trend (for detail see Oettinger (1999)).

The best transfer functions from 14 sites were selected for 2D inversion and forward modelling. The final model with the best data fit is shown in Fig. 4. NW of the granulites, a good conductor at 1 km depth, with *c.* 10 km lateral extent and *c.* 10 Ω m resistivity, outlines the lower margin of a Permo-Carboniferous sedimentary basin. With the exception of this near-surface conductor, the upper crust is highly resistive (500 Ω m) with very high values in particular in the Erzgebirge (10 000 Ω m), which coincides with the area of Palaeozoic thrusting. The upper crust does not show a macroscopic electric anisotropy as found further south in the KTB region of the northern zone of Erbendorf–Vohenstrauß (see enclosure; Eisel & Haak 1999).

A good conductor linked to the shear zone between the exposed granulite body and its schist mantle in the upper crust was not observed. A series of alternative models was calculated to test the reliability of the measured data. A hypothetical conductor was introduced along the margins of the granulitic core from the surface to 10 km depth, according to the seismic reflection data (Fig. 2). Comparison of real data with synthetic models based on different conductance values and dip angles of the critical structure suggested that only a very slight conductivity contrast would fit the measured data. Thus, the ductile shear zone between the granulites and its schist mantle is not a good conductor.

The most prominent feature in the resistivity model is the highly conductive zone at 18–20 km depth with a resistivity of *c.* 5 Ω m. To test the depth significance of this conductor, the model response for different depths was calculated. With respect to the confidence limits of the measured data, the upper boundary of the conductor lies between 15 and 22 km depth (Oettinger 1999). The crustal conductor terminates between the SE margin of the Saxonian granulites and the NW margin of the Erzgebirge. This remarkable lateral limitation is strongly confirmed by the data. Figure 5 shows the model response of a hypothetical model in which the conductor continues underneath the Erzgebirge. The comparison of the hypothetical and measured long-period (10–1600 s) induction arrows for three sites near the assumed margin of the deep-crustal conductor yields large discrepancies between theoretical and measured data.

The induction arrows along profile 95A for periods larger than 100 s mainly point to the SE, which is the direction of lower conductivity. This result is further confirmed by other profiles (see compilation by Oettinger (1999) and references therein), suggesting that the deep electric structures continue at depth along-strike to the Franconian Line. The deep crustal conductor below the SGM may extend at shallower depth to the SW, as shown from modelling and borehole results in the KTB area in the northern zone

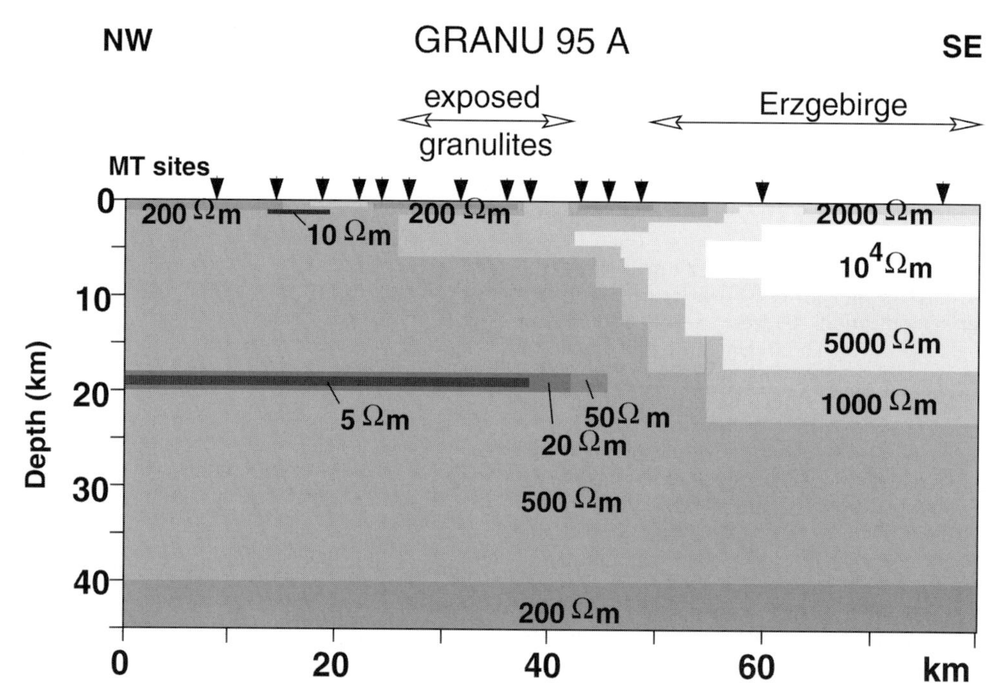

Fig. 4. 2D resistivity model along profile GRANU 95A. NW of the exposed granulites, a good conductor at 1 km depth with *c.* 10 km lateral extent outlines the lower margin of a Permo-Carboniferous sedimentary basin. In the SE part of the profile, the highly conductive zone at 18–20 km depth is not evident. In general, the upper crust is highly resistive (500 Ω m) with maximum values in the Erzgebirge (10 kΩ m).

of Erbendorf–Vohenstrauß (see enclosure; ELEKTB Group 1997; Eisel & Haak 1999).

Gravity database

Three Bouguer gravity datasets were merged in this study for comparison of modelling and real data. The first is a 2 km × 2 km grid of Bouguer anomalies (BA), based on more than 600 000 gravity observations in the area of Saxony (Germany) and covering both the SGM and the Erzgebirge. The second is a grid with the same spacing in the area of the Czech Republic. Because of the lack of original point data, gravity data were also taken from the gravity map of the Brandenburg, Mecklenburg-Vorpommern, Sachsen and Sachsen-Anhalt districts (Sächsisches Landesamt für Umwelt und Geologie 1996). The third set consists of *c.* 1000 gravity observations for the Münchberg Gneiss Massif (MGM) in the area of Franconia (Plaumann 1986). Bouguer anomalies range from -600 μm s^{-2} near the Eibenstock Granites to 400 μm s^{-2} in the Lausitz area (Fig. 6). In the SGM and MGM most of the gravity values vary around 0 μm s^{-2}, in the Erzgebirge between -200 and -600 μm s^{-2}.

To compare the results of gravity modelling with observed field data along lines 95A and 9502, two density models were compiled. The seismic refraction velocity field (Enderle *et al.* 1998) was first interpreted petrophysically, and general density–velocity relationships were used (Conrad *et al.* 1994; Sobolev & Andrey 1994). The density–velocity gradients $D(V_p)$ vary significantly with rock composition and the grade of metamorphism. Using both modal rock compositions and experimental data for elastic modules of single crystals, Sobolev & Andrey (1994) presented synthetic phase diagrams, which consist of densities, elastic-wave velocities under specific pressure and temperature conditions for different anhydrous magmatic rocks:

$$D(V_p) = 0{\cdot}446V_p - 0{\cdot}074 \quad \text{for } V_p = 6{\cdot}05\text{–}6{\cdot}95\ km\ s^{-1}$$

$$D(V_p) = 0{\cdot}487V_p - 0{\cdot}359 \quad \text{for } V_p = 6{\cdot}95\text{–}7{\cdot}80\ km\ s^{-1}.$$

For further 3D modelling the density–velocity values according to Müller (1995) were used,

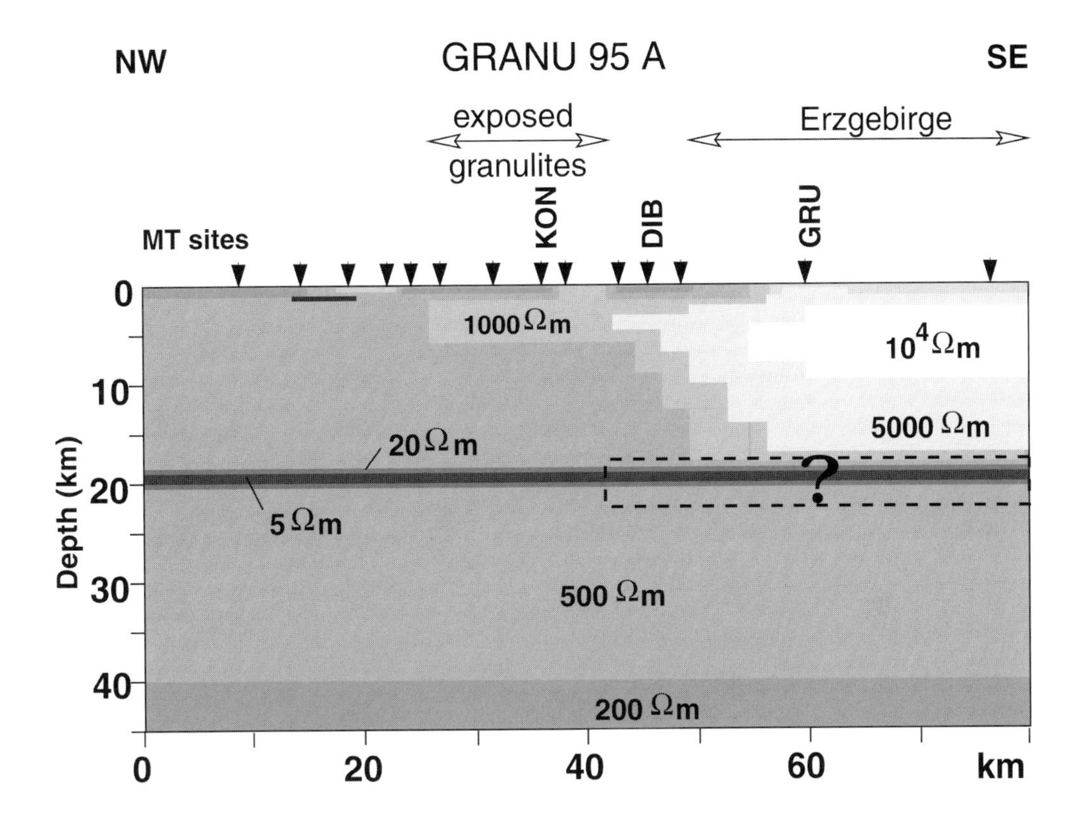

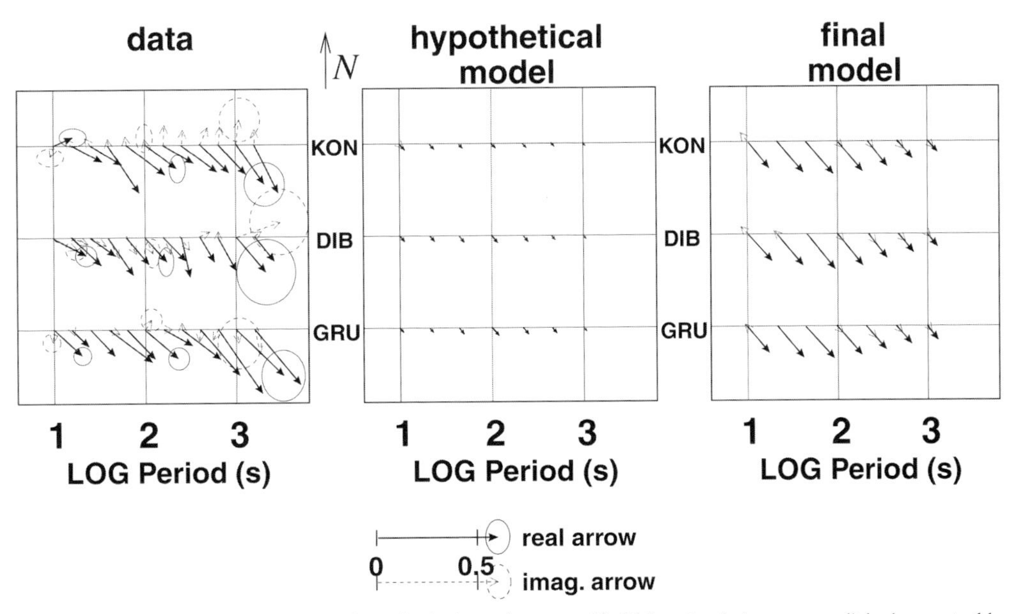

Fig. 5. Resistivity model with a laterally unlimited conductor at 18–20 km depth (upper panel) is demonstrably inconsistent with the data. A clear misfit of induction arrows is obvious (lower panel).

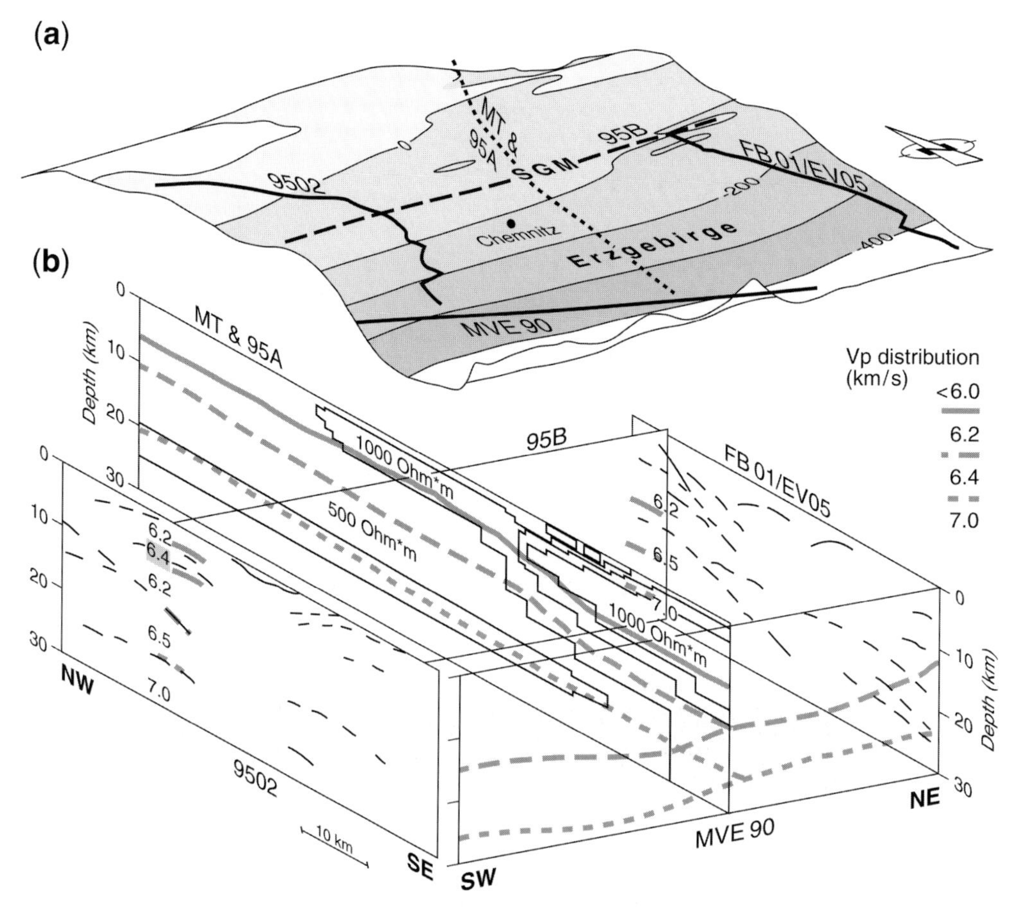

Fig. 6. (**a**) Gravity map of the homogenized 2 km × 2 km grid (Bouguer anomaly, 100 μm s^{-2} contour lines) and (**b**) integrated block model of seismic and magnetotelluric datasets in the area of interest (see Fig. 1 for orientation). Bouguer anomaly ranges from −600 μm s^{-2} (near the Eibenstock Granites) to 400 μm s^{-2} in the Lausitz area. In the Saxonian Granulite and Münchberg Gneiss Massifs most of the gravity values are around 0 μm s^{-2}, in the Erzgebirge between −200 and −600 μm s^{-2}.

determined on rock samples of the eastern Saxo-Thuringian Zone (Table 1).

In the first model (Fig. 7), which was used as a starting model, the measured gravity anomaly along profile 95A varies significantly from the density distribution inferred from seismic refraction data. This velocity model is characterized by a high-velocity upper-crustal layer (V_p = 6.4 km s^{-1}) which dips towards the SE from 5 to 9 km depth beneath the SE margin of the SGM. This layer causes local anomalies in the resulting Bouguer field between +100 μm s^{-2} in the SGM and *c.* − 150 μm s^{-2} in the Erzgebirge area (Fig. 7). The best fit and at the same time the most simple model accounts for this deviation by introducing a high-density body (3000 kg m^{-3}) at the SE end of the profile between 6 and 12 km depth.

The second model (Fig. 8) uses the velocity field from line 95A, the available geological and well log data (Krentz *et al.* 1997), and the seismic reflection constraints presented above. All datasets were projected onto line 9502, and were

Table 1. *Velocity–density values for the eastern Saxo-Thuringian Belt (after Müller 1995)*

V_p (km s^{-1})	Density (kg m^{-3})	Rock type
<6.0	2620	Granites
6.0–6.2	2700	Gneiss
6.2	2740	Pyroxene-free granulites
6.4	2840	Pyroxene granulites
6.5–7.0	2900	60% metabasite
7.0–8.0	3000	70% metabasite
>8.0	3250	Mantle

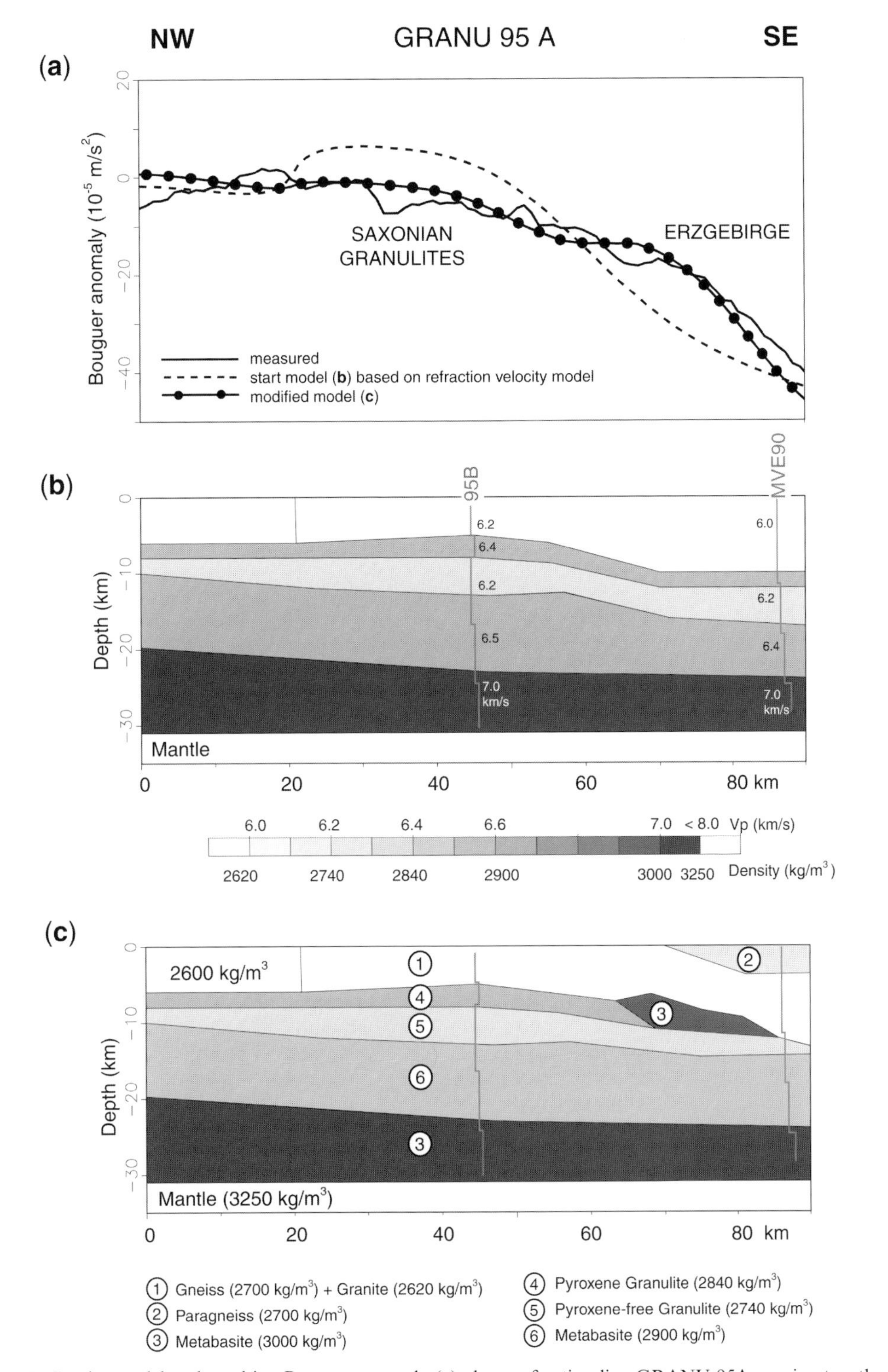

Fig. 7. Gravity model and resulting Bouguer anomaly (**a**) along refraction line GRANU 95A, serving together with crossing profiles 95B (Enderle *et al.* 1998) and MVE-90 (DEKORP Research Group 1994) as a starting model (**b**). The modified and best-fit model (**c**) is characterized by a sheet of metabasic rocks in the Erzgebirge area.

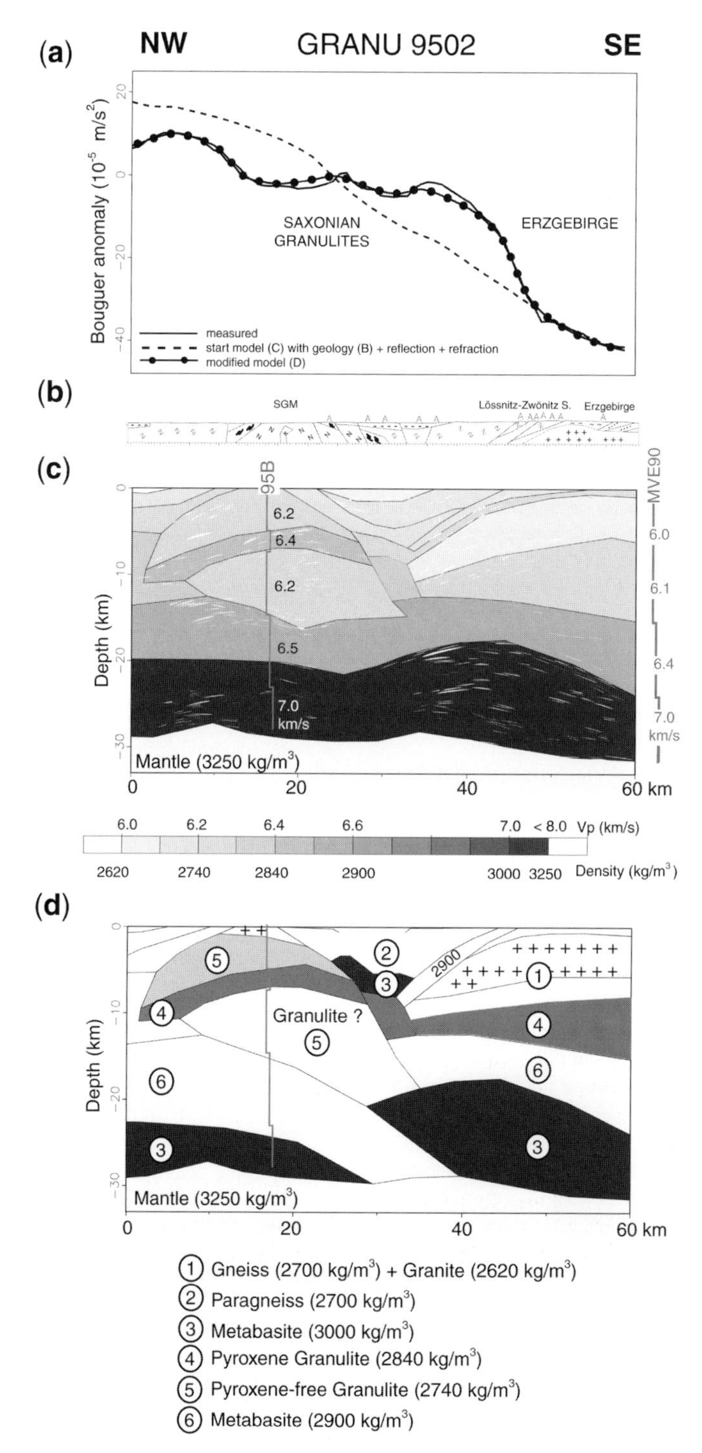

Fig. 8. Gravity model and resulting Bouguer anomaly (**a**) along reflection line 9502. In contrast to Fig. 7, structural information from surface geology (**b**) and seismic reflection survey (**c**) define additional boundary conditions; (**d**) shows best modification of simplified starting model. SE of the exposed granulites a sheet of metabasic rocks is necessary in the upper crust between 3 and 8 km depth (km 25–35).

translated into a density distribution, yielding a more detailed starting model. The differences obtained between measured and calculated anomalies are as high as 150 μm s^{-2} (Fig. 8). For a better fit, a sheet of metabasic rock is necessary SE of the SGM in the upper crust between 3 and 8 km depth, and a low-density middle–lower-crustal body of approximately 10 km thickness is required (2900 kg m^{-3}). It deepens from 15 to 30 km depth from the NW end of the profile towards the SE (Choi *et al.* 1998). The domed structures in the SGM area are expected to be composed of pyroxene granulite or pyroxene-free granulites down to *c.* 8 km depth, which extend as pyroxene granulite further below the Erzgebirge between 10 and 15 km depth (Fig. 8). Representation of deeper parts of body 6 by material from body 3 would be allowed from seismic reflection survey, but would worsen the fit (Choi *et al.* 1998). The strong Bouguer anomaly gradient towards the Erzgebirge is caused by granites in the upper 5 km.

Geophysical features of the eastern Saxo-Thuringian Zone

The combined seismic reflection and refraction data show high velocities of 6.4 km s^{-1} at shallow depths (*c.* 5 km) below or within the exposed granulites along reflection line 9502 and refraction line 95A. The velocity boundary coincides with a highly reflective zone and prominent reflections (Fig. 9). SE of the granulite exposure, the high-velocity upper-crustal layer and the related reflectivity pattern seem to either disappear or to step down.

In the NW half of profile 9502, the reflectivity pattern changes below 15 km from dome-like structures below the SGM to a more structureless image of the middle–lower crust, whereas moderately dipping structures are imaged in the SE half of the line (Fig. 9). This depth and the extent of the structure roughly coincide with the depth and extent of the good conductor modelled from MT data along line 95A (Fig. 9). The overall upper–middle-crustal conductivity also changes drastically between the SGM and the Erzgebirge. More regional phenomena are the general decrease from almost zero to −450 μm s^{-2} Bouguer gravity from NW to SE, and the change of induction arrow direction south of the SGM and the Münchberg Massif.

The deep crustal conductor below the SGM correlates either with a zone of weak reflectivity or (with regard to the depth confidence limits of 15–22 km) with the seismic reflectors that form the boundaries of this zone (Fig. 9). In the Rhenish Massif, a good conductor at similar depths (15 km) is interpreted as a tectonic shear zone, and electronic conductance as a result of graphite is assumed (ERCEUGT Group 1992). Structural modelling corroborated this interpretation (Behrmann *et al.* 1991). As there are strong objections against fluids in the lower crust (Yardley & Valley 1997), we favour an interpretation of the conductor below the SGM as a graphitic shear zone. The conductor and its lateral limitation may also represent the top of the Rheno-Hercynian plate assumed at similar depths (Oncken 1997), or it may correlate with the basal detachment of the thrust wedge that displaced the granulites at a later stage (Schäfer *et al.* this volume).

The seismic response from the lowermost crust indicates a complex reflective structure. Its character may be related to a change in the scale of heterogeneities near the crust–mantle boundary, possibly because of an intercalation of thin high- and low-velocity layers (Enderle *et al.* 1998). A classification of the lower crust in terms of reflectivity patterns as underthrust Rheno-Hercynian or Saxo-Thuringian basement is ambiguous.

In summary, all methods independently observe a segmentation of crustal architecture. All upper-crustal structures either end between the SGM and the Erzgebirge or show different characteristics (Fig. 9). Lower-crustal structures are more similar, but also indicate a slightly different nature north and south of the above boundary, with an important boundary identified from gravity modelling and reflectivity patterns. A similar, albeit less distinctive segmentation is observed in the profile 9501. Hence, a partly different origin and development of the upper–middle-crustal units imaged on the NW and SE halves of the NW–SE-striking profiles has to be envisaged.

Geological interpretation

The Saxo-Thuringian Belt developed from two successive thrust events that overprinted the earlier structures in the Carboniferous sequence (Schäfer *et al.* this volume). First, an older thrust wedge developed during the Saxo-Thuringian–Teplá–Barrandian collision and migrated northwestwards during Early Carboniferous time until *c.* 320 Ma, reaching as far as the exposed Saxonian granulites. A second thrust belt to the northwest developed as an upper-plate retro-wedge during the Saxo-Thuringian–Rheno-Hercynian collision to the north. It migrated SE-wards mainly at around 310 Ma as far south

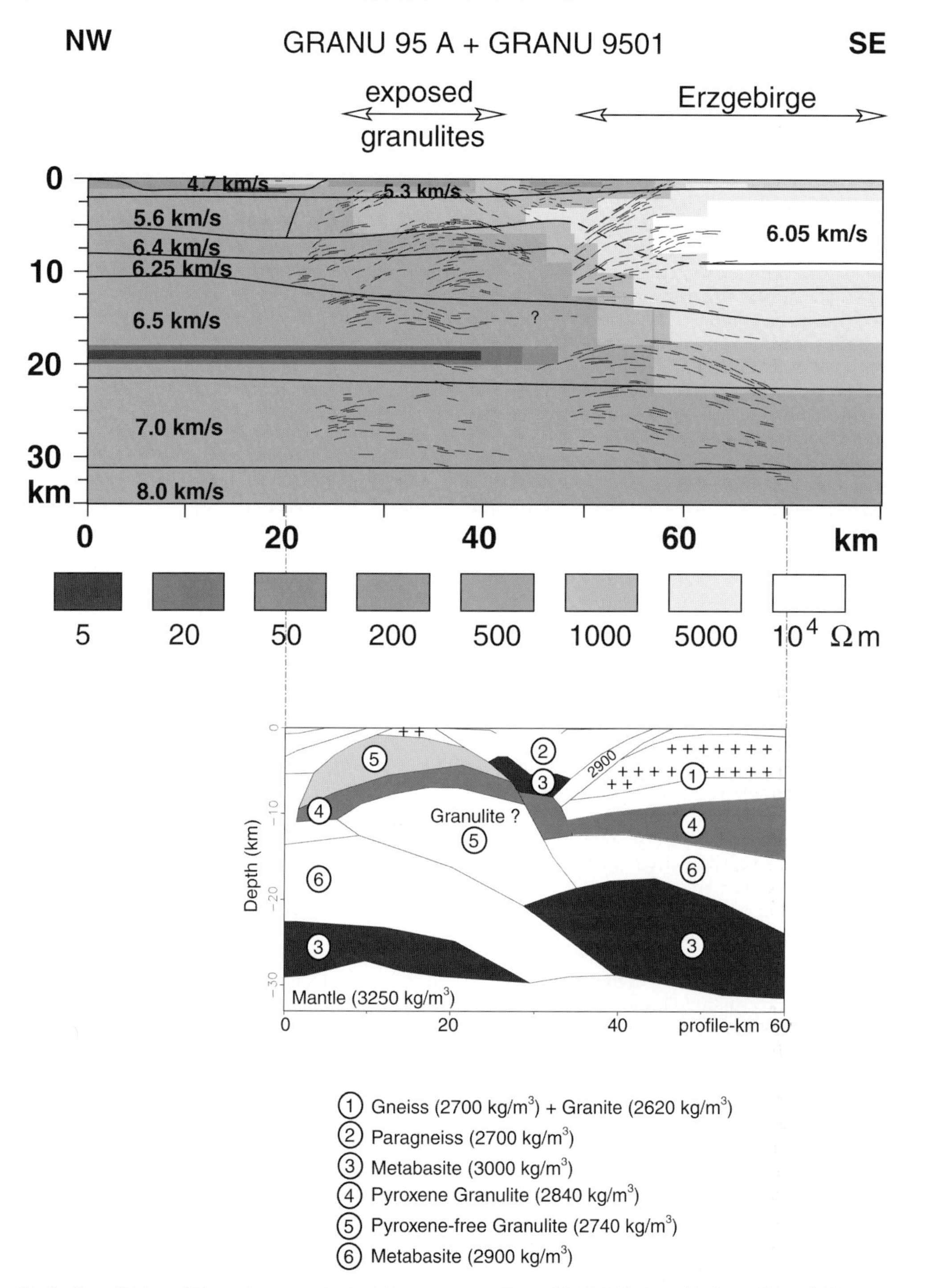

Fig. 9. Compilation of the various geophysical datasets along lines GRANU 95A and 9502, offset *c.* 25 km towards the SW (see Fig. 1 for location). The upper panel shows the combined MT and refraction velocity model, with the line drawing of line 9502 projected into it. The corresponding gravity model is shown below. All geophysical structures mark a change between the Saxonian Granulite Massif and the Erzgebirge extending to 15 km depth, indicating a different origin and development of the two HP units.

as the northern flank of the Fichtelgebirge–Erzgebirge area. Although Schäfer *et al.* (this volume) interpreted these structures in the vicinity of the seismic reflection profiles DEKORP '85-4N and MVE-90, the NE-ward continuation of both wedges is readily identified in the reflection lines 9501 and 9502 (Fig. 10).

The earlier Teplá–Barrandian thrust wedge is imaged as a stack of antiform-type and foreland-dipping duplexes involving the Palaeozoic Saxo-Thuringian sedimentary sequence (Fig. 10, km 30–57, light grey, square-dotted lines) as well as the overthrust basement nappes of the Erzgebirge and Frankenberg Units, an equivalent to the Münchberg Nappes further west (Fig. 10, km 30–57). Field observations clearly identify folding of these uppermost nappes above the subsequently stacked, flat to NW-dipping Palaeozoic thrust sheets. The underlying basal detachment is at around 16 km depth in the SE and shallows to the NW. The detachments of the basement nappes and of the Palaeozoic duplexes rise above surface along the southern flank of the Saxonian Granulite Dome (Fig. 10, km 20–25) and probably do not re-enter on the exposed northern flank (Fig. 10, km 0–15). Further west it can be observed that these detachments are cut off by the Göttengrün Fault on the southern flank of the Berga Anticline and are offset above the erosional surface.

The Göttengrün Fault forms part of the later SE-ward propagating retro-arc system. Its southernmost exposed features are the NW-dipping thrusts in the Lössnitz–Zwönitz Zone on the northern flank of the Erzgebirge and of the thrust system south of Plauen. The basal detachment of this system is rooted in a weakly NW-dipping level at some 15–17 km depth and also involves the basement below the Palaeozoic sediments, in contrast to the earlier thrust wedge (Fig. 10, km 0–25). Under the granulite dome, a backthrust rooted at the base of the ramp apparently helped in uplifting the dome at a late stage (Fig. 10, km 0–25, 10–17 km depth). This is also apparent from minor fault bend folding of the Viséan to Upper Carboniferous piggyback sedimentary basin (Hainichen Basin) between the Erzgebirge and the Saxonian granulites that unconformably overlies the earlier wedge. Displacement and shortening of the later wedge system were estimated to be only minor at its front (Schäfer *et al.* this volume), therefore largely retaining the pre-shortening structure in this area. The faults near the southern deformation front mainly reactivated the earlier crustal fabric and are graphitic near the surface, as revealed by mining in the Erzgebirge (Hösel *et al.* 1997). They might link up with the mid-crustal

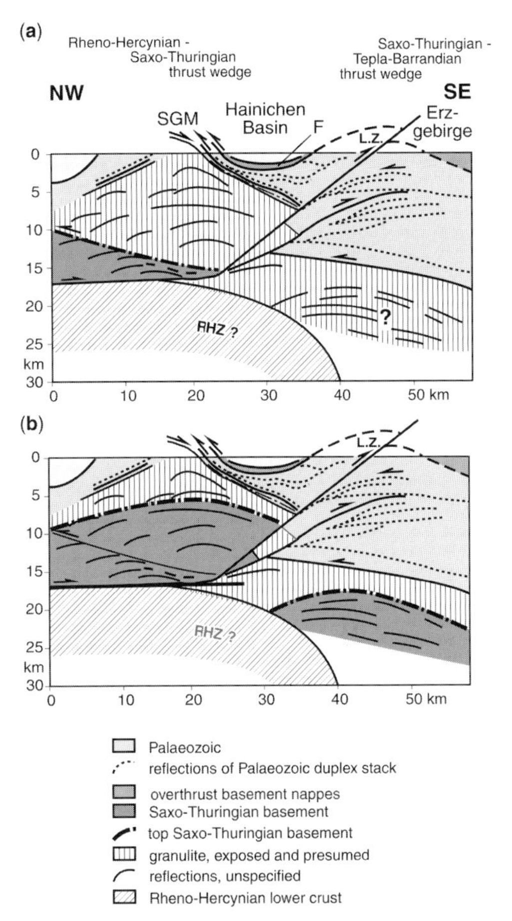

Fig. 10. Interpretation of the seismic reflection line GRANU 9502, crossing the Saxonian Granulite Massif (SGM) with a NW–SE trend. The development of two thrust wedges determines the reflection seismic structures in the upper 15 km of the crust. The depth extent of the exposed granulites, and thus the presence of Saxo-Thuringian basement, is disputable. Towards the SE, the granulites either extend down to Moho level (**a**), and the basement of the Palaeozoic cover would be largely lost, or the granulites continue at mid-crustal levels (**b**). F, Frankenberg; L.Z., Lössnitz–Zwönitz Zone; RHZ, Rheno-Hercynian lower crust.

good conductor. The downward-bending velocity layer in the refraction model between the Erzgebirge and Granulite dome is therefore preferably interpreted as offset by this fault zone rather than as a flexural structure.

Distribution of high-grade rocks

The reflection seismic data suggest that high-grade rocks are underlying the SGM antiform, and extend in the shallow subsurface towards the

SW. Lateral changes of reflective patterns, in the refraction velocity model, and in the gravity field suggest important changes of lithology along-strike. The main common features of the granulite axis are the lateral persistence of the dome structure and increased *P*-wave velocities combined with considerable reflectivity in the upper crust (Fig. 9). These observations are best explained by an updoming of high-grade metamorphic rocks of varying lithologies beneath the low-grade Palaeozoic sequences of the Saxo-Thuringian Belt.

It is not possible to trace the reflective and high-velocity layer in the upper–middle crust towards the SE into the Erzgebirge antiform, probably because of stacking and refolding of the rocks during the development of the Saxo-Thuringian–Teplá–Barrandian orogenic wedge in this region. The important change in crustal structure at the NW margin of the Fichtelgebirge–Erzgebirge Antiform is visible not only in all reflection profiles, but also in refraction profile 95A, and by the termination of the mid-crustal conductor at the NW margin of the Erzgebirge (Fig. 9).

On the basis of surface geology and seismic structures, the exposed granulite with its schist mantle shows a dome-like internal structure (Fig. 10, km 0–35) with its flanks reaching some 7–8 km depth. These are partly outlined by high-density bodies that may be associated with exposed ultramafic rocks found in the roof of the granulites. The depth extent of the seismically transparent granulites is under debate: it may encompass the entire stack of dome-shaped reflections, implying an interlayering of felsic and mafic granulites, down to the basal retro-wedge detachment at some 15–17 km depth (Fig. 10a, km 5–30), or, alternatively, it may be confined to the reflective high-velocity layer at some 7 km depth (Fig. 10b, km 5–30). If the exposed granulites were restricted to the transparent domains in the uppermost crust, then the reflectivity below the high-velocity boundary and below the backthrust between 10 and 16 km depth would also be interpreted as Saxo-Thuringian basement underlying the Palaeozoic sequence (Fig. 10b, km 0–25, dark grey). The latter alternative correlates better with the geometry proposed by gravity modelling (Fig. 9).

From seismic data in the SE half of the profile it is unclear whether the dome-shaped reflections below the detachment at 18–25 km depth may represent granulite (Fig. 10, km 30–55). From gravity modelling, the more weakly reflective NW part (km 0–30) is suggested to dip SE-wards under the lower-crustal reflection stacks in the SE (km 30–55). This would suggest some crustal stacking in the lower crust that is independent of the upper-crustal features. Crustal-scale balancing by Oncken (1997) has suggested that the underthrust Rheno-Hercynian lower crust should reach about as far as the Erzgebirge, which would agree well with this feature and the observed limit of the good mid-crustal conductor (Fig. 9).

Structures in profile 9501 are less pronounced, because of nearby granites, but still allow correlation (compare Fig. 3). Pro- and retro-wedge structures can be interpreted as in profile 9502. There is, however, no unambiguous evidence for the existence of granulites at shallow depth, although the reflectivity between 3 and 5 km depth correlates with an increased refraction velocity. The entire structural level of profile 9501 seems to be *c.* 2–3 km higher than in the East Erzgebirge, and the basal detachment is located at *c.* 12 km depth. This may be caused by the NW–SE-trending Gera–Jachimov Fault, a mainly Late Cretaceous reverse fault that separates the two profiles. A reflective lower crust is observed throughout the entire line again with similar features (Fig. 3).

Discussion of proposed geodynamic models

In summary, two alternative geometric interpretations for the occurrence of the HP granulites are supported by the geophysical data together with geological field evidence and map analysis. Either a thin (3–7 km thick) seismically transparent sheet of granulites separates the Palaeozoic sediments from their continental basement (Fig. 10b), or the granulite layer is substantially thicker and encompasses the upper-crustal rhomb-shaped unit (15 km thick) in the NW and the lower crust in the SE (Fig. 10a). In the latter case, the continental basement of the Palaeozoic sediments would have been largely removed in the area surveyed. From the granulite emplacement age (340 Ma) this process must have occurred before the formation of the two thrust wedges. From the established geometry, emplacement has probably occurred from the south or from below rather than from the north. Gravity modelling results corroborate this image and also indicate segmentation of the lower crust, suggesting the ending of a crustal unit dipping from the north. This is interpreted as the southward end of the underthrust Rheno-Hercynian lower crust, but may also hint at a downward continuation of granulitic material to the base of the crust.

Granulite emplacement has been explained by a variety of models that foresee different material

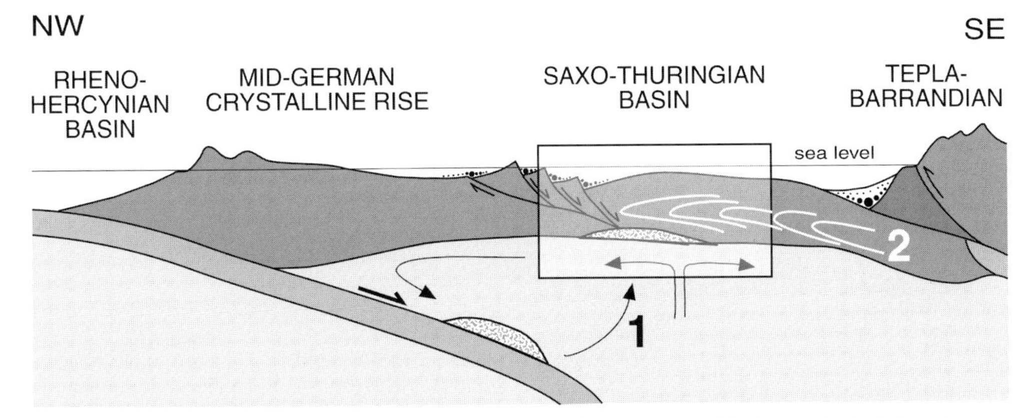

Fig. 11. Conceptual models of granulite emplacement that are in agreement with the geophysical data: 1, subduction erosion (Oncken 1998); 2, injection model (DEKORP & Orogenic Processes Working Groups 1999).

paths through the crust, and should potentially be imaged in the geophysical data. Several of the earlier suggested models have been rejected on mainly geological, geochemical and geochronological grounds. The classical metamorphic core complex model (e.g. Franke 1993; Reinhardt & Kleemann 1994) involving vertical uprise of Pre-Palaeozoic basement is incongruous with the irreconcilable observations of continuous marine sedimentation during uplift of granulitic material (Sebastian *et al.* 1993), which should have had at least 18 kbar (some 60 km) thickness (Rötzler 1992). A good conductor at mid-crustal depth may also suggest that at least part of the granulite exhumation arose from horizontal transport and not from vertical uplift through the entire crust.

Underplated and exhumed Rheno-Hercynian lower crust underthrust from the north is in conflict with the different lead isotope composition of the granulites (Molzahn *et al.* 1998) and with the fact that underthrusting started after collision at *c.* 335 Ma (Oncken 1997), which post-dates granulite emplacement and cooling. Buoyant rise of the SE-ward subducted Saxo-Thuringian lithosphere as recently suggested from analogue and numerical modelling of exhumation of HP rocks (Chemenda *et al.* 1995; Hynes *et al.* 1996) is in conflict with the required kinematic pattern at the top and base of the granulites and their position below sediments of the lower-plate foreland basin instead of below the upper-plate fore-arc. This mechanism, however, fits well the observations of the Erzgebirge HP rocks.

Two models are left (Fig. 11) that may be tested against the geophysical evidence presented. Both integrate age determinations, kinematic constraints and isotopic data. The first is that the granulites may have formed through subduction erosion of parts of the leading edge of the Saxo-Thuringian Plate in the north and their subsequent underplating under the Saxo-Thuringian Basin, after buoyantly rising through the upper-plate asthenosphere (Oncken 1998). Further emplacement into the upper crust during continuing marine sedimentation would require substantial crustal extension before shortening. The second model, the so-called injection model, transports Saxo-Thuringian lower-crustal material subducted to the SE below the Teplá–Barrandian Unit back northwards by intracrustal plug flow, with contemporaneous extension of the Saxo-Thuringian upper crust to accommodate marine sedimentation (DEKORP & Orogenic Processes Working Groups 1999). This model has additionally been tested by thermo-mechanical modelling (Henk this volume), corroborating the general feasibility of this mechanism under the condition of an overheated crust.

Both models (Fig. 11) may be seen to agree with the two interpretations of the geophysical data discussed here (Fig. 10). The subduction erosion model is probably consistent only with the interpretation that suggests large volumes of HP granulites in the crust down to the Moho (Fig. 10a). Ultramafic bodies, identified as derived from a hot mantle (Werner 1981), are consistently associated with the HP granulites in the field and are suggested to continue towards depth from the interpreted high densities and high velocities in the crust. This would support a mantle contact of the granulites on their path which agrees with the interpretation and a subduction erosion origin. The injection model is compatible with both interpretations of the geophysical constraints (Fig. 10).

Conclusions

Geophysical modelling of seismic reflection and refraction, gravity and MT data integrating geological observations in the area of the Saxonian Granulite Massif and Erzgebirge leads to the following conclusions:

(1) All geophysical structures mark a structural change between the Saxonian Granulite Massif and the Erzgebirge down to 15 km depth, indicating a different origin and development of the two HP units (Fig. 9). Interpretation of the upper-crustal structure from field geology and seismic reflection data clearly shows two subsequently developed thrust wedges: an earlier one involving the Saxo-Thuringian Palaeozoic sediments and propagating from the south to the north transporting the allochthonous Erzgebirge Nappes in piggyback fashion; a later one propagating from the north, which undercuts and uplifts the Saxonian granulites and overthrusts the northern flank of the Erzgebirge antiformal stack. The detachments of both wedges root at mid-crustal depth between the upper and lower crust, which also differs from north to south (Fig. 10).

(2) The dome-like structure expressed by the seismic reflection data beneath the SGM (Fig. 2) and modelled in the gravity model (Fig. 8) is either composed of layered felsic to pyroxene granulites down to mid-crustal depth or indicates a thin sheet of felsic HP granulites between the Saxo-Thuringian Palaeozoic sedimentary sequence and its continental basement. The continuation of the granulites in the subsurface is to the southeast in both cases, either at mid-crustal levels or connected to the Moho, thereby prohibiting an origin from the north (Fig. 10).

(3) The geophysical results may be seen to support two models of granulite emplacement: the first involves buoyant rise of Saxo-Thuringian material through the asthenospheric mantle after tectonic erosion from the upper plate in the subduction zone to the north; the second involves NW-ward intrusion of hot and weak material by intracrustal plug flow from the Saxo-Thuringian subduction zone in the SE (Fig. 11). Both models envisage crustal extension for the final emplacement into the upper crust.

This manuscript has benefited from discussions with numerous colleagues, among them H. Kämpf and H.-J. Förster. The thorough reviews of H. Thybo, D. Tanner and an anonymous reviewer helped to improve the manuscript. We gratefully acknowledge funding of our research, which was provided by Deutsche Forschungsgemeinschaft through the Special Research Programme 'Orogenic Processes' (projects Go380/17-1, -2; Ha1210/19; On7/9-1; Pr74/17-3, -4). M. Stiller is thanked for help during processing, and the Geophysical Instrument Pool Potsdam for technical support during the field campaigns.

References

AHRENDT, H., FRANZKE, H. J., MARHEINE, D., SCHWAB, M. & WEMMER, K. 1996. Zum Alter der Metamorphose in der Wippraer Zone/Harz—Ergebnisse von K/Ar-Altersdatierungen an schwachmetamorphen Sedimenten. *Zeitschrift der Deutschen Geologischen Gesellschaft*, **147**(1), 39–56.

ANSORGE, J., BLUNDELL, D. & MUELLER, S. 1992. Europe's lithosphere—seismic structure. *In*: BLUNDELL, D., FREEMAN, R. & MUELLER, S. (eds) *A Continent Revealed: the European Geotraverse*. Cambridge University Press, Cambridge, 33–70.

BEHRMANN, J., DROZDZEWSKI, G., HEINRICHS, T., HUCH, M., MEYER, W. & ONCKEN, O. 1991. Crustal-scale balanced cross sections through the Variscan fold belt, Germany: the central EGT-segment. *Tectonophysics*, **196**, 1–21.

CHEMENDA, A. I., MATTAUER, M., MALAVIEILLE, J. & BOKUN, A. N. 1995. A mechanism for syncollisional rock exhumation and associated normal faulting: results from physical modelling. *Earth and Planetary Science Letters*, **132**, 225–232.

CHOI, S., GÖTZE, H.-J. & KRENTZ, O. 1998. Gravity stripping of Palaeozoic crustal domains—an example from the Erzgebirge. *Terra Nostra*, **98**(2), 38–41.

CONRAD, W., HAUPT, M. & BÖLSCHE, J. 1994. Interpretation des tiefenseismischen Regionalprofils EV01-EV02/1978–80 Vogtland–Erzgebirge–Lausitz (Adorf-Gutzen) mit Hilfe von Gravimetrie und Magnetik. *Zeitschrift für Geologische Wissenschaften*, **22**, 603–615.

DEKORP RESEARCH GROUP 1988. Results of DEKORP4/KTB Oberpfalz deep seismic reflection investigations. *Journal of Geophysics*, **62**, 69–101.

—— 1994. The deep reflection seismic profiles DEKORP3/MVE-90. *Zeitschrift für Geologische Wissenschaften*, **22**, 623–825.

DEKORP & OROGENIC PROCESSES WORKING GROUPS 1999. Structure of the Saxonian granulites: geological and geophysical constraints on the exhumation of high-pressure/high-temperature rocks in the mid-European Variscan belt. *Tectonics*, **18**(5), 756–773.

EISEL, M. & HAAK, V. 1999. Macro-anisotropy of the electrical conductivity of the crust: a magnetotelluric study from the German Continental Deep

Drilling site (KTB). *Geophysical Journal International*, **136**, 109–122.

ELEKTB GROUP 1997. KTB and the electrical conductivity of the crust. *Journal of Geophysical Research*, **102**, 18289–18306.

ENDERLE, U., SCHUSTER, K., PRODEHL, C., SCHULZE, A. & BRIBACH, J. 1998. The refraction seismic experiment GRANU 95 in the Saxothuringian zone, SE-Germany. *Geophysical Journal International*, **133**, 245–259.

ERCEUGT GROUP 1992. An electrical resistivity crustal section from the Alps to the Baltic Sea (central segment of the EGT). *Tectonophysics*, **207**, 123–139.

FRANKE, W. 1984. Variszischer Deckenbau im Raume der Münchberger Gneismasse, abgeleitet aus der Fazies, Deformation und Metamorphose im umgebenden Paläozoikum. *In*: DALLMEYER, R. D. (ed.) *Terranes in the Circum-Atlantic Palaeozoic Orogens. Geotektonische Forschungen*, **68**, 1–253.

—— 1989. Tectonostratigraphic units in the Variscan belt of central Europe. Geological Society of America, Special Papers, **230**, 67–90.

—— 1993. The Saxonian granulites: a metamorphic core complex. *Geologische Rundschau*, **82**, 505–515.

—— & STEIN, E. 2000. Exhumation of high-grade rocks in the Saxo-Thuringian Belt: geological constraints and geodynamic concepts. *This volume*.

HENK, A. 2000. Foreland-directed lower-crustal flow and its implications for the exhumation of high-pressure–high-temperature rocks. *This volume*.

HÖSEL, G., TISCHENDORF, G. & WASTERNACK, J. 1997. Erläuterungen zur Karte 'Mineralische Rohstoffe Erzgebirge–Vogtland 1:100 000'. *Bergbau in Sachsen*, **3**, 7–104.

HYNES, A., ARKANI-HAMED, J. & GREILING, R. 1996. Subduction of continental margins and the uplift of high-pressure metamorphic rocks. *Earth and Planetary Science Letters*, **140**, 13–25.

KRENTZ, O., LEONHARDT, D. & BERGER, H.-J. 1997. Einbindung von Tiefenbohrungen in die Interpretation des tiefenseismischen Profils Granu 9502. *Terra Nostra*, **97**(5), 94–97.

MATTE, P. 1991. Accretionary history and crustal evolution of the Variscan Belt in Western Europe. *Tectonophysics*, **196**(3–4), 309–337.

MINGRAM, B. 1996. *Geochemische Signaturen der Metasedimente des erzgebirgischen Krustenstapels*. Scientific Technical Report, GFZ Potsdam, **STR96/04**.

MOLZAHN, M., ONCKEN, O. & REISCHMANN, T. 1998. Isotopengeochemische Charakterisierung von Krustenblöcken innerhalb der Varisziden Mitteleuropas. *Terra Nostra*, **98**(2), 108–109.

MÜLLER, H. J. 1995. Modelling the lower crust by simulation of *in situ* conditions: an example from the Saxonian Erzgebirge. *Physics of the Earth and Planetary Interiors*, **92**, 3–15.

OETTINGER, G. 1999. *Magnetotellurische Messungen im Sächsischen Granulitgebirge: Separation von Nutz- und Störsignalen und Verteilung der elektrischen Leitfähigkeit*. PhD thesis, FU Berlin.

ONCKEN, O. 1997. Transformation of a magmatic arc and an orogenic root during oblique collision and its consequences for the evolution of the European Variscides (Mid-German Crystalline Rise). *Geologische Rundschau*, **86**, 2–20.

—— 1998. Evidence for precollisional subduction erosion in ancient collisional belts: the case of the Mid-European Variscides. *Geology*, **26**(12), 1075–1078.

PLAUMANN, H. 1986. Die Schwerekarte der Oberpfalz und ihre Bezüge zu Strukturen der oberen Erdkruste. *Geologisches Jahrbuch*, **33**, 5–13.

REINHARDT, J. & KLEEMANN, U. 1994. Extensional unroofing of granulite lower crust and related low-pressure, high-temperature metamorphism in the Saxonian Granulite Massif, Germany. *Tectonophysics*, **238**, 71–94.

RÖTZLER, J. 1992. Zur Petrogenese im Sächsischen Granulitgebirge—Die pyroxenfreien Granulite und die Metapelite. *Geotektonische Forschungen*, **77**, 1–100.

SÄCHSISCHES LANDESAMT FÜR UMWELT UND GEOLOGIE 1996. *Schwerekarte der Länder Brandenburg, Mecklenburg-Vorpommern, Sachsen, Sachsen-Anhalt. Geoprofil*, **6**.

SCHÄFER, F., ONCKEN, O., KEMNITZ, H. & ROMER, R. L. 2000. Upper-plate deformation during collisional orogeny; a case study from the German Variscides (Saxo-Thuringian Zone). *This volume*.

SCHMÄDICKE, E., MEZGER, K., COSCA, M. A. & OKRUSCH, M. 1995. Variscan Sm–Nd and Ar–Ar ages of eclogite facies rocks from the Erzgebirge, Bohemian Massif. *Journal of Metamorphic Geology*, **13**, 537–552.

SCHULZE, A. & BORMANN, P. 1990. Deep seismic sounding in eastern Germany. *In*: FREEMAN, R., GIESE, P. & MUELLER, S. (eds) *The European Geotraverse: Integrative Studies*. European Science Foundation, Strasbourg, 109–114.

SEBASTIAN, U., FACHMANN, S., RÖTZLER, K. & HOFMANN, J. 1993. Postcompressive structural development in a Variscan reversed and reduced metamorphic sequence. *Terra Abstracts*, **5**(2), 31.

SOBOLEV, S. V. & ANDREY, Y. B. 1994. Modelling of mineralogical composition, density and elastic wave velocities in anhydrous magmatic rocks. *Surveys in Geophysics*, **15**, 515–544.

VAVRA, G. & REINHARDT, J. 1999. Low-resistance of granulite zircons to post-climax alteration and Pb-loss—New strategies for dating high-grade metamorphic processes. *Terra Nostra*, **98**(2), 149–151.

VON QUADT, A. 1993. The Saxonian Granulite Massif: new aspects from geochronological studies. *Geologische Rundschau*, **82**, 516–530.

WEBER, K. & BEHR, H. J. 1983. Geodynamic interpretation of Mid-European Variscides; case studies in the Variscan Belt of Europe and the Damara Belt in Namibia. *In*: MARTIN, H. & EDER, F. W. (eds) *Intracontinental Fold Belts*. Springer, Berlin, 427–469.

WERNER, C. D. 1981. Outline of the evolution of the magmatism in the G.D.R. *Guidebook PK IX 2, Potsdam/Freiberg*, 17–68.

WERNER, O. & LIPPOLT, H. J. 2000. White-mica $^{40}Ar/^{39}Ar$ ages of Erzgebirge metamorphic rocks: simulating the chronological results by a model of Variscan crustal imbrication. *This volume*.

—— & REICH, S. 1997. $^{40}Ar/^{39}Ar$-Abkühlalter von Gesteinen mit unterschiedlicher *P–T*-Entwicklung aus dem Schiefermantel des Sächsischen Granulitgebirges. *Terra Nostra*, **97**(5), 196–198.

YARDLEY, B. W. D. & VALLEY, J. W. 1997. The petrologic cause for a dry lower crust. *Journal of Geophysical Research*, **102**, 12173–12185.

White mica $^{40}Ar/^{39}Ar$ ages of Erzgebirge metamorphic rocks: simulating the chronological results by a model of Variscan crustal imbrication

OLAF WERNER & HANS J. LIPPOLT

Laboratorium für Geochronologie, Ruprecht-Karls-Universität, INF 234, D-69120 Heidelberg, Germany (e-mail: olaf.werner@urz.uni-heidelberg.de)

Abstract: In the Erzgebirge, which is part of the Saxo-Thuringian Mid-European Variscides, an $^{40}Ar/^{39}Ar$ study was performed on white mica and hornblende separates from 68 metamorphic rocks of varying lithologies. Two groups of late Early Carboniferous cooling ages of 340 ± 2 Ma and 329.7 ± 1.5 Ma are distinguished by evaluating spectra shapes. A third group of rocks yielded intermediate argon ages, presumably related either to thermal rejuvenation or to neoformation of white mica during a reheating process 330 Ma ago. A model that explains rejuvenation is proposed. The observed age patterns and the shape of the age spectra cannot be simply explained by assuming a single uplift and exhumation process set in motion by extensive forces. Chronological and structural arguments favour the idea that Variscan crustal imbrication took place during the mineral cooling phase. A tectonic process that intercalated, at about 20 km depth, a unit of cool rocks into a much hotter environment suitably models the chronological results. This is shown by a simulation of the age record. Temperature profiles are calculated for both intercalation and subsequent uplift models. Rocks in these profiles, which preserved their 340 Ma signature, are distinguished from those heated 330 Ma ago by contact with adjacent hotter rocks and reset isotopically to varying degrees. The agreement between analytical results and modelled ages favours the crustal imbrication hypothesis as explanation for the age distribution throughout all lithologies.

The Erzgebirge in Central Europe consists mainly of Pre- to Early Variscan metamorphic rock units and Variscan postkinematic granitic plutons. It is an element of the Saxo-Thuringian Zone of the Mid-European Variscides (Kossmat 1927).

Since 1959, isotopic dating has been undertaken by several research groups to refine the geological history of the Erzgebirge. Mainly K–Ar techniques were applied (summarized by Haake (1972) and Lorenz & Pilot (1994)). K–Ar results determined on whole-rock samples, mineral concentrates and separates gave a range of 450–280 Ma for metamorphic rocks and 380–280 Ma for plutonic rocks. Rb–Sr whole-rock and K–Ar mica studies by Gerstenberger (1989) and Gerstenberger *et al.* (1995) yielded granite ages between 320 and 310 Ma.

More recent investigations included Sm–Nd, U–Pb and $^{207}Pb/^{206}Pb$(rad) techniques (Schmädicke *et al.* 1995; Kröner & Willner 1998) to obtain the age range of high-pressure metamorphism. Additionally, to recognize metamorphic evolution steps, Tikhomirova *et al.* (1995) measured mica Rb–Sr ages between 330 and 320 Ma for the eastern part of the Erzgebirge. There are three reasons why a further, comprehensive argon age study may advance our knowledge of the Erzgebirge basement. First, the relatively wide spread of published data is astonishing, although a plausible geological model has been developed (Lorenz & Pilot 1994). Second, the success of argon chronometry of rocks of the Dora–Maira massif in the Western Alps is encouraging, because numerous potassium–argon investigations had shown that white mica had partially retained its isotopic inventory even through high-pressure metamorphism (e.g. Monié 1990; Scaillet 1996). Third, the geological setting of the various metamorphic units of the Erzgebirge suggested that cooling patterns might exist from which information on the tectonic evolution could be inferred.

Therefore extensive $^{40}Ar/^{39}Ar$ studies (total fusion and step-degassing experiments) on white mica and other minerals from the Erzgebirge have been performed. They narrowed the age

From: FRANKE, W., HAAK, V., ONCKEN, O. & TANNER, D. (eds). *Orogenic Processes: Quantification and Modelling in the Variscan Belt.* Geological Society, London, Special Publications, **179**, 323–336. 0305-8719/00/$15.00

ranges for the metamorphic and magmatic rocks between 342 and 330 Ma and between 320 and 310 Ma, respectively (Werner & Lippolt 1998, in prep.). Only few exceptions, which do not impair the white mica ages of the metamorphic rocks, were observed. The high precision of the $^{40}Ar/^{39}Ar$ technique provided isotopic dates with uncertainties of only 1.2–2.5 Ma. It therefore became possible to decipher differences between the cooling ages of distinct metamorphic units as well as between grain-size fractions. The age differences between the metamorphic units cannot be explained by uniform uplift of the whole area. They suggest, however, that a detailed tectonic model worked out by structural geologists (e.g. Nega *et al.* 1997; Konopàsek & Schulmann 1999), which assumes crustal-stacking processes at greater depth in the Erzgebirge crust during Carboniferous time, should be considered.

In the following, we summarize our analytical evidence and present model calculations. The models describe the thermal evolution of the investigated part of the crust after the supposed imbrication event, and demonstrate consistently how these temperatures may have influenced the evolution of the present cooling-age patterns. The agreement of analytical record and simulated model ages may serve as circumstantial evidence for Carboniferous stacking processes in the Erzgebirge crust.

Geological and analytical setting

Geological situation

The Erzgebirge is located to the north of the Bohemian Massif and adjoins the Saxonian Granulite Massif to the north (see enclosure at back). Together with the Fichtelgebirge, it has an antiform structure exposing rocks of high metamorphic grade in the core and of lower grade in the cover (e.g. Pietzsch 1954; Lorenz & Hoth 1990). However, the present regional coexistence of the metamorphic rocks has been complicated by later tectonic processes. Figure 1 shows the location of the main rock units. The core of the gneiss dome, made up of red and grey gneisses, is overlain by a series of high-pressure metamorphic rocks (gneisses and mica schists) with intercalations of granulite and eclogite relics (demonstrating metamorphic inversion). In the last-mentioned rocks high- and low-temperature units are distinguished (gneiss–eclogite and mica schist–eclogite units, respectively; Willner *et al.* 1997). The gneisses of the mica schist–eclogite unit are intercalated with eclogite relics, whereas the mica schists are free from eclogite relics. The mica schists are covered by rocks of greenschist to anchimetamorphic facies (garnet-bearing phyllites, phyllites, slates). In Fig. 2 the probable stratigraphic relationship of all these lithologies is shown. The contacts between the core and the cover are formed by ductile shear zones. The original metamorphic position of the granulite and eclogite relics in the gneisses of the high-pressure rock unit is unknown. Granitic rocks mainly are exposed in the SW and the NE of the dome structure. A detailed geological summary has been given by Franke & Stein (this volume).

Tectonic models and age prediction

Several tectonic models were proposed to explain the observed rock coexistence in the exposures of the Erzgebirge. Mainly extensional scenarios, caused either by primary rifting or by orogenic collapse, were assumed (Kroner & Voigt 1992; Franke 1993; Krohe 1996; Willner *et al.* 1997). Other workers (Nega *et al.* 1997; Konopàsek & Schulmann 1999) argued for models of orogenic collision, thrusting Moldanubian rocks over subducted Saxo-Thuringian domains.

The tectonic models can be used to predict cooling-age patterns as a basis for comparisons with measured isotopic cooling ages. Higher ages imply that cooling below a critical crustal temperature occurred earlier than in a rock with lower ages. A model of compressional stacking or nappe formation would imply lower cooling ages for the metamorphic cover because the hot cover rocks would be pushed onto a cooler core and therefore would cool later. Higher cooling ages of the metamorphic cover rocks could be indicative of a model of primary rifting caused by uniform thermal gradients and slow uplift. The orogenic collapse model demands uniform cooling ages because of fast uplift and cooling.

When modelling an imbrication process, in this case two shear zones have to be postulated. The lower one could be the transition of the red and grey gneiss unit into the gneiss-eclogite unit. Petrological investigations on retrograde mylonites of this transition zone indicate 8 kbar and 600 °C as boundary conditions of shearing (Rötzler 1995; Willner *et al.* 1997). The existence of an upper shear zone is suggested by the contrasting metamorphic grades between the gneiss-eclogite unit and the mica schist–eclogite unit (Sebastian & Kroner 1992).

Analytical evidence

Taking into account that isotopic mineral ages of the Erzgebirge cover only a narrow range and

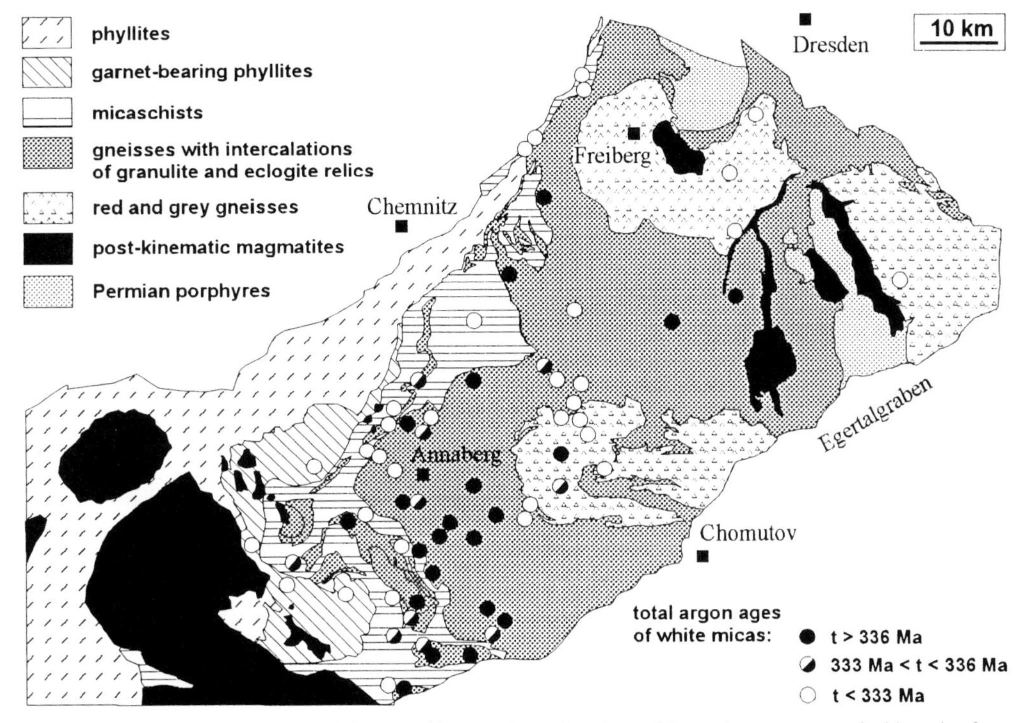

Fig. 1. Simplified geological map of the Erzgebirge: main rock units and isotopic age groups of white mica from metamorphic rocks.

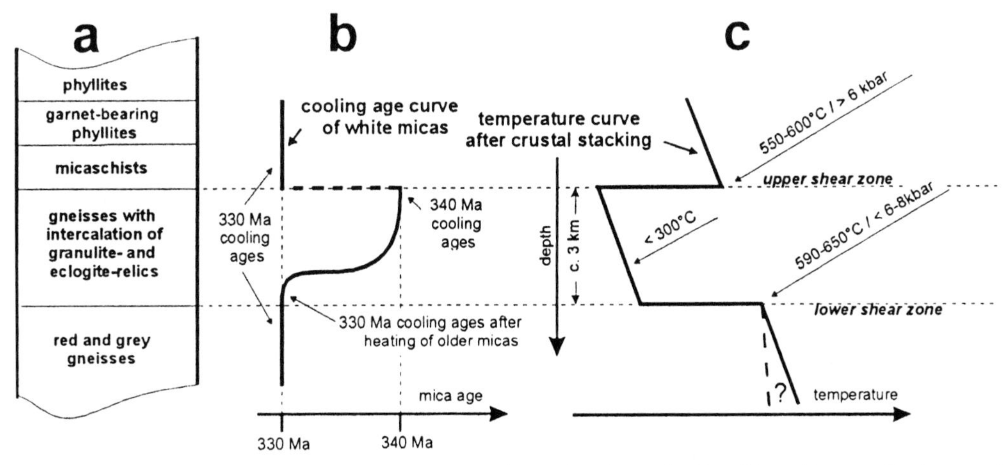

Fig. 2. (**a**) Schematic stratigraphic section of metamorphic rock units of the Erzgebirge; (**b**) generalized curve of cooling ages (in Ma) of white mica; (**c**) hypothetical temperature profile 330 Ma ago after imbrication process.

may be distinguished only by measurements with high accuracy, we consider here, besides our own data (Werner & Lippolt 2000*a*, *b*), mainly zircon dates measured by Kröner & Willner (1998) and Rb–Sr ages determined by Tikhomirova *et al.* (1995). The Pb-evaporation zircon age of 341.2 ± 0.3 Ma (1σ mean error) from granulite facies metamorphic rock is thought to mark the peak of high-pressure metamorphism (Kröner & Willner 1998). Further evaluations of the analytical accuracy of these measurements have been given by Franke & Stein (this volume), who stated an age of 340 ± 4 Ma (1σ error). This is a mineral formation age by definition and a upper age limit for regional uplift. $^{40}Ar/^{39}Ar$ (and Rb–Sr) dates, however, are cooling ages, insofar as their closure temperatures are lower than the mineral formation temperatures.

With a few exceptions, the $^{40}Ar/^{39}Ar$ cooling ages of *c.* 80 white mica and two hornblende separates are below 342 Ma and above 330 Ma. The argon age results will be documented in detail elsewhere (Werner & Lippolt, in prep.). They are based on both total fusion and step-degassing analyses (age spectra with both integrated and plateau step ages), accompanied by multiple analyses and grain-size investigations. Apparently, within 10 Ma during late Early Carboniferous time, all the investigated metamorphic rocks cooled below the relevant closure temperatures.

Two groups of late Early Carboniferous cooling ages of 340 ± 2 Ma (1σ accuracy) and 329.7 ± 1.5 Ma (1σ accuracy) are distinguished by evaluating spectra shapes (Werner & Lippolt, in prep.). The analytical error of the $^{40}Ar(rad)/^{39}Ar$ ratio of both monitors and samples is *c.* 0.1% (1σ error). The poorer error is always higher. The less age precision even of samples with a high grade of homogeneity is mainly caused by the gradient of neutron intensity during activation. It is in a range between 0.5 and 2.5 Ma (1σ precision). The accuracy additionally considers the age error of the monitor (*c.* 0.3%).

A third group of rocks yielded intermediate argon ages, presumably related to thermal rejuvenation or neoformation of white mica during a reheating process 330 Ma ago.

Figure 3a–f shows six typical spectra. Four white mica and two hornblende separates are demonstrated. All measured white mica spectra belong to one of the four spectra groups. Two white mica spectral types show unquestionable plateaux (Fig. 3a and c), the other two show regular step age distributions (Fig. 3d and e) between 342 and about 330 Ma. The latter are interpretable with respect to their cooling ages, argon diffusion losses and neoformation of white mica 330 Ma ago (Werner & Lippolt, in prep.). The highest plateau ages of about 340 Ma were found for phengite (Fig. 3a) and hornblende (Fig. 3b) from the high-pressure gneisses and their granulitic and eclogitic intercalations. This age coincides with the previously mentioned zircon age and thus suggests that the host rocks cooled very fast. Plateau ages of about 330 Ma (Fig. 3c) were found for phengite and muscovite from red and grey gneisses, high-pressure mica schists and garnet-bearing phyllites. The micas with type e spectra show, besides their cooling ages at about 330 Ma, both a remnant (at high-temperature steps) and a potentially extraneous excess argon component (at low-temperature steps).

With a two-stage temperature history (cooling at 340 Ma and reheating at 330 Ma with partial argon loss), the measured total fusion $^{40}Ar/^{39}Ar$ ages carry essentially the same information as the age spectra. The partial argon loss 330 Ma ago (in per cent) equals the [(measured total argon age − 330 Ma)/10 Ma] × 100. The hornblende spectrum type f may be explained in a similar manner. The micas with type d spectrum prove the existence of mixtures with 330 Ma muscovite and older phengite (Werner & Lippolt, in prep.).

The regional distribution of white mica age groups is shown in Fig. 1. The lowest cooling ages of coarse-grained metamorphic white mica amount to 329.7 ± 0.5 Ma (precision), ± 1.5 Ma (accuracy). The garnet-bearing phyllites, the mica schists and the red and grey gneisses belong to this age group. Apparent contradictions to this assignment have been explained elsewhere (Werner & Lippolt, in prep.). For all rocks of this age group, excess argon, grain-size effects and thermal overprint by post-kinematic granites can be excluded. The argon age dates for the 340 Ma rock units (gneisses with intercalations of granulite and eclogite relics) show a relationship to both the lithostratigraphic position of the rocks and the size of the measured mica fraction. The coarse phengites (>1 mm) from high positions in the section yielded undisturbed age dates between 341 and 339 Ma with only negligible additional argon components. The finer fractions yielded mainly disturbed spectra with lower total argon ages. The grain-size relation can be reasonably explained by a heating event at about 330 Ma which influenced argon retention as a function of grain size.

The exceptions with regard to the mentioned narrow band of isotope ages between 340 and 330 Ma are two hornblendes (from low-temperature mica schist–eclogite unit) in the West Erzgebirge and two white mica separates (from high-temperature gneiss–eclogite unit) in the East Erzgebirge. Provided that their argon is a remnant of an older component and not of extraneous origin, they have cooling ages of about 380–360 Ma. Taking into account the shape of spectra and the temperature of metamorphism, it is probable that the hornblendes could have partially preserved their isotopic inventory even through high-pressure metamorphism at 340 Ma or the reheating process at 330 Ma ago, but this is not possible for the white mica. The shape of the white mica spectra show that 340 or 330 Ma ago rejuvenation and neoformation of white mica influenced the isotopic system. The consequences

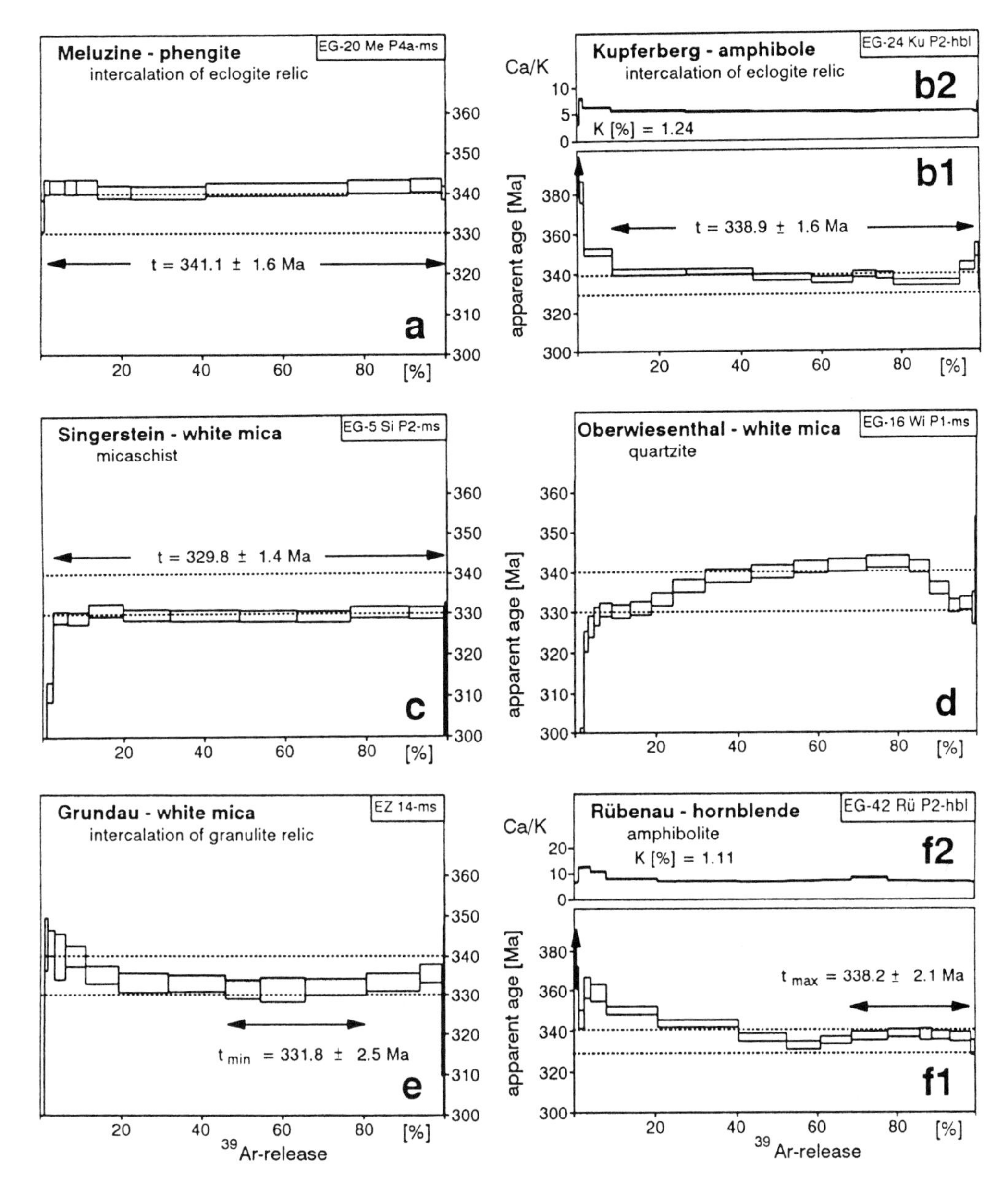

Fig. 3. Examples of $^{40}Ar/^{39}Ar$ spectra of white mica (**a**, **b**, **d**, **e**) and hornblende separates (**b1**, **f1**, together with Ca/K spectra **b2**, **f2**) of metamorphic rocks from the Erzgebirge. All measured spectra belong to one of these spectral types. (**a**, **b**) Plateau-type spectra of phengites or hornblende with ages of 340 Ma; (**c**) plateau spectra with 330 Ma ages; (**d**) disturbed spectra of phengite–muscovite mixtures; (**e**, **f1**) spectra of white mica or hornblende disturbed by resetting 330 Ma ago. Errors: 1σ precision; analytical error of $^{40}Ar(rad)/^{39}Ar$ ratio *c*. 0.1% (1σ error).

with regard to the tectonic model are addressed in the section Inferences.

In addition, biotites from metamorphic rocks were also dated using Rb–Sr and K–Ar methods. They yielded lower age values than the white mica, possibly because of alterations, which occurred during pluton intrusion and hydrothermal mineralization. It can be excluded with certainty that the observed age differences indicate slow geological cooling.

In this work the relatively wide spread of published data of cooling ages is reduced to a narrow band of a few million years. The majority of white micas obviously did not retain their isotopic inventory through high-pressure metamorphism as observed for examples from Dora–Maira.

Ages, lithologies and the crustal imbrication hypothesis

The mineral cooling ages of the investigated rocks are summarized in a cooling-age curve in Fig. 2b, by which observed cooling ages and rock units of the schematic metamorphic profile of the Erzgebirge are related. Rock units with cooling ages of about 340 or 330 Ma and rocks with apparent intermediate ages are distinguished. The core (gneisses with intercalations) of the demonstration is mainly based on data shown in Fig. 4. Disregarding neoformation of white mica 330 Ma ago by evaluating the spectra shapes, Fig. 2 shows cooling ages and intermediate ages related to thermal rejuvenation during a reheating process presumably 330 Ma ago as well.

Figure 2b demonstrates that the intercalation of a rock unit with a cooling age of about 340 Ma may be assumed. This unit now has cooling ages of about 340 Ma in the upper part, whereas in the lower part the cooling ages decrease to 330 Ma. All other rocks of the reconstructed profile have cooling ages of only 330 Ma. The derived cooling-age profile therefore requires a crustal imbrication event 330 Ma ago. Neoformation of 330 Ma white mica in rocks, originally cooled 340 Ma ago, also indicates the time of new shearing and heating. Because arguments from structural geology are also in favour of crustal imbrication in the West Erzgebirge (e.g. Nega *et al.* 1997; Konopàsek & Schulmann 1999), it is likely that cool rocks (i.e. below the closure temperatures of white mica) were enclosed in a crustal segment with rocks at higher temperatures. Subsequent heating of the rocks to temperatures above the Ar closure temperature partly may have opened the Ar isotope systems of the white mica, above all in the lower parts of the intercalated unit. Subsequently, the

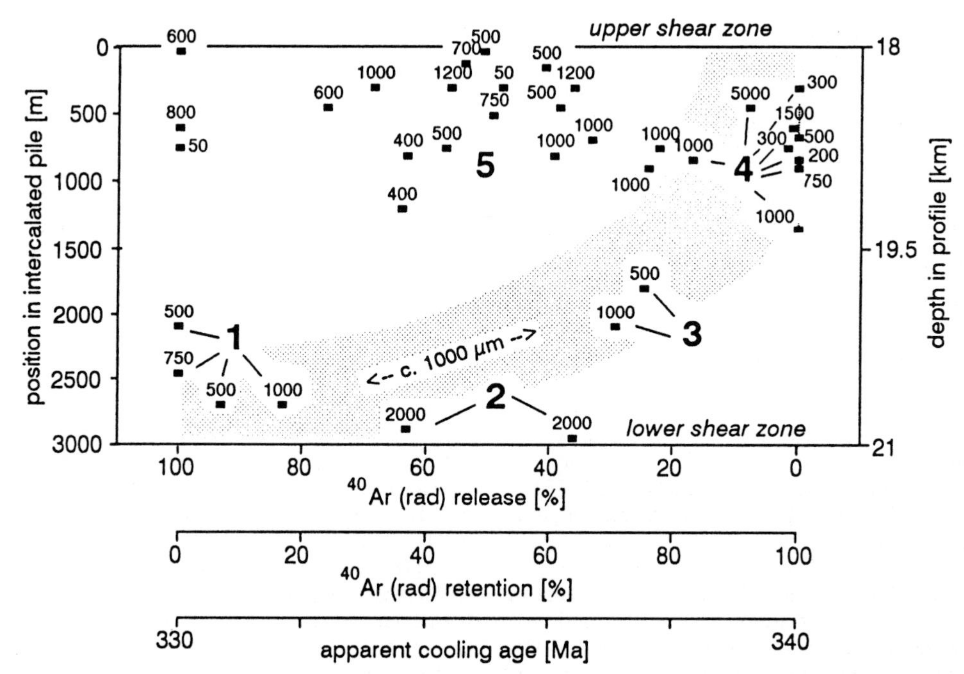

Fig. 4. Position of investigated rocks with white mica in the crustal profile (right axis) and imbricated rock pile (left axis), in relation to the apparent cooling ages (on the lowermost abscissa). The cooling ages are transformed into hypothetical ^{40}Ar(rad) retention and release values (per cent; on the other two abscissae) based on the assumption that 10 Ma old rocks were partially degassed 330 Ma ago. The large numbers denote age groups (see below), the small ones mica grain sizes (in µm), determined on thin section. The shaded area corresponds to the cooling-age curve of Fig. 2b. Five white mica types are distinguished: 1, completely reset 330 Ma ages of minerals near the lower shear zone; 2, partially reset ages of coarse-grained minerals near the lower shear zone; 3, partially reset ages of minerals in the centre of the intercalated rock pile; 4, preserved 340 Ma ages of coarse- and fine-grained minerals near the upper shear zone; 5, age mixtures by neoformation of white mica.

stacked pile was uplifted, cooled and exhumed. It is noteworthy that later extensive movements, which may have occurred, are not recorded in the cooling-age data. Alternative models (e.g. shear heating metamorphism) that could also explain the age pattern will be discussed in detail elsewhere (Werner & Lippolt, in prep.).

Within the three lithological units that display the 330 Ma ages, an internal reproducibility of within $< \pm 0.5$ Ma is obtained, demonstrating that the rocks cooled at the same time. The original thickness of the three units presumably was of the order of 6 km. The identical cooling ages of about 330 Ma for the rocks of this pile indicate that the pertinent regional uplift rate was at least 10 km Ma^{-1}. Fast uplift and cooling of the rocks is also indicated by the results on different grain sizes of white mica separates from several 330 Ma rocks. To show the consistency of this tectonic hypothesis with the measured age record a simulation of the argon age pattern is performed, achieved in two steps by evaluation of the crustal thermal evolution during the event and by calculating the argon retention of the minerals. The simulation has to check whether the physical factors of the tectonic model allow the existence of the two predominant age groups at 340 and 330 Ma and whether the proposed interpretation of the spectra with intermediate total argon ages is substantiated.

Petrological constraints on the imbrication model

The highest formation pressures have been measured for the gneisses and their intercalations of granulite and eclogite relics. In the literature there are still different views on the evolution steps of most of these gneisses, but for eclogites maximal temperatures of 700–750 °C and pressures of 25 kbar were determined (Willner *et al.* 1997). Some orthogneisses reached 700–750 °C and 16–20 kbar. The garnet-bearing phyllites which overlie the mica schists are characterized by 460 °C and 9 kbar (Rötzler *et al.* 1998).

Modelling the derived imbrication process we have to argue as follows. The red and grey gneisses are mainly orthogneisses, which were formed at medium temperatures and pressures of 650–590 °C and 8–6 kbar (Rötzler 1995). Amphibolites of this unit do not show any indications of earlier high-pressure metamorphism. For the gneisses below the assumed lower shear zone formation temperature and pressures of 620 °C and 7 kbar have been estimated, which implies that the lower shear zone may have been active at a depth of not more than 21 km.

For the mica schists above the gneiss unit with intercalations of eclogite and granulite relics, temperatures of formation of 550–600 °C and pressures of 14–16 and 8 kbar have been determined (Rötzler *et al.* 1998). Given isothermal uplift, temperatures close to the shear plane at the time of thrusting were probably about 575 °C. The pressure was >5 kbar, because during decompression the grade of metamorphism remained above the sillimanite–andalusite facies boundary. The upper shear zone probably was at a depth of about 18 km. In this case the thermal gradient was 30 K km^{-1}.

The original thickness of the rock pile displaying the 340 Ma and intermediate ages was about 3 km. At the time of thrusting it probably extended from a depth of 21 km to 18 km. The temperature within this pile, before thrusting, may have been as low as 100 °C. Exact knowledge of this temperature, however, is insignificant for the modelling because it does not influence the results.

Model and age simulation

Temperature evolution of tectonic model variants

The cover rocks of the stack were mica schists and garnet-bearing phyllites, the substratum red and grey gneisses. The assumed temperature gradient of 30 K km^{-1} and the petrological data allow us to derive a temperature profile in the crust after the shearing and imbrication process (Fig. 2c). The model implies sudden and sharp temperature changes of about 300 and 250 °C across the shear zones after the event. This simple model cannot consider details of the relative position of mica schists and gneisses in the mica schist–eclogite unit. But it explains the present day, partially inverse, metamorphic constellation in the Ergebirge profile. Immediately after stacking adjustment of the incongruous temperatures by conductive heat transfer and isostatic uplift were initiated. The cooler rocks were in parts heated above the closure temperature of argon in white mica, but not long enough to reset the K–Ar clock in the white mica everywhere and completely. The cooling ages of the phengites of the cooler rocks at that time, which were of the order of 10 Ma (corresponding to 340 Ma today), were reset differently on a regional scale. Formation of new white mica at that time also has to be taken into consideration. In the lower regions of the heated intercalated pile, the temperature must have reached about 500 °C, which is proved by reset hornblende ages.

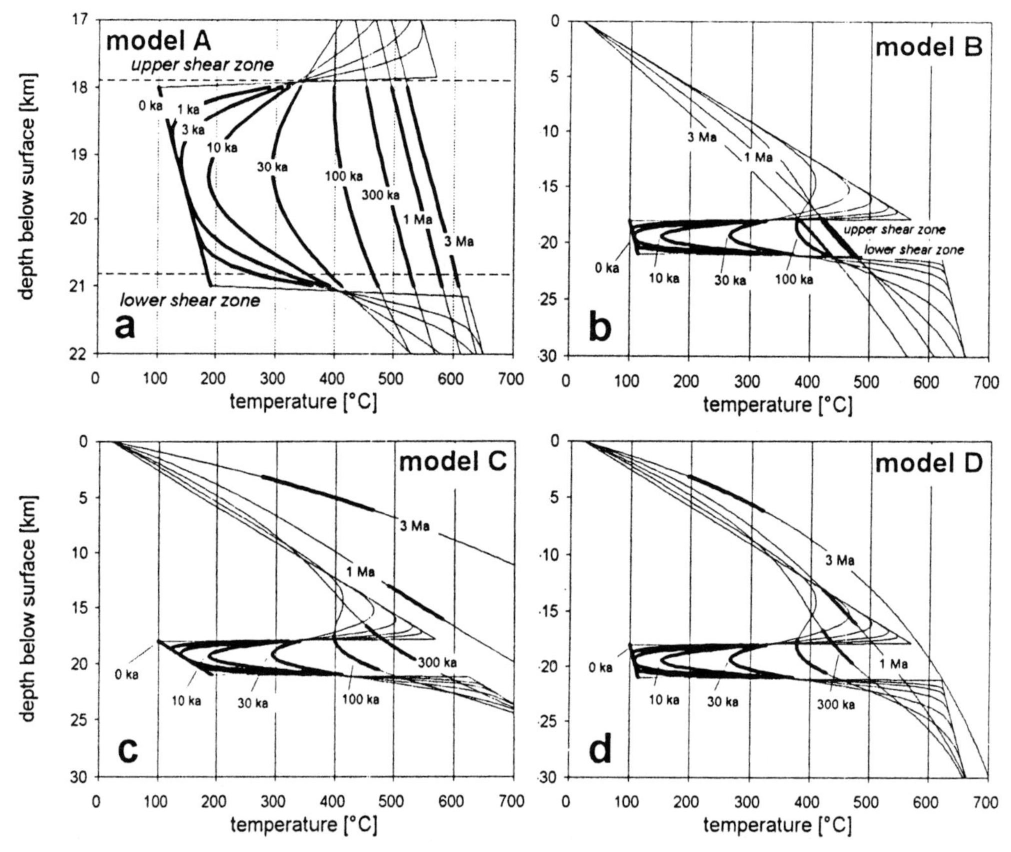

Fig. 5. Diagrams of model temperature evolution (models A–D) in a cool intercalated rock pile and the under- and overlying hotter regions. (Note the different scale of (**a**).) Model parameters are listed in Table 1. The diagrams show the temperatures in the depth profiles at intervals after the intercalation process (from zero time to 3 Ma). The bold lines mark the position of the intercalated rock pile. The evolution of the temperatures at the shear zones is illustrated in Fig. 6.

In later phases of the process further compression produced doming and folding, which together produced the present-day tectonic relation of the metamorphic rocks in the West Erzgebirge. Continuing exhumation probably removed about 20 km of the profile since 330 Ma.

The conductive heat exchange was calculated numerically by an approach using equation (4) of Carslaw & Jaeger (1959, Chapter I, Section 1.6):

$$dT/dt = (\lambda/\rho c)\Delta T$$

where T is temperature, t is time, λ is thermal heat conductivity, ρ is density, c is specific heat and Δ is the Laplace operator. The starting conditions were as in Fig. 2c. In addition, it was assumed that temperatures at the surface and at the mantle–crust boundary remained constant. Convective heat transfer was not considered, but was compensated mathematically by the assumption of a relatively high thermometric conductivity ($\lambda/\rho c$) of 45 m^2 a^{-1}. The developed computer program calculates the temperature profiles at various times after the intercalation in the newly formed pile of metamorphic rocks, at first only for heat transfer (models A and B), then for an uplifting crust (models C and D, rate 5 km Ma^{-1}.). The temperature profile evolution of the four model variants is given in Fig. 5a–d. The model parameters are summarized in Table 1. For the four model variants Fig. 6a and b shows the changes of the temperatures at the two shear zones with time. The maximum temperatures for model C reached within the shear zones are shown additionally as a function of the assumed uplift rates in Fig. 7.

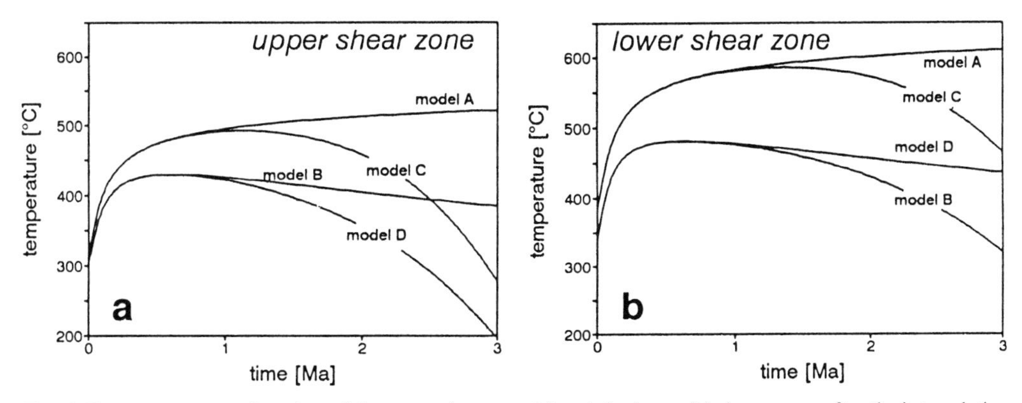

Fig. 6. Temperatures as a function of time near the upper (**a**) and the lower (**b**) shear zones after the intercalation process for the four variants of the intercalation model, described in Fig. 5 and Table 1. In models C and D temperatures in the intercalated pile at first rise and then decrease rapidly as a result of assumed uplift.

Table 1. *Alternative model boundaries for the calculations of temperature in the Erzgebirge crust after the assumed tectonic imbrication process*

Model	Uplift rate (km Ma^{-1})	Thermal gradient: Above upper shear zone (K km^{-1})	Below lower shear zone (K km^{-1})
A	0	30	30
B	0	30	5
C	5	30	30
D	5	30	5

Physical variables for diffusion rate considerations

Minerals that are used for K–Ar dating are known to lose their radiogenic argon when they are heated above a critical temperature range for a significant period of time both in nature and in laboratory experiments. The ^{40}Ar(rad) losses are a function of the size of the crystal domains from which argon migration occurs. These domains may be the observed mineral grains or smaller regions within them. In general, volume diffusion is thought to be responsible for the losses, but also other mechanisms may prevail (e.g. Harrison *et al.* 1985; Sletten & Onstott 1998; Villa 1998). In the following model, it is assumed that the white mica of the Erzgebirge rocks retained or lost ^{40}Ar(rad) in a way that is described by the diffusion theory of rare gases in solids (Flügge & Ziemens 1939; Hauffe 1966). The diffusion rates are functions of activation energy, pre-exponential factor, temperature, time, grain size and Ar concentration profile (see Fechtig & Kalbitzer 1966). These factors also control Ar retention, and together with the cooling rate, the Ar closure temperatures (Dodson 1986).

Temperatures and heating times for the simulation of Ar age adjustments are supplied by the crustal temperature models A–D. The specific mineral parameters (muscovite) were taken from Robbins (1972) and Lippolt *et al.* (1991). As activation energy a value of 180 kJ mol^{-1} and as pre-expontential D_0 log −3.5 were adopted. This is compatible with closure temperatures in the range between 300 and 400 °C. For phengite slightly differing values were used, which were derived from the evaluation of step-degassing experiments on muscovite and phengite (Werner 1998). The pre-exponential factor is nearly identical but the activation energy is 13 kJ mol^{-1} higher, which results in higher closure temperatures for phengite (about 20 K above those of muscovite). Wijbrans & McDougall (1986) suggested a 50 K higher closure temperature for phengite. For hornblende 250 kJ mol^{-1} and log −1.05 were used, respectively (Harrison 1981; Baldwin *et al.* 1990).

Isotopic record

On the basis of the temporal evolutions of the temperature (models A–D), partial ^{40}Ar(rad) losses from white mica and hornblende were calculated. Before the imbrication process these minerals were in a state below the relevant closure temperatures. The calculations were performed for different grain sizes and different positions in the rock profile of the intercalated rock pile. Partial ^{40}Ar(rad) losses were plotted as functions of three parametres: possible degassing times (Fig. 8a and b) for all four models, position in the rock pile for models B and C (Fig. 9a and b) and assumed uplift rates for models C and D

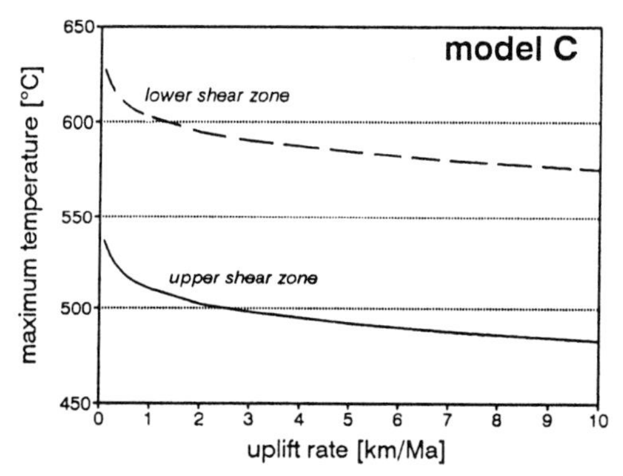

Fig. 7. The maximum temperatures near the lower and upper shear zones after the imbrication process for model C as a function of the uplift rate. Dotted line, at lower shear zone; continuous line, at upper shear zone. For a rate of 5 km Ma^{-1}, 480 °C and 570 °C are obtained. Model D is independent of the uplift rate; the maximum temperatures are 420 °C and 470 °C, respectively.

(Fig. 10a and b). Figure 4 shows the model-adjusted analytical data that are to be simulated. The apparent partial argon losses and the retentions (both in per cent) of the dated minerals from the intercalated gneisses are calculated. It is assumed that they had a K–Ar age of 10 Ma (*c.* 340–330 Ma) at the time of thrusting; 340 Ma is the present cooling age of the intercalated gneisses and 330 Ma is the time of thermal overprinting. The data of the framing red and grey gneisses and mica schists are not presented here (but are given by Werner *et al.*, in prep.), because these rocks cooled 330 Ma ago in a simple single-stage process. They probably did not experience the same cooling history as the gneisses. The idea is that from the comparison of Fig. 4 and Fig. 9a and b arguments can be derived that favour one of the models A–D.

Modelling results

Crustal thermal evolution

Two further factors also influence the temperature profiles (Fig. 5): the thermal gradient below the lower shear zone and the uplift rate. The latter was also calculated by the vertical distance of the 330 Ma rock units (see section above: Ages, lithologies and crustal imbrication hypothesis). The four models differ with respect to the assumptions regarding thermal gradients and the uplift rate of the whole rock stack (Table 1). Model A (Fig. 5a) uses only a unique gradient and neglects uplift. Model B (Fig. 5b) uses, below the lower shear zone, a smaller gradient (of 5 K km^{-1}), which leads to a larger negative slope than in Fig. 5a. This value complies with the assumption that the crust was 60 km thick and reached a temperature of 800 °C at the boundary to the mantle. Models C and D (Fig. 5c and d) consider additionally an uplift rate of 5 km Ma^{-1}.

It is evident that models A and C result in unrealistically high temperatures below 30 km. An additional heat source would be needed to reach these temperatures. Such a heat source could probably have been the granitic plutons below the metamorphic basement in the Erzgebirge. They were emplaced 320–310 Ma ago (Gerstenberger *et al.* 1995; Werner & Lippolt 1998). Heating and melting processes, however, already could have taken place during crustal stacking 330 Ma ago (Gerdes *et al.* 1998).

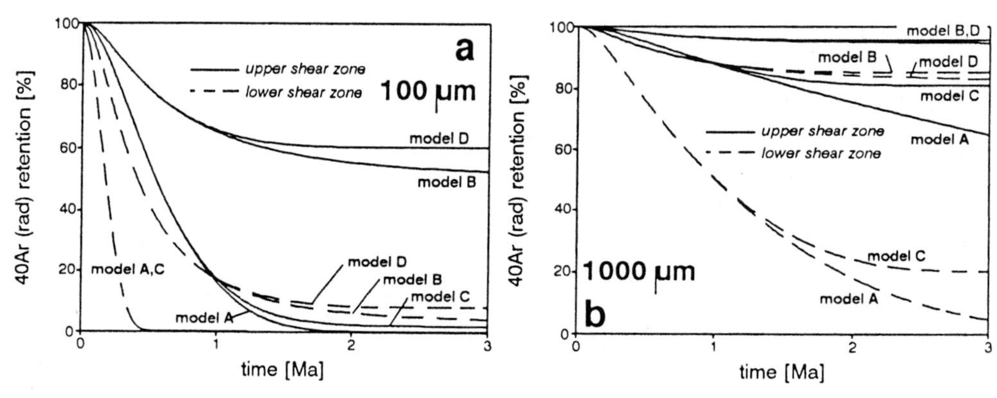

Fig. 8. The calculated partial ^{40}Ar(rad) retentions (in per cent) from 1000 µm (**a**) and 100 µm (**b**) separates after the assumed intercalation at the lower and upper shear zones bordering the cooler rock pile. The curves denoted the four models. Losses decrease from models D to model B, C and A.

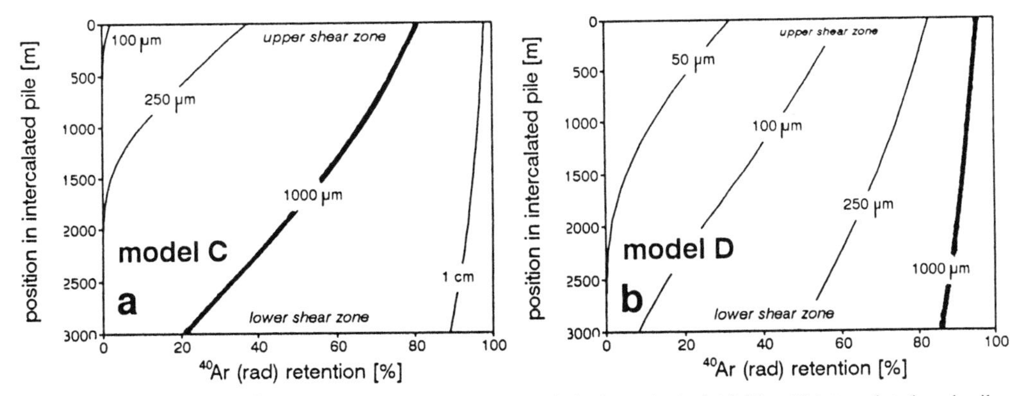

Fig. 9. Calculated partial ^{40}Ar(rad) retention (in per cent) of the hypothetical 10 Ma old intercalated rock pile, based on models C (**a**) and D (**b**). Model C predicts large losses for the grain size 1000 µm (25–80%, depending on the position in the intercalated rock pile), whereas model B losses are only between 7 and 18%, independent of position. The model D curves for grain sizes 250–1000 µm fit rather well the cooling-age curves in Figs 2 and 4.

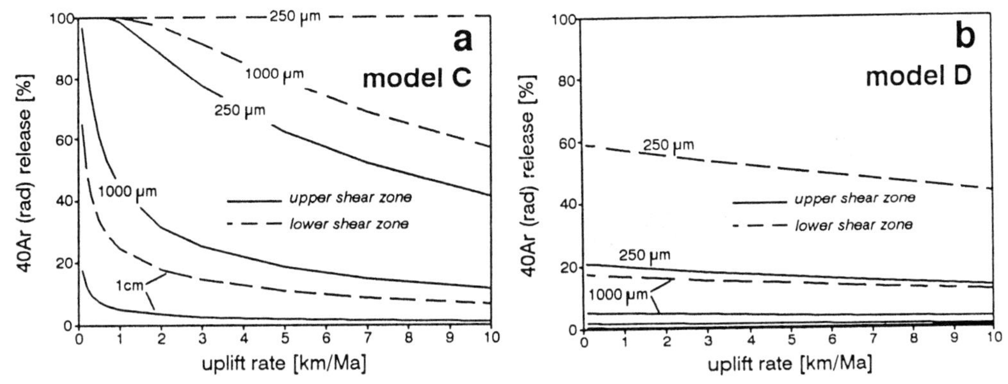

Fig. 10. The calculated partial ^{40}Ar(rad) losses for models C (**a**) and D (**b**) in the regions of the lower and the upper shear zones as a function of the uplift rate. Continuous lines, at upper shear zone; dotted lines, at lower shear zone. Model C predicts for an uplift rate of 3 km Ma^{-1} an argon loss of about 25% at the upper shear zone, during isostatic adjustment after crustal imbrication, but only of about 10% for 5 km Ma^{-1}. (**b**) demonstrates that uplift is of minor importance for model D.

Model A yields, 1 Ma after the imbrication, depending on the position in the profile, temperatures of about 500–580 °C for the initially cool rocks (temperature below argon closure). These estimates are above the Ar retention range. According to models B, C and D after 1 Ma, temperatures of 420–470 °C, 480–570 °C and 420–470 °C, respectively, would be reached. They still are above the Ar retention range. They are, however, as a result of the assumed lower gradient below the lower shear zone and the assumed uplift, lower than the temperatures predicted by model A. For model C the maximum temperatures that may be reached depend distinctly on the actual uplift rate (Fig. 7). For a rate of 10 km Ma^{-1} the temperature range would be 470–550 °C. For model D, which combines the additional conditions considered by models B and C, the uplift rate is of negligible influence.

All the calculations therefore demonstrate that the heating of the cold intercalated rock unit must obviously have been much shorter than 1 Ma because the temperatures in the intercalated rock pile increase very rapidly above 400 °C (Fig. 6a and b). Otherwise all K–Ar ages would be around 330 Ma. The heating time during which the radiogenic argon would have been partially preserved from being degassed is probably *c*. 300 ka.

Evolution of ^{40}Ar(rad) storage

Figure 8a and b shows that the ^{40}Ar(rad) retention, independent of the choice of the models, strongly depends on the grain size of the mineral separates. The 100 μm grains at the lower shear zone lose nearly 100% ^{40}Ar(rad) within 0.3 Ma in models A and C, whereas they still retain 20% ^{40}Ar(rad) after 1 Ma in models B and D. In models A and C the 1 mm grains lose within 1 Ma about 50% of their argon, but only 10% in models B and D. It follows that the small grain-size separates, close to the lower shear zone, should generally be completely reset. Figure 9 not only demonstrates the strong dependence of the Ar losses on grain size (between 50 μm and 1 cm) for models B and C, but also shows that there should be large differences in Ar retention for mica in the transition from the region of the lower to the upper shear zone. Figure 10 illustrates how strongly the uplift rate influences the Ar storage in model C and how insignificantly in model D.

Comparison of analytical and simulation evidence

Model A can be excluded because relic 340 Ma cooling ages would have been completely reset 330 Ma ago. Also, model B, with a very low uplift rate, does not work realistically because the granites intruded about 10 Ma later at a high structural level (Förster & Tischendorf 1994). Consequently the uplift rate minimum is 1 km Ma^{-1}. Therefore other variants have to be taken into consideration.

A comparison of Fig. 4 and Fig. 9a and b favours arguments for a model with rapid uplift and a large thermal gradient below the lower shear zone, caused maybe by an additional heat source below the shear zone (model C). Crustal thinning and melting processes may be the trigger. Large Ar retention differences for differing grain sizes below 1000 μm of white mica from the region of the lower to the upper shear zone favour this model. Models B and D, with low thermal gradients below the lower shear zone, call for a more homogeneous Ar retention. Supporting observations are as follows. First, the relatively fine-grained white micas (up to 1000 μm) from close to the lower shear zone are severely reset, evidenced by the last degassing steps of the pertinent age spectra (see Fig. 3e), which do indeed show relic ^{40}Ar(rad) components. However, two coarse-grained white micas (*c*. 200 μm) have 50% Ar loss. In addition, close to the upper shear zone also fine-grained white micas have undisturbed 340 Ma spectra. Second, the largest phengites of the sample show the highest Ar ages, indicating that there are distinct grain-size effects. A competing mechanism to produce disturbed spectra (see Fig. 3d) and apparent partial Ar loss would be neoformation of muscovite (Werner & Lippolt, in prep.).

In the case of model C, an uplift rate of 3 km Ma^{-1} would exclude Ar losses from 1000 μm phengite close to the upper shear zone (Fig. 10). An uplift rate of 5 km Ma^{-1} even would exclude gas losses larger than 10%. From independent thermal modelling calculations (see above) it was indicated that the whole rock pile was uplifted at a minimal uplift rate of 10 km Ma^{-1}. It therefore may be assumed that this minimum uplift rate should also be applied here.

Model C is additionally supported by the hornblende results. The hornblende (Fig. 3f1)) must have lost *c*. 60% of its radiogenic argon. Extraneous argon components were incorporated during the mineral closing phase at about 330 Ma (lower steps in spectrum f1, Fig. 3). The sample was located in the region of the lower shear zone. For the 500 μm grain size of this sample a theoretical Ar(rad) loss of 30% is calculated. On the other hand, for the 500 μm grain size of the sample from the upper shear zone level with 340 Ma plateau age spectra (Fig. 3b) the theoretical Ar(rad) loss is negligible.

Inferences

The narrow accordance of the 330 Ma mica argon ages above and below the intercalated pile may be taken as evidence that exhumation of most of the crustal profile was very fast. With an uplift rate of 5–10 km Ma^{-1} an exhumation period of 1–2 Ma is calculated. Diffusion rate considerations show that, given the initial and boundary conditions of model C, an uplift rate of 5 km Ma^{-1} is needed to preserve 340 Ma cooling ages in the present-day upper crust. The intercalation process or, in other words, the emplacement of the rocks that are now coexisting, occurred about 332 Ma ago. Phengites from the high-pressure gneisses as well as phengites and amphiboles from basic inclusions yield cooling ages of 340 Ma $\pm$ 2 Ma (accuracy), identical to the Pb-evaporation zircon data of 340 $\pm$ 4 Ma. This finding suggests fast geological cooling. It caused the zircons to date the peak of the high-pressure event, and the phengite and amphiboles to indicate ages of its waning stages. Together with their gneissic host rocks, the granulites and eclogites were reheated 330 Ma ago.

Differing cooling ages within the mica schist–eclogite unit contradict a common evolution and metamorphism of this petrological unit as is postulated in the literature (Willner *et al.* 1997; Rötzler *et al.* 1998). Mica schists yield ages of about 330 Ma ± 1.5 Ma (accuracy), whereas the gneisses and their intercalations of eclogite relics show cooling ages of 340 ± 2 Ma (accuracy). Accompanied by partial thermal resetting of the isotopic systems and neoformation of white mica, these gneisses were imbricated with the mica schists and subsequently folded not before 330 Ma ago.

The majority of our data mainly was received for the West Erzgebirge. The dates for the Central and East Erzgebirge possibly indicate that the evolution in the whole region was more complex. Probably, in addition to 340 Ma rock units, 380–360 Ma rocks were also included in the late Early Carboniferous crustal imbrication process.

Summary and conclusions

In the Erzgebirge basement, white micas from metamorphic rocks demonstrate argon ages between 340 and 330 Ma. Two age groups at 340 Ma and 330 Ma and a third group with intermediate ages are distinguished. To explain the meaning of the argon age behaviour and regional distribution, a tectonic imbrication scenario was developed, which simultaneously explains the metamorphic inversion of red and grey gneisses, and granulite and eclogite relics.

It was assumed that thrust tectonics intercalated a relatively cool, about 3 km thick pile of gneisses into hotter rocks at a crustal depth of about 20 km. Conductive heat flow when hot and cold segments of the crust became intercalated and cooling during isostatic adjustment (uplift) were considered. The evolution of temperatures in the intercalated rock pile was calculated for various sets of boundary conditions. The thermal rejuvenation of Ar ages existing 330 Ma ago was estimated as a function of rock position in the pile and the grain size of the minerals. Comparison of the measured ages and the predicted Ar losses from the white mica favours a imbrication model in a crustal region where a high thermal gradient prevailed, caused by additional heat sources. This tectonic event probably was accompanied by melting processes in the lower crust leading to pluton intrusions at higher structural levels 10–20 Ma later. The analytical data presented here are not interpretable by a single phase extension model. Although the imbrication model presented here convincingly explains the cooling-age patterns, it cannot be excluded that there might be another tectonic model, that does the same.

The Deutsche Forschungsgemeinschaft (DFG, Bonn–Bad Godesberg) supported this study by a grant from the priority programme 'Orogenic processes: Quantification and Simulation'. We thank W. Franke (Giessen), co-ordinator of this programme, for his encouragement. O.W. gratefully remembers fruitful discussions with M. Nega (Frankfurt). J. Hess (Heidelberg) supported the studies. We acknowledge the help by the neutron irradiation service groups of the Geesthacht and Jülich reactors.

References

Baldwin, A. L., Harrison, T. M. & Fitzgerald, J. D. 1990. Diffusion of ^{40}Ar in metamorphic hornblende. *Contributions to Mineralogy and Petrology*, **105**, 691–803.

Carslaw, H. J. & Jaeger, J. C. 1959. *Conduction of Heat in Solids*. Clarendon Press, Oxford.

Dodson, M. H. 1986. Closure profiles in cooling systems. *Material Science Forum*, **7**, 145–154.

Fechtig, H. & Kalbitzer, S. 1966. The diffusion of argon in potassium bearing solids. *In*: Schaeffer, O. A. & Zähringer, J. (eds) *Potassium Argon Dating*. Springer, Berlin, 68–107.

Flügge, S. & Ziemens, K. G. 1939. Die Bestimmung von Korngrößen und von Diffusionskonstanten aus dem Emanierungsvermögen. *Zeitschrift der Physikalischen Chemie*, **42**, 179–220.

Förster, H. J. & Tischendorf, G. 1994. The western Erzgebirge–Vogtland granites: implications to the Hercynian magmatism in the Erzgebirge–Fichtelgebirge anticlinorium. *In*: Seltmann, R., Kämpf, H. & Möller, P. (eds) *Metallogeny of Collisional Orogens*. Czech Geological Survey, Prague, 35–48.

Franke, W. 1993. The Saxonian Granulites: a metamorphic core complex? *Geologische Rundschau*, **82**, 505–515.

—— & Stein, E. 2000. Exhumation of high-grade rocks in the Saxo-Thuringian Belt: geological constraints and geodynamic concepts. *This volume*.

Gerdes, A., Wörner, G. & Henk, A. 1998. Geochemical and thermal constraints on the origin of granites in the Southern Moldanubian zone. *Terra Nostra*, **98**(2), 58–61.

Gerstenberger, H. 1989. Autometasomatic Rb enrichments in highly evolved granites causing lowered Rb–Sr isochron intercepts. *Earth and Planetary Science Letters*, **93**(1), 65–75.

——, Haase, G. & Wemmer, K. 1995. Isotope systematics of the Variscan postkinematic granites in the Erzgebirge (E. Germany). *Terra Nostra*, **95**(7), 36–41.

Haake, R. 1972. Zur Altersstellung granitoider Gesteine im Erzgebirge. *Geologie*, **21**, 641–671.

Harrison, T. M. 1981. Diffusion of ^{40}Ar in hornblende. *Contributions to Mineralogy and Petrology*, **78**, 324–331.

——, Duncan, I. & McDougall, I. 1985. Diffusion of ^{40}Ar in biotite: temperature, pressure and compositional effects. *Geochimica et Cosmochimica Acta*, **49**, 2461–2468.

Hauffe, K. 1966. *Reaktionen in und an festen Stoffen*. Springer, Berlin, 372–603.

KONOPÀSEK, J. & SCHULMANN, K. 1999. Tectonic evolution of the central part of the Kruzne Hory Mts (Erzgebirge) in the Czech Republic (Saxothuringian Domain of the Bohemian Massif). *Terra Nostra*, **99**(1), 121–122.

KOSSMAT, F. 1927. Gliederung des variscischen Gebirges. *Abhandlungen der Sächsischen Geologischen Landesanstalt*, **1**, 1–39.

KROHE, A. 1996. Variscan tectonics of central Europe: postaccretionary intraplate deformation of weak continental lithosphere. *Tectonics*, **15**(6), 1364–1388.

KRÖNER, A., WILLNER, A. P. 1998. Time of formation and peak of Variscan HP–HT metamorphism of quartz–feldspar rocks in the central Erzgebirge, Saxony, Germany. *Contributions to Mineralogy and Petrology*, **132**, 1–20.

KRONER, U. & VOIGT, T. 1992. Postkollisionale Extension—ein Modell für die Entwicklung am Nordrand der Böhmischen Masse. *Frankfurter Geowissenschaftliche Arbeiten*, **A11**, 104–107.

LIPPOLT, H. J., VOGEL, H., HESS, J. C. & METZ, P. 1991. Ar diffusion in micas under experimental hydrothermal conditions. *Terra Abstracts*, **3**, 13.

LORENZ, W. & HOTH, K. 1990. Lithostratigraphie im Erzgebirge—Konzeption, Entwicklung, Probleme, Perspektiven. *Abhandlungen des Staatlichen Museums für Mineralogie und Geologie, Dresden*, **7**, 7–35.

—— & PILOT, J. 1994. K/Ar-Alter und thermische Anomalien in erzgebirgischen Metamorphiten. *Zeitschrift geologischer Wissenschaften*, **22**(3–4), 391–396.

MONIÉ, P. 1990. Preservation of Hercynian $^{40}Ar/^{39}Ar$ ages through high-pressure low-temperature Alpine metamorphism in the Western Alps. *European Journal of Mineralogy*, **2**, 343–361.

NEGA, M., KRUHL, J. H., KRENTZ, O. & LEONHARDT, D. 1997. Gefügeprägung und Exhumierung durch Kontinent-Kollision: die spätvariscische Kinematik im westlichen Erzgebirge. *Terra Nostra*, **97**(5), 109–115.

PIETZSCH, K. 1954. Die Gneise des Sächsischen Erzgebirges. *Geologie*, **3**, 391–412.

ROBBINS, G. A. 1972. *Radiogenic argon diffusion in muscovite under hydrothermal conditions*. Master Thesis, Brown University, Providence, RI, USA.

RÖTZLER, K. 1995. *Die P–T-Entwicklung der Metamorphose des Mittel- und Westerzgebirges*. Scientific Technical Report, GeoForschungsZentrum Potsdam, **95/14**, 220S.

——, SCHUMACHER, R., MARESCH, W. V. & WILLNER, A. P. 1998. Characterization and geodynamic implications of contrasting metamorphism evolution in juxtaposed high-pressure units of the Western Erzgebirge (Saxony, Germany). *European Journal of Mineralogy*, **10**, 261–280.

SCAILLET, S. 1996. Excess ^{40}Ar transport scale and mechanism in high-pressure phengites: a case study from an eclogitized metabasite in the Dora–Maira nappe, western Alps. *Geochimica et Cosmochimica Acta*, **60**(6), 1075–1090.

SCHMÄDICKE, E., MEZGER, K., COSCA, M. A. & OKRUSCH, M. 1995. Variscan Sm–Nd and Ar–Ar ages of eclogite facies rocks from the Erzgebirge, Bohemian Massif. *Journal of Metamorphic Geology*, **13**, 537–552.

SEBASTIAN, U. & KRONER, U. 1992. Scherzonenentwicklung kontra Intrusionskontakt—eine Fallstudie im mittleren Erzgebirge (Sachsen). *Zentralblatt Geologie Paläontologie*, **7(8), 785–790.**

SLETTEN, V. W. & ONSTOTT, T. C. 1998. The effect of the instability of muscovite during in vacuo heating on $^{40}Ar/^{39}Ar$ step-heating spectra. *Geochimica et Cosmochimica Acta*, **62**(1), 123–141.

TIKHOMIROVA, M., BERGER, H. J. & KOCH, E. A. 1995. Altersdatierungen an Osterzgebirgsgneisen. *Terra Nostra*, **95**(8), 134.

VILLA, I. M. 1998. Isotopic closure. *Terra Nova*, **10**, 42–47.

WERNER, O. 1998. *K–Ar und Rb–Sr-Chronologie spätvariscischer Krustenkonvergenz—Bilanzierung des Wärme-und Stofftransportes im Erzgebirge*. Inaugural dissertation, Universität Heidelberg.

—— & LIPPOLT, H. J. 1998. Datierung von postkinematischen Intrusionsphasen des Erzgebirges: thermische und hydrothermale Überprägung der Nebengesteine. *Terra Nostra*, **98**(2), 160–163.

WILLNER, A. P., RÖTZLER, K. & MARESCH, W. V. 1997. Pressure–temperature and fluid evolution of quartzo-feldspathic metamorphic rocks with a relic high-pressure, granulite-facies history from the Central Erzgebirge (Saxony, Gemany). *Journal of Petrology*, **38**(3), 307–336.

WIJBRANS, J. R. & MCDOUGALL, I. 1986. $^{40}Ar/^{39}Ar$ dating of white-micas from an alpine high pressure metamorphic belt on Naxos (Greece): the resetting of the argon isotope system. *Contributions to Mineralogy and Petrology*, **93**, 187–194.

Exhumation of high-grade rocks in the Saxo-Thuringian Belt: geological constraints and geodynamic concepts

W. FRANKE[1] & E. STEIN[2]

[1]*Institut für Geowissenschaften der Universität, D-35390 Giessen, Germany (e-mail: wolfgang.franke@geo.uni-giessen.de)*

[2]*Institut für Mineralogie, Technische Universität Darmstadt, D-64287 Darmstadt, Germany*

Abstract: The Saxo-Thuringian Belt on the northern flank of the European Variscides resulted from SE-ward subduction of a Cambro-Ordovician rift basin under the Teplá–Barrandian (Bohemian) margin. It contains ultra-high-pressure (UHP) metamorphic rocks, which are now exposed in different tectonic settings, and were exhumed in different modes, under different thermal regimes. Eclogites of the upper allochthon (tectonic klippen of Münchberg, Wildenfels and Frankenberg) originated from early Devonian (*c.* 400 Ma) subduction. Cooling ages in the klippen, combined with clast spectra and ages of detrital minerals in the foreland flysch record exhumation in Famennian time. The high-pressure (HP) rocks rose in a narrow corridor along the suture zone, were retrogressed under amphibolite facies conditions, and rapidly recycled into the foreland flysch. Rocks of the lower allochthon are exposed in dome structures emerging from under the relative autochthon. The HP Saxonian Granulites were formed at *c.* 20 kbar/1050 °C, and the Erzgebirge contains diamond-bearing quartzo-feldspathic rocks, associated with eclogites. Peak temperatures of the lower allochthon were attained at *c.* 340 Ma. Immediately afterwards, these rocks were emplaced under the floor of the relative autochthon, which, at this time, was a foreland basin receiving clastic sediments from the orogen adjacent to the SE. Emplacement of the lower allochthon is most clearly documented in the Saxonian Granulite Dome, in which HP granulites are juxtaposed against low-pressure–high-temperature (LP–HT) rocks of the hanging wall. The interface is a zone of HT shear, which cuts out $\geqslant$60 km of crustal thickness. In more internal parts of the belt (Erzgebirge), the newly emplaced lower allochthon was subsequently reworked by thrusting and polyphase refolding. Emplacement of the hot, low-viscosity lower allochthon was probably driven by buoyancy and the hydraulic gradient between the crustal root to the SE and the lower crust of the foreland. Unlike the earlier HP rocks exposed in the upper allochthon, the 340 Ma HP rocks of the lower allochthon were thermally softened, and, instead of piercing their cover, intruded into the foreland. Therefore, the 340 Ma rocks do not appear in the clastic record of the flysch. The two contrasting mechanisms of exhumation observed in the upper and lower allochthon are apparently due to different thermal regimes.

The Saxo-Thuringian Belt is situated on the northern flank of the European Variscan Orogen. It represents the hinterland of the Rheno-Hercynian Orogen adjacent to the NW, and, at the same time, the foreland to the Teplá–Barrandian leading edge impinging from the SE (see foldout map, and Franke (this volume)). The Saxo-Thuringian is the most complex part of the central European Variscides (Matte *et al.* 1990). It contains, in a narrow outcrop belt, late Proterozoic through to Carboniferous sedimentary and volcanic rocks in diverse facies (Franke this volume), but also an extremely large spectrum of metamorphic grades. The metamorphic terranes comprise 'world record' rocks, such as the Saxonian Granulites (the world type locality of the metamorphic facies) with metamorphic temperatures in excess of 1000 °C, and diamond-bearing eclogite-facies rocks. These lithologies are fascinating as such, because exhumation of high-pressure rocks has become a topic of international interest and controversy during the last decade (e.g. Platt 1986; Lister & Baldwin 1993; Chemenda *et al.* 1995). Important issues include the partitioning of erosion and tectonic extension during exhumation, and the causes of extension (wedge-controlled corner flow v. gravitational collapse v. active rifting). In addition, the Saxo-Thuringian high-grade rocks occur in a very unusual tectonic position:

From: FRANKE, W., HAAK, V., ONCKEN, O. & TANNER, D. (eds). *Orogenic Processes: Quantification and Modelling in the Variscan Belt.* Geological Society, London, Special Publications, **179**, 337–354. 0305-8719/00/$15.00

they occupy the cores of antiformal structures that belong to the foreland part of the Saxo-Thuringian orogenic belt, and their ages of metamorphism coincide with the age of sedimentation of the overlying marine flysch sediments. This paradoxical situation requires a novel approach to the problem of exhumation, which is discussed in this paper. A wealth of geological and palaeontological data acquired during past decades, together with new isotopic ages (Werner & Lippolt this volume) and geophysical studies (Krawczyk *et al.* this volume) provide tight constraints on the exhumation processes and even permit numerical modelling (Henk this volume).

Regional geological setting

In the Saxo-Thuringian Belt, it is possible to distinguish three structural levels (Fig. 1): an upper allochthon (tectonic klippen), a relative autochthon and a lower allochthon, which will be shown to have been emplaced under the relative autochthon (high-grade rocks contained in the domal structures of the Saxonian Granulites and the Erzgebirge).

Upper allochthon and relative autochthon: geological events

The tectonic klippen of Münchberg, Wildenfels and Frankenberg (Figs 1 and 2) are contained in a late-tectonic synform (Vogtland Synform). They are derived from the Teplá–Barrandian margin adjacent to the SE (Behr *et al.* 1982; Franke 1984*a*, *b*), and were thrust over what is now the Fichtelgebirge–Erzgebirge Antiform. These klippen expose very similar tectonic sequences, of which the Münchberg is the largest and most differentiated. It consists of up to eight thrust sheets, whose metamorphic and stratigraphic order is inverted (not overturned!) by thrusting. The tectonic sequence (Fig. 2) comprises metamorphic facies from eclogite to very low grade, and protolith ages ranging from Late Proterozoic to Early Carboniferous time (see recent surveys by Falk *et al.* (1995)). The ages of metamorphism are likewise inverted: eclogitic rocks of the upper thrust sheet derived from *c.* 495 Ma mid-ocean ridge (MOR)-type mafic rocks were formed at *c.* 395 Ma, and underwent a medium-pressure (MP) overprint at 380 Ma. K–Ar and Ar–Ar cooling ages of amphiboles and micas from all metamorphic thrust sheets range between *c.* 380 and 365 Ma (summary by Franke *et al.* (1995)). Similar ages have recently been obtained for the Frankenberg Klippe (Roetzler *et al.* 1999). With the exception of one small occurrence of granulite facies in the Münchberg Klippe (Kleinschrodt & Gayk 1999), the high-grade rocks of the upper allochthon have not been heated beyond 700 °C, and retrogression occurred under amphibolite facies conditions.

The klippen were derived from a position to the SE (Behr *et al.* 1982). However, ductile mineral lineations are consistently oriented SW–NE (Schwan 1974) and asymmetric clasts in the metamorphic rocks invariably indicate transport top to the SW (Franke *et al.* 1992*a*). These findings suggest a regime of (gross) SE-ward subduction of the Saxo-Thuringian passive margin under the NW margin of the Teplá–Barrandian (Bohemian terrane), at least partly combined with dextral transpression (Franke this volume).

Rocks with similar metamorphic ages occur to the SE of the Fichtelgebirge Antiform, in the Zone of Erbendorf–Vohenstrauß (ZEV), in the Mariánské Lázně mafic complex and the Teplá–Barrandian adjacent to the SE (enclosure at back). Most conspicuous are eclogites derived from lower Ordovician MOR-type rocks, with early Devonian eclogite facies metamorphism, often with a granulite facies overprint (see reviews by Franke *et al.* (1995) and Medaris *et al.* (1995)). They are bounded against the main, less metamorphosed part of the Teplá–Barrandian by ductile normal faults, for which K–Ar and Ar–Ar hornblende ages range between 390 and 370 Ma (Fig. 2; Zulauf *et al.* 1997).

The thrust sheets were emplaced on Devonian and lower Carboniferous sequences of the Saxo-Thuringian foreland, the relative autochthon. It consists of very low grade sedimentary and volcanic rocks, which record early Palaeozoic rifting with shelf sedimentation on a Cadomian continental basement and a (hemi-)pelagic stage from Late Ordovician to Tournaisian time, with a second pulse of rifting in Frasnian time. The latter is documented in widespread basaltic intraplate volcanism, and coarse-grained clastic sediments derived from emergent hosts between extensional half-grabens (see reviews by Franke (1993) and Falk *et al.* (1995)).

NW-ward approach of the Teplá–Barrandian tectonic front is heralded by synorogenic clastic sediments, which extend well beyond the outcrop of the Saxonian Granulites, up to the south-eastern margin of the Mid-German Crystalline High (see enclosure). The oldest flysch (of Early Famennian age: Greiling 1961; Franke *et al.* 1992*b*) occurs in the relative autochthon at the SE flank of the Fichtelgebirge Antiform. Petrological

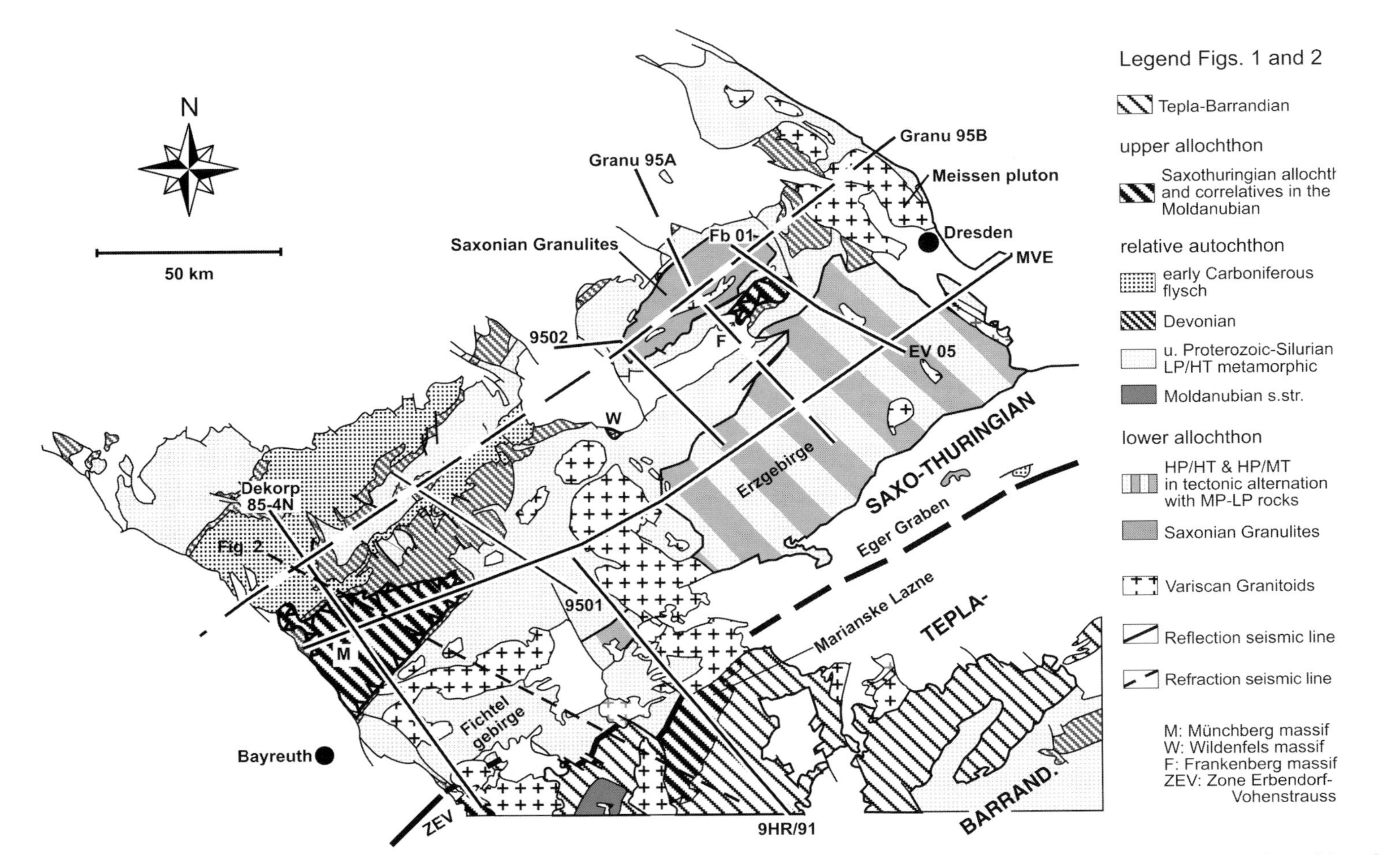

Fig. 1. Geological map of the Saxo-Thuringian Belt between the SW margin of the Bohemian Massif (Franconian Line) and the Elbe Fault Zone, with the position of seismic lines.

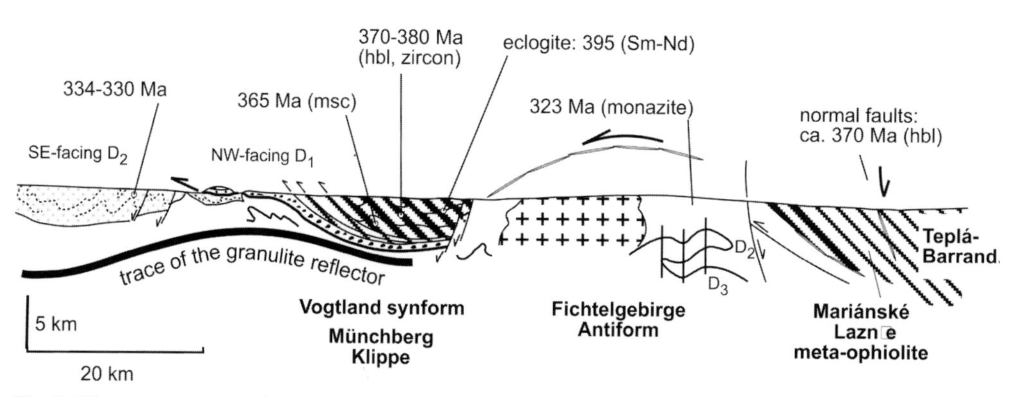

Fig. 2. Diagrammatic tectonic cross-section through the western part of the Saxo-Thuringian Belt, with ages of metamorphism (see Fig. 1 for location; compilation of Franke *et al.* (1995)).

and isotopic analyses of clasts and detrital minerals demonstrate that the greywacke turbidites were fed by the advancing Münchberg thrust stack, and that the tectonic pile was already largely assembled in Famennian time (Neuroth 1997; Schäfer *et al.* 1997). In earliest Viséan time, the clastic front arrived in the relative autochthon of the northwestern foreland.

More proximal stratigraphic equivalents of the autochthonous flysch are preserved in the lowermost thrust sheet of the Münchberg stack (Franke 1984*a*, *b*). The original area of deposition of these proximal flysch sediments must have been to the north of the present-day Fichtelgebirge–Erzgebirge Antiform or beyond, with the source area even further to the SE. Consequently, in Viséan time (*c*. 340–330 Ma), all of the present-day Saxo-Thuringian Belt, from the NW margin of the Teplá–Barrandian to the SE margin of the Mid-German Crystalline High, must have belonged to the foreland basin.

It is very important to note that the Saxo-Thuringian flysch contains detrital micas and zircons, whose ages of 380–370 Ma match those of the overriding crystalline thrust sheets. In addition, there is a record of older events such as 500 Ma magmatism and Cadomian metamorphism (Neuroth 1997; Schäfer *et al.* 1997). However, there is no detrital record of rocks formed around 340 Ma.

The Palaeozoic sequences of the relative autochthon underwent polyphase contraction and very low to amphibolite grade metamorphism in late Viséan time. A first tectonic increment is represented by W–NW-facing folds and the grossly westward emplacement of the Münchberg and related klippen. D1 dies out toward the northwestern, external part of the relative autochthon. Further toward the NW, retro-thrusting at the SE flank of the Mid-German Crystalline High has produced SE-facing structures (Schäfer *et al.* this volume). To the SE, these folds overprint the earlier, NW-directed structures, and have consequently been labelled as D2 (Fig. 2; Franke 1984*a*, *b*). A SE-facing D2 deformation is also present to the SE of the Münchberg Klippe in the Fichtelgebirge Antiform (Stein 1988), where it correlates with peak low-pressure (LP) metamorphism of greenschist to lower amphibolite grade. The latter has yielded K–Ar hornblende ages between 332 ± 6 and 330 ± 4 Ma, and muscovite ages between 330 ± 2.5 and 325.5 ± 2.5 Ma (Kreuzer *et al.* 1989). Concordant monazite dated at 323 ± 3 Ma is possibly more reliable (Teufel 1988). Open D3 folds, formed in a semi-brittle regime, are responsible for the major antiforms and synforms, which control the large-scale outcrop pattern. Like the Erzgebirge (see below), the Fichtelgebirge Antiform is flanked by late normal faults, which possibly correlate with the formation of Permo-Carboniferous basins (Franke et al., 1992*a*).

Constraints on the termination of sedimentation

The Saxo-Thuringian paradox consists in the coincidence of the age of flysch sedimentation in the relative autochthon and high-grade metamorphism of rocks now contained in the underlying lower allochthon. It is therefore necessary to critically review the palaeontological and isotopic database. As documented by brachiopods and foraminifera, flysch sedimentation in the relative autochthon continued at least to the start of Late Viséan time (V2b–3a boundary, Gandl & Mansourian 1978). The fossils are derived from a shallow-water realm, and occur as clasts in mixed siliciclastic–carbonate turbidites; this sedimentation age is therefore a minimum. In addition, the fossiliferous levels are concordantly overlain by at least 1000 m of non-fossiliferous greywackes, shales and some conglomerates. For these reasons, Gandl (1998, pp. 28 and 182) has proposed that the upper part of the flysch extends into the Go β zone of the goniatite chronology, that is, the Viséan 3b zone of the Belgian zonation. However, this is not backed up by palaeontological findings, and in a flysch basin 1000 m of sediment may be accumulated within a very short period of time. We therefore choose the middle–upper Viséan (V2b–V3a) boundary as a conservative estimate for the end of marine sedimentation.

Roberts *et al.* (1995) have reported SHRIMP zircon ages from volcanic ash layers in Eastern Australia. The Kurra Keith (dated at about the Tournaisian–Viséan boundary) has yielded 342 ± 3.2 Ma, the Neerong (*c.* Viséan 2a) 338.6 ± 3.8 Ma, and the Martins Creek (*c.* Viséan 3a) 332.3 ± 2.2 Ma. It must be noted that the error bars on these data represent only the statistical variation of ages obtained from different grains, but do not account for the analytical error, which is much larger for SHRIMP measurements. Also, the volcanic layers are ignimbrites erupted in a continental setting. Correlation with marine beds, biostratigraphic dating of these marine beds and their correlation with Europe might also enlarge the error. It is therefore important that Trapp *et al.* (1998) have reported conventional, single-grain U–Pb ages from felsic tuffs in the NW Harz Mts. One bed in the *Gnathodus texanus* conodont zone yielded a very precise age of 334 ± 1 Ma (2σ), and a bed from the basal *Gnathodus bilineatus* zone gave a discordant age of 327 ± 5 Ma. In the correlation chart of Braun & Gursky (1991), the *texanus* zone corresponds to the V1 zone. Hence, the youngest proven sediments here (V2–V3 boundary) must be clearly younger than 334 ± 1 Ma. The early *bilineatus* zone correlates with the late V3a–early V3b zones, but this level has not been detected in the flysch.

The end of sedimentation is also constrained by the unconformable onlap of coal-bearing, intramontane clastic sediments on deformed flysch of the Frankenberg Klippe. These beds have yielded a rich macroflora of latest Viséan age (see discussion by Franke (1984*b*)). From a tuff in beds dated as Namurian A, Lippolt *et al.* (1994) have obtained a sanidine age of 325 ± 4 Ma, so that the latest Viséan age is slightly older. A somewhat earlier time limit is defined by the emplacement of the post-tectonic Meissen Pluton in the Elbe Zone, which intrudes folded Palaeozoic rocks and has been dated by Wenzel *et al.* (1997) at 330 ± 2 Ma (Ar–Ar, amphibole). Hence, the youngest part of marine sedimentation, as well as subsequent deformation and metamorphism in the Saxo-Thuringian relative autochthon, must have taken place between 334 ± 1 Ma and 330 ± 2 Ma. The younger ages in the Fichtelgebirge reflect a late tectonometamorphic event, which is restricted to the SW margin of the Bohemian Massif and transsects the main collisional structures (Franke this volume).

Lower allochthon

As a result of late Variscan transverse faulting, the lower allochthon is exposed only in the NE (Figs 1 and 3), in the Saxonian Granulites (Fig. 4) and the Erzgebirge (Figs 5 and 6). However, recently acquired seismic data have revealed that the dome structure of the granulites can be traced toward the SW as far as the northwestern foreland of the Münchberg Klippe (Fig. 7; Franke 1993; DEKORP & Orogenic Processes Working Groups 1999; Krawczyk *et al.* this volume). It has not been possible to identify equivalents of the Erzgebirge high-pressure (HP) rocks in the Fichtelgebirge as yet. Such rocks are possibly hidden at deeper levels, but alternative interpretations are possible (Stein & DEKORP Research Group 1998). The Fichtelgebirge sequence have therefore not been incorporated into Fig. 6.

The Saxonian Granulites are exposed in a brachy-anticlinal structure, overlain by a 'schist mantle' of greenschist and very low grade rocks of Late Proterozoic to Early Carboniferous age (Figs 3 and 4). The granulites are mainly felsic, and were probably derived from magmatic rocks with some sedimentary intercalations. In addition, there are lenses of mafic–ultramafic pyroxene granulites and garnet peridotites. According to Roetzler *et al.* (1998), mafic and felsic granulites were subjected to peak metamorphism at 20–25 kbar and 1000–1060 °C, followed by near-isothermal decompression (Roetzler 1992; Reinhardt & Kleemann 1994).

Main granulite facies metamorphism appears to be well dated at *c.* 340 Ma by numerous zircon studies (von Quadt 1993; Baumann *et al.* 1996; Kröner *et al.* 1998; Vavra *et al.* 1999) (Fig. 4). A poorly constrained eclogite stage around 380 Ma (Sm–Nd mineral–whole-rock (WR) isochron;

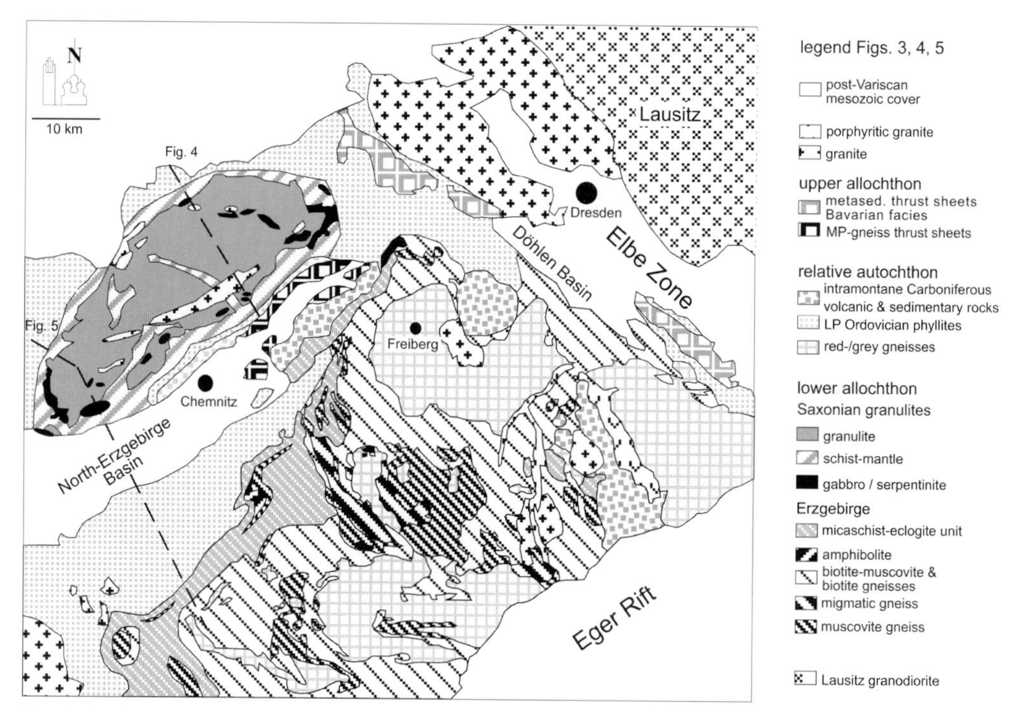

Fig. 3. Geological map of the northeastern part of the Saxo-Thuringian Belt (after Sebastian (1995) and Geol. Übersichtskarte des Freistaates Sachsen 1:400 000). The tectono-metamorphic foliation is roughly parallel to the contours of the tectonic units.

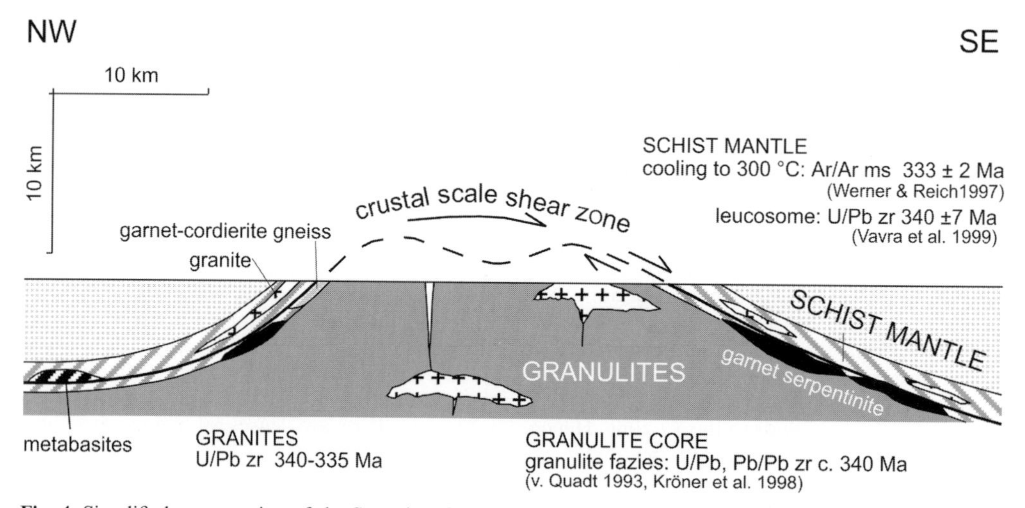

Fig. 4. Simplified cross-section of the Saxonian Granulite Antiform (see Fig. 3 for location; after Reinhardt & Kleemann (1994)) with isotopic ages.

von Quadt 1993) was later questioned by Baumann *et al.* (1996) and Kröner *et al.* (1998).

Most of the sedimentary and volcanic rocks of the schist mantle exhibit prograde low-pressure metamorphism effected by the rising hot granulite core. Metamorphic grade in the schist mantle increases rapidly downwards from very low over greenschist grade up to the formation of leucosome in the roof of the granulite. The contact between core and mantle is a ductile shear zone with consistent sense of shear top to the SE (Kroner 1995). Reich (1996) has also

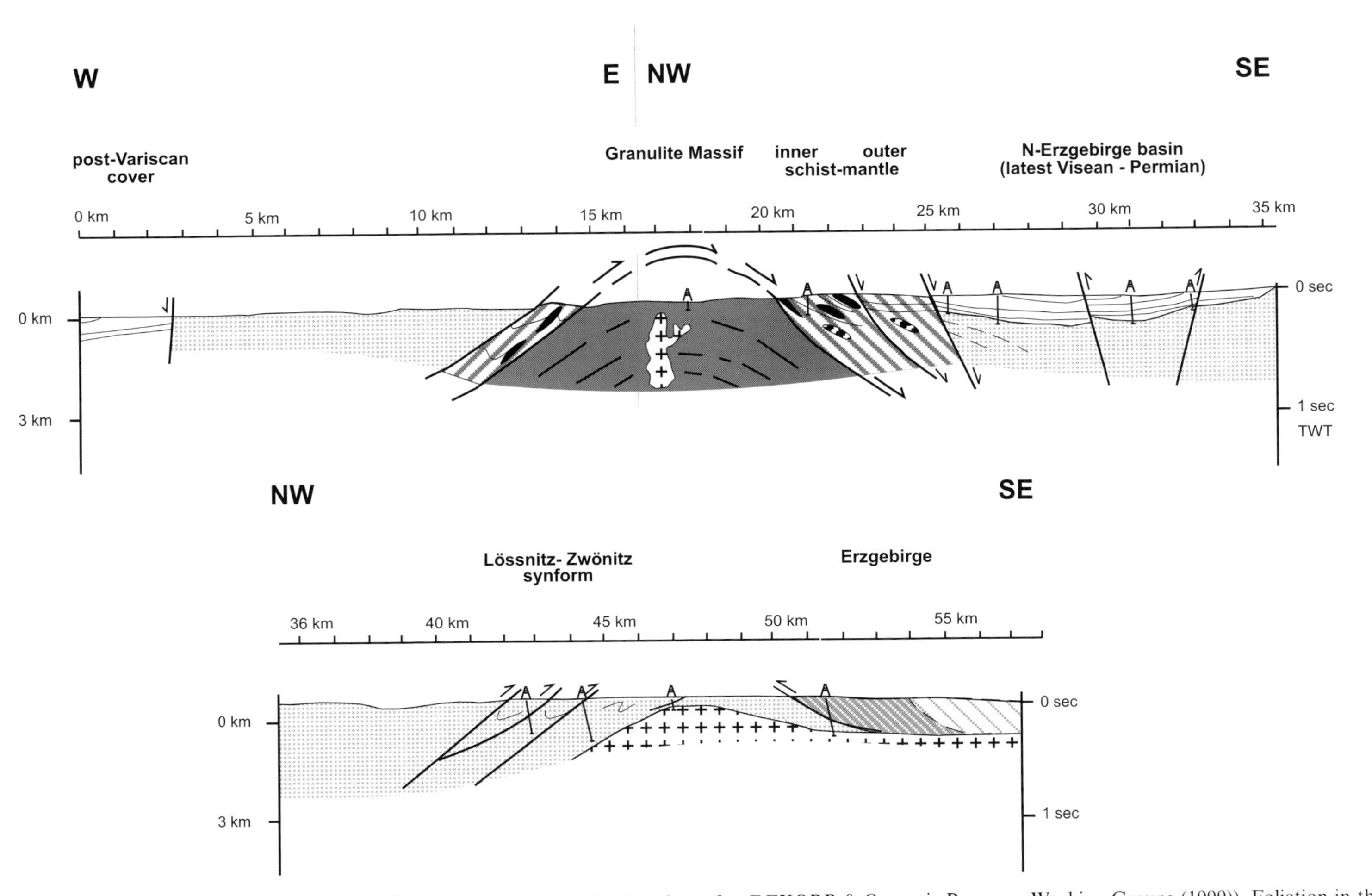

Fig. 5. Tectonic section along seismic profile DEKORP 9502 (see Fig. 3 for location, after DEKORP & Orogenic Processes Working Groups (1999)). Foliation in the Saxonian Granulite is taken from seismic reflection profile.

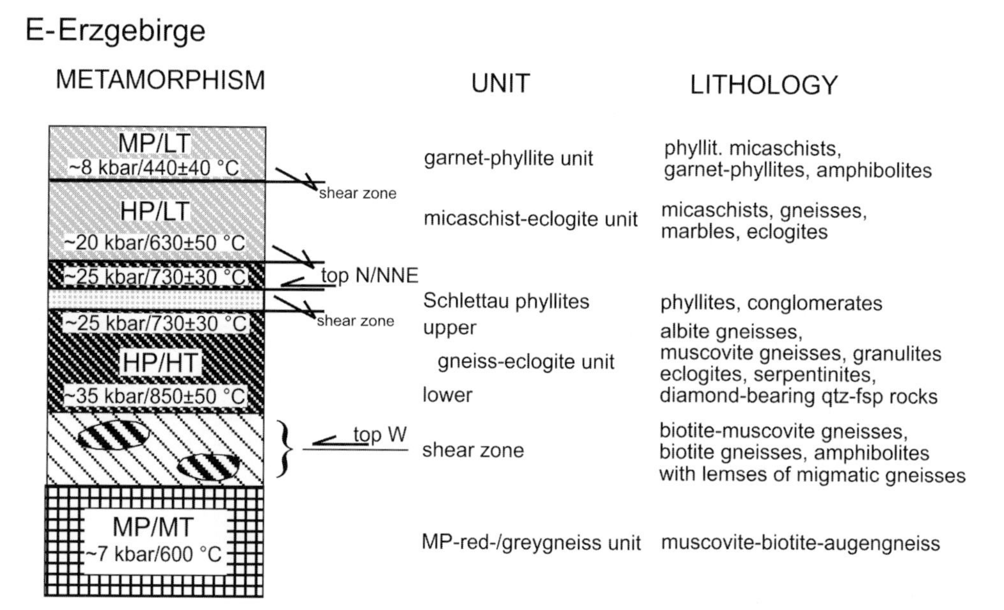

Fig. 6. Tectonic stratigraphy of the Erzgebirge Antiform. *P–T* data from Schmädicke (1994) and Roetzler (1995).

described relics of an earlier medium-pressure evolution, whose age and tectonic setting are unknown.

The granulites and the schist mantle were intruded by late- to post-tectonic granites. The Berbersdorf Granite in the southeastern schist mantle has been dated at 338 ± 5 Ma (U–Pb zircon; von Quadt 1993). A granite in the northwestern schist mantle gave an Pb–Pb zircon evaporation age of 333 ± 1 Ma (Kröner *et al.* 1998). Ar–Ar cooling ages on muscovite and biotite from the schist mantle cluster within the narrow range of 333 ± 3 Ma (Werner & Reich 1997). Zircons from LP leucosomes at the base of the schist mantle have yielded U–Pb ages of 340 ± 7 Ma (Vavra *et al.* 1999). As metamorphism in the schist mantle was brought about by the rising granulites, these ages imply that the granulites were emplaced at *c.* 340 Ma and already cooled, together with the schist mantle, before 333 ± 3 Ma (Fig. 4). Some concordant U–Pb monazite ages from the granulites range between 315 and 303 Ma (Baumann *et al.* 1998). These numbers post-date the mica ages and cannot be related to any known geological event. The position of the data points on the concordia diagram is possibly caused by excess thorium (Romer, pers. comm.).

The protolith age and original site of granulite formation are unknown. Kröner *et al.* (1998) have reported U–Pb ages of zircon cores of *c.* 1700 and 485–470 Ma. The latter age group is widespread in the Saxo-Thuringian relative autochthon (Linnemann *et al.* this volume). Molzahn *et al.* (1998) found that the Sr, Nd and Pb isotopic signature of the granulites clearly differs from that of the Rheno-Hercynian basement, but is comparable with that of the northern Mid-German Crystalline High. Samples from the Saxo-Thuringian and Teplá–Barrandian remain to be analysed.

The Erzgebirge Dome, SE of the Vogtland Synform, exposes a tectonic alternation of HP–MT (medium-temperature), HP–HT (high-temperature), MP–MT and even LP–LT (low-temperature) units (Figs 5 and 6; Schmädicke *et al.* 1992; Roetzler 1995; Mingram 1996; Willner *et al.* 1997; Nega 1998; Nega *et al.* 1998; Massonne, pers. comm.). It is important to note that the tectonic sequence shows a twofold inversion of metamorphic grades, with eclogites overriding amphibolite-grade rocks. This implies that, before thrusting, the overburden of the eclogites had already been reduced to a thickness corresponding to the metamorphic pressure of the interleaved lower-grade rocks, that is, that exhumation pre-dates stacking. The thrust faults responsible for the metamorphic inversion are poorly exposed, but reveal tectonic transport either to the west, or to the N–NNE (Fig. 6, observations by E.S.). The fault surfaces, which tectonically omit important parts of the barometric profile, are not exposed. After the stacking event, the tectonic sequence underwent polyphase refolding (Nega 1998; Nega *et al.* 1998; Konopásek & Schulmann 1999). A late phase of

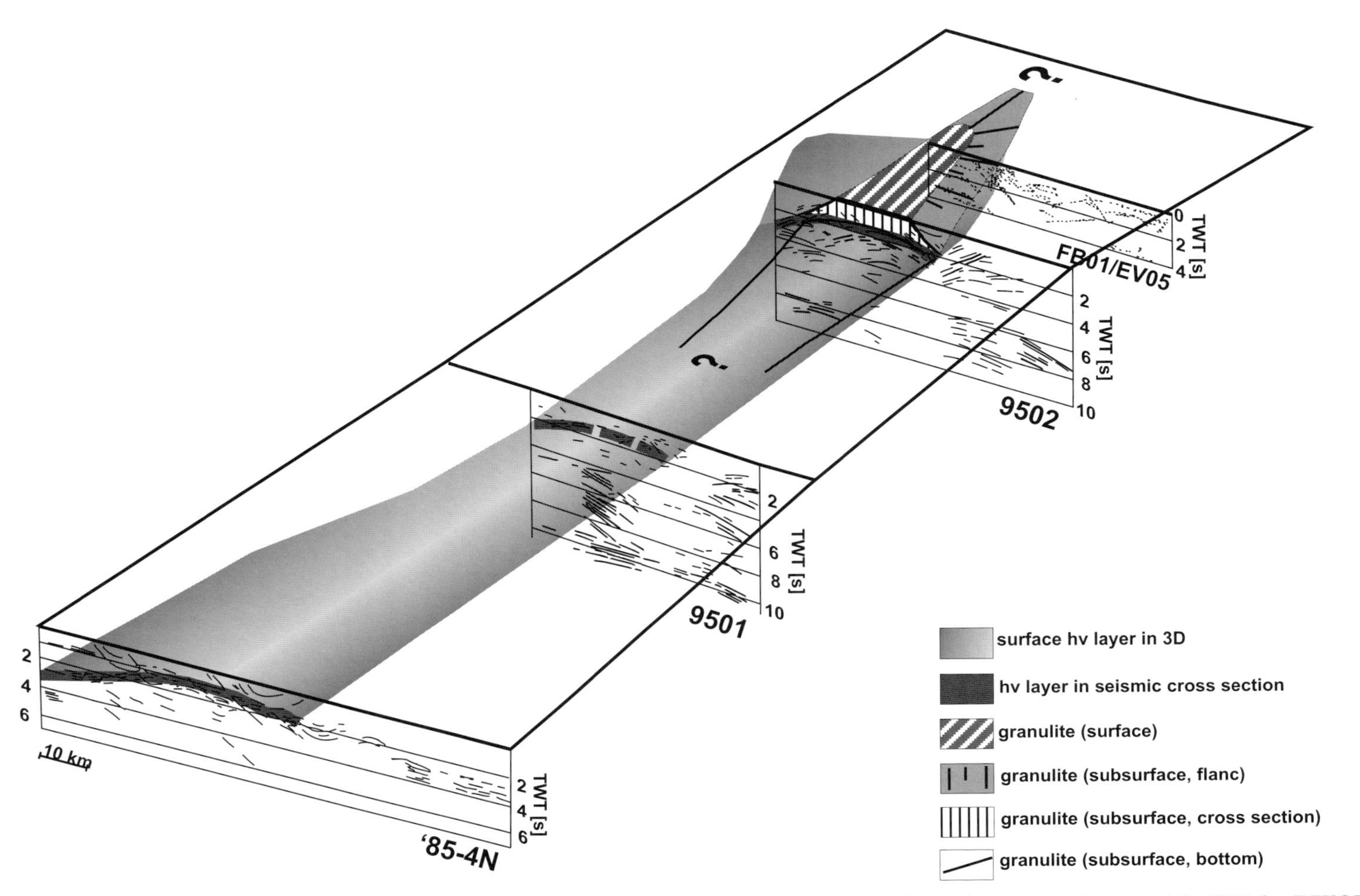

Fig. 7. 3D reconstruction of the refractor and main reflector depicting the dome structure of the Saxonian Granulites and its continuation toward the SW (after DEKORP & Orogenic Processes Working Groups (1999)). hv, high velocity.

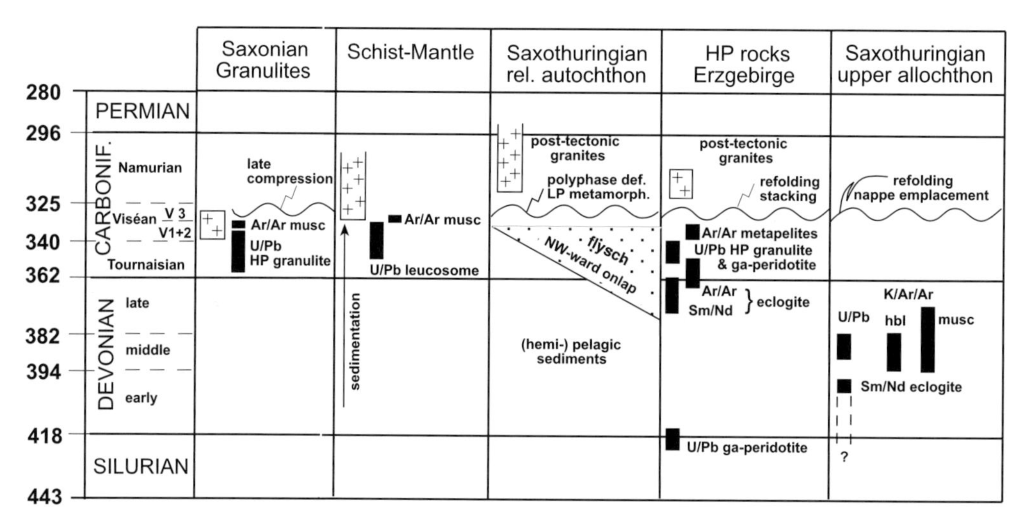

Fig. 8. Timing of geological events in different parts of the Saxo-Thuringian Belt.

extension has effected relative uplift of the Fichtelgebirge–Erzgebirge, and the formation of Permo-Carboniferous intramontane basins on both flanks. In the SW part of the Czech Erzgebirge and in the adjacent Eger Graben (see enclosure), there are also extensive outcrops of HP granulites (Kotkova *et al.* 1996), whose relative tectonic position is uncertain, so that they are not included in Fig. 6.

Isotopic data from the high-grade thrust sheets in the Erzgebirge (Fig. 8) show some scatter. Werner (1998) has reported a few Ar–Ar hornblende ages of *c.* 370 Ma. Schmädicke *et al.* (1995) suggested HP metamorphism at 360 ± 7 Ma (Sm–Nd grt–cpx–WR isochron), and cooling through the closing temperature of phengite at 348 ± 2 and 355 ± 2 Ma, Ar–Ar). However, eclogites have yielded Sm–Nd WR–garnet–clinopyroxene isochrons between 343 ± 1.8 and 335 ± 4 Ma, and zircon ages of 339 ± 1.6 Ma (conventional U–Pb, concordant) and between 344 ± 6 and 338 ± 5 Ma (laser inductively coupled plasma mass spectrometry (ICP-MS), von Quadt & Günther 1999). Pb–Pb evaporation of zircon from eclogite yielded 346.5 ± 8 and 341.5 ± 10 Ma (Köhler & Tichomirova 1998), and 340 ± 8 Ma (Kröner & Willner 1995, 1998). Similar data were also obtained from the HP granulites of the Eger Graben: Kotkova *et al.* (1996) have reported U–Pb multigrain data on zircon and rutile of 348 ± 10 and 346 ± 14 Ma, and Pb–Pb evaporation ages on zircon of 339 ± 1.5 Ma. Ar–Ar ages of orthogneisses from the same area range between 347 ± 3 and 344 ± 3 Ma (biotite) and 344 ± 3 and 341 ± 4 Ma (muscovite; Zulauf *et al.* 1998). Ages $\geqslant 350$ Ma from HP rocks may or may not indicate older increments of metamorphism (see the discussion by Kröner & Willner (1998)). Although there are some inconsistencies also among the younger ages, it appears that the Erzgebirge HP rocks were affected by a very important metamorphic event around 340 Ma.

This event is most clearly depicted in widespread Ar–Ar ages on white mica at *c.* 340 Ma from the country rocks of the eclogites, and *c.* 330 Ma in the lower-grade units (Werner *et al.* 1996, 1997, 1998; Werner & Lippolt this volume). This latter age group is also recorded in Rb–Sr muscovite–WR ages of 330–320 Ma from the lowest tectonic unit (Freiberg gneisses, MP–MT; Tichomirova & Belyatski 1995). Contrary to Kröner & Willner (1998), the younger age group does not represent a later increment of one coherent *P–T* path, but, instead, indicates cooling after the thrusting, which emplaced the largely exhumed HP rocks over the lower-P rocks (Werner & Lippolt this volume). This thrusting episode is also responsible for the final emplacement of the upper allochthon.

Some zircon data reveal the pre-340 Ma history of the HP rocks. From garnet peridotite in borehole T7 in the Eger Graben (enclosure), SHRIMP dating on zoned zircon revealed two stages of HP, one at 342 ± 5 and one at 424 ± 4 Ma (Gebauer 1991). The older age is referred to a HP melting event. This is similar to Sm–Nd ages from the Winklarn eclogite klippe, south of the ZEV, along-strike in the northwestern corner of the Teplá–Barrandian (see enclosure; von Quadt & Gebauer 1993). The eclogites are contained in an association of felsic and mafic magmatic rocks, quartzites, metapelites and carbonates, which clearly represent

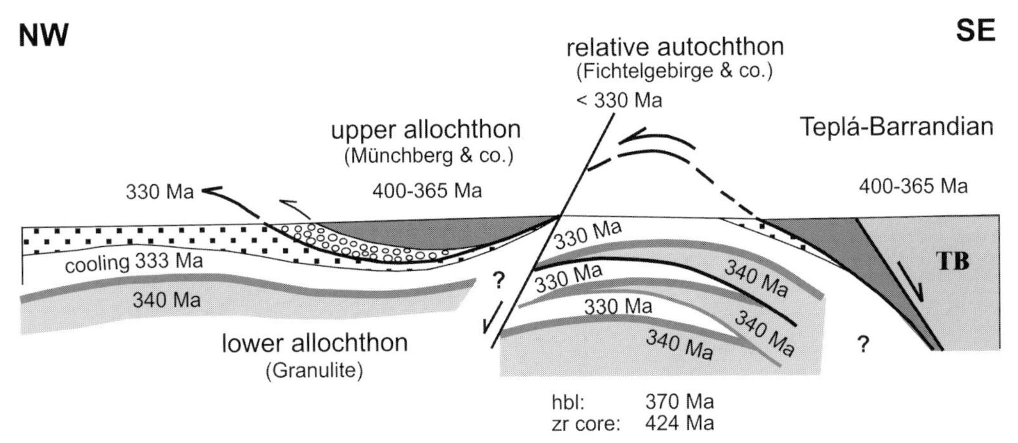

Fig. 9. Diagrammatic section across the eastern part of the Saxo-Thuringian Belt.

continental crust. On the basis of the lithological sequence and its geochemical characteristics, Mingram & Rötzler (1995) concluded that these rocks represent metamorphosed equivalents of the much less metamorphosed sequences in the relative autochthon further to the NW. This is backed up by protolith ages of some of the HP rocks. Kröner & Willner (1995) interpreted U–Pb zircon ages around 480 Ma as dating magmatic crystallization of felsic granitoid and volcanic protoliths in the gneiss–eclogite unit. von Quadt & Günther (1999) obtained 490 ± 8 Ma (SHRIMP) and 495 ± 8 Ma (laser ICP-MS) from magmatic zircons in one eclogite. A magmatic episode of 500–470 Ma is widespread in the Saxo-Thuringian autochthon as well as in the upper allochthon (Franke *et al.* 1995; Linnemann *et al.* this volume).

The assemblage of HP rocks in the Erzgebirge is much more complex than that of the Saxonian Granulites: it represents a refolded thrust stack, and includes HP rocks with a clearly lower geothermal gradient. However, rocks in both areas have registered an important metamorphic event at *c.* 340 Ma, which corresponds to some deeper stage of burial history, if not to peak pressure. Hence, it is plausible to look for a common explanation for the origin and exhumation of all high-grade rocks in the lower allochthon. The main difference between the two areas lies in the much more complex tectonic evolution of the Erzgebirge, which is probably due to its more internal position in the Saxo-Thuringian Belt. The Erzgebirge was affected by the late early Carboniferous stacking, which also emplaced the Münchberg and related klippen, whereas the exhumation geometry of the Saxonian Granulites remained more or less intact (Fig. 9). The tectonic relationships between the Granulites and the Erzgebirge have been discussed in more detail by Krawczyk *et al.* (this volume).

It is obvious that the formation of the Saxonian HP granulites at *c.* 340 Ma is incompatible with marine sedimentation on the hanging wall, which continued beyond 334 ± 1 Ma. Hence, the granulites must represent a lower allochthon, which was emplaced under the relative autochthon. The relative timing of granulite emplacement and marine sedimentation is not clearly resolved. Taken at face value, sedimentation outlasts emplacement of the granulites. However, if one makes full use of the dating error towards a younger age of granulite emplacement, leucosome formation at the base of the schist mantle might have occurred at $340 - 7 = 333$ Ma, and subsequent cooling at $333 - 3 = 330$ Ma, that is, later than the youngest proven age of sedimentation at $334 + 1 = 335$ Ma. It therefore remains uncertain whether emplacement of the Saxonian Granulites (and their equivalents in the Erzgebirge) occurred under the floor of the foreland basin while sedimentation and subsidence were still going on, or a little later. The same considerations apply to the time-equivalent metamorphic rocks in the Erzgebirge. Nevertheless, it is surprising to observe that, in spite of the granulite emplacement, the flysch sediments and even parts of the overlying upper allochthon have not been removed by uplift and erosion.

Geodynamic models

Upper allochthon

The formation and exhumation of the upper allochthon can be explained with available

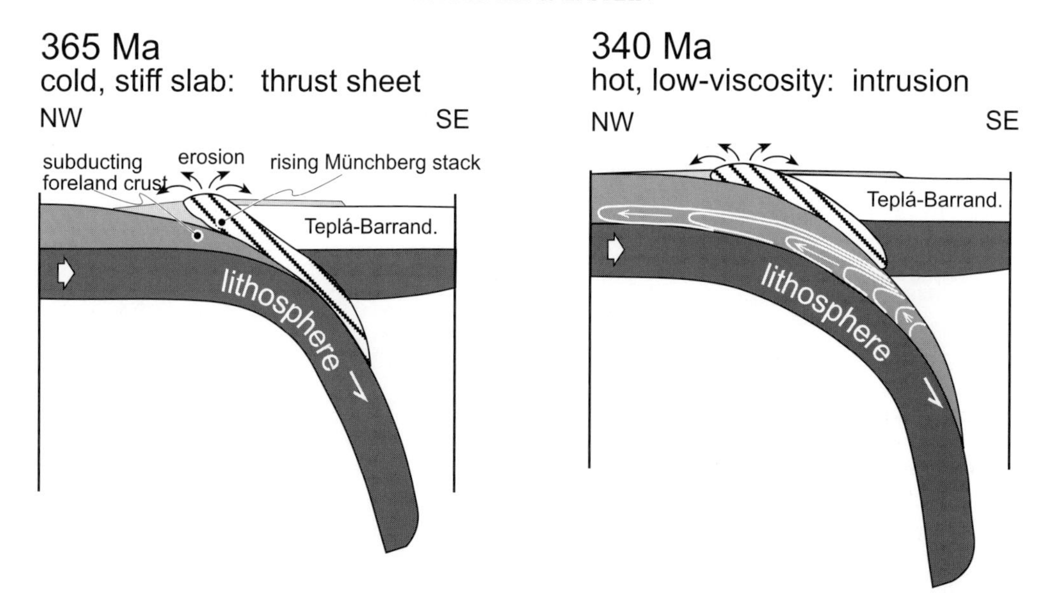

Fig. 10. Geodynamic models for two phases of exhumation of HP rocks in the Saxo-Thuringian Belt.

tectonic models. The upper allochthon is a narrow, yet laterally continuous assemblage of rocks, which is bounded at the rear by ductile normal faults. The K–Ar age of 397 ± 5 Ma obtained from synkinematic hornblende on one of these faults (Zulauf *et al.* 1997) falls in the same range as K–Ar ages of synkinematic muscovites and hornblendes from the Münchberg thrust stack, the bulk of which is constrained between *c.* 400 and 360 Ma (Kreuzer *et al.* 1989). Most of the metamorphic thrust sheets are composed of continental material. To explain these features, we adopt the scenario of Chemenda *et al.* (1995), who proposed subduction, followed by buoyant rise, of continental crust (Fig. 10). One feature typical of this model is the coexistence of normal and reverse faulting, which is also observed in the Saxo-Thuringian Belt (Figs 2 and 10). In the Saxo-Thuringian case, buoyancy as the driving body force is overprinted by plate margin forces (dextral strike-slip component). Erosional unloading (an essential prerequisite in the analogue model of Chemenda *et al.*) is well documented in the upper Devonian–lower Carboniferous flysch.

Lower allochthon

The evolution of the lower allochthon is much more complex. Any geological model for the Saxonian Granulites and the Erzgebirge has to explain the 'Saxo-Thuringian paradox': an uninterrupted sequence of lower Palaeozoic to Upper Viséan rocks, whose deposition continued at least to 333 Ma, is underlain not by a pre-Variscan crystalline basement but by HP granulite and eclogite facies rocks formed at *c.* 340 Ma. In an Alpine analogue, this would translate into a Tertiary age of HP metamorphism in the External Massifs (= Erzgebirge), and the Saxonian Granulites would occupy a position NW of the Pré-Alpes Klippe (= Münchberg Klippe), for example, under the Folded Jura!

Previous models for the lower allochthon

In a first reaction to the isotopic ages, Franke (1993) proposed that the zircon ages do not record the time of HP metamorphism, but isotopic resetting of Proterozoic or older HP rocks, possibly in a back-arc setting at the rear of the Rheno-Hercynian subduction zone (Franke 1994, 1995). In fact, zircon growth has been reported from lower-crustal rocks in areas where the crust has been thinned and heated (Ivrea Zone: Vavra *et al.* 1994; xenoliths from underneath the French Massif Central: Costa & Rey 1995; Red Sea Rift: Bosch & Bruguier 1998). On the other hand, there is as yet no positive isotopic evidence of a Proterozoic HP event in the Saxo-Thuringian Belt. Also, a thermal event strong enough to affect the U–Pb system in zircon in such a situation should also have brought about a pervasive re-equilibration of metamorphic parageneses and fabrics at lower pressures, a feature not recorded in the high-grade rocks exposed at the surface.

The metamorphic event at *c*. 340 Ma has been confirmed by many researchers and various methods. It is relatively unimportant whether this age indicates time of peak pressure or peak temperature. In any case, the HP rocks of the basement and the low-grade Palaeozoic cover must have been brought together tectonically. The ductile detachment in the roof of the Saxonian Granulites has juxtaposed rocks with 17–22 kbar difference in peak metamorphic pressure. With regard to this metamorphic break, as well as the dome-like structure, most workers have agreed with Franke (1989, 1993) that the Saxonian Granulites represent a prime example of a metamorphic core complex. As the cutoff angles in both the hanging wall and the footwall are very low, the enormous reduction of the crustal section would require dramatic extensional displacements. However, timing and geodynamic causes of extension are still hotly debated.

Several workers (e.g. Kroner 1995; Krohe 1996; Willner *et al.* 1997) have proposed gravitational collapse of previously thickened Saxo-Thuringian crust. However, this model is not compatible with the observation that the Saxo-Thuringian Belt was a foreland basin with marine clastic sedimentation until at least *c*. 334 Ma.

As high-grade metamorphism at ⩾340 Ma cannot have occurred under the floor of the subsiding Saxo-Thuringian foreland basin, and HP metamorphism in the well-known upper allochthon is significantly older, Franke (1995) and Franke & Oncken (1996*a*, *b*) have considered tectonic underplating of rocks, originally formed by subduction of Rheno-Hercynian (Avalonian) crust towards the E–SE under the Mid-German Crystalline High. This concept is plausible, as balancing studies of Oncken (1997) suggest that ⩾200 km of thinned lower continental Rheno-Hercynian crust must have been underplated below the nascent Mid-German Crystalline High, and possibly beyond. However, the isotopic signatures of the granulites reported by Molzahn *et al.* (1998) do not support correspondence to Rheno-Hercynian crust.

Hydraulic intrusion model

With regards to the deficiencies of the previously mentioned concepts, we are left with the derivation of the 340 Ma rocks from the SE, that is, from the northwestern, active margin of the Bohemian terrane, which overrode the Saxo-Thuringian foreland. The upper allochthon and the flysch derived from this margin do not contain any record of a 340 Ma event, but the 340 Ma overprint on *c*. 420 Ma garnet peridotite from a borehole in the Eger Graben (Gebauer 1991) hints at the existence of such rocks at depth. Reinhardt & Kleemann (1994) have proposed derivation of the high grade rocks from the SE in a gravitational collapse model. Rocks of relatively low density and low viscosity, originally positioned in the root of the orogen, are suggested to have risen diapirically and to have been emplaced in the upper crust of the northwestern foreland. A diapir model had already been proposed by Weber & Behr (1983). Undoubtedly, the reduction of an orogenic root requires some degree of viscous, buoyant displacement of material away from the root and upwards (e.g. numerical models of Govers & Woertel (1993)). However, the density contrasts cannot be expected to be high enough to cause emplacement into the upper crust, especially as the western part of the shallow basement in the Saxo-Thuringian Zone correlates with a positive gravity anomaly (Franke 1993). In addition, diapiric rise of lower-crustal material in a compressional foreland situation should cause uplift, which is not compatible with continuing sedimentation.

As suggested by DEKORP & Orogenic Processes Working Groups (1999), the inconsistencies of the diapir model may be overridden, if the rise of the granulites were produced by the hydraulic gradient between the orogenic root to the SE and the lower crust of the foreland. This gradient may have caused injection of material into a channel bounded by the lithospheric mantle below and the brittle–ductile boundary above (Fig. 10). Injection must have been fast enough to prevent metamorphic equilibration of the granulites at shallower depths. This model is supported by the predominantly felsic nature of the lower allochthon, which implies low bulk viscosity. Eclogites and associated serpentinites form only small lenses, often embedded in metagranitoids or even marbles. These mafic rocks have probably been entrained by the low-viscosity felsic materials. The intrusion model is backed up by the observation that debris of the 340 Ma rocks is missing in the flysch, whereas evidence for older events such as 380 Ma metamorphism is abundant. Apparently the 340 Ma rocks never reached the surface.

As shown by Reinhardt & Kleemann (1994), emplacement of the low-viscosity material in the foreland creates a space problem. Either the lower crust of the foreland has to be pushed aside, or the granulites were intruded like a magmatic sill, which should have raised its roof. The difference between these alternatives lies in the localization of the materials displaced from the lower crust of the relative autochthon. In either case, it will be difficult to maintain marine sedimentation on top. Therefore the hydraulic injection model required extension in the foreland, which compensated for the intrusion of granulite, and thus leaves space for sediment and water on top. Extension could have been provided by back-arc spreading at the rear of the Rheno-Hercynian subduction zone. The dramatic reduction of the crustal profile can be satisfactorily explained by the intrusion model on its own. In such a model, extension is only a late increment of exhumation, and the Granulite Dome cannot therefore be classed as a classical metamorphic core complex.

In the southeastern part of the Saxo-Thuringian Belt, the previously exhumed rocks underwent renewed stacking and multiphase folding, at *c.* 330 Ma, during the NW-ward advance of the Teplá–Barrandian orogenic wedge. This implies that the intrusion of the lower allochthon proceeded faster than the advance of the orogenic front. As the NW-driving D1 deformation dies out towards the NW, the more external area occupied by the Saxonian Granulite dome and its continuation towards the SW has been little affected, so that the roof of the intrusion has largely been preserved.

Low-viscosity flow of lower-crustal materials has also been proposed in several recent publications. Intrusive nappe transport driven by pure shear has been suggested for the Alps (Merle & Guillier 1989; Escher & Beaumont 1997), but also for extensional environments (e.g. Block & Royden 1990; Bird & Kong 1994; Kaufmann & Royden 1994; Hopper & Buck 1996; MacCready *et al.* 1997; Wernicke & Getty 1997). Our concept of low-viscosity intrusion into the foreland crust has been tested by numerical modelling by Henk (this volume).

Discussion and conclusions

The upper and lower allochthon of the Saxo-Thuringian Belt are probably derived from the same subduction zone, but were exhumed at different times and in different ways. Metamorphic rocks of the upper allochthon, formed and cooled between 395 and 365 Ma, were uplifted and exhumed as a narrow slice between the Teplá–Barrandian hinterland and the Saxo-Thuringian foreland. They had already reached the surface in Famennian time, and first fed, and then overrode the foreland basin adjacent to the NW. The lower allochthon, formed at 340 Ma or little earlier, did not apparently reach the surface, but was intruded into the lower crust of the foreland.

These different modes of exhumation are probably related to the different peak temperatures of the metamorphic rocks, which produced different rheologies. The upper allochthon has a granulite overprint only in its internal, southeastern part (Mariánské Lázně), which, according to the model of Chemenda *et al.* (1995), could have been in contact with the mantle. Its northwestern part (contained in the Münchberg and other klippen) was not heated beyond 700 °C. Hence, the upper allochthon can be regarded as a relatively stiff slab, which pierced its roof and was available for erosion. The Saxonian Granulites and their equivalents in the Erzgebirge have attained much higher temperatures (in excess of 1000 °C) and, consequently, lower viscosities. Therefore, they were unable to break their cover and were spread laterally under the rigid lid of the orogen and the foreland crust.

The divergent thermal regimes (relatively cool during late Devonian time and hot during early Carboniferous time are recognizable in the entire central Variscides. From *c.* 340 Ma onwards, all processes appear to have proceeded at high temperatures. The origin of this late orogenic high-temperature regime is a matter of debate, which we leave to other contributions in this volume (see Henk *et al.* (this volume), O'Brien (this volume)).

From the scenario depicted in Fig. 10, it might be expected that the lithospheric mantle was completely delaminated from the crust, or even broke off. In fact, delamination has been proposed in such a setting by Chemenda *et al.* (1997). However, these events should cause significant heating, and, in the latter case, rise of primitive mantle melts (eg. Henk *et al.* this volume), both of which have not been recorded.

The northwestward advance of the orogenic front lagged behind the granulite intrusion. In the southeastern, internal part of the Saxo-Thuringian Belt (Fichtelgebirge–Erzgebirge Antiform), it produced a tectonic alternation of rocks derived from the top of the lower allochthon and from its cover, which underwent polyphase refolding. We propose that the Fichtelgebirge–Erzgebirge Antiform is a late compressive feature, and not an extensional core complex as has been suggested by Kroner (1995) and Kröner & Willner (1998). However, the geometry of the post-340 Ma structures remains to be resolved in detail. Normal faulting along the flanks of the Antiform occurred during late Carboniferous–Permian time and is coeval with the formation of intramontane basins to the NW and the SE.

Although we are still far from completely understanding the process of its emplacement, it is safe to state that the lower allochthon of the Saxo-Thuringian Belt represents a fascinating addition to the assortment of exhumation scenarios.

The authors gratefully acknowledge helpful suggestions and constructive criticism by M. Sausson and O. Werner. Facts and views presented in this paper were assembled during research projects funded by the DEKORP project (Pr 74/17-5//Schw617/2-2//Schu824/6-4 und Pr 74/17-6//Ha 1210/19-6).

References

BAUMANN, N., PILOT, J., TODT, W., WERNER, C.-D. & WILLIAMS, I. S. 1998. Konventionelle U/Pb- und SHRIMP-Daten zum Sächsischen Granulitgebirge: Überlegungen zur Hebungsgeschichte. *Terra Nostra*, **98**(2), 20–22.

——, ——, WERNER, C.-D., TODT, W. & HOFMANN, A. W. 1996. Neue Datierungen im Sächsischen Granulitgebirge. *Terra Nostra*, **96**(2), 16–17.

BEHR, H.-J., ENGEL, W. & FRANKE, W. 1982. Variscan wildflysch and nappe tectonics in the Saxothuringian Zone (Northeast Bavaria, West Germany). *American Journal of Science*, **282**, 1438–1470.

BIRD, P. & KONG, X. 1994. Computer simulations of California tectonics confirm very low strength of major faults. *Geological Society of America Bulletin*, **106**, 159–174.

BLOCK, L. & ROYDEN, L. H. 1990. Core complex geometries and regional scale flow in the lower crust. *Tectonics*, **9**, 557–567.

BOSCH, D. & BRUGUIER, O. 1998. An early Miocene age for a high-temperature event in gneisses from Zabargad Island (Red Sea, Egypt): mantle diapirism? *Terra Nova*, **10**(5), 274–279.

BRAUN, A. & GURSKY, H. J. 1991. Kieselige Sedimentgesteine des Unter-Karbons im Rhenohercynikum—eine Bestandsaufnahme. *Geologica et Palaeontologica*, **25**, 57–77.

CHEMENDA, A. I., MATTAUER, M., MALAVIEILLE, J. & BOKUN, A. L. 1995. A mechanism for syncollisional rock exhumation and associated normal faulting: results from a physical modelling. *Earth and Planeatary Science Letters*, **132**, 225–232.

——, YANG, R. K., HSIEH, C. H. & GROHOLSKY, A. N. 1997. Evolutionary model for the Taiwan collision based on physical modelling. *Tectonophysics*, **274**, 253–274.

COSTA, S. & REY, P. 1995. Lower crustal rejuvenation and growth during post-thickening collapse: insights from a crustal cross section through a Variscan metamorphic core complex. *Geology*, **23**, 905–908.

DEKORP & OROGENIC PROCESSES WORKING GROUPS 1999. Structure of the Saxonian Granulites—geological and geophysical constraints on the exhumation of HP/HT-rocks. *Tectonics*, **18**, 756–773.

ESCHER, A. & BEAUMONT, C. 1997. Formation, burial and exhumation of basement nappes at crustal scale: a geometric model based on the Western Swiss–Italian Alps. *Journal of Structural Geology*, **19**, 955–974.

FALK, F., FRANKE, W. & KURZE, M. 1995. Saxothuringian Basin: autochthon and nonmetamorphic nappe units: stratigraphy. *In*: DALLMEYER, D., FRANKE, W. & WEBER, K. (eds) *Pre-Permian Geology of Central and Western Europe*. Springer, Berlin, 219–234.

FRANKE, W. 1984*a*. Variszischer Deckenbau im Raume der Münchberger Gneismasse, abgeleitet aus der Fazies, Deformation und Metamorphose im ungebenden Paläozoikum. *Geotektonische Forschungen*, **68**, 1–253.

—— 1984*b*. Late events in the tectonic history of the Saxothuringian Zone. *In*: HUTTON, D. W. H., SANDERSON, D. J. (eds) *Variscan Tectonics of the North Atlantic Region*. Blackwell, Oxford, 33–45.

—— 1989. Tectonostratigraphic units in the Variscan belt of central Europe. *In*: DALLMEYER, R. D. (ed.) *Terranes in the Circum-Atlantic Palaeozoic Orogens*. Geological Society of America, Special Papers, **230**, 67–90.

—— 1993. The Saxonian granulites: a metamorphic core complex. *Geologische Rundschau*, **82**, 505–515.

—— 1994. Exhumation of Saxothuringian HP rocks: alternative concepts. *Terra Nostra*, **94**(3), 41–43.

—— 1995. Exhumation of the Saxonian Granulites: geological, petrological and isotopic constraints. *Terra Nostra*, **95**(8), 22–24.

—— 2000. The mid-European Segment of the Variscides: tectono-stratigraphic units, terrane boundaries and plate evolution. *This volume*.

—— & ONCKEN, O. 1996*a*. Auswege aus dem saxothuringischen Paradoxon. *Terra Nostra*, **96**(2), 55–57.

—— & —— 1996*b*. Crustal structure of the Saxo-Thuringian Belt: a layered cake of terranes. *Annales Geophysicae*, **14**, C86.

——, BEHRMANN, J. & MOEHRMANN, H. 1992*a*. Zur Deformationsgeschichte des Kristallins im Münchberger Deckenstapel. *KTB Report*, **92**(4), 225–240.

——, DALLMEYER, D. & WEBER, K. 1995. Geodynamic evolution. *In*: DALLMEYER, D., FRANKE, W. & WEBER, K. (eds) *Pre-Permian Geology of Central and Western Europe*. Springer, Heidelberg, 579–593.

——, PRÖSSL, K. F. & SCHWARZ, J. 1992*b*. Devonische Grauwacken im Erbendorfer Paläozoikum—Alter, tektonische Stellung und geotektonische Bedeutung. *KTB Report*, **92**(4), 213–224.

GANDL, J. 1998. Neue Daten zum jüngeren Paläozoikum NE-Bayerns und angrenzender Gebiete—Fazies-Entwicklung und geotektonische Konsequenzen. *Geologica Bavarica*, **103**, 19–273.

—— & MANSOURIAN, E. 1978. Neue Daten zur Entwicklung des Unter-Karbons im Frankenwald (NE-Bayern). *Zeitschrift der Deutschen Geologischen Gesellschaft*, **129**, 99–108.

GEBAUER, D. 1991. Two Palaeozoic high-pressure events in garnet-peridotite of northern Bohemia, Czechoslovakia. *Terra abstracts*, **5**.

GOVERS, R. & WOERTEL, M. J. R. 1993. Initiation of asymmetric extension in continental lithosphere. *Tectonophysics*, **223**, 75–96.

GREILING, L. 1961. Devon im Oberpfälzer Wald. *Senckenbergiana Lethaea*, **42**, 265–271.

HENK, A. 2000. Foreland-directed lower-crustal flow and its implications for the exhumation of high-pressure–high-temperature rocks. *This volume*.

——, VON BLANCKENBURG, F., FINGER, F., SCHALTEGGER, U. & ZULAUF, G. 2000. Syn-convergent high-temperature metamorphism and magmatism in the Variscides; a discussion of potential heat sources. *This volume*.

HOPPER, J. R. & BUCK, W. R. 1996. The effect of lower crustal flow on continental extension and passive margin formation. *Journal of Geophysical Research*, **101**, 20175–20194.

KAUFMANN, P. S. & ROYDEN, L. H. 1994. Lower crustal flow in an extensional setting: constraints from the Halloran Hill region, eastern Mojave Desert, California. *Journal of Geophysical Research*, **100**, 17587–17602.

KLEINSCHRODT, R. & GAYK, T. 1999. Analogies and synchronism in the exhumation of HP units in the Variscan Orogen: a comparison of the Münchberg Complex and Cabo Ortegal Complex. *Terra Nostra*, **99**(1), 119.

KÖHLER, R. & TICHOMIROVA, M. 1998. Geochronologie und Geochemie von amphibolitfaziellen Metabasiteinschaltungen des Osterzgebirges. *Terra Nostra*, **98**(2), 91–94.

KONOPÁSEK, J. & SCHULMANN, K. 1999. Tectonic evolution of the central part of the Kruzne Hory Mts. (Erzgebirge) in the Czech Republic (Saxothuringian domain of the Bohemian Massif). *Terra Nostra*, **99**(1), 120–121.

KOTKOVA, J., KRÖNER, A., TODT, W. & FIALA, J. 1996. Zircon dating of North Bohemian granulites, Czech Republic: further evidence for the Lower Carboniferous high-pressure event in the Bohemian Massif. *Geologische Rundschau*, **85**, 154–161.

KRAWCZYK, C. M., STEIN, E., CHOI, S. *et al.* 2000. Geophysical constraints on exhumation mechanisms of high-pressure rocks; the Saxo-Thuringian case between the Franconian Line and Elbe Zone. *This volume*.

KREUZER, H., SEIDEL, E., SCHÜSSLER, U., OKRUSCH, M., LENZ, K.-L. & RASCHKA, H. 1989. K–Ar geochronology of different tectonic units at the northwestern margin of the Bohemian Massif. *Tectonophysics*, **157**, 149–178.

KRÖNER, A. & WILLNER, A. P. 1995. Magmatische und metamorphe Zirkonalter für Quarz–Feldspat-Gesteine der Gneis–Eklogit-Einheit des Erzgebirges. *Terra Nostra*, **95**(8), 112.

—— & —— 1998. Time of deformation and peak of Variscan HP–HT metamorphism of quartz–feldspar rocks in the central Erzgebirge, Saxony, Germany. *Contributions to Mineralogy and Petrology*, **132**, 1–20.

——, JAECKEL, P., REISCHMANN, T. & KRONER, U. 1998. Further evidence for an early Carboniferous (*c.* 340 Ma) age of high-grade metamorphism in the Saxonian granulite complex. *Geologische Rundschau*, **86**, 751–766.

KROHE, A. 1996. Variscan tectonics of Central Europe: postaccretionary intraplate deformation of weak continental lithosphere. *Tectonics*, **15**, 1364–1388.

KRONER, U. 1995. Postkollisionale Extension am Nordrand der Böhmischen Masse—die Exhumierung des Sächsischen Granulitgebirges. *Freiberger Forschungshefte*, **C457**, 1–114.

LINNEMANN, U., GEHMLICH, M., TICHOMIROVA, M. *et al.* 2000. From Cadomian subduction to Early Palaeozoic rifting: The evolution of Saxo-Thuringia at the margin of Gondwana in the light of single zircon geochronology and basin development (Central European Variscides, Germany). *This volume*.

LIPPOLT, H. J., HESS, J. C. & GOLL, M. 1994. Quantitative Erfassung des Einsetzens und der Dauer des älteren subsequenten Vulkanismus im Thüringer Wald (Gehren Schichten). *Terra Nostra*, **94**(3), 73–74.

LISTER, G. S. & BALDWIN, S. L. 1993. Plutonism and the origin of metamorphic core complexes. *Geology*, **21**, 607–610.

MACCREADY, T., SNOKE, A. W., WRIGHT, J. E. & HOWARD, K. A. 1997. Mid-crustal flow during Tertiary extension in the Ruby Mountains core complex, Nevada. *Geological Society of America Bulletin*, **109**, 1576–1594.

MATTE, P., MALUSKI, H., RAJLICH, P. & FRANKE, W. 1990. Terrane boundaries in the Bohemian massif: result of large-scale shearing. *Tectonophysics*, **177**, 151–170.

MEDARIS, L. G., JR, JELÍNEK, E. & MÍSAR, Z. 1995. Czech eclogites: terrane settings and implications for Variscan tectonic evolution of the Bohemian Massif. *European Journal of Mineralogy*, **7**, 7–28.

MERLE, O. & GUILLIER, B. 1989 The building of the Central Swiss Alps: an experimental approach. *Tectonophysics*, **165**, 41–56.

MINGRAM, B. 1996. *Geochemische Signaturen des erzgebirgischen Krustenstapels*. Scientific Technical Report, GeoForschungsZentrum Potsdam, **STR96**.

—— & RÖTZLER, K. 1995. Deckenstapelung altpaläozoischer Einheiten auf proterozoischem Basement im Erzgebirge. *Terra Nostra*, **95**(8), 117.

MOLZAHN, M., ONCKEN, O. & REISCHMANN, T. 1998. Isotopengeochemische Charakterisierung von Krustenblöcken innerhalb der Varisziden Mitteleuropas. *Terra Nostra*, **98**(2), 108–109.

NEGA, M. 1998. *Von der Subduktion bis zur Exhumierung: Entwicklung eines Kollisionsorogens und Exhumierungsmechanismen am Beispiel des westlichen Erzgebirges, mitteleuropäische Varisziden*. PhD thesis, TU München.

——, KRUHL, J., KRENTZ, O. & LEONHARDT, D. 1998. Polyphase Krustenstapelung und Exhumierung von HP-Gesteinen durch Kontinent-Kollision im westlichen Erzgebirge. *Terra Nostra*, **98**(2), 109–112.

NEUROTH, H. 1997. K/Ar-Datierungen an detritischen Muskoviten—'Sicherungskopien' orogener Prozesse am Beispiel der Varisziden. *Göttinger Arbeiten zur Geologie und Paläontologie*, **72**, 1–114.

O'BRIEN, P. J. 2000. The fundamental Variscan problem: high-temperature metamorphism at different depths and high-pressure metamorphism at different temperatures. *This volume.*

ONCKEN, O. 1997. Transformation of a magmatic arc and orogenic root during the oblique collision and its consequences for the evolution of the European Variscides (Mid-German Crystalline Rise). *Geologische Rundschau*, **86**, 2–21.

PLATT, J. P. 1986. Dynamics of orogenic wedges and the uplift of high-pressure metamorphic rocks. *Geological Society of America Bulletin*, **97**, 1037–1053.

REICH, S. 1996. *Erzgebirgsrandzone—Schiefermantel des Sächsischen Graulitgebirges: ein petrologischer Vergleich.* PhD thesis, TU Darmstadt.

REINHARDT, J. & KLEEMANN, U. 1994. Extensional unroofing of granulite lower crust and related low-pressure, high-temperature metamorphism in the Saxonian Granulite Massif, Germany. *Tectonophysics*, **238**, 71–94.

ROBERTS, J., CLAOUE-LONG, J., JONES, P. J. & FOSTER-CLINTON, B. 1995. SHRIMP zircon age control of Gondwanan sequences in Late Carboniferous and Early Permian Australia. *In*: DUNAY, R. E. & HAILWOOD, E. A. (eds) *Non-biostratigraphical Methods of Dating and Correlation.* Geological Society, London, Special Publications, **89**, 145–174.

ROETZLER, J. 1992. Zur Petrogenese im Sächsischen Granulitgebirge—die pyroxenfreien Granulite und die Metapelite. *Geotektonische Forschungen*, **77**, 1–100.

——, CARSWELL, D. A., GERSTENBERGER, H. & HAASE, G. 1999. Transitional blueschist–epidote amphibolite facies metamorphism in the Frankenberg massif, Germany, and geotectonic implications. *Journal of Metamorphic Geology*, **16**, 109–125.

——, HAGEN, B. & HOERNES, S. 1998. Ultrahigh-temperature high-pressure metamorphism in the Saxonian granulite massif, Germany. *Terra Nostra*, **98**(2), 130–131.

ROETZLER, K. 1995. *Die PT-Entwicklung der Metamorphite im Mittel- und Westerzgebirge.* Scientific Technical Report, GeoForschungsZentrum Potsdam, **STR95/14**.

SCHÄFER, F., ONKEN, O., KEMNITZ, H. & ROMER, R. L. 2000. Upper-plate deformation during collisional orogeny; a case study from the German Variscides (Saxo-Thuringian Zone). *This volume.*

SCHÄFER, J., NEUROTH, H., AHRENDT, H., DÖRR, W. & FRANKE, W. 1997. Accretion and exhumation at a Variscan active margin, recorded in the Saxothuringian flysch. *Geologische Rundschau*, **86**, 599–611.

SCHMÄDICKE, E. 1994. Die Eklogite des Erzgebirges. *Freiberger Forschungshefte*, **C456**, 1–338.

——, MEZGER, K., COSCA, M. A. & OKRUSCH, M. 1995. Variscan Sm–Nd and Ar–Ar ages of eclogite facies rocks from the Erzgebirge, Bohemian Massif. *Journal of Metamorphic Geology*, **13**, 537–552.

——, OKRUSCH, M. & SCHMIDT, W. 1992. Eclogite-facies rocks in the Saxonian Erzgebirge, Germany: high pressure metamorphism under contrasting *P–T* conditions. *Contributions to Mineralogy and Petrology*, **110**, 226–241.

SCHWAN, W. 1974. Die Sächsischen Zwischengebirge und Vergleiche mit der Münchberger Gneismasse und anderen analogen Kristallinvorkommen im Saxothuringikum. *Erlanger Geologische Abhandlungen*, **99**, 189.

SEBASTIAN, U. 1995. Die Strukturentwicklung des spätorogenen Erzgebirgsaufstiegs in der Flöhazone. Ein weiterer Beitrag zur postkollisionalen Extension am Nordrand der Böhmischen Masse. *Freiberger Forschungshefte*, **C461**, 1–114.

STEIN, E. 1988. Die strukturgeologische Entwicklung im Übergangsbereich Saxothuringikum/Moldanubikum in NE-Bayern. *Geologica Bavarica*, **92**, 5–131.

—— & DEKORP RESEARCH GROUP 1998. Geologische Randbedingungen für die Krustenstruktur im Saxothuringikum. *Terra Nostra*, **98**(2), 138–143.

TEUFEL, S. 1988. Vergleichende U–Pb- und Rb–Sr-Altersbestimmungen an Gesteinen des Übergangsbereiches Saxothuringikum/Moldanubikum, NE-Bayern. *Göttinger Arbeiten zur Geologie und Paläontologie*, **35**, 1–87.

TICHOMIROVA, M. & BELYATSKI, B. V. 1995. Evidence of Variscan metamorphism in the Eastern Erzgebirge. *Terra Nostra*, **95**(7), 133–136.

TRAPP, E., ZELLMER, H., BAUMANN, A., MEZGER, K. & WACHENDORF, H. 1998. Die Kieselgesteins-Fazies des Unterkarbons im Harz—Biostratigraphie, U–Pb-Einzelzirkon-Alter, Petrographie und Sedimentologie. *Terra Nostra*, **98**(3), V 368.

VAVRA, G., GEBAUER, D. & SCHMID, R. 1994. Zircon growth and recrystallization during granulite facies metamorphism in the Ivrea Zone (Southern Alps): a combined cathodoluminescence and ion microprobe study. *Journal of the Czech Geological Society*, **39**, 114–115.

——, REINHARDT, J. & PIDGEON, R. T. 1999. Lower Carboniferous (340 Ma) HP/HT-metamorphism and rapid exhumation of the Saxonian Granulite Massif. *Terra Nostra*, **99**(1), 203.

VON QUADT, A. 1993. The Saxonian Granulite Massif: new aspects from geochronological studies. *Geologische Rundschau*, **82**, 516–530.

—— & Gebauer, D. 1993. Sm–Nd and U–Pb dating of eclogites and granulites from the Oberpfalz, NE Bavaria, Germany. *Chemical Geology*, **109**, 317–339.

—— & Günther, D. 1999. Evolution of Cambrian eclogitic rocks in the Erzgebirge: a conventional and LA-ICP-MS U–Pb zircon and Sm–Nd study. *Terra Nostra*, **99**(1), 164.

WEBER, K. & BEHR, H.-J. 1983. Geodynamic interpretation of the Mid-European Variscides. *In*: MARTIN, H. & EDER, F. W. (eds) *Intracontinental Fold Belts. Case Studies in the Variscan Belt of*

Europe and the Damara Belt in Namibia. Springer, Heidelberg, 427–469.

WENZEL, TH., MERTZ, D. F., OBERHÄNSLI, R., BECHER, T. & RENNO, P. R. 1997. Age, geodynamic setting, and mantle enrichment processes of a K-rich intrusion from the Meissen massif (northern Bohemian massif) and implications for related occurrences from the mid-European Hercynian. *Geologische Rundschau*, **86**, 556–570.

WERNER, O. 1998. K–Ar- und Rb–Sr-Chronolgie spätvariscischer Krustenkonvergenz—Bilanzierung des Wärme- und Stofftransportes im Erzgebirge. PhD thesis, Heidelberg.

—— & LIPPOLT, H. J. 2000. White-mica $^{40}Ar/^{39}Ar$ ages of Erzgebirge metamorphic rocks: simulating the chronological results by a model of Variscan crustal imbrication. *This volume*.

—— & REICH, S. 1997. $^{40}Ar/^{39}Ar$ Abkühlalter von Gesteinen mit unterschiedlicher *P–T*-Entwicklung aus dem Schiefermantel des Sächsischen Granulitgebirges. *Terra Nostra*, **97**(5), 196–198.

——, HESS, J. C. & LIPPOLT, H. J. 1996. Das variszische Abkühlungsmuster des Westerzgebirges—Ergenbisse isotopischer Mineraldatierungen. *Terra Nostra*, **96**(2), 192–196.

WERNER, O., LIPPOLT, H. J. & HESS, J. C. 1997. $^{40}Ar/^{39}Ar$ and Rb–Sr-investigations of the cooling-history of metamorphic and plutonic rocks in the Erzgebirge (Mid-European Saxothuringian). *Terra Nova, Abstract Supplement 1*, **9**, 106.

WERNER, O., NEGA, M., KRUHL, J. H. & LIPPOLT, H. J. 1998. Krustenstapelung und Teilheraushebung im Erzgebirge um 340 Ma: Glimmerchronologie und struktureller Befund der Metapsammopelite der Fuchsleithe. *Terra Nostra*, **98**(2), 163–166.

WERNICKE, B. & GETTY, B. 1997. Intracrustal subduction and gravity currents in the deep crust: Sm–Nd, Ar–Ar, and thermobarometric constraints from the Skagit Gneiss Complex, Washington. *Geological Society of America Bulletin*, **109**, 1149–1166.

WILLNER, A. P., RÖTZLER, K. & MARESCH, W. V. 1997. Pressure–temperature and fluid evolution of quartzo-feldspathic metamorphic rocks with a relic high-pressure, granulite-facies history from the Central Erzgebirge (Saxony, Germany). *Journal of Petrology*, **38**, 307–336.

ZULAUF, G., BUES, C., DÖRR, W., FIALA, J., KOTKOVA, J., SCHEUVENS, D. & VEJNAR, Z. 1998. Extrusion tectonics due to thermal softening of a thickened crustal root: the Bohemian Massif in Lower Carboniferous times. *Terra Nostra*, **98**(2), 177–180.

——, DÖRR, W. & VEJNAR, Z. 1997. 13 km normal displacement along the West Bohemian shear zone—evidence for dramatic gravitational collapse in the Bohemian Massif (Central European Variscides). *Terra Nova, Abstract Supplement 1*, **9**, 107.

Foreland-directed lower-crustal flow and its implications for the exhumation of high-pressure–high-temperature rocks

ANDREAS HENK

Institut für Geologie, Universität Würzburg, Pleicherwall 1, D-97070 Würzburg, Germany (e-mail: a.henk@mail.uni-wuerzburg.de)

Abstract: Thermo-mechanical finite element models are used to study foreland-directed lower-crustal flow as a potential process to transport high-pressure–high-temperature (HP–HT) rocks from a continental collision zone to areas that have never experienced crustal thickening or deep burial. The numerical simulations show that lower-crustal rocks can indeed flow over substantial horizontal and vertical distances, provided a thermal anomaly reducing lithospheric strength exists in the foreland. As lower-crustal flow is a fast process compared with conductive heat transport and occurs subparallel to the isotherms, the resulting *PT* path of the rocks exhumed will be characterized by near-isothermal decompression. Modelling results are applied to the Saxonian Granulite Massif in Eastern Germany, where HP–HT granulites have been exhumed beneath a marine sedimentary basin during the Variscan Orogeny.

Strength profiles through the continental lithosphere image the lower crust as a region of very low strength, particularly in areas of high heat flow. High temperatures can reduce the viscosity of the lower crust to such an extent that it behaves almost like a fluid on geological time scales (Wernicke 1990) and flows according to lateral pressure gradients (e.g. Turcotte & Schubert 1982). Such a regional-scale lower-crustal flow has been suggested, for example, to explain the formation of metamorphic core complexes (Fig. 1; Block & Royden 1990; Wernicke 1990; Kaufman & Royden 1994) and passive continental margins (Hopper & Buck 1996). In compressional settings, plateau collapse (Bird 1991; Royden 1996) and surface deformation (Royden *et al.* 1997) have been attributed to lower-crustal flow. The high mobility of the lower crust in contrast to the upper crust and the upper mantle has important implications on how strain is partitioned within the lithosphere during extensional or compressional deformation. As the lower crust tends to flow towards areas of minimum lithostatic head, it will tend to thicken in regions where the upper crust is thinnest and vice versa (Fig. 1). In spite of these lateral variations, the Moho will have a flat relief, as the lower crust acts as a hydraulic reservoir compensating for the differences in upper-crustal thinning (e.g. Block & Royden 1990; Wernicke 1990).

The high temperatures required for substantial lower-crustal flow can result from a variety of processes. In extensional settings they may be related to active or passive upwelling of the asthenosphere beneath thinned crust, whereas in areas of thickened crust the increased radiogenic heat production in the crustal stack and/or the increased basal heat flow (i.e. caused by delamination or convective removal of the thermal boundary layer) are important. In any case, lower-crustal flow will be spatially restricted to the region of elevated temperatures. For a continental collision zone, this will essentially be the internal part of the orogen where the crust was thickened and/or the mantle lithosphere has been removed.

The present paper addresses the particular requirements that are necessary to allow lower-crustal flow to areas outside the orogen. Such an outward lower-crustal flow from beneath an orogen towards the foreland is potentially a very efficient process to transport high-pressure rocks beneath areas where the upper-crustal rocks have never been involved in crustal thickening or continental subduction. However, it obviously requires a special physical state of the foreland lithosphere and involves pronounced strain partitioning and stress field perturbations, both in the vertical and horizontal direction. To gain a quantitative insight into the requirements for and the processes during

From: Franke, W., Haak, V., Oncken, O. & Tanner, D. (eds).
Orogenic Processes: Quantification and Modelling in the Variscan Belt.
Geological Society, London, Special Publications, **179**, 355–368. 0305-8719/00/$15.00

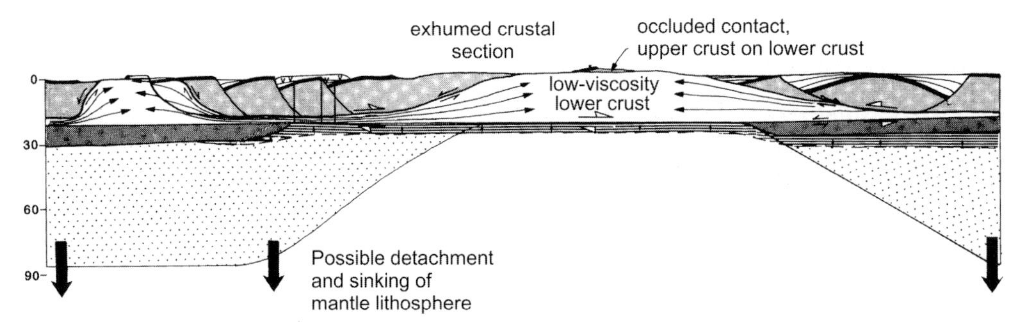

Fig. 1. Schematic model of lower-crustal flow in response to pressure gradients resulting from lateral variations in lithostatic head, which in turn are a consequence of heterogeneous extension of the upper crust (simplified after Wernicke (1990)). Because of its very low viscosity, the lower crust can flow towards regions of thinned upper crust and lower lithostatic pressure. Consequently, the lower crust is thickest where the upper crust has been thinned the most and vice versa. The Moho in the entire area remains subhorizontal, as the fluid-like lower crust can provide for a kind of intracrustal isostatic balance.

foreland-directed lower-crustal flow, a numerical modelling approach is used. Model predictions are compared with field data from the Saxonian Granulite Massif, where high-pressure–high-temperature (HP–HT) granulites have been exhumed beneath a marine sedimentary basin.

Some general considerations concerning foreland-directed lower-crustal flow

During the final convergence stage, the interiors of continental collision zones are typically characterized by high crustal temperatures in response to radiogenic heating and/or increased basal heat flow. Consequently, the lower crust in the orogenic realm has a very low strength and can flow according to lateral pressure gradients resulting from variations in the lithostatic pressure at the top of the lower crust (Bird 1991). Although there may be minor variations in lithostatic head within the internal part of the orogen resulting from differential thickening or plateau collapse, the major difference is between the orogen and the foreland. However, although a suitable pressure gradient exists, the cooler and consequently stronger foreland crust prevents any outward flow of hot lower-crustal rocks from the orogen. This force balance between the hot lower crust in the internal zone of the orogen and its cooler counterpart in the foreland is illustrated by Fig. 2, which shows lithospheric strength profiles deduced from a 2D finite-element (FE) model of orogenic evolution (see below for detailed description of the modelling approach). Because of the excess potential energy stored in the thickened crust the orogen is in a tensile state of stress (Artyushkov 1973), which may ultimately lead to a gravitational collapse if it exceeds the strength of the lithosphere and the compressive plate boundary forces. The hot lower crust in the orogen has a very low strength and viscosity, respectively, but the cool lower crust in the foreland has sufficient compressive strength to prevent any outward flow. Consequently, exhumation of deeply buried rocks by lower-crustal flow will be restricted to the internal zone of the orogen only. Foreland-directed lower-crustal flow will be possible only if two basic requirements are met: in addition to the lateral pressure gradient driving lower-crustal flow, the foreland crust must have a strongly reduced strength, allowing the lower crust to escape from beneath the orogen.

The first requirement is already fulfilled by the crustal thickness variations across a collision zone, as the lithostatic pressure at the top of the lower crust is greater beneath the orogen than beneath the foreland. It should be emphasized that this does not necessarily imply that the entire overburden above the lower-crustal channel is formed by upper-crustal rocks. A suitable pressure gradient would also exist if the lower crust had detached from the upper crust and was subducted beneath a wedge of thickened crust and mantle lithosphere. The second prerequisite for outward lower-crustal flow, a weak lower crust in the foreland, could already be caused by compositional variations between the orogen and its foreland resulting in low foreland viscosities (Fig. 3). However, because of the strong temperature dependence of lower-crustal rock rheologies, geodynamic scenarios providing elevated temperatures are far more effective in reducing lithospheric strength. Processes unrelated to orogeny, such as rifting or the ascent of a mantle plume, could heat and weaken the foreland crust significantly. During plate convergence, two subduction zones acting concomitantly

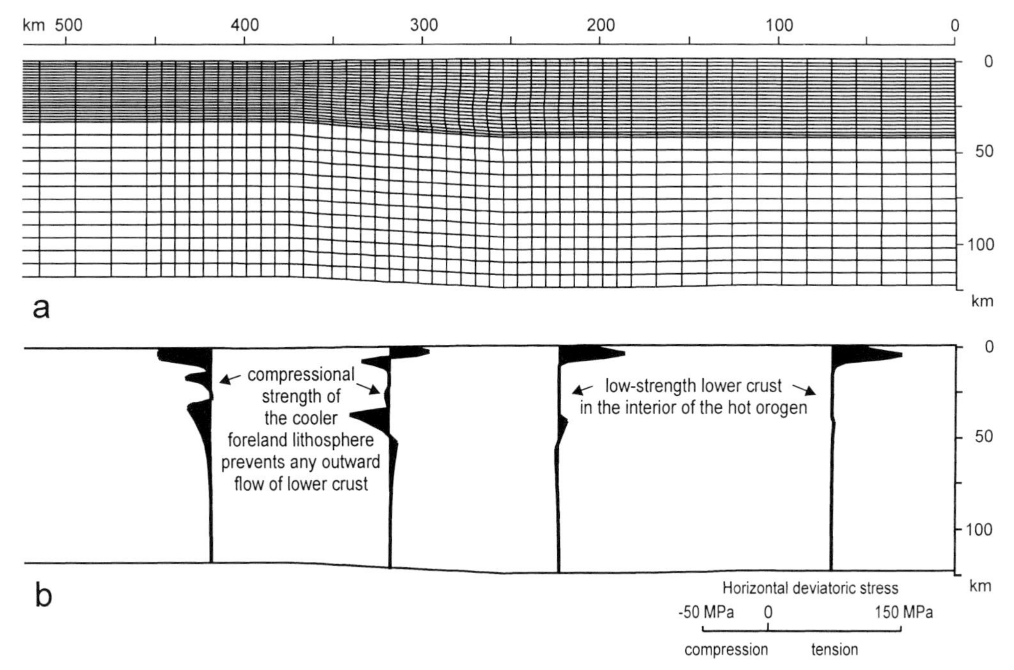

Fig. 2. Horizontal deviatoric stress profiles after 15 Ma taken from a 2D lithosphere model (see Fig. 4) assuming elevated temperatures in the internal zone of the orogen (Moho temperature 870 °C) and no thermal anomaly in the foreland (Moho temperature 620 °C). The excess potential energy stored in the thickened crust leads to extensional stresses in the orogen, but is balanced by compression in the foreland. The cool and consequently strong foreland crust prevents any outward flow of lower-crustal rocks from beneath the orogen although it has only a very low strength.

could promote foreland-directed lower-crustal flow. In this case, the foreland of one subduction zone is located above the back-arc mantle plume of the other leading to abnormally high foreland temperatures (Fig. 3).

Modelling approach

A fully dynamic modelling approach was chosen to study lithospheric deformation during foreland-directed lower-crustal flow. The numerical simulations utilize the finite-element software ANSYS® (Ansys Inc., Houston, TX, USA). It is assumed that the physical properties and geometry of the model lithosphere are relatively constant along-strike, so that the orogen can be adequately described in two dimensions. It is fully appreciated that such a plane-strain assumption is a gross simplification of the complexity of real orogens and orogenic processes, but it is considered as an important basis for future 3D models.

Conceptually, the model lithosphere contains a region of thickened lithosphere representing the internal zone of the orogen, a transition zone equivalent to the external fold and thrust belt and a region unaffected by orogenic processes, the foreland or reference lithosphere (Fig. 4; Braun & Beaumont 1989). The model lithosphere is divided into three compositional layers representing upper crust, lower crust and mantle lithosphere. The mechanical and thermal properties applied to these three layers are summarized in Table 1. At the right side of the model representing the orogen's interior only vertical movements are allowed, whereas at the left side displacements or forces can be applied to simulate extension in the foreland of the orogen.

An important consequence of the given crustal thickness variations is lateral differences in the potential energy of the lithosphere and in the lithostatic head acting on the lower crust. The excess potential energy stored in the thickened crust can result in extension (i.e. gravitational collapse) provided it is sufficient to exceed the strength of the lithosphere (e.g. Artyushkov 1973; Dewey 1988). Likewise, the difference in lithostatic pressure acting on the lower crust between the orogen and the foreland provides the driving force for outward lower-crustal flow.

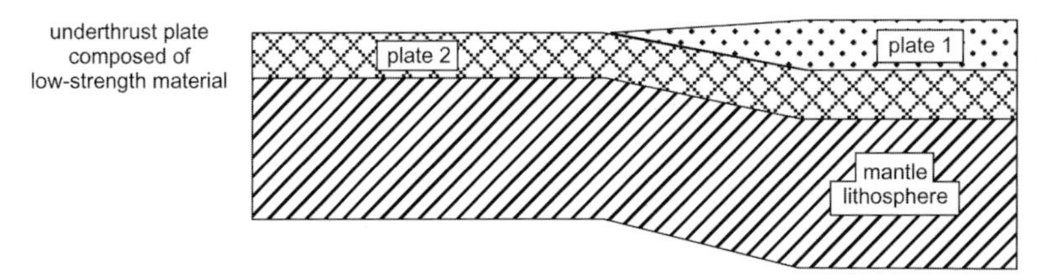

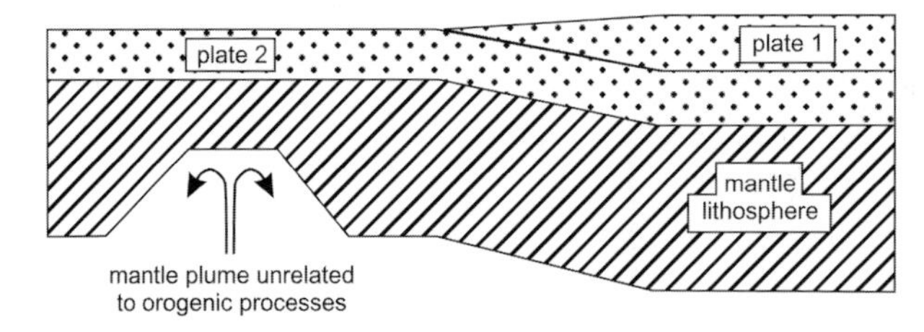

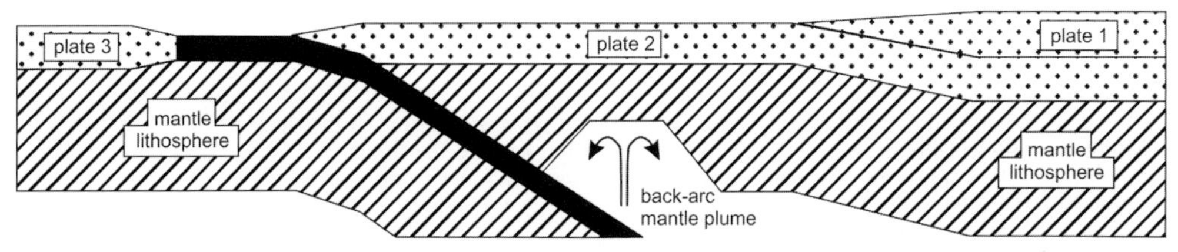

Fig. 3. Possible geodynamic scenarios that can result in a reduced strength of the foreland lithosphere.

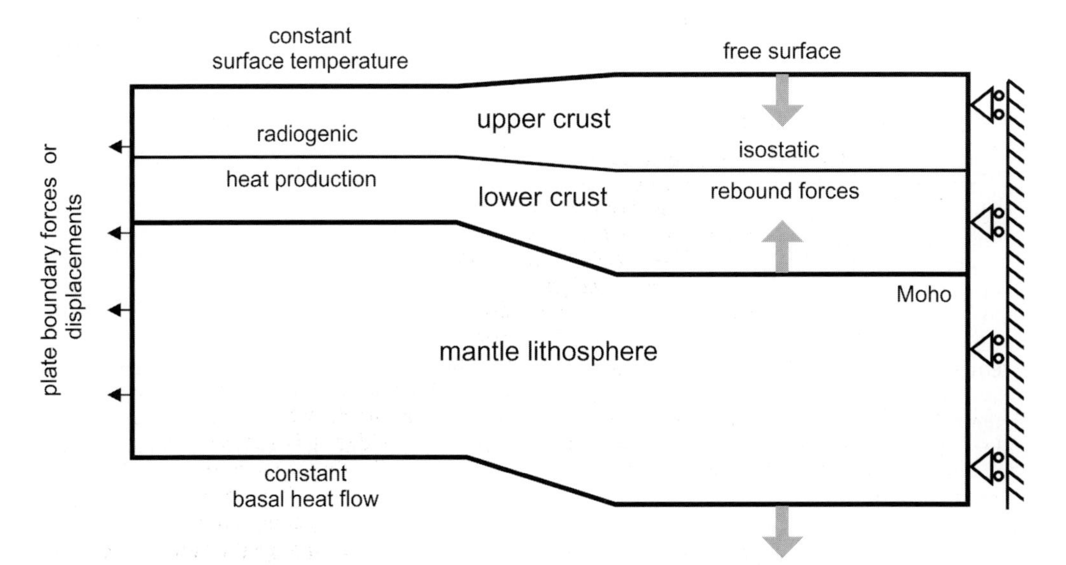

Fig. 4. Modelling concept.

Table 1. *Material parameters used for thermal and mechanical calculations. As modelling results are compared to field data from the Variscan orogen, material parameters typical for the Central European lithosphere were used whenever possible*

Variable	Symbol	Value	Reference
Thermal lithosphere properties			
Surface temperature (K)	T_s	273	
Asthenosphere temperature (K)	T_a	1623	
Lithospheric basal heat flow (W m^{-2})	q^*	0.026	
Upper crust			
Thermal conductivity (W m^{-1} K^{-1})	k	3–1.3	(1), Zoth & Hänel (1988)
Specific heat (m^2 s^{-2} K^{-1})	C	1.3×10^3	
Radiogenic heat production (W m^{-3})	H	2.3×10^{-6}	(2), Cermák (1995)
Lower crust			
Thermal conductivity (W m^{-1} K^{-1})	k	2.5–1.7	(1), Zoth & Hänel (1988)
Specific heat (m^2 s^{-2} K^{-1})	C	1.3×10^3	
Radiogenic heat production (W m^{-3})	H	0.52×10^{-6}	(2), Cermák (1995)
Upper mantle			
Thermal conductivity (W m^{-1} K^{-1})	k	3–4	(1), Zoth & Hänel (1988)
Specific heat (m^2 s^{-2} K^{-1})	C	1.3×10^3	
Radiogenic heat production (W m^{-3})	H	0.02×10^{-6}	(2), Cermák (1995)
Mechanical lithosphere properties			
Reference continent thickness (m)		3×10^4	
Reference lithosphere thickness (m)		1.17×10^5	
Poisson's ratio	ν	0.25	
Upper crust			
Density at 273 K (kg m^{-3})	ρ	2800	
Young's Modulus (Pa)	E	0.5×10^{11}	
Tensile strength gradient (Pa m^{-1})		1.3×10^4	(3)
Strain rate coefficient (Pa^{-n} s^{-1})	a_0	1.633×10^{-26}	Paterson & Luan (1990)
Activation constant (K)	Q/R	16238	
Power-law stress exponent	n	3.1	
Lower crust			
Density at 273 K (kg m^{-3})	ρ	3000	
Young's Modulus (Pa)	E	0.8×10^{11}	
Strain rate coefficient (Pa^{-n} s^{-1})	a_0	2.009×10^{-21}	Wilks & Carter (1990)
Activation constant (K)	Q/R	29228	
Power-law stress exponent	n	3.1	
Upper mantle			
Density at 273 K (kg m^{-3})	ρ	3400	
Young's Modulus (Pa)	E	1.5×10^{11}	
Strain rate coefficient (Pa^{-n} s^{-1})	a_0	5.248×10^{-25}	Chopra & Paterson (1981)
Activation constant (K)	Q/R	59899	
Power-law stress exponent	n	4.48	

(1) Temperature dependent; (2) corrected for reduction in the last 300 Ma; (3) based on a frictional coefficient of 0.75 and a hydrostatic pore fluid pressure.

Numerical model

The strong temperature dependence of the lithospheric rheologies and the advective heat transport in relation to lithospheric deformation requires a coupled thermo-mechanical approach. Thermo-mechanical coupling is achieved by solving the thermal and mechanical equations successively for two geometrically identical FE grids. The calculations are coupled via displacements, and via temperatures and thermal stresses, which result from thermal expansion. Depending on the rate of deformation, coupling may be necessary every 100 ka–1 Ma. Not only this field coupling, but also the temperature dependence of some of the thermal parameters and the material nonlinearities require an iterative solution. Time stepping at intervals of 20 ka and the Newton–Raphson procedure are used, in which the stiffness matrix is updated at every equilibrium iteration.

Calculation of the lithospheric temperature field is based on 2D heat transport by conduction and advection, taking into account, amongst other things, temperature-dependent thermal conductivities and lithology dependent radiogenic heat production. Boundary conditions are a constant surface temperature of 0 °C and a constant basal heat flow. This allows the base of the lithosphere, as defined by the 1350 °C isotherm, to migrate vertically during thermal relaxation. No lateral heat flow is allowed through the sides of the model.

For the mechanical calculations, the two most important mechanisms for lithospheric deformation, brittle failure and viscous flow, are considered (e.g. Ranalli 1995). Brittle deformation by fracture and frictional sliding on existing faults occurs, if the differential stresses ($\sigma_1 - \sigma_3$) reach the yield stress σ_y as defined by the Coulomb–Navier criterion (e.g. Jaeger & Cook 1979; Ranalli 1995). In a simplified form, this relationship can be given as

$$\sigma_y(z) = Ag\rho z(1 - \lambda) \tag{1}$$

(Sibson 1977; Ranalli 1995), where A is a variable that depends on the coefficient of friction and the predominant type of faulting, g is gravitational acceleration, ρ is the density of the rocks overlying a depth z, and λ is the ratio of pore pressure to lithostatic pressure.

Deformation in the ductile domain is assumed to be governed by thermally activated power-law creep (e.g. Kohlstedt *et al.* 1995; Ranalli 1995). The non-linear relationship between strain rate $\dot{\varepsilon}$, differential stress $\sigma_1 - \sigma_3$ and temperature T is given by

$$\dot{\varepsilon} = a_0(\sigma_1 - \sigma_3)^n \exp\left(\frac{-Q}{RT}\right) \tag{2}$$

where R is the universal gas constant and a_0 (strain rate coefficient), n (power-law stress exponent) and Q (activation energy) are material properties derived from laboratory experiments.

In the numerical model, the two rheological laws outlined above are combined to calculate the strength at any point in the lithosphere. At differential stresses below the yield stress, lithospheric rocks are assumed to behave as incompressible viscous fluids deforming by steady-state creep according to equation (2). If the stresses reach the yield stress as defined by equation (1), rocks are assumed to deform according to an ideal plastic flow law. This rheology is designed to describe the permanent strain caused by brittle failure, but does not explicitly describe frictional sliding and fault movement. The rheological law for each element is not fixed but is continuously updated depending on the actual temperature, pressure and strain rate. Thus the strength of the model lithosphere varies not only laterally and vertically, but also with time.

The boundary conditions for the mechanical calculations permit only vertical displacements at the right wall and no tilt at the model's left side. Displacement constraints can be applied on the foreland side of the model lithosphere to simulate extension or compression. Isostatic rebound forces act on the density interfaces at the top and bottom of the model as well as at the base of the crust.

Initial geometry and physical state of the model lithosphere

The numerical model comprises 2108 four-node isoparametric coupled field elements, which describe a section of 754 km length from the internal part of an orogen towards the undeformed foreland (Fig. 5). For the reference lithosphere, crustal and total thicknesses of 30 km and 117 km, respectively, were selected. The orogen has a crust of 50 km thickness and a total lithospheric thickness of 130 km. As high lower-crustal temperatures are difficult to reconcile with a thickened mantle lithosphere (Bird 1991), it is assumed that part of the upper mantle had already detached, leaving a thickness of 80 km beneath the orogen. Any additional delamination or convective erosion of the mantle lithosphere would increase the basal heat flow and raise lower-crustal temperatures even further.

Within the orogenic domain, lower-crustal temperatures were chosen such that typical HT granulite facies metamorphic conditions of 700–900 °C (Bohlen 1987; Sandiford 1989) existed. The foreland lithosphere possesses a thermal anomaly caused by a prescribed mantle plume. The anomaly has a diameter of 180 km and Moho temperatures above the plume head reach 870 °C. Inherited in the assumption of a mantle plume heating the foreland crust is localized extension. For a back-arc setting behind an active subduction zone, for example, extension could be caused by the active injection of a mantle diapir rising from the downgoing slab, viscous drag of flow in the mantle wedge above the subduction zone or retreat of the descending plate (subduction rollback; Moores & Twiss 1995). This is supported by the observation that all modern subduction systems with active

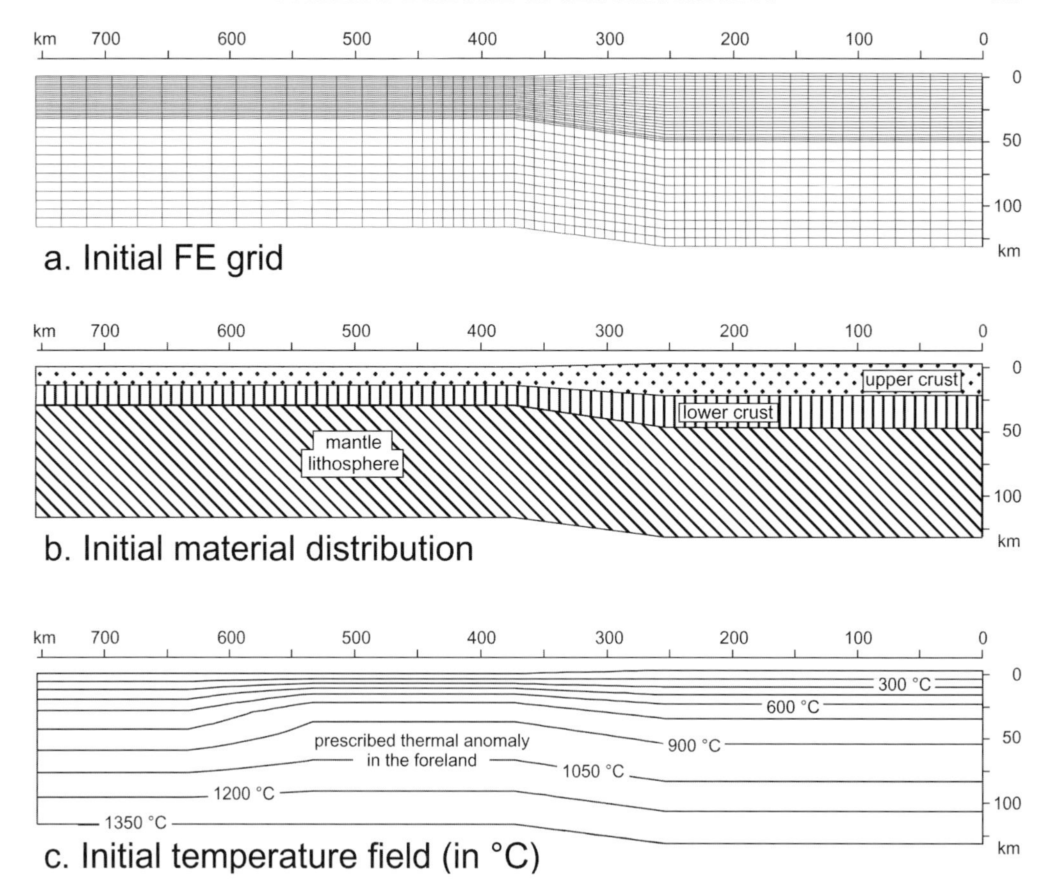

Fig. 5. Initial FE grid (**a**), material distribution (**b**) and temperature field (**c**).

back-arc basins show a seaward migration of the trench axis in a hotspot reference frame (Taylor 1995). The numerical model accounts for this process by a displacement boundary condition at the left side of the model arbitrarily set to 5 mm a^{-1}.

Modelling results

The numerical experiment shows that lateral extrusion from areas of thickened crust towards the foreland is indeed possible and can occur over substantial distances. This is illustrated by the deformed FE grid after 12 Ma shown in Fig. 6a. Several of the initially rectangular elements are strongly deformed and the distortion of initially vertical lines through the lithosphere now documents the large differential movements between upper crust, lower crust and mantle lithosphere. The crust has developed a 'pinch-and-swell' style (Braun & Beaumont 1989) of deformation similar to the pattern shown schematically in Fig. 1. Lower-crustal necks are located below upper-crustal swells and vice versa. This strongly resembles the geometry of metamorphic domes and core complexes, respectively.

Maximum horizontal displacement relative to the initial position amounts to 97 km (Fig. 6b). Vertical displacement in the crust reaches depths up to 10 km and affects predominantly lower-crustal rocks in the transition zone between orogen and foreland (Fig. 6c). This is equivalent to a decompression of up to 4 kbar, as the surface topography in the area decreases as well. As lower-crustal flow is a fast process compared with conductive heat transport and the lower-crustal channel runs subparallel to the isotherms, the resulting *PT* path of the rocks exhumed will be characterized by near-isothermal decompression.

The displacement pattern in the lower crust can be best described as a kind of plug flow with maximum displacements in the lowermost crust, where temperatures are highest and viscosities are minimal. The large differential movements

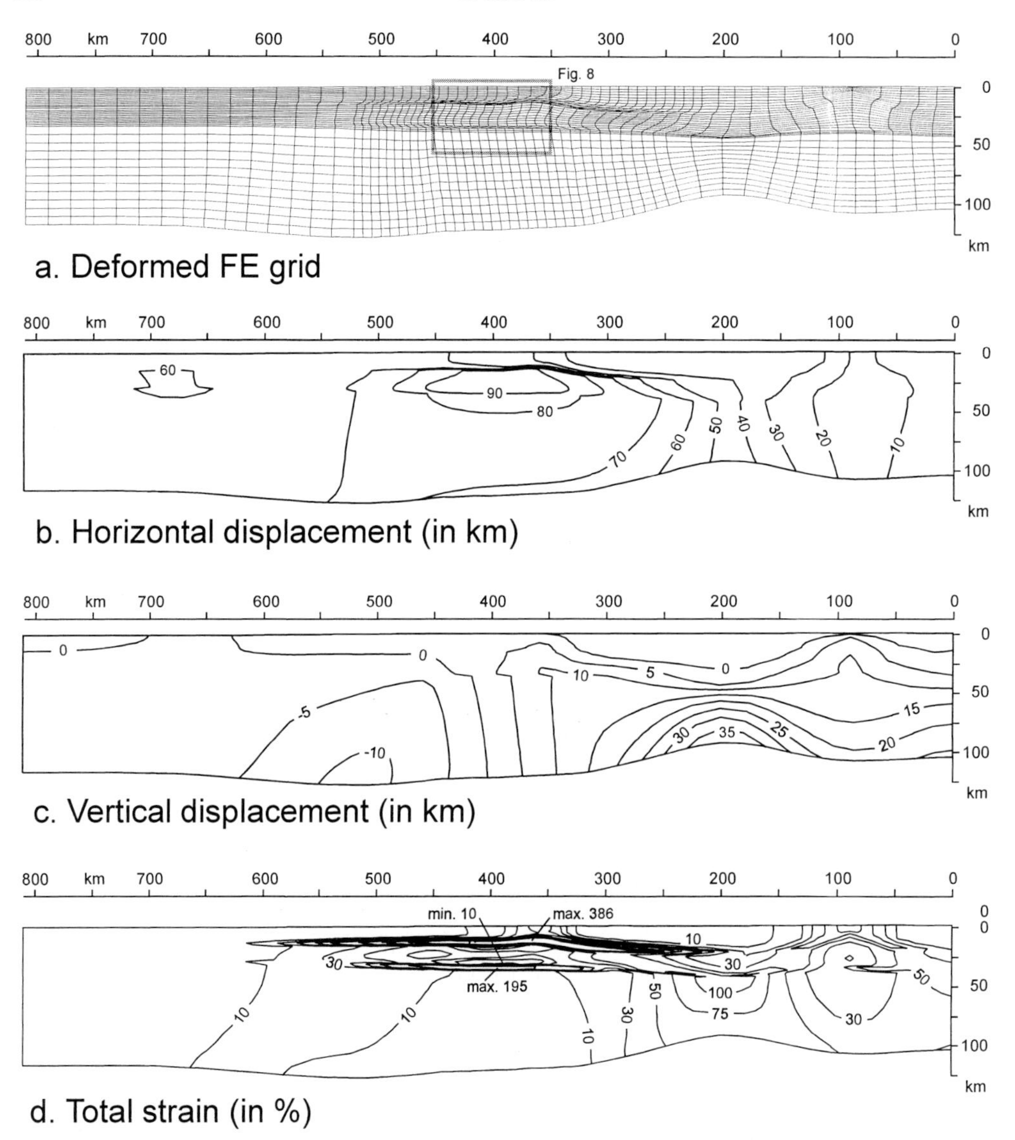

Fig. 6. Modelling results after 12 Ma assuming 5 mm a^{-1} displacement on the foreland side of the model. Deformed FE grid (**a**), horizontal displacement (**b**), vertical displacement (**c**) and total strain (**d**).

lead to a pronounced strain partitioning between upper crust, lower crust and mantle lithosphere (Fig. 6d). Foreland-directed extrusion results in high-strain zones at the top and base of the lower crust, whereas the bulk of the lower crust experiences little internal deformation. The two strain zones define a lower-crustal channel with low-strength rock bordered by higher strengths in the lowermost upper crust and uppermost mantle lithosphere, respectively. The most dominant feature is the upper strain zone, which shows deformation up to several hundred per cent. The strain zone at the base of the crust is less pronounced because the mantle lithosphere has also a low strength and can partly follow the foreland-directed movement of the lower crust. The relative displacement between upper and lower crust is equivalent to the kinematics of a low-angle normal fault, whereas the distortion of the elements at the transition between lower crust and mantle lithosphere resembles a thrust fault. However, it should be noted that in spite of the strong distortion and asymmetric displacement this modelling approach is still a variation on

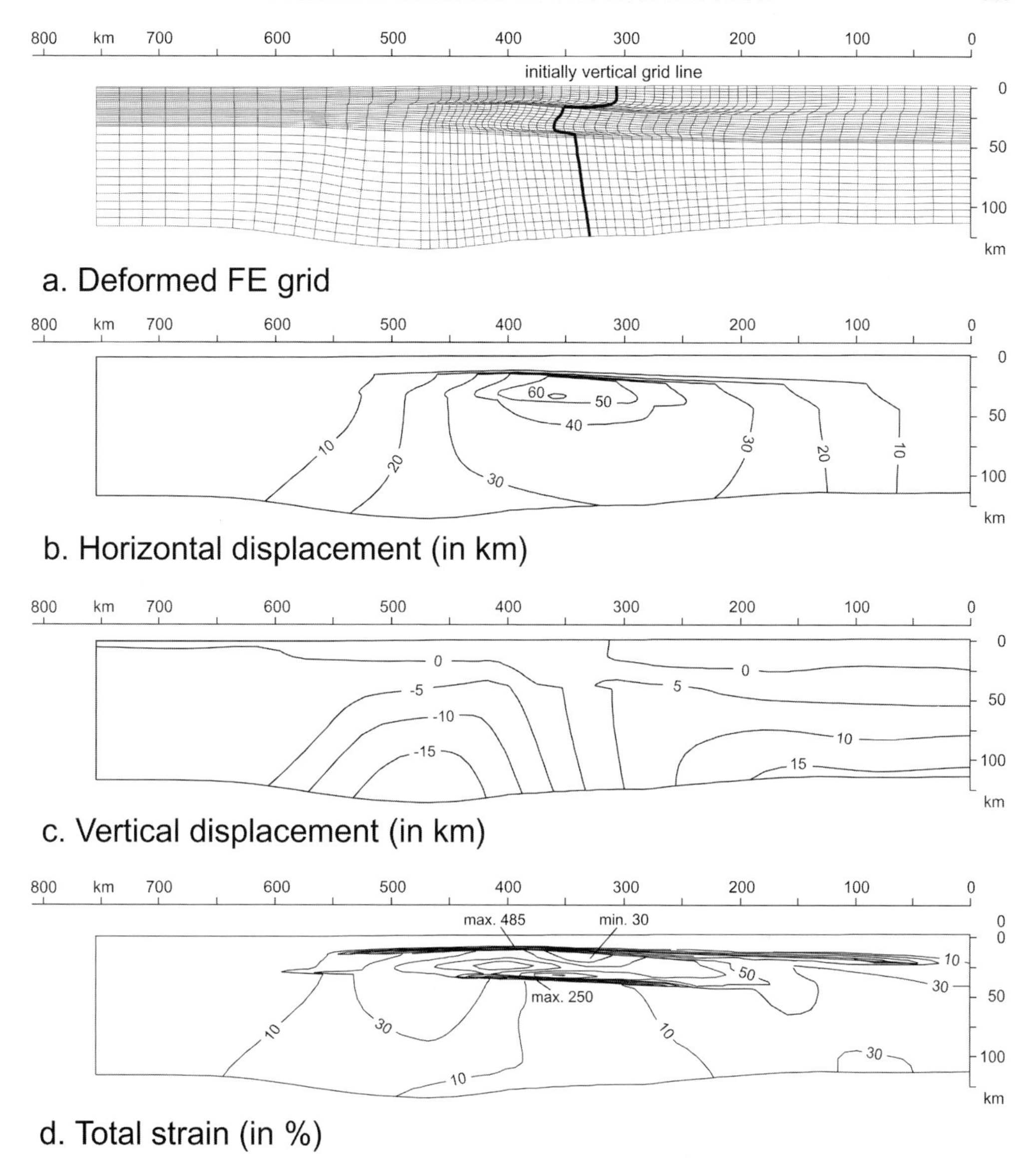

Fig. 7. Modelling results after 12 Ma assuming fixed side-walls of the model. Deformed FE grid (**a**), horizontal displacement (**b**), vertical displacement (**c**) and total strain (**d**).

pure-shear extension that varies both laterally and vertically. It does not account for any differential movements along a discrete fault zone and exhumation by erosion.

Model runs with fixed side-walls show that the displacement boundary condition of 5 mm a^{-1} applied on the foreland side of the model is not necessary to allow substantial foreland-directed lower-crustal flow. Even then, lower-crustal flow over more than 60 km occurs, inflating and thickening the foreland crust and mantle lithosphere (Fig. 7a–c). As the brittle upper crust is essentially chucked between the fixed side-walls, deformation at the top and base of the outward flowing lower crust is even greater than in the previous example and reaches almost 500% (Fig. 7d). This variation in the boundary conditions shows that foreland-directed lower-crustal flow can occur even without the far-field extensional forces caused by the moving side-wall of

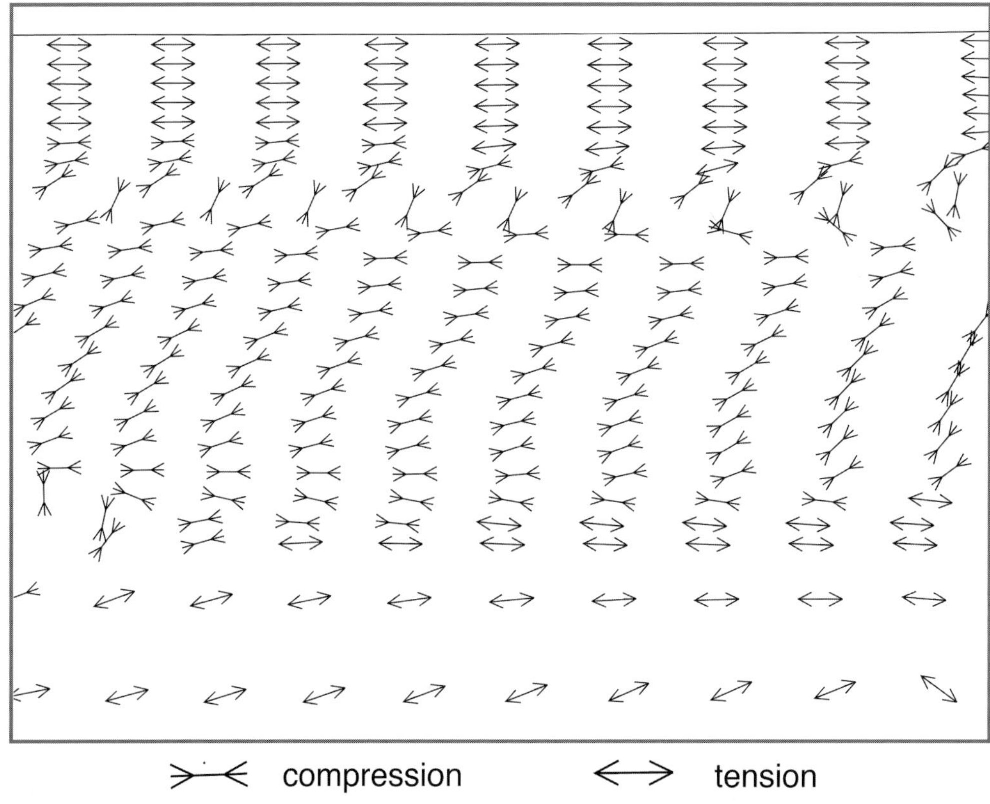

Fig. 8. Orientation of the least principal stress axis σ_3 in the vicinity of the lower-crustal channel (see Fig. 6a for location).

the numerical model and assumed back-arc spreading, respectively.

Another important observation is the strong stress field perturbations predicted by the numerical model (Fig. 8). As a result of the prescribed slow extension, stress fields in the upper crust are rather uniform and characterized by a horizontal tensional σ_3. Stress fields near the boundary between upper and lower crust are highly variable. Within the lower crust σ_3 orientations vary between horizontal and vertical but are always compressional in the area shown as extrusion displaces material. In the mantle lithosphere stress fields are again similar to those in the upper crust. Considering these vertical but also lateral variations in stress and strain, it will obviously be very difficult to extrapolate outcrop data derived from a certain crustal level to the entire crustal section.

Application to the Saxonian Granulite Massif

The Saxonian Granulite Massif (SGM) in East Germany is located within the Saxo-Thuringian Zone of the Variscan orogen (see Franke & Stein (this volume) and Krawczyk *et al.* (this volume) for comprehensive discussions of the regional geology and geophysics). Whereas most of the Variscan granulites are related to metamorphic thrust sheets, the SGM represents a tectonic window and has been interpreted as a metamorphic core complex (Franke 1993; Kroner 1995). Its dome-like structure is characterized by a core of granulitic rocks and a cover of low-grade metasediments, the so-called Schist Mantle. Both units are separated by a major normal fault with top-to-the-SE sense of displacement (Reinhardt & Kleemann 1994; Kroner 1995).

Important constraints for the evolution of the SGM are provided by petrological and geochronological data which are summarized in Fig. 9. Although the exhumation stage from a depth equivalent to about 9 kbar to the surface is rather well documented (Rötzler *et al.* 1994; Reinhardt *et al.* 1998), the maximum metamorphic conditions experienced by the granulites are still a matter of debate. According to Rötzler (1992), felsic granulites yielding 11.5 ± 1 kbar

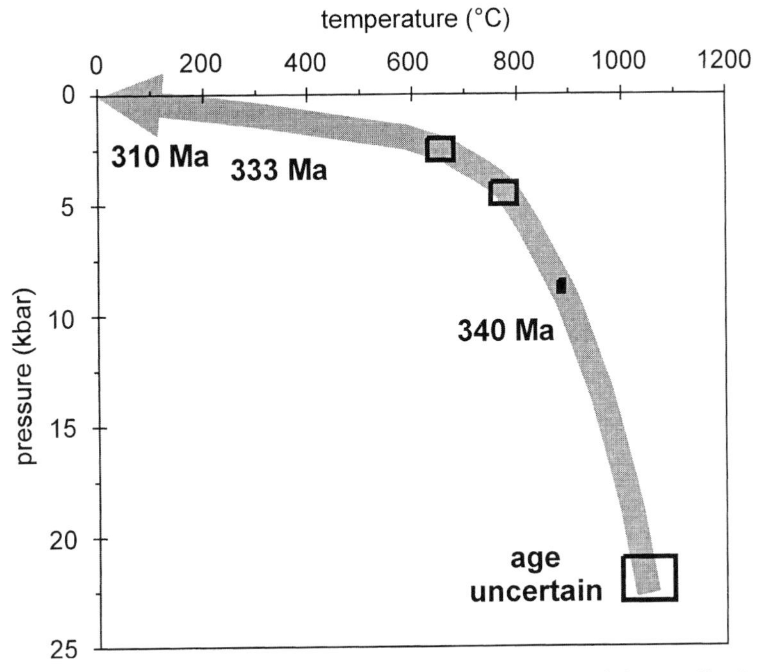

Fig. 9. Pressure–temperature–time (*P–T–t*) describing the exhumation history of the granulites in the Saxonian Granulite Massif (data after Rötzler *et al.* (1994), Werner & Reich (1997), Kröner *et al.* (1998) and Reinhardt *et al.* (1998)).

and 780–850 °C provide only a minimum estimate of the peak *PT* conditions. Instead, Rötzler *et al.* (1994) suggested that the 20 kbar and 1050 °C *PT* estimates they derived for the mafic granulites are representative for the entire SGM. The time of peak metamorphism may be documented by several U–Pb zircon ages, which cluster around 340 Ma (von Quadt 1993; Kröner *et al.* 1998). At this time, the sediments in the present hanging wall record continuous marine sedimentation (Franke & Stein this volume). Consequently, the granulites cannot have achieved this HP–HT metamorphism at the base of a thickened crust vertically beneath their present position.

Exhumation of the SGM from maximum burial to a depth of about 10 km may have occurred in one step, as recent age data from Kröner *et al.* (1998) give no convincing evidence for an ultra-high-pressure (UHP) event at about 380 Ma followed by a phase of intermediate pressure metamorphism at about 340 Ma, as was previously suggested by von Quadt (1993) and Rötzler *et al.* (1994). The *PT* evolution of the granulites shows near-isothermal decompression. The heat advected by the uprising granulites caused a strong thermal overprint of the meta-sediments in the hanging wall of the detachment. Their metamorphic evolution is characterized by isobaric heating and cooling following granulite emplacement and thermal re-equilibration, respectively. Peak metamorphic conditions of the Schist Mantle are 2–3 kbar and 680 °C (Reinhardt *et al.* 1998). Exhumation can be bracketed between about 340 Ma, the age of high-grade metamorphism (e.g. Kröner *et al.* 1998) and 333 ± 2 Ma (Werner & Reich 1997). The latter age represents muscovite and biotite $^{40}Ar/^{39}Ar$ cooling ages, which document the time when the Schist Mantle had already cooled below 300 °C following granulite emplacement to shallow crustal levels. Exhumation was entirely due to tectonic denudation by normal faulting and ductile flow, as is indicated by coeval marine sedimentation in the upper plate. Only the final uplift from about 10 km depth to the surface involved erosion, as is documented by pebbles derived from the Schist Mantle and granulites that occur in adjacent molasse basins (Franke 1993; Kroner 1995).

The evolution of the SGM points to a process that can rapidly exhume hot lower-crustal rocks from depths of at least 45 km without any major heat loss and emplace them at depth of about 10 km beneath upper-crustal rocks recording contemporaneous marine sedimentation. A geodynamic scenario in which the granulites were exhumed by gravitational orogen collapse as was suggested by Kroner (1995) can be ruled out, as the contemporaneous marine sedimentation is not compatible with overthickened crust in the area of the present SGM. Instead, the hypothesis is put forward that the granulites achieved maximum metamorphic conditions in a collision zone to the south and reached their present position by foreland-directed lower-crustal flow. This builds on a qualitative exhumation model proposed by Reinhardt & Kleemann (1994, fig. 18d), who assumed a crustal-scale low-angle detachment dipping from the foreland beneath the orogen. Franke & Stein (this volume) have further elaborated on this idea.

The potential mobility of the lower crust in the area of the SGM can be assessed using some simple calculations based on channel flow algorithms (Turcotte & Schubert 1982; Block & Royden 1990). Assuming a granulitic rheology, strain rates of 10^{-13}–10^{-14} s^{-1} and temperatures of 900–1050 °C as indicated by the petrological data, effective viscosities of the order of 10^{18}–10^{19} Pa s can be expected. This is in the range of, or even a magnitude lower than, present-day viscosities of the asthenosphere. Assuming a lateral pressure gradient of about 4700 Pa m^{-1} (difference in lithostatic head between crustal sections of 25 and 60 km thickness over a distance of 200 km) and a channel of 10 km thickness, lower-crustal rocks would be able to flow at velocities of 0.16–1.95 m a^{-1}.

Two basic requirements have to be fulfilled to allow foreland-directed lower-crustal flow: a lateral pressure gradient and a thermal anomaly in the foreland. Both requirements are met in case of the SGM, which is located between two collision zones, the Rheno-Hercynian–Saxo-Thuringian to the north and Saxo-Thuringian–Moldanubian to the south. Along both zones, subduction of oceanic crust and continental collision occurred contemporaneously throughout most of the Variscan convergence history. The SGM was therefore in a foreland position relative to the Saxo-Thuringian–Moldanubian collision zone, but in a back-arc position relative to the Rheno-Hercynian–Saxo-Thuringian convergence zone. I suggest that heating of the lower crust in the area of the present SGM occurred in a back-arc regime related to subduction of the Rheno-Hercynian ocean towards the south, beneath the Mid-German Crystalline High and the Saxo-Thuringian basin. The effect of this back-arc spreading and mantle upwelling may be documented by prominent extensional uplift and basaltic volcanism occurring along the axis of the Saxo-Thuringian basin and later site of the SGM in Frasnian times (Adorf, 370–365 Ma; Franke 1993; Franke & Stein this volume).

As all prerequisites are fulfilled, foreland-directed lower-crustal flow could provide an explanation of how HP–HT granulites can be exhumed beneath a marine sedimentary basin, as follows. The granulites represent subducted lower crust of the Saxo-Thuringian terrane. They achieved maximum metamorphic pressures in the Saxo-Thuringian–Moldanubian collision zone to the south. Subsequently, following the lateral pressure gradient induced by crustal thickening, they flowed towards the foreland. There, lithospheric strength was strongly reduced by the thermal effects of the back-arc spreading and mantle upwelling, which had been formed as a result of subduction of a small oceanic basin between the Rheno-Hercynian and Saxo-Thuringian zone. This process of foreland-directed lower-crustal flow can explain the near-isothermal exhumation of the granulites from maximum burial in the collision zone to lower-crustal levels beneath the foreland. The final exhumation stage from a depth equivalent to about 8 kbar to the surface is not yet included in the thermo-mechanical model but has already been studied quantitatively using a 2D thermal-kinematic modelling approach (Henk 1995). Modelling results show that subsequent exhumation at moderate rates of about 3 mm a^{-1} can explain both the continuing near-isothermal exhumation of the granulites as well as the thermal overprint of the hanging-wall rocks.

The exhumation model outlined above acknowledges the special geodynamic setting of the SGM at the northern flank of the Variscan orogen, between two subduction zones, which provides also the unusually high lower-crustal temperatures required for substantial foreland-directed lower-crustal flow. It is consistent with published petrological and geochronological data for the granulites and Schist Mantle, as well as the sedimentological record in the hanging wall. Further field evidence may be difficult to furnish, as exhumation during the early stages occurred almost isothermally at high temperatures (above 800 °C). Thus, it is no surprise that cooling ages from minerals with various closure temperatures yield similar ages. Additional precise age data for the exhumation stage between 20 and 10 kbar is still required but may be very difficult to obtain.

Conclusions

The numerical simulations show that lower-crustal flow from an orogen towards the foreland is physically possible and can provide a very efficient process for the rapid exhumation of HT rocks provided a thermal anomaly exists in the foreland lithosphere. Raised temperatures and reduced lower-crustal strengths, respectively, can occur in a variety of geodynamic settings, for example, a tandem of subduction zones or the ascent of a mantle plume unrelated to orogenic processes. If this prerequisite is fulfilled, lower-crustal flow driven by lateral variations in lithostatic head can substantially displace HP–HT rocks both horizontally and vertically. The numerical experiments show that for a crust of 50 km thickness (by no means an extreme value regarding the crustal thickness of modern orogens) the lower crust can flow over 100 km towards the foreland within less than 12 Ma. The thermal anomaly caused by the back-arc rifting and mantle upwelling provides a kind of casting mould localizing foreland-directed lower-crustal flow. As lower-crustal flow is a fast process compared with conductive heat transport and the lower-crustal channel runs subparallel to the isotherms, the resulting *PT* path will be characterized by near-isothermal decompression.

The large differential movements inherited in foreland-directed lower-crustal flow also cause pronounced strain partitioning and strong stress field perturbations within the lithosphere. High strain zones develop at the top and base of the lower-crustal channel, whereas the internal part is only slightly deformed. The related kinematics are equivalent to a low-angle detachment at the upper and a thrust fault at the lower high-strain zone. Calculated stress fields are not only different in upper crust, lower crust and mantle lithosphere, but also vary laterally within each unit. Thus, the general lesson that can be learned from the numerical experiments is that contemporaneous deformation at different levels within the lithosphere can be strongly heterogeneous, even in two dimensions and thus most certainly in three dimensions. Consequently, it may be inappropriate to extrapolate palaeo-stress and -strain data obtained from a certain outcrop level to the entire crust.

If applied to the SGM in East Germany, the concept of foreland-directed lower-crustal flow can explain several field observations. It provides not only an exhumation process consistent with first-order physical principles (e.g. heat transfer and temperature-dependent and strain rate dependent rock rheology) but can also explain emplacement of HP–HT rocks beneath a marine sedimentary basin. Model predictions are consistent with the geological, petrological and geochronological data at present available but more information, particularly on the exact time and pressure of peak metamorphism and the early exhumation history, is urgently required to place further constraints on the exhumation process. Nevertheless, from the present knowledge, thermo-mechanical modelling proves to be a very valuable tool to gain quantitative insights into complex orogenic processes.

W. Franke, O. Oncken, E. Stein and C. Krawczyk are thanked for interesting discussions. J. M. Lardeaux, D. Tanner and S. Wdowinski are thanked for helpful comments on an earlier version of the manuscript. Financial support by the Deutsche Forschungsgemeinschaft as part of the programme 'Orogene Prozesse—ihre Simulation und Quantifizierung am Beispiel der Varisciden' is gratefully acknowledged.

References

ARTYUSHKOV, E. V. 1973. Stresses in the lithosphere caused by crustal thickness inhomogeneities. *Journal of Geophysical Research*, **78**, 7675–7708.

BIRD, P. 1991. Lateral extrusion from under high topography, in the isostatic limit. *Journal of Geophysical Research*, **96**, 10275–10286.

BLOCK, L. & ROYDEN, L. H. 1990. Core complex geometries and regional scale flow in the lower crust. *Tectonics*, **9**, 557–567.

BOHLEN, S. R. 1987. Pressure–temperature time paths and a tectonic model for the evolution of granulites. *Journal of Geology*, **95**, 617–632.

BRAUN, J. & BEAUMONT, C. 1989. Contrasting styles of lithospheric extension: implications for differences between the Basin and Range Province and rifted continental margins. *In*: TANKARD, A. J. & BALKWILL, H. (eds) *Extensional Tectonics and Stratigraphy of the North Atlantic Margins*. American Association of Petroleum Geologists, Memoirs, **46**, 53–80.

CERMÁK, V. 1995. A geothermal model of the central segment of the European Geotraverse. *Tectonophysics*, **244**, 51–55.

CHOPRA, P. N. & PATERSON, M. S. 1981. The experimental deformation of dunite. *Tectonophysics*, **78**, 453–473.

DEWEY, J. F. 1988. Extensional collapse of orogens. *Tectonics*, **7**, 1123–1139.

FRANKE, W. 1993. The Saxonian granulites: a metamorphic core complex? *Geologische Rundschau*, **82**, 505–515.

—— & STEIN, E. 2000. Exhumation of high-grade rocks in the Saxo-Thuringian Belt: geological constraints and geodynamic concepts. *This volume*.

HENK, A. 1995. Variscan exhumation of the Saxonian Granulite Massif—preliminary results from two-dimensional thermal–kinematic modeling. *Terra Nostra*, **95**(8), 29–37.

HOPPER, J. R. & BUCK, W. R. 1996. The effect of lower crustal flow on continental extension and passive margin formation. *Journal of Geophysical Research*, **101**, 20175–20194.

JAEGER, J. C. & COOK, N. G. W. 1979. *Fundamentals of Rock Mechanics*, Chapman and Hall, London.

KAUFMAN, P. S. & ROYDEN, L. H. 1994. Lower crustal flow in an extensional setting: constraints from the Halloran Hills region, eastern Mojave Desert, California. *Journal of Geophysical Research*, **99**, 15723–15739.

KOHLSTEDT, D. L., EVANS, B. & MACKWELL, S. J. 1995. Strength of the lithosphere: constraints imposed by laboratory experiments. *Journal of Geophysical Research*, **100**, 17587–17602.

KRAWCYZK, C. M., STEIN, E., CHOI, S. *et al.* 2000. Geophysical constraints on exhumation mechanisms of high-pressure rocks; the Saxo-Thuringian case between the Franconian Line and Elbe Zone. *This volume*.

KRÖNER, A., JAECKEL, P., REISCHMANN, T. & KRONER, U. 1998. Further evidence for an early Carboniferous (~ 340 Ma) age of high-grade metamorphism in the Saxonian granulite complex. *Geologische Rundschau*, **86**, 751–766.

KRONER, U. 1995. *Postkollisionale Extension am Nordrand der Böhmischen Masse—Die Exhumierung des Sächsischen Granulitgebirges*. Freiberger Forschungshefte, **C457**.

MOORES, E. M. & TWISS, R. J. 1995. *Tectonics*. W. H. Freeman, New York.

PATERSON, M. S. & LUAN, F. C. 1990. Quartzite rheology under geological conditions. *In*: KNIPE, R. J. & RUTTER, E. H. (eds) *Deformation Mechanisms, Rheology and Tectonics*. Geological Society, London, Special Publications, **54**, 299–307.

RANALLI, G. 1995. *Rheology of the Earth*, Chapman and Hall, London.

REINHARDT, J. & KLEEMANN, U. 1994. Extensional unroofing of granulitic lower crust and related low-pressure, high temperature metamorphism in the Saxonian Granulite Massif, Germany. *Tectonophysics*, **238**, 71–94.

——, VAVRA, G., REICH, S., WERNER, O. & HORN, A. 1998. Petrological, geochronological and tectonic aspects of the Saxonian Granulite Massif: getting to the core of a metamorphic complex. *Terra Nostra*, **98**(2), 123–126.

RÖTZLER, J. 1992. Zur Petrogenese im Sächsischen Granulitgebirge, die pyroxenfreien Granulite und die Metapelite. *Geotektonische Forschungen*, **77**, 1–100.

——, BUDZINSKI, H. & BUDZINSKI, G. 1994. Evidence for early Variscan very high temperature (>1000 °C)–high pressure metamorphism in the Saxonian granulite massif. *Terra Nostra*, **94**(3), 92–95.

ROYDEN, L. 1996. Coupling and decoupling of crust and mantle in convergent orogens; implications for strain partitioning in the crust. *Journal of Geophysical Research*, **101**, 17679–17705.

——, BURCHFIELD, B. C., KING, R. W., WANG, E., CHEN, Z., SHEN, F. & LIU, Y. 1997. Surface deformation and lower crustal flow in eastern Tibet. *Science*, **276**, 788–790.

SANDIFORD, M. 1989. Horizontal structures in granulite terrains: a record of mountain building or mountain collapse. *Geology*, **17**, 449–452.

SIBSON, R. H. 1977. Fault rocks and fault mechanism. *Journal of the Geological Society, London*, **133**, 191–213.

TAYLOR, B. 1995. Preface. *In*: TAYLOR, B. (ed.) *Backarc Basins—Tectonics and Magmatism*. Plenum, New York.

TURCOTTE, D. L. & SCHUBERT, G. 1982. *Geodynamics: Applications of Continuum Physics to Geological Problems*. John Wiley, New York.

VON QUADT, A. 1993. The Saxonian granulite massif: new aspects from geochronological studies. *Geologische Rundschau*, **82**, 516–530.

WERNER, O. & REICH, S. 1997. $^{40}Ar/^{39}Ar$-Abkühlalter von Gesteinen mit unterschiedlicher *P–T*-Entwicklung aus dem Schiefermantel des Sächsischen Granulitgebirges. *Terra Nostra*, **97**(5), 196–198.

WERNICKE, B. 1990. The fluid crustal layer and its implications for continental dynamics. *In*: SALISBURY, M. H. & FOUNTAIN, D. M. (eds) *Exposed Cross-Sections of the Continental Crust*, Kluwer, Dordrecht, 509–544.

WILKS, K. R. & CARTER, N. L. 1990. Rheology of some continental lower crustal rocks. *Tectonophysics*, **182**, 57–77.

ZOTH, G. & HÄNEL, R. 1988. Appendix. *In*: HÄNEL, R., RYBACH, L. & STEGENA, L. (eds) *Handbook of Terrestrial Heat Flow Density Determination*. Kluwer, Dordrecht, 449–466.

The fundamental Variscan problem: high-temperature metamorphism at different depths and high-pressure metamorphism at different temperatures

PATRICK J. O'BRIEN

Bayerisches Geoinstitut, Universität Bayreuth, D-95440 Bayreuth, Germany
(e-mail: Patrick.Obrien@uni-bayreuth.de)

Abstract: The evolution of the crystalline internal zone of the European Variscides (i.e. Moldanubian and Saxo-Thuringian) is best understood within a framework of two distinct subduction stages. An early, pre-Late Devonian (older than 380 Ma), subduction stage is recorded in medium-temperature eclogites and blueschists derived from low-pressure basaltic and gabbroic protoliths now found as minor relics in amphibolite facies meta-ophiolite or gneiss–metabasite nappe complexes. A second subduction and exhumation event produced further nappe complexes containing different types of mantle peridotites, along with their enclosed pyroxenites and high-temperature eclogites, associated with large volumes of high-*T*–high-*P* (900–1000 °C, 15–20 kbar) felsic granulites. Abundant geochronological evidence points to a Carboniferous age (*c.* 340 Ma) for the high-*P*–high-*T* metamorphism as well as an extremely rapid exhumation because the fault-bounded, granulite–peridotite-bearing tectonic units are also cut by late Variscan granitic plutons (315–325 Ma). The massive heat energy for the characteristic, and most widespread feature of the Variscan event, the low-*P*–high-*T* metamorphism (750–800 °C, 4–6 kbar) and voluminous granitoid magmatism (325–305 Ma), comes from three sources. An internal heat component comes from imbrication of crust with upper-crustal radiogenic heat production potential in the region parallel to the subduction zone; an external mantle heat component is undoubtedly contributing to the transformation of crust taken to mantle depths (i.e. the granulites); and a heat component advected to the middle and lower crust seems inescapable if the hot granulite–peridotite complexes were exhumed and cooled as rapidly as petrological and geochronological evidence seems to suggest. Major mantle delamination and asthenospheric upwelling as a cause of heating in Early Carboniferous times is not supported by geochemical, geophysical or petrological–geochronological studies, although slab break-off probably did occur.

The European Variscides, part of a *c.* 8000 km long orogenic belt caused essentially by the late Palaeozoic collision of Laurasia and Gondwana and intervening microplates (Matte 1991), can be followed from Portugal to Poland in a series of basement blocks protruding through Mesozoic and Cenozoic cover (Fig. 1). Further Variscan basement occurs within the Alpine chain. A subdivision of the orogenic belt into linear zones with differing characteristic lithostratigraphic, metamorphic and deformation styles has existed since the pioneering work of Kossmat (1927) whereby the high-grade, crystalline core of the orogen (Fig. 1) comprises the Saxo-Thuringian Zone and Moldanubian Region (subdivided, in the Bohemian Massif, into the Moldanubian and Teplá–Barrandian; Fig. 2). This traditional view (Kossmat 1927; Chaloupský 1989) distinguishes the fault-bounded Teplá–Barrandian in the core of the Bohemian Massif (Fig. 2), with its pre-Mid-Cambrian Barrovian metamorphism and apparently continuous Cambrian to Middle Devonian sedimentary sequence (Barrandian Basin, around Prague), from the polymetamorphic, monotonous, high-grade, migmatitic paragneisses of the Moldanubian in the south and the monometamorphic, mostly low-grade, biostratigraphically defined, upper Proterozoic to Palaeozoic volcano-sedimentary rocks of the Saxo-Thuringian to the north (see also Zulauf *et al.* 1997).

A first-order characteristic of the Variscan chain is the abundance of granitoid magmatism and associated low-pressure (*P*)–high-temperature (*T*) regional metamorphism. This basic feature, along with an apparent absence of ophiolites, led Zwart (1967) to separate a supposed ensialic, high-*T*, Variscan-style

From: FRANKE, W., HAAK, V., ONCKEN, O. & TANNER, D. (eds).
Orogenic Processes: Quantification and Modelling in the Variscan Belt.
Geological Society, London, Special Publications, **179**, 369–386. 0305-8719/00/$15.00

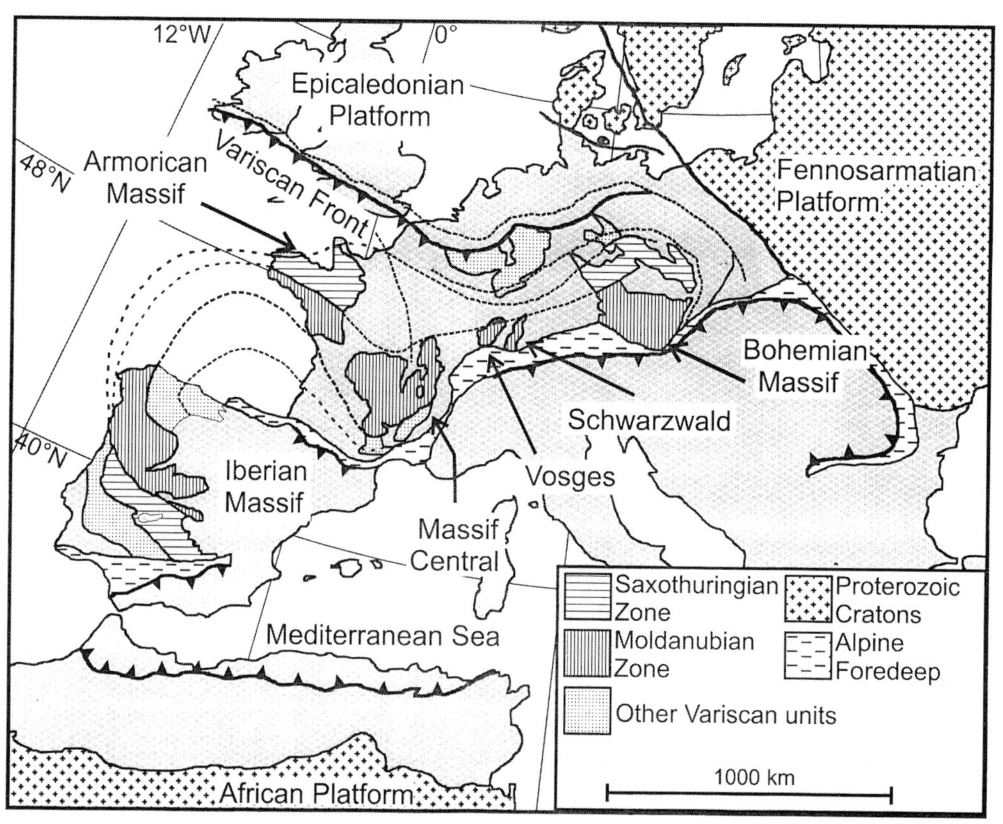

Fig. 1. Location of the main Variscan basement blocks in Europe with the high-grade Saxo-Thuringian and Moldanubian Zones highlighted.

evolution from that of the Alps, where high-*P*, subduction zone metamorphism, ophiolites, abundant nappes and scarce granites typified a collisional orogen. The presence of high-*P* rocks has been known for a long time in the Variscides (eclogites were described even before the name 'eclogite' existed (Goldfuss & Bischof 1817) and the term 'granulite' was actually coined for the dominant rock type of the Saxonian Granulitgebirge (Weiss 1803)) but the recognition that these rocks were products of a Palaeozoic orogeny emerged only after systematic geochronological investigations had been undertaken (e.g. Gebauer & Grünenfelder 1979; Van Breemen *et al.* 1982; Gebauer 1986; Kröner *et al.* 1988; Aftalion *et al.* 1989). Likewise, allochthonous nappe models, disregarded for decades despite early studies (Suess 1913), were resurrected (Ries & Shackleton 1971; Burg & Matte 1978; Tollmann 1982; Behr *et al.* 1984) and mafic–ultramafic complexes with the geochemical properties of oceanic crust were slowly recognized (Gebauer & Grünenfelder 1982; Pin & Vielzeuf 1983; Jelínek *et al.* 1984; Bernard-Griffiths & Cornichet 1985; Bernard-Griffiths *et al.* 1985; Pin 1990). In addition, palaeomagnetic (Van der Voo *et al.* 1980) and palaeontological (Cocks & Fortey 1982) data suggested the existence of a wide oceanic domain, between northern (Armorican) and southern (Gondwanan) Variscan units, during Early Palaeozoic time. Subsequent research (e.g. Tait *et al.* 1997, this volume) points to the late Ordovician period as the time of maximum separation. Thus, the Variscides have all the features of a collision orogen; they are just harder to recognize. Study of such an eroded mountain belt is extremely important for understanding modern, active, high mountain chains such as the Alps or Himalayas because it documents the thermal and mechanical processes leading to orogenic collapse and the return of the crust to a stable thickness, that is, the future events expected in the high mountain belts. With respect to the Alps, these Variscan rocks also provide important information because much of the basement involved in the Alpine event was consolidated during the Variscan event and thus

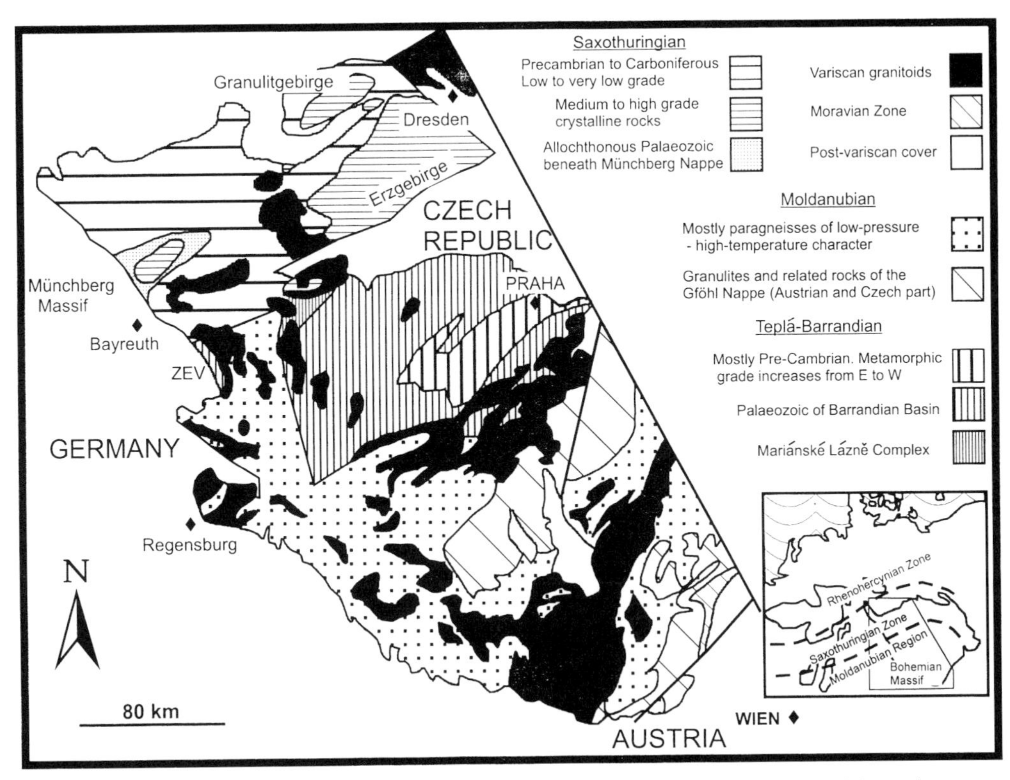

Fig. 2. Simplified geological map of part of the Bohemian Massif showing the distribution of the main metamorphic units and the abundant granites. ZEV, Zone of Erbendorf–Vohenstrauss.

metamorphism as a result of this older orogenic stage must first be deciphered before any effects caused solely by the younger Alpine event can be evaluated.

In recent years, the rapid accumulation of new knowledge on structural, metamorphic and geochronological aspects of the Variscan evolution has led to the emergence of more complex, dynamic models to explain its evolution. Such present-day models consider: (1) the location of major tectonic boundaries marked by rift or ophiolite sequences; (2) the presence of nappes containing remnants of high-P rocks in both Moldanubian and Saxo-Thuringian domains; (3) the timing, and regional extent, of distinct metamorphic events; (4) the temporal thermal and baric development of individual tectonic units; (5) the importance of structural level, with respect to survival of older features, during the final, high-T, Early Carboniferous stage of the evolution; (6) the magnitude of Late Carboniferous disruption caused by late strike-slip movements (Arthaud & Matte 1977; Matte & Burg 1981; Behr *et al.* 1984; Burg *et al.* 1984; Franke 1989; Pin 1990; Matte *et al.* 1990; Matte 1991; O'Brien & Carswell 1993; Krohe 1998). There is still considerable dispute, however, over the cause of the one feature that is so characteristic: the high-T–low-P metamorphism and associated magmatism. Undoubtedly this simple, first-order feature requires a large thermal anomaly. However, to understand the magmatic and metamorphic evolution, it is necessary to ascertain the magnitude and timing of the thermal anomaly on a lithospheric scale. In other words we need to determine what temperatures were reached, at what time and at what depth. The aim of this review, concentrating on examples from the Bohemian Massif, is to document the distinguishing features of both high- and low-P rock units, the duration and timing of specific stages in the monophase, multiphase or polyphase metamorphic evolution of these rocks, and to ascertain the applicability and suitability of previously presented tectonometamorphic models with particular emphasis on the possible factors responsible for the development of high temperatures.

The characteristic rock types of the Variscides

High-grade rocks in the Variscides occur predominantly in the Moldanubian, but it is important to understand, for the large-scale evolution, that comparable high-grade rocks appear as tectonically emplaced, allochthonous bodies within lower-grade series of the Saxo-Thuringian. Three fundamentally different tectonometamorphic units can be recognized: (1) low-*P* series; (2) series containing high-*P*–low- to medium-*T* rocks and (3) units preserving remnants of a high-*P*–high-*T* evolution. In the following, taking the examples from the Bohemian Massif, each of these basic units will be addressed in turn and the characteristic features outlined.

Low-P units

The Moldanubian Zone is dominated by a monotonous series mostly composed of metagreywackes but with minor acid and mafic orthogneiss, metaquartzites and lenses of calc-silicate rock (e.g. Fiala 1995; Fuchs 1995). Distinctly subordinate are local, more variegated units which contain alternations of amphibolite, marble, calc-silicate rock, quartzite and graphitic schist within paragneisses. A late Proterozoic age is supposed for the majority of the series although, locally, biostratigraphically proven Palaeozoic (up to Silurian) series structurally overlie the monotonous paragneisses in the Bohemian Massif (Pflug & Reitz 1987; Stettner 1990). The rocks of this series show the archetypal features of low-*P*–high-*T* metamorphism, being commonly migmatitic and containing sillimanite, K-feldspar and, when bulk chemistry permits, cordierite. The high degree of melting of biotite, absence of muscovite, common presence of spinel and occasional orthopyroxene-bearing leucosomes all indicate that temperatures of at least 750–800 °C were reached (Jones & Brown 1990; Montel *et al.* 1992; Linner 1996; Kalt *et al.* 1999). These surprisingly high temperatures are supported by the development of low-*P*–high-*T* pyroxene hornfels overprints (Fe-rich olivine–pyroxene–plagioclase) in meta-eclogites (O'Brien 1997*b*, unpublished data) from several locations in the Moldanubian. Despite these extremes of temperature it is rather surprising that some of the low-*P* paragneisses still preserve relics of garnet and/or staurolite and/or kyanite pointing to an earlier metamorphism of medium-*P* character (Jones & Brown 1990; Tanner *et al.* 1993; Linner 1996; Reche *et al.* 1998). Interestingly, the preservation of strong compositional gradients in some large relict garnets and the extreme reaction of armoured inclusions of medium-*P* relics when exposed by fluid influx along fractures (sapphirine, corundum–quartz and spinel–quartz assemblages are documented from kyanite and staurolite breakdown) strongly suggest only a short-lived exposure to extreme temperatures (Schuster & O'Brien unpublished data). This is what would be expected for contact rather than regional metamorphism, and the fast cooling rates implied in some multi-method geochronological studies (e.g. Brown & Dallmeyer 1996) would certainly be consistent with such a model.

The low-*P*–high-*T* metamorphism is of Carboniferous (335–325 Ma) age as constrained by U–Pb (monazite, zircon), and Rb–Sr, Ar–Ar and K–Ar (mica, amphibole, feldspar and whole-rock) data (e.g. Dallmeyer *et al.* 1992; Friedl *et al.* 1993, 1994; Kalt *et al.* 1994, 1998), with the younger ages (i.e. around 325 Ma) found predominantly in the SW margin of the Bohemian Massif. However, there is no consensus on the age of the medium-*P* event indicated by the kyanite and staurolite relics. This is a rather difficult problem to solve. In the SW Teplá–Barrandian, close to its border with the Moldanubian, a regional metamorphism leading to kyanite and staurolite growth is clearly at least of Cambrian age, as regional isograds are overprinted by pyroxene-hornfels facies assemblages (Vejnar 1966) as a result of a gabbro intrusion dated at 530 Ma (U–Pb zircon, Gebauer & Grünenfelder 1982). An early Palaeozoic medium-*P* event would fit with the simple, apparently mono-metamorphic history recorded in the biostratigraphically proven Palaeozoic succession of the Moldanubian, but it should also be pointed out that second generations of kyanite and staurolite occur in the Teplá–Barrandian, post-dating the contact metamorphism (Cháb & Zácek 1994), and the same minerals also appear in complexes yielding Devonian cooling ages that represent shallow tectonic units unaffected by the Early Carboniferous event (e.g. Stettner 1990; Kleemann 1991). The age of this medium-*P* stage is an important constraint on tectonic models for the low-*P*–high-*T* metamorphism. If the medium-*P* stage is of Carboniferous age, the *P*–*T* path involves significant decompression (either isothermal or with increasing *T*) whereas an older medium-*P* event allows the Carboniferous regional metamorphism to be explained by near-isobaric heating, that is, regional contact metamorphism: two completely different scenarios.

Also belonging to the low-*P* unit are parts of the Saxo-Thuringian (for example, in Bavaria and some areas of the northern Bohemian Massif). Here, the unit is composed of a biostratigraphically well-defined (in its very low-grade metamorphic parts), upper Proterozoic to Carboniferous parauthochthonous, passive margin, volcano-sedimentary sequence, showing a strong Variscan deformation (von Gaertner *et al.* 1968). Generally, greenschist to very low grade assemblages are found, but the peak of the monophase metamorphism, in deeply eroded parts, reaches the lower amphibolite facies with staurolite and garnet present (Mielke *et al.* 1979).

*High-P–low- to medium-*T *units*

Isolated areas of higher-grade gneissic rocks, surrounded by lower-grade units, have long been the centre of controversy in the Variscides. The classic example, the Münchberg Gneiss Massif in NE Bavaria (Fig. 2), where a unit of hornblende schist, mica schist, augen gneiss, corona gabbro, granulite and eclogite rests on phyllites, has attracted geological attention for over two centuries. The earliest nappe models for the Variscides were formulated here (Suess 1913) only to be widely rejected for basement-uplift models (e.g. Cloos 1927) before returning to favour in the 1980s (e.g. Behr *et al.* 1984). The complex is actually a composite stack of five nappe units with lowest grade (sub-greenschist) at the base and highest grade (eclogite) at the top of the pile, and was emplaced onto very low grade rocks of the Saxo-Thuringian. It is extremely well preserved (although poorly exposed) and so its temporal structural and metamorphic evolution can be used as a yardstick against which other eclogite-bearing nappes can be compared (Franke 1984).

The Münchberg nappe complex, from base to top comprises: (1) slivers of very low grade Ordovician to Lower Carboniferous volcanic rocks and sediments; (2) greenschist facies calc-alkaline basalts, quartz phyllites and minor serpentinites; (3) sub-alkaline, tholeiitic, metabasaltic garnet- or epidote-bearing amphibolites at amphibolite facies or retrograded along shear zones to greenschist facies; (4) metaclastic rocks with deformed granitic intrusions (augen-gneisses), gabbro and granodiorite showing upper amphibolite facies garnet corona growth, and minor serpentinites; (5) banded amphibolites and hornblende gneisses, acid gneisses and some eclogites (e.g. Franke 1984; Okrusch *et al.* 1989). Protolith ages for the eclogite, gabbro and augen-gneiss are around 500 Ma and K–Ar cooling ages from the amphibolite and greenschist facies overprint fall between 380 and 360 Ma (Gebauer & Grünenfelder 1979; Söllner *et al.* 1981; Kreuzer *et al.* 1989).

Münchberg Massif eclogites, often well preserved, are typically either darker, kyanite-free types with mid-ocean ridge basalt (MORB)-like chemistry or kyanite- and/or zoisite-bearing, high-alumina metabasalts or gabbros (Matthes *et al.* 1975; Stosch & Lugmair 1990). Growth from former low-*P*, plagioclase-bearing protoliths is evidenced by abundant kyanite and zoisite inclusions, cluster-like aggregate form, and prograde zoning of garnet as is known from corona gabbros. The distinctly high d*P*-d*T* for eclogite formation, reflected by peak conditions of around 20–25 kbar and 650 °C (Franz *et al.* 1986; Klemd 1989; O'Brien 1993), favours a subduction environment, as do breakdown reactions during decompression, which do not exceed the temperatures of the amphibolite facies. The age of eclogite metamorphism was deduced to be around 380–395 Ma (Gebauer & Grünenfelder 1979; Stosch & Lugmair 1990), although amphiboles and mica from a clearly retrogressive (amphibolite) stage also yield 380 Ma ages (Kreuzer *et al.* 1989).

South of the Münchberg Massif, generally less well-preserved eclogites, with a slightly higher-temperature (up to granulite facies) overprint, occur in probably allochthonous gneiss–metabasite complexes sitting at high structural levels within both the Moldanubian (Zone of Erbendorf–Vohenstrauss: O'Brien *et al.* 1997*a*) and Teplá–Barrandian (Mariánské Lázne Complex, Beard *et al.* 1995). Strong lithological, geochemical and geochronological similarities exist between these two bodies and the Münchberg Massif and suggest the former existence of a much more extensive nappe sheet. In the Massif Central of France, numerous eclogite-bearing nappe complexes comparable with the Münchberg Massif have been identified (Bouchardon *et al.* 1989; Ledru *et al.* 1989) with again a bimodal acid and basic rock association (so-called leptyno-amphibolite complexes) being a common characteristic. In the Massif Central it is clearly demonstrated that these high-*P* units have been exhumed before Carboniferous time as they are overlain by Devonian sedimentary sequences (Bouchardon *et al.* 1989). This fundamental pattern, the kyanite-bearing and kyanite-free types, similarity of bulk chemistry and protolith age, presence in ophiolite or gneiss–metabasite-complexes, clear prograde evolution from low-*P* protoliths, pre-Carboniferous age of high-*P* metamorphism, multi-step breakdown via amphibolite or granulite facies stages dependent on regional position, and association dominantly with spinel peridotite, is repeated throughout the Variscides (see review by O'Brien *et al.* (1990)). Whether in the Bohemian Massif, Massif Central, Armorican Massif or Iberian Massif (Paquette 1987; Bouchardon *et al.* 1989; Gil Ibarguchi *et al.* 1990; O'Brien *et al.* 1990; Peucat *et al.* 1990; Beard *et al.* 1995; O'Brien 1997*a*), these basic features are repeated and point to a fundamental aspect of the Variscan evolution, that is, rift and/or marginal basin crust was involved in a pre-Carboniferous subduction event and exhumed, in nappe complexes, to shallower crustal levels, before *c.* 380 Ma (see Pin 1990; Matte 1991). The connection with a low-*T* subduction environment is strongly supported by the rare presence of blueschists and low-*T* eclogites in scattered parts of the Variscan belt (Schermerhorn & Kotsch 1984; Gil Ibarguchi & Ortega Girones 1985; Peucat 1986; Guiraud *et al.* 1987; Kryza *et al.* 1990; Bakun-Czubarow 1998).

*High-P–high-*T *units*

The second type of high-*P* rock association comprises dominantly high-*P* granulites that enclose lenses and masses of garnet peridotite, high-*T* eclogite and pyroxenite. This association is also widely distributed. In the Bohemian Massif alone, significant volumes of high-*P*–high-*T* rocks are preserved in the Granulitgebirge and Erzgebirge (Germany), in South Bohemia (Czech Republic) and Lower Austria, as well as smaller bodies in the Polish Sudetes (Scharbert & Kurat 1974; Vrána & Jakeš 1982; Fiala *et al.* 1987; Vrána 1989; Medaris & Carswell 1990; Rötzler 1992; Beard *et al.* 1992; Carswell & O'Brien 1993; Medaris *et al.* 1995*b*; Kotková *et al.* 1996; Kryza *et al.* 1996; O'Brien *et al.* 1997*b*; Willner *et al.* 1997; O'Brien 1999). Not all granulite units are tectonically uppermost. The Saxonian Granulitgebirge actually forms a gneissic dome or core complex that protrudes through low-grade sediments of the Saxo-Thuringian (Reinhardt & Kleemann 1994; Franke & Stein this volume). The gneissic high-grade core of the Erzgebirge has also been explained in this way (Krohe 1998). Although often heavily retrogressed, garnet peridotite-bearing granulite bodies are known in the Schwarzwald, Vosges and Massif Central (Lasnier 1971; Bonhomme & Fluck 1981; Fabriès & Latouche 1988; Bouchardon *et al.* 1989; Kalt *et al.* 1995; Altherr & Kalt 1996; Kalt & Altherr 1996) as well as in Variscan basement blocks of the Alps (e.g. Hinterlechner-Ravnik 1971; Obata & Morten 1987).

A good example of one of these high-*P*–high-*T* units is the Gföhl Nappe complex (Fig. 2) found in the southern Bohemian Massif (Fiala 1995; Fuchs 1995). Fragments of this composite unit can be recognized over most of the southern Bohemian Massif, on either side of the Moldanubian Pluton. Tectonically uppermost are acid granulites and these overlie migmatitic, granitic orthogneisses known as the Gföhl gneiss. More variegated, amphibolite-rich series occur below the granulites and below the Gföhl gneiss. Whereas the granulite bodies contain infolded lenses of garnet peridotites, pyroxenite and high-*T* eclogite, serpentinized peridotites in the variegated series underlying the Gföhl gneiss show no evidence that they ever contained garnet and, in addition, magmatic textures are still preserved in some of the metagabbros. In contrast, some of the mafic bodies enclosed in, or directly underlying the granulites, show evidence, in the form of high-*P*–high-*T* garnet- and clinopyroxene-bearing assemblages, of having reached conditions comparable with those experienced by the granulites. This is clear evidence that not all rocks of the Gföhl nappe had the same metamorphic evolution; a feature noted also in Massif Central and Iberian Massif nappes, where comparable garnet peridotites and lower-grade ophiolitic peridotites occur in different levels of the thrust stack (Mercier *et al.* 1982; Peucat *et al.* 1990).

The typical Variscan felsic granulites are light coloured, are of medium- to high-K calc-alkaline granitic composition, dominantly composed of quartz and feldspars (mesoperthite is characteristic) with minor kyanite and garnet, and characterized by a strong mylonitic fabric that formed during exhumation at lower grade (e.g. Pin & Vielzeuf 1983; Fiala *et al.* 1987; Urban 1992; Vellmer 1992). Geochemical evidence for major melt extraction is lacking as, apart from low Cs, and variable U and Th, the rocks are essentially undepleted (see Vellmer 1992). Subordinate layers with more intermediate compositions are clinopyroxene-bearing varieties with (probably) secondary orthopyroxene. Although it has been often stated, based on the granitic bulk composition, that the granulites represent metamorphosed rhyolites, more likely is a derivation by high-*P*–high-*T* partial melting processes (Vrána & Jakeš 1982; Vrána 1989; Kotková & Harley 1999). In this respect the poikiloblastic form of garnet with many large mesoperthite inclusions is interesting: a characteristic garnet form documented from migmatite belts world-wide where incongruent melting reactions such as biotite + plagioclase + aluminosilicate + quartz = garnet + K-feldspar + melt (biotite dehydration melting) have taken place (e.g. Waters & Whales 1984; Le Breton & Thompson 1988). Also, garnet coronas around idiomorphic feldspar grains have been noted in some of these rocks, clearly a magmatic rather than a metamorphic (solid-state) feature.

Deformation and retrogression have led to variable degrees of breakdown of original, high-*T* ternary feldspars to mesoperthite (in mylonitic zones a two-feldspar + quartz matrix wraps mesoperthite augen) and growth of biotite. These two features have caused difficulty in recovering the peak metamorphic conditions, as (1) it is necessary to reconstitute the mesoperthite to its original pre-exsolution composition to apply geothermobarometric methods, and (2) the growth of secondary biotite may significantly modify primary garnet composition. However, the simple coexistence of kyanite and K-feldspar indicates high-*P* conditions (see Scharbert & Kurat 1974) and, where care has been taken to reconstitute the original feldspar composition, geothermobarometric results for the high-*P* granulites suggest extreme conditions in the realm 850–1000 °C and 15–20 kbar (Carswell & O'Brien 1993; Kotková *et al.* 1996; Kryza *et al.* 1996; O'Brien *et al.* 1997; Willner *et al.* 1997*b*; O'Brien 1999). More robust geothermobarometry in the rare intermediate and mafic granulites (e.g. Kryza *et al.* 1996; O'Brien *et al.* 1997) fully confirms these values as well as allowing a second, medium-*P* granulite facies stage to be distinguished in the granulite complex, this second phase being more prominent in some areas (e.g. Petrakakis 1997). In some granulites, for example in South Bohemia, a later low-*P*–high-*T* stage is also indicated by the occurrence of cordierite and partial melt patches in the acid granulites.

Most geochronological studies of the granulites of the Bohemian Massif, especially zircon and monazite dating, have yielded, contrary to earlier expectations, Palaeozoic, and in most cases Carboniferous, ages (Van Breemen *et al.* 1982; Schenk & Todt 1983; Aftalion *et al.* 1989; Carswell & Jamtveit 1990; von Quadt 1993; Wendt *et al.* 1994; Kröner & Willner 1995; Kotková *et al.* 1996; Becker 1997; O'Brien *et al.* 1997*b*; Kröner *et al.* 1998). With few exceptions, an age of 340 Ma is repeatedly obtained. This value even appears in samples where clear petrographic evidence for high-, medium- and low-*P* reactions is visible: rounded, multifaceted, metamorphic zircons, regardless of petrographic location (controlled by *in situ* measurements of zircons extracted from thin section) yield the same age as newly grown, prismatic zircons in low-*P* cordierite-bearing melt pods in the same rocks (Kröner *et al.* 1996). There is no correlation with zircon grain size. Thus, diffusive resetting of the isotopic clock, such that the ages represent the time at which a post-peak temperature 'blocking temperature' was reached, can be ruled out. The only remaining explanation is that the ages record the metamorphism and thus the whole of the granulite high-*P* to low-*P* evolution at high *T* took place within the error of the measured age (i.e. 340 ± 1 Ma). This is consistent with other geochronological methods that show that the high-*T* exhumation process ended by 340–330 Ma (e.g. Dallmeyer *et al.* 1992; Costa *et al.* 1993).

A characteristic feature of the felsic granulite complexes is the occurrence of pods and lenses of garnet peridotite, which themselves contain layers of various types of pyroxenite and high-*T* eclogites (Medaris & Carswell 1990; Beard *et al.* 1992; Kalt *et al.* 1995; Medaris *et al.* 1995*a*; Altherr & Kalt 1996; Becker 1996*a*, *b*; Kalt & Altherr 1996). In contrast to the eclogites described above, this variety shows higher equilibration temperatures, quartz- and kyanite-bearing types are extremely rare, and garnet lacks growth zoning (Dudek & Fediuková 1974; O'Brien 1997*a*). These rocks have been interpreted as products of crystal accumulation during reaction between melts and the host peridotites at high *P* (Beard *et al.* 1992; Medaris *et al.* 1995*a*; Becker 1996*a*, *b*). In contrast to the low- to medium-*T* eclogites therefore, this group did not undergo prograde metamorphism from low-*P* protoliths but were extracted from the deep mantle along with their host peridotites. The peridotites, intensively investigated in recent years, are chemically and petrologically inhomogeneous rocks, but fall essentially into three groups (Medaris & Carswell 1990; Medaris *et al.* 1995*a*, *b*; Becker 1996*a*, *b*; Brueckner & Medaris 1998). The dominant group of garnet peridotites, formed at 30–50 kbar, represents subcontinental mantle depleted by melt extraction but enriched in an upper-crustal isotopic component, most probably caused by metasomatism intimately linked to the production of the melts that produced the pyroxenites and eclogites (Medaris *et al.* 1995*a*, Becker 1996*b*). These melts probably originated in the metasomatized mantle wedge above subducted lithosphere, and, on the basis of Sr and Nd isotope ratios and mineral isochrons, this melting may have occurred shortly before the peridotites and enclosed rocks were incorporated into the crust (Brueckner & Medaris 1998). Equilibration conditions for this group of garnet peridotites correspond to a geotherm typical for a lithosphere of 200 km thickness. In contrast, the hot spinel peridotite at Mohelno, SE Czech

Republic, contains garnet only along its margins, that is, it cooled into the garnet stability field. Equilibration conditions for this latter type of peridotite, 20–30 kbar and 1000–1300 °C, are more typical for the hotter geotherm of a thinner (100 km) lithosphere and thus this peridotite has more in common with hot, upwelling asthenosphere. A final, minor component, is suggested to be subducted mafic material accreted to the lithospheric mantle (Becker 1996*a*). As for the granulites, diffusion modelling studies show that these rocks must have undergone extremely fast exhumation and cooling from peak conditions (Medaris *et al.* 1990).

Discussion

Summarizing the above information allows construction of the simplified *P–T* diagram for the Variscan evolution as shown in Fig. 3. This distinguishes an initial, pre-Carboniferous, subduction stage that produced the blueschists and eclogites that are now widely distributed in gneiss–metabasite (or leptynite–amphibolite) units. These rocks of the high-*P*–low- to medium-*T* subduction complexes were already exhumed by late Devonian times, as evidenced by sediments of this age overlying the complexes or the presence of high-*P* detritus in sediments of this age (e.g. Schäfer *et al.* 1997). The second major stage, the Early Carboniferous phase that truly characterizes the Variscan Orogeny, involves two components. The high-*P*–high-*T* rocks, judging from the geochronological and diffusion modelling results, were very rapidly exhumed and sometimes emplaced at high structural levels before granite intrusion, as the granites transect nappe boundaries. The rocks of the low-*P* series, including any of the earlier high-*P*–low- to medium-*T* rocks undergoing a second metamorphism, show a *P–T* evolution apparently convergent with that of the high-*P*–high-*T* units. The timing of the high-*P*–high-*T* granulite metamorphism at 340 Ma and the regional low-*P* metamorphism shortly followed by massive granite intrusion in the period 335–315 Ma, strongly suggests a correlation between the lithosphere-scale processes leading to their joint formation. Thus the fundamental Variscan problem, high-temperature metamorphism at different depths (i.e. pressures), and high-pressure metamorphism at different temperatures, can be explained by a two-stage development. The remaining problem is the cause of the high temperatures.

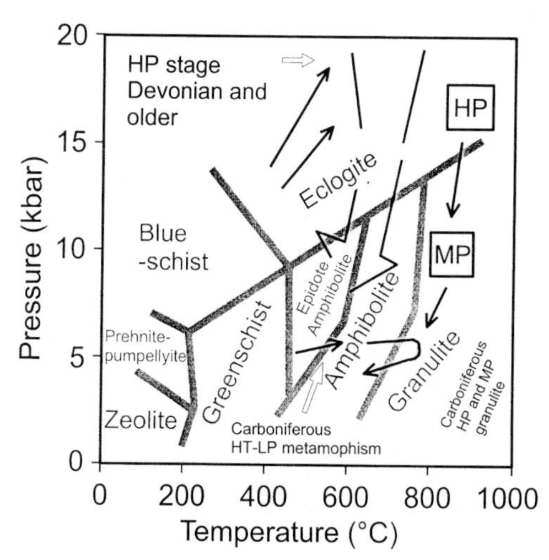

Fig. 3. Schematic pressure–temperature evolution diagram for the metamorphic rocks of the Variscides with emphasis on the two-stage nature and the different thermal character of each stage.

The heat sources for regional crustal metamorphism may be internal, external or advected; that is, internal radiogenic heat production from the decay of isotopes of Th, U and K, external heat from the underlying mantle, or heat advected into the crust as a result of magma emplacement or tectonic juxtaposition of hot and cold units.

Internal heat source

An internal source of heat, by the stacking of fertile upper-crustal lithologies over a *c.* 20 Ma period, has been proposed to explain the production of the voluminous granitoids (330–320 Ma, Friedl *et al.* 1993) of the southern Bohemian Massif (Gerdes *et al.* 1996). Geochemical and isotopic data point to a high degree of biotite dehydration melting of pelites and greywackes, that is, melting of crustal material, with only a very minor (*c.* 5% of the total volume of granite) mantle component, which itself indicates only a small degree of partial melting of lithospheric mantle (Gerdes *et al.* 1998, this volume). The granites intruded into crust that was already hot and so a simple, massive, mantle-derived heat source, regardless of type, could provide the heat at deeper levels required for melting but could not explain why regional low-*P* metamorphism, at shallower levels of the same crust, was earlier than granite intrusion. The internal heat production model, although compatible with the features of the South Bohemian Batholith (Gerdes *et al.* 1998, this volume), cannot explain the formation of the felsic granulites. In this model, production of 1000 °C temperatures to produce granulites at

340 Ma, temperatures higher than those needed for biotite dehydration melting to produce the granites (intrusion at 325–315 Ma), could not be reached early enough in the process such that the granulites could be exhumed to high structural levels before being themselves cut by the granites. In addition, there are other large intrusive complexes in the Variscides (e.g. the Central Bohemian Pluton) where a much larger mantle component is present and so, clearly, some modification to the above model is required.

External heat source

Models of continent–continent collision leading to crustal thickening often neglect the lithospheric mantle (and dense lower crust) of the colliding plates. These are, however, the volumetrically most important parts of the colliding plates and their accommodation in the overall collision model may play an important role in the syn- to post collisional development of an orogen. Thermal and mechanical instability of the thickened mantle lithosphere below the crustal stack has been predicted to result in tectonic removal of the mantle lithosphere and its replacement by asthenosphere: a process known as delamination (Bird 1978; Houseman *et al.* 1981). One of the possible consequences of delamination may be melt production in the mantle (Houseman *et al.* 1981; Platt & England 1993): for depths less than *c.* 50 km, hot, upwelling asthenosphere will be above the dry peridotite solidus and is thus capable of producing large volumes of basaltic magma (McKenzie & Bickle 1988). Underplating of these magmas at, and within, the base of the crust and the combination of conductive (hot mantle) and advected (basalt magma) heat would probably cause high-*T* metamorphism and massive melting in the crust and thus the production of granites.

In such a scenario, mixing of mantle-derived and crust-derived melts is highly probable and thus delamination is ruled out at least for the south Bohemian granites (Gerdes *et al.* 1996, 1998). Although a component of underplating has been proposed for the Variscan crust (e.g. Pin & Vielzeuf 1983) this represents only the result of late orogenic extension of late Carboniferous to Permian age and so cannot be the cause of the metamorphism and magmatism in Early Carboniferous time. Likewise, the enormous volumes of basalt needed to balance the degree of crustal melting required by the measured granite production (see modelling of Gerdes *et al.* (1998)) should be visible on geophysical profiles: they are not! Certainly, the high-*T* granulites could be formed in such an environment, but these rocks are not really comparable with lower-crustal rocks and, in addition, formed at depths above 50 km, that is, at conditions where upwelling mantle would be below its melting point. An alternative model would be total removal of the thickened lithospheric root below crust >50 km thick and its replacement by asthenosphere, which would then act as a conductive heat source; a scenario that would provide only minor melts of enriched, metasomatized mantle. This could explain the mix of asthenospheric and lithospheric mantle components within the granulite complexes but, again, the lack of a lower-crustal signature in the granulites, the difficulty in this model of forming granite in the upper-crustal lithologies by conductive heat transport alone, and the necessity to emplace the granulites at high structural levels before granite emplacement, makes this scenario improbable.

A significant pointer to the validity of possible delamination models is the present nature of the lithospheric mantle. Assuming mantle detachment or delamination did occur at the end of the Variscan Orogeny, what would be the consequences for the post-Variscan mantle and how could such an event be recognized? To test possible models it is clearly important to determine the petrological and geochemical evolution of the European lithospheric mantle. There are numerous bodies of mantle-derived ultramafic rock exposed throughout the Moldanubian and Saxo-Thuringian Zones of the Variscides (e.g. Medaris & Carswell 1990; Brueckner & Medaris 1998). These record features of the mantle environment before the Permo-Carboniferous collapse of the orogen. In addition, a picture of the post-Variscan mantle can be obtained from peridotite and pyroxenite xenoliths entrained in Tertiary basalts (Stosch & Lugmair 1986; Hegner *et al.* 1995; Zagana *et al.* 1997), as well as from investigation of mantle-derived melts (Wörner *et al.* 1986; Wilson & Downes 1991; Hegner *et al.* 1998). Many of the mantle rocks within the granulite complexes show isotopic evidence for metasomatic enrichment shortly before entrainment (Medaris *et al.* 1995*a*; Becker 1996*a*, *b*; Brueckner & Medaris 1998): a contamination reflected in the chemistry of lamprophyres intruded in the initial post-collisional stage (Hegner *et al.* 1998). The geochemical evidence from magmas derived from the post-Variscan lithosphere is that a source depleted in major elements, but metasomatically enriched in certain trace components, is present below central Europe (Wörner *et al.* 1986; Wilson & Downes 1991).

These two features together seem to rule out the possibility that some kind of large-scale delamination process took place in Carboniferous time. Geophysical results from the European Geotraverse (EGT) project show a well-defined Moho and, in stark contrast to the thick lithosphere of the old, stable, cratonic region in Scandinavia to the north and the deep lithospheric root below the orogenically active Alps to the south, a fairly shallow lithosphere throughout the Variscides (Blundell *et al.* 1992). This feature appears, however, to be the product of a young asthenospheric upwelling event rather than being evidence for a renewal of the Variscan lower lithosphere in Carboniferous time (e.g. Hoernle *et al.* 1995). Xenoliths of mantle peridotite and pyroxenite have been identified within Tertiary basalts both from the internal (Moldanubian, Saxo-Thuringian) and external (Rheno-Hercynian) parts of the orogen but it is rather surprising that, despite the abundance of garnet peridotite in the metamorphic complexes, the peridotites of the xenolith suite are almost exclusively spinel bearing (Stosch & Lugmair 1986; Hegner *et al.* 1995; Zagana *et al.* 1997). This suggests selective sampling by the young volcanic rocks, post-Variscan metamorphic recrystallization of possible former garnet to pyroxenes + spinel, or that the Variscan garnet peridotites formed in a particular geotectonic setting.

Now there is a problem. There is clear evidence for the presence of very hot, asthenospheric mantle in the granulite bodies. If the hot mantle did not come to the crust (i.e. by delamination) then an obvious alternative is that the crust went to the mantle! The high determined pressures for metamorphism in the granulite complexes (pseudomorphs after possible coesite have been reported by Schmädicke (1991) and Bakun-Czubarow (1991)), the broadly granitic, upper-crustal composition of the dominant, felsic granulites, and the interdigitation of mantle garnet peridotites can all be explained by deep subduction of thinned, continental crust. The large volumes of essentially granitic material in the granulite complexes would mean that, regardless of mineralogy at high-*P* conditions, the bodies would always be less dense than the mantle and thus would be forced up to normal crustal levels by buoyancy forces. A possible variant on such a model is slab break-off during the subduction phase (e.g. Davies & von Blankenburg 1995). In such a process, asthenospheric mantle penetrates a weak point of the subducting slab, thus detaching the dense, relatively cold, mantle root, and perhaps even mafic lower crust. The result would be juxtaposition of hot mantle and subducted rock, a sudden added buoyancy as the dense down-dragging force of the root was removed, and melting of low melting-temperature (metasomatized) components of the mantle. Such a process is certainly compatible with the presence of different types of mantle rocks in the granulite complexes and perhaps also with the minor volumes of shoshonitic to ultrapotassic rocks (e.g. Wenzel *et al.* 1997) found throughout the high-grade core of the Variscides.

Advected heat sources

A important consequence of transportation of the hot, dry granulite complexes back to shallow lithospheric levels, and a simple conclusion reached purely from the short time-scale of the event (i.e. fast cooling), is that these rocks would provide a source of advected heat. Whether by thrusting into mid-crustal levels or brought to shallower levels by tectonic unroofing, the heat from the hot granulite complexes would have added to the heat budget of a crust already heated as a result of crustal imbrication and stacking in the zones away from the subduction environment. Models of extensional tectonics leading to exhumation of crystalline rocks in core complexes have been proposed for many parts of the Variscides (e.g. Echtler & Malavieille 1990; Malavieille *et al.* 1990; Krohe 1998). One of the consequences of nappe tectonics followed by extension is that rocks of tectonically overlying, high-*P* series are suddenly juxtaposed against younger, hot rock units, undergoing decompression. It could be that the sort of short-lived, thermal pulse indicated by the formation of anomalous pyroxene hornfels assemblages (see above), in rocks still preserving abundant evidence of earlier events, could be attributed to such an advected heat component. The timing of the high-*P*–high-*T* and low-*P*–high-*T* events so close together in time and space cannot be a coincidence.

Large-scale picture and conclusions

The large-scale structure of the Variscides, deciphered during the last decades, shows a significant divergence with nappe complexes thrust from internal towards external zones as a result of two major subduction and obduction processes spanning a time period of 100–150 Ma (e.g. Burg & Matte 1978; Matte & Burg 1981; Behr *et al.* 1984; Matte 1991; Quesada 1991). The high-*P* rocks formed during each major subduction episode are significantly different in character (see Pin 1990). The older event has a

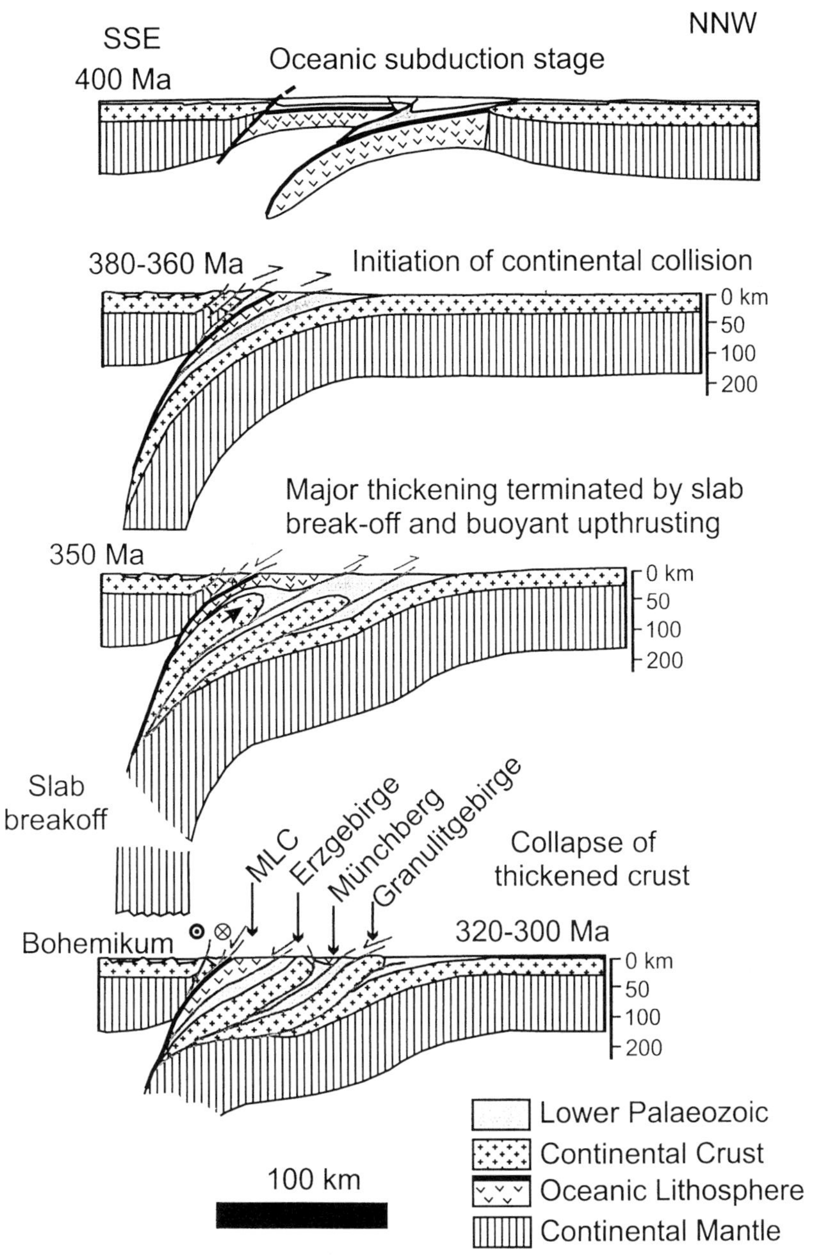

Fig. 4. A simplified subduction–collision model to explain the typical Variscan high-P metamorphism at different temperatures for the example of the northern Bohemian Massif (redrawn after Matte (1998)).

typical subduction zone low- to moderate-T character, in which blueschists were formed, and it most probably represents, on the basis of the common gneiss–metabasite (or leptynite–amphibolite) association, true oceanic crust subduction (and obduction) and marginal basin closure without the influence of large continental plates. The second, and more major, subduction event led to subduction of rocks of the continental margin to considerable depth, as well as major crustal imbrication, as final closure of inter- and intra-plate sutures occurred. The second event is the one that led to the typical low-*P*–high-*T* Variscan metamorphism and magmatism. A model that permits the features of this tectonometamorphism to be explained (deep subduction of crustal material, presence of a variety of mantle rocks within dominantly crustal complexes, and at the same time an opportunity to develop high temperatures and granites in crust without a mantle influence) has been refined from a series of precursor models by Matte (1998). This version of the model (Fig. 4) was constructed (Matte 1998) to explain the Saxonian Granulitgebirge but could just as easily be used to explain the evolution of the Gföhl Nappe, ophiolitic series accompanying the granulites, and low-*P* Monotonous Series gneisses in the southern Bohemian Massif. In the model (Fig. 4), although oceanic subduction with its colder thermal regime operates at an early stage, as thinned continental crust is subducted and the crust away from the subduction zone is compressed and stacked, a different thermal environment develops. The model also accommodates the observation that, even within the complexes with high-grade rocks, there are mafic–ultramafic, ophiolite-like lenses that show no evidence of having experienced the high-grade history; that is, the nappe complexes are composite bodies. Slab detachment in the subduction zone led to incorporation of asthenospheric mantle into subducted crust, removed the dense root, and initiated buoyancy-driven, rapid exhumation of the subducted rocks. The granulite-bearing nappe complexes were exhumed during continued collision, by extension at high structural levels, in a manner perhaps comparable with present-day exhumation of High Himalayan high-*T*–low-*P* rocks, enclosing minor coesite-bearing eclogite relics (O'Brien *et al.* 1999), below the low-grade to unmetamorphosed Tethyan series (e.g. Hodges *et al.* 1993).

Certainly we have expanded our knowledge of the Variscides considerably since the time of Zwart's 1967 paper. The separation of complex polyphase and multistage metamorphic events in this subduction–collision orogen is now better understood, and the lessons learned in the Variscides can well be used to help decipher and predict the evolution of modern, active, collision belts such as the Alps and Himalayas.

I would like to thank the reviewers, especially H. Becker and W. Franke, for their comments and suggestions. Numerous fruitful discussions with W. Franke, A. Henk, O. Oncken, G. Wörner and G. Zulauf have also contributed to the ideas and models presented here although all failings and misrepresentations are obviously the author's responsibility. Generous financial support from the Deutsche Forschungsgemeinschaft within the priority programme 'Orogene Prozesse' is gratefully acknowledged.

References

Aftalion, M., Bowes, D. R. & Vrána, S. 1989. Early Carboniferous U–Pb zircon age for garnetiferous perpotassic granulites, Blanský les massif, Czechoslovakia. *Neues Jahrbuch für Mineralogie, Monatshefte*, **4**, 145–152.

Altherr, R. & Kalt, A. 1996. Metamorphic evolution of ultrahigh-pressure garnet peridotites from the Variscan Vosges Mts. (France). *Chemical Geology*, **134**, 27–47.

Arthaud, F. & Matte, P. 1977. Late Paleozoic strike-slip faulting in southern Europe and northern Africa: result of a right-lateral shear zone between the Appalachians and the Urals. *Geological Society of America Bulletin*, **88**, 1305–1320.

Bakun-Czubarow, N. 1991. On the possibility of occurrence of quartz pseudomorphs after coesite in the eclogite–granulite rock series of the Zlote Mountains in the Sudetes (SW Poland) *Archiwum Mineralogiczne*, **XLVII**, 5–16.

—— 1998. Ilmenite-bearing eclogites of the West Sudetes—their geochemistry and mineral chemistry. *Archiwum Mineralogiczne*, **LI**, 29–99.

Beard, B. L., Medaris, L. G. Jr, Johnson, C. M., Brueckner, H. K. & Mísař, Z. 1992. Petrogenesis of Variscan high-temperature Group A eclogites from the Moldanubian Zone of the Bohemian Massif, Czechoslovakia. *Contributions to Mineralogy and Petrology*, **111**, 468–483.

——, ——, ——, Jelínek, E., Tonika, J. & Riciputi, L. R. 1995. Geochronology and geochemistry of eclogites from the Mariánské Lázné Complex, Czech Republic: implications for Variscan orogenesis. *Geologische Rundschau*, **84**, 552–567.

Becker, H. 1996*a*. Crustal trace element and isotopic signatures in garnet pyroxenites from garnet peridotite massifs from lower Austria. *Journal of Petrology*, **37**, 785–810.

—— 1996*b*. Geochemistry of garnet peridotite massifs from lower Austria and the composition of the deep lithosphere beneath a Palaeozoic convergent margin. *Chemical Geology*, **134**, 49–65.

—— 1997. Sm–Nd garnet ages and cooling history of high-temperature garnet peridotite massifs and high-pressure granulites from lower Austria. *Contributions to Mineralogy and Petrology*, **127**, 224–236.

Behr, H.-J., Engel, W., Franke, W., Giese, P. & Weber, K. 1984. The Variscan belt in central Europe: main structures, geodynamic implications, open questions. *Tectonophysics*, **109**, 15–40.

Bernard-Griffiths, J. & Cornichet, J. 1985. Origin of eclogites from southern Brittany, France: a Sm–Nd isotopic and REE study. *Chemical Geology*, **52**, 185–201.

——, Peucat, J. J., Cornichet, J., Iglesias, M. & Gil-Ibarguchi, I. 1985. U–Pb, Nd isotope and REE geochemistry in eclogites from the Cabo

Ortegal complex, Galicia, Spain: an example of REE immobility conserving MORB-like patterns during high-grade metamorphism. *Chemical Geology*, **52**, 217–225.

BIRD, P. 1978. Initiation of intra-continental subduction in the Himalaya. *Journal of Geophysical Research*, **83**, 4975–4987.

BLUNDELL, D., FREEMAN, R. & MUELLER, S. 1992. *A Continent Revealed. The European Geotraverse*. Cambridge University Press, Cambridge.

BONHOMME, M. G. & FLUCK, P. 1981. Nouvelles données isotopiques Rb–Sr sur les granulites des Vosges—âge protérozoique terminal de la série volcanique calc-alcaline et âge acadien du métamorphisme régional. *Comptes Rendus de l'Académie des Sciences*, **293**, 771–774.

BOUCHARDON, J.-L., SANTALLIER, D., BRIAND, B., MÉNOT, R.-P. & PIBOULE, M. 1989. Eclogites in the French Palaeozoic Orogen: geodynamic significance. *Tectonophysics*, **169**: 317–332.

BROWN, M. & DALLMEYER, R. D. 1996. Rapid Variscan exhumation and the role of magma in core complex formation: southern Brittany metamorphic belt, France. *Journal of Metamorphic Geology*, **14**, 361–379.

BRUECKNER, H. K. & MEDARIS, L. G. JR 1998. A tale of two orogens: the contrasting *P–T–t* history and geochemical evolution of mantle in high- and ultrahigh-pressure metamorphic terranes of the Norwegian Caledonides and the Czech Variscides. *Schweizerische Mineralogische und Petrographische Mitteilungen*, **78**, 293–307.

BURG, J. P. & MATTE, P. 1978. A cross section through the French Massif Central and the scope of its Variscan geodynamic evolution. *Zeitschrift der Deutschen Geologischen Gesellschaft*, **129**, 429–460.

——, LEYRELOUP, A., MARCHAND, J. & MATTE, P. 1984. Inverted metamorphic zonation and large-scale thrusting in the Variscan belt: an example in the French Massif Central. *In*: HUTTON, D. H. W. & SANDERSON, D. J. (eds) *Variscan Tectonics of the North Atlantic Region*. Geological Society, London, Special Publications, **14**, 47–61.

CARSWELL, D. A. & JAMTVEIT, B. 1990. Variscan Sm–Nd ages for the high-pressure metamorphism in the Moldanubian Zone of the Bohemian Massif in Lower Austria. *Neues Jahrbuch für Mineralogie, Abhandlungen*, **162**, 69–78.

—— & O'BRIEN, P. J. 1993. Thermobarometry and geotectonic significance of high pressure granulites: examples from the Moldanubian Zone of the Bohemian Massif in Lower Austria. *Journal of Petrology*, **34**, 427–459.

CHÁB, J. & ZÁCEK, V. 1994. Metamorphism of the Teplá Crystalline Complex, *KTB Report*, **94-3**, 33–37.

CHALOUPSKÝ, J. 1989. Major tectonostratigraphic units of the Bohemian Massif. *In*: DALLMEYER, R. D. (ed.) *Terranes in the Circum-Atlantic Palaeozoic Orogens*. Geological Society of America, Special Papers, **230**, 101–114.

CLOOS, H. 1927. Zur Frage des Deckenbaus in Schlesien und im Fichtelgebirge. *Geologische Rundschau*, **18**, 221–225.

COCKS, L. R. M. & FORTEY, R. A. 1982. Faunal evidence for oceanic separation in the Palaeozoic of Britain. *Journal of the Geological Society, London*, **139**, 465–478.

COSTA, S., MALUSKI, H. & LARDEAUX, J.-M. 1993. ^{40}Ar–^{39}Ar chronology of Variscan tectono-metamorphic events in an exhumed crustal nappe: the Monts du Lyonnais complex (Massif Central, France). *Chemical Geology*, **105**, 339–359.

DALLMEYER, R. D., NEUBAUER, F. & HÖCK, V. 1992. Chronology of late Palaeozoic tectonothermal activity in the southeastern Bohemian Massif, Austria (Moldanubian and Moravo-Silesian zones): $^{40}Ar/^{39}Ar$ mineral age controls. *Tectonophysics*, **210**, 135–153.

DAVIES, J. H. & VON BLANCKENBURG, F. 1995. Slab breakoff: a model of lithospheric detachment and its test in the magmatism and deformation of collisional orogens. *Earth and Planetary Science Letters*, **129**, 85–102.

DUDEK, A. & FEDIUKOVÁ, E. 1974. Eclogites of the Bohemian Moldanubicum. *Neues Jahrbuch für Mineralogie, Abhandlungen*, **121**, 127–159.

ECHTLER, H. & MALAVIEILLE, J. 1990. Extensional tectonics, basement uplift and Stephano-Permian collapse basin in a late Variscan metamorphic core complex (Montagne Noire, Southern Massif Central). *Tectonophysics*, **177**, 125–138.

FABRIÈS, J. & LATOUCHE, L. 1988. Granulite facies conditions in the Sainte Marie aux Mines 'Varied Group' (Central Vosges, France). *Terra Cognita*, **8**, 249.

FIALA, J. 1995. General characteristics of the Moldanubian Zone. *In*: DALLMEYER, R. D., FRANKE, W. & WEBER, K. (eds) *Pre-Permian Geology of Central and Eastern Europe*, Springer, Berlin, 417–418.

——, MATEJOVSKÁ, O. & VANKOVÁ, V. B. 1987. Moldanubian granulites: source material and petrogenetic considerations. *Neues Jahrbuch für Mineralogie, Abhandlungen*, **157**, 133–165.

FRANKE, W. 1984. Variszischer Deckenbau im Raume der Münchberger Gneismasse—abgeleitet aus der Fazies, Deformation und Metamorphose im umgebenden Paläozoikum. *Geotektonische Forschungen*, **68**, 1–253.

—— 1989. Tectonostratigraphic units in the Variscan belt of central Europe. *In*: DALLMEYER, R. D. (ed.) *Terranes in the Circum-Atlantic Palaeozoic Orogens*. Geological Society of America, Special Papers, **230**, 67–90.

—— & STEIN, E. 2000. Exhumation of high-grade rocks in the Saxo-Thuringian Belt; geological constraints and geodynamic concepts. *This volume*.

FRANZ, G., THOMAS, S. & SMITH, D. C. 1986. High pressure phengite decomposition in the Weissenstein eclogite, Münchberger Gneiss Massif, Germany. *Contributions to Mineralogy and Petrology*, **92**, 71–85.

FRIEDL, G., VON QUADT, A. & FINGER, F. 1994. 340 Ma U/Pb-Monazitalter aus dem niederösterreichen Moldanubikum und ihre geologische Bedeutung. *Terra Nostra*, **94**(3), 43–46.

——, ——, OCHSNER, A. & FINGER, F. 1993. Timing of the Variscan orogeny in the Southern Bohemian Massif (NE Austria) deduced from new U–Pb zircon and monazite dating. *Terra Nova*, **5**, 235–236.

FUCHS, G. 1995. The Austrian part of the Moldanubicum. *In*: DALLMEYER, R. D., FRANKE, W. & WEBER, K. (eds) *Pre-Permian Geology of Central and Eastern Europe*, Springer, Berlin, 422–426.

GEBAUER, D. 1986. The development of the continental crust of the European Variscides since the Archean based on radiometric data. *In*: FREEMAN, R., MUELLER, S. & GIESE, P. (eds) *Proceedings of the Third Workshop on the European Geotraverse (EGT) Project. The Central Segment*. European Science Foundation, Strasbourg, 15–23.

—— & GRÜNENFELDER, M. 1979. U–Pb zircon and Rb–Sr mineral dating of eclogites and their country rocks. Example: Münchberg Gneiss Massif, northeast Bavaria. *Earth and Planetary Science Letters*, **42**, 35–44.

—— & —— 1982. Geological development of the Hercynian belt of Europe based on age and origin of high-grade and high-pressure mafic and ultramafic rocks. *Fifth International Conference on Geochronology, Cosmochronology, Isotope Geology*, Nikko National Park, Japan, 111–112.

GERDES, A., WÖRNER, G. & FINGER, F. 1996. Mantelquellen in variscischen Granitoiden? *Terra Nostra*, **96**(2), 75–78.

——, —— & —— 2000. Hybrids, magma mixing and enriched mantle melts in post-collisional Variscan granitoids: the Rastenberg pluton, Austria. *This volume*.

——, —— & HENK, A. 1998. Geochemical and thermal constraints on the origin of granites in the Southern Moldanubian zone. *Terra Nostra*, **98**(2), 58–61.

GIL IBARGUCHI, J. I. & ORTEGA GIRONES, E. 1985. Petrology, structure and geotectonic implications of glaucophane-bearing eclogites and related rocks from the Malpica–Tuy unit, Galicia, Northwest Spain. *Chemical Geology*, **50**, 145–162.

——, MENDIA, M., GIRARDEAU, J. & PEUCAT, J. J. 1990. Petrology of eclogites and clinopyroxene–garnet metabasites from the Cabo Ortegal complex (northwest Spain). *Lithos*, **25**, 133–162.

GOLDFUSS, A. & BISCHOF, G. 1817. *Physikalisch-statistische Beschreibung des Fichtelgebirges*. Nürnberg.

GUIRAUD, M., BURG, J. P. & POWELL, R. 1987. Evidence for a Variscan suture in the Vendée, France: a petrological study of blueschist gacies rocks from Bois de Cené. *Journal of Metamorphic Geology*, **5**, 225–237.

HEGNER, E., KÖLBL-EBERT, M. & LOESCHKE, J. 1998. Post-collisional Variscan lamprophyres (Black Forest, Germany): $^{40}Ar/^{39}Ar$ phlogopite dating, Nd, Pb, Sr isotope, and trace element characteristics. *Lithos*, **45**, 395–411.

——, WALTER, H. J. & SATIR, M. 1995. Pb–Sr–Nd isotopic compositions and trace element geochemistry of megacrysts and melilitites from the Tertiary Urach volcanic field: source compositions of small volume melts under SW Germany. *Contributions to Mineralogy and Petrology*, **122**, 322–335.

HINTERLECHNER-RAVNIK, A. 1971. The metamorphic rocks of Pohorje (English summary). *Geologija Razprave Porocila*, **14**, 187–226

HODGES, K. V., BURCHFIEL, B. C., ROYDEN, L. H., CHEN, Z. & LIEU, Y. 1993. The metamorphic signature of contemporaneous extension and shortening in the central Himalayan orogen: data from the Nyalam transect, southern Tibet. *Journal of Metamorphic Geology*, **11**, 721–737.

HOERNLE, K., ZHANG, Y.-S. & GRAHAM, D. 1995. Seismic and geochemical evidence for large scale mantle upwelling beneath the eastern Atlantic and western and central Europe. *Nature*, **374**, 34–39.

HOUSEMAN, G. A., MCKENZIE, D. P. & MOLNAR, P. 1981. Convective instability of a thickened boundary layer and its relevance for the thermal evolution of continental convergent belts. *Journal of Geophysical Research*, **86**, 6115–6132.

JELÍNEK, E., PACESOVÁ, M., MÍSAŘ, Z., MARTINEC, P. & WEISS, Z. 1984. Geochemistry of a dismembered metaophiolite complex, Letovice, Czechoslovakia. *Transactions of the Royal Society of Edinburgh: Earth Sciences*, **75**, 37–48.

JONES, K. & BROWN, M. 1990. High-temperature 'clockwise' *P–T* paths and melting in the development of regional migmatites: an example from southern Brittany, France. *Journal of Metamorphic Geology*, **8**, 551–578.

KALT, A. & ALTHERR, R. 1996. Metamorphic evolution of garnet–spinel peridotites from the Variscan Schwarzwald (F.R.G.). *Geologische Rundschau*, **85**, 211–224.

——, —— & HANEL, M. 1995. Contrasting *P–T* conditions recorded in ultramafic high-pressure rocks from the Variscan Schwarzwald (F.R.G.). *Contributions to Mineralogy and Petrology*, **121**, 45–60.

——, BERGER, A. & BLÜMEL, P. 1999. Metamorphic evolution of cordierite-bearing migmatites from the Bayerische Wald (Variscan Belt, Germany). *Journal of Petrology*, **40**, 601–627.

——, CORFU, F. & WIJBRANS, J. 1998. Time calibration of a *P–T* path from a Variscan high-temperature low-pressure metamorphic complex (Bayerische Wald, Germany). *Terra Nostra*, **98**(2), 80–81.

——, GRAUERT, B. & BAUMANN, A. 1994. Rb–Sr and U–Pb isotope studies on migmatites from the Schwarzwald (Germany): constraints on isotopic resetting during Variscan high-temperature metamorphism. *Journal of Metamorphic Geology*, **12**, 667–680.

KLEEMANN, U. 1991. *Die* P–T–t–d-*Entwicklung im Grenzbereich zwischen der Zone von Erbendorf–Vohenstrauss (ZEV) und dem Moldanubikum in der Oberpfalz/NE-Bayern*. PhD thesis, University of Bochum.

KLEMD, R. 1989. *P–T* conditions and fluid inclusion characteristics of retrograded eclogites, Münchberger Gneiss Massif, Germany. *Contributions to Mineralogy and Petrology*, **102**, 221–229.

KOSSMAT, F. 1927. Gliederung des variszischen Gebirgsbaues. *Abhandlungen des sächsischen geologischen Landesamtes*, **1**, 1–39.

KOTKOVÁ, J. & HARLEY, S. 1999. Formation and evolution of high-pressure leucogranulites: experimental constraints and unresolved issues. *Physics and Chemistry of the Earth*, **24**, 299–304.

——, KRÖNER, A., TODT, W. & FIALA, J. 1996. Zircon dating of North Bohemian granulites, Czech Republic: further evidence for the Lower Carboniferous high-pressure event in the Bohemian Massif. *Geologische Rundschau*, **85**, 154–161.

KREUZER, H., SEIDEL, E., SCHÜSSLER, U., OKRUSCH, M., LENZ, K.-L. & RASCHKA, H. 1989. K–Ar geochronology of different tectonic units at the northwestern margin of the Bohemian Massif. *Tectonophysics*, **157**, 149–178.

KROHE, A. 1998. Extending a thickened crustal bulge: towards a new geodynamic evolution model of the Paleozoic NW Bohemian Massif, German Continental Deep Drilling site (SE Germany). *Earth-Science Reviews*, **44**, 95–145.

KRÖNER, A. & WILLNER, A. P. 1995. Magmatische und metamorphe Zirkonalter für Quarz–Feldspat-Gesteine der Gneis–Eklogit-Einheit des Erzgebirges. *Terra Nostra*, **95**(8), 112.

——, JAECKEL, P., REISCHMANN, T. & KRONER, U. 1998. Further evidence for an early Carboniferous (~340 Ma) age of high-grade metamorphism in the Saxonian granulite complex. *Geologische Rundschau*, **86**, 751–766.

——, O'BRIEN, P. J., PIDGEON, R. T. & FIALA, J. 1996. SHRIMP zircon ages for HP–HT granulites from southern Bohemia. *Terra Nostra*, **96**(2), 131–132.

——, WENDT, I., LIEW, T. C. *et al.* 1988. U–Pb zircon and Sm–Nd model ages of high-grade Moldanubian metasediments, Bohemian Massif, Czechoslovakia. *Contributions to Mineralogy and Petrology*, **99**, 257–266.

KRYZA, R., MUSZYNSKI, A. & VIELZEUF, D. 1990. Glaucophane-bearing assemblages overprinted by greenschist-facies metamorphism in the Variscan Kaczawa complex, Sudetes, Poland. *Journal of Metamorphic Geology*, **8**, 345–355.

——, PIN, C. & VIELZEUF, D. 1996. High-pressure granulites from the Sudetes (south-west Poland): evidence of crustal subduction and collisional thickening in the Variscan Belt. *Journal of Metamorphic Geology*, **14**, 531–546.

LASNIER, B. 1971. Les péridotites et pyroxénolites à grenat des Bois de Feuilles (Monts du Lyonnais, France). *Contributions to Mineralogy and Petrology*, **34**, 29–42.

LE BRETON, N. & THOMPSON, A. B. 1988. Fluid-absent (dehydration) melting of biotite in metapelites in the early stages of crustal anatexis. *Contributions to Mineralogy and Petrology*, **99**, 226–237.

LEDRU, P., LARDEAUX, J.-M., SANTALLIER, D. *et al.* 1989. Où sont les nappes dans le Massif central français? *Bullétin de la Societé Géologique de la France*, **8**(5), 605–618.

LINNER, M. 1996. Metamorphism and partial melting of paragneisses of the Monotonous Group, SE Moldanubicum (Austria). *Mineralogy and Petrology*, **58**, 215–234.

MCKENZIE, D. P. & BICKLE, M. J. 1988. The volume and composition of melt generated by extension of the lithosphere. *Journal of Petrology*, **29**, 625–667.

MALAVIEILLE, J., GUIHOT, P., COSTA, S., LARDEAUX, J. M. & GARDIEN, V. 1990. Collapse of thickened Variscan crust in the French Massif Central: Mont Pilat extensonal shear zone and St. Etienne Carboniferous basin. *Tectonphysics*, **177**, 139–149.

MATTE, P. 1991. Accretionary history and crustal evolution of the Variscan Belt in western Europe. *Tectonophysics*, **196**, 309–337.

—— 1998. Continental subduction and exhumation of HP rocks in Palaeozoic orogenic belts: Uralides and Variscides. *Geologiska Föreningens i Stockholm Förhandlingar*, **120**, 209–222.

—— & BURG, J. P. 1981. Sutures, thrusts and nappes in the Variscan Arc of western Europe: plate tectonic implications. *In*: MCCLAY, K. R. & PRICE, N. J. (eds) *Thrust and Nappe Tectonics*. Geological Society, London, Special Publications, **9**, 353–358.

——, MALUSKI, H., RAJLICH, P. & FRANKE, W. 1990. Terrane boundaries in the Bohemian Massif: Results of large scale Variscan shearing. *Tectonophysics*, **177**, 151–170.

MATTHES, S., RICHTER, P. & SCHMIDT, K. 1975. Die Eklogitvorkommen des kristallinen Grundgebirges in NE-Bayerns, IX. Petrographie, Geochemie und Petrogenese der Eklogite des Münchberger Gneisgebietes. *Neues Jahrbuch für Mineralogie, Abhandlungen*, **126**, 45–86.

MEDARIS, L. G. JR & CARSWELL, D. A. 1990. Petrogenesis of Mg–Cr garnet peridotites in European metamorphic belts. *In*: CARSWELL, D. A. (ed.) *Eclogite Facies Rocks*. Blackie, Glasgow, 260–290.

——, BEARD, B. L., JOHNSON, C. M., VALLEY, J. W., SPICUZZA, M. J., JELÍNEK, E. & MÍSAŘ, Z. 1995*a*. Garnet pyroxenite and eclogite in the Bohemian Massif: geochemical evidence for Variscan recycling of subducted lithosphere. *Geologische Rundschau*, **84**, 489–505.

——, JELÍNEK, E. & MÍSAŘ, Z. 1995*b*. Czech eclogites: terrane settings and implications for Variscan tectonic evolution of the Bohemian Massif. *European Journal of Mineralogy*, **7**, 7–28.

——, WANG, H. F., MÍSAŘ, Z. & JELÍNEK, E. 1990. Thermobarometry, diffusion modelling and cooling rates of crustal garnet peridotites: two examples from the Moldanubian Zone of the Bohemian Massif. *Lithos*, **25**, 189–202.

MERCIER, J.-C. C., POZZO DI BORGO, M., FRISON, J. Y. & GIRARDEAU, J. 1982. Les associations basiques et ultrabasiques du Bas-Limousin, restes d'un complexe ophiolitique démembré d'une fraicheur remarquable. *9ème Réunion Annales Science de la Terre, Paris*, 430.

MIELKE, H., BLÜMEL, P. & LANGER, K. 1979. Regional low-pressure metamorphism of low and medium grade in metapelites and psammites of the Fichtelgebirge area, NE Bavaria. *Neues Jahrbuch für Mineralogie, Abhandlungen*, **137**, 83–112.

MONTEL, J. M., MARIGNAC, C., BARBEY, P. & PICHAVANT, M. 1992. Thermobarometry and granite genesis: the Hercynian low-*P*, high-*T* Velay anatectic dome (French Massif Central). *Journal of Metamorphic Geology*, **10**, 1–15.

OBATA, M. & MORTEN, L. 1987. Transformation of spinel lherzolite to garnet lherzolite in ultramafic lenses of the Austridic Crystalline Complex, Northern Italy. *Journal of Petrology*, **28**, 599–623.

O'BRIEN, P. J. 1993. Partially retrograded eclogites of the Münchberg Massif, Germany: records of a multistage Variscan uplift history in the Bohemian Massif. *Journal of Metamorphic Geology*, **11**, 241–260.

—— 1997*a*. Garnet zoning and reaction textures in overprinted eclogites, Bohemian Massif, European Variscides: a record of their thermal history during exhumation. *Lithos*, **41**, 119–133

—— 1997*b*. Granulite facies overprints of eclogites: short-lived events deduced from diffusion modelling. *In*: QIAN, X., YOU, Z., JAHN, B-M., HALLS, H. C. (eds) *Precambrian Geology and Metamorphic Petrology. Proceedings, 30th International Geological Congress*, **17**, Part II, 157–171.

—— 1999. Asymmetric zoning profiles in garnet from HP–HT granulite and implications for volume and grain boundary diffusion. *Mineralogical Magazine*, **63**, 227–238.

—— & CARSWELL, D. A. 1993. Tectonometamorphic evolution of the Bohemian Massif: evidence from high pressure metamorphic rocks. *Geologische Rundschau*, **82**, 531–555.

——, —— & GEBAUER, D. 1990. Eclogite formation and distribution in the European Variscides. *In*: CARSWELL, D. A. (ed.) *Eclogite Facies Rocks*. Blackie, Glasgow, 204–224.

——, DUYSTER, J., GRAUERT, B., SCHREYER, W., STÖCKHERT, B. & WEBER, K. 1997*a*. Crustal evolution of the KTB drill site: from oldest relics to the late Hercynian granites. *Journal of Geophysical Research*, **102**, 18203–18220.

——, KRÖNER, A., JAEKEL, P., HEGNER, E., ZELAZNIEWICZ, A. & KRYZA, R. 1997*b*. Petrological and isotopic studies on Palaeozoic high pressure granulites, Góry Sowie (Owl) Mts, Polish Sudetes. *Journal of Petrology*, **38**, 433–456.

——, ZOTOV, N., LAW, R., KHAN, M. A. & JAN, M. Q. 1999. Coesite in eclogite from the Upper Kaghan Valley, Pakistan: a first record and implications. *Terra Nostra*, **99**(2), 109–111.

OKRUSCH, M., SEIDEL, E., SCHÜSSLER, U. & RICHTER, P. 1989. Geochemical characteristics of metabasites in different tectonic units of the northeast Bavarian basement. *In*: EMMERMANN, R. & WOHLENBERG, J. (eds) *The Continental Deep Drilling Program (KTB)*. Springer Verlag, Berlin, 67–79

PAQUETTE, J. L. 1987. Comportement des systèmes isotopiques U–Pb et Sm–Nd dans le métamorphisme éclogitique. Chaîne hercynienne et chaîne alpine. Thèse, Rennes. *Mémoires, Centre Armoricain d'Études Structurales de Socles*, **14**, 1–189.

PETRAKAKIS, K. 1997. Evolution of Moldanubian rocks in Austria: review and synthesis. *Journal of Metamorphic Geology*, **15**, 203–222.

PEUCAT, J. J. 1986. Rb–Sr and U–Pb dating of the blueschists of the Île de Groix (France). In: Evans, B. W. & Brown, E. H. (eds) *Blueschists and Eclogites*. Geological Society of America, Memoirs, **164**, 229–238.

——, BERNARD-GRIFFITHS, J., GIL IBARGUCHI, J. I., DALLMEYER, R. D., MENOT, R. P., CORNICHET, J. & IGLESIAS PONCE DE LEON, M. 1990. Geochemical and geochronological cross section of the deep Variscan crust: the Cabo Ortegal high-pressure nappe (northwestern Spain). *Tectonophysics*, **177**, 263–292.

PFLUG, H. D. & REITZ, E. 1987. Palynology in metamorphic rocks: indication of early land plants. *Naturwissenschaften*, **74**, 386–387.

PIN, C. 1990. Variscan oceans: ages, origins and geodynamic implications inferred from geochemical and radiometric data. *Tectonophysics*, **177**, 215–227.

—— & VIELZEUF, D. 1983. Granulites and related rocks in Variscan median Europe: a dualistic interpretation. *Tectonophysics*, **93**, 47–74.

PLATT, J. P. & ENGLAND, P. C. 1993. Convective removal of lithosphere beneath mountain belts: thermal and mechanical consequences. *American Journal of Science*, **293**, 307–336.

QUESADA, C. 1991. Geological constraints on the Paleozoic tectonic evolution of tectonostratigraphic terranes in the Iberian Massif. *Tectonophysics*, **185**, 225–245.

RECHE, J., MARTÍNEZ, F. J. & ARBOLEYA, M. L. 1998. Low- to medium-pressure Variscan metamorphism in Galicia (NW Spain). *In*: TRELOAR, P. J. & O'BRIEN, P. J. (eds) *What Drives Metamorphism and Metamorphic Reactions?* Geological Society, London, Special Publications, **138**, 61–79.

REINHARDT, J. & KLEEMANN, U. 1994. Extensional unroofing of granulitic lower crust and related low-pressure high-temperature metamorphism in the Saxonian Granulite Massif, Germany. *Tectonophysics*, **238**, 71–94.

RIES, A. C. & SHACKLETON, R. M. 1971. Catazonal complexes of North-West Spain and North Portugal, remnants of a Hercynian thrust plate. *Nature, Physical Science*, **234**, 65–68.

RÖTZLER, J. 1992. Zur Petrogenese im sächsischen Granulitgebirge. Die pyroxenfreien Granulite und die Metapelite. *Geotektonische Forschungen*, **77**, 1–100.

SCHÄFER, J., NEUROTH, H., AHRENDT, H., DÖRR, W. & FRANKE, W. 1997. Accretion and exhumation at a Variscan active margin, recorded in the Saxo-Thuringian flysch. *Geologische Rundschau*, **86**, 599–611.

SCHARBERT, H. G. & KURAT, G. 1974. Distribution of some elements between ferromagnesian minerals in some Moldanubian granulite facies rocks. *Tschermaks Mineralogische und Petrographische Mitteilungen*, **21**, 110–134.

SCHENK, V. & TODT, W. 1983. U–Pb Datierungen an Zirkon und Monazit der Granulit im Moldanubikum Niederösterreichs. *Fortschritte der Mineralogie*, **61**, 190–191.

SCHERMERHORN, L. J. G. & KOTSCH, S. 1984. First occurrence of lawsonite in Portugal and tectonic implications. *Communicaçoes dos Serviços Geologicos de Portugal*, **70**, 23–29.

SCHMÄDICKE, E. 1991. Quartz pseudomorphs after coesite in eclogites from the Saxonian Erzgebirge. *European Journal of Mineralogy*, **3**, 231–238.

SÖLLNER, F., KÖHLER, H. & MÜLLER-SOHNIUS, D. 1981. Rb/Sr-Altersbestimmungen an Gesteinen der Münchberger Gneismasse (MM), NE-Bayern. Teil 1, Gesamtgesteinsdatierungen. *Neues Jahrbuch für Mineralogie, Abhandlungen*, **141**, 90–112.

STETTNER, G. 1990. *KTB Umfeldgeologie*. Bayerisches Geologisches Landesamt, München.

STOSCH, H.-G. & LUGMAIR, G. W. 1986. Trace element and Sr and Nd isotope geochemistry of peridotite xenoliths from the Eifel (West Germany) and their bearing on the evolution of the subcontinental mantle. *Earth and Planetary Science Letters*, **80**, 281–298.

—— & —— 1990. Geochemistry and evolution of MORB-type eclogites from the Münchberg Massif, southern Germany. *Earth and Planetary Science Letters*, **99**, 230–249.

SUESS, F. E. 1913. Vorläufige Mitteilung über die Münchberger Deckscholle. *Anzeiger der K Akademie der Wissenschaften, Mathematische-naturwissenschafliche Klasse*, **50**, 255–258.

TAIT, J. A., BACHTADSE, V., FRANKE, W., SOFFEL, H. C. 1997. Geodynamic evolution of the European Variscan fold belt: palaeomagnetic and geological constraints. *Geologische Rundschau*, **86**, 585–598.

——, SCHÄTZ, M., BACHTADSE, V. & SOFFEL, H. 2000. Palaeomagnetism and Palaeozoic palaeogeography of Gondwana and European terranes. *This volume*.

TANNER, D. C., SCHUSTER, J., BEHRMANN, J. & O'BRIEN, P. J. 1993. New clues to the Moldanubian puzzle: structural and petrological observations from the Waldmünchen area, eastern Bavaria. *KTB Report*, **93-2**, 97–102.

TOLLMANN, A. 1982. Großräumiger variszischer Deckenbau im Moldanubikum und neue Gedanken zum Variszikum Europas. *Geotektonische Forschungen*, **64**, 1–91.

URBAN, M. 1992. Kinematics of the Variscan thrusting in the Eastern Moldanubicum (Bohemian Massif, Czechoslovakia): evidence from the Namest granulite massif. *Tectonophysics*, **201**, 371–391.

VAN BREEMEN, O., AFTALION, M., BOWES, D. R., DUDEK, A., MÍSAŘ, Z., POVONDRA, P. & VRÁNA, S. 1982. Geochronological studies of the Bohemian Massif, Czechoslovakia, and their significance in the evolution of Central Europe. *Transactions of the Royal Society of Edinburgh: Earth Sciences*, **73**, 89–108.

VAN DER VOO, R., BRIDEN, J. C. & DUFF, A. 1980. Late Precambrian and Palaeozoic palaeomagnetism of the Atlantic-bordering continents. *In*: COGNÉ, J. & SLANSKY, M. (eds) *Géologie de l'Europe*. Mémoires du BRGM, **108**, 203–212.

VEJNAR, Z. 1966. The petrogenetic interpretation of kyanite, sillimanite and andalusite in the southwestern Bohemia crystalline complexes. *Neues Jahrbuch für Mineralogie, Abhandlungen*, **104**, 172–189.

VELLMER, C. 1992. *Stoffbestand und Petrogenese von Granuliten und granitischen Gesteinen der südlichen Böhmischen Masse in Niederösterreich*. PhD thesis, University of Göttingen.

VON GAERTNER, H. R., VON HORSTIG, G., STETTNER, G. & WURM, A. 1968. Saxo-Thuringikum in Bavaria. *In*: *Guide to Excursion C34, International Geological Congress, 23rd session, Prague*, Bundesanstalt für Bodenforschung, Hannover, 1–160.

VON QUADT, A. 1993. The Saxonian Granulite Massif: new aspects from geochronological studies. *Geologische Rundschau*, **82**, 516–530.

VRÁNA, S. 1989. Perpotassic granulites from southern Bohemia. A new rock-type derived from partial melting of crustal rocks under upper mantle conditions. *Contributions to Mineralogy and Petrology*, **103**, 510–522.

—— & JAKEŠ, P. 1982. Orthopyroxene and two-pyroxene granulites from a segment of charnockitic crust in southern Bohemia. *Bulletin of the Geological Survey, Prague*, **57**, 129–143.

WATERS, D. J. & WHALES, C. J. 1984. Dehydration melting and the granulite transition in metapelites from southern Namaqualand, S. Africa. *Contributions to Mineralogy and Petrology*, **88**, 269–275.

WENDT, J. I., KRÖNER, A., FIALA, J. & TODT, W. 1994. U–Pb zircon and Sm–Nd dating of Moldanubian HP/HT granulites from South Bohemia. *Journal of the Geological Society, London*, **151**, 83–90.

WEISS, C. S. 1803. Über die Gebirgsart des sächsichen Erzgebirges, welche unter dem Namen Weiss-Stein neuerlich bekannt gemacht worden ist. *Neue Schriften Gesellschaft Naturforschender Freunde*, **4**, 342–366.

WENZEL, T., MERTZ, D. F., OBERHÄNSLI, R., BECKER, T. & RENNE, P. R. 1997. Age, geodynamic setting, and mantle enrichment processes of a K-rich intrusion from the Meissen Massif (northern Bohemian Massif) and implications for related occurrences from the mid-European Hercynian. *Geologische Rundschau*, **86**, 556–570.

WILLNER, A. P., RÖTZLER, K. & MARESCH, W. V. 1997. Pressure–temperature and fluid evolution of quartzo-feldspathic metamorphic rocks with a relic high-pressure, granulite-facies history from the Central Erzgebirge (Saxony, Germany). *Journal of Petrology*, **38**, 307–336.

WILSON, M. & DOWNES, H. 1991. Tertiary–Quaternary extension-related alkaline magmatism in Western and central Europe. *Journal of Petrology*, **32**, 811–849.

WÖRNER, G., ZINDLER, A., STAUDIGEL, H. & SCHMINKE, H. U. 1986. Sr, Nd, and Pb isotope geochemistry of Tertiary and Quaternary alkaline volcanics from West Germany. *Earth and Planetary Science Letters*, **79**, 107–119.

ZAGANA, N. A., DOWNES, H., THIRLWALL, M. F. & HEGNER, E. 1997. Relationship between deformation, equilibrium temperatures, REE and radiogenic isotopes in mantle xenoliths (Ray Pic, Massif Central, France): an example of plume–lithosphere interaction? *Contributions to Mineralogy and Petrology*, **127**, 187–203.

ZULAUF, G., DÖRR, W., FIALA, J. & VEJNAR, Z. 1997. Late Cadomian crustal tilting and Cambrian transtension in the Teplá–Barrandian unit (Bohemian Massif, Central European Variscides). *Geologische Rundschau*, **86**, 571–584.

ZWART, H. J. 1967. The duality of orogenic belts. *Geologie en Mijnbouw*, **46**, 283–309.

Syn-convergent high-temperature metamorphism and magmatism in the Variscides: a discussion of potential heat sources

A. HENK[1], F. von BLANCKENBURG[2], F. FINGER[3], U. SCHALTEGGER[4] & G. ZULAUF[5]

[1]*Institut für Geologie, Universität Würzburg, Pleicherwall 1, D-97070 Würzburg, Germany (e-mail: andreas.henk@mail.uni-wuerzburg.de)*

[2]*Mineralogisch-Petrographisches Institut, Abt. für Isotopengeologie, Universität Bern, Erlachstr. 9a, CH-3012 Bern, Switzerland*

[3]*Institut für Mineralogie, Universität Salzburg, Hellbrunnerstr. 34, A-5020 Salzburg, Austria*

[4]*Institut für Isotopengeochemie und Mineralische Rohstoffe, ETH-Zentrum NO, Sonneggstr. 5, CH-8092 Zürich, Switzerland*

[5]*Institut für Geologie und Mineralogie, Universität Erlangen–Nürnberg, Schloßgarten 5, D-91054 Erlangen, Germany*

Abstract: A period of pervasive high-temperature metamorphism and igneous activity from 340 to 325 Ma is a well-established characteristic of the Variscan Orogen of Central Europe. During this stage, the internal zone of the orogen was virtually soaked by granitic to granodioritic magmas. Petrological data point to temperatures of 600–850 °C at upper- to mid-crustal levels. These elevated temperatures occurred during the final convergence stage and may be comparable with similar processes inferred from geophysical evidence for the present-day Tibetan Plateau, in both regional extent and significance for the orogen's evolution. We review various geodynamic scenarios that may have provided the heat for melting and metamorphism, and compare model predictions with field data from the Variscides. All lines of evidence point to a geodynamic scenario that led to thickening of the continental crust with increased internal radiogenic heating, but without simultaneous thickening of the mantle lithosphere. Possible mechanisms include convective removal of the thermal boundary layer, delamination of part of the lithospheric mantle, and subduction of the mantle lithosphere of the downgoing plate. However, with the present stage of knowledge it is virtually impossible to single out one of these three mechanisms, as their geological consequences are so similar.

The Paleozoic Variscan Belt of Europe is a collisional orogen that has been pervasively affected by late- and post-convergent processes, in particular high-temperature (HT) metamorphism, magmatism and formation of extensional basins (e.g. Matte 1991; Franke *et al.* 1995). If compared with recent collision zones, these processes are probably matched only by the Himalayan Orogen and the Tibetan Plateau in regional extent and significance. Therefore it is no surprise that the Variscan Orogen has often been likened to the Tibetan Plateau, or the Basin and Range Province (e.g. Menard & Molnar 1988). Seismic surveys across part of the India–Asia collision zone have suggested that southern Tibet is at present underlain by a partially molten middle crust (Nelson *et al.* 1996). As such, the Variscides offer the opportunity to study the geological consequences of these processes at a deeply exhumed crustal level, at which in the Himalayas only geophysical data are available.

In recent years, a major effort of European researchers has been focused on refining our knowledge on the evolution of the European Variscan Orogen (e.g. this volume). One of the outstanding results is the remarkably narrow time window within which an HT event imprinted both the evolution of the Variscan Orogen as well as the crustal structure of most of Central Europe. This pervasive HT event is characterized not only by its intensity, documented by HT–LP metamorphism and

From: Franke, W., Haak, V., Oncken, O. & Tanner, D. (eds). *Orogenic Processes: Quantification and Modelling in the Variscan Belt.* Geological Society, London, Special Publications, **179**, 387–399. 0305-8719/00/$15.00

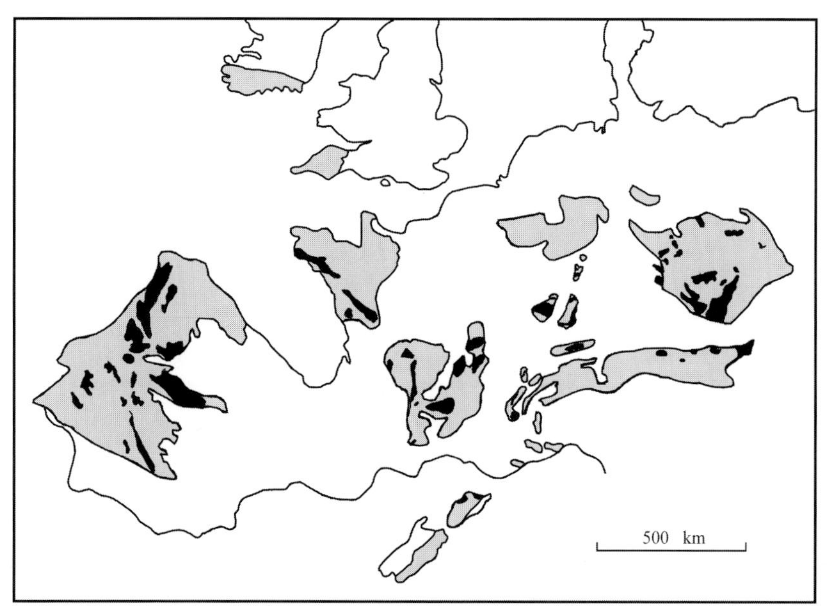

Fig. 1. Distribution of granitic to granodioritic rocks in the Variscan Orogen with intrusion ages between about 340 and 325 Ma (compiled after Franke (1989), Martinez *et al.* (1990), Neubauer & von Raumer (1993), Carron *et al.* (1994) and Ledru *et al.* (1994)).

large-scale plutonic activity, but also by its large areal extent: the thermal peak occurred near-synchronously throughout the orogen, that is, over a distance of 2000 km between Bohemia and Spain (Fig. 1). Starting locally as early as 350 Ma high temperatures documented by metamorphism and magmatism peaked between 340 and 325 Ma. Thus, the HT event was still occurring during the convergence stage and post-dates both the end of the oceanic subduction stage and the onset of continental collision in the Variscan Internides by about 20–30 Ma (Franke 1989).

The topic of this paper is a discussion of geodynamic scenarios that may have provided the heat required for the widespread magmatism and HT metamorphism observed during the final convergence stage of the Variscan Orogen (i.e. between 340 and 325 Ma). In principle, four potential heat sources are envisaged: (1) radiogenic production from the decay of U, Th, and K in thickened crustal rocks; (2) advection of heat by exhumation of deeply buried hot rocks; (3) advection of heat by intrusion of mantle-derived melts; (4) conduction of heat from a hotter than normal mantle. For the case of the Variscan Orogen, tectonic settings that have been suggested to account for one or more of these processes are continental rifting and strike-slip movements (Wickham & Oxburgh 1986), large-scale extension by gravitational orogenic collapse (Dewey 1988; Ménard & Molnar 1988), slab detachment (Brown & Dallmeyer 1996), extension triggered by loss of a deep lithospheric mantle root (Inger 1994; Schaltegger 1997; Zulauf *et al.* 1998*a*), heat advected into the middle and lower crust by rising slivers of hot granulite (O'Brien this volume) and internal radiogenic heating within the thickened crust (Gerdes *et al.* 2000).

Here we will briefly summarize the spatial and temporal distribution of HT metamorphism and associated magmatism in the Variscides. The integrated analysis of geological, geochemical, petrological and geochronological data will be used to review and discriminate between various geodynamic scenarios that could provide the heat required. It should be noted that this discussion concentrates on only the heat sources for syn-convergent metamorphism and magmatism. The geodynamic processes during the younger, post-convergent evolution (in Stephanian–Permian time) are not considered here but have been treated quantitatively by Henk (1999).

Temporal and spatial distribution of late-convergent HT metamorphism and magmatism

Radiometric ages of Variscan magmatic rocks scatter over a wide range between 360 and

270 Ma. However, most of this scatter is only apparent and is caused by analytical bias, such as rejuvenation of Rb–Sr and K–Ar systems of whole rocks and minerals by pervasive fluid percolation during post-orogenic times. Recent age data based on more robust isotopic systems (e.g. Finger *et al.* 1997; Schaltegger 1997) show that magmatism occurred in several pulses, which are separated by periods without magmatic activity. Although local magmatism (Central Bohemian Batholith, Dörr *et al.* 1996; Holub *et al.* 1997; NW Bohemian Massif, Siebel *et al.* 1997) is documented from 350 Ma onwards, certainly the most voluminous magmatic episode occurred during Late Viséan and Early Namurian time (i.e. between 340 and 325 Ma), during the final stage of convergence in the Variscides. Numerous plutons between Bohemia and Spain give evidence for the intensity and large areal extent of this melting event (Fig. 1). During this stage, the internal zone of the orogen was virtually soaked by granitic to granodioritic magmas, which typically represent more than 30% of the rocks exposed. Metamorphic conditions of the related HT event vary from place to place but typically range between 600 and 850 °C at 3–6 kbar (Dallmeyer *et al.* (1995) and references therein). To discriminate between the magmas being either the source or the product of the elevated crustal temperatures, the relative timing between magmatic intrusions and HT metamorphism of the country rocks is of critical importance.

Relative timing of granite magmatism and HT metamorphism

Because of the close spatial and temporal coincidence it is often difficult to decide if magmatism preceded metamorphism or vice versa. Locally, it can be shown that the crust was already hot or even cooling after peak metamorphism when the intrusions took place. For example, geochronological and petrological data for the South Bohemian Batholith indicate formation as several large magma batches between 330 and 320 Ma (Friedl *et al.* 1996), shortly after the thermal peak of regional metamorphism (Büttner & Kruhl 1997). A similar evolution with HT metamorphism preceding granite plutonism has been reported from the fore-Sudetic Block (Kryza 1995). In the Armorican Massif, contact aureoles are observed in rocks that previously experienced high-grade metamorphism; this feature points to major cooling before granite emplacement (Ballèvre *et al.* 1994). Rolet *et al.* (1994) described granites from the same area that frequently cut or enclose high-grade, anatectic gneisses, indicating high temperatures already before the granite intrusions. Rolet *et al.* concluded that the granites appear to be the products of the temperature anomaly and not the cause. However, there are also examples where different relationships are observed. For example, magmatism at *c.* 340 Ma and 330–328 Ma, respectively, pre-dates HT granulite-facies metamorphism in the southern Vosges, but post-dates it in the central Vosges (Schaltegger 1997). Likewise, there are some HT migmatites ('Schlierengranite') in the area of the South Bohemian Batholith, that seem to be generated by the granite intrusions (Finger & Clemens 1995).

The observation that, at least locally, granite magma intruded after the thermal peak in their country rocks argues against the assumption that in these cases the ascending magma itself provided the heat for HT metamorphism. In addition, it is difficult to reconcile this with an increased basal heat flow into the crust, for example, as a result of magmatic underplating as sole heat source for HT metamorphism. The resulting temperature rise could certainly generate granitic magmas by partial melting of the lower crust, but heat transport into the upper crust by these melts would be much more rapid than slow conductive heat transport, thereby intruding before the temperature peak.

Relative timing of granite magmatism, exhumation and extension

Several datasets indicate that continental collision in the Variscides was accompanied by substantial exhumation of deep crustal levels by erosion and tectonic denudation, particularly during the final convergence stage. In the Black Forest, for example, petrological and geochronological data suggest that a decrease in pressure of at least 8 kbar, equivalent to about 25 km of exhumation (Flöttmann 1988; Kalt 1991), was related to convergent retrograde shear zones between about 334 and 332 Ma (Lippolt *et al.* 1994; Schaltegger 2000), that is, still during overall convergence and essentially before granite intrusion. In southern Bohemia peak metamorphic conditions for the country rocks not far from the South Bohemian Batholith are in the range 5–9 kbar and 650–800 °C but pressures decreased by up to 5 kbar before the granites were emplaced at *c.* 2.5–5 kbar (*c.* 9–18 km depth; Büttner & Kruhl 1997). In the central part of the Bohemian Massif, high-pressure granulites were exhumed along the

Bohemian Shear Zone during Early Carboniferous elevator-style sinking of the supracrustal Teplá–Barrandian Unit into extruding deep-seated melt-bearing rocks of the Moldanubian Unit (Zulauf *et al.* 1998*b*; Bues & Zulauf 2000; Scheuvens & Zulauf 2000). The Moldanubian unit includes ultra-high-pressure rocks (26–28 kbar, 830 °C; Kotková *et al.* 1997), where the peak of metamorphism has recently been dated at *c.* 345 Ma using U–Pb on monazite (Doerr, pers. comm.). Pebbles of exhumed and eroded lower-crustal rocks of the Moldanubian Unit are part of Upper Viséan conglomerates in the Moravo-Silesian Unit further to the south (e.g. Dvorak 1973; Fiala & Patocka 1994). The high-pressure granulites of the North Bohemian Shear Zone were formed at *c.* 347 Ma. Subsequently, they were cooled and possibly also exhumed very rapidly, as is indicated by $^{39}Ar/^{40}Ar$ age data for biotite and muscovite yielding ages between 341 and 347 Ma (Zulauf *et al.* (1998*b*) and references therein). Rolet *et al.* (1994) described a syn-collisional metamorphic evolution with progressive pressure decrease and temperature rise from the Armorican Massif (see also Brown & Dallmeyer (1996)) and, similarly, Santallier *et al.* (1994) reported nappe stacking under decreasing pressures from the French Massif Central. For part of the Variscan Internides (Saxo-Thuringian Zone), a combination of petrological data, isotopic ages and thermal modelling (Henk 1995) suggest exhumation rates of several mm a^{-1} pointing to tectonic denudation as the dominant exhumation mechanism. The local palaeogeography and contemporaneous sedimentation do not support long-term erosion at such rates. Similarly, syn-convergent extension is indicated by extensional movements along the high-angle normal faults of the Bohemian Shear Zone which were active in Early Carboniferous time (*c.* 350–320 Ma; Zulauf *et al.* 1998*a*).

Localized extension is also documented by sedimentary basin formation in Viséan time, but was apparently restricted to transtensional settings within an oblique collision zone (Ziegler 1990; Schaltegger 1997). However, several basins in the Variscan realm show a continuous marine sedimentary record from Late Devonian to Viséan time (Eisbacher *et al.* 1989; Franke 1989; Ziegler 1990), which is difficult to reconcile with a Himalaya-like thickened crust and surface topography, which would be required to drive wholesale lithospheric extension by gravitational collapse.

The close spatial and temporal relationship between Viséan magmatism, rapid exhumation of metamorphic complexes, and local sedimentary basin formation point to a prominent role of syn-convergent extension. The question then is whether extension caused decompressional melting or whether HT metamorphism and magmatism weakened the crust and thus triggered extension and exhumation. From the relative timing of exhumation and magmatism outlined above it appears that, at least locally, extension preceded magmatism. However, regardless of the precise sequence of events, as soon as partial melting and intrusion of magma had occurred, the integrated strength of the crust would have been strongly reduced, thereby enhancing any further extension. In addition, a change in the dominant deformation mechanism in quartz from dislocation creep to solid-state diffusion creep has been invoked for the low viscosity of the lower crust at temperatures above 800 °C (Zulauf *et al.* 1998*a*).

Convergence in the internal zone of the Variscan Orogen lasted until about 330 Ma in the Black Forest area, as is indicated by undeformed granites intruding shear zones (Schaltegger 2000). For the French Massif Central, Ledru *et al.* (1994) reported that convergence and crustal stacking lasted until about 320 Ma. The end of convergence in the most external parts of the orogen (Ruhr Basin) is documented by the youngest sediments still affected by compression, which are Westphalian D sediments, about 305 Ma in age (Ziegler 1990). The subsequent modification in the plate boundary stresses, enhanced by gravitational forces, resulted in wholesale extension of the Variscan Orogen (Henk 1999). This phase of post-convergent lithospheric extension is well documented by widespread Stephanian–Early Permian basin formation and magmatism (Ziegler 1990; Burg *et al.* 1994).

Geochemical evidence and granite typologies

The large volume of Viséan S-type and high-K_2O I-type granites in the Moldanubian central zone are mainly products of intracrustal anatexis (Gerdes *et al.* 1998; Ploquin & Stussi 1994). They are generally characterized by a crustal chemical and isotope signature, as opposed to the pre-Viséan Variscan granites, which include several primitive calc-alkaline I-type magmas (Janoušek *et al.* 1995; Finger *et al.* 1997, and references therein). They are often associated with migmatites (diatexites), some of which may have been newly produced in the contact aureoles of the granites, but others appear to be slightly older than the granites (Finger & Clemens 1995).

In the Southern Bohemian Batholith, high formation temperatures of almost 900 °C, at probable pressures of 5–7 kbar, have been inferred for some of the Viséan granite magmas (e.g. the prominent Weinsberg Granite). According to Finger & Clemens (1995), magma formation occurred as a result of fluid-absent dehydration melting of high-grade paragneisses. Gerdes *et al.* (1998) have inferred from geochemical calculations that the source gneisses experienced typically between 30 and 50% partial melting.

Judging from recent precise U–Pb zircon and monazite ages for granites from the Southern Bohemian Batholith (Friedl *et al.* 1996) and the Black Forest (Schaltegger 1997), an important pulse of Moldanubian crustal melting occurred close to 330 Ma, with the production of a large melt volume in a short time-span. After this almost catastrophic crustal melting event, the rate of melt production apparently decreased until, at *c.* 310–290 Ma, the Viséan granites were in turn intruded in a number of places by late-Variscan I-type granodiorites (Finger *et al.* 1997). This later high heat flow event is distinct from the Viséan one and will be not further discussed in this paper.

It is important to note that the granite magmas that intruded the Saxo-Thuringian Unit and the Mid-German Crystalline Rise at *c.* 330 Ma are different from the Moldanubian Viséan granites. They include many calc-alkaline I-type granodiorites, often interpreted as subduction related (Reischmann & Anthes 1986). However, these rocks may have received their I-type characteristics simply through remagmatization of older arc-type crust in a non-subduction-related regime of high heat flow, or may have formed as a consequence of thermal reactivation of an older subduction zone below the orogen (Finger *et al.* 1997).

The contributions of lithospheric mantle melts to the Viséan plutonism seem to vary from place to place, although overall they were probably minor compared with the amounts of crustal magmas. For example, in the case of the large Southern Bohemian Batholith, it has been considered unlikely that enriched mantle melts made up more than 5% of the total melt budget (Gerdes *et al.* 1998). The lithospheric mantle source appears to be present in most plutons of this age, however. Coeval mafic magma batches of lithospheric mantle origin have been identified in numerous granites (Wenzel *et al.* 1997; Altherr *et al.* 1999*a*, *b*). Locally, these mafic magmas intimately mingled with the felsic crustal melts, producing sometimes almost homogeneous hybrids (Holub 1997; Gerdes *et al.* this volume). In the Vosges, the *c.* 340 Ma magmatic suite may have been mainly extracted from a lithospheric mantle source, with the melt products undergoing variable crustal contamination during ascent. Also, for some Viséan granites in the Massif Central, a mixing process between lithospheric mantle melts and crustal melts has been invoked, based mainly on isotope chemical studies (Downes & Duthou 1988; Turpin *et al.* 1988). However, it is important to note that although there is evidence for melts derived from the lithospheric mantle, any indications for depleted asthenospheric mantle magmas are absent in the Viséan period.

Potential heat sources

A number of processes have been suggested to explain high temperatures in orogenic settings: subduction zone melting, slab break-off (Davies & von Blanckenburg 1995), convective removal of a thickened thermal boundary layer (Houseman *et al.* 1981), delamination of the mantle lithosphere (Bird 1979), ascent of a mantle plume, thickening of the radiogenic heat producing layers (England & Thompson 1984) and lithospheric extension. Here we will briefly review each scenario and subsequently discuss its potential as a heat source for Variscan syn-collisional HT metamorphism and magmatism.

Subduction zone melting

At active continental margins melting of the mantle lithosphere of the overriding plate can be induced by fluids released from the downgoing slab (Fig. 2a). Subduction zone melting can generate large volumes of calc-alkaline magmas, and by assimilation of country rocks and differentiation, a wide range of acidic to intermediate magmatic rocks can form, most of which show an I-type signature. Accretion of subducted material to the accretionary wedge will lead to build-up of a thick crustal sequence, which may thin by extension (e.g. Platt 1986).

In the case of the Variscan Orogen, such activity might have led to the pre-Viséan 'Cordilleran'-type granitoids (e.g. Košler *et al.* 1993; Košler & McFarrow 1994; Finger *et al.* 1997) but the Viséan magmatism seems not to be compatible with closure of oceanic basins within the Variscan domain, which was completed by the end of Devonian time (Franke 1989). Additionally, the widespread and nonlinear extent of granitoids would argue against subduction magmatism. Finger & Steyrer (1990) discussed the possibility of a late Palaeozoic subduction zone along the southern flank of the Variscides, based on the presence of a distinct belt of Variscan-I-type granites extending

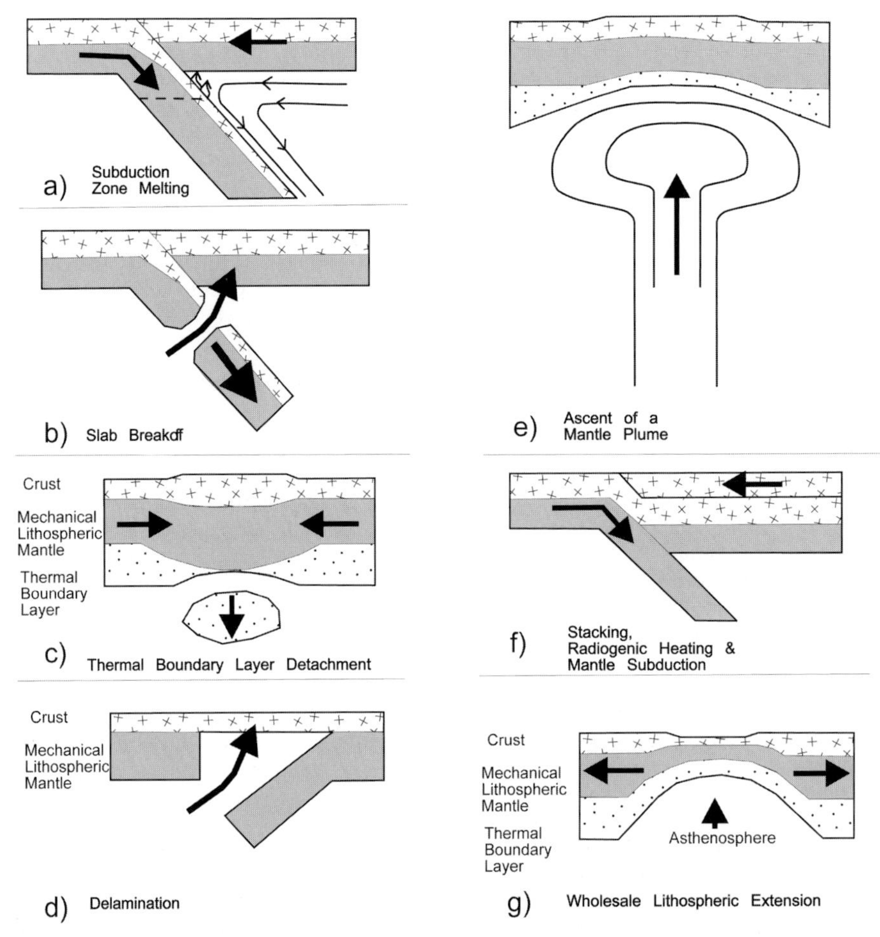

Fig. 2. Conceptual models accounting for high-temperature metamorphism and magmatism in orogenic settings. (**a**) Subduction zone melting; (**b**) slab break-off; (**c**) thermal boundary layer detachment; (**d**) delamination; (**e**) ascent of a mantle plume; (**f**) crustal stacking, radiogenic heating and mantle subduction; (**g**) wholesale lithospheric extension. (See text for further discussion.)

through the Alps. However, it is equally possible that the I-type signature of these granites reflects the geochemical composition of an older lower crust and yields no information on the Viséan geodynamic setting and source of the heat, respectively.

Slab break-off

Slab break-off describes the detachment of subducted oceanic lithosphere from continental lithosphere during continental collision (Fig. 2b; Davies & von Blanckenburg 1995). Slab break-off may occur after the closure of an oceanic basin if light continental lithosphere attempts to follow the dense oceanic lithosphere into the subduction zone. The change in buoyancy forces and the contrast in the vertically integrated strength between the two types of lithosphere may result in the detachment of the downgoing oceanic slab from the more buoyant continental lithosphere. This will bring hot asthenospheric mantle into direct contact with the lithospheric mantle of the overriding plate. Although the asthenospheric mantle is unlikely to melt by decompression, as the pressure decrease is mostly not shallower than the 50 km required for decompression melting of dry peridotite (McKenzie & Bickle 1988), the overriding litho-spheric mantle with its lower solidus will melt because it is heated from below. The localized nature of mantle and induced crustal melting above the break-off point will result in a linear trace of igneous activity parallel to the strike of the orogen. The change in potential energy of the thickened orogen after break-off will result in uplift and, as a consequence, enhanced erosion, extension and associated metamorphism. Of all the models discussed here, slab break-off is the only one that presents a consistent explanation for exhumation of high-pressure rocks from mantle depths (Davies & von Blanckenburg 1998).

Slab break-off might have been important for the exhumation of eclogite facies rocks in the Variscan Orogeny (O'Brien this volume), and might have made a minor contribution to early collisional magmatism. Slab break-off is also a plausible mechanism for certain locations, for example, in the Southern Brittany Metamorphic Belt (Brown & Dallmeyer 1996). O'Brien (this volume) has postulated that the rise of hot granulites from mantle depths into the middle and lower crust after slab break-off has sufficient potential to carry the required amount of advected heat for the production of the HT metamorphism. This mechanism would require simultaneous, and very widespread, closure of basins with concomitant break-off of several slabs. Although slab break-off is a plausible model for the linear array of Tertiary plutons of the Alps (von Blanckenburg & Davies 1995), the hypotheses that the process made more than a minor, and mostly local, contribution to the Variscan magmatism and HT metamorphism between 340 and 325 Ma would require further testing and thermal modelling because the expressions have been so widespread, nonlinear, and much more pervasive than is predicted by slab break-off.

Convective removal of a thickened thermal boundary layer

The thermal boundary layer (TBL) comprises the lower part of the lithosphere, which has a lower strength as compared with the strong upper part, the mechanical boundary layer. Thus, although being weak, the TBL is part of the moving plates and transfers heat from the convecting asthenosphere to the overlying layers by conduction. During continental collision, both the mechanical and the thermal boundary layer thicken, and the cooler and denser mantle root is pressed into the hotter and less dense asthenosphere. Because of the lateral density and temperature contrasts, the TBL may become unstable, and may be removed from the upper parts of the lithosphere into the asthenospheric mantle by convection (Fig. 2c). According to numerical experiments of Houseman *et al.* (1981), the time-scale of TBL detachment depends primarily on the assumed mantle viscosity and varies between one and several tens of million years. As such, TBL removal can be catastrophic, after a long build-up of lithospheric thickness, or it can be a continuous process during thickening. In the latter case, the crust will thicken whereas the mantle part remains at constant thickness.

An obvious consequence of TBL detachment is a pronounced modification of the thermal regime in the continental collision zone. The hot asthenospheric mantle replacing the detached TBL heats the bottom part of the mechanical lithosphere, thereby producing mantle melts. Their ascent and intrusion at or near the Moho will advect heat into the crust. Without magmatic underplating, temperatures in the crust would still rise conductively, and may cause crustal melting as the heat flow at the base of the lithosphere is increased. An immediate consequence of the replacement of the TBL by less dense asthenosphere is rapid upward movement of the mechanical lithosphere and surface, respectively, by as much as 1–3 km (England & Houseman 1989; Platt & England 1993). Such a plateau uplift may be difficult to detect directly in fossil orogens, as palaeotopographical information is usually lacking. However, it also increases the excess potential energy stored in the thickened crust, which may lead to a gravitational collapse of the orogen (Dewey 1988). Thus, similar to the present situation in the Himalaya, the orogen may convert into a state of local horizontal (i.e. orogen-parallel) extension in spite of the continuing overall convergence. This will leave a clear signal in the geological record, because crustal extension may be documented not only by the corresponding rock fabrics, but also by rapid exhumation of metamorphic rocks, sedimentary basin formation and decompression melting.

Detachment of the TBL is a mechanism by which syn- to post-collisional magmatism of mantle and crustal origin can be produced concomitantly with thermotectonic metamorphism and also over a wide areal extent. Most observations from the Early Carboniferous evolution of the Variscides seem to be compatible with this scenario.

Delamination

Delamination is similar to TBL removal. However, it is postulated that the entire lithosphere peels off the overlying crust (Fig. 2d; Bird 1979; Schott & Schmeling 1998). This will lead to very extensive lower-crustal magmatism, high-T metamorphism and plateau uplift. If the entire lithospheric mantle delaminates, no lithospheric mantle melts can be produced. Such melts may be produced if only parts of the mantle delaminate. In this case, the overall consequences would be the same as for the TBL detachment removal scenario described above.

Ascent of a mantle plume

Mantle plumes consist of hot material rising by thermal buoyancy through the upper mantle to impinge on the overriding lithosphere (Fig. 2e). This results in temperatures of the asthenosphere in excess by a few hundreds of degrees of the usual upper-mantle potential temperature of 1280 °C (McKenzie & Bickle 1988). Because such a temperature increase would allow for dry peridotite to melt, this is the only mechanism (of those discussed here) that might result in melting of asthenospheric mantle with enriched mid-ocean ridge basalt- to ocean-island basalt–HIMU-like isotopic compositions. In addition, heating of the overriding lithospheric mantle (with its lower solidus) will lead to widespread production of basalts with lithospheric mantle-like compositions. If concomitant with crustal thickening, the rise of basaltic melts will produce thermal metamorphism with anti-clockwise *PT* paths and crustal melting. The areal extent might range from hot-spot tracks (Hawaii) to flood basalts (Deccan Traps). It is possible that the thermal buoyancy of a plume results in widespread uplift of the overriding crust, followed by extension. Such a scenario has been postulated for the Cenozoic magmatism of Central Europe (Wilson & Downes 1991).

For the Carboniferous magmatism, although not impossible, it appears to be too much of a coincidence that widespread continental thickening should be followed immediately by the rise of a mantle plume with the same areal extent. Also, the limited basaltic magmatism and absence of voluminous asthenospheric magmatism in the Variscides argues against active mantle upwelling as a major heat source.

Thickening of the radiogenic heat producing layers

In contrast to the geodynamic scenarios discussed above, a higher surface heat flow can result not only from an increased mantle input but also from an increased contribution from the decay of radiogenic elements in areas of thickened continental crust (Fig. 2f). Such heat-producing elements (U, Th and K) are preferentially concentrated in the upper crust and immature sediments. Consequently, the contribution of radiogenic heat and the equilibrium surface heat flow will increase significantly when upper-crustal rocks are stacked and thickened during continental collision. One- and two-dimensional thermal models indicate that crustal thickening can indeed result in significant steepening of geotherms and crustal anatexis, in particular if lithological units with a high radiogenic heat production are involved and fertile protoliths occur in the crustal section (e.g. England & Thompson 1984, 1986; Patiño Douce *et al.* 1990; Thompson & Connolly 1995).

However, as Patiño Douce & McCarthy (1998) pointed out, in thickened crust underlain by a lid of thickened lithospheric mantle, temperatures are unlikely to exceed 800–850 °C, because in this case the mantle acts as a buffer between crustal and asthenospheric temperatures. Under such conditions, only muscovite-rich pelites can undergo dehydration melting, yielding peraluminous leucogranites. Thus, radiogenic heating is particularly effective if, during convergence, the mantle lithosphere is not thickened by the same degree as the crust. This can be achieved by continuous convective removal of the TBL, delamination of part of the lithospheric mantle, or subduction of the mantle lithosphere of the downgoing plate.

For part of the Variscides, numerical modelling by Gerdes *et al.* (2000) showed that radiogenic heating is indeed capable of generating high crustal temperatures within the time frame available. Using a 2D thermal–kinematic model of crustal stacking and material characteristics (e.g. radiogenic heat production) typical for the area of the South Bohemian Batholith, their model predictions are in good agreement with the observed time lag between maximum temperature in high-*P* rocks, the thermal peak in amphibolite-facies country rocks, and the timing and volume of granite magmatism. In this study the lithospheric mantle thickness was held constant by assuming subduction of the mantle part of the downgoing slab. Radiogenic heating might be the dominant heat source in parts of the Variscan Orogen. The widespread intrusion of I-type granitoids, and also the observed partial melting of lithospheric mantle, however, may suggest a heat input from below that is higher than that expected from a double thickness lithospheric mantle, which would act as a thermal buffer and strongly delay thermal equilibration. Hence a process capable of reducing the thickness of the lithospheric mantle is required.

Lithospheric extension

Extension of the lithosphere is an attractive explanation for post-convergent magmatism (Fig. 2g). Such thinning might be induced by a change in plate boundary stresses, leading to stretching of the entire lithosphere by external forces. Alternatively, crustal and mantle lithospheric thinning can result from the excess potential energy stored in the previously thickened orogen. Such a gravitational collapse is often postulated to be a consequence of the loss of the lithospheric root, by convective removal of a TBL, delamination or slab break-off (Dewey 1988; Platt & England 1993). Magmatic consequences are melting of dry asthenospheric peridotite, if stretching resulted in lithospheric thicknesses of less than 50 km (McKenzie & Bickle 1988), such as is the case at mid-ocean ridges. In contrast, the enriched and hydrous sections of the lithospheric mantle will melt by decompression because their solidus is lower. Whether, and how much, crustal melts are being produced depends on the time-scales of the process: if extension immediately follows rapid crustal thickening, geotherms will only be restored to their initial state and no melting will occur. If however, the time between thickening and extension allows for considerable relaxation of geotherms by internal heat production and conduction, extensive crustal melting can take place by subsequent decompression during extension. England & Thompson (1986) have modelled intra-crustal melting by extension and have concluded that a low-heat-flow regime will lead to S-type magmatism, with minor I-type magmatism. In contrast, a higher thermal regime (likely to be associated with a higher mantle heat flux) will lead to higher rates of melting. This will allow for more melting of intermediate and mafic lithologies, and hence more I-type compositions. The abundance of I-type granites will increase as the metamorphism progresses.

The Viséan evolution of Europe has many features compatible with lithospheric extension: the progression from S-type to I-type with time (at least in the Alpine external massifs and the Tauern window (Finger *et al.* 1997)), the present but not dominant mantle components in resulting granites, the preceding or simultaneous HT metamorphism, resulting from decompression reactions and heating during decompression, and the formation of basins during the final stage of HT activity. Although there is no doubt that extension strongly affected the Variscan Orogen, one should carefully distinguish between syn-convergent and post-convergent extension. Syn-convergent extension would be driven by local gravitational instabilities induced by topographical relief in an area of lithospheric shortening and crustal stacking. Thus, it is confined to upper-crustal levels, whereas the lower parts of the lithosphere are still under compression (e.g. Central Vosges, Schaltegger 1997). In contrast, post-convergent extension applies to the entire orogenic lithosphere and results in extension at all levels. From the amount of syn-convergent exhumation described above, it appears that a substantial part of the total crustal extension documented in the Variscides had already occurred during the final convergence stage. Therefore, this syn-convergent thinning must have contributed significantly to the overall heat budget. Wholesale lithospheric extension, driven by a change in the external stress field, however, was a later feature. It affected the orogen from 305 Ma onwards (Henk 1999) and finally restored the original crustal thickness of 30 km.

Discussion and conclusions

Of the mechanisms described above, some appear to be unlikely or at best minor players in providing a sufficiently strong heat source for the specific late-convergent evolution of the Variscides. For example, for subduction magmatism the geodynamic setting, timing and spatial distribution seem inappropriate. Slab break-off must have contributed to a minor extent after closure of oceanic basins, but the overall extent of magmatism was too widespread, too synchronous and not sufficiently linear, and thus appears to require an additional cause. One such possibility has been brought forward by Chemenda *et al.* (2000) for the Himalaya–Tibet system: that is, slab break-off immediately followed by delamination. A plume, although possible, appears too fortuitous a possibility. Furthermore, signs of major basaltic magmatism and of advected heat by melts (in the form of counter-clockwise *PT* paths) are absent. Wholesale lithospheric extension driven by a change in external boundary stresses was a later phenomenon, and can be excluded too. TBL removal, delamination of parts of the mantle lithosphere, subduction of the mantle lithosphere of the downgoing plate, radiogenic heating and syn-convergent lithospheric extension as a result of any of these processes deserve further investigation.

In this regard it is important to note that the former three models all have one assumption in common: this is that heat flow into the base of the crust is at least constant or increases, which is achieved by thickening the crust only without thickening (possibly even with thinning) of the mantle part of the lithosphere. In thermal models where the whole lithosphere is uniformly thickened, temperatures in the crust are much lower than in crust-only thickening models (England & Thompson 1984) because the thickened mantle lithosphere acts as a buffer between crustal and asthenospheric temperatures. As such, it appears that all three models have the similarity that in their first-order physical characteristics there is a reduction in the relative thickness of the mantle lithosphere compared with the thickened crust.

Present-day mantle thicknesses provide no information on the Variscan processes, because these have been affected by the Cenozoic extension in Europe (Menzies & Bodinier 1993), but there is some controversial evidence that indeed parts, but not all, of the mantle were removed during the Variscan Orogeny. First, Turpin *et al.* (1988) have measured very uniform enriched isotopic compositions from K-rich lamprophyres in Viséan granites throughout Europe. In contrast, Cenozoic volcanic rocks of Central European extensional regions are much less enriched isotopically (Wilson & Downes 1991). This might simply imply that different lithospheric mantle reservoirs have been tapped in different events, or that the enriched mantle pockets have been depleted by melting in the Variscides. However, it may also hint at the removal of some of the enriched parts of the mantle during the Variscan cycle. Second, Menzies & Bodinier (1993) have observed that whereas Variscan potassic magmas have tapped lithospheric sources, spinel lherzolite facies mantle xenoliths in Cenozoic volcanic rocks from Variscan terranes tap widespread depleted sources, and speculated that these are derived from young (post-Variscan) mantle. A wholesale delamination of the mantle part beneath the Variscan Orogen seems unlikely in view of the fact that lithospheric mantle melts have been produced during Early Carboniferous time (e.g. Turpin *et al.* 1988). Although it is difficult to envisage what should stop this process once it has been initiated, the overall differences between delamination and TBL removal are not all that significant, if only part of a vertical mantle sequence is delaminated.

Increased radiogenic heating will always contribute to the thermal budget in the crustal stack and can be the dominant cause for HT metamorphism and magmatism if the initial lithology and the location of fertile protoliths in the crustal profile are favourable. This seems to be the case at least for parts of the Variscides (e.g. Gerdes *et al.* 2000) but widespread I-type magmatism and substantial volumes of lithospheric mantle melts may suggest an additional heat contribution from below. Geodynamic scenarios that fit the field observations and might account for an increased basal heat flow and a reduced mantle lithosphere thickness are convective removal of the TBL, delamination of part of the lithospheric mantle, or subduction of the mantle lithosphere of the downgoing plate. Whether the last is actually possible is disputed and depends on the actual density of the lithospheric mantle. However, because the first-order physical characteristics of all three processes are similar, the geological consequences are similar too: in each case, the increased basal heat flow from the thinned mantle section, in combination with radiogenic heating, will support metamorphism and intracrustal melting. After removal of parts of the lithospheric root, the orogen will have obtained excess potential energy. In that case, the heat generated by both basal heat flow and radiogenic heating will weaken the crust and trigger extension, which in turn enhances further decompression melting and heating. In addition, lithospheric mantle melts will be produced to varying degrees. The rather instantaneous occurrence and close interaction of these processes is indeed what characterized the evolution of the Variscan orogen from 340 Ma onwards.

Unfortunately, at the present stage of knowledge it seems virtually impossible to establish which of the three theoretical mechanisms actually reduced the thickness of the mantle lithosphere and acted in conjunction with radiogenic heating in the Variscan Orogen, because their expressions in the geological record are so similar. It may well turn out that this remains subject for further speculation, as cause and consequence follow so closely in time and space.

This review resulted from various research projects that have been carried out within the frame of the 'Schwerpunktprogramm: Orogene Prozesse—Quantifizierung und Simulation am Beispiel der Varisciden'. Financial support of the Deutsche Forschungsgemeinschaft is gratefully acknowledged. G. Bergantz, M. Brown, W. Franke and D. Tanner are thanked for reviewing an earlier version of the manuscript.

References

ALTHERR, R., HENES-KLAIBER, U., HEGNER, E., SATIR, M. & LANGER, C. 1999*a*. Plutonism in the Variscan Odenwald (Germany): from subduction to collision. *International Journal of Earth Sciences*, **88**, 422–443.

——, HENJES-KUNST, F., LANGER, C. & OTTO, J. 1999*b*. Interaction between crustal-derived felsic and mantle-derived mafic magmas in the Oberkirch Pluton (European Variscides, Schwarzwald, Germany). *Contributions to Mineralogy and Petrology*, **137**, 304–322

BALLÉVRE, M., MARCHAND, J., GODARD, F., GOUJOU, J. C. & WYNS, R. 1994. Eo-Hercynian events in the Armorican Massif. *In*: KEPPIE, J. D. (ed.) *Pre-Mesozoic Geology in France and Related Areas*. Springer, Berlin, 183–194.

BIRD, P. 1979. Continental delamination and the Colorado Plateau. *Journal of Geophysical Research*, **84**, 7561–7571.

BROWN, M. & DALLMEYER, R. D. 1996. Rapid Variscan exhumation and the role of magma in core complex formation; southern Brittany metamorphic belt, France. *Journal of Metamorphic Geology*, **14**, 361–379.

BUES, C. & ZULAUF, G. 2000. Microstructural evolution and geologic significance of garnet pyriclasites in the Hoher–Bogen shear zone (Bohemian massif, Germany). *International Journal of Earth Science*, **88**, 803–813.

BURG, J. P., VAN DEN DRIESSCHE, J. & BRUN, J. P. 1994. Syn- to post-thickening extension in the Variscan Belt of Western Europe: modes and structural consequences. *Géologie de la France*, **3**, 33–51.

BÜTTNER, S. & KRUHL, J. 1997. The evolution of a late-Variscan high-*T*/low-*P* region: the south-eastern margin of the Bohemian Massif. *Geologische Rundschau*, **86**, 21–38.

CARRON, J. P., LEGUEN DE KERNEIZON, M. & NACHIT, H. 1994. Variscan granites from Brittany. *In*: KEPPIE, J. D. (ed.) *Pre-Mesozoic Geology in France and Related Areas*. Springer, Berlin, 231–239.

CHEMENDA, A. I., BURG, J. P. & MATTAUER, M. 2000. Evolutionary model of the Himalaya–Tibet system: geopoem based on new modeling, geological and geophysical data. *Earth and Planetary Science Letters*, **174**, 397–409.

DALLMEYER, R. D., FRANKE, W. & WEBER, K. (eds) 1995. *Pre-Permian Geology of Central and Eastern Europe*. Springer, Berlin.

DAVIES, J. H. & VON BLANCKENBURG, F. 1995. Slab breakoff: a model of lithosphere detachment and its test in the magmatism and deformation of collisional orogens. *Earth and Planetary Science Letters*, **129**, 85–102.

—— & —— 1998. Thermal controls on slab breakoff and the rise of high-pressure rocks during continental collisions. *In*: HACKER, B. R. & LIOU, J. G. (eds) *When Continents Collide: Geodynamics and Geochemistry of Ultrahigh-Pressure Rocks*. Kluwer, Dordrecht, 97–115.

DEWEY, J. F. 1988. Extensional collapse of orogens. *Tectonics*, **7**, 1123–1139.

DÖRR, W., ZULAUF, G., SCHASTOCK, J., SCHEUVENS, D., VEJNAR, Z., WEMMER, K. & AHRENDT, H. 1996. The Teplá–Barrandian/Moldanubian s.str. boundary: preliminary geochronological results of fault-related plutons. *Terra Nostra*, **96**(2), 34–38.

DOWNES, H. & DUTHOU, J. L. 1988. Isotopic and trace-element arguments for the lower-crustal origin of Hercynian granitoids and Pre-Hercynian orthogneisses, Massif Central (France). *Chemical Geology*, **68**, 291–308.

DVORAK, J. 1973. Synsedimentary tectonics of the Paleozoic of the Drahany Upland (Sudeticum, Moravia, Czechoslovakia). *Tectonophysics*, **17**, 359–391.

EISBACHER, G. H., LÜSCHEN, E. & WICKERT, F. 1989. Crustal-scale thrusting and extension in the Hercynian Schwarzwald and Vosges, Central Europe. *Tectonics*, **8**, 1–21.

ENGLAND, P. C. & HOUSEMAN, G. A. 1989. Extension during continental convergence, with application to the Tibetan Plateau. *Journal of Geophysical Research*, **94**, 17561–17579.

—— & THOMPSON, A. B. 1984. Pressure–temperature–time paths of regional metamorphism. Part I: Heat transfer during the evolution of regions of thickened continental crust. *Journal of Petrology*, **25**, 894–928.

—— & —— 1986. Some thermal and tectonic models for crustal melting in continental collision zones. *In*: COWARD, M. P. & RIES, A. C. (eds) *Collision Tectonics*. Geological Society, London, Special Publications, **19**, 83–94.

FIALA, J. & PATOCKA, F. 1994. The evolution of Variscan terranes of the Moldanubian Region, Bohemian Massif. *In*: HIRSCHMAN, G. & HARMS, U. (eds) *Beiträge zur Geologie und Petrologie der KTB—Lokation und ihres Umfeldes*. KTB Report, **94-3**, 1–8.

FINGER, F. & CLEMENS, J. D. 1995. Migmatization and 'secondary' granitic magmas: effects of emplacement and crystallization of 'primary' granitoids in Southern Bohemia, Austria. *Contributions to Mineralogy and Petrology*, **120**, 311–326.

—— & STEYRER, H. P. 1990. I-type granitoids as indicators of a late Paleozoic convergent ocean–continent margin along the southern flank of the central European Variscan orogen. *Geology*, **18**, 1207–1210.

——, ROBERTS, M. P., HAUNSCHMID, B., SCHERMAIER, A. & STEYRER, H. P. 1997. Variscan granitoids of central Europe: their typology, potential sources and tectonothermal relations. *Mineralogy and Petrology*, **61**, 67–96.

FLÖTTMANN, T. 1988. *Strukturentwicklung*, P–T-*Pfade und Deformationsprozesse im Zentralschwarzwälder Gneiskomplex*. Frankfurter geowissenschaftliche Arbeiten, **A6**.

FRANKE, W. 1989. Tectonostratigraphic units in the Variscan belt of central Europe. *In*: DALLMEYER, R. D. (ed.) *Terranes in the Circum-Atlantic Palaeozoic Orogens*. Geological Society of America, Special Papers, **230**, 67–90.

——, DALLMEYER, R. D. & WEBER, K. 1995. Geodynamic evolution. *In*: DALLMEYER, R. D., FRANKE, W. & WEBER, K. (eds) *Pre-Permian Geology of Central and Eastern Europe*. Springer, Berlin, 579–593.

FRIEDL, G., VON QUADT, A. & FINGER, F. 1996. Timing der Intrusionstätigkeit im Südböhmischen Batholith. *TSK Salzburg, Abstract Volume*, 127.

GERDES, A., WÖRNER, G. & FINGER, F. 1998. Late-orogenic magmatism in the southern Bohemian

Massif; geochemical and isotopic constraints on possible sources and magma evolution. *Acta Universitatis Carolinae*, **42**(1), 41–45.

——, —— & —— 2000. Hybrids, magma mixing and enriched mantle melts in post-collisional Variscan granitoids: the Rastenberg pluton, Austria. *This volume*.

——, —— & HENK, A. 2000. Post-collisional granite generation and HT–LP metamorphism by radiogenic heating: the Variscan South Bohemian Batholith. *Journal of the Geological Society, London*, **157**, 577–587.

HENK, A. 1995. Late Variscan exhumation histories of the southern Rhenohercynian Zone and western Mid-German Crystalline Rise: results from thermal modeling. *Geologische Rundschau*, **84**, 578–590.

—— 1999. Did the Variscides collapse or were they torn apart? A quantitative evaluation of the driving forces for postconvergent extension in Central Europe. *Tectonics*, **18**, 774–792.

HOLUB, F. V. 1997. Ultrapotassic plutonic rocks of the durbachite series in the Bohemian Massif: petrology, geochemistry and petrogenetic interpretation. *Sborník Geologických Ved, Geologie, Mineralogie*, **31**, 5–26.

——, COCHERIE, A. & ROSSI, Ph. 1997. Radiometric dating of granitic rocks from the Central Bohemian Plutonic Complex (Czech Republic): constraints on the chronology of thermal and tectonic events along the Moldanubian–Barrandian boundary. *Comptes Rendus de l'Académie des Sciences, Sciences de la Terre et des Planètes*, **325**, 19–26.

HOUSEMAN, G. A., MCKENZIE, D. P. & MOLNAR, P. 1981. Convective instability of a thickened boundary layer and its relevance for the thermal evolution of continental convergent belts. *Journal of Geophysical Research*, **86**, 6115–6132.

INGER, S. 1994. Magma genesis associated with extension in orogenic belts: examples from the Himalaya and Tibet. *Tectonophysics*, **238**, 183–197.

KALT, A. 1991. *Isotopengeologische Untersuchungen an Metabasiten des Schwarzwaldes und ihren Rahmengesteinen*. Freiburger Geowissenschaftliche Beiträge, **3**.

JANOUŠEK, V., ROGERS, G. & BOWES, D. R. 1995. Sr–Nd isotopic constraints on the petrogenesis of the Central Bohemian Pluton, Czech Republic. *Geologische Rundschau*, **84**, 520–534.

KOŠLER, J. & MCFARROW, C. M. 1994. Mid–late Devonian arc-type magmatism in the Bohemian Massif: Sr and Nd isotope and trace element evidence from the Staré Sedlo and Mirotice gneiss Complex, Czech Republik. *Journal of the Czech Geological Society*, **39**(1), 56–58.

——, AFTALION, M. & BOWES, D. R. 1993. Mid–late Devonian activity in the Bohemian Massif: U–Pb zircon isotopic evidence from the Staré Sedlo and Mirotice gneiss complexes, Czech Republic. *Neues Jahrbuch für Mineralogie Monatschafte*, **1993**(H.9), 417–431.

KOTKOVÁ, J., HARLEY, S. L. & FIŠERA, M. 1997. A vestige of very high-pressure (*c*. 28 kbar) metamorphism in the Variscan Bohemian Massif, Czech Republic. *European Journal of Mineralogy*, **1997**(9), 1017–1033.

KRYZA, R. 1995. Igneous activity. *In*: DALLMEYER, R. D., FRANKE, W. & WEBER, K. (eds) *Pre-Permian Geology of Central and Eastern Europe*, Springer, Berlin, 341–350.

LEDRU, P., COSTA, S. & ECHTLER, H. 1994. Structure. *In*: KEPPIE, J. D. (ed.) *Pre-Mesozoic Geology in France and Related Areas*. Springer, Berlin, 305–323.

LIPPOLT, H. J., HRADETZKY, H. & HAUTMANN, S. 1994. K–Ar dating of amphibole-bearing rocks in the Schwarzwald, SW Germany: I. $^{40}Ar/^{39}Ar$ age constraints in Hercynian HT metamorphism. *Neues Jahrbuch für Mineralogie, Monatshefte*, **10**, 433–448.

MARTINEZ, F. J., CORRETGE, L. G. & SUAREZ, O. 1990. Distribution, characteristics and evolution of metamorphism. *In*: DALLMEYER, R. D. & MARTINEZ GARCIA, E. (eds) *Pre-Mesozoic Geology of Iberia*. Springer, Berlin, 207–211.

MATTE, P. 1991. Accretionary history and crustal evolution of the Variscan belt in Western Europe. *Tectonophysics*, **196**, 309–337.

MCKENZIE, D. & BICKLE, M. J. 1988. The volume and composition of melt generated by extension of the lithosphere. *Journal of Petrology*, **29**, 625–679.

MÉNARD, G. & MOLNAR, P. 1988. Collapse of a Hercynian Tibetan Plateau into a late Paleozoic European Basin and Range province. *Nature*, **334**, 235–237.

MENZIES, M. A. & BODINIER, J. L. 1993. Growth of the European lithospheric mantle—dependence of upper-mantle peridotite facies and chemical heterogeneity on tectonics and age. *Physics of the Earth and Planetary Interiors*, **79**, 219–240.

NELSON, K. D., ZHAO, W., BROWN, L. D. *et al.* 1996. Partially molten middle crust beneath southern Tibet: synthesis of project INDEPTH results. *Science*, **274**, 1684–1688.

NEUBAUER, F. & VON RAUMER, J. 1993. The Alpine Basement—linkage between Variscides and east-Mediterranean mountain belts. In: von Raumer, J. & Neubauer, F. (eds.) *Pre-Mesozoic Geology in the Alps*. Springer, Berlin, 641–663.

O'BRIEN, P. 2000. The fundamental Variscan problem: high-temperature metamorphism at different depths and high-pressure metamorphism at different temperatures. *This volume*.

PATIÑO-DOUCE, A. E. & MCCARTHY, T. C. 1998. Melting of crustal rocks during continental collision and subduction. *In*: HACKER, B. R. & LIOU, J. G. (eds) *When Continents Collide: Geodynamics and Geochemistry of Ultrahigh-Pressure Rocks*. Kluwer, Dordrecht, 27–55.

——, HUMPHREYS, D. & JOHNSTON, A. D. 1990. Anatexis and metamorphism in tectonically thickened continental crust exemplified by the Sevier hinterland, western North America. *Earth and Planetary Science Letters*, **97**, 290–315.

PLATT, J. P. 1986. Dynamics of orogenic wedges and the uplift of high-pressure metamorphic rocks. *Geological Society of America Bulletin*, **97**, 1037–1053.

—— & ENGLAND, P. C. 1993. Convective removal of lithosphere beneath mountain belts: thermal and mechanical consequences. *American Journal of Science*, **293**, 307–336.

PLOQUIN, A. & STUSSI, J. M. 1994. Felsic plutonism and volcanism in the Massif Central. *In*: KEPPIE, J. D. (ed.) *Pre-Mesozoic Geology in France and Related Areas*. Springer, Berlin, 363–378.

REISCHMANN, T., & ANTHES, G. 1996. Geochronology of the mid-German crystalline rise west of the river Rhine. *Geologische Rundschau*, **85**, 761–774.

ROLET, J., GRESSELIN, F., JEGOUZO, P., LEDRU, P. & WYNS, R. 1994. Intracontinental Hercynian events in the Armorican Massif. *In*: KEPPIE, J. D. (ed.) *Pre-Mesozoic Geology in France and Related Areas*. Springer, Berlin, 195–219.

SANTALLIER, D. S., LARDEAUX, J. M., MARCHAND, J. M. & MARIGNAC, C. 1994. Metamorphism. *In*: KEPPIE, J. D. (ed.) *Pre-Mesozoic Geology in France and Related Areas*. Springer, Berlin, 324–340.

SCHALTEGGER, U. 1997. Magma pulses in the Central Variscan Belt: episodic melt generation and emplacement during lithospheric thinning. *Terra Nova*, **9**, 242–245.

—— 2000. U–Pb geochronology of the Southern Black Forest batholith (central Variscan Belt): timing of exhumation and granite emplacement. *International Journal of Earth Science*, **88**, 814–828.

——, FANNING, G., MAURIN, J. C., SCHULMANN, K. & GEBAUER, D. 1999. Growth, annealing and recrystallization of zircon and preservation of monazite in high-grade metamorphism: conventional and in-situ U–Pb isotope, cathodoluminescence and microchemical evidence. *Contributions in Mineralogy and Petrology*, **134**, 186–201.

SCHEUVENS, D. & ZULAUF, G. 2000. Exhumation, strain localization, and emplacement of granitoids along the western part of the Central Bohemian shear zone (central European Variscides, Czech Republic). *International Journal of Earth Science*, in press.

SCHOTT, B. & SCHMELING, H. 1998. Delamination and detachment of a lithospheric root. *Tectonophysics*, **296**, 225–247.

SIEBEL, W., RASCHKA, H., IRBER, W., KREUZER, H., LENZ, K. H., HÖHNDORF, A. & WENDT, I. 1997. Early Palaeozoic acid magmatism in the Saxo-Thuringian Belt; new insights from a geochemical and isotopic study of orthogneisses and metavolcanic rocks from the Fichtelgebirge, SE Germany. *Journal of Petrology*, **38**, 203–230.

THOMPSON, A. B. & CONNOLLY, J. A. D. 1995. Melting of the continental crust: some thermal and petrological constraints on anatexis in continental collision zones and other tectonic settings. *Journal of Geophysical Research*, **100**, 15565–15579.

TURPIN, L., VELDE, D. & PINTE, G. 1988. Geochemical comparison between minettes and kersantites from the Western European Hercynian orogen: trace element and Pb–Sr–Nd isotope constraints on their origin. *Earth and Planetary Science Letters*, **87**, 73–86.

VON BLANCKENBURG, F. & DAVIES, J. H. 1995. Slab breakoff: a model for syncollisional magmatism and tectonics in the Alps. *Tectonics*, **14**, 120–131.

WENZEL, T. H., MERTZ, D. F., OBERHÄNSLI, R., BECKER, T. & RENNE, P. R. 1997. Age, geodynamic setting, and mantle enrichment processes of a K-rich intrusion from the Meissen massif (northern Bohemian massif) and implications for related occurrences from the mid-European Hercynian. *Geologische Rundschau*, **86**, 556–570.

WICKHAM, S. M. & OXBURGH, E. R. 1986. A rifted tectonic setting for Hercynian high-thermal gradient metamorphism in the Pyrenees. *Tectonophysics*, **129**, 53–69.

WILSON, M. & DOWNES, H. 1991. Tertiary–Quaternary extension-related alkaline magmatism in western and central Europe. *Journal of Petrology*, **32**, 811–849.

ZIEGLER, P. A. 1990. *Geological Atlas of Western and Central Europe*. Shell Internationale Petroleum Maatschappij, The Hague.

ZULAUF, G., BUES, C., DÖRR, W., FIALA, J., KOTKOVA, J., SCHEUVENS, D. & VEJNAR, Z. 1998*a*. Extrusion tectonics due to thermal softening of a thickened crustal root: The Bohemian Massif in Lower Carboniferous times. *Terra Nostra*, **98**(2), 177–180.

——, FIALA, J., KOTKOVA, J. & MALUSKI, H. 1998*b*. Rheologischer Kollaps eines unterkarbonischen Hochplateaus in der Böhmischen Masse: die Rolle der Nordböhmischen Scherzone. *Freiberger Forschungsheft*, **C471**, 261–263.

The Variscan lower continental crust: evidence for crustal delamination from geochemical and petrophysical investigations

A. WITTENBERG[1], C. VELLMER[2], H. KERN[3] & K. MENGEL[4]

[1]*Institut für Mineralogie, Universität Hannover, Welfengarten 1, 30167 Hannover, Germany (e-mail: Wittenberg@mineralogie.uni-hannover.de)*

[2]*Mineralogisch-Petrographisches Institut, Universität Hamburg, Grindelallee 48, D-20146 Hamburg, Germnay*

[3]*Institut für Geowissenschaften, Universität Kiel, Olshausenstr. 40, 24098 Kiel, Germany*

[4]*Institut für Mineralogie und Mineralische Rohstoffe, Technische Universität Clausthal, Adolph-Roemer-Straße 2A, 38678 Clausthal-Zellerfeld, Germany*

Abstract: Seismic observations along the European Geotraverse (EGT) Central Segment indicate an overall felsic composition of the Variscan crust, including 15 vol. % of mafic lower-crustal (MLC) rocks. As the Variscan continental crust is largely post-Archaean, its bulk composition should be basaltic andesitic rather than felsic (tonalitic), because primitive post-Archaean crust was formed by mantle extracted basaltic andesite magmas. The observed proportions of the Variscan continental crustal layers do not match the requirement of a mafic overall crust. Combining laboratory-derived, *in situ* seismic data (V_P) with the refraction seismic data and the evidence from geological and geochemical investigations suggests that large amounts of MLC rocks are missing in the present-day lower crust. A mass balance calculation of the basaltic andesite bulk crust, considering a refined compilation of the Variscan MLC, gives a volume proportion of 37 vol. % for the felsic and 63 vol. % for the mafic components, for primitive crust prior to collision tectonics and orogenic root delamination. If large parts of this MLC have undergone eclogite-facies metamorphism at the orogenic root, large amounts of former basaltic–gabbroic and MLC cumulates could have been subtracted from the crust by its subsequent delamination processes which results in the currently observed proportion of the continental crust of the Variscan Orogen with a predominance of felsic rocks.

The Variscan orogenic belt structure was a focus of the European Geotraverse Project (EGT Central Segment) and is seismically well studied. Wedepohl (1995) suggested that the European continental crust had a 40 km overall crustal thickness, including 7.2 km mafic lower crust (MLC). This observation is in good agreement with the world-wide observed average crustal thickness of 37–41 km. It is widely accepted that the lower crust at the crust–mantle boundary is of mafic composition (e.g. Griffin & O'Reilly 1987; O'Reilly & Griffin 1987; Rudnick & Taylor 1987, Bohlen & Metzger 1989; Mengel *et al.* 1991). In comparison with the thickness of the continental crust in older orogens the Moho of young orogens is located at depths between 50 and 80 km (e.g. Alps, Himalayas). There is still debate about the relevant subduction processes and the amounts of continental material subducted into the mantle; in particular, because the subducted material cannot be detected seismically. To explain the very different depths of the crust–mantle boundary in orogenic belts, we discuss the possibility of crustal delamination by eclogite-facies overprint of MLC (e.g. Rudnick & Fountain 1995). Such transformations from gabbroic rocks to eclogitic rocks need at least 1.6 GPa lithostatic pressure corresponding to a tectonic overburden of >50 km.

During continent–continent collision the thickness of the continental crust increases markedly and the lower- and lowermost-crustal parts are metamorphosed with the result that the MLC rocks are transformed into eclogites. As eclogite exhibits almost the same elastic properties as lithospheric mantle peridotite, the rock types are seismically not distinguishable from each other. If delamination of the MLC is a

From: FRANKE, W., HAAK, V., ONCKEN, O. & TANNER, D. (eds). *Orogenic Processes: Quantification and Modelling in the Variscan Belt.* Geological Society, London, Special Publications, **179**, 401–414. 0305-8719/00/$15.00

major process, the rest of the continental crust should become more felsic with time.

We will discuss aspects of delamination of MLC from an assumed primitive continental crust having bulk composition of basaltic to basaltic andesites. As an example we use the Moldanubian part of the Variscan orogenic belt crossed by the EGT. Model calculations for continental crust composition reveal a lack of an Eu anomaly (Eu/Eu*) in the bulk continental crust (e.g. Rudnick & Fountain 1995). Compilations based on the observed rock types in the Variscan belt, however, indicate a prominent negaive Eu anomaly for the total crust. The negative Eu anomaly could be compensated by missing MLC components containing plagioclase-rich residual and plagioclase-dominated cumulate gabbroic units.

Starting from the assumption that the post-Archaean bulk continental crust equals basaltic to basaltic andesite composition, we will calculate the mass of residual MLC produced by subsequent intra-crustal differentiation, which has formed about 20 km thick felsic component. In a second step we calculate the mass of residual MLC required for the formation of seismically observed felsic crust. This calculated mass of MLC is compared with the present-day abundance of MLC in the Variscan fold belt.

Geological setting

The Mid- to Late Palaeozoic Variscan Orogen was the result of a continent–continent collision involving parts of Laurasia and Gondwana (Matte 1991). Widespread granulite-facies metamorphism took place until Late Carboniferous time (Pin & Vielzeuf 1983). In general, the high-grade metamorphic rocks of the Variscan Orogen are younger than 480 Ma and represent a major part of the Pan-African evolution (Van Breemen *et al.* 1982; Gebauer 1983, 1986; Kröner *et al.* 1988).

The Central Segment of the Variscan belt is partly covered by Mesozoic to Cenozoic sediments; in other places a series of isolated massifs are exposed, such as the Iberian Peninsula in the west and the Bohemian Massif in the east (see enclosure at back). The Bohemian Massif comprises large blocks of plutonic and metamorphic rocks, which underwent polyphase deformation and metamorphism during Palaeozoic time. The southeastern part of the Bohemian Massif is dominated by Late Palaeozoic intrusive rocks ('Southern Bohemian Plutons'). High-pressure granulites and associated garnet-bearing meta-peridotites form the top of a high-grade nappe stack in the Moldanubian Zone of the southern Bohemian Massif. Detailed geochronological studies of eclogite-facies rocks of the Bohemian Massif have been given by Schmädicke *et al.* (1995). The Moldanubian is divided into four units: the Moldanubian *sensu stricto*, the Gföhler Nappe, the Kutná Hora–Svratka Complex and the Teplá–Barrandian Unit (Zoubek *et al.* 1988). All subdivisions contain eclogite-facies rocks of former mafic precursors (Dudek & Fediuková 1974; Fediuková 1989).

The Lower Austrian Waldviertel is part of the Gföhl Unit and is dominated by felsic to mafic granulites and orthogneisses (Gföhler Nappe). Geochrological data indicate that granulite-facies metamorphism is related to crustal stacking during Carboniferous collision (Van Breemen *et al.* 1982; Kröner *et al.* 1988; Aftalion *et al.* 1989; Wendt *et al.* 1994). These studies give metamorphism ages of 345 Ma for granulites of the Gföhler Gneiss Unit (Van Breemen *et al.* 1982). Sm–Nd isotope data of high-pressure rocks of the Waldviertel prove the Late Devonian–Early Carboniferous age of garnet pyroxenites (344 Ma) and garnet peridotites (370 Ma) (Carswell & Jamtveit 1990). Late extensional collapse, low-pressure–high-temperature metamorphism and extensive plutonism affected all Moldanubian nappes during Late Carboniferous time to various degrees. $^{40}Ar/^{39}Ar$ ages (*c.* 325–340 Ma) obtained from hornblende and micas of amphibolite-facies gneisses constrain the cooling history of the southern Bohemian Massif (Dallmeyer *et al.* 1992). U–Pb zircon and monazite ages yield 314–318 Ma as intrusion ages of 'post-metamorphic' plutons (granitoids) from southern Bohemia (Friedl *et al.* 1992). The Kutná Hora–Svratka Complex is characterized by muscovite–biotite high-grade gneisses with intercalated eclogite-facies units and is thought to represent a higher structural level than the Moldanubian in the south (Beard *et al.* 1992).

A much smaller complex at the northeastern boundary of the Saxo-Thuringian–Moldanubian Zones is the nappe unit of the Münchberg Massif (see Variscan map at back of volume). The Wei*ß*enstein and Oberkotzau eclogites are some of the best studied high-pressure rocks of this region. The eclogite-facies metamorphism is dated to 395–380 Ma by Sm–Nd and Rb–Sr mineral isochrons (Stosch & Lugmair 1987, 1990). The eclogites-facies rocks are interlayered with paragneisses, which were interpreted by Gebauer (1990) as rift- or back-arc-basin precursors, rather than as a continental margin setting.

The uplifted sections of the crust–mantle boundary in the Ivrea–Verbano Zone represent a window into the processes occurring in the lower continental crust. The Ivrea–Verbano Zone contains uplifted sections of former lower continental crust and subcontinental upper mantle. The lower-crustal sections in the Val Sesia and Val Mastalone as well as in the Val Strona comprise granulite-facies rocks of mafic and felsic composition (Mehnert 1975; Sighinolfi & Gorgoni 1978; Zingg 1980). The meta-gabbroic rocks in the Val Sesia range from ultramafic pyroxene-dominated gabbros to more plagioclase-dominated gabbros and diorites (Rivalenti *et al.* 1984). The intrusion of a 6 km thick gabbroic body into pre-existing lower crust caused granulite-facies metamorphism at around 0.8–1.0 GPa and 700–900°C.

Seismic observations

The European continental crust has been studied in detail by seismic refraction survey in the framework of EGT. In principle the crust can seismically be subdivided into three major layers, comprising upper crust ($V_P \leqslant 6.2$ km s^{-1}), middle crust of mainly felsic composition (V_P6.1–6.5 km s^{-1}) including a thin velocity inversion at around 10 km depth, and lowermost crust mainly of mafic composition with V_P between 6.7 and 7.0 km s^{-1}.

The northern part of the EGT profile exhibits an average Moho depth of about 40 km. The seismic structure of the Variscan Central Segment is shown in Fig. 1a (after Prodehl & Giese 1990, p. 82, fig. 3). The Moho Discontinuity in this part of the Variscan Zone is observed at 29 km depth, and the crustal *P*-wave velocities are not higher than 7.0 km s^{-1}. Although MLC rocks such as granulites and meta-gabbros with *P*-wave velocities between 6.9 and 7.5 km s^{-1} are known from outcrops, the seismic profile of the EGT Central Segment clearly points to a scarceness of such rocks (Fig. 1b).

Although the last orogenic event dates back to Devonian–Carboniferous times (Franke 1989), orogenic roots are not visible beneath the two prominent collision zones (Moldanubian–Saxo-Thuringian, Saxo-Thuringian–Rheno-Hercynian). The absence of orogenic roots and the relatively shallow Moho depth is not in accordance with observations of younger

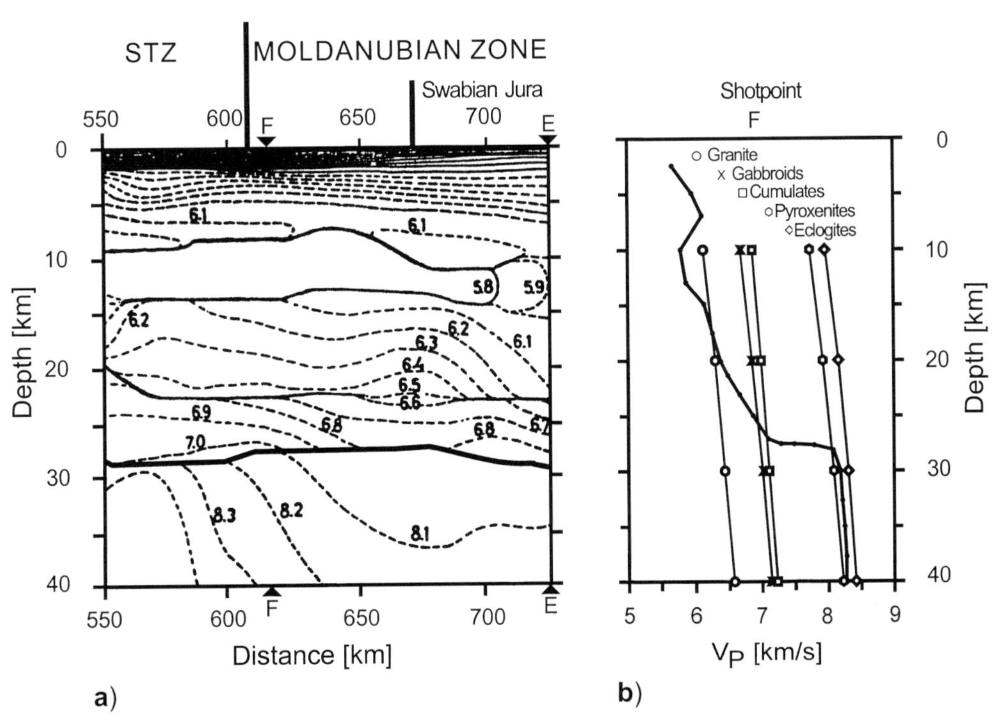

Fig. 1. Seismic profile (**a**) of the studied part of the Variscan Orogen modified after Prodehl & Giese (1990, p. 82, fig. 3). The letters indicate the shotpoints from the 1986 studies in the frame of the EGT project. STZ, Saxo-Thuringian Zone. The *P*-wave velocity profile is given at shotpoint F in (**b**), where the relation between depth and V_P is presented for the rock types discussed here.

orogenic belts. The Alpine Orogen reveals a double Moho in its southern part and crustal roots tending to have a depth of 50–70 km in the central and northern part (Pfiffner 1992, p. 188, fig. 6-27). Among the most important results of the EGT Alpine Section is the observation of a lower-crustal segment, which has been removed from the thickened orogenic root by southward-directed delamination into lithospheric mantle.

On the basis of these observations it is tempting to suggest tectonic erosion of the Variscan collision-induced crustal roots, which results in a flattened and shallow Moho depths observed today. Tectonic erosion implies that mafic sections of the lowermost continental crust are detached from the rest of the crust upon transformation under eclogites-facies conditions. Such a process would convert the MLC or parts of it into dense ($\rho > 3.2$ g cm^{-3}) rocks with a high *P*-wave velocity ($V_P > 7.8$–8.1 km s^{-1}) corresponding to the velocities of the lithospheric mantle.

Samples, analytical and experimental techniques

A set of 17 samples of the Variscan lower crust were analysed for their elastic properties (V_S, V_P), their major element composition and mineralogy. They include three plagioclase-dominated cumulates, two pyroxenites, one granitoid, four gabbros and five eclogites from the Münchberger Mass (Germany), the Lower Austrian Waldviertel and Kutna Horá (Czech), respectively and, in addition, two gabbros from the Ivrea–Verbano Zone (northern Italy). Sample numbers, rock types, locations and modal compositions of the rocks are presented in Table 1, and chemical analyses of major oxides including their CIPW normative minerals and trace element compositions of these 17 samples can be obtained from the Society Library or the British Library Document Supply Centre, Boston Spa, Wetherby, West Yorkshire LS23 7BQ, UK as Supplemetary Publication No. SUP 18153 (7 pages).

The velocity and density of the 17 samples were measured at simulated *in situ* conditions using specimens free of macroscopically visible fractures and secondary alterations.

The seismic velocity measurements were carried out on sample cubes (43 mm length) in a multi-anvil pressure apparatus. Compressional-wave (V_P) and shear-wave (V_S) velocities were measured on oven-dried rocks using the pulse transmission technique. Compressional and shear waves were generated by means of 2 MHz and 1 MHz lead titanite zirconate (PTZ) transducers, respectively. The sample reference system corresponded to fabric elements (normal to foliation (z), perpendicular to lineation (y) and parallel to lineation (x)). Splitting of shear waves was obtained by two sets of orientated transducers with perpendicular planes (for details, see Kern *et al.* (1997)). Length and volume changes of the sample cubes, caused by changes of principal stress and temperature, were obtained by the piston displacement. Densities for each sample were calculated from the masses and the measured volumes of the cubes. The cumulate error in V_P and V_S measurements is estimated to be less than 1%.

Whereas the measurements of the elastic properties of Variscan lower continental rocks were restricted to a set of 17 representative samples, a reference set of 36 mafic rocks was investigated geochemically to obtain a more detailed geochemical database for the various types of lower-crustal rocks of the Variscan crust. These results, together with a compilation of otherwise analysed mafic xenoliths and surface outcrops, give insights into the chemical composition of the Variscan lowermost continental crust and will be discussed in the following sections.

Results

Petrophysical data

We measured *P*- and *S*-wave velocities simultaneously in the three orthogonal directions of the sample cubes, first at increasing confining pressures (to 600 MPa) and then at increasing temperature (to 600°C) at 600 MPa confining pressure. The intrinsic pressure and temperature derivatives of velocities and the reference velocities, V_0, derived from the regression of the linear segments of the velocity *v.* pressure (300–600 MPa) and the velocity *v.* temperature curves (20–500°C) are available as a Supplimentary Publication, see p. xxx. They allow us to extrapolate seismic velocities for any *P*–*T* condition within the stability field of the constituent assemblage of rock-forming minerals.

Coupling of the laboratory seismic *in situ* data of relevant crustal rock types with those determined by geophysical field experiments allowed us to constrain the composition of the *in situ* deep crust. Using the experimentally determined pressure and temperature derivatives and the reference velocities (V_0) we calculated the *in situ* velocities at various depths (z) according to

$$V_{(z)} = V_{(0)} + (\mathrm{d}V/\mathrm{d}P)_P + (\mathrm{d}V/\mathrm{d}T)_T \quad (1)$$

where $P = \rho g z$ and $T = (\mathrm{d}T/\mathrm{d}z)z$.

Table 1. *Description of rocks; sample number, rock type, location and modal composition by least-squares fits or point counting of 17 samples of the Variscan lower crust*Title

Sample	Lithology	grt	px	amph	plg	ep	zo	rt	mica	Kfsp	qtz	chl	sm	op
ρ (g cm^{-3})		4.2	3.4	3.3	2.7	3.4	3.2	4.7	2.8	2.6	2.7	2.8	2.5	4.5
Waldviertel (Lower Austria)														
Se-D	plag cumulate			30	70 (An_{65})			<1						
Se-I	plag cumulate			26	57 (An_{88})	<1	16							
Se-H	plag cumulate		2		90 (An_{85})	3						5		
SW 2761	grt pyroxenite	11	72		≪1			≪1					17	≪1
DW 2898	grt pyroxenite	40	43		≪1		17	≪1	≪1					≪1
DW 2889	qtz diorite			14	44				26	3	13			≪1
go 3185	gabbro		≪1	30	57				10		3			≪1
go 3186	gabbro		≪1	38	52				7		3			≪1
go 3187	gabbro			19	52				18	≪1	11			≪1
go 3188	gabbro			17	48				25	≪1	10			≪1
Kutná Hora (Czech Republic)														
91027	eclogite	22	15	32	26	<1		4	<1					<1
91028	eclogite	29	33	30	6			1	<1					1
91031	eclogite	23	42	31	3	<1		<1	<1					<1
Münchberger Massif (Southern Germany)														
3141	eclogite	44	40				3	2	<1		11			
E-11	eclogite	35	44	6				2	1		7		4	
Ivrea Zone (Western Alps)														
GD	hbl gabbro		1	73	22									3
GH	grt gabbro	23	10	40	26				<1	<1				<1

grt, garnet; px, pyroxene; amph, amphibole; plg, plagioclase; ep, epidote; zo, zoisite; rt, rutile; Kfsp, alkali feldspar; gtz, quartz; chl, chlorite; sm, symplectite; op, opaque.

Table 2. *Experimentally derived* in situ *seismic properties of the different rock types for various depths (averages)*

	V_P (km s^{-1})			
Depth (km):	10	20	30	40
Pressure (MPa):	300	580	850	1120
Temperature (°C):	288	511	669	781
Diorite (1)	6.12	6.29	6.43	6.57
Gabbroids (6)	6.90	7.06	7.20	7.35
Plag cumulate (3)	6.86	6.99	7.11	7.23
Pyroxenite (2)	7.15	7.34	7.49	7.63
Eclogite (5)	7.73	7.92	8.08	8.22

Number of samples in parentheses.

Because there are no indications from the seismic experiments for velocity anisotropy in the crust of the EGT Central Segment we used the averages of the velocities measured in the three orthogonal directions, x, y, and z, for the calculation. The *in situ* P-wave velocities of the various rock types calculated for four depth levels of the Variscan crust are listed in Table 2. The calculation is based on crustal geotherms of the EGT Central Segment (shotpoint F) as reported by Cermak (1995) and refers to the velocity averages of the various lithologies. Lithostatic pressure was calculated using an average crustal density of 2.75 g cm^{-3} to 10 km depth, 2.80 g cm^{-3} to 20 km depth and 2.85 g cm^{-3} at depths >20 km. The experimentally derived *in situ* velocities represent the intrinsic velocities for depths >5 km where the effect of microcracks is almost eliminated. Figure 1b compares the experimentally derived *in situ* velocities of the various rock types collected from the Variscan lower crust with those obtained from seismic refraction profiling. From the comparison it is clear that a mixture of felsic and mafic granulites as well as meta-gabbros matches the lowermost crust very well and that eclogites are not a major component of lower crust in the EGT Central Segment.

Chemical composition of the Variscan mafic lower-crustal reference suite and implications for an estimated bulk composition

The reference suite of the lower continental crust studied here includes 10 plagioclase-dominated meta-cumulates, 17 meta-basalts and nine pyroxene-dominated meta-cumulates from the Bohemian Massif and the Ivrea–Verbano Zone. The mean chemical composition of major and trace elements of the MLC reference suite are available as a Supplementary Publication, see p. xxx. The samples have been grouped according to their major elements, Cr, Ni and Eu anomalies into high-grade rocks of either meta-basaltic or meta-cumulate composition. The meta-cumulates are subdivided into pyroxene-dominated and plagioclase-dominated samples.

The meta-basaltic granulites have a mean *mg*-number of 54 and Cr and Ni contents of 221 ppm and 72 ppm, respectively, and the Eu anomaly is 0.92. The pyroxene-dominated granulites have a higher *mg*-number of 73 and higher Cr (576 ppm) and Ni (126 ppm) contents. The Eu anomaly of 1.3 reflects plagioclase accumulation in the pyroxene-dominated magmatic assemblage. The plagioclase-dominated granulites are characterized by very low Mg ($MgO = 1.8$ wt %), very high Al (Al_2O_3 up to 33 wt %) and Ca ($CaO = 16$ wt %) contents and a mean Eu anomaly of 1.6, whereas the Cr and Ni values are much lower compared with the meta-basalts and pyroxene-dominated cumulates. This reflects the predominance of plagioclase over pyroxene in the cumulate rocks.

In Figure 2 the chondrite-normalized rare earth element (REE) pattern of these three groups are presented. The meta-basaltic granulites and the meta-plagioclase cumulates are light REE (LREE) enriched, whereas the pyroxene-dominated granulites are generally LREE depleted. The type of basaltic rock that gave rise to the meta-basaltic rocks cannot be assigned easily to a certain plate tectonic environment; for example there is no distinct depletion of high field strength elements over Ba, Th and La.

The compilation of otherwise investigated MLC of the central European Variscan region is presented in Table 3, together with information on lower-crustal compositions from worldwide sampling. The compiled database for the Variscan lower mafic crust includes a total of 206 samples (for references see footnote to Table 3), 147 of which are of cumulate character (both pyroxene and plagioclase dominated) and 59 refer to meta-basaltic bulk compositions. A problem arises of how to combine the two contrasting types of meta-mafic rocks to give an overall representative composition for MLC. It does not seem justified to estimate a mean composition merely on the basis of the number of samples analysed, as this would arbitrarily give preference to one or the other type. Unfortunately, there is no evidence from the EGT seismic refraction studies to give preference to one of the two types, as they largely have identical P-wave velocities (Mengel *et al.* 1991). Thus, as a model, we assume an equal proportion of both types in the lower-crustal section discussed here.

Table 3. *Compilation of the mafic lower continental crust*

	Meta-Basaltes (M)				Cumulates (C)				C : M = 1 :			LCC–R&P	
	mean	median	$\pm\sigma$	N	mean	median	$\pm\sigma$	N	mean	median	$\pm\sigma$	mean	median
SiO_2	46.8	47.3	0.4	59	47.3	47.8	0.2	147	47.1	47.6	0.3	50.5	49.4
TiO_2	1.55	1.62	0.09	59	1.08	0.76	0.07	147	1.32	1.19	0.08	1.04	0.91
Al_2O_3	15.7	15.4	0.4	59	17.3	16.2	0.2	147	16.5	15.8	0.3	16.5	16.7
FeO^T	11.7	11.5	0.3	59	8.5	8.9	0.2	147	10.1	10.2	0.2	9.0	8.8
MnO	0.22	0.21	0.01	59	0.15	0.16	0.004	146	0.19	0.19	0.01	0.15	0.14
MgO	8.3	7.9	0.4	59	9.5	8.7	0.4	147	8.9	8.3	0.4	7.7	7.3
CaO	11.1	11.1	0.3	59	12.4	12.0	0.2	147	11.7	11.5	0.2	9.8	10.4
Na_2O	2.14	2.18	0.1	59	2.01	2.10	0.07	147	2.08	2.14	0.1	2.39	2.50
K_2O	0.35	0.24	0.07	59	0.22	0.12	0.02	147	0.29	0.18	0.04	0.79	0.40
P_2O_5	0.22	0.18	0.03	59	0.16	0.07	0.02	145	0.19	0.13	0.02	0.23	0.13
H_2O^T	0.77	0.34	0.1	30	0.75	0.51	0.08	80	0.76	0.43	0.1	n.g.	n.g.
CO_2	0.13	0.05	0.03	16	0.25	0.14	0.05	54	0.19	0.10	0.04	n.g.	n.g.
LOI	1.5	1.5	0.8	2	1.0	0.7	0.2	11	1.2	1.1	0.5	n.g.	n.g.
Σ	100.5	99.5			100.6	98.2			100.5	98.9		98.1	96.7
Mg No.	55.8	55.1			66.6	63.5			61.1	59.2		60.3	59.7
Li	14	14	1	26	14	10	2	61	14	12	2	6	
Rb	10	4	5	21	5	1	1	64	8	3	3	17	5
Cs	0.4	0.3	0.1	12	0.6	0.5	0.1	26	0.5	0.4	0.09	1.04	0.14
Sr	268	187	39	56	299	242	21	127	284	215	30	422	338
Ba	118	76	23	42	206	50	78	92	162	63	51	521	269
Y	32	29	2	45	16	14	1	94	24	22	2	20	18
Zr	87	81	8	50	70	54	9	107	78	68	8	89	223
Hf	3.2	2.8	0.3	43	1.7	1.4	0.2	105	2.5	2.1	0.3	2.5	1.5
Nb	9	7	1	31	8	5	0.9	58	9	6	1	11	5
Ta	0.6	0.4	0.1	38	0.6	0.3	0	75	0.6	0.4	0.09	0.6	n.g.
Th	2.9	0.9	1	17	2.1	0.5	0.6	45	2.5	0.7	0.9	1.59	0.3
Sc	42	42	1	48	35	35	1	122	38	38	1.2	34	35
V	303	308	10	40	209	220	11	88	256	264	11	214	193
Cr	234	207	19	58	523	278	52	141	379	243	36	308	148
Co	45	46	1	45	48	45	2	92	46	46	2	51	75
Ni	92	74	8	46	200	152	16	111	146	113	12	113	83
Zn	81	81	7	23	51	40	5	41	66	61	6	90	82
La	10.6	6.9	1	59	6.7	4.5	0.7	145	8.6	5.7	1.0	13	9.3
Ce	25.6	18.1	3	59	15.9	12.0	1	140	20.7	15.1	2	27.8	19.7
Pr	1.5	1/3	0.3	14	1.5	1.0	0.3	24	1.5	1.1	0.3		1.2
Nd	19.1	13.8	3	32	9.8	6.3	1	85	14.5	10.1	1.8	10	9.3
Sm	4.8	4.5	0.3	59	2.7	2.3	0.2	146	3.7	3.4	0.3	3.52	3
Eu	1.41	1.38	0.08	59	1.12	0.92	0.07	146	1.27	1.15	0.07	1.24	1.2
Gd	5.55	5.41	0.3	59	1.85	1.24	0.2	46	3.7	3.32	0.3	3.58	3.26
Tb	0.96	0.97	0.05	56	0.52	0.48	0.03	130	0.74	0.72	0.04	0.58	0.54
Dy	5.53	5.18	0.9	14	1.73	1.22	0.3	33	3.63	3.2	0.6	3.86	3.77
Ho	1.06	0.81	0.2	14	0.33	0.18	0.06	24	0.69	0.49	0	0.77	0.62
Er	3.47	3.59	0.6	17	1.06	0.64	0.2	37	2.27	2.12	0.4	2.22	1.76
Tm	0.39	0.31	0.08	14	0.18	0.11	0.04	25	0.28	0.21	0.06	n.g.	n.g.
Yb	3.12	2.9	0.2	59	1.61	1.4	0.09	145	2.36	2.15	0	1.9	1.46
Lu	0.46	0.44	0.03	55	0.25	0.23	0.02	135	0.36	0.33	0.02	0.28	0.22
Eu/Eu*	0.84	0.85			1.55	1.69			1.05	1.05		1.07	1.17

The data sets are based on Variscan samples from outcrops of central Europe and include data from numerous xenoliths ($N = 206$, see references). Samples of cumulate-dominated mafic lower continental crust as Cr > 500 ppm, Ni > 200 ppm and/or Eu/Eu* > 1. Samples of meta-basaltic lower continental crust as Cr < 500 ppm, Ni < 200 ppm. The composition of the lower continental crust by Rudnick & Presper (1990) is given in the last two columns. $\pm$s = standard deviation of the mean. n.g., not given.

Data sources: Northern Hessian Depression (Germany)—xenoliths $N = 42$ ($N_{\text{cumulates}} = 34$, $N_{\text{meta-basalts}} = 8$), Mengel (1990). Eifel (Germany)—xenoliths $N = 15$ ($N_{\text{cumulates}} = 10$; $N_{\text{meta-basalts}} = 5$), Stosch *et al.* (1990); Mengel (1990). Helsburger Gangschar (Germany)—xenoliths $N = 8$ ($N_{\text{cumulates}} = 8$), Mengel (1990). Münchberg Mass (Germany) $N = 23$ ($N_{\text{cumulates}} = 17$, $N_{\text{meta-basalts}} = 6$), Stosch & Lugmuir (1990) and this study. Bohemian Massif and Lower Austria $N = 67$ ($N_{\text{cumulates}} = 48$, $N_{\text{meta-basalts}} = 19$), Beard *et al.* (1992); Vellmer (1992); Becker (1996) and this study. Massif Central (France) $N = 12$ ($N_{\text{cumulates}} = 5$, $N_{\text{meta-basalts}} = 7$), Downes *et al.* (1990); Paquette *et al.* (1995). Ivrea-Verbano Zone (Italy) $N = 39$ ($N_{\text{cumulates}} = 25$, $N_{\text{meta-basalts}} = 14$), Dostal & Capredi (1979); Pin & Sills (1986); Mengel (1990); Mazzuchelli *et al.* (1992); Wittenberg (1997) and this study.

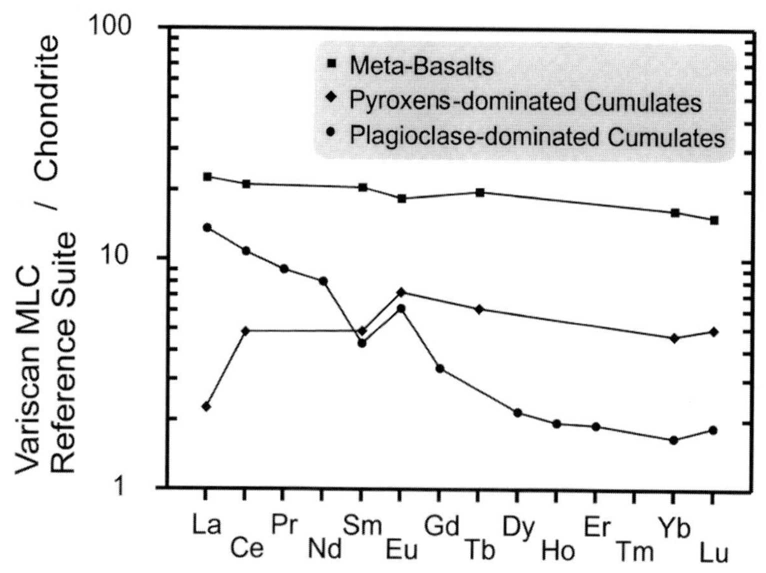

Fig. 2. REE pattern of the Variscan mafic lower continental crustal rocks. Chondrite-normalized values accoridng to Boynton (1984). The meta-basalts show a flat MORB-type pattern, whereas the pyroxene-dominated cumulates show a depletion in LREE relative to HREE, and the plagioclase-dominated cumulates are LREE enriched. Both cumulates show positive Eu anomaly.

The assumption of a 1 : 1 proportion of meta-cumulate relative to meta-basaltic lower-crustal rocks yields a composition that is very close to the lower-crustal model compositions compiled by Rudnick & Presper (1990) and Rudnick & Fountain (1995) in terms of Al, Fe, Mn, Mg, Ca and Na. Our estimate yields lower SiO_2 values; this might be explained by the presence of granulite samples, which have been identified by Mengel & Hoefs (1990) as former spilites, on the basis of their Li–SiO_2–$\delta^{18}O$ signatures. In addition, the lower SiO_2 values may reflect previous tonalitic to granitic melt extractions. The mean values of the compilation for the Variscan MLC are for many trace elements close to the mean values reported by Rudnick & Presper (1990) (Table 5). However, their Sr and Ba values are significantly higher, whereas their V, Cr and Ni values are slightly lower compared with the compilation presented here. The lower V, Cr and Ni contents in the Rudnick & Presper (1990) compilation are probably due to higher feldspar and lower pyroxene abundance in the suite of rocks. The REE pattern of the mean composition of the MLC components (meta-basalts, meta-cumulates) and the calculated bulk composition of MLC is given in Fig. 3. Compared with the global averages, the LREE are slightly lower and the heavy REE (HREE) are slightly higher for the calculated Variscan mafic lower crust.

Assumption: crustal growths by mantle magmatism

The Variscan crust of Central Europe is distinctly dominated by post-Archaean rocks with respect to Nd model ages (e.g. Liew & Hofmann 1988) and U–Pb zircon discordia upper intersects of less than 2.5 Ga. Thus, Archaean components are largely missing in the Central European crust along the EGT Central Segment.

The modes of post-Archaean crustal growth are still a matter of debate. Some models postulate a tonalitic to andesitic bulk composition. The following mass balance is based on the fact that the overall input of magmas from the mantle is largely basaltic, not andesitic and certainly not tonalitic. Taylor & McLennan (1985 and subsequent papers) have postulated that the continental crust grows by andesitic magma additions resulting in a low *mg*-number of 50, low Cr and Ni contents of 55 and 30 ppm, respectively, and a slight negative Eu anomaly (0.92). Kay & Kay (1986, 1991) have convincingly shown that the input from the mantle should be even tholeiitic rather than andesitic. Here we focus on the *mg*-number, Cr and Ni arguments which rule out that andesitic melts originated from the depleted mantle. In contrast to Taylor & McLennan (1985) we propose that the input from the mantle has a bulk composition of basaltic andesite with *mg*-number > 55

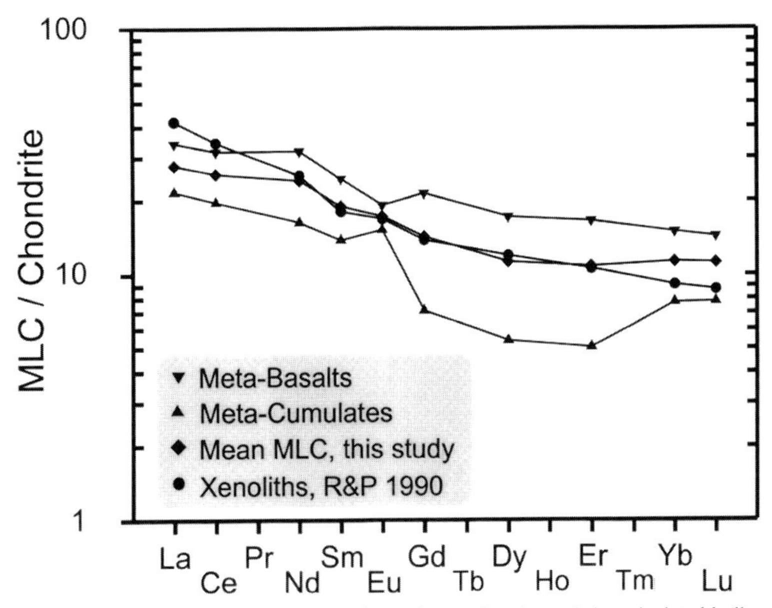

Fig. 3. REE pattern of the Variscan mafic lower continental crustal rocks and the calculated bulk composition of the mafic lower crust. Chondrite-normalized values according to Boynton (1984). Additionally, the mafic lower crust as observed by Rudnick & Presper (1990) on world-wide xenoliths is given in the last two colums of Table 3. (For further description, see text.)

corresponding to MgO values between 5 and 11 wt %. Cr and Ni in this type of magma vary in the range 43–669 ppm and 25–312 ppm, respectively, indicating almost primitive melt compositions (for mean values see Table 4). A compilation of basaltic andesites from world-wide sampling is included in Table 4 (for references, see the footnote to Table 4).

Consequences for bulk crustal evolution

On the assumption that post-Archaean crustal growth was accomplished by an overall input of basaltic andesites it is tempting to calculate the mass proportions of the felsic upper and middle and the mafic lower continental crust that should match a bulk basaltic andesite composition. The model proposed here explains the observed overall felsic composition of the post-Archaean crust in central Europe.

$$\text{basaltic andesite} = a \text{ upper crust} + b \text{ middle crust} + c \text{ mafic lower crust} \quad (2)$$

where a, b and c are the mass proportions of the crustal sections. Because the densities of the respective rocks vary over a small range between 2.8 and 3.0 g cm^{-3}, as a first approximation, the mass proportions can be taken as volume proportions.

For the calculations we used the averages reported by Wedepohl (1995) for the upper continental crust (UCC), by Rudnick & Presper (1990) for middle continental crust (MCC) and the calculated composition for the MLC of the Variscan Orogen are available as a Supplementary Publication, see p. xxx. All data were normalized to phosphorus- and volatile-free major element sums. In a first step, the mass balance is calculated by means of least-squares fits for major elements of a simplified three-layer continental crust. The results are given in Table 4 and plotted in Fig. 4.

The calculated proportions of felsic upper and felsic lower to mafic lower crust are 0.29 and 0.09 to 0.62, respectively. For a total crustal thickness of 30 km, these proportions would result in a thickness of 11.4 km for upper and felsic lower crust to 18.6 km for MLC.

If this model is true, the mass proportions for felsic upper to felsic middle and MLC should match the composition of basaltic andesites with respect to trace elements. To check this assumption we have plotted trace element concentrations in average basaltic andesites over the calculated trace element contents in the calculated post-Archaean continental crust (0.38 felsic relative to 0.62 mafic units). Table 4 summarizes the results

Table 4. *Mass balance of the bulk continental crust*

	Continental crust									
	UC		MC		MLC		BCC			
Mass proportions (%)	29	+	9	+	62	=	100	$BA_{comp.}$	$\pm\sigma$	n
SiO_2	66.9		62.5		48.0		54.8	54.8	0.2	35
TiO_2	0.54		0.72		1.34		1.05	0.85	0.03	35
Al_2O_3	15.08		15.98		16.8		16.23	16.51	0.3	35
FeO_T	4.09		6.6		10.32		8.18	7.87	0.2	35
MnO	0.07		0.1		0.19		0.15	0.15	0	35
MgO	2.31		3.5		9.06		6.60	7.06	0.3	35
CaO	4.25		5.26		11.93		9.10	8.81	0.2	35
Na_2O	3.57		3.3		2.12		2.65	2.96	0.08	35
K_2O	3.19		2.07		0.29		1.29	1.05	0.1	35
Σ	100.0		100.0		100.0		100.0	100.0		
Mg No.	50.2		48.6		61.0		59.0	61.5		
Rb	110		62		8		42	17	2	24
Cs	5.8		2.4		0.5		2.2	0.8	0.1	15
Sr	316		281		284		293	393	29	25
Ba	668		402		162		330	213	21	24
Y	21		22		24		23	18	1	25
Zr	237		125		78		128	82	6	25
Hf	5.8		4		2.5		3.6	2.1	0.2	18
Nb	26		8		9		14	3.7	0.5	20
Ta	1.5		0.6		0.6		0.9	0.2	0.02	3
Th	10.3		6.1		2.5		5.1	2.2	0.3	26
U	2.5		1.6		0.4		1.1	0.5	0.1	11
Sc	7		22		38		28	28	1	27
V	53		118		256		185	214	11	20
Cr	35		83		379		252	241	26	31
Co	12		25		46		34	24	1	22
Ni	19		33		146		99	104	15	31
Zn	52		70		66		62	78	3	16
Pb	17		15		8		11	6	0.9	14
La	32.2		17		8.6		16.2	9.2	0.6	31
Ce	65.7		54		20.7		37	21	1	31
Pr	6.3		5.8		1.5		3.3	1.6	0.4	5
Nd	25.9		24		14.4		18.6	13.0	0.9	25
Sm	4.7		4.4		3.7		4.1	3.0	0.2	35
Eu	1.0		1.5		1.3		1.2	1.0	0.04	29
Gd	2.8		4		3.7		3.5	2.8	0.4	8
Tb	0.5		0.6		0.7		0.6	0.5	0.02	28
Dy	2.9		3.8		3.6		3.4	2.9	0.3	10
Ho	0.6		0.8		0.7		0.7	0.6	0.09	5
Yb	1.5		2.3		2.4		2.1	1.8	0.07	33
Lu	0.3		0.4		0.4		0.4	0.3	0.01	26
Eu/Eu*	0.80		1.09		1.05		1.00	1.01		

BCC, bulk continental crust, calculated as 29% UC + 9% MC + 62% MLC = 100% BCC, where UC is upper continental crust (Wedepohl 1995), MC is middle crust (Rudnick & Fountain 1995) and MLC as compiled in Table 3 and given in colum C : M = 1 : 1 median therein. (For BA_{comp}, see references.) $\pm\sigma$ is standard deviation of the mean.

Data sources: Chile ($n = 10$), Thorpe *et al.* (1976), Hickey-Vargas *et al.* (1989), López-Escobar *et al.* (1995), Feeley *et al.* (1988); Aleutians ($n = 3$), Singer *et al.* (1992); South Kamchatka ($n = 3$), Bindeman & Bailey (1994); Japan ($n = 8$), Ishizaka & Carlson (1983); Kurile Islands ($n = 2$), Bailey *et al.* (1989); Philippines ($n = 4$), Sajona *et al.* (1994); Papua New Guinea ($n = 3$), Woodhead & Johnson (1993); South Africa ($n = 2$), Jochum *et al.* (1991).

of the mass balance calculations. The majority of trace elements show a good agreement between the compilation of basaltic andesites and the calculated bulk crust. The elements Rb, Ba, Ce and Zr are higher in the calculated bulk continental crust compared with the compiled basaltic andesite dataset, whereas Sr and V are relatively depleted.

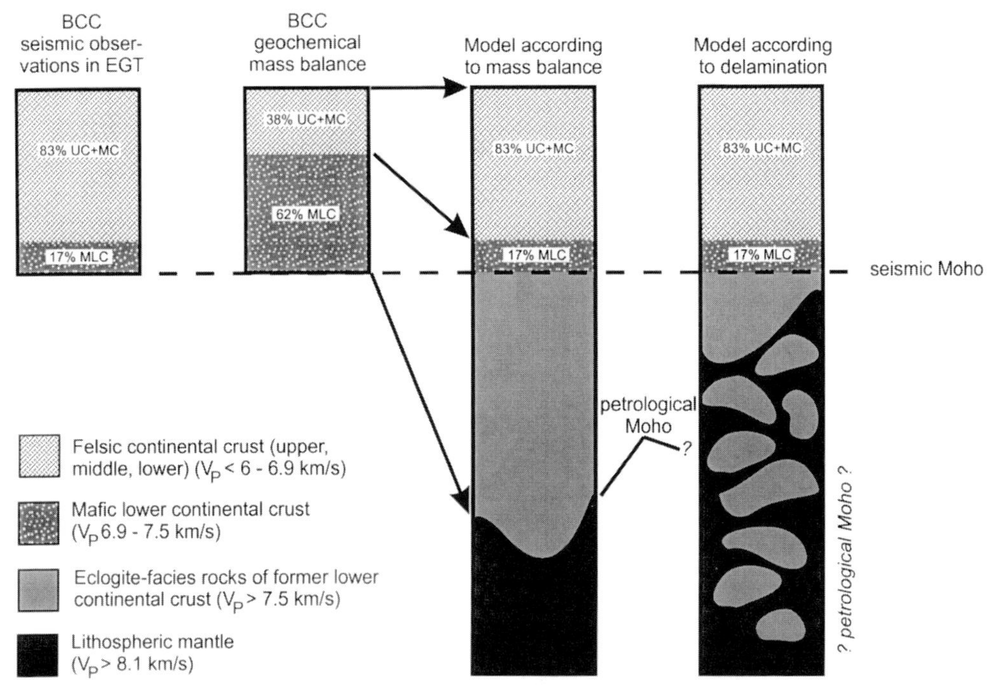

Fig. 4. Outer left column: distribution of principal crustal units as observed in the EGT Central Segment. Inner left column: calculated by mass balance on the assumption that bulk continental crust equals basaltic andesites. Inner right column: distribution of principal central units according to mass balances in accordance with seismic observations. The outer right column describes lower continental crust in eclogites facies residing in the upper mantle. BCC, bulk continental crust; UC, upper felsic continental crust; MC, middle to lower felsic continental crust; MLC, mafic lower continental crust. (For further explanation, see text).

Discussion

An inspection of Table 3 shows discrepancies between trace element data of the assumed basaltic andesite crust and the bulk crust calculated from the data of felsic and mafic units. This may be exemplified by a closer look at the Sr and Ba data. The standard deviation of the mean in basaltic andesites is ± 29 ppm for Sr and ± 21 ppm for Ba. The variation of Sr and Ba in meta-basaltic and meta-cumulate rocks is ± 39 and ± 21 for Sr and ± 23 and ± 78 for Ba, respectively (see Table 3). The observed discrepancies for Ba and Sr between basaltic andesites and calculated bulk crust can thus be assigned to the large diversities of these trace elements in either dataset. In addition, it can not be excluded at the present state of the investigation that early intra-crustal differentiation processes have not been accounted for correctly. Although these uncertainties are implicit in our model, some conclusions for the evolution of the Variscan continental crust can be drawn.

The model presented here for thin felsic upper and felsic lower and thick mafic lower continental crust is in strong contrast to the seismic observations made for the EGT Central Segment. Ansorge *et al.* (1992) observed a thickness of around 20 km for rocks with *P*-wave velocities <6.9 km s^{-1}. The results of our mass balance for the proportion of the MLC are not in agreement with the observations of the present-day continental crust, as the calculated overall crustal thickness must have reached at least about 53 km before delamination. Hence, according to the calculated proportions of 38 vol. % felsic crust and 62 vol. % MLC an initial thickness of 20–33 km of felsic crust and 33–54 km of MLC has to be assumed for the EGT Central Segment. These findings are in agreement with the observed continental crust thickness at orogenic collision zones.

The implications of the presented mass balance are surprising and, at a first glance, seem to be totally incompatible with the observed 7 km thickness of MLC in the EGT Central Segment. However, it has to be taken into account that the conversion of MLC rocks into eclogites under high-pressure metamorphism in collision zones gives rise to *P*-wave velocities that cannot be distinguished from those of upper-mantle peridotites (Mengel & Kern 1992). This is true not only for meta-basaltic compositions but also for plagioclase-dominated meta-cumulates. As a consequence, the roots of collision type orogens residing in the subcontinental mantle cannot be detected by seismic refraction studies. Furthermore, eclogite-facies mafic rocks of a former lower crust have higher densities than average subcontinental peridotites (3.5 g cm^{-3} relative to 3.3 g cm^{-3}). For the transformation of MLC rocks into eclogite-facies rocks with densities of >3.5 g cm^{-3} pressures up to 1.5 GPa are needed, corresponding to about 53 km depth (Kay & Kay 1991). Under favourable tectonic conditions high-density mafic units of orogenic roots may thus be delaminated from the base of the crust and recycled in the upper mantle.

Processes that affect the roots of the lower continental crust and the crust–mantle boundary during collision-type tectonics are not well understood. Among the few examples studied so far is the Alpine orogenic belt. Detailed seismic studies of the Alps give evidence for a 'subducted' lower crust as a result of the collision tectonics (Pfiffner 1992, p. 188. fig. 6-27). Continuous 'subduction' and delamination of lower-crustal units being converted into eclogite-facies rocks by pressure increase will result in a successive loss of MLC. By this process, the volume of the felsic upper crust progressively increases relative to that of the MLC, resulting in the observed overall tonalitic bulk composition along the EGT Central Segment. One of the consequences of the relative enrichment of the felsic crustal units is the observation of a negative Eu anomaly. The currently observed value of Eu/Eu* is 0.86, which implies that the excess Eu has been removed by delamination of residual granulites and/or plagioclase-dominated meta-cumulates into the mantle.

Similar results have recently been obtained by Gao *et al.* (1998), who also invoked massive crustal roots of eclogite-facies rocks. That example for the crust of central East China demonstrates that delamination of MLC causes bulk upper- and middle-crust compositions to become increasingly felsic during each orogenic cycle.

The authors are indebted to D. Tanner, R. Kay and P. Sachs for constructive criticism.

References

AFTALION, M., BOWES, D. R. & VRÁNA, S. 1989. Early Carboniferous U–Pb zircon age for garnetiferous, perpotassic granulites. Blansky les massif, Czecheslovakia. *Neues Jahrbuch für Mineralogie, Monatshefte*, **4**, 145–152.

ANSORGE, J., BLUNDELL, D. & MUELLER, S. 1992. Europe's lithosphere—seismic structure. *In*: BLUNDELL, D., FREEMAN, R. & MUELLER, S. (eds) *A Continent Revealed. The European Geotraverse*. Cambridge University Press, Cambridge, 33–70.

BAILEY, J. C., FROLOVA, T. I. & BURIKOVA, I. A. 1989. Mineralogy, geochemistry and petrogenesis of Kurile island-arc basalts. *Contributions to Mineralogy and Petrology*, **102**, 265–280.

BEARD, B. L., MEDRARIS, L. G., JOHNSON, C. M., BRUECKNER, H. K. & MISAR, Z. 1992. Petrogenesis of Variscan high-temperature group A eclogites from the Moldanubian Zone of the Bohemian Massif, Czechoslovakia. *Contributions to Mineralogy and Petrology*, **111**, 468–483.

BECKER, H. 1996. Crustal trace elements and isotope signatures in garnet pyroxenites from garnet peridotite massifs from lower crusts. *Journal of Petrology*, **37**, 785–811.

BINDEMAN, I. N. & BAILEY, J. C. 1994. A model of reverse differentiation at Dikii Greben Volcano, Kamchatka: progressive basic magma vesiculation in a silicic magma chamber. *Contributions to Mineralogy and Petrology*, **117**, 263–278.

BOHLEN, S. R. & METZGER, K. 1989. Origin of granulite terrains and the formation of the lowermost continental crust. *Science*, **244**, 326-329.

BOYNTON, W. V. 1984. Geochemistry of rare earth elements: meteorite studies. *In*: HENDERSON, P. (ed.) *Rare Earth Element Geochemistry*. Elsevier, Amsterdam, 63–114.

CARSWELL, D. A. & JAMTVEIT, B. 1990. Variscan Sm–Nd ages for the high-pressure metamorphism in the Moldanubian Zone of the Bohemian Massif, Lower Austria. *Neues Jahrbuch für Mineralogie, Abhandlungen*, **162**, 69–78.

CERMAK, V. 1995. A geothermal model of the Central segment of the European Geotraverse. *Tectonophysics*, **244**, 51–55.

DALLMEYER, R. D., NEUBAUER, F. & HÖCK, V. 1992. Chronology of late Paleozoic tectonothermal activity in the southeastern Bohemian Massif, Austria (Moldanubian and Moravo-Silesian zones): $^{40}Ar/^{39}Ar$ mineral age controls. *Tectonophysics*, **210**, 135–153.

DOSTAL, J. & CAPREDI, S. 1979. Rare earth elements in high-grade metamorphic rocks from the Western Alps. *Lithos*, **12**, 41–49.

DOWNES, H., DUPUY, C. & LEYRELOUP, A. F. 1990. Crustal evolution of the Hercynian belt of Western Europe. Evidence from lower-crustal granulite xenoliths (French Massif Central). *Chemical Geology*, **83**, 209–231.

DUDEK, A. & FEDIUKOVÁ, E. 1974. Eclogites of the Bohemian Moldanubicum. *Neues Jahrbuch für Mineralogie, Abhandlungen*, **21**, 127–159.

FEDIUKOVÁ, E. 1989. Eclogites in Czechoslovakia. *Krystalinikum*, **20**, 27–48.

FEELEY, T. C., DUNGAN, M. A. & FREY, F. A. 1998. Geochemical constraints on the origin of mafic and silicic magmas, Cordón El Guadal,

Tatara–San Pedro Complex, central Chile. *Contributions to Mineralogy and Petrology*, **131**, 393–411.

FRANKE, W. 1989. Tectonostratigraphic units in the Variscan belt of central Europe. *In*: DALLMEYER, R. D. (ed.) *Terranes in the Circum-Atlantic Palaeozoic Orogens*. Geological Society of America, Special Papers, **230**, 67–90.

FRIEDL, G., VON QUADT, A., OCHSNER, A. & FINGER, F. 1992. Timing of the Variscan Orogeny in the southern Bohemian Massif (NE-Austria) deduced from new U–Pb zircon and monazite dating. *Terra Abstracts*, **5**, *Supplement 1*, 235–236.

GAO, S., ZHANG, B.-R., JIN, Z.-M., KERN, H., LUO, T.-C. & ZHAO, Z.-D. 1998. How mafic is the lower continental crust? *Earth and Planetary Science Letters*, **161**, 101–117.

GEBAUER, D. 1983. FIEC feedback sections on Hercynides and isotopes. *Terra Cognita*, **3**, 320–323.

—— 1986. The development of the continental crust of the European Herzynides since the Archaean based on radiometric data. *In*: FREEMAN, R., MUELLER, S. & GIESE, P. (eds) *The Central Segment ESF/ESRC. Proceedings of the Third Workshop on the European Geotraverse (EGT) Project, Strasbourg*, 15–23.

—— 1990. Isotopic systems—geochronology of eclogites. *In*: CARSWELL, D. A. (ed.) *Eclogite Facies Rocks*. Blackie, Glasgow, 141–179.

GRIFFIN, W. L. & O'REILLY, S. Y. 1987. Is the continental Moho the crust–mantle boundary? *Geology*, **15**, 241–244.

HICKEY-VARGAS, R., ROA, H. M., ESCOBAR, L. L. & FREY, F. A. 1989. Geochemical variations in Andean basaltic and silicic lavas from the Villarrica–Lanin volcanic chain (39.5° S). An evaluation of source heterogeneity, fractional crystallization and crustal assimilation. *Contributions to Mineralogy and Petrology*, **103**, 361–386.

ISHIZAKA, K. & CARLSON, R. W. 1983. Nd–Sr systematics of the Setouchi volcanic rocks, southwest Japan: a clue to the origin of orogenic andesites. *Earth and Planetary Science Letters*, **64**, 327–340.

JOCHUM, K. P., ARNDT, N. T. & HOFMANN, A. W. 1991. Nd–Th–La in komatiites and basalts: constraints on komatiite petrogenesis and mantle evolution. *Earth and Planetary Science Letters*, **107**, 272–291.

KAY, R. W. & KAY, S. M. 1986. Petrology and geochemistry of the lower continental crust. An overview. *In*: DAWSON, J. B., CARSWELL, D. A., HALL, J. & WEDEPOHL, K. H. (eds) *The Nature of the Lower Continental Crust*. Geological Society, London, Special Publications, **24**, 147–159.

—— & —— 1991. Creation and destruction of lower continental crust. *Geologische Rundschau*, **80**, 259–278.

KERN, H., LIU, B. & POPP, T. 1997. Relationship between anisotropy of P- and S-wave velocities and anisotropy of attenuation in serpentine and amphibolite. *Journal of Geophysical Research*, **102**, 3051–3065.

KRÖNER, A., WENDT, I., LIEW, T. C. *et al.* 1988. U–Pb zircon and Sm–Nd model ages of high grade Moldanubian metasediments, Bohemian Massif, Czechoslovakia. *Contributions to Mineralogy and Petrology*, **99**, 257–266.

LIEW, T. C. & HOFMANN, A. W. 1988. Precambrian crustal components, plutonic associations, plate environment of the Hercynian Fold Belt of central Europe: indications from a Nd and Sr isotopic study. *Contributions to Mineralogy and Petrology*, **98**, 129–138.

LOOK, G., STOSCH, H.-G. & SECK, H. A. 1990. Granulites facies lower-crustal xenoliths from the Eifel, West Germany: petrological and geochemical aspects. *Contributions to Mineralogy and Petrology*, **105**, 25–41.

LÓPEZ-ESCOBAR, L., PARADA, M. A., HICKEY-VARGAS, R., FREY, F. A., KEMPTON, P. D. & MORENO, H. 1995. Calbuco Volcano and minor eruptive centers distributed along the Liqinñe-Ofqui Fault Zone, Chile (41°–42° S): contrasting origin of andesitic and basaltic magma in the Southern Volcanic Zone of the Andes. *Contributions to Mineralogy and Petrology*, **119**, 345–361.

MATTE, P. 1991. Accretioning history and crustal evolution of the Variscan Belt in western Europe. *Tectonophysics*, **196**, 309–337.

MAZZUCCHELLI, M., RIVALENTI, G., VANNUCCI, R., BOTTAZZI, P., OTTOLONI, L., HOFMANN, A. W. & PARENTI, M. 1992. Primary positive Eu anomaly in clinopyroxenes of low-crust gabbroic rocks. *Geochimica et Cosmochimica Acta*, **56**, 2363–2370.

MEHNERT, K. R. 1975. The Ivrea Zone—a model for the deep crust. *Neues Jahrbuch für Mineralogie, Abhandungen*, **125**, 156–199.

MENGEL, K. 1990. Crustal xenoliths from Tertiary volcanics of the Northern Hessian Depression. *Contributions to Mineralogy and Petrology*, **104**, 8–26.

—— & HOEFS, J. 1990. Li–$\delta^{18}O$–SiO_2 systematics in volcanic rocks and mafic lower-crustal granulite xenoliths. *Earth and Planetary Science Letters*, **101**, 42–53.

—— & KERN, H. 1992. Evolution of the petrological and seismic Moho—implications for the continental crust–mantle boundary. *Terra Nova*, **4**, 109–116.

——, SACHS, P. M., STOSCH, H. G., WÖRNER, G. & LOOCK, G. 1991. Crustal xenoliths from Cenozoic volcanic field of West Germany: implications for structure and composition of the continental crust. *Tectonophysics*, **195**, 271–289.

O'REILLY, S. Y. & GRIFFIN, W. L. 1987. The composition of the lower crust and the nature of the continental Moho—xenolith evidence. *In*: NIXON, P. H. (ed.) *Mantle Xenoliths*. Wiley, Chichester, 413–430.

PAQUETTE, J. L., MONCHOUX, P. & COUTURIER, M. 1995. Geochemical and isotopic study of norite-eclogite transition in the European Variscan belt; implications for U–Pb zircon systematics in metabasic rocks. *Geochimica et Cosmochimica Acta*, **59**, 1611–1622.

PFIFFNER, A. 1992. Deep-reaching geodynamic processes in the Alps. *In*: BLUNDELL, D., FREEMAN, R. & MUELLER, S. (eds) *A Continent Revealed. The European Geotraverse*. Cambridge University Press, Cambridge, 180–190.

PIN, C. & SILLS, J. D., 1986. Petrogenesis of layered gabbros and ultramafic rocks from the Val Sesia, The Ivrea Zone, northwest Italy. *In*: DAWSON, J. B., CARSWELL, D. A., HALL, J. & WEDEPOHL, K. H. (eds) *The Nature of the Lower Continental Crust*. Geological Society, London, Special Publications, **24**, 231–250.

—— & VIELZEUF, D. 1983. Granulites and related rocks in Variscan median Europe: a dualistic interpretation. *Tectonophysics*, **93**, 47–74.

PRODEHL, C. & GIESE, P. 1990. Seismic investigations around the EGT in central Europe. *In*: FREEMAN, R., MUELLER, S. & GIESE, P. (eds) *The European Geotraverse: Integrative Studies*. European Science Foundation, Strasbourg, 77–97.

RIVALENTI, G., ROSSI, A., SIENA, F. & SINIGOI, S. 1984. The layered series of the Ivrea–Verbano igneous complex, Western Alps, Italy. *Tschermarks Mineralogische und Petrographische Mitteilungen*, **33**, 77–99.

RUDNICK, R. L. & FOUNTAIN, D. M. 1995. Nature and composition of the continental crust. A longer crustal perspective. *Reviews of Geophysics*, **33**, 267–309.

—— & PRESPER, T. 1990. Geochemistry of intermediate- to high-pressure granulites. *In*: VIELZEUF, D. & VIDAL, P. (eds) *Granulites and Crustal Evolution*. Kluwer, Dordrecht, 523–550.

—— & TAYLOR, S. R. 1987. The composition and petrogenesis of the lower crust: a xenolith study. *Journal of Geophysical Research*, **92**, 13981–14005.

SAJONA, F. G., BELLON, H., MAURY, R. C., PUBELLIER, M., COTTEN, J. & RANGIN, C. 1994. Magmatic response to abrupt changes in geodynamic settings: Pliocene–Quaternary calc-alkaline and Nd-enriched lavas from Mindanao (Philippines). *Tectonophysics*, **237**, 47–72.

SCHMÄDICKE, E., MEZGER, K., COSCA, M. A. & OKRUSCH, M. 1995. Variscan Sm–Nd and Ar–Ar ages of eclogite facies rocks from Erzgebirge, Bohemian Massif. *Journal of Metamorphic Geology*, **13**, 537–552.

SIGHINOLFI, G. P. & GORGONI, C. 1978. Chemical evolution of high-grade metamorphic rocks—anatexis and remotion of material from granulite terrains. *Chemical Geology*, **22**, 157–176.

SINGER, B. A., MYERS, J. D. & FROST, C. D. 1992. Mid-Pleistocene lavas from the Seguam volcanic center, central Aleutian arc: closed-system fractional crystallization of a basalt to rhyodacite eruptive suite. *Contributions to Mineralogy and Petrology*, **110**, 87–112.

STOSCH, H.-G. & LUGMAIR, G. W. 1987. Geochronology and geochemistry of eclogites from the Münchberger Gneiss Massif, FRG. *Terra Cognita*, **7**, 163.

—— & —— 1990. Geochemistry and evolution of MORB-type eclogites from the Münchberg Massif, southern Germany. *Earth and Planetary Science Letters*, **99**, 230–249.

——, —— & SECK, H. A., 1986. Geochemistry of granulite-facies lower-crustal xenoliths: implications for the geological history of the lower continental crust beneath the Eifel, West Germany. *In*: DAWSON, J. B., CARSWELL, D. A., HALL, J. & WEDEPOHL, K. J. (eds) *The Nature of the Lower Continental Crust*. Geologcial Society, London, Special Publications, **24**, 309–317.

TAYLOR, S. R. & MCLENNAN, S. M. 1985. *The Continental Crust. Its Composition and Evolution*. Blackwell, Oxford.

THORPE, R. S., POTTS, P. J. & FRANCIS, P. W. 1976. Rare earth data and petrogenesis of andesite from the North Chilean Andes. *Contributions to Mineralogy and Petrology*, **54**, 65–78.

VAN BREEMEN, O., AFLALION, M., BROWES, D. R., DUDEK, A., MISAR, Z., POVONDRA, P. & VRÀNA, S. 1982. Geochronology studies of the Bohemian Massif, Czechoslovakia, and their significance in the evolution of Central Europe. *Transactions of the Royal Society of Edinburgh: Earth Sciences*, **73**, 89–108.

VELLMER, C. 1992. *Stoffbestand und Petrogenese von Granuliten und granitischen Gesteinen der südlichen Böhmischen Masse in Niederösterreich*. Ph D thesis, Universität Göttingen.

WEDEPOHL, K. H. 1995. The composition of the continental crust. *Geochimica et Cosmochimica Acta*, **59**, 1217–1232.

WENDT, I., KRÖNER, A., TODT, W. & FIALA, J. 1994. U–Pb-zircon and Sm–Nd dating of Moldanubian high-*P*, high-*T* granulites from south Bohemia. *Journal of the Geological Society, London*, **151**, 83–90.

WITTENBERG, A. 1997. *Geochemische und isotopengeochemische Untersuchungen an mafischen Gesteinen der kontinentalen Unterkruste und deren Hochdruckäquivalenten (Grospydite)*. PhD thesis, Universität Hannover.

WOODHEAD, J. D. & JOHNSON, R. W. 1993. Isotopic and trace-element profiles across the New Britain island arc, Papua New Guinea. *Contributions to Mineralogy and Petrology*, **113**, 479–491.

ZINGG, A. 1980. Regional metamorphism in the Ivrea Zone (Southern Alps, N-Italy). Field and microscopic investigations. *Schweizerische Mineralogische und Petrographische Mitteilungen*, **60**, 153–173.

ZOUBEK, V., FIALA, J., VANKOVA, V., MACHART, J. & STETTNER, G. 1988. Moldanubian region. *In*: ZOUBEK, V., COGNE, J., KOZHOUKHAROV, D. & KRAUTER, H. G. (eds) *Precambrian in Younger Fold Belts*. Wiley, Chichester, 183–267.

Hybrids, magma mixing and enriched mantle melts in post-collisional Variscan granitoids: the Rastenberg Pluton, Austria

A. GERDES[1,3,4], G. WÖRNER[1] & F. FINGER[2]

[1]*Geochemisches Institut, Goldschmidtstr. 1, 37077 Göttingen, Germany*

[2] *Institut für Mineralogie, Hellbrunnerstr. 34, 5020 Salzburg, Austria*

[3] *Departamento de Mineralogía y Petrología, Campus Fuentenueva, 18002 Granada, Spain*

[4] *Present address: NERC Isotope Geosciences Laboratory, Keyworth, Nottingham NG12 5GG, UK (e-mail: ager@bgs.ce.uk)*

Abstract: The composite Rastenberg Pluton in the South Bohemian Massif preserves an example of generation of relatively homogeneous granitoid hybrids by mixing of mafic and felsic magmas. Metaluminous melagranites and quartz monzonites, the main lithologies of the pluton, are interpreted to be hybrids. In contrast, a slightly peraluminous biotite granodiorite is considered to be a lower-crustal melt. In addition, abundant mafic ultrapotassic enclaves, country-rock lamprophyres and quartz monzodiorite bodies represent distinct lithospheric mantle-derived magmas, which were only slightly modified by fractionation and/or magma mixing. Almost continuous linear chemical and isotope correlations joining the enclaves, quartz monzonites, melagranites and granodiorites indicate the importance of a mixing process in the generation of these rocks. Incompatible elements decrease with increasing silica and moderate negative Eu anomalies disappear towards the granodioritic endmember. Pb, Sr and Nd isotopes show typical crustal values in all granitoids but are more radiogenic in the ultrapotassic endmember. In the Rastenberg Pluton, the interaction of mantle- and crustal-derived magmas has produced unequivocal petrographic, chemical and isotopic evidence for mixing and mingling. Because similar features are lacking in most other Variscan plutons, we suggest that mantle magmas were not substantially involved in their genesis. Nevertheless, small-volumes of hybrids, involving variably enriched mantle-derived melts, do crop out locally throughout the central Variscides. In view of the generally strongly enriched nature of these small volume hybrid magmas, we suggest that voluminous mantle melting and large-scale magmatic underplating are unlikely to have occurred.

A fundamental question in recent granite research is whether mafic mantle magmas and felsic crustal melts can mix to form homogeneous hybrids that broadly fit the chemical composition of I-type granites (e.g. Castro *et al.* 1991). Some researchers believe that magma mixing is essential to granite generation in general and consider that even large, homogeneous granodioritic to granitic intrusions could represent well-mixed hybrids (e.g. Collins 1998).

A problematic point in the magma-mixing discussion is that the processes usually occur at deep unexposed crustal levels and cannot be directly observed. Unequivocal geological evidence for chemical or physical homogenization (mixing) between contrasting magmas has been presented for only relatively few granitoids and on small scales (e.g. Frost & Mahood 1987; Zorpi *et al.* 1991; Neves & Vauchez 1995; D'Lemos 1996). In many cases, the hybrid nature of granitoids is merely assumed, either because of the presence of mafic enclaves (Barbarin & Didier 1991) or on the basis of chemical and isotopic modelling (e.g. DePaolo 1981; Keay *et al.* 1997). However, other magmatic processes could also cause linear elemental co-variation in magmatic suites (Wall *et al.* 1987), and mafic enclaves in a pluton do not automatically prove a hybrid nature for the host granitoid. On the other hand, from fluid dynamic arguments, hybrids are expected to be strongly heterogeneous on a larger scale, because of high viscosity contrasts, slow rates of chemical diffusion, fast cooling rates and limited convection in plutonic systems (Campbell & Turner 1986; Oldenburg *et al.* 1989; Van de Laan & Wyllie 1993). This would imply that homogeneous felsic granitoids are unlikely to be hybrids of mantle-derived and crustal melts.

To constrain better the mechanism of magma mixing and to evaluate its overall importance in granite petrology, there is a particular need for

From: FRANKE, W., HAAK, V., ONCKEN, O. & TANNER, D. (eds).
Orogenic Processes: Quantification and Modelling in the Variscan Belt.
Geological Society, London, Special Publications, **179**, 415–431. 0305-8719/00/$15.00

detailed studies on hybrid granitoids. The identification of such hybrids also has important implications for unravelling the thermal evolution of large granitoid provinces. As post-collisional magmatism and widespread high-T metamorphism are striking features of the Central Variscan Belt, the characterization and quantification of juvenile mantle input will be critical for the evaluation of lithospheric orogenic tectonothermal models (see Henk *et al.* this volume).

We studied a series of plutons in the Central Variscan Belt of the Moldanubian Zone, the genesis of most of which can be explained in terms of crustal melting plus fractional crystallization (Finger & Clemens 1995). Other plutons contain rocks that were identified as potential hybrids between mafic and felsic magmas (Gerdes 1997; Holub 1997). One of these is the Rastenberg Pluton (RbP) in Lower Austria. Using major and trace elements as well as Sr, Nd, Pb, O and H isotopes, we identify the relation between the different rock types and characterize likely magma sources. Some of the observed geochemical co-variations of the RbP rocks are highly unusual but indicative of a mixing process. On the basis of RbP example, we go on to discuss the possibility of generating homogeneous hybrids and the role of mantle melts in the genesis of the late-Variscan granitoids.

Geological background

The Rastenberg Pluton (RbP) is located on the eastern margin of the large South Bohemian Batholith (Fig. 1), which is one of the major post-collisional plutonic complexes in the Variscan Orogen. There is general agreement that the magma sources were predominantly crustal (Finger *et al.* 1997), although the local occurrence of K-rich mantle-derived magmas and their interaction with felsic melts has been reported from various places throughout the Variscan

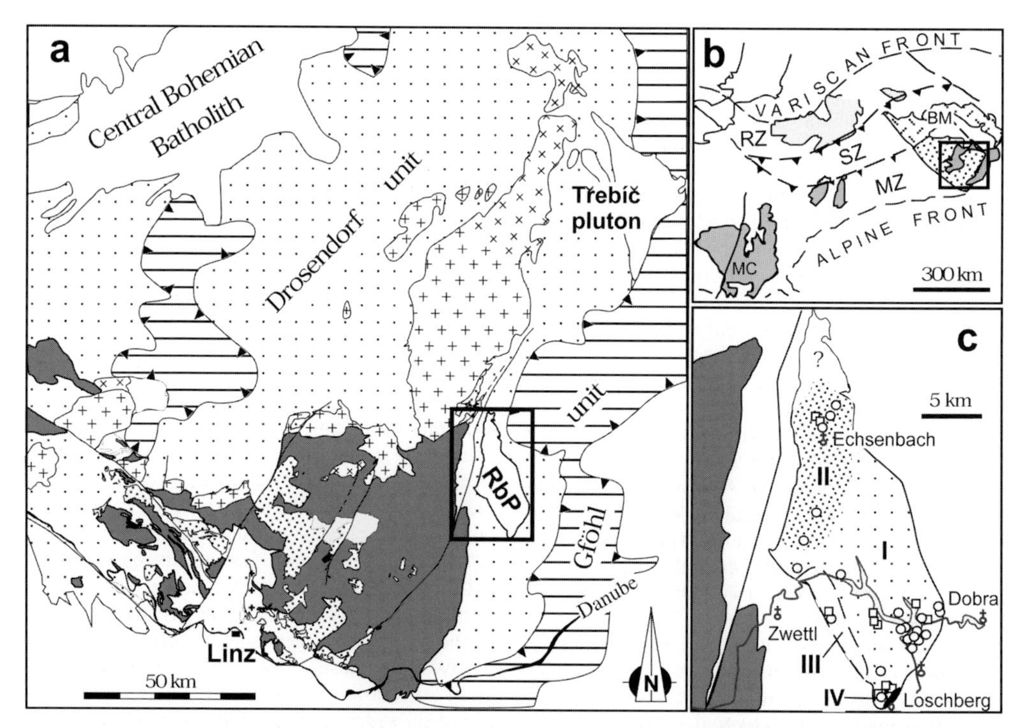

Fig. 1. Geological sketch map of the South Bohemian Batholith. (**a**) Regional distribution of the major granitoids; + and ×, Eisgarn granites; shaded area, Weinsberg granites; stippled area, Mauthausen–Freistadt granitoids (after Gerdes *et al.* (1999) and references therein). The Rastenberg Pluton (RbP) is located at the eastern border of the South Bohemian Batholith. (**b**) The Bohemian Massif (BM) in the context of the Variscan Belt; MZ, SZ and RZ, Moldanubian, Saxo-Thuringian and Theno-Hercynian Zones; MC, Massif Central. (**c**) Sample locations (○, samples from this study; □, samples from Vellmer (1992)) and distribution of rock types in the Rastenberg Pluton (modified after Nickel (1950) and Exner (1969)). I, main type melagranite; II, Echsenbach quartz monzonite; III, Kleehof granodiorite; IV, Loschberg quartz monzodiorite.

Orogen (e.g. Cocherie *et al.* 1994; Holub 1997; Wenzel *et al.* 1997; Bea *et al.* 1999). However, there is still considerable debate concerning the heat source that triggered crustal melting at that time (Gerdes *et al.* 2000; Henk *et al.* this volume).

The RbP was emplaced at *c.* 330 Ma (Vellmer 1992; Friedl *et al.* 1993; Klötzli & Parrish 1996), i.e. at a similar time to most other southern Bohemian granitoids. On the basis of inherited zircon ages, Klötzli & Parrish (1996) have postulated the reworking of Cadomian and Proterozoic to Archean crustal components in the main granite type. The pluton intruded, with partly discordant contacts, into high-grade metamorphic gneisses (Drosendorf unit, Fig. 1) of the Moldanubian nappe pile at depths of *c.* 10–12 km, as estimated from contact metamorphic mineral parageneses (Büttner & Kruhl 1997). The regional distribution of mineral cooling ages (Dallmeyer *et al.* 1992) suggests that the pluton was emplaced during the exhumation stage of the Variscan Orogeny.

Petrographic features and field relations

Five main lithologies are distinguished macroscopically in the RbP (Fig. 1c): (1) The dominant facies, which comprises up *c.* 60% of the outcrop, is a hornblende–biotite melagranite with an isotropic fabric, irregular texture and megacrysts of idiomorphic K-feldspar; (2) hornblende–biotite quartz monzonites crop out in the northern part of the pluton (Echsenbach type; Nickel 1950) as dark varieties of the main melagranite; (3) dark, mostly fine-grained, igneous enclaves are abundant in both the melagranites and quartz monzonites; (4) a distinctive type of medium-grained quartz monzodiorite (Loschberg type) forms isolated bodies (> 500 m in diameter) in the southern corner of the pluton; (5) finally, K-feldspar-phyric biotite granodiorites are present at the southwestern margin of the pluton (Kleehof type; Exner 1969) and as small dyke fragments in the melagranites.

The main melagranite and the Echsenbach quartz monzonite show a continuous compositional range. They consist of characteristic K-feldspar megacrysts up to 80 mm long in a medium- to coarse-grained matrix of plagioclase, quartz, K-feldspar, hornblende and biotite. The K-feldspar megacrysts, which are generally larger and more abundant in the melagranites, in places show magmatic flow orientation and frequently have concentric rings of small epitactic plagioclase, biotite and amphibole inclusions (Frasl 1954). The plagioclase is generally euhedral and zoned with oligoclase margins and cores of up to An_{50} in the melagranites and up to An_{70} in the

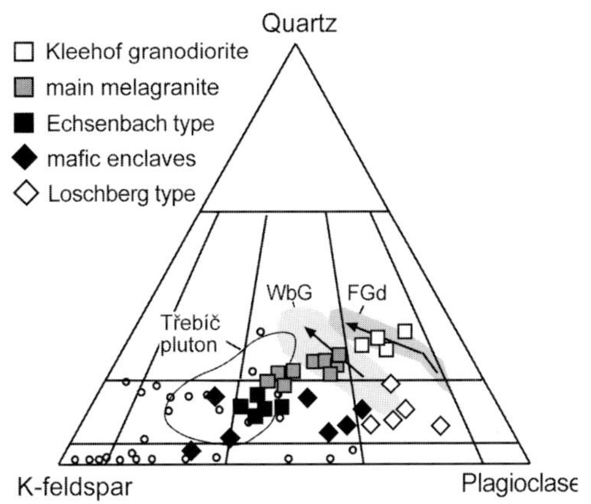

Fig. 2. Classification of the rocks of the Rastenberg Pluton in the Streckeisen Q–P–F diagram. Model contents were calculates from whole-rock composition and representative mineral analyses using the program PETMIX (LeMaitre 1979). Eleven RbP samples are from Vellmer (1992). In addition, field of ultrapotassic granitoids (durbachites) from the Třebíč Pluton (Scharbert & Veselá 1990), co-genetic enclaves (○) of ultrapotassic granitoids from South and Central Bohemia (Holub 1977) and 'typical' granitic trends from the Freistadt granodiorite (FGd) and Weinsberg-granites (WbG) are shown (Gerdes 1997).

quartz monzonites. Estimates of modal contents are difficult in such coarse-grained and porphyritic rocks, and were therefore calculated from the chemical data (see caption to Fig. 2). The plagioclase/K-feldspar ratio ranges from 1.4 to 0.6 and generally decreases in the melagranites and quartz monzonites (Fig. 2) with decreasing amount of feldspars (*c.* 57–53 and *c.* 51–45 vol. %, respectively) and quartz (*c.* 20–13 and 10–7 vol. %, respectively). The biotite and amphibole contents vary from *c.* 13–17 and 7–16 vol. % in the melagranites to *c.* 17–21 and 21–27 vol. % in the quartz monzonites. Biotite is rich in titanium and, like the amphiboles, relatively Al poor and Mg rich. Amphiboles are typically zoned with actinolitic hornblende in the core and an actinolite rim. Relicts of clinopyroxene are present in cores of amphiboles or as inclusions in the feldspars. In both the melagranite and the quartz monzonites, clots of fine-grained biotite and actinolite are abundant; these could be pseudomorphs after large primocrysts of clinopyroxene and/or hornblende. Accessory minerals include acicular apatite, euhedral sphene, sulphides, zircon, epidote, chromite and in more felsic varieties also monazite.

Mafic enclaves with various dimensions, shapes, compositions and textures are very common in the main melagranites and the

Echsenbach quartz monzonites. They range from a few centimetres to several metres in diameter and are of (quartz) syenitic, quartz monzonitic to quartz monzodioritic composition (Fig. 2). Grain size varies from fine grained to medium grained. The enclaves' igneous textures are variably porphyritic to equigranular and generally undeformed. Most striking are the inclusions of K-feldspar megacrysts, frequently corroded and more abundant near contacts. These large feldspars are apparently derived from interaction with the host granitoids. The modal composition of the enclaves is variable with 12–28 vol. % plagioclase, 15–30 vol. % K-feldspar, 2–9 vol. % quartz, 18–35 vol. % biotite and 25–45 vol. % hornblende $\pm$ clinopyroxene $\pm$ olivine. K-feldspar is abundant in the matrix and is mostly hypidiomorphic. Plagioclase is zoned, with oligoclase rim and cores sometimes as calcic as An_{80}. Clinopyroxene primocrysts occur as corroded crystals, and abundant amphibole and biotite clusters are probably pseudomorphic after clinopyroxene. Clots of fine-grained, pale green actinolites with biotite rims ('pilites') were interpreted as pseudomorphs after olivine (Nickel 1950; Holub 1977). Fresh olivine, however, is rare (Nickel 1950).

The shapes of the mafic enclaves are strongly elongated to well rounded. The contacts to their host are irregular to cuspate, varying from sharply to poorly defined and diffuse. Their distribution in the melagranites and quartz monzonites is irregular and makes up <5 up to *c.* 20% of the rock volume. Enclaves are also present as schlieren, disaggregated pillows or disrupted dykes, or are stretched and folded together with their host. Relicts of the melagranites are found enclosed in the enclaves, and host and enclaves repeatedly show concordant magmatic foliations. The temperature contrasts between enclave and granitoid magma were probably only moderate, because chilled margins are only rarely present. Instead, progressive to total dispersion of the enclaves as microfragments in melagranites and quartz monzonite is very common. Agglomerates of idiomorphic, zoned K-feldspar megacrysts (10–60 mm) with interstitial amphiboles (up to 20 mm) occasionally appear near the mafic enclaves. All these observations suggest the coexistence of mafic enclaves, main melagranite and Echsenbach quartz monzonite before crystallization as contemporaneous magmas and intense interaction between these contrasting magmas.

The Loschberg quartz monzodiorite (quartz <5 vol. %) is medium grained and usually has an equigranular texture. Magnesio-hornblende (*c.* 40–45 vol. %) and plagioclase (*c.* 25 vol. %) dominate the mineralogy, with abundant relicts of clinopyroxene (*c.* 2–5 vol. %) and brown hornblende. K-feldspar (*c.* 6–11 vol. %) and biotite (*c.* 14–21 vol. %) are less common than in the mafic enclaves. The margins with the melagranites are not well exposed, but where they are seen, they are usually sharp. Except for some isolated large K-feldspars near the host contact, no evidence for an interaction with the melagranitic magma exists. Fine-grained dykes with a composition similar to the Loschberg rocks crosscut the main melagranites. The presence of these suggests that the quartz monzodioritic Loschberg magma intruded later than the enclave magma, that is after most of the pluton was nearly solidified.

The Kleehof biotite granodiorites represent the most quartz-rich (*c.* 23–27 vol. %) and felsic variety of the pluton. They are finer grained and less porphyritic than the main melagranites. Plagioclase forms euhedral and moderately zoned crystals (*c.* An_{36-16}) and is much more frequent than K-feldspar (*c.* 38–47 vol. % and *c.* 9–18 vol. %, respectively). Biotite contents vary from *c.* 12–17 vol. %, whereas amphibole appears only in small amounts (0–1 vol. %). Enclaves are very rare.

We note an unusual petrographic variation from syenitic–monzonitic rocks to melagranites and granodiorites (Fig. 2): amounts of mafic minerals (Ci *c.* 60 to 15) are decreasing whereas the plagioclase/K-feldspar ratio (0.38–5.2) increases, so that plagioclase become more abundant in felsic melagranites and Kleehof granodiorites.

Samples and analytical procedures

Twenty fresh, representative RbP samples were collected from quarries and roadside exposures (Fig. 1c). Two minettes crosscutting the metamorphic units *c.* 30 km east of the RbP were also sampled. Samples of 1–8 kg of each mafic enclave and 10–35 kg of each porphyritic rock were prepared for analysis at Geochemisches Insitut, Universität Göttingen. Accuracy, precision and sample homogeneity were controlled by repeated analyses of samples, international reference material and in-house standards (see Gerdes (1997) for details). Major, minor and trace element were determined under routine conditions by X-ray fluorescence (XRF) on glass fusion beads using a Phillips PW 1480 instrument (see Hartmann & Wedepohl 1993) and by inductively coupled plasma mass spectrometry (ICP-MS) techniques on a VG PlasmaQuad2 Instrument (see Gao *et al.* 1999). H_2O was determined by gravimetry. The precision of major and minor elements (XRF) was $\pm 0.4–1 > 4$ rel. % and for trace elements <10 rel. %, except HF, Ta, Tm and Lu (<14 rel. %). Rock powders (100 mg) were dissolved in pressurized

Teflon reaction vessels for ICP-MS with super-pure HF and H_2SO_4, and for Rb–Sr and Sm–Nd isotopes with doubly distilled HF and HNO_3. Ion exchange technique was used for Pb, Rb–Sr and rare earth element (REE) separation, and HDEHP-Kel-F columns for Nd and Sm. Enriched ^{87}Rb-^{84}Sr and ^{149}Sm-^{150}Nd spikes were added before dissolution. For Pb-isotope analyses hand-picked K-feldspar samples (*c.* 100 mg, 60–125 μm) were treated with hot 7 N HNO_3 plus 6 N HCl overnight, leached in steps in 10–30% HF, and then dissolved in concentrated HF. This leaching removed radiogenic Pb from the decay of U and Th after crystallization (e.g. Ludwig & Silver 1977). Analytical blanks were <200, <800 and <300 pg for Pb, Sr and Nd, respectively, representing <0.007% of the dissolved Sr, Nd, and Pb. Isotope ratios were measured on Finnigan MAT 262 RPQ mass spectrometer with multi-collectors operating in static mode. Mass fractionation of Sr and Nd isotopes was linearly corrected using $^{86}Sr/^{88}Sr = 0.1194$ and $^{146}Nd/^{144}Nd = 0.7219$. Pb-isotope ratios were corrected by 0.134% a.m.u.$^{-1}$ based on repeated measurements of NBS-981 ($n = 18$). The precision was <0.02% a.m.u.$^{-1}$ (1σ). Frequent measurements of the La Jolla ($n = 42$) and NBS-987 ($n = 51$) yielded $^{143}Nd/^{144}Nd = 0.511851 \pm 5$ (1σ) and $^{86}Sr/^{88}Sr = 0.710244 \pm 11$ (1σ). Eight analyses of BCR-1 gave $^{147}Sm/^{144}Nd$ of 0.1379 ± 6 (1σ) and $^{143}Nd/^{144}Nd$ of 0.512637 ± 6 (1σ). $\delta^{18}O$ ratios were determined on quartz and δD on biotite separates on a Finnigan MAT 251 gas mass spectrometer following the methods described by Clayton & Mayeda (1963) and Simon (1988). Quartz was chosen, because of its resistance to post-magmatic alteration (e.g. Hoefs 1996). $\delta^{18}O$ whole-rock values were then calculated from experimental quartz–mineral partition factors and modal composition (see caption to Fig. 8, below).

Geochemistry

Major and trace elements

The RbP rocks cover a large SiO_2 range from 49 to 68% and show almost linear correlations between all major and most trace elements (Figs 3–5; Table 1).

The mafic enclaves are strongly metaluminous and define the low-SiO_2 endmember on the RbP trends. They are characterized by 6–10% MgO, 4.4–6.6% K_2O, *mg*-numbers (mol % Mg/(Mg + Fe^{2+})) usually of 66–71 and negative Eu anomalies (Eu/Eu* *c.* 0.62–0.70). K_2O/Na_2O ratios >2 define them as being ultrapotassic (Foley *et al.* 1987). Echsenbach quartz monzonites are also ultrapotassic and metaluminous but have a slightly less mafic composition (53–58% SiO_2, 4.9–5.5% MgO and *mg*-numbers of 63–66). The mafic enclaves have a greater compositional spread, and element contents for both rock types often overlap. Both are enriched in incompatible elements (light REE (LREE) *c.* 220–480 ppm, Rb *c.* 250 ppm, Th 40–60 ppm, U 11–21 ppm) and compatible elements (e.g. Cr > 230 ppm). $(La/Yb)_n$ ratios vary from 8–31. The enrichment of incompatible elements in the ultrapotassic rocks and their relative abundances is similar to that of K-rich calc-alkaline lamprophyres (Fig. 5). Silica-poor, ultrapotassic minettes from RbP country rocks and high-K Loschberg quartz monzodiorites are also characterized by a primitive, metaluminous composition with high Mg, Ca, Cr and *mg*-numbers. However, the minettes have no Eu anomaly and many elements are even more enriched than in the mafic enclaves, whereas Al, Na, U, Th and heavy REE (HREE) concentrations are lower. Quartz monzodiorites also have high LREE but compared with mafic enclaves their Eu/Eu* is 0.85–0.95; K, Rb and Zr are lower and Ca, Na and Sr contents are higher.

The slightly peraluminous Kleehof granodiorites (66–68% SiO_2) represent the felsic endmember of the RbP trends. They are characterized by *c.* 1.5% MgO, K_2O/Na_2O of 1 and *mg*-numbers of *c.* 50. Sr and Ba contents are high at low K and Rb contents and relatively high $(Gd/Yb)_n$ (3.4–3.7), $(La/Yb)_n$ (35–46) and Th/U (5.3–6.6) ratios. There is no Eu anomaly. Such compositions are characteristic of granodiorites and trondhjemites that are generally identified as lower-crustal melts (Rudnick & Taylor 1986).

The main melagranites are intermediate between the mafic ultrapotassic and granodioritic endmembers (Figs 3–5). They are weakly metaluminous and moderately silicic with 60–65% SiO_2, *mg*-numbers of 56–64, 100–241 ppm Cr, 807–1048 ppm Sr and Eu/Eu* of 0.78–0.87.

RbP rocks thus define highly unusual trends that crosscut common melting or differentiation paths (Figs 3 and 4): only elements that are compatible in feldspar (Sr, Pb, Al and Na) and the Eu anomaly correlate positively with SiO_2; all other elements correlate negatively. The $(La/Yb)_n$ ratio increases with increasing SiO_2, because the HREE decrease faster than the LREE. Correlations not only exist between different rock types but also within each group (e.g. in the main melagranites). For the range of *c.* 55–68% SiO_2 and 5.5–1.3% MgO, the trends are particularly well defined. Most elements show correlation coefficients with SiO_2 and MgO better than 0.9 (Pearson product coefficient), except for U, Pb, Nb, K (>0.8), Ba and Th (<0.8). Whereas the silicic endmember is clearly identified as a granodiorite, increased scatter (e.g. for Ba, K, Al and LREE) at the mafic end of the trends suggests the existence of variable enriched mafic components in the RbP. Fractionation and/or diffusion processes during

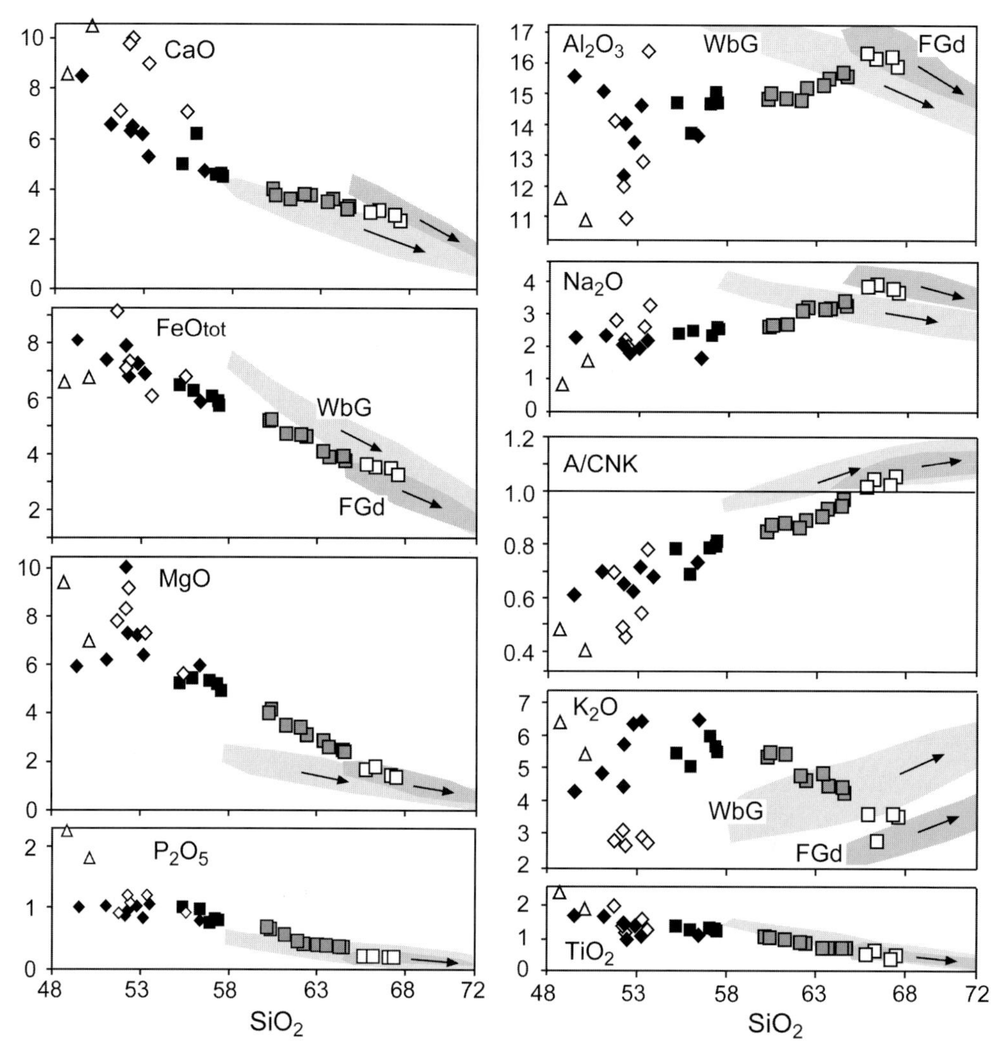

Fig. 3. Major element variations (in wt %) of the RbP rocks and two minettes (△). Symbols and abbreviations as in Fig. 2. Eleven analyses are from Vellmer (1992). 'Typical' granitic trends (FGd and WbG) are show for comparison.

enclave–host interaction, however, could also explain some compositional variation in the mafic enclaves.

Pb, Sr, Nd, O and H isotopes

The isotope data of the RbP rocks also display well-defined correlations, but again with highly unusual trends. The Pb, Sr and Nd isotopes are more radiogenic and the Rb/Sr and Sm/Nd ratios higher in the mafic ultrapotassic than in the felsic samples (Table 1). Pb isotopes of leached K-feldspars show a range of $^{207}Pb/^{204}Pb$ and $^{208}Pb/^{204}Pb$ at less variable $^{206}Pb/^{204}Pb$ and form a steep array at lower $^{206}Pb/^{204}Pb$ compared with 'normal' Variscan granites (Fig. 6). The range in the small-volume RbP is larger than that of all other South Bohemian granitoids together. Linear arrays between mafic ultrapotassic and granodioritic RbP endmembers are also observed in the $^{87}Sr/^{86}Sr$ v. $^{87}Rb/^{86}Sr$ and $^{143}Nd/^{144}Nd$ v. $^{147}Sm/^{144}Nd$ diagrams. These correlations give dates with large errors (446 ± 39 Ma, 433 ± 86 Ma; Gerdes 1997) that cannot be interpreted as ages but instead are indicative of a mixing process. Scharbert & Veselá (1990) reported a comparable correlation of $^{87}Sr/^{86}Sr$ v. $^{87}Rb/^{86}Sr$ from similar

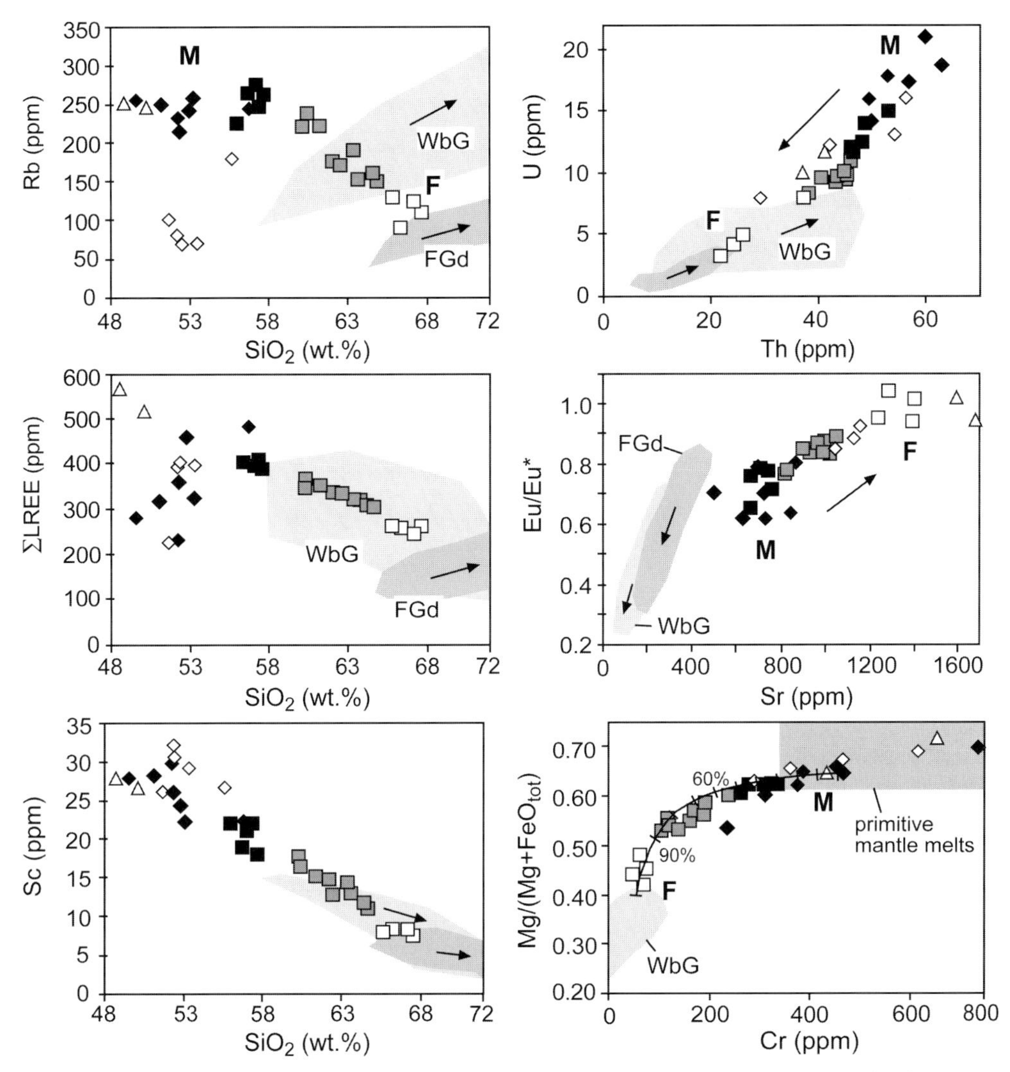

Fig. 4. Selected trace element (in ppm) variations of the RbP rocks. Symbols and abbreviations as in Figs 2 and 3. M, ultrapotassic endmember; F, felsic endmember. Eleven analyses are from Vellmer (1992). It should be noted that these trace elements usually are highly sensitive to fractionation processes. Cr v. *mg*-number shows mixing curve (with 10% ticks) between crustal and mantle-derived melts.

ultrapotassic rocks of the northern Třebíč Pluton. The ultrapotassic, melagranitic and granodioritic RbP rocks show typical crustal εNd_{330} values of −4.6 to −5.2 and $^{87}Sr/^{86}Sr_{330}$ = 0.707–0.709. These ranges overlap the fields from granites and quartz diorites from South Bohemia (Fig. 7) and form well-defined positive correlation between initial Nd- and Sr-isotope composition, which are highly unusual. Compared to major and trace elements, the Sr- and Nd-isotopic variations are relatively small. Loschberg quartz monzodiorites have a slightly less evolved isotopic composition compared with the ultrapotassic rocks with $^{87}Sr/^{86}Sr_{330}$ of 0.706–0.707 and εNd_{330} of −4.2, and Pb isotopes less radiogenic than the granodioritic endmember. The RbP minette has a $^{87}Sr/^{86}Sr_{330}$ of *c.* 0.707 and εNd_{330} of −2.6, and falls into the field of Variscan lamprophyres (Turpin *et al.* 1988).

Oxygen isotope ratios for the RbP rocks vary from +8 to +11‰ (relative to SMOW (Standard Mean Ocean Water)) and show linear positive correlations with silica content (Fig. 8). Two quartz monzonites with low $\delta^{18}O$ were interpreted by Vellmer (1992) to result from local

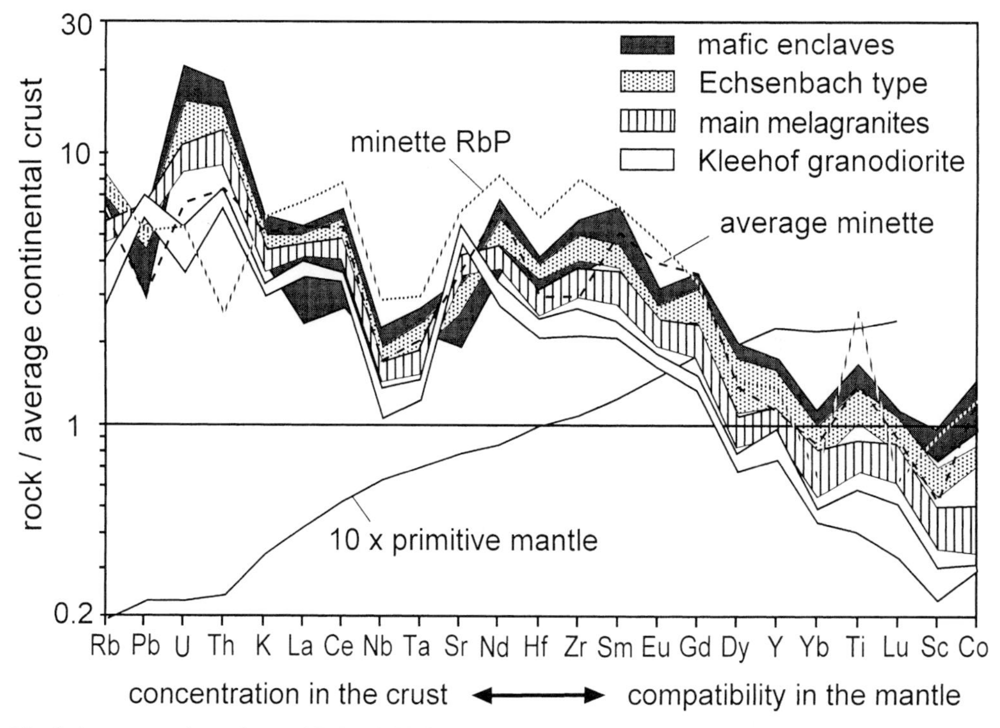

Fig. 5. Average continental-crust (Taylor & McLennan 1985) normalized multi-element diagram for the different rock groups of the RbP and a minette dyke. The patterns illustrate the co-variation of the rocks and the increasing crustal signature from the Kleehof towards the mafic enclaves. The average composition of K-rich calc-alkaline lamprophyres granodiorites (average minette, Rock 1991) and the primitive mantle composition (Hofmann 1988) are shown for comparison. Sequence of the elements is from Hofmann (1988).

high-temperature alteration. However, the δD values of RbP biotite and amphibole separates generally vary between −59 to −83‰ and −84 to −89‰, respectively (Table 1; Vellmer 1992) and fall into the range of unaltered magmatic rocks (Hoefs 1996). As hydrogen isotopes are more sensitive indicators for fluid–rock interaction than O-isotopes, a significant post-magmatic alteration by a meteoric fluid is excluded.

Discussion

Many host–enclave studies have shown complex, scattered variations with uncorrelated spreads of data points (e.g. Chen *et al.* 1990; Flinders & Clemens 1996) or, for example bell-shaped trends of certain elements in Harker diagrams (Poli & Tommasini 1991). By contrast, the RbP produced simple chemical and isotopic co-variations between enclaves and their hosts, suggesting, therefore, a relatively simple, single-stage process. None of the enclaves described here have textures, or chemical composition of cumulates or restites. Holub (1997) has shown that curvilinear trends of Mg and Cr *v.* *mg*-number from Central Bohemian ultrapotassic granitoids, similar to that of the RbP (Fig. 4), are unlike common fractionation trends and are best explained by magma mixing. Restite unmixing or fractional crystallization also should not cause linear negative correlation of elements such as K, Rb, Li, Th and U, or the positive correlation of Sr, Na, Al and Eu/Eu* with SiO_2. In addition, restite unmixing and fractional crystallization would produce identical isotopic signatures between enclaves and their hosts. Furthermore, field observations suggest the intrusion of various mafic to felsic magmas and their intense interaction. Each of the RbP rock types have distinct chemical and isotopic compositions. Together they form linear elemental and isotopic co-variations, suggesting successive mixing and mingling of two contrasting magmas as the fundamental petrogenetic process in the RbP. However, at the mafic end there is some compositional scatter, which could be due to limited host–enclave interaction, differentiation

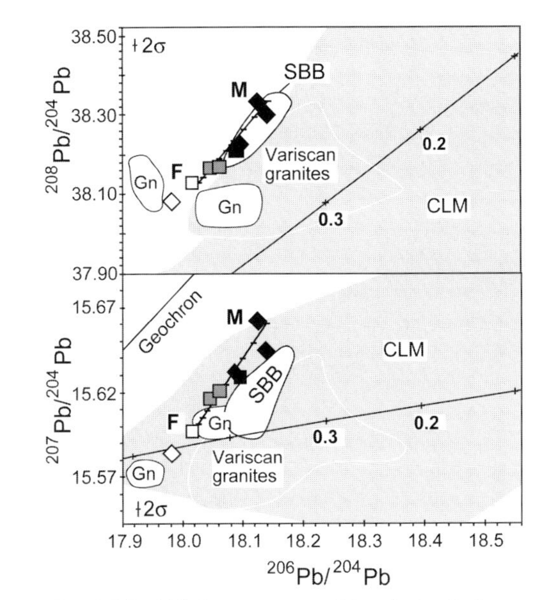

Fig. 6. Initial Pb isotope composition for leached K-feldspare samples from the RbP with mixing line (10% ticks) between ultrapotassic (M) and granodioritic endmember (F). Symbols as in Fig. 2. Fields of Variscan granites (Schwarzwald and Bohemian Massif), gneisses (Gn), lamprophyres (CLM) and granites from South Bohemia (SBB) are shown for comparison (Kober & Lippolt 1985; Turpin *et al.* 1988; Gerdes 1997, and references therein). Pb evolution curve (0.1 Ga ticks) is from Stacey & Kramers (1975).

and/or variable composition of the mafic ultrapotassic endmember.

Mafic, ultrapotassic endmember(s)

The ultrapotassic mafic enclaves have *mg*-numbers and transition element concentrations comparable with primitive mantle melts, which have undergone only limited fractionation (e.g. Le Roex *et al.* 1983). However, the strong enrichment of large ion lithophile elements (LILE), LREE and crustal-like isotope compositions indicate a clear geochemical affinity to K-rich calc-alkaline lamprophyres (minettes, Fig. 5) and lamproites. High concentration of these elements and primitive bulk chemistry exclude crustal contamination as the cause of this enrichment. The genesis of such K-rich magmas is generally explained by melting of a metasomatically enriched mantle source (Fraser *et al.* 1985; Foley *et al.* 1987; Rock 1991; Nelson 1992).

Rb and K_2O do not fall on the extrapolated linear RbP trends (Figs 3 and 4). This suggests that the ultrapotassic RbP endmember was even more enriched in these elements (K_2O *c.* 6.7%, Rb *c.* 350 ppm). Ultrapotassic granitoids (durbachites) with a very similar chemical composition to the RbP enclaves but with such higher Rb and K_2O contents are very common in the northern Třebíč Pluton (Holub 1997) and are likely candidates for the ultrapotassic endmember in the RbP.

Chemical and isotopic data from peridotites and pyroxenites from various localities (Fig. 7) in the Variscides indicate that an enriched, heterogeneous subcontinental lithospheric mantle existed beneath the Variscan Orogen. For instance, in the Gföhl unit in Lower Austria, 40 km SE from the RbP, high-Mg pyroxenites were found, which equilibrated under mantle *P*–*T*-conditions. Therefore formation by assimilation is excluded, and Becker (1996) interpreted these rocks as cumulates of LREE- and LILE-enriched lithospheric melts. The calculated composition of melts in equilibration with these cumulates is very similar to that of ultrapotassic RbP enclaves, with respect to Tb/Yb, La/Ta, Zr/Sm, Sr/Nd, La/Yb, $^{87}Sr/^{86}Sr_{330}$ and $^{143}Nd/^{144}Nd_{330}$ ratios as well as Cr and Ni contents. Becker (1996) also postulated a moderate negative Eu anomaly for these magmas, again comparable with that of the ultrapotassic mafic rocks in the RbP. Geochemical and isotopic data therefore suggest that the ultrapotassic mafic endmember of the RbP was generated from a metasomatically enriched lithospheric mantle. The strongly crustal-like isotope signatures require long-term enrichment of the lithospheric mantle source (Fraser *et al.* 1985). The generation of high $^{207}Pb/^{204}Pb$ and $^{208}Pb/^{204}Pb$ ratios at relatively low $^{206}Pb/^{204}Pb$ imply a multi-stage evolution of the old enriched mantle source over >1 Ga, similar to that of other K-rich mantle magmas (Fraser *et al.* 1985; Nelson 1992). Sr and Nd isotope signatures can be explained by a less complex enrichment process, for example, by adding a sediment component (3–5%) to the mantle during Late Proterozoic to Early Palaeozoic subduction. Loschberg quartz monzodiorites, RbP minettes, quartz monzodiorites from South Bohemia and the ultrapotassic rocks from Central Bohemia and the Třebíč Pluton have a distinct, variably enriched chemical and isotopic composition (Fig. 7 and references given in the caption). This suggests heterogeneity in the Variscan lithospheric mantle on a small scale. Variably enriched mantle-derived mafic igneous rocks with crustal-like isotope composition seem to be a characteristic feature of Variscan magmatism (Cocherie *et al.* 1994; Holub 1997; Wenzel *et al.* 1997; Bea *et al.* 1999; Fig. 7).

Table 1. *Chemical and isotopic composition of representative RbP rocks*

	12/54	13/54	10/54	18/54	23/54	7/54	35/92	17/54	78/94	23/92*	20/94
Sample:	RbP melagranites			Echsenb.	Kleehof	Mafic enclaves			Loschberg type		Minette
SiO_2 (wt%)	64.5	62.5	62.1	57.5	65.8	52.8	52.3	52.2	52.4	55.6	48.7
TiO_2	0.64	0.78	0.80	1.16	0.52	1.14	0.90	0.95	1.11	1.18	2.40
Al_2O_3	15.5	15.2	14.8	14.8	16.3	13.5	14.1	12.4	11.0	15.4	11.6
Fe_2O_3	0.70	0.81	0.94	0.98	0.54	0.59	0.60	0.78	0.89	0.78	0.92
FeO	3.32	3.90	3.85	4.85	3.14	6.71	6.26	7.20	6.52	5.38	5.75
MnO	0.07	0.08	0.09	0.10	0.07	0.18	0.14	0.13	0.14	0.14	0.13
MgO	2.50	3.14	3.45	4.92	1.66	7.24	7.29	10.01	9.15	5.63	9.42
CaO	3.33	3.76	3.82	4.53	3.23	6.52	6.72	6.52	10.02	7.09	8.58
Na_2O	3.33	3.22	3.19	2.54	3.85	1.42	2.16	1.79	2.00	2.32	0.83
K_2O	4.49	4.55	4.64	5.42	3.58	6.38	5.74	4.36	2.56	2.63	6.39
H_2O	0.83	0.97	0.95	1.19	0.79	1.54	1.89	1.71	1.77	1.88	3.73
P_2O_5	0.37	0.44	0.45	0.78	0.22	1.03	0.93	0.84	0.88	0.94	2.23
Total	99.6	99.6	99.9	98.7	99.7	99.1	99.0	99.0	98.4	99.0	100.6
Li (ppm)	35	38	43	46	30	54	45	37	19		
Sc	12	13	15	18	7	24	26	30	31	27	28
V	71	97	98	126	54	150	119	172	196	140	140
Cr	138	164	188	280	77	465	452	739	615	287	655
Co	14.0	15.0	14.5	25.4	9.8	31.0	31.0	43.5	41.5	26.5	36.5
Ni	41	43	50	79	25	149	166	178	162	77	269
Zn	70	81	85	96	61	118	79	93	92	79	90
Rb	163	169	176	259	129	241	210	225	70	182	251
Sr	977	1048	1028	706	1397	729	702	507	1045	797	1592
Y	20	22	24	28	20	28	28	27	33	28	30
Zr	307	352	353	472	271	405	534	355	156	482	809
Nb	17	18	18	19	15	20	24	21	22	35	32
Ba	2020	2238	2197	2220	2610	3710	1970	1473	2680	2294	6110
La	69	74	74	75	61	86	57	36	78	58	106

Ce	146	158	155	186	120	207	144	92	190	130	261
Pr	16.2	18.0	17.8	23.9	14.7	26.4	20.1	12.9	22.6		33.0
Nd	60	65	71	90	54	108	89	59	88		133
Sm	9.8	11.4	12.0	16.1	8.5	17.9	18.6	12.5	14.9		22.4
Eu	2.24	2.63	2.59	3.14	1.99	2.72	3.55	2.22	3.29		5.25
Gd	6.3	7.4	7.7	9.2	4.9	9.9	10.1	7.5	9.5		11.2
Tb	0.65	0.79	0.83	1.00	0.58	1.07	1.06	0.85	1.15		1.21
Dy	3.4	4.1	4.2	5.2	2.7	5.2	5.2	4.5	5.7		5.5
Ho	0.55	0.73	0.77	0.90	0.47	0.90	0.90	0.81	1.04		0.91
Er	1.54	1.89	2.01	2.35	1.27	2.21	2.28	2.12	2.69		2.10
Tm	0.21	0.27	0.29	0.31	0.18	0.31	0.32	0.31	0.37		0.26
Yb	1.37	1.63	1.83	2.06	1.17	1.93	1.91	1.85	2.22		1.46
Lu	0.19	0.24	0.27	0.31	0.16	0.29	0.28	0.27	0.32		0.19
Ta	1.1	1.2	1.1	1.3	1.0	1.3	1.7	1.6	1.6		1.9
Pb	51	45	50	38	46	34	27	24	29	33	41
Th	38	41	43	46	26	60	63	39	42	58	37
U	8.5	9.8	9.3	10.8	4.9	21.0	18.8	11.3	12.3		10.0
Eu/Eu*		0.88	0.87	0.82	0.79	0.95	0.62	0.79	0.70	0.85	1.01
$^{206}Pb/^{204}Pb$ Kfs		18.011	18.045	18.045	18.096	18.093	18.137	18.125	17.982		
$^{207}Pb/^{204}Pb$ Kfs		15.595	15.618	15.621	15.631	15.630	15.645	15.663	15.584		
$^{208}Pb/^{204}Pb$ Kfs		38.115	38.173	38.168	38.223	38.206	38.311	38.335	38.080		
$\delta^{18}O$ (‰) quartz		11.5	11.5		11.4	11.4		11.3	11.0		
$\delta^{18}O$ (‰) WR_{calc}		10.0	9.6		9.1			8.3	8.1		
δ D (‰) biotite		−65.1	−79.3		−58.9	−81.2		−75.4			
H_2O^+ (%) biotite		3.70	2.93		2.98	3.02		3.80		(L12/54*)	
$^{87}Rb/^{86}Sr$		0.4481	0.4530	1.0599	0.2873	0.9451	0.8657	1.2644	0.1996	1.9652	0.3956
$^{87}Sr/^{86}Sr$†		0.70943	0.70968	0.71322	0.70866	0.71275	0.71267	0.71500	0.70765	0.71552	0.70883
$^{87}Sr/^{86}Sr_{330Ma}$		0.70733	0.70755	0.70824	0.70731	0.70831	0.70860	0.70906	0.70671	0.70629	0.70697
$^{147}Sm/^{144}Nd$		0.1041	0.1006	0.1109	0.0896	0.1021	0.1146	0.1303	0.1023	0.1239	0.1045
$^{143}Nd/^{144}Nd$†		0.512184	0.512174	0.512207	0.512142	0.512187	0.512226	0.512252	0.512215	0.512264	0.512306
ϵNd_{330Ma}		−5.0	−5.0	−4.8	−5.2	−4.8	−4.6	−4.7	−4.3	−4.2	−2.6

*Sample 23/92 was analysed by XRF only; isotope analyses are from a similar dyke, which crosscuts 12/54. †$2\sigma_m$ of $^{87}Sr/^{86}Sr$ and $^{143}Nd/^{144}Nd$ was less than ±0.0027 rel. % and ±0.0014 rel.%, respectively. (See text.)

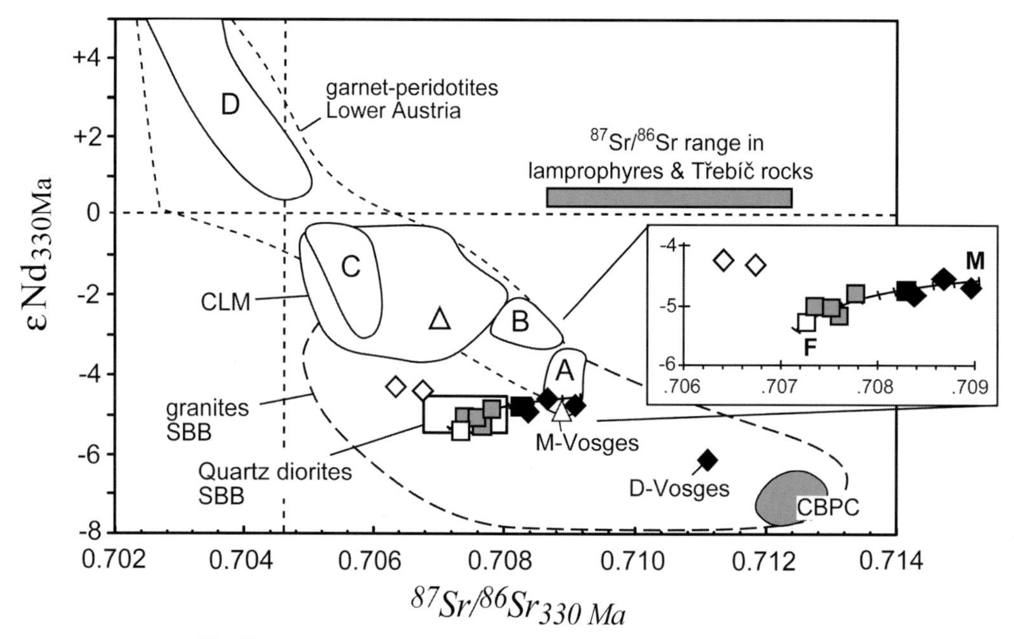

Fig. 7. εNd *v.* initial $^{87}Sr/^{86}Sr$ for the RbP rocks. The rocks define a mixing curve (10% ticks) with unusual positive correlation from the felsic (F) to the ultrapotassic endmember (M). Symbols as in Figs 2 and 3. One melagranite analysis is from Liew *et al.* (1989). The composition of the granites, quartz monzodiorites, lamprophyres and Třebíč granitoids (only $^{87}Sr/^{86}Sr$) from South Bohemia (SBB) is outlined (Liew *et al.* 1989; Scharbert & Veselá 1990; Vellmer 1992; Gerdes 1997). Variable isotopically enriched mafic rocks suggest the heterogeneity of the Variscan lithospheric mantle. A–D, fields of garnet peridotites and various pyroxenites with and without Eu anomaly (Becker 1996). CLM and M-Vosges, field of lamprophyres of the western Variscan Belt and minette from the Vosges (Turpin *et al.* 1988); D-Vosges, durbachite of the Des Cretes Pluton (Langer *et al.* 1995); CBPC, minettes and durbachites from Central Bohemia (Janoušek *et al.* 1995).

Crustal endmember

The Kleehof biotite granodiorites as the felsic endmember on the RbP trend have relatively low initial Sr- and Pb-isotope ratios and low K, Rb, Th and U contents, comparable with some SiO_2-poor varieties from the Mauthausen granite group of South Bohemia (Vellmer & Wedepohl 1994; Gerdes *et al.* 1998). Pb, Sr, Nd and O isotopes overlap with fields defined by Variscan granites, gneisses and felsic to intermediate granulites (Kober & Lippolt 1985; Downes & Duthou 1988; Downes *et al.* 1991; Vellmer 1992; Gerdes 1997). High Sr concentrations and the missing Eu anomaly suggest an intermediate felsic, plagioclase-rich meta-igneous source, and high proportions of plagioclase in the melt as a result of high water activity and/or melting temperature (Conrad *et al.* 1988). Elevated $(La/Yb)_n$ but moderate $(Gd/Yb)_n$ and low Nd/Th are consistent with amphibole and/or clinopyroxene as important residual minerals. The Kleehof granodiorites are similar in composition to experimental melts produced by *c.* 20–30% vapour-absent partial melting of tonalitic protoliths (Singh & Johannes 1996). In conclusion, the most likely origin for these magmas is melting of a moderately felsic lower crust.

Magma mixing

Progressive hybridization of two contrasting magmas is proposed to have produced the linear RbP trends in major elements, trace elements and isotopes. Such a process was also postulated by Holub (1997) to explain comparable chemical trends in ultrapotassic rocks from Central Bohemia and the Třebíč Pluton. Holub's proposed felsic endmember, however, had a granitic composition with high K_2O and Rb and low Sr contents. The composition of the ultrapotassic endmember is similar to that of the RbP and the amount of mafic rocks seems to be slightly higher (Holub 1997).

A binary mixing process for the RbP is also supported by the co-variations in the typical isochron diagrams, so-called *pseudochrons*, and the excellent correlation ($r > 0.95$; Fig. 9) of 1/Sr

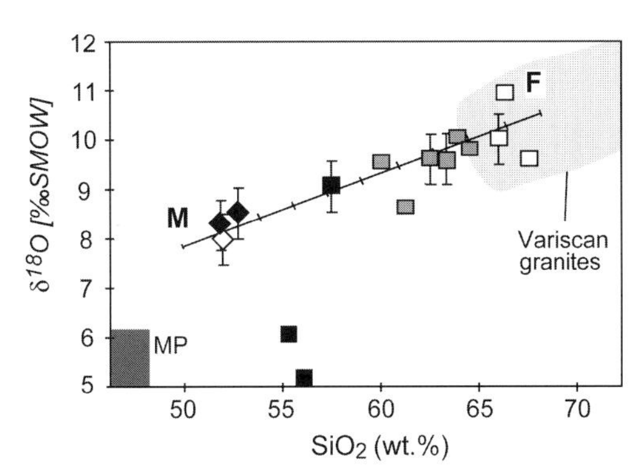

Fig. 8. $\delta^{18}O$ *v.* SiO_2 diagram for the rocks of the RbP with mixing line (10% ticks) between two assumed endmembers (50% SiO_2, 7.8‰ $\delta^{18}O$ and 68% SiO_2, 10.4‰ $\delta^{18}O$). Symbols as in Fig. 2. $\delta^{18}O$ whole-rock values (squares with error bars) were calculated from experimental quartz–mineral fractionation factors (equilibration temperature $\pm$ 100 °C; Bottinga & Javoy 1975; Chiba *et al.* 1989), measured $\delta^{18}O$ of quartz, and modal content. Calculated values shown good agreement with $\delta^{18}O$ whole-rock measurements (squares) of unaltered samples at similar SiO_2 content (Vellmer 1992). The fields of Variscan granites (Downes & Duthou 1988; Vellmer 1992; Gerdes 1997) and mantle peridotites (MP, from Downes *et al.* 1991) are shown for comparison. Low $\delta^{18}O$ of two samples suggest high-temperature alteration by meteoric water (Vellmer 1992).

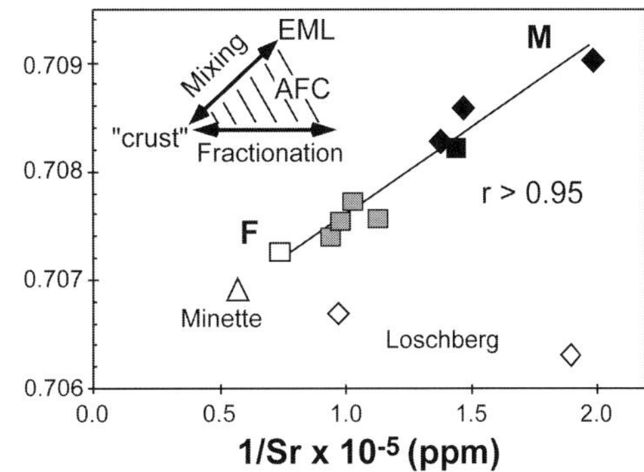

Fig. 9. $^{87}Sr/^{86}Sr_{330}$ *v.* 1/Sr plot for the RbP rocks. Symbols as in Fig. 2. The linear correlation ($r > 0.95$) between Kleehof biotite granodiorites (F) and ultrapotassic rocks (M) indicates mixing of isotopically contrasting magmas. AFC, assimilation and fractional crystallization; ELM, enriched lithospheric mantle.

with initial $^{87}Sr/^{86}Sr$ ratio. Despite the fact that many trace elements (Fig. 4) are highly sensitive to fractionation and assimilation, these processes were obviously unable in the present case to affect the mixing trends. Initial Nd-, Pb- and Sr-isotope compositions were not re-equilibrated even at the metre scale (see sample 17/54 and 10/54 in Table 1) during enclave–host interaction. On the other hand, the main melagranites and the Echsenbach quartz monzonites represent macroscopically relatively homogeneous hybrids. Minor chemical variation within these groups and the presence of large amounts of almost completely digested enclaves show, nevertheless, that hybridization was incomplete and failed to produce chemical and isotopic homogeneity. On a regional scale, the Kleehof granodiorites, main melagranites and Echsenbach quartz monzonites all form individual intrusions of distinct composition that suggest only limited interaction between these magmas during ascent and emplacement. Melagranitic and quartz monzonitic hybrids thus probably formed at deeper crustal level by mixing between ultrapotassic and granodioritic magma in distinct proportions (see Figs 3–9). The difficulties of forming large-scale homogeneous hybrids were outlined in the introduction (e.g. Huppert *et al.* 1984; Campbell & Turner 1986; Frost & Mahood 1987). However, the RbP case shows that it may be possible to generate smaller-scale quasi-homogeneous hybrids in the lower crust where viscosity and viscosity contrast are low and slow cooling is combined with stronger convection.

Implication for Variscan granite genesis

In most previous studies, large contributions of mantle magmas in Variscan batholiths have been considered to be unlikely, because of the crustal isotope signatures of almost all granitoid types (Liew *et al.* 1989; Pin & Duthou 1990; Vellmer & Wedepohl 1994; Downes *et al.* 1997; Finger *et al.* 1997; Bea *et al.* 1999). However, this argument rules out only depleted mantle components, and the presence of enriched mantle melts could possibly be masked because of their more evolved, crust-like isotope signatures (e.g. Janoušek *et al.* 1995). On this basis, it might be argued that the mantle contributions in the Variscan plutons, provided from mantle-derived parental melts or hybrid magmas, could have been much more important than previously assumed (Langer *et al.* 1995).

The melagranitic hybrid of the RbP has a chemical composition that is clearly different from the common Variscan granitoid spectrum (e.g. Downes *et al.* 1997; Gerdes *et al.* 1998; Bea *et al.* 1999). The relative small RbP is characterized by large compositional heterogeneity from mafic to felsic melt compositions and trends different from that expected from igneous fractionation. The observation of only incomplete small-scale mixing forming the RbP is in accordance with experimental and theoretical studies (e.g. Campbell & Turner 1986; Frost & Mahood, 1987) and many field observations (e.g. Zorbi *et al.* 1991; D'Lemos 1992; Neves & Vauchez 1995), which rule out large-scale homogenization of felsic and mafic magmas. Homogeneous hybrids are therefore unlikely to form, and we suggest that the dominantly felsic Variscan granites and granodiorites cannot be seen as such hybrids.

Our observations in the RbP therefore indicate that, where mixing between mantle-derived magmas and crustal melts did occur, it produced clear field and compositional evidence. Plutons with such clear evidence of mingling and hybridization are documented only for small volumes and at limited occurrences in the Variscan Orogen compared with the large monotonous crustal-derived granites. Therefore, on the whole, the overall contribution of mantle-derived magmas to Variscan plutonism must be small. On the basis of the documented mixing trends and proportions of rock types, a contribution of around 40 vol. % of lithospheric mantle-derived magma is estimated for the RbP. However, the large South Bohemian Batholith (*c.* 80 000 km^3) is represented mainly by homogeneous crustal-derived granites (Vellmer & Wedepohl 1994; Gerdes *et al.* 1998) and the total amount of mantle melts is probably <4 vol. %.

Viséan K-rich to ultrapotassic mantle magmas, as in the RbP, suggest small degrees of partial melting of a hydrated phlogopite-bearing lithospheric mantle (e.g. Foley 1992). Contemporaneous melting of this mantle and deep tonalitic crust was probably caused by conductive heating associated with thermal re-equilibration of thickened lithosphere during post-collisional extension (Gerdes *et al.* 2000). The presence of small degree enriched mantle melts in the Central Variscides and the apparent absence of mantle input in the genesis of most post-collisional plutons is inconsistent with large-scale mantle melting. We can therefore conclude that magmatic underplating was probably not the main heat source for the extensive Late Variscan granite generation.

Summary and conclusions

Field and petrographic evidence together with linear elemental and isotopic trends of the RbP rocks suggest the progressive hybridization of ultrapotassic lithospheric mantle magma and a felsic crustal melt. The ultrapotassic mafic endmember(s) are characterized by lower Sr, Pb, Al and Na contents and generally high compatible *and* incompatible element concentrations. Crustal-like trace element and isotope compositions and primitive bulk chemistry cannot be explained by crustal contamination and suggest the generation by small-degree partial melting of an enriched lithospheric mantle. Mafic rocks with variable enriched compositions suggest a heterogeneous lithospheric mantle source, which was possibly contaminated by crustal components during pre-Variscan subduction. The granodioritic crustal endmember with low K and Rb but high Sr, fractionated REE patterns and no Eu anomaly can be explained by fluid-absent partial melting of a tonalitic lower crust.

Hybridization did not proceed to complete compositional homogenization. However, the predominance of melagranites and quartz monzonites in the RbP suggests that mixing of contrasting magmas at similar proportions resulted in individual magma batches (<10 km^3) of similar and nearly homogeneous composition. Mixing and progressive hybridization probably occurred at depth followed by incomplete homogenization at higher crustal levels during ascent and emplacement.

Financial support was provided by the DFG, SPP 'Orogene Prozesse' (Wo 362/12-1,2,3) and the Austrian FWF (Grant 9434 to F. Finger). This paper benefited from the comments of H. Downes, M. Roberts, V. Janoušek and J. H. Scarrow. B. Haunschmid is thanked for providing us with some unpublished mineral analyses.

References

Barbarin, B. & Didier, J. 1991. Review of the main hypotheses proposed for the genesis and evolution of mafic microgranular enclaves. *In*: Didier, J. & Barbarin, B. (eds) *Enclaves and Granite Petrology*, **13**, Elsevier, Amsterdam, 367–373.

Bea, F., Montero, P. & Molina, J. F. 1999. Mafic precursors, peraluminous granitoids and late lamprophyres in the Avila batholith. A model for the generation of Variscan batholiths in Iberia. *Journal of Geology*, **107**, 399–419.

Becker, H. 1996. Crustal trace element and isotopic signatures in garnet pyroxenites from garnet peridotite massifs from Lower Austria. *Journal of Petrology*, **37**, 785–810.

Bottinga, Y. & Javoy, M. 1975. Oxygen isotope partitioning among the minerals in igneous and metamorphic rocks. *Reviews of Geophysics and Space Physics*, **13**, 401–418.

Büttner, S. & Kruhl, J. 1997. The evolution of a late-Variscan high-*T*/low-*P* region: the south-eastern margin of the Bohemian Massif. *Geologische Rundschau*, **86**, 21–38.

CAMPBELL, I. H. & TURNER, J. S. 1986. The influence of viscosity in fountains in magma chambers. *Journal of Petrology*, **27**, 1–30.

CASTRO, A., MORENO-VENTAS, I. & DE LA ROSA, J. D. 1991. H-type (hybrid) granitoids: a proposed revision of the granite-type classification and nomenclature. *Earth-Science Review*, **31**, 237–253.

CHEN, Y. D., PRICE, R. C. & WHITE, A. J. R. 1990. Mafic inclusions from the Glenbog and Blue Gum Granite Suites, Southeastern Australia. *Journal of Geophysical Research*, **95**, 17757–17785.

CHIBA, H., CHACKO, T., CLAYTON, R. N. & GOLDSMITH, J. R. 1989. Oxygen isotope fractionation involving diopside, forsterite, magnetite and calcite: application to geothermometry. *Geochimica et Cosmochimica Acta*, **53**, 2985–2995.

CLAYTON, R. N. & MAYEDA, T. K. 1963. The use of bromine pentafluoride in the extraction of oxygen from oxides and silcates for isotopic analysis. *Geochimica et Cosmochimica Acta*, **27**, 43–52.

COCHERIE, A., ROSSI, P., FOUILLAC, A. M. & VIDAL, P. 1994. Crust and mantle contributions to granite genesis—an example from the Variscan batholith of Corsica, France, studied by trace-element and Nd–Sr–O-isotope systematics. *Chemical Geology*, **115**, 173–211.

COLLINS, W. J. 1998. Evaluation of petrogenetic models for Lachlan Fold Belt granitoids: implications for crustal architecture and tectonic models. *Australian Journal of Earth Science*, **45**(4), 483–500.

CONRAD, W. K., NICHOLLS, I. A. & WALL, V. J. 1988. Water-saturated and -undersaturated melting of metaluminous and peraluminous crustal compositions at 10 kb: evidence for the origin of silicic magmas in the Taupo Volcanic Zone; New Zealand, and other occurrences. *Journal of Petrology*, **29**, 765–803.

DALLMEYER, R. D., NEUBAUER, F. & HÖCK, V. 1992. Chronology of late Paleozoic tectonothermal activity in the southeastern Bohemian Massif, Austria (Moldanubian and Moravo-Silesian zones): $^{40}Ar/^{39}Ar$ mineral age controls. *Tectonophysics*, **210**, 135–153.

DEPAOLO, D. J. 1981. A Nd and Sr isotopic study of the Mesozoic calcalkaline granitic batholiths of the Sierra Nevada and Peninsular Ranges, California. *Journal of Geophysical Research*, **86**(B11), 10470–10488.

D'LEMOS, R. S. 1996. Mixing between granitic and dioritic crystal mushes, Guernsey, Channel Islands, UK. *Lithos*, **38**, 233–257.

DOWNES, H. & DUTHOU, J.-L. 1988. Isotopic and trace-element arguments for the lower-crustal origin of Hercynian granitoids and pre-Hercynian orthogneisses, Massif Central (France). *Chemical Geology*, **68**, 291–308.

——, KEMPTON, P. D., BRIOT, D., HARMON, R. S. & LEYRELOUP, A. F. 1991. Pb and O isotope systematics in granulite facies xenoliths, French Massif Central: Implications for crustal processes. *Earth and Planetary Science Letters*, **102**, 342-357.

——, SHAW, A., WILLIAMSON, B. J. & THIRLWALL, M. F. 1997. Sr, Nd and Pb isotopic evidence for the lower crustal origin of Hercynian granodiorites and monzogranites, Massif Central, France. *Chemical Geology*, **136**, 99–122.

EXNER, C. 1969. Zur Rastenberger Granittektonik im Bereich der Kampkraftwerke (Südliche Böhmische Masse). *Mitteilungen der Geologischen Gesellschaft Wien*, **61**, 6–39.

FINGER, F. & CLEMENS, J. D. 1995. Migmatization and 'secondary' granitic magmas: effects of emplacement and crystallization of 'primary' granitoids in Southern Bohemia, Austria. *Contributions to Mineralogy and Petrology*, **120**, 311–326.

——, ROBERTS, M. P., HAUNSCHMID, B., SCHERMAIER, A. & STEYRER, H. P. 1997. Variscan granitoids of central Europe: their typology, potential sources and tectonothermal relations. *Mineralogy and Petrology*, **61**, 67–96.

FLINDERS, J. & CLEMENS, J. D. 1996. Non-linear dynamics, chaos, complexity and enclaves in granitoid magmas. *Transactions of the Royal Society of Edinburgh: Earth Sciences*, **87**, 217–233.

FOLEY, S. 1992. Petrological characterization of the source components of potassic magmas: geochemical and experimental constraints. *Lithos*, **28**, 187–204.

FOLEY, S. F., VENTURELLI, D. H., GREEN, D. H. & TOSCANI, L. 1987. The ultrapotassic rocks: characteristics, classification, and constraints for petrogenetic models. *Earth Science Reviews*, **24**, 81–134.

FRASL, G. 1954. Anzeichen schmelzflussigen und hochterperierten Wachstums an den grossen Kalifeldspaten einiger Pophyrgranite, Porphyrgranitgneise und Augen gneise Osterreichs. *Jahrbuch der Geologischen Bundesanstalt*, **97**, 71–137.

FRASER, K. J., HAWKESWORTH, C. J., ERLANK, A. J., MITCHELL, R. H. & SCOTT-SMITH, B. H. 1985. Sr, Nd and Pb isotope and minor element geochemistry of lamproites and kimberlites. *Earth and Planetary Science Letters*, **76**, 57–70.

FRIEDL, G., VON QUADT, A., OCHSNER, A. & FINGER, F. 1993. Timing of the Variscan Orogeny in the Southern Bohemian Massif (NE-Austria) deduced from new U–Pb zircon and monazite dating. EUG 5 Strasbourg, *Terra Nova*, **5**, 235.

FROST, T. P. & MAHOOD, G. A. 1987. Field, chemical, and physical constraints on mafic-felsic magma interaction in the Lamarck Granodiorite, Sierra Nevada, California. *Geological Society of America Bulletin*, **99**, 272–291.

GAO, S., LING, W., QUI, Y., LAIN, Z., HARTMANN, G. & SIMON, K. 1999. Contrasting geochemical and Sm–Nd isotopic compositions of Archean metasediments from Kongling high-grade terrain of the Yangtze craton: evidence for cratonic evolution and redistribution of REE during crustal anatexis. *Geochimica et Cosmochimica Acta*, **63**, 2071–2088.

GERDES, A. 1997. *Geochemische und thermische Modelle zur Frage der spätorogenen Granitgenese am Beispiel des Südböhmischen Batholiths:*

Basaltische Underplating oder Krustenstapelung. PhD thesis, Georg-August University, Göttingen.
——, Wörner, G. & Finger, F. 1998. Late-orogenic magmatism in South Bohemia—geochemical and isotopic constraints on possible sources and magma evolution. *Acta Universitatis Carolinae, Geologica*, **42**(1), 41–45.
——, —— & Henk, A. 2000. Post-collisional granite generation and HT–LP metamorphism by radiogenic heating: the example from the Variscan South Bohemian Batholith. *Journal of the Geological Society, London*, **157**, 577–587.
Hartmann, G. & Wedepohl, K. H. 1993. The composition of peridotite tectonites from the Ivrea Complex, northern Italy: residues from melt extraction. *Geochimica et Cosmochimica Acta*, **57**, 1761–1782.
Henk, A., von Blanckenburg, F., Finger, F., Schaltegger, U. & Zulauf, G. 2000. Syn-convergent high-temperature metamorphism and magnatism in the Variscides; a discussion of potential heat sources. *This volume*.
Hoefs, J. 1996. Stable Isotope Geochemistry. Springer, Berlin.
Hofmann, A. W. 1988. Chemical differentiation of the earth: the relation between mantle, continental crust, and ocean crust. *Earth and Planetary Science Letters*, **90**, 297–324.
Holub, F. V. 1977. Petrology of inclusions as a key to petrogenesis of the durbachitic rocks from Czechoslovakia. *Tschermaks Mineralogische und Petrolographische Mitteilungen*, **24**, 133–150.
—— 1997. Ultrapotassic plutonic rocks of the durbachite series in the Bohemian Massif: petrology, geochemistry and petrogenetic interpretation. *Sborník Geologickych Ved, Lozisková Geologie*, **31**, 5–26.
Huppert, H. E., Sparks, R. S. & Turner, J. S. 1984. Some effects of viscosity on the dynamics of replenished magma chambers. *Journal of Geophysical Research*, **B89**, 6857–6877.
Janoušek, V., Rogers, G. & Bowes, D. R. 1995. Sr–Nd isotopic constraints on the petrogenesis of the Central Bohemian Pluton, Czech Republik. *Geologische Rundschau*, **84**, 520–534.
Keay, S., Collins, W. J. & McCulloch, M. T. 1997. A three-component Sr–Nd isotopic mixing model for granitoid genesis, Lachlan fold belt, eastern Australia. *Geology*, **25**(4), 307–310.
Klötzli, U. S. & Parrish, R. R. 1996. Zircon U/Pb and Pb/Pb geochronology of the Rastenberg granodiorite, South Bohemian Massif, Austria. *Mineralogy and Petrology*, **58**, 197–214.
Kober, B. & Lippolt, H. J. 1985. Pre-Hercynian mantle lead transfer to basement rocks as indicated by lead isotopes of the Schwarzwald crystalline, SW-Germany. *Contributions to Mineralogy and Petrology*, **90**, 162–171.
Langer, C., Hegner, E., Altherr, R., Satir, M. & Henjes-Kunst, F. 1995. Carboniferous granitoids from the Odenwald, the Schwarzwald and the Vosges—constraints on magma sources. *Terra Nostra*, **95**(8), 114.
LeMaitre, R. W. 1979. A new generalized petrological mixing model. *Contributions to Mineralogy and Petrology*, **71**, 133–137.
Le Roex, A. P., Dick, H. J. B., Erlank, A. J., Raid, A. M., Frey, F. A. & Hart, S. R. 1983. Geochemistry, mineralogy and petrogenesis of lavas erupted along the Southwest Indian Ridge between the Bouvet triple junction and 11 degrees east. *Journal of Petrology*, **24**, 267–318.
Liew, T. C., Finger, F. & Höck, V. 1989. The Moldanubian granitoid plutons of Austria: chemical and isotopic studies bearing on their environmental setting. *Chemical Geology*, **76**, 41–55.
Ludwig, K. R. & Silver, L. T. 1977. Lead-isotope inhomogeneity in Precambrian igneous K-feldspars. *Geochimica et Cosmochimica Acta*, **41**, 1457–1471.
Nelson, D. R. 1992. Isotopic characteristics of potassic rocks: evidence for the involvement of subducted sediments in magma genesis. *Lithos*, **28**, 403–420.
Neves, S. P. & Vauchez, A. 1995. Successive mixing and mingling of magmas in a plutonic complex of Northeast Brazil. *Lithos*, **34**, 275–299.
Nickel, E. 1950. Das Mischgestein vom Typus Echsenbach (Niederösterreich) und seine Stellung im Rastenberger Tiefenkörper. *Neues Jahrbuch für Mineralogie, Monatshefte*, **81**, 273–314.
Oldenburg, C. M., Spera, F. J. & Sewell, G. 1989. Dynamic mixing in magma bodies: theory, simulations, and implications. *Journal of Geophysical Research*, **94**(B7), 9215–9236.
Pin, C. & Duthou, J. L. 1990. Sources of Hercynian granitoids from the French Massif Central: inferences from Nd isotopes and consequenses for crustal evolution. *Chemical Geology*, **83**, 281–296.
Poli, G. E. & Tommasini, S. 1991. Model for the origin and significance of microgranular enclaves in calc-alkaline granitoids. *Journal of Petrology*, **32**, 657–666.
Rock, N. M. S. 1991. *Lamprophyres*. Blackie, Glasglow.
Rudnick, R. L. & Taylor, S. R. 1986. Geochemical constraints on the origin of Archean tonalitic–trondhjemitic rocks and implications for lower continental crustal composition. *In*: Dawson, B. J., Carswell, D. A., Hall, J. & Wedepohl, K. H. (eds) *The Nature of the Lower Continental Crust*. Geological Society, London, Special Publications, **24**, 179–191.
Scharbert, S. & Veselá, M. 1990. Rb–Sr systematics of intrusive rocks from the Moldanubicum around Jihlava. *In*: Minarikova, D. & Lobitzer, H. (eds) *Thirty Years of Geological Cooperation between Austria and Czechoslovakia*. Federal Geological Survey, Vienna, 262–272.
Simon, K. 1988. *Wasser/Gestein Wechselwirkungen in der Granitserie des südöstlichen Schwarzwaldes*. PhD thesis, Georg-August University, Göttingen.
Singh, J. & Johannes, W. 1996. Dehydration melting of tonalites. Part II. Composition of melts and solids. *Contributions to Mineralogy and Petrology*, **125**, 26–44.

STACEY, J. S. & KRAMERS, J. D. 1975. Approximation of terrestrial lead isotope evolution by a two-stage model. *Earth and Planetary Science Letters*, **26**, 207–221.

TAYLOR, S. R. & MCLENNAN, S. M. 1985. *The Continental Crust: its Composition and Evolution*. Blackwell, Oxford.

TURPIN, L., VELDE, D. & PINTE, G. 1988. Geochemical comparison between minettes and kersantites from the Western European Hercynian orogen: trace element and Pb–Sr–Nd isotope constraints on their origin. *Earth and Planetary Science Letters*, **87**, 73–86.

VAN DER LAAN, S. & WYLLIE, P. J. 1993. Experimental interaction of granitic and basaltic magmas and implications of mafic enclaves. *Journal of Petrology*, **34**, 491–517.

VELLMER, C. 1992. *Stoffbestand und Petrogenese von Granuliten und granitischen Gesteinen der südlichen Böhmischen Masse in Niederösterreich*. PhD thesis, Georg-August University, Göttingen.

—— & WEDEPOHL, K. H. 1994. Geochemical characterization and origin of granitoids from the South Bohemian batholith in Lower Austria. *Contributions to Mineralogy and Petrology*, **118**, 13–32.

WALL, V. J., CLEMENS, J. D. & CLARKE, D. B. 1987. Models for granitoid evolution and source compositions. *Journal of Geology*, **95**, 731–749.

WENZEL, T., MERTZ, D. F., OBERHANSLI, R., BECKER, T. & RENNE, P. R. 1997. Age, geodynamic setting, and mantle enrichment processes of a K-rich intrusion from the Meissen massif (northern Bohemian massif) and implications for related occurrences from the mid-European Hercynian. *Geologische Rundschau*, **86**(3), 556–570.

ZORPI, M. J., COULON, C. & ORSINI, J. B. 1991. Hybridisation between felsic and mafic magmas in calc-alkaline granitoids—a case-study in northern Sardinia, Italy. *Chemical Geology*, **92**, 45–86.

Geochemistry and provenance of Devono-Carboniferous volcano-sedimentary sequences from the Southern Vosges Basin and the geodynamic implications for the western Moldanubian Zone

JÜRGEN EISELE[1,3], RALF GERTISSER[2] & MICHAEL MONTENARI[1]

[1]*Geologisches Institut der Universität Freiburg, Albertstr. 23 B, D-79104 Freiburg i. Br., Germany (e-mail: monte@perm.geologie.uni-freiburg.de)*

[2]*Institut für Mineralogie, Petrologie und Geochemie der Universität Freiburg i. Br., Albertstr. 23 B, D-79104 Freiburg i. Br., Germany*

[3]*Present address: Max-Planck-Institut für Chemie, Postfach 3060, 55020 Mainz, Germany*

Abstract: The Southern Vosges host a volcano-sedimentary basin subdivided into a Lower Unit of Late Devonian age containing marine greywackes associated with bimodal volcanism, a marine Middle Unit (of Early Carboniferous age) and a terrestrial Upper Unit (of Late Carboniferous age), both accompanied by volcanism of intermediate to evolved composition. Petrographically and geochemically, sediments from the Lower Unit are characterized by mafic volcanic, sedimentary and metamorphic components and a positive Cr–Ni anomaly. A negative Nb anomaly weakens in the Middle and Upper Units. These units are dominated by intermediate to felsic volcanic detritus, and exhibit an enrichment in incompatible trace elements and disappearance of the Cr–Ni anomaly. The facies and composition of the volcano-sedimentary sequences indicate that an extensional basin existed in Late Devonian time (*c.* 360 Ma). Subduction-related volcanism and successive closure of the basin occurred in the Lower Carboniferous (*c.* 330 to *c.* 325 Ma). Sedimentation graded from flysch into molasse in Late Carboniferous time, where terrestrial conditions indicate that continental collision had occurred. The volcanism reported here coincides with arc magmatism in the Mid-German Crystalline High and closure of the Rheno-Hercynian Ocean, which was possibly responsible for the geodynamic evolution of the Southern Vosges Basin.

During the last few years the Mid-European Central Variscan Belt has been the subject of intensive investigations. Whereas the crustal, magmatic and sedimentological evolution of the Rheno-Hercynian and the Saxo-Thuringian Zones is now relatively well understood (see overviews by Franke & Oncken (1990), Franke (1992, this volume) and Oncken (1998), only a few models for the metamorphic evolution of the Moldanubian Schwarzwald (e.g. Kalt *et al.* 1994; Kalt & Altherr 1996) and the Vosges Mountains (Altherr & Kalt 1996) have been so far established. Detailed geochemical investigations have been carried out for the Lower Carboniferous sedimentary sequences of the Badenweiler–Lenzkirch Zone (BLZ) in the southern Schwarzwald (Güldenpfennig & Loeschke 1991; Güldenpfennig 1997), which lead to a plate tectonic model for this zone (Loeschke *et al.* 1998).

However, little attention has been paid to the sedimentary sequences of the Southern Vosges Basin (for reviews, see Maass (1988) and Schaltegger *et al.* (1996)). Despite the relatively good biostratigraphic control of the successions preserved within the basin, little work has been done to determine the plate tectonic evolution of this basin. With this aim, the geochemical composition of sedimentary rocks with well-known biostratigraphic position was studied to determine the provenance and to elucidate the geodynamic setting and evolution of the Southern Vosges Basin.

Sedimentary rock petrography and geochemistry is one of the most important and widely used tools in the determination of sandstone and greywacke provenance. Many recent studies have shown that the compositions of ancient greywackes and modern deep-sea turbidites are closely related to the provenance and tectonic setting of sedimentary basins (e.g. Dickinson & Suczek 1979; Dickinson & Valloni 1980; Maynard *et al.* 1982; Taylor & McLennan

From: FRANKE, W., HAAK, V., ONCKEN, O. & TANNER, D. (eds). *Orogenic Processes: Quantification and Modelling in the Variscan Belt.* Geological Society, London, Special Publications, **179**, 433–444. 0305-8719/00/$15.00

1985; McLennan *et al.* 1990; Floyd *et al.* 1991; McLennan & Taylor 1991).

In magmatic arc settings, immature sandstone and greywacke geochemistry generally provide a compositional record of convergent plate interaction. Although the direct study of ancient arcs is often hindered by denudation and fragmentary exposure, the rapid redeposition of volcanoclastic rocks into sedimentary basins flanking the volcanic arcs leads to the most complete record of the magmatic activity along active plate margins. The petrographical and geochemical analysis of these deposits, therefore, has a great potential for identifying the provenance of the sediments, the evolution of their source rocks and geodynamic setting.

Geological setting

Within the Central Variscan Belt, the Vosges Mountains are subdivided into areas belonging to the Saxo-Thuringian and the Moldanubian Zones (see Fig. 1 and enclosure at back). The Saxo-Thuringian Zone of the Northern Vosges Mountains consists of Lower Palaeozoic metasediments and the Devono-Carboniferous volcano-sedimentary sequences of the Vallée de la Bruche. It is separated from the Moldanubian Zone in the central and Southern Vosges Mountains by the Lalaye–Lubine Fault (see enclosure). The Central Vosges Mountains consist of high-grade metamorphic sequences, which have been intruded by plutons of mainly granitic composition. They are bordered to the south by the Southern Vosges, which host a volcano-sedimentary basin. The southern margin of this basin is formed by terrestrial and fluvial deposits of Permian age. Within the basin, three lithologically and stratigraphically different units can be distinguished (Maass 1988), as follows.

The Lower Unit forms the base of the sequence with monotonous Upper Devonian (Famennian) marine shales (the so-called 'Treh-shales') and greywackes of turbiditic origin

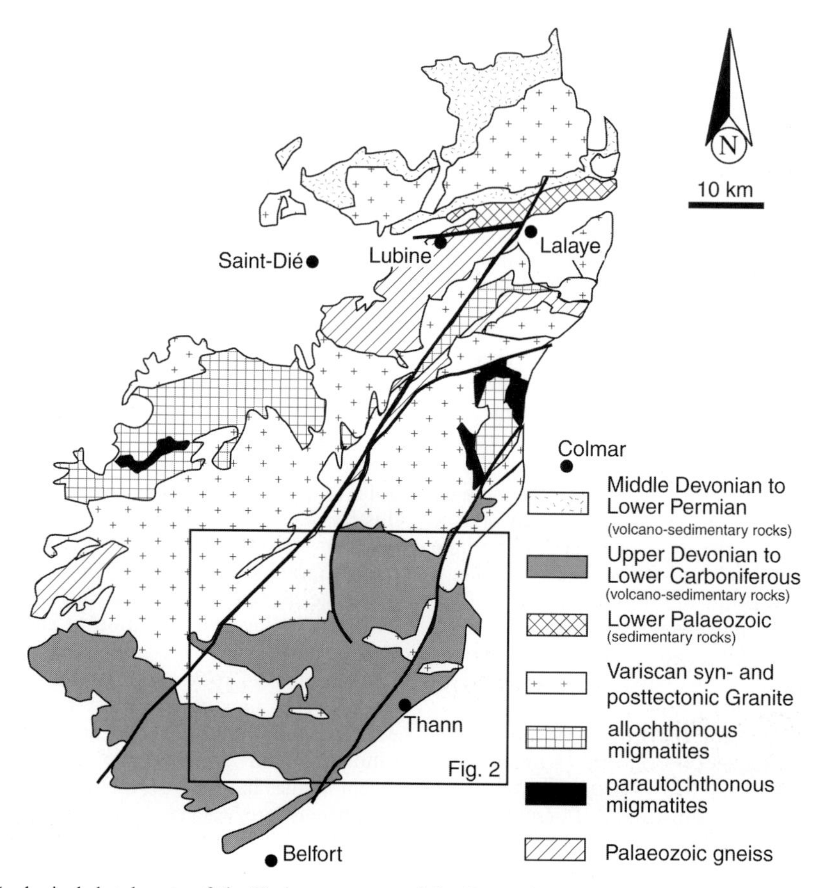

Fig. 1. Geological sketch map of the Variscan aspects of the Vosges Mountains (redrawn and modified after Schaltegger *et al.* (1996)).

(Maass & Stoppel 1982; Fig. 2). The sediments are associated with bimodal volcanic rocks such as low-K rhyolites and basalts (Schaltegger *et al.* 1996), and have been interpreted differently as oceanic arc tholeiites (Lefèvre *et al.* 1994) and continental tholeiites (Bébien & Gagny 1978).

The sequences of the Middle Unit, which also comprise marine turbiditic shales and greywackes, yield fossils of Early (Coulon *et al.* 1975) to Late Viséan age (Doubinger & Rauscher 1966; Hahn *et al.* 1981; Vogt 1981; Maass 1988). Synsedimentary volcanic activity of andesitic to rhyolitic composition has been documented by the occurrence of 'pepperites' in places where lavas intruded wet sediments (Maass 1992). Olistostromes (Schneider *et al.* 1990), slump-structures with intercalated conglomerates and breccias that occur within the Middle Unit can be interpreted as submarine slope deposits (Krecher 1997). They probably document tectonic unrest and a high submarine relief during Late Viséan times.

The Upper Unit constitutes the final Variscan sedimentation event. It contains predominantly terrigenous sequences of ?Namurian to Westphalian age (Doubinger & Rauscher 1966; Traiser *et al.* 1998). The base of this succession is composed of a possibly brackish, coal- and plant-fragment-bearing greywacke–pelite sequence (Schneider 1990), with intercalated pyroclastic rocks (rhyodacitic to rhyolitic) of ignimbritic origin. These deposits are followed by a rapid change from a brackish to a continuous terrestrial environment, documented by several facies associations, such as river-channel fills, fluvial sandbars and floodplain sediments (Maass & Schneider 1995) and lake deposits, which are interrupted by more or less structureless high-energy deposits of arkosic arenites and primary pyroclastic and volcano-clastic deposits.

Methods

Four stratigraphic sections of 12–210 m thickness were measured and sampled within the three units of the Southern Vosges Basin. The structureless, sometimes graded sand fraction from the lowest turbidite unit and some pelites were sampled at 2–20 m intervals from the stratigraphic sections (see Tables 1 and 2 for lithology). Mineralogical composition of the rocks was determined by thin-section petrography. Additionally, quartz–feldspar–lithic-fragment (QFL) values (after Dickinson & Suczek 1979) of 26 representative samples were determined by point-counting (400–500 points per thin section). X-ray fluorescence (XRF) analyses were performed at the Institut für Mineralogie, Petrologie und Geochemie, Universität Freiburg on a Philips PW-1450/20 XRF spectrometer. Before analysis, fresh interior slabs of the samples were pulverized in an agate mill and 1.000 g and 4.000 g weighed for analyses of major and trace elements, respectively. The sample material for major elements was prepared as a glass bead from powdered samples mixed with dilithium tetraborate and lithium metaborate, and samples for trace element measurements were prepared for analysis as pressed powder disc. The geochemical data are shown in Tables 1 and 2. Analytical error is about 0.1 wt% for major elements

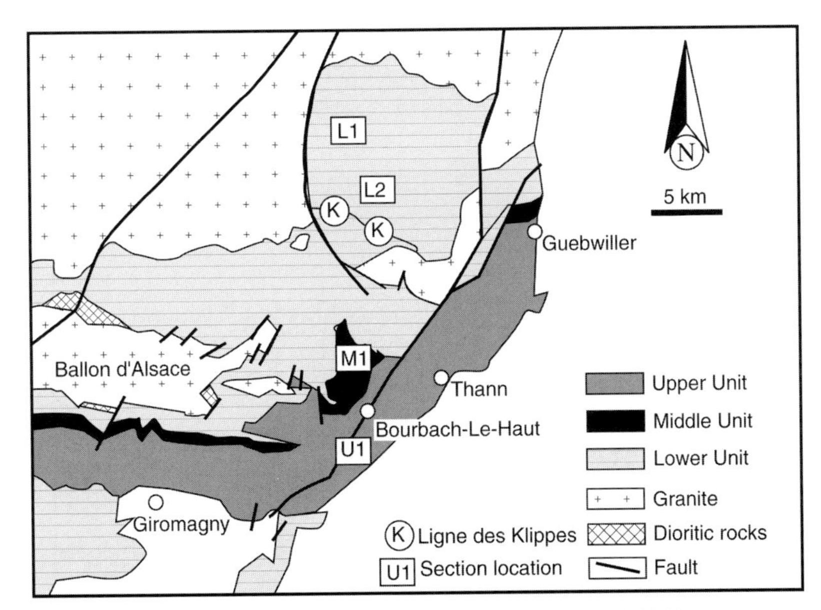

Fig. 2. Geological map of the Southern Vosges Basin (after Schaltegger *et al.* (1996)).

Table 1. *Major and trace element composition of Lower Unit 1 (L1) and Lower Unit 2 (L2)*

Sample:	L1-1	L1-2	L1-3	L1-4	L1-5	L1-7	L1-8	L1-9	L1-10	L1-11	L1-12	L1-14	L1-15	L2-2	L2-3	L2-4	L2-5	L2-6	L2-7	L2-8	L2-10	L2-11
Rock type*:	P	Gl	Gl	Gl	Gl	Gl	Gl	Gf	P	Gf	Gf	Gf	Gf	Gl	Gf	Gf	Gf	Aa	Gl	Al	Gf	Gl
Major elements (wt %)																						
SiO_2	60.88	65.13	66.35	63.17	65.05	64.22	63.00	64.61	59.46	65.33	64.17	63.43	64.61	58.93	59.96	73.19	70.63	74.25	72.04	70.20	65.07	63.83
TiO_2	0.74	1.05	1.02	1.27	0.92	0.82	0.97	1.09	0.90	0.93	1.00	1.03	0.86	2.29	0.91	0.56	0.70	0.57	0.63	0.66	1.01	1.17
Al_2O_3	18.84	15.28	14.45	14.93	15.19	15.80	16.09	15.22	17.93	15.25	15.19	15.61	15.49	14.92	18.38	13.47	14.59	12.54	14.11	13.95	15.09	15.50
$Fe_2O_3^{tot}$	6.14	7.16	7.08	7.82	6.70	7.18	8.29	7.29	9.13	7.07	7.04	8.57	7.74	11.50	7.88	3.61	4.37	3.55	3.74	4.25	7.12	8.23
MnO	0.04	0.09	0.08	0.10	0.08	0.08	0.09	0.09	0.07	0.10	0.10	0.12	0.09	0.08	0.06	0.06	0.05	0.06	0.05	0.06	0.09	0.10
MgO	3.05	3.57	3.11	3.94	3.29	3.54	4.22	3.31	4.19	3.59	3.76	3.55	3.51	4.97	3.11	1.63	1.97	2.14	1.92	2.71	3.45	3.77
CaO	1.69	3.77	2.30	4.42	3.08	3.19	0.69	2.80	1.20	3.74	4.21	2.92	2.60	0.98	0.40	1.03	0.69	0.88	1.83	1.17	2.52	2.79
Na_2O	2.66	3.11	3.08	3.35	3.21	3.35	2.96	3.61	1.72	3.36	3.46	3.49	3.45	3.22	3.02	4.76	3.90	3.82	4.45	3.56	3.88	3.65
K_2O	5.30	1.67	1.90	1.22	1.78	1.67	2.37	1.38	4.51	1.01	0.94	1.68	1.40	1.23	3.18	1.47	2.52	2.06	1.75	2.56	1.17	1.46
P_2O_5	0.23	0.12	0.13	0.15	0.11	0.11	0.12	0.14	0.13	0.12	0.12	0.13	0.11	0.24	0.18	0.15	0.16	0.16	0.16	0.16	0.13	0.14
Total	99.57	100.95	99.50	100.37	99.41	99.96	98.80	99.54	99.24	100.50	99.99	100.53	99.86	98.36	97.08	99.93	99.58	100.03	100.68	99.28	99.53	100.64
Trace elements (ppm)																						
Cu	29	21	116	13	22	20	21	18	13	19	15	15	29	42	36	14	10	10	25	13	21	18
Zn	70	89	83	85	79	78	114	72	92	76	74	90	88	107	115	46	70	51	55	76	75	90
Ga	11	9	8	8	8	7	12	8	15	8	8	10	9	11	18	5	8	4	8	7	8	9
Rb	195	64	67	53	65	60	118	66	210	31	29	70	51	48	108	38	65	44	74	62	30	44
Sr	163	209	196	251	200	202	120	213	91	196	215	251	195	230	162	309	249	324	352	370	377	317
Y	30	27	28	31	27	27	22	29	24	27	28	27	28	59	25	29	30	28	26	28	30	32
Zr	200	149	160	171	141	140	163	166	171	142	142	140	155	208	184	190	190	189	197	216	148	153
Nb	11	4	4	4	3	4	5	4	7	2	3	4	6	4	11	4	7	3	4	5	3	5
Ba	1054	316	457	230	230	332	361	254	366	196	246	201	253	261	630	490	561	618	284	818	398	309
La	50	21	28	27	20	17	20	27	28	20	21	25	15	27	29	36	37	33	24	32	20	19
Pb	24	19	19	13	16	16	16	15	14	15	16	17	18	13	24	18	21	43	17	23	17	17
Th	20	12	14	11	11	14	11	14	14	13	12	10	13	10	19	14	16	16	17	16	13	11
Sc	5	9	10	10	10	12	9	8	9	10	11	10	9	9	6	5	5	6	5	7	9	7
V	89	166	188	222	193	152	189	205	129	180	187	199	169	314	136	68	97	78	77	91	187	217
Cr	135	154	197	205	191	203	168	170	143	170	181	202	157	229	155	113	173	211	180	144	156	191
Ni	72	50	57	48	45	49	58	47	91	43	44	51	50	72	103	59	68	81	78	80	42	47

*Samples classified as: A, arenite; G, greywacke; P, pelite. Further subdivision into: a, arkosic; f, feldspathic; l, lithic (after Pettijohn *et al.* (1987).

Table 2. *Major and trace element composition of Middle Unit 1 (M1) and Upper Unit (U1)*

Sample:	M1-1	M1-2	M1-3	M1-4	M1-5	M1-6	M1-7	M1-8	M1-9	M1-10	M1-11	M1-11A	M1-12	M1-13	M1-14	M1-15	M1-16	M1-17	M1-18	M1-19	M1-20	U1-3	U1-4	U1-5	U1-7	U1-8
Rock type*:	Al	P	Al	Aa	P	Al	Gl	Al	Aa	Al	Al	Al	Al	Al	Al	Al	Al	Al	P	Al	Al	Al	Al	Al	Al	P
Major elements (wt %)																										
SiO_2	62.67	60.77	64.23	68.18	67.11	61.19	66.23	65.06	66.19	64.36	67.03	67.75	64.33	67.79	64.81	60.60	65.39	65.32	61.22	62.60	64.50	71.62	70.07	66.34	66.24	73.06
TiO_2	0.53	0.78	0.81	0.68	0.74	0.72	0.66	0.74	0.65	0.75	0.62	0.76	0.90	0.74	0.83	1.03	0.86	0.83	0.87	1.31	0.83	0.43	0.48	0.54	0.61	0.43
Al_2O_3	17.10	19.21	18.22	16.65	16.66	18.58	16.10	17.12	16.20	16.90	17.32	17.15	17.20	17.31	17.30	18.33	16.85	17.45	18.75	17.47	17.33	14.18	14.09	15.00	15.61	13.74
$Fe_2O_3^{tot}$	3.73	5.12	5.00	3.53	5.16	6.04	6.04	5.75	5.11	5.25	3.65	3.81	5.21	4.02	4.99	6.30	5.59	4.77	6.90	6.54	5.27	3.57	4.46	5.35	4.33	2.31
MnO	0.09	0.06	0.10	0.08	0.08	0.11	0.16	0.16	0.20	0.14	0.10	0.12	0.23	0.12	0.16	0.25	0.17	0.13	0.07	0.18	0.14	0.04	0.06	0.07	0.08	0.03
MgO	1.08	1.38	1.43	1.08	1.42	2.00	1.64	1.48	1.43	1.45	0.96	0.89	1.62	0.92	1.33	1.65	1.73	1.36	1.86	2.26	1.58	1.56	2.29	2.84	2.49	1.18
CaO	3.80	1.01	1.59	2.10	1.90	2.61	2.02	1.92	1.79	1.72	1.62	0.99	1.45	1.38	1.34	2.03	1.63	1.59	1.37	2.11	1.94	0.41	1.01	0.94	1.38	1.07
Na_2O	4.53	2.31	4.93	4.82	2.40	3.80	3.42	3.43	5.85	5.29	5.62	5.05	5.35	4.10	4.68	5.33	4.90	5.13	2.07	4.88	5.47	3.86	3.41	3.46	3.42	4.72
K_2O	2.68	4.66	2.50	1.83	3.34	3.10	2.74	3.28	1.44	3.41	2.41	2.45	2.01	2.71	3.50	3.03	2.10	3.06	4.19	1.68	1.47	4.32	3.70	4.66	5.17	3.29
P_2O_5	0.14	0.15	0.21	0.14	0.14	0.20	0.14	0.15	0.19	0.19	0.17	0.23	0.19	0.14	0.21	0.31	0.15	0.23	0.26	0.17	0.24	0.15	0.15	0.21	0.26	0.24
Total	96.35	95.45	99.02	99.09	98.95	98.35	99.15	99.09	99.05	99.46	99.50	99.20	98.49	99.23	99.15	98.86	99.37	99.87	97.56	99.20	98.77	100.14	99.72	99.41	99.59	100.07
Trace elements (ppm)																										
Cu	7	17	8	9	15	13	11	9	11	5	8	11	5	9	8	11	5	7	6	5	10	10	16	10	19	4
Zn	41	81	66	56	93	55	98	67	160	65	226	102	71	83	62	77	66	68	90	89	74	54	67	76	78	49
Ga	6	12	8	5	8	12	8	9	8	7	7	8	8	7	7	11	9	9	14	10	8	5	7	8	14	6
Rb	94	189	89	66	133	107	111	127	41	84	88	86	72	110	106	86	69	98	218	64	59	158	142	187	233	149
Sr	398	238	393	461	349	389	292	317	379	339	447	396	347	376	359	456	438	395	135	360	402	196	268	239	276	224
Y	29	49	34	30	38	30	41	32	37	35	34	44	37	34	43	45	34	37	58	32	57	22	22	21	40	28
Zr	183	356	219	197	299	208	262	236	201	222	205	255	205	216	261	253	193	217	312	190	267	169	168	190	253	200
Nb	5	13	7	6	9	8	8	8	5	7	6	8	6	6	8	11	5	6	13	6	9	6	8	8	15	10
Ba	1269	2500	2192	1272	1725	1587	1362	1499	1182	2051	2543	2641	1690	1278	2602	3093	1380	2140	1335	920	1708	1161	952	1074	1056	757
La	45	64	38	42	46	27	48	41	48	54	35	56	46	33	51	55	32	46	60	27	44	29	29	45	57	50
Pb	28	40	45	34	50	53	58	41	44	56	58	51	28	29	37	25	21	33	33	20	22	37	31	35	40	71
Th	24	46	27	26	33	31	32	33	27	31	26	31	26	26	32	34	20	27	37	19	33	29	28	34	33	29
Sc	6	6	5	7	5	6	6	6	6	8	5	5	7	7	5	5	8	7	8	7	7	5	4	5	3	3
V	50	70	80	57	67	89	58	63	90	82	69	65	76	63	91	117	90	82	109	140	92	72	84	101	69	43
Cr	36	15	34	9	19	30	14	13	30	18	3	1	15	12	11	38	23	20	47	41	23	63	74	90	74	21
Ni	6	16	9	3	15	9	12	8	9	6	7	9	7	10	12	16	6	8	46	13	13	27	34	37	47	15

*Samples classified as: A, arenite; G, greywacke; P, pelite. Further subdivision into: a, arkosic; f, feldspathic; l, lithic (after Pettijohn *et al.* (1987).

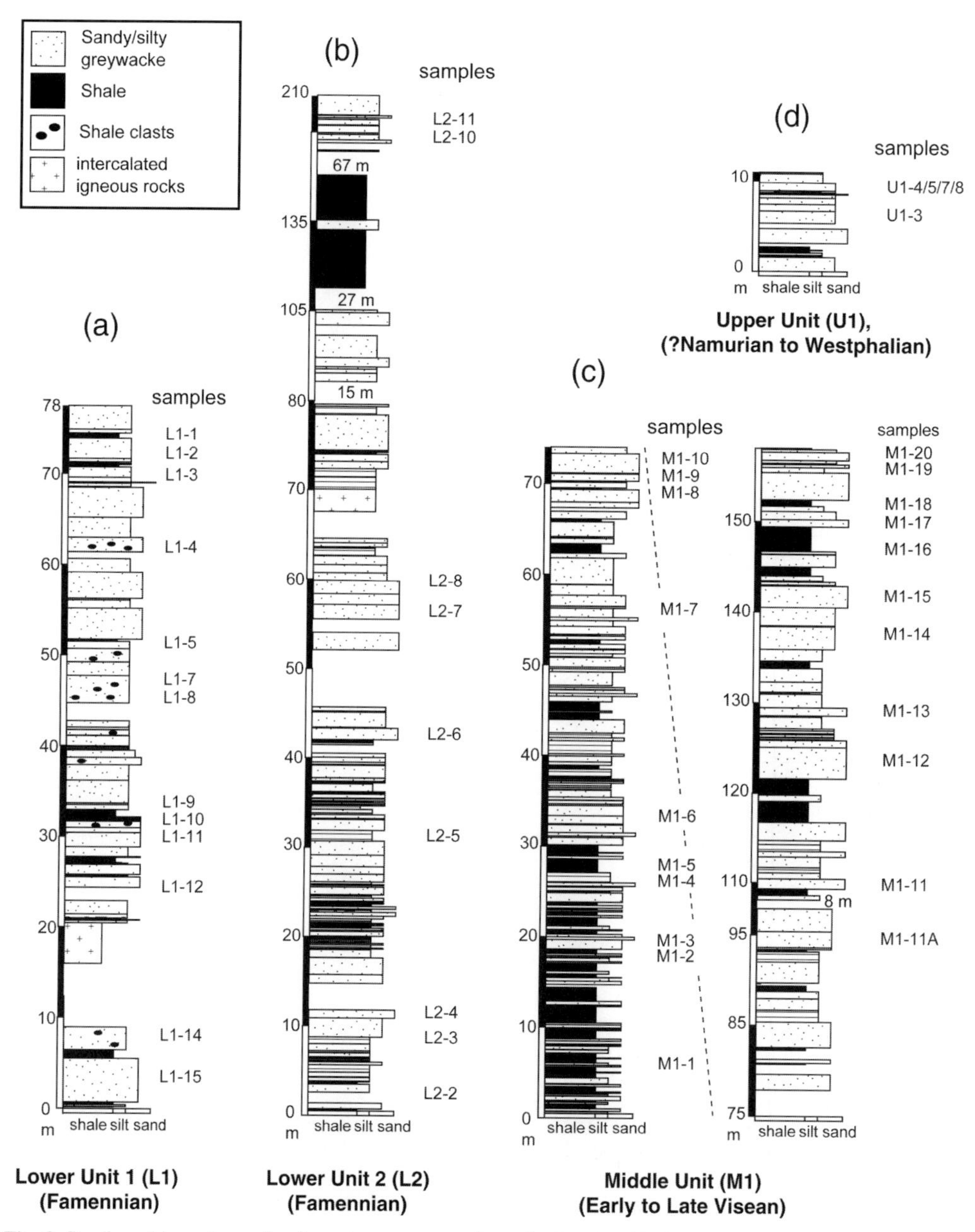

Fig. 3. Stratigraphic sections of sedimentary sequences from the Lower, Middle and Upper Units of the Southern Vosges Basin. Sections were sampled where indicated. (**a**) Lower Unit (L1); (**b**) Lower Unit (L2); (**c**) Middle Unit (M1); (**d**) Upper Unit (U1).

and varies from 1 to 10 ppm for trace elements, depending on the specific element.

Results

The first section taken from the Lower Unit (L1) of Famennian age is about 80 m thick (Fig. 3a). It is dominated by structureless greywacke beds of 0.5–5 m thickness, with only minor intercalations of pelitic rocks. The section probably represents fan deposits of turbidity currents. The second section from the Lower Unit (L2) is about 210 m thick with several outcrop gaps (Fig. 3b). It shows thinner beds of sand grain size and a

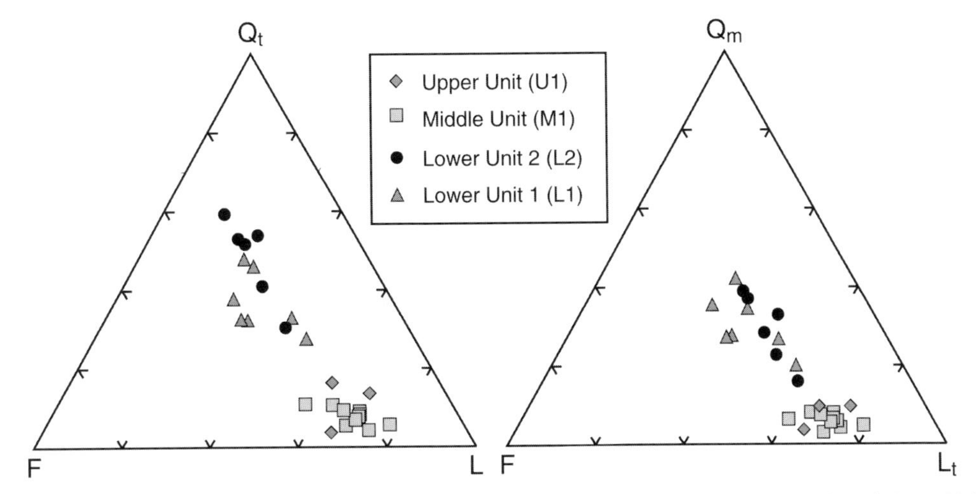

Fig. 4. QFL values (plots after Dickinson & Suczek (1979)) of samples from the Lower, Middle and Upper Units ($n = 26$). Q_m, monocrystalline quartz; Q_t, total quartz (monocrystalline and polycrystalline); F, Feldspar; L, unstable lithic fragments (volcanic, sedimentary and metamorphic); L_t; total unstable lithic fragments (L and polycrystalline quartz).

higher amount of pelites, with respect to L1, indicating more distal turbidite fan deposition. Section M1 from the Middle Unit (Early to Late Viséan age) shows a *c.* 150 m coarsening- and thickening-upward cycle with increasing volcanic detritus to the top (Fig. 3c). A narrow outcrop belt of the Upper Unit (U1, ?Namurian to Westphalian age) shows medium-bedded rocks of sand to pelite grain-size which probably represent lake deposits (Fig. 3d).

Petrography

The mean grain size of the clastic particles ranges from medium to coarse sand in L1 and L2, except for the sampled pelites. Matrix contents range from about 5% to 40%. Most of the samples are lithic or feldspathic greywackes (classification after Pettijohn *et al.* (1987); see Table 1). Qualitative thin-section petrography indicates input of quartz, feldspar and sedimentary clasts along with subordinate mafic volcanic fragments and minerals of igneous origin (olivine, hornblende, biotite) in the lower part of L1. At the top of L1 polycrystalline quartz and gneissic lithic fragments occur. Sedimentation in L2 is similar to L1, with increased input of polycrystalline quartz and felsic igneous lithic fragments in samples L2-6 to L2-8. The grain size in the psammites from M1 and U1 is mainly coarse sand to fine gravel and a pelitic matrix is absent in most of the rocks, which can mainly be classified as lithic arenites (see Table 2). In M1 and U1, clastic components consist mainly of intermediate to felsic volcanic lithic fragments and feldspar.

These petrographical findings are corroborated by the point-counting analyses (Fig. 4). The QFL values show input of quartz, feldspar and lithic fragments in L1 and L2. Input of polycrystalline quartz in some samples of L2 leads to higher values in total quartz (Q_t) than in L1. M1 and U1 are dominated by rock fragments, containing subordinate feldspar and minor quartz. There is a trend towards less detrital quartz from the Lower Unit to M1 and U1.

Geochemistry

There are major geochemical variations between the sampled sections that reflect the change in clastic input. Figure 5 shows the variation of selected elements and element ratios with stratigraphic height. The main variations occur between the sampled sections, not within them. It should be noted that there was an unknown amount of sediment accumulation between the sampled sections, which represent isolated outcrops. The geochemical trends do not, to a first order, vary with matrix content or grain size (e.g. Cr in M1, Th and Ba in L1; Fig. 5). Additionally, sediments with similar grain size and matrix content from the different sections exhibit the same geochemical variations recorded for the different rock types. Consequently, the elemental differences between the sections are not simply a

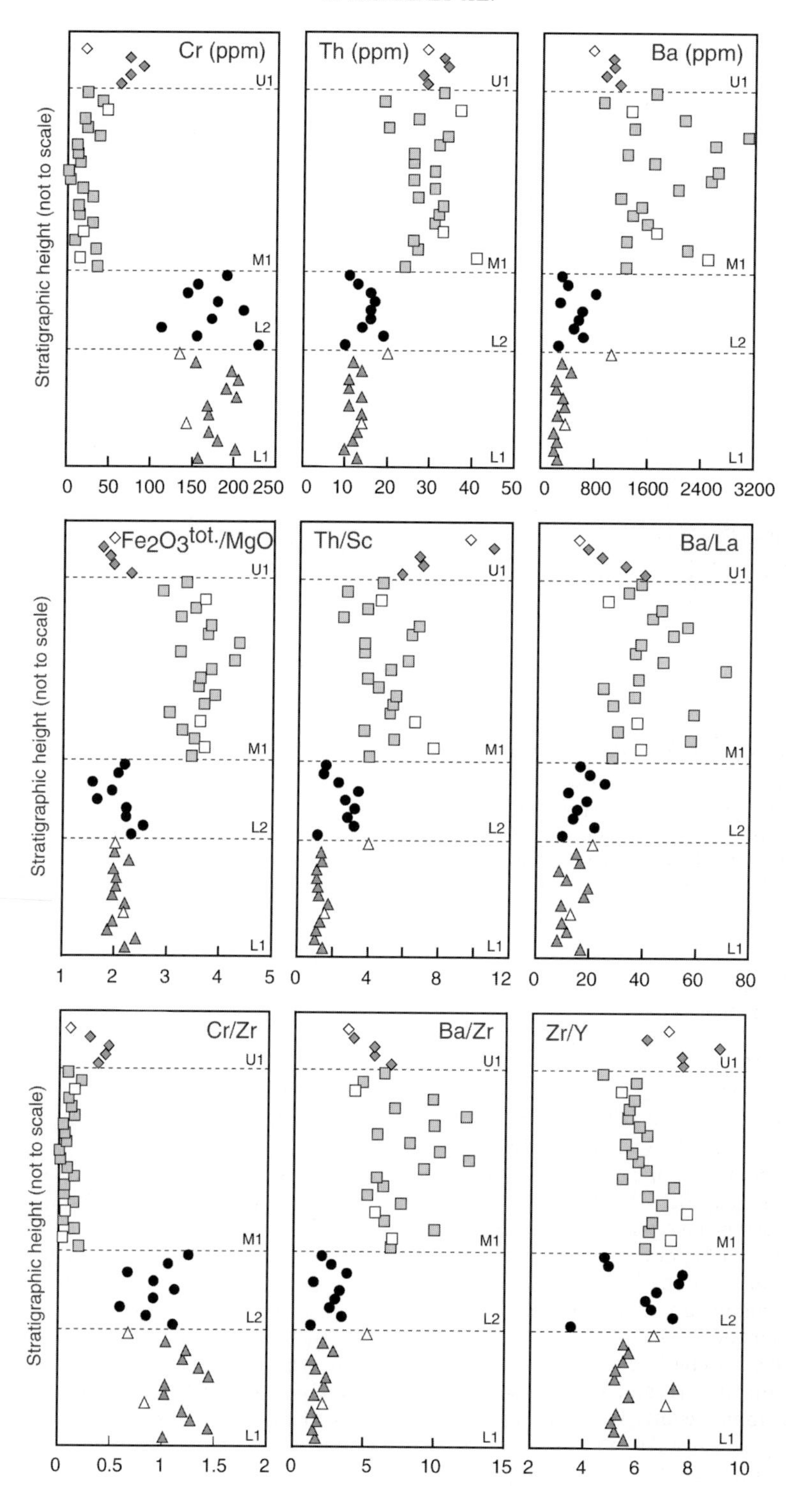

Fig. 5. Geochemical variation of trace elements and element ratios from the Lower, Middle and Upper Units with stratigraphic height. Symbols as in Fig. 4; open symbols, pelites.

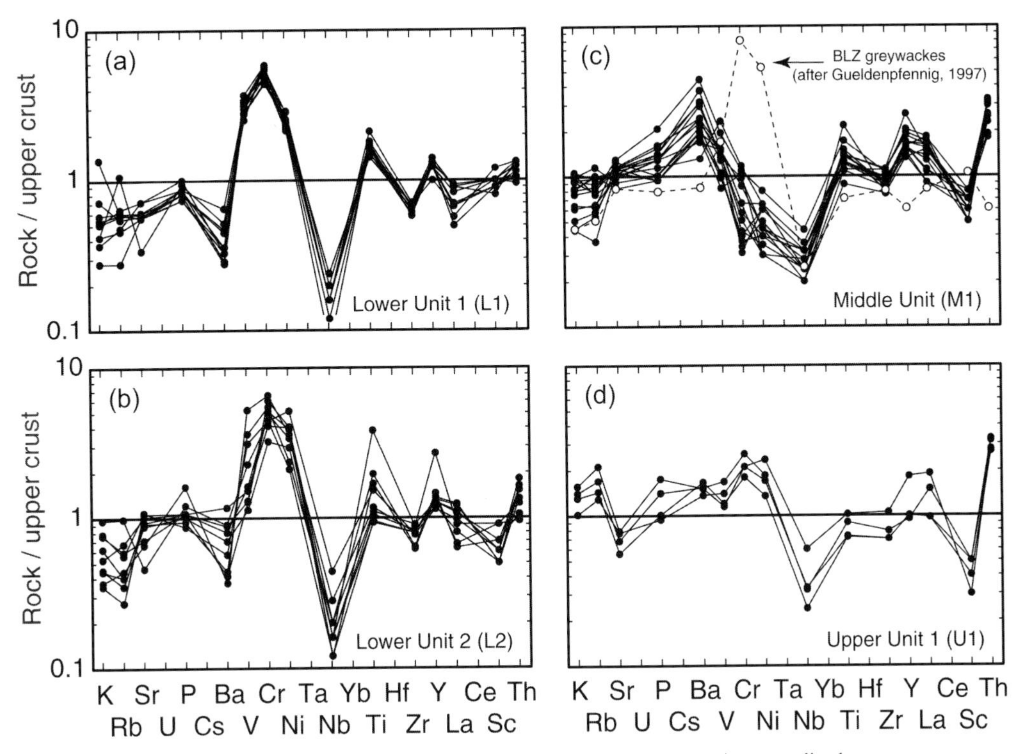

Fig. 6. Multi-element diagrams for the greywackes of the Southern Vosges Basin normalized to average upper continental crust (Taylor & McLennan 1985). Plotting order of the elements according to Floyd *et al.* (1991). Pelitic sediments are excluded. (**a**) L1; (**b**) L2; (**c**) M1 and average geochemical signature of Lower Carboniferous greywackes from the Badenweiler–Lenzkirch Zone (BLZ), southern Schwarzwald, after Güldenpfennig (1997), shown as dashed line; (**d**) U1.

function of grain size but record the geochemical composition of the clastic input.

Prominent features (Fig. 5) are high Cr values and high Cr/Zr ratios in the Lower Unit, that disappear in M1 and U1, as a result of the cessation of mafic input. The element ratio of $Fe_2O_3^{tot}/MgO$ is low in L1 and L2, but high in M1. Furthermore, an increase of incompatible trace elements (e.g. Ba, Th) from L1 and L2 to M1 and U1 can be recognized. The element ratio of Th/Sc (and La/Sc, not shown) also reflects an enrichment in incompatible trace elements in M1. Furthermore, an enrichment of large-ion lithophile elements over light rare earth elements and high field-strength elements (e.g. Ba/La, Ba/Zr), occurs in M1 and O1. These features indicate a more evolved volcanic sediment source. The geochemical composition of samples from M1 is consistent with andesitic to rhyodacitic volcanic source rocks, inferred from SiO_2 *v.* Zr/TiO_2 (plot after Floyd & Winchester (1978)), which is not shown. The Zr/Y ratio shows only minor variations (slightly higher in U1) indicating similar behaviour of these two elements throughout the sequences and therefore no major input of garnet- or apatite-bearing continental basement sources.

In Fig. 6, the analysed trace elements and some major elements are shown, normalized to average upper crust (Taylor & McLennan 1985) and plotted in the order given by Floyd *et al.* (1991). As only XRF measurements were performed, some elements (U, Cs, Ta, Yb, Hf, Ce) are missing in this multi-element plot.

Again, the strong positive Cr, Ni, V and Ti anomalies in L1 and L2 are evident (Fig. 6a and b), indicating a mafic input that disappears in M1 (Fig. 6c). A negative Nb anomaly, which is characteristic for an arc source, occurs in all sampled sections. This anomaly is most pronounced in L1 and L2 and becomes weaker in M1 and U1, accompanied by Th enrichment as a result of more evolved volcanic components in these samples. There is a positive Ba anomaly in

M1, which is controlled by feldspar. Figure 6c also shows the average trace element composition from Lower Carboniferous greywackes from the BLZ in the southern Schwarzwald (Güldenpfennig 1997). They are a time-equivalent of the greywackes of the Middle Unit but show marked differences in trace element composition. The main difference is the pronounced positive Cr–Ni anomaly in the BLZ. These elements are depleted in M1. The Cr–Ni anomalies were interpreted as ophiolitic input by Güldenpfennig (1997), which is absent in M1 of the Southern Vosges Basin. Also, Zr and La, indicative of evolved volcanic input, are lower in the greywackes from the southern Schwarzwald than in the samples from M1.

Discussion and conclusions

The composition of the volcanic rock fragments and detrital minerals as well as the geochemical composition of the sedimentary rocks from the Southern Vosges Basin exhibit consistent variations with stratigraphic height. These variations correlate well with compositional changes in synsedimentary volcanism (Maass 1988). Volcanic input in the Lower Unit of Famennian age is dominated by mafic volcanic rock fragments and detrital minerals, which cause a positive Cr–Ni anomaly. Input of intermediate to felsic volcanic rock fragments in the Middle and Upper Units of Viséan and ?Namurian–Westphalian age shows up in the whole-rock geochemistry as enrichment in incompatible elements, such as Th, Ba and Zr. The negative Nb anomaly reveals dominance of an arc source. The marked difference in geochemical composition of the greywackes from the Middle Unit compared with their time-equivalent, Lower Carboniferous greywackes from the southern Schwarzwald indicates ultramafic input in the BLZ (Güldenpfennig 1997) and evolved volcanic input in M1. Therefore, distinct sediment sources for the two sample sites existed in the Early Carboniferous time.

Zircons dated from volcanic rocks in the Southern Vosges Basin yielded ages of 345 ± 2 Ma and 340 ± 2 Ma (Schaltegger *et al.* 1996). The volcanic input into the biostratigraphically dated sediments recorded here, as well as the synsedimentary volcanic rocks (Maass 1988, 1992), extends this phase of igneous activity. Our results indicate that mafic volcanism occurred in Late Devonian time (Famennian, *c.* 360 Ma) giving way to intermediate to felsic volcanism in Late Viséan to Namurian times (*c.* 330 to *c.* 325 Ma; time-scale after Gradstein & Ogg (1996)).

The petrographical and geochemical data combined with previously published tectonic and sedimentological field observations (Maass 1988; Schneider 1990; Maass & Schneider 1995) allow us to develop a tentative model for the origin and evolution of the Southern Vosges Basin.

The Southern Vosges Basin hosts sedimentary deposits of marine to terrestrial origin with an overall thickness of several thousand metres (Maass & Schneider 1995). In general, the sequences of the Lower Unit are characterized by thick greywacke beds (Fig. 3). The 'Treh-shales' from the northern part of the basin indicate deep-water conditions, at least locally. The input of mafic volcanic, sedimentary and metamorphic rock fragments associated with bimodal volcanic rocks (Schaltegger *et al.* 1996) could represent a phase of extension, subsidence and sedimentation on thinned continental crust. The slope deposits of the Middle Unit (Krecher 1997) and the increase of arc volcanic detritus can be interpreted as a phase of tectonic unrest and subduction-related volcanism associated with compression in Viséan times. The onset of continental collision probably led to successive closure of the basin and ultimately to the terrestrial conditions of the Upper Unit in ?Namurian to Westphalian times.

For an interpretation of the geodynamic setting of the Southern Vosges Basin, knowledge of time and location of the subduction activity in the Central Variscan Belt is essential. This question, however, is still a matter of debate. Loeschke *et al.* (1998) inferred accretionary prism sedimentation and favoured northward subduction in the BLZ, adjacent to the Southern Vosges. In the area studied here, evidence for mélange-type deformation is weak. However, it is striking that the timing of subduction-related volcanism and sedimentation in the Southern Vosges Basin coincides with arc magmatism in the Mid-German Crystalline High and the closure of the Rheno-Hercynian oceanic basin (see Franke this volume), which therefore could have been responsible for the igneous activity in the Vosges Mountains. We therefore favour an extensional back-arc setting on thinned continental crust for the Southern Vosges in Late Devonian to Early Carboniferous time, related to a magmatic arc to the north of the basin. The onset of convergence and continental collision in Early to Mid-Carboniferous time (see Franke this volume) could have transformed the basin into a retro-arc foreland basin with sedimentation, synchronous igneous activity and flysch grading into molasse in the Upper Unit (in Late Carboniferous time).

We thank J. Otto for making access to the XRF spectrometer possible, and S. Dachnowsky for help with sample preparation. A. Wamsler assisted us in the field and during sample preparation. The Deutsche Forschungsgemeinschaft (DFG) supported this project financially (Le 690/4-1). We thank P. Floyd, P. Jakeš and D. Tanner for their constructive reviews.

References

ALTHERR, R. & KALT, A. 1996. Metamorphic evolution of ultrahigh-pressure garnet peridotites from the Variscan Vosges Mts (France). *Chemical Geology*, **134**(1–3), 27–47.

BÉBIEN, J. & GAGNY, C. 1978. Le plutonisme viséen des Vosges méridionales: un nouvel exemple de combinaison magmatique entre roches tholéiitiques et calco-alcalines. *Comptes Rendus de l'Académie des Sciences, Série II,* **286/II**, 1045–1048.

COULON, M., FOURQUIN, C., PAICHELER, J. C. & HEDDEBAUT, C. 1975. Mise au point sur l'âge des faunes de Bourbach-le-Haut et sur la chronologie des différentes séries du Culm des Vosges du Sud. *Sciences Géologiques Bulletin*, **28**(2), 141–148.

DICKINSON, W. R. & SUCZEK, C. A. 1979. Plate tectonics and sandstone compositions. *AAPG Bulletin*, **63**, 2164–2182.

—— & VALLONI, R. 1980. Plate settings and provenance of sands in modern ocean basins. *Geology*, **8**, 82–85.

DOUBINGER, J. & RAUSCHER, R. 1966. Spores du viséen marine de Bourbach-le-Haut dans les Vosges du sud. *Pollen et Spores*, **8**(2), 361–405.

FLOYD, P. A. & WINCHESTER, J. A. 1978. Identification and discrimination of altered and metamorphosed volcanic rocks using immobile elements. *Chemical Geology*, **21**, 291–306.

——, SHAIL, R., LEVERIDGE, B. E. & FRANKE, W. 1991. Geochemistry and provenance of Rhenohercynian synorogenic sandstones: implications for tectonic environment discrimination. *In*: MORTON, A. C., TODD, S. P. & HOUGHTON, P. D. W. (eds) *Developments in Sedimentary Provenance Studies*. Geological Society, London, Special Publications, **57**, 173–188.

FRANKE, W. 1992. Phanerozoic structures and events in central Europe. *In*: BLUNDELL, D., FREEMAN, R. & MUELLER, S. (eds) *A Continent Revealed: the European Geotraverse*. Cambridge University Press, Cambridge, 164–180.

—— 2000 The mid-European segment of the Variscides: tectono-stratigraphic units, terrane boundaries and plate evolution. *This volume*.

—— & ONCKEN, O. 1990. Geodynamic evolution of the North–Central Variscides—a comic strip. *In*: FREEMAN, R., GIESE, P. & MUELLER, S. (eds) *The European Geotraverse: Integrative Studies*. European Science Foundation, Strasbourg, 187–194.

GRADSTEIN, F. M. & OGG, J. 1996. A Phanerozoic time scale. *Episodes*, **19**(1–2), 3–5.

GÜLDENPFENNIG, M. 1997. Geologische Neuaufnahme der Zone von Badenweiler–Lenzkirch (Südschwarzwald) unter besonderer Berücksichtigung unterkarbonischer Vulkanite und Grauwacken. *Tübinger Geowissenschaftliche Arbeiten*, **A32**, 1–120.

—— & LOESCHKE, J. 1991. Petrographie und Geochemie unterkarbonischer Grauwacken und Vulkanite der Zone von Badenweiler–Lenzkirch in der Umgebung von Präg (Südschwarzwald). *Jahreshefte des Geologischen Landesamts Baden-Württemberg*, **33**, 5–32.

HAHN, G., HAHN, R. & MAASS, R. 1981. Trilobiten aus dem Unterkarbon der S-Vogesen. *Oberrheinische Geologische Abhandlungen*, **30**, 1–26.

KALT, A. & ALTHERR, R. 1996. Metamorphic evolution of garnet–spinel peridotites from the Variscan Schwarzwald (Germany). *Geologische Rundschau*, **85**, 211–224.

——, GRAUERT, B. & BAUMANN, A. 1994. Rb–Sr and U–Pb isotope studies on migmatites from the Schwarzwald (Germany): constraints on isotopic resetting during Variscan high-temperature metamorphism. *Journal of Metamorphic Geology*, **12**, 667–680.

KRECHER, M. 1997. Viséan slope sedimentation in the Southern Vosges (NE France). *Terra Nova*, **9**, Abstract Supplement 1, 264.

LEFÈVRE, C., LAKHRISSI, M. & SCHNEIDER, J. L. 1994. Les affinités magmatiques du volcanisme dinantien des Vosges méridionales (France); approche géochimique et interprétation. *Comptes Rendus de l'Académie des Sciences, Série II*, **319**, 79–86.

LOESCHKE, J., GÜLDENPFENNIG, M., HANN, H. P. & SAWATZKI, G. 1998. Die Zone von Badenweiler-Lenzkirch (Schwarzwald): eine variskische Suturzone. *Zeitschrift der Deutschen Geologischen Gesellschaft*, **149**(1), 197–212.

MAASS, R. 1988. Die Südvogesen in variszischer Zeit. *Neues Jahrbuch für Geologie und Paläontologie Monatshefte*, **10**, 611–638.

MAASS, R. 1992. Peperite im Unterkarbon der Südvogesen. *Jahreshefte Geologisches Landesamt Baden-Württemberg*, **34**, 312–237.

—— & SCHNEIDER, J. L. 1995. Die südlichen Vogesen. *Jahresberichte und Mitteilungen des Oberrheinischen Geologischen Vereins Neue Folge*, **77**, 139–153.

—— & STOPPEL, D. 1982. Nachweis von Oberdevon bei Markstein (Bl. Munster, Südvogesen). *Zeitschrift der Deutschen Geologischen Gesellschaft*, **133**, 403–408.

MAYNARD, J. B., VALLONI, R. & YU, H. 1982. Composition of modern deep sea sands from arc-related basins. *In*: LEGGETT, J. K. (ed) *Trench Forearc Geology*. Geological Society, London, Special Publications, **10**, 551–561.

MCLENNAN, S. M. & TAYLOR, S. R. 1991. Sedimentary rocks and crustal evolution: tectonic setting and secular trends. *Journal of Geology*, **99**, 1–21.

——, ——, MCCULLOCH, M. T. & Maynard, J. B. 1990. Geochemical and Nd–Sr isotopic composition of deep-sea turbidites: crustal evolution and plate tectonic associations. *Geochimica et Cosmochimica Acta*, **54**, 2015–2050.

ONCKEN, O. 1998. Evidence for precollisional subduction erosion in ancient collisional belts: the case of the Mid-European Variscides. *Geology*, **26**(12), 1075–1078.

PETTIJOHN, F. J., POTTER, P. E. & SIEVER, R. 1987. *Sand and Sandstone*. Springer, New York.

SCHALTEGGER, U., SCHNEIDER, J. L., MAURIN, J. C. & CORFU, F. 1996. Precise U–Pb chronometry of 345–340 Ma old magmatism related to syn-convergence extension in the Southern Vosges (Central Variscan Belt). *Earth and Planetary Science Letters*, **144**, 403–419.

SCHNEIDER, J. L. 1990. *Enregistrement de la dynamique varisque dans les Bassins volcano-sédimentaires dévono-dinantiens: exemple des Vosges du Sud (zone moldanubienne)*. PhD thesis, Strasbourg University.

——, HASSENFORDER, B. & PAICHELER, J. C. 1990. Une ou plusieurs 'Ligne des Klippes' dans les Vosges du Sud (France)? Nouvelles données sur la nature des 'klippes' et leur signification dans la dynamique varisque. *Comptes Rendus de l'Académie des Sciences, Série II*, **311**, 1221–1226.

TAYLOR, S. R. & MCLENNAN, S. M. 1985. *The Continental Crust: its Composition and Evolution*. Blackwell, Oxford.

TRAISER, C., MONTENARI, M. & SPECK, T. 1998. The Upper Unit of the Southern Vosges Basin (Central Variscan Belt): a genetic model and a stratigraphic discussion. *Terra Nostra*, **98**(3), 189–190.

VOGT, C. 1981. Benthonische Klein-Foraminiferen aus dem Unterkarbon der Südvogesen. *Neues Jahrbuch für Geologie und Paläontologie, Monatshefte*, **6**, 363–384.

Index

Note: Page numbers in *italic* refer to illustrations, those in **bold** type refer to tables.